Handbook of Thick Film Hybrid Microelectronics

Other McGraw-Hill Handbooks of Interest

Handbook of Thick Film Hybrid Microelectronics

A PRACTICAL SOURCEBOOK FOR DESIGNERS, FABRICATORS, AND USERS

Charles A. Harper editor-in-chief

Westinghouse Electric Corporation
Baltimore, Maryland

McGRAW-HILL BOOK COMPANY

New York   St. Louis   San Francisco   Düsseldorf   Johannesburg
Kuala Lumpur   London   Mexico   Montreal   New Delhi
Panama   Paris   São Paulo   Singapore
Sydney   Tokyo   Toronto

Library of Congress Cataloging in Publication Data

Harper, Charles A.
    Handbook of thick film hybrid microelectronics.

    1.   Microelectronics—Handbooks, manuals, etc.
    2.   Integrated circuits—Handbooks, manuals, etc.
    I.   Title.
    TK7874.H39   621.381'71   74-2460
    ISBN 0–07–026680–8

1234567890KPKP7987654

*The editors for this book were Harold B. Crawford and
Lila M. Gardner, the designer was Naomi Auerbach, and
its production was supervised by Teresa F. Leaden.
It was set in Caledonia by Monotype Composition Company, Inc.*

It was printed and bound by The Kingsport Press.

Contents

Contributors

CLARK, RICHARD J. *Defense Electronics Division, General Electric Company:* CHAPTER 9. PACKAGING AND INTERCONNECTION OF ASSEMBLED CIRCUITS

COTÉ, RENÉ E. *Integrated Circuits, Incorporated:* CHAPTER 3. FACILITIES, EQUIPMENT, AND MANUFACTURING OPERATIONS FOR CIRCUIT DEPOSITION AND TESTING

HARPER, CHARLES A. *Systems Development Division, Westinghouse Electric Corporation:* CHAPTER 4. SUBSTRATES FOR THICK FILM CIRCUITS

HICKS, WILLIAM T. *Electronic Products Division, E. I. du Pont de Nemours & Co.:* CHAPTER 5. CONDUCTOR MATERIALS, PROCESSING, AND CONTROLS

HILGERS, JOHN L. *Centralab Electronics Division, Globe-Union, Incorporated:* CHAPTER 1. CIRCUIT DESIGN, COMPONENT SELECTION, AND LAYOUT GENERATION

LINDEN, ALBERT E. *Industrial Division, Honeywell, Incorporated:* CHAPTER 8. COMPONENT-ATTACHMENT TECHNIQUES AND EQUIPMENT

PELNER, JOSEPH J., JR. *Consulting Engineer:* CHAPTER 1. CIRCUIT DESIGN, COMPONENT SELECTION, AND LAYOUT GENERATION

STALNECKER, STEWART G. *Microcircuit Engineering Corporation:* CHAPTER 2. PHOTOFABRICATION OPERATIONS, SCREENS, AND MASKS

ULRICH, DONALD R. *Space Sciences Laboratory, General Electric Company:* CHAPTER 7. DIELECTRIC MATERIALS, PROCESSING, AND CONTROLS

WAITE, GILBERT C. *Information Systems Department, Honeywell, Incorporated:* CHAPTER 6. RESISTOR MATERIALS, PROCESSING, AND CONTROLS

Preface

The maturing of modern electronic design and the forces of economics and schedules have led to the inevitable growth and prominence of thick film hybrid microelectronics. From time to time in the past there have been debates on pure thick film designs, pure thin film designs, or total semiconductor circuit designs. However, it soon became obvious that hybrid designs offered the broadest range and most practical set of design opportunities. It also became apparent that deposited thick film circuitry had many advantages from the viewpoints of design flexibility, cost of electronic designs, shorter and simpler fabrication cycles, and capital equipment requirements. Hence, the great growth of thick film hybrid microelectronics.

In spite of the great recent growth in thick film hybrid microelectronics, large numbers of electronic designers continue to be entering this design area for the first time. At the same time, design engineers already involved in thick film hybrid microelectronics require a broader base of knowledge in the technology and a ready desk-top reference sourcebook. It is this need which dictated the preparation and publication of this *Handbook of Thick Film Hybrid Microelectronics*. The complete premise around which this Handbook has been written is that it should be a practical sourcebook for designers, fabricators, and users of thick film hybrid microelectronics. As the reader will observe, it is heavily oriented towards the electronic design engineer but, in addition, has a wealth of data and guidelines for fabricators and users as well.

A review of the coverage and scope will clearly show that this Handbook is by far the most thorough and complete sourcebook in its field.

The largest and most extensive treatment available today on circuit design, component selection and layout generation is presented in the first chapter, which includes nearly 200 data-packed figures and tables, and lists over 100 invaluable data sources. A wealth of equally invaluable data and guidelines are presented in separate, individual chapters on deposited conductors, deposited resistors, deposited dielectrics and capacitors, and substrates. Further thorough and separate chapter treatments are devoted to packaging, component attachment, and interconnection. In addition, separate chapters covering selection, installation and utilization of facilities and equipment, as well as chapter coverage on screens, masks and photofabrication techniques are sufficiently thorough to guide any operation from initial conception through volume operations.

In addition to the broad presentation of data and guidelines, several other features of this Handbook should be noted. First, an extremely thorough glossary of terms is included. Such definitions can frequently be invaluable. Next, the reader will find trade-off comparison considerations spread generously throughout the book. These will be very useful in making decisions among various possible approaches to given problem areas. Last, every attempt has been made to provide a very complete and thoroughly cross-referenced index. It is suggested that each reader acquaint himself with this index.

Length and coverage of a Handbook of this magnitude are necessarily measured compromises. Inevitably, varying degrees of shortages and excesses will exist, depending on the needs of the individual reader. Then too, the time required to complete such a major work as this necessarily demands that some of the most recent data may not be fully covered. Further, in spite of the tremendous efforts involved, some errors or omissions may exist. While every effort has been made to minimize such shortcomings, it is my greatest desire to improve each successive edition. Toward this end, any and all reader comments will be welcomed and appreciated.

Charles A. Harper

Handbook of Thick Film Hybrid Microelectronics

Chapter **1**

Circuit Design, Component Selection, and Layout Generation

JOHN L. HILGERS

Centralab Electronics Division,
Globe-Union Inc., Milwaukee, Wisconsin

and

JOSEPH J. PELNER, JR.

Consulting Engineer, Milwaukee, Wisconsin

INTRODUCTION

Comparison of Microcircuit Technologies

The development and growth of microcircuit technologies were precipitated when the aerospace industry presented a need for electronic circuitry with increased functional capability per unit volume, reduced weight, and improved reliability and environmental stability. These requirements could not be achieved with conventional, exclusively discrete electronic components and printed-circuit boards or circuit-card assemblies. Attempts to miniaturize discrete versions of circuit functions generally failed or resulted in assemblies which were not only costly and difficult to assemble but fell short of reliability requirements. As a result, new avenues within the field of microminiaturization were sought to meet the desired objectives. Three major technologies evolved: (1) thick film technology, (2) thin film technology, and (3) monolithic integrated circuit technology including medium-scale integration (MSI) and large-scale integration (LSI).   The general characteristics of these technologies are compared in Table 1.

The thick and thin film technologies are best suited for low-volume custom circuits, complex arrays, and applications requiring medium to high power, high voltage, or stable, tight-toleranced passive components.   Thick film has the added advantage of low product cost, low design cost, and low capital investment.   Thin films offer the highest component stability.

Monolithic integrated circuits fill the need for high-volume, low-power circuits where wider passive-component tolerances and temperature coefficients are acceptable. They are generally used whenever the required performance can be achieved and the quantity required justifies the high design and tooling costs.

TABLE 1   Comparison of Microcircuit Technologies

Parameter	Thick-film hybrid circuits	Thin-film hybrid circuits	Monolithic circuits
Performance	High	High	Limited
Design flexibility, digital	Medium	Medium	High
Analog	High	High	Low
Parasitics	Low	Low	High
Resistors, maximum sheet resistivity	High	Low	Lowest
Temperature coefficient of resistance	Low	Lowest	High
Tolerance	Low	Lowest	High
Power dissipation	High	Medium	Low
Frequency limit	Medium	High	Medium
Voltage swing	High	High	Low
Size	Small	Small	Smallest
Package density	Medium	Medium	High
Reliability	High	High	Highest
Circuit development time	1 month	2 months	3–6 months
1:1 design transfer from bench	Yes	Yes	No
Turnaround time for design change	2 weeks	1 month	2 months
Part cost, low quantity	Medium	High	Impractical
High quantity	Medium	Medium	Low
Cost of developing one circuit	Low	Medium	High
Capital outlay	Low	Medium	High
Production setup and tooling costs	Low	Medium	High

Definition of Thick Film

The term *thick film* has gained acceptance as the preferred generic description for that field of microelectronics in which specially formulated pastes are applied and fired onto a ceramic substrate in a definite pattern and sequence to produce a set of individual components, such as resistors and capacitors, or a complete functional circuit.   The pastes are usually applied using a silk-screen method.   The high-temperature firing matures the thick film elements and bonds them integrally to the ceramic substrate.

Typically, the thickness of a thick film element will be 0.5 to 1 mil or more.   This distinguishes it from thin film technology, where conductor thicknesses are generally in the neighborhood of 30 Å.

When active devices such as diodes and transistors are attached to a thick film network, the resulting product is known as a *thick film hybrid circuit.*

History of Thick Film Technology

Basic thick film technology actually dates back to World War II.   About 1943 the military was seeking a method of miniaturizing the electronics portion of a mortar proximity fuse.   The idea evolved of depositing silver conductive and carbon resistive inks on a ceramic substrate and was successfully implemented. Most of these early circuits were passive *RC* networks, but some were more sophisticated and contained miniature vacuum tubes.

After the war this thick film technique was applied to consumer products. Because they proved to be more reliable and less costly than equivalent discrete-component types, thick film networks soon became widely used in such products as radios, television sets, and electronic organs. Most of the circuits were custom-designed to meet specific customer requirements.

With the introduction of the transistor in the late 1940s, full advantage of the miniaturization offered with thick film techniques was finally realized. Complete functional circuits for consumer and industrial applications became practical. At first, devices packaged in the standard TO-5 cans were used. Later the smaller, less expensive epoxy-encased transistors were employed. These devices are still widely used in many thick film hybrid circuits today.

The birth of the monolithic integrated circuit in the late fifties posed a threat to thick film technology. Optimistic technologists predicted that monolithic circuits could be constructed to perform all electronic functions. However, it soon became apparent that monolithic circuits could not do everything. Certain limitations with regard to power, voltage, current, speed, cost, and even design exist. As a result a new emphasis was placed on thick film technology. Advances in thick film materials technology evolved, providing higher-stability resistor pastes with wider resistivity ranges and lower temperature coefficients. Significant improvements were achieved in dielectric materials for thick film capacitors, crossovers, and multilayer insulation glazes. Process and material improvements made possible fine-line screening and the attachment of semiconductors in chip form directly to a thick film network, thus greatly improving component density. More sophisticated assembly techniques and production equipment, which increased yields, improved quality, and reduced costs, were developed. The net result is thick film as we know it today, a highly versatile and flexible technology.

Advantages of Thick Film Technology

Some of the many advantages of thick film technology are listed below, grouped into the four major categories of performance, flexibility, reliability, and economy.

Performance

1. A proved conventional circuit constructed from discrete components can be converted into a more reliable thick film hybrid circuit with substantial savings in size, weight, and volume without extensive changes in circuit design.

2. Thick film construction reduces the parasitic capacitive coupling between components and minimizes lead lengths, which in turn minimizes lead resistance and inductance. This provides improved high-speed and high-frequency performance.

3. The high thermal conductivity of the substrate materials typically utilized for component mounting minimizes the thermal gradient between components. Improved and repeatable thermal tracking and stability at high temperatures result.

4. Close thermal tracking between thick film resistors fabricated from paste with the same resistivity can readily be achieved.

5. Resistors can be functionally adjusted to compensate for parametric variations of other components to optimize total circuit performance.

6. Resistor power ratings are high.

7. Thick film construction provides high dielectric isolation. This is an advantage for high-frequency, high-voltage, and radiation environments.

8. Low-cost capacitors can be produced using the ceramic substrate as the dielectric.

9. Close resistor tolerances and ratio matching are possible.

Flexibility

1. Circuits can be designed and breadboarded using conventional techniques.

2. Existing designs can be readily transferred on a direct one-to-one conversion basis.

3. Circuits can be custom-designed to satisfy specific needs or to achieve maximum benefit from the inherent performance advantages of thick film technology.

4. Circuits can be tailored to meet a wide variety of packaging requirements and system interconnection schemes.

5. Fast turnaround times are possible.

6. Prototype or evaluation units can be readily assembled.

7. Design changes can be accomplished with minimal time and effort.

8. Transfer from the prototype stage to production is rapid.

9. A wide variety of patterns can be achieved with little variation in the manufacturing process.

10. The designer may choose from a wide selection of active and passive components in packaged and unpackaged (chip) form.

11. The designer can specify and closely control passive-component parameters.

12. A wide assortment of component assembly and attachment techniques are possible.

Reliability

1. The high reliability of thick film circuits has been demonstrated.

2. The increased reliability is due primarily to the reduction in the number of physical interconnections, the most common failure mode of discrete components.

3. Replacement of a large percentage of the circuits' solder connections by chemically bonded material interfaces has reduced their susceptibility to wiring errors and shock, vibration, and acceleration damage.

4. The close bond between thick film resistors and the high-thermal-conducting substrates minimizes localized heating and hot spots within the resistive elements.

Economy

1. Development costs are low.

2. Since processes, materials, and equipment are commercially available, costly development programs are not required.

3. Initial investment in capital equipment and personnel training is relatively low.

4. Mask and production tooling are minimal and relatively inexpensive.

5. Thick film processes are ideally suited for mass-production techniques.

6. Prototype or evaluation units can be assembled at minimal cost.

7. Circuit changes can be readily accomplished with minimal time, effort, and cost.

8. Procurement of outside components is reduced.

9. Incoming inspection is reduced.

10. Inventory control and production-line operation orders are reduced.

11. Final-product assembly time is significantly reduced.

Application of Thick Film

Thick film is an extremely versatile and flexible technology, as Fig. 1 clearly illustrates. Resistors and capacitors in any of a wide variety of combinations, values, and characteristics can be constructed with the basic thick film materials and substrates. Thick film can even be employed in high-power and high-voltage applications. Note in Fig. 1 the thick film resistive voltage divider capable of handling 25,000 V.

The ability to attach discrete active and passive devices either in encased or chip form further enhances the versatility of thick film. Complete functional circuits from simple amplifiers to complex arrays containing many chip-type monolithic integrated circuits are possible in thick film form. Thick film technology also offers many advantages at high frequencies and has been utilized in circuits operating in the gigahertz range.

In general, thick film technology is employed in any application where limited quantities or quick turnaround renders monolithic or thin film circuits uneconomical; where the circuit power, voltage, frequency, or passive-device tolerance or temperature-coefficient requirements exceed the capability of monolithic technology; where existing designs are to be retained; or where a reduction in size, cost, or assembly time or an improvement in reliability over conventional discrete devices is desired.

Although the majority of thick film hybrid circuits at present are custom-designed for specific applications, this technology can be used to produce standard off-the-shelf functional circuits as well.

Because of the great capabilities and versatility of thick film networks, the electronics industry is increasingly turning to them for solutions to problems that have not been solved by conventional, monolithic, thin film, or any other known techniques. Today thick film networks can be found in a host of consumer, commercial, industrial, and even military products. Table 2 lists but a few typical examples. As the electronics field continues to grow and improvements in thick film materials and manufacturing techniques continue to evolve, increased application of thick film technology into new product areas is certain to ensue.

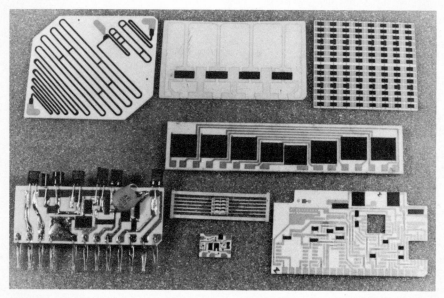

Fig. 1 Example of the versatility of thick film technology. (*Centralab Division, Globe-Union Inc.*)

Thick Film Design Procedures[1]

The specific design approach to be employed in converting a given circuit into thick film form is dictated by a combination of end-use functional and environmental performance requirements and a set of size, weight, package, thermal, reliability, and economic objectives. The design transition from a conventional schematic to the final thick film module is an involved process. The design engineer must continuously consider many alternate solutions to problems as the design progresses. He must perform his work under conditions which differ from those to which he has become accustomed. Above all, he must ensure that his design remains within the framework of compatible elements. This requires close coordination between the design engineer and the process engineers and a knowledge of thick film materials and processes and of the variety of discrete components that can be attached to thick film networks.

A diagrammatic representation of a generalized step-by-step process for the complete design of a thick film hybrid circuit is shown in Fig. 2. Central to all design phases are the functional requirements and the constraints of the hybrid technology. The arrows pointing outward from the requirements indicate the central importance of this item and the need to get through the technological constraints in order to even consider hybridizing the function. The smaller return arrows indicate that in many practical situations, each step in the design cycle may force a modification of the

TABLE 2   Typical Applications of Thick Film Circuits

Consumer	Commercial and industrial	Military
Television	Computers	Proximity fuses
Radio	Computer peripherals	Fire-control computers
Phonographs	Communication equipment	Missiles
Electronic organs	Desk calculators	Helmet radios
Garage-door openers	Safety equipment	Sonar systems
Power tools	Signaling equipment	Communication systems
Automotive products	Emergency lighting	Navigation systems
Timepieces	Medical instruments	Guidance systems
Appliances	Control systems	Radar systems
Toys	Test equipment	Infrared detectors

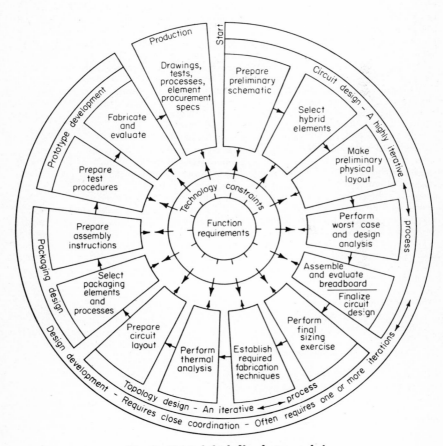

Fig. 2   Typical thick film design cycle.[1]

basic function requirements. These central functions are implicit in each step of the design process from preparation of the preliminary schematic through finalization of the production documentation. In general, the typical existing technology capabilities and constraints are given in this chapter. Clearly, it is extremely important that the design engineer be fully familiar with the technological capabilities and constraints of the materials and production facilities available to him.

In this generalized procedure the complete cycle is divided into five major phases: circuit design, topology design, packaging design, prototype development, and production. Further divisions are made within each major phase to complete the generalized step-by-step procedure.

It is recognized at the outset that the indicated steps are not independent. On the contrary, they are so interdependent that the design-development cycle can be viewed in this context: each step considers in successively greater detail all the remaining steps. There is also, as indicated, an interplay and iterative activity within and between the major phases. Thus, while the first indicated step concerns itself principally with developing an electrically functioning schematic, it also considers in a more cursory manner the remaining circuit-design steps. The major phase of the design-development cycle is arranged somewhat arbitrarily in accordance with current design-activity practice. Thus, while no attempt is made to assign specific tasks to specific specialists, there is a logical break between circuit, topology, and packaging design.

The following paragraphs summarize each step in the design process. The remainder of this chapter presents detailed information relative to the various phases of the design cycle up to prototype development. The organization of the material does not exactly follow the design sequence as shown in Fig. 2; rather, closely interactive and similar design phases are logically integrated into several major sections. Information relative to prototype development, thick film processes, and production techniques will be found in other chapters of this handbook.

Functional requirements   The starting point in the design cycle is assumed to occur after a system has been partitioned and stated function requirements can with confidence be considered hybridizable into a single hybrid package or a decision is reached to convert an existing discrete circuit into a thick film hybrid circuit on a one-to-one direct conversion basis. The basic decision to hybridize must result from a consideration of the total requirements and available technologies—discrete, monolithic, and hybrid. In other words, the systems partitioner must have performed a preliminary feasibility study to assure that hybridization is the best overall approach.

The characteristics of the function should be completely specified. In most practical cases they will be idealized to start with and modified somewhat in the final product by cost, design, and manufacturing constraints. The following is a general list of the requirements which should be considered:

General Requirements

1. Applicable specifications
2. Size
3. Weight
4. Package envelope and terminal configuration
5. Cost restrictions
6. Reliability allotment or goal
7. Qualification-test requirements
8. Acceptance-test requirements

Environmental Requirements

1. Ambient nominal operating temperature
2. Ambient operating-temperature extremes
3. Storage-temperature range
4. Required shelf life
5. Vibration level
6. Humidity requirements
7. Shock level

8. Acceleration level
9. Derating criteria
10. Thermal-shock criteria
11. Hermeticity
12. Altitude

Electrical Requirements

1. Performance description
2. Complete definition of input and output signals
3. Available power and decoupling
4. Response time
5. Dynamic and thermal stability
6. Timing information
7. Interface definitions

Technology constraints   The capabilities and limitations of the technology must be considered at each step of the design process.   Cost, size, weight, performance, process, manufacturing, and patent limitations are among the prime considerations. Each step of the design procedure generally describes those factors appropriate to it.

Circuit design   This phase of the cycle starts with the given function requirements and carries the design through circuit design and worst-case, preliminary thermal, performance, and reliability analyses.   In addition the circuit is breadboarded using the intended hybrid elements, and a physical layout is prepared which approximates the final hybrid configuration to the highest degree practical.

Preparation of Preliminary Schematic.   Utilizing the given function requirements, a preliminary circuit is designed and a schematic is prepared.   In this step cursory consideration must be given to circuit partitioning and worst-case and thermal analysis.   Thick film circuit-design considerations are treated in greater detail later.

Selection of Hybrid Elements.   The Component Selection and Application section of this chapter contains procedures, curves, and tables for selecting hybrid circuit elements to fit the preliminary schematic requirements.   It is at this step that the first confirmation of whether the circuit is hybridizable in a single package normally becomes apparent.

Preliminary Physical Layout.   This step of the circuit-design phase is intended to confirm that the size, weight, and volume requirements of the function can be met with the projected design.   This step is discussed in the Layout Generation section of this chapter.   Thermal and packaging considerations are of significance at this step. Some changes in the preliminary schematic may be anticipated at this point.

Worst-Case Analysis and Design Analysis.   Analysis of the extreme values of circuit-element parameters and worst-case combinations thereof are covered in this phase.   Design analysis includes an assessment of the design capability to meet specified reliability goals and a comprehensive thermal analysis.   Further refinements in the circuit design will often result from the analysis.   Worst-case design is discussed later in this chapter.

Breadboard Preparation and Evaluation and Circuit-Design Finalization.   Confirmation of the suitability of the proposed circuit design is best accomplished by assembly and checkout of a circuit breadboard which simulates to the highest degree practicable the final production device.   New hybrid designs should be breadboarded to verify satisfactory circuit performance and to establish test procedures and test equipment.   The degree of confidence in circuit performance depends upon the degree to which the breadboard simulates the final circuit layout and the thoroughness with which the breadboard is checked out.   For dc or low-frequency applications, a discrete-component form of breadboard will normally suffice.   For high-frequency applications, where breadboard parasitics are a problem, consideration should be given to using hybrid circuit elements similar to those chosen for the preliminary layout. Where element placement, spacing, or interconnections may be critical, consideration should be given to fabricating the breadboard using the preliminary-feasibility layout. In high-frequency applications, several layouts are frequently required before the desired performance is attained.

Topology design   This phase of the design cycle covers the actual conversion of the design schematic into hybrid hardware. The first step in this phase is to perform a detailed sizing exercise to optimize such items as RC time constants, resistor configurations, and line widths. The exact fabrication techniques, conductor composition, and resistor characteristics are selected to meet the design requirements.

Final Sizing Exercise.   At this state (the circuit has been breadboarded and the schematic finalized) it is necessary to perform a final sizing exercise to optimize the physical design. The exact size, configuration, and interconnection requirements should be established for each add-on or attached device. Further information on this subject is provided in the Package Selection section of this chapter.

Fabrication Techniques.   Almost simultaneously with the final sizing exercise, the exact substrate, interconnect, and element-deposition process must be established. The techniques chosen will depend upon the function requirements and the existing capabilities and constraints of the available fabrication facilities. This subject is thoroughly treated in other chapters of the handbook.

Thermal Analysis.   Preliminary thermal analyses will have been performed at several points in the circuit-design cycle. This should give high confidence that major thermal problems will not be encountered; however, optimized thermal management to provide nearly uniform temperature on the entire substrate is possible only at this stage of the design cycle. A detailed discussion of thermal analysis is given later in this chapter.

Circuit Layout.   With the completion of the previous three tasks the actual and final layout of the circuit can be accomplished. The Layout Generation section of this chapter provides procedures and guidelines concerned with the details of the topological design. These layout rules and processes must of necessity be tailored to a particular fabrication process. The data given have general application and are considered typical of existing industry capability. It is particularly important at this point to coordinate the final layout with the probable packaging and assembly arrangement.

Packaging design   On completion of the substrate designs, the remaining package elements and processes must be selected. With these decisions made, assembly and packaging drawings can be prepared. Packages are briefly discussed later in this chapter and are treated in detail in Chap. 9.

Packaging Elements and Processes.   In addition to the package itself, which will be largely dictated by the system and function requirements, other elements (including interconnect wire, die-bond materials, and processes for bonding wires and chips and sealing the package) must be chosen. These decisions are based largely on the cost and reliability requirements of the function and on the fabrication techniques available. Procedures and guidelines for making these decisions are given in subsequent chapters.

Assembly Instructions.   At this point, all the hybrid-circuit elements, including the substrate designs, are ready for assembly into a hybrid circuit. Because the probable final packaging arrangement has already been considered in some detail, interconnection problems resulting from a physical placement of the hybrid element should be minimized. The assembly instructions should (1) cover drawings and instructions to orient and install all attached components and drawings and instructions to orient and install the substrates in the package (exploded and/or cutaway drawings should be prepared for multilayered or stacked-substrate arrays); (2) clearly identify each of the external pin connections by pin number and function; (3) show routings of all wire bonds, indicating wire materials, diameters, and types of bonds; (4) call out substrate-bonding materials and processes; and (5) specify any trimming processes required after assembly; and (6) specify package-sealing processes.

Prototype development   Fabrication and evaluation of completed prototype circuits constitutes the final step in the standard development cycle. The results of this evaluation determine whether the design should be committed to production or recycled through the design-development phase to correct deficiencies.

Preparation of Test Procedures.   As with all other steps of the design cycle, the acceptance and evaluation tests for the device should have been considered in a preliminary fashion throughout the design cycle. A recurring consideration is the need to allow pin connections for testing otherwise inaccessible circuit nodes. However, the major effort in preparing detailed evaluation tests normally occurs after the device design is finalized.

Prototype Fabrication and Evaluation. Ideally, the prototype devices should be fabricated exactly as required by the final design. The degree to which this is done depends upon the prototype facilities available and the need to prove out all details of the design. Substrate layout and design must always be accomplished, whereas final and exact packaging may not be necessary. Most electrical testing can be accomplished without a final seal, and the exact specified header may be unnecessary. Electrical performance to the function requirements is of primary concern in assessment of prototype devices. Mechanical and environmental tests are usually of concern only with new design, material, processing, or packaging techniques. Thermal evaluation and hot-spot determinations can be made if required.

Production When the device design has been proved out by evaluation of the prototype devices, additional effort is usually required to successfully implement the production article. Of primary concern are production assembly and process documentation, in-process and final acceptance-test procedures, lot qualification and/or screening tests, and procurement specifications.

THICK FILM CIRCUIT-DESIGN CONSIDERATIONS

A normal thick film design cycle should begin with the basic circuit-design phase; for it is here that the capabilities and limitations of thick film technology must first be considered if overall optimum designs are to be achieved. The general stability of the circuit, the compatibility of components to thick film capabilities, the value and tolerance of components, and the short- and long-term behavior of components all greatly influence the performance, yield, quality, reliability, and manufacturability of the final thick film product.

To adequately design for thick film, a circuit designer must have a basic understanding of the capabilities and limitations of thick film technology, a working knowledge of proper circuit partitioning, and the ability to perform a worst-case circuit analysis. It is the purpose of this section to present background information on these basic points.

Circuit-Design Guidelines

Unlike monolithic integrated circuits, which are passive-component-limited and require special design techniques to accomplish certain circuit functions, thick film microcircuits can be fabricated directly from conventional-circuit designs. An example of this capability is illustrated in Fig. 3. The significant advantages of this ability are minimal design costs and reduced design and end-product lead times. It should not be construed, however, as meaning that all circuits can be directly converted to thick film form. As with all technologies, certain physical and economic limitations do exist. System and circuit designers contemplating the use of thick film techniques should be cognizant of these limitations and plan their designs accordingly.

Listed below in the form of design guidelines are the typical capabilities and limitations of thick film technology which are applicable to the circuit-design phase. If these guidelines are followed in this basic design phase, many problems can be eliminated and much time can be saved later in the design cycle.

1. It is not necessary to use Electronics Industries Association (EIA) standard resistor values. Thick film resistors can be screened and adjusted to any specific value.

2. Resistor tolerances should be kept as broad as possible. The cost of a thick film resistor is inversely proportional to its tolerance requirement. Tolerances below ±1 percent should be avoided. Sometimes tolerance requirements can be broadened if the recommendations of the preceding guideline are followed. For example, if a resistor must be held in the range between 1,155 and 945 Ω, it should be specified as 1,050 ±10 percent rather than 1,000 ±5 percent. The latter value is the closest EIA standard with a tolerance that will assure that the desired value range will be achieved.

3. Resistors greater than 10 MΩ occupy significant substrate area and should be avoided.

4. Capacitors greater than 1 μF require considerable substrate area and should be avoided. Capacitance values up to 220 μF can be obtained in relatively small physical sizes with solid-tantalum capacitors; however, care should be taken in utilizing these devices as they exhibit certain performance and process limitations. See the Component Selection section of this chapter for more details. Utilization of direct-coupling techniques and large-value resistors in *RC* timing applications will help reduce capacitor requirements.

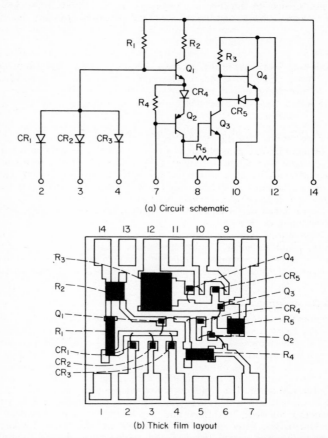

(a) Circuit schematic

(b) Thick film layout

Fig. 3   Direct conversion from circuit schematic to thick film layout.

5. There is no room in thick film circuits for large inductive-type components. However, many miniature and microminiature inductors and transformers are available for certain applications. Information on these components is provided in the Component Selection section.

6. Circuits should be designed whenever possible with readily available active devices. A minimum number of device types should be used. Specify devices by JEDEC part number and state which parameters can be relaxed or must be tightened for the application.

7. The maximum number of active devices on a single substrate should be limited to around 10. Economical yields above this figure may be difficult to achieve, particularly in chip hybrid circuits, where chip and wire bonding techniques are used.

8. Although high-power and high-voltage resistors can readily be fabricated in thick film form, they require considerable substrate area and therefore should be avoided in miniature hybrid circuits.

9. Monolithic integrated circuits are a complement to thick film technology and should be used freely in thick film designs.

10. If possible, the pin or lead assignments should be left to the thick film designer, as this provides maximum layout flexibility and minimizes fabrication complexity.

It should be noted that capabilities beyond these guidelines exist and improvements in technology are a continuing matter. Therefore, circuit designers should consult with thick film designers when requirements conflict or extend beyond these stated guidelines.

Circuit Partitioning

Often an electronic circuit is too large and complex to be economically and reliably fabricated on a single substrate or in a single package. In such cases, it is necessary to divide or partition the circuit into smaller functional parts. Any circuit can be partitioned in a number of different ways. The approach used often depends on the specific optimization desired, e.g., size, economy, producibility, or reliability. It is usually difficult to achieve all desired objectives; e.g., what may be optimum in size may not be readily producible, or what may be optimum economically may not be thermally acceptable. A practical division which is a reasonable compromise of all major design objectives can be obtained if the following guidelines are observed:

1. Divide the overall circuit into smaller, distinct circuit functions. This will assure better end performance of the complete circuit and simplify the specification and test of the individual circuit elements.

2. Keep the component count and density fairly equal. This will permit utilization of standard fabrication and packaging techniques.

3. Maintain the number of active devices within a circuit to a level commensurate with acceptable yields for the attachment process to be used.

4. Spread heat-dissipating elements. Maintain the power dissipation within each circuit fairly equal and below the maximum acceptable level of the package type selected.

5. Be practical with regard to physically large components. Leave them outside the package where necessary.

6. Take advantage of integrated-circuit and MOS devices for increased component and circuit density.

7. Consider the critical nature of components with respect to specific circuit functions and keep elements involved within the same subdivision.

Worst-Case Design [2]

A worst-case analysis involves the observation of the performance of an electronic circuit while the values of the components constituting the circuit are varied over their known tolerance limits, temperature limits, end-of-life values, etc. Depending on the method of analysis employed, information such as the sensitivity of circuit output parameters with respect to each individual component, allowable component tolerance ranges, circuit reliability, operating life, maximum operating temperature range, and expected production yields can be obtained.

All electronic circuits to be fabricated in thick film form should be given a thorough worst-case analysis. This is necessary to establish the proper component values and tolerances and to assure, especially when high-volume production is anticipated, that all circuit-performance parameters will be met when the specified components are randomly selected and assembled. Although a worst-case analysis of a circuit may seem time-consuming and costly, it is economically justifiable in comparison to the high costs experienced when problems are encountered and changes made during a production run. Furthermore, worst-case analysis can be employed as a cost-savings and yield-improvement tool. For instance, sensitivity and tolerance-limit data can be used to investigate the possibility of broadening resistor tolerances and thus reducing adjustment costs. Parameter-variation studies on circuits with components exhibiting wide parameter spreads such as transistors will assist in setting appropriate component tolerance limits so that high yields of the end product to specified performance requirements will be realized.

A worst-case analysis is generally accomplished either by empirical or computer-aided methods.

Empirical method   In the empirical method of worst-case analysis, the analysis is performed on an actual breadboard of the circuit. Physical components with known high and low tolerance limits are substituted into the circuit, and performance data are taken over the desired temperature range. This is perhaps the oldest method of worst-case analysis and also the most time-consuming. However, it is still used quite often when other methods are not available or cannot readily be used. If the proper range and combination of component variables are introduced into the circuit, this method of analysis is suitably accurate for most applications.

Computer-aided methods   The digital computer, with its ability to perform a high volume of calculations in an extremely short time, has become a very useful tool for worst-case analysis. Many computer-aided analysis programs based on a variety of analysis techniques have been written and are available in direct form or included in large general circuit-analysis programs. Four common computer-aided worst-case analysis techniques are the parameter-variation method, Mandex worst-case method, moment method, and Monte Carlo method.

Parameter-Variation Method. One or two input or component parameters are varied in discrete steps within their expected variation limits while all other component values are maintained at nominal. The computer analyzes the circuit and records the effect of the parameter variation on a selected output. Other components or component pairs are then varied and the output changes recorded. The resulting data consist of a complete listing of the performance variations of the selected circuit output with respect to the variations anticipated for each component. These data can be analyzed or plotted to determine the overall effects on the circuit performance or to select component-tolerance limits to assure that the desired performance will be obtained.

Mandex Worst-Case Analysis Method. The computer is used to analyze the circuit in question and to determine the partial derivatives and corresponding sensitivities of each output with respect to all component parameters when set at their end-point limits. After the partials are known, the computer determines from the sign and magnitude of each partial the worst-case condition for each output parameter of interest. The worst-case output parameters are printed out by the computer. A quick analysis of these results will pinpoint potential component-tolerance problems and indicate areas for redesign.

Moment Method. This method utilizes statistical input data such as mean value, variance, and possibly parameter correlation coefficient for each of the circuit components. The computer determines the partial derivatives of each output with respect to each input and performs the necessary statistical calculations. Output data are often in the form of the percentage of effect each component has on the output parameters and the mean value and standard deviation of each output value determined from the statistical-input parameters.

Monte Carlo Method. This predicts the circuit output-parameter frequency distribution as expected from the random selection and assembly of component parts in a production lot. To perform this type of analysis, complete frequency-distribution data on all component variables must be fed into the computer. The computer randomly selects the component parameter values from the input frequency-distribution data and solves the circuit equations for the desired output parameters. The calculation process is repeated until sufficient data to construct the output frequency distribution are obtained. The distribution can be compared with performance specifications to determine if circuit redesign is required.

The method of analysis used depends greatly on the input data available and the output information desired. Input data for the parameter variation and Mandex methods are quite simple and usually readily available. Considerable input data normally acquired through extensive test programs are needed for the moment and Monte Carlo methods. Output data such as partials and sensitivities offered by the moment and Mandex methods are extremely useful in locating critical components and areas that require redesigning. Because of its simplicity, the Mandex method is becoming the most popular.

Computer-aided Design

Utilization of the digital computer for design and analysis purposes is by no means limited to worst-case analysis. Thick film designers, especially those having access to a timeshare computer terminal, will find this high-speed computational tool extremely useful in performing many of the iterative and complex tasks normally encountered in thick film design. Typical are circuit design and analysis, layout design and mask generation, cost analysis and generation, thermal analysis, and miscellaneous routine calculations.

Circuit design and analysis Many excellent general-purpose circuit-analysis programs are readily available. Most can perform dc, ac, and transient analysis. Table 3 lists common programs along with their capabilities and characteristics. These programs are very easy to use since no programming experience is required and the input formats are quite simple. Inputting is generally of a form which describes the circuit topology and the values of the circuit components. A host of output data can be obtained, including node voltages, branch currents, power dissipations, sensitivities, and frequency response. Many of these large circuit-analysis programs such as ECAP include worst-case analysis, thus eliminating the need for separate programs of this type. The programs can be run in batch mode or on a timeshare computer. Operating with cathode-ray-tube (CRT) displays, plotters, and similar peripherals is possible with some of the programs.

In computer-aided circuit-analysis programs all active components such as transistors and diodes must be represented by equivalent circuits or models. Each program requires its own unique model type, which can be as simple as the ECAP diode and transistor models or as complex as the NET I models shown in Fig. 4. An exception is the Sceptre program, which will accept any model type, thus leaving the choice and utilization to the discretion of the user. Virtually no models exist for the more exotic semiconductor types, including field-effect transistors. The accuracy of any circuit-analysis program is highly dependent on the quality and completeness of the values assigned to the model parameters. Unfortunately model-parameter data cannot be obtained from manufacturers' data sheets and usually must be derived through test and statistical evaluation of actual devices. Parameter data for specific models of some device types can be found in the literature.[5,6]

Figure 5 and Table 4 illustrate the relative simplicity of initiating a computer-aided circuit-analysis program. Figure 5a shows a simple amplifier in normal schematic form. Figure 5b depicts the same circuit in the equivalent form needed for an ECAP dc analysis. Note the use of models for the transistors and the identification of all nodes and branches. The simple input format required to obtain a complete dc analysis of the circuit is provided in Table 4. Here the topological nature of describing the circuit by identifying each element by branch, nodes, and value is clearly illustrated.

Because of the ease in utilizing these programs and the speed at which a high volume of circuit performance data can be obtained, computer-aided circuit design is finding wide application in the industry, especially in the design of high-reliability circuits and high-volume integrated and hybrid circuits. Further information on the capabilities and applications of these programs can be found in the many articles that have appeared in the literature and the various users' manuals which have been prepared for each program type.[7-9]

In addition to the large programs, many smaller programs exist, most of which are special programs written by designers for specific applications. Some have appeared in the literature,[10-12] but most have been written for specific design problems and are not generally known. Any circuit designer with a knowledge of computer programming will find many applications for the computer in solving day-to-day circuit-design problems.

Layout design and mask generation Perhaps the most repetitive and time-consuming task in microelectronic design is the preparation of the final layout and

TABLE 3 Characteristics of Available Computer-aided Circuit-Design Programs

	ECAP	NET-1	Predict	Sceptre	Calahan	Circus	Nasap
General-purpose	Yes	Yes	Yes	Yes	No	No	Yes
Special-purpose	No	No	No	No	Yes	Yes	No
Application:							
Circuit analysis	Yes	Yes	Yes	Yes	Yes	Yes	Yes
Circuit synthesis	Yes	Yes	Yes	Yes	Yes	Yes	Yes
Device design	Yes	Yes	Yes	Yes	Yes	Yes	Yes
Input features:							
Special coding required	Yes	Yes	Yes	Yes	No	Yes	Yes
Sequential node numbering	Yes	Yes	No	...	Yes	Yes	Yes
Random sequence of input	Limited	Yes	Yes	Yes	No	Yes	No
Voltage or current sources	Yes (with modification)	Yes (with modification)	Yes	Yes	Current sources only	Yes	Yes
Tabular input for signal sources	Yes	Yes	Yes	Yes	Yes	Yes	No
Analytic description of branch elements	Yes	Yes	Yes	Yes	No	Yes	No
Analytic descriptions of signal sources	No	No	Yes	Yes	Yes (available with modifications)	Yes	No
Automatic modification of input (repeated runs)	Yes	Yes	No	Yes	Yes	Yes	Yes
Type of input:							
Batch cards	Yes	Yes	Yes	Yes	Yes	Yes	Yes
Light pen	Yes	Yes	No	Yes	No	No	No
Modeling capabilities:							
Built-in models	No	Yes (transistors, diodes)	No	Yes	No	No	No
Allows small-signal models	Yes	No (convenient)	Yes	Yes	Yes	Yes	Yes
Allows large signal models	No (inconvenient)	Yes (built-in)	Yes	Yes	No	Yes	No
Output options:							
Transfer functions	No	No	No	No	Yes	No	Yes
Pole-zero locations	No	No	No	No	Yes	No	Yes
Symbolic expression for time response	No	No	No	No	Yes	...	Yes
Time-response output	Selected variables	Node voltages and currents	Branch voltages and currents	Selected variables	Output node	Yes	Yes
Steady-state solution	Separate analysis	Yes	Separate analysis	Yes	No	Yes	Yes

Programming features:	Fortran II and IV	Madcap and FAD	Fortran II and FAD	Fortran IV	Fortran II and IV	Fortran IV	Fortran IV
Program language	Fortran II and IV	Madcap and FAD	Fortran II and FAD	Fortran IV	Fortran II and IV	Fortran IV	Fortran IV
Recommended memory capacity	32,000	32,000	32,000	32,000	32,000	Overlayed in 175,000	48,000
Network formulation:							
Topological or matrix	Both	Matrix	Matrix	Matrix	Topological	Matrix	Topological
Primary integration routine	Implicit numerical relation	Predictor-corrector	Runge-Kutta	Runge-Kutta, predictor-corrector	Runge-Kutta, inverse Laplace		
Machine it can be used on	Univac 1108, IBM 7094, CDC 3600	IBM 7094	IBM 7094	IBM 7094	IBM 7094, CDC 3600, CDC 6600	IBM 360, IBM 7094	Spectra 70, B 5500, IBM 7094, 360, Univac 1108, GE 635, 645, Sigma-10
Unconventional features	Plotting routine (transient)	None	Cal Comp, CRT plots	Printer	Cal Comp, CRT plots	Printer plotting, radiation responses	Symbolic transfer function

SOURCE: Reprinted from *Electronics*, Apr. 30, 1970, copyright McGraw-Hill, Inc., 1970.

artwork. Methods of using the digital computer to perform this tedious task have been devised and are now fully operational. Although most of these methods were originally developed for monolithic integrated-circuit layout and mask generation, several have been modified to handle thick film networks as well. The simplest method utilizes the digital computer in conjunction with a drum or flat-bed plotter to generate the final artwork only. Input data to the computer consist of *XY* coordinate information describing the configuration to be drawn. Inputting is usually accomplished with paper tape, magnetic tape, or punched cards. The computer converts the input data to the appropriate language and commands required by the plotter. Digital plotters such as the Cal Comp* and Complot† types are normally used. By following the instructions from the computer, the digital plotter produces

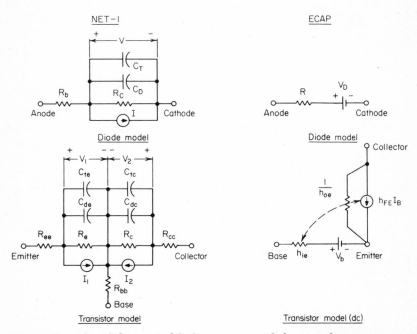

Fig. 4 Examples of device models for computer-aided circuit-design programs.[4]

a complete scaled drawing or mask of the desired layout. Typical methods of recording with the plotter for artwork generation include ink pen on a plastic film, scribing of Rubylith‡ or light beam on a photosensitive plate. The computer can be programmed to produce the artwork to any desired scale. Scaling factors for thick film layouts normally range between 5× and 50×.

The ultimate in computer-aided layout consists of a system in which the computer through direct man-machine interface is used to prepare the layout as well as the final artwork. In this type of system, input data are provided through a keyboard terminal or a CRT with a light pen or both. Design equations, constants, and rules relative to resistor-size calculations, conductor line widths and spacings, interconnect patterns, etc., are part of the basic operational computer program. A layout is initiated by inputting specific information about the circuit to be drawn. This information normally includes a description of the circuit topology, the approximate placement of components, the number and sizes of lead pads, and the values and charac-

* Trademark of California Computer Products, Inc.
† Trademark of Houston Instrument.
‡ Trademark of Ulano Products Co., Inc.

teristics of the thick film components. Once these data are entered, an initial layout is automatically generated and recorded on a plotter or CRT display. Since the computer lacks intelligence and reasoning power and can only perform the tasks it is programmed to do, it is highly unlikely that this initial layout will be the desired optimum design. The designer must exercise his superior ability at this point and determine what changes are necessary to improve the layout. The modifications

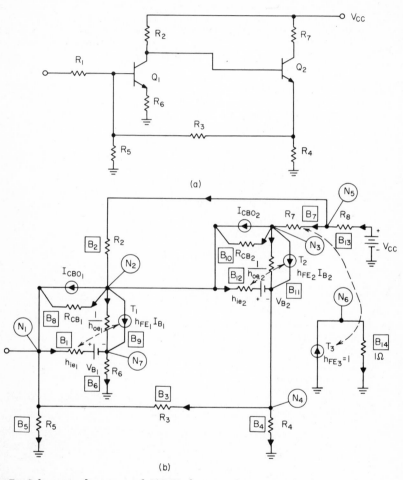

(a)

(b)

Fig. 5 Schematic diagram and ECAP dc equivalent network of a simple two-stage amplifier.

might include increasing or decreasing pad sizes, line widths, substrate size, etc., or shifting, moving, or rotating components. Rerouting of the interconnect patterns may also be necessary. Simple instructions entered through the keyboard or with a light pen on a CRT display initiate the changes and establish a revised layout. Several iterations of this type are normally required before an acceptable design is achieved. Once the layout is finalized, the computer, upon command, will generate the individual mask drawings for each of the screening operations needed to fabricate the thick film circuit. Again some form of digital plotter is normally used to obtain this final output.

Any number of variations in programming techniques and plotting methods between the two extremes described above are possible. Some programs are capable of simultaneously planning and plotting conductor patterns in several distinct layers. Other programs can simultaneously design complete thick film layouts on both sides of a single substrate. It is also possible to create a library of interconnect patterns for various types of attached components. These patterns can be recalled and plotted through a simple input command, thus eliminating the need for detailed input instructions each time a specific pattern is required.

Computer-aided layout techniques were initially expected to provide considerable savings in both time and money over manual design methods. However, it has been

TABLE 4   Example of ECAP Input Format

	ECAP INPUT EXAMPLE
	DC ANALYSIS
B1	$N(1,7), R = 4000, E = -.47$
B2	$N(5,2), R = 33.3E3$
B3	$N(4,1), R = 86.1E3$
B4	$N(4,0), R = 150$
B5	$N(1,0), R = 16.5E3$
B6	$N(7,0), R = 26.1$
B7	$N(5,3), R = 145$
B8	$N(2,1), R = 2.5E5, I = 2E-6$
B9	$N(2,7), R = 5.5E3$
B10	$N(3,2), R = 1.6E10, I = 7.5E-9$
B11	$N(3,4), R = 2.5E3$
B12	$N(2,4), R = 380, E = -.32$
B13	$N(0,5), R = 0.1, E = 120$
B14	$N(6,0), R = 1$
T1	$B(1,9), BETA = 500$
T2	$B(12,11), BETA = 64$
T3	$B(7,14), BETA = 1$
	SENSITIVITIES
	PRINT, NV, CA, CV, BA, BV, BP, SE
	EXECUTE

KEY: NV = node voltages
CA = element currents
CV = element voltages
BA = branch currents
BV = branch voltages
BP = element power loss
SE = partial derivatives and sensitivities of each node voltage with respect to each branch element

found that the most significant savings are realized in the time factor only. The cost savings gained through labor reduction are essentially offset by the cost of the peripheral equipment and computer time needed to perform the plotting task. In a highly competitive business, time savings alone is a very important factor. Other less apparent advantages of computer-aided layout include design consistency, layout accuracy, and error reduction.

Over the past few years many articles on computer-aided layout and drafting techniques have appeared in the literature. Further information on this subject can be found in these articles.[13-15] Also some complete programs are available and can be procured at a moderate cost.[16]

Cost generation and analysis The generation of cost data on thick film networks is another tedious task easily performed by the computer. Since thick film circuits are fabricated by simply combining a group of purchased parts, materials, and manufacturing operations, the preparation of a program to perform this task is quite straightforward. One common method is to generate within the computer a cost-data file on all purchased parts and all possible fabrication steps and to create a program that accesses this file, performs the appropriate calculations, and outputs the desired information. A list of the various parts and operations needed to fabricate a specific circuit is all that is required to initiate the program. These input data can be in an abbreviated (one-word) descriptive form. The output may include a tabulated list of items such as the cost of each operation, the accumulative cost, the anticipated yields, and, of course, the total cost. Programs of this type can also be expanded to provide output information such as routings, bills of material, and other documentation needed for final production of the part.

Thermal analysis   The digital computer has also been successful in the thermal evaluation of thick film and other types of microelectronic circuits. In one method of computer-aided thermal analysis, a thick film network is subdivided into numerous two-dimensional rectangles. A temperature node is established at the center of each rectangle. Basic heat-balance equations, including boundary and other effects, are written for each node and entered into the computer. Through a trial-and-error process, the computer assigns temperature values to each of the nodes and solves the heat-balance equation until an equilibrium condition is reached. The final temperature values for each of the nodes are then recorded. With a digital plotter and an appropriate computer program, a thermal-profile plot of the complete thick film network can be obtained.

Other less complex methods of computer-aided thermal analysis are also possible. One such method employs the electrothermal-analog techniques described in the section on Thermal Design and Analysis.

Miscellaneous routine calculations   Many small mathematical computations turn up frequently in the normal process of designing thick film networks. The digital computer, particularly a timeshare type, performs many of these tasks effectively. Typical examples are listed below.

1. Calculation of thick film resistor sizes
2. Calculation of required substrate areas
3. Statistical analysis of data
4. Generation of design tables
5. Curve fitting and curve plotting
6. Preparation of reliability estimates
7. Performance of an analysis-of-variance study

In addition, the computer can be used to accomplish many of the larger tasks normally encountered during the engineering and production of thick film circuits. Some typical applications include:

1. Project planning and control
2. Information retrieval
3. Process control
4. Automated testing
5. Inventory control

COMPONENT SELECTION AND APPLICATION

A large assortment of microdiscrete active and passive components in chip and miniature packaged form are available for application in thick film hybrid networks. These components are chiefly used to provide the component functions and performance capabilities unattainable with normal thick film techniques. A basic working knowledge of the availability, physical characteristics, and electrical properties of these devices and their relationship to thick film components is essential if optimum cost, performance, and reliability are to be achieved in a thick film design. This section is intended to provide this information for the basic component forms, namely, resistors, capacitors, inductors, and active semiconductor devices. Critical performance parameters of the various component forms are discussed, and (where applicable) advantages, disadvantages, and potential areas of application are indicated. Since it is not possible to cover in detail every available microdiscrete component, and since many new types are continually being introduced, the designer should consult manufacturers' data sheets and electronic trade journals for complete product information and new product trends.

Resistors

In thick film networks, the primary method of obtaining the electronic resistance function is obviously with thick film resistors formed by screen printing and firing resistive pastes directly onto a ceramic substrate. Since the design, fabrication, and performance of thick film resistors is thoroughly covered in other sections and chapters of this handbook, it should suffice to include here a summary of their typical value ranges and performance characteristics. This information (Table 5) is based on capabilities readily obtainable with commercially available cermet paste formulations. Although thick film resistors offer a wide value range and excellent stability, conditions can arise which may preclude their use. For example:

1. Very low resistor values, typically less than 10 Ω
2. Very high resistor values, typically greater than 5 MΩ
3. Any resistor with very tight tolerance or stability requirements
4. Replacement of an out-of-specification thick film resistor on a complex part
5. Elimination of a screening operation when only one resistor of a certain resistive formulation is required
6. Insufficient substrate area to fit all needed thick film components

TABLE 5   Typical Thick Film Resistor Capabilities

Resistance range	50 Ω to 10 MΩ
Sheet resistivity range	10 Ω/sq to 1 MΩ/sq
Resistance tolerance	Down to ±0.5%
Resistance ratio	Down to ±0.5%
Temperature coefficient (−55 to +125°C):	
100 Ω/sq to 100 KΩ/sq	±100 ppm/°C
10 Ω/sq and 1 MΩ/sq	±200 ppm/°C
Resistance-ratio tracking	±50 ppm/°C
Voltage capability	500 V/in.
Voltage coefficient	20 ppm/(V)(in.)
Noise:	
100 Ω/sq	−20 dB
100 KΩ/sq	0 dB
Power density	40 to 50 W/in.²
Short-term overload (2.5 times rated voltage)	$\Delta R < 0.5\%$
Long-term stability (1,000 h at rated power density)	$\Delta R < 0.5\%$
Temperature range	−55 to +175°C

TABLE 6   Checklist of Critical Resistor Performance Characteristics

Value	Temperature coefficient
Tolerance	Noise
Maximum power capability	Frequency-response effects
Maximum operating voltage	Short-term stability
Insulation resistance	Long-term stability
Maximum operating temperature	

When conditions of this type are encountered, other resistor forms, such as chip resistors and miniature encased resistors, can be used. They are quite small and can readily be soldered or otherwise bonded to conductor patterns on the circuit substrate. Detailed information regarding the configuration, size, and performance characteristics of these resistor types is provided below.

The choice between a thick film resistor and an attached discrete resistor depends on the performance parameters needed to fulfill the specific circuit function. Normally thick film resistors can be employed exclusively, but when exceptions do occur, the designer should be able to recognize them and make the appropriate substitutions. To assist in evaluating and selecting the proper resistor type, a checklist of key resistor performance parameters is included (Table 6).

Chip resistors[17]   Chip resistors are available in a wide assortment of values, types, and sizes.   The more common types are composed of thick or thin resistive films on tiny, chip-type ceramic substrates, as shown in Fig. 6.   They can be obtained in a number of termination materials, the material depending upon the assembly technique employed.   When wire-bonding methods are to be used, gold terminations are recommended.   Pt-Au, Pd-Ag, and other noble-metal compositions can be obtained for solder assembly.

The chip size is primarily dictated by power dissipation.   No size standards exist,

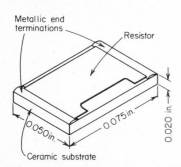

Fig. 6   Physical characteristics of a typical thick film chip resistor.

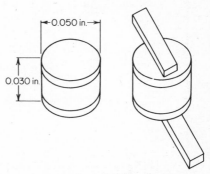

Fig. 7   Typical configurations of miniature pellet-type resistor forms.[18]

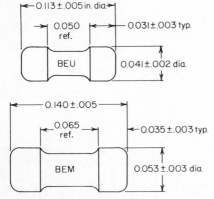

Fig. 8   Outline dimensions of miniature metal-film resistors.[19]

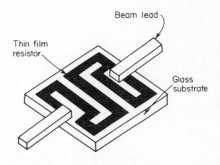

Chip size: 50×50×1.75 mils typical
Beam size: 5×20×0.3 mils typical

Fig. 9   Typical configuration of a thin film beam-lead resistor chip.[20]

but physical dimensions usually range from 0.035 by 0.035 in. to 0.100 by 0.100 in. The 0.035-in. chip will dissipate about 35 mW and the 0.100-in. size about 500 mW. Other miniature resistor chips, such as tubular-type metal-film resistors and pellet-type cermet and carbon-film resistors, are also available for inclusion in thick film networks.   The physical size and outline of these two resistor types are shown in Figs. 7 and 8.   Thin film resistors in beam-lead form (Fig. 9) are also available.

The electrical properties of chip resistors vary from supplier to supplier and with the basic resistive materials used.   Thin film resistors primarily span the low-value range and exhibit excellent stability, particularly at the very low resistor values. Thick film chip resistors cover a wide resistor range, up to 1,000 MΩ, and exhibit varying degrees of stability.   Table 7 summarizes the size and performance characteristics of chip resistors offered by a number of typical suppliers.

TABLE 7 Typical Characteristics and Suppliers of Chip Resistors[21]

Company	Size, in.	Resistance tolerance, %	Temperature coefficient, ppm/°C	Resistance values	Termination
ASC Microelectronics	0.050 × 0.050 × 0.012	1, 2, 5, 10, 20	−150	10 Ω to 1 MΩ	Au, Ag
	0.050 × 0.040 × 0.012	1, 2, 5, 10, 20	−350	10 Ω to 15 MΩ	Au, Ag
Airco-Speer:					
Thick film	0.050 × 0.050 × 0.020	1, 2, 5, 10	<100	1 Ω to 100 MΩ	Wraparound solderable Metal-ceramic monolithic substrate
	0.050 × 0.100 × 0.020				
	0.050 × 0.150 × 0.020				
Thin film	0.025 × 0.025	0.1, 1, 5	25	100 Ω to 500 kΩ	Au or Al pad
	0.050 × 0.050				
American Components	0.050 × 0.050 × 0.025	5, 10, 20	0 to ±150	5 Ω to 50 kΩ	Fired Ag, solder tinned
	0.100 × 0.050 × 0.025			10 Ω to 150 kΩ	
	0.300 × 0.050 × 0.025			25 Ω to 300 kΩ	
	0.400 × 0.100 × 0.025			50 Ω to 500 kΩ	
Angstrom Precision	0.040 × 0.040 × 0.025	±10	±200 to ±25	10 to 4,600 Ω	Solder pad
CTS	0.075 × 0.050 × 0.030	±10 standard (±5, 2, and 1 available)	±300 (lower available)	200 Ω to 350 kΩ	End-around (Au, Pt-Au, Ag)
	0.090 × 0.050 × 0.030				
Caddock Electronics	0.047 × 0.047 × 0.014	±1	±80	10 Ω to 400 kΩ	Metallized pad
	0.195 × 0.095 × 0.016	±1	±80	0.4 to 7.5 MΩ	
California Microcircuit	0.050 × 0.100	±20, ±15, ±10, ±5, ±1	200	10 Ω to 1 MΩ	Topside or endaround
CAL-R	0.050 × 0.040 × 0.012	20, 10, 5, 2, 1	±300, 150, ±50 available (matched resistors with tracking to ±10 ppm available) 100, 50, 25 (special: 10, 5, 0)	10 Ω to 5.6 MΩ	Pd-Ag, Pt-Au
	0.075 × 0.050 × 0.016				
	0.100 × 0.050 × 0.016				
Crownover Electronics	0.055 × 0.040 × 0.015	±10, 5, 2, 1	0 ±200	10 Ω to 1 MΩ	Au, Ag, or Pd combinations
	0.080 × 0.040 × 0.015				
	0.100 × 0.040 × 0.040				
Dale Electronics:					
Thick film	0.050 × 0.050 × 0.020	±1, ±2, ±5, ±10, ±20	0 ±25, 50, 100	100 Ω to 500 kΩ (10 Ω to 5 MΩ by special order)	60-40 Ag termination band; Au, Pd-Au, Pt-Au, or Ag also available
	0.050 × 0.075 × 0.020				
	0.050 × 0.100 × 0.020				
	0.050 × 0.150 × 0.020				
Thin film	0.050 × 0.050 × 0.017	±1, ±2, ±5, ±10, ±20	0 ±25, 50, 100	25 Ω to 100 kΩ (1 Ω to 1 MΩ by special order)	Au bonding pads with Au backing
	0.050 × 0.100 × 0.017				
	0.050 × 0.150 × 0.017				

Manufacturer	Size	Tolerance	TC	Resistance range	Termination
Dickson Electronics	0.025 × 0.025 × 0.010	±10 standard	±50	30 Ω to 3.9 MΩ	Sil-backed and Al pads
	0.035 × 0.035 × 0.010	(±5 and 1 available)	±50	3.9 to 10 MΩ	
	0.050 × 0.050 × 0.010	(±5 and 1 available)	±50	10 to 20 MΩ	
Electra/Midland	0.035 × 0.050 × 0.010	±20, ±10, ±5	±100	20 Ω to 1 MΩ	Au pads
	0.050 × 0.050 × 0.010	±2, ±1			
	0.050 × 0.075 × 0.010				
	0.050 × 0.100 × 0.010				
	0.050 × 0.150 × 0.010				
Eltec Instruments	0.100 × 0.050 × 0.012	±25	−0.3%/°C	100 to 100,000 MΩ	Au
Mepco	0.050 × 0.050 × 0.015	±10 (±5, 2 available)	±200	56 Ω to 100 kΩ	Pads
Monolithic Dielectrics	0.050 × 0.040 × 0.012	20, 10, 5, 2, 1	300, 150, 50	10 Ω to 5 MΩ	Ag Pt-Ag
	0.075 × 0.050 × 0.016				
Motorola	0.050 × 0.050 × 0.015	±10	±50	5 Ω to 10 kΩ	Beam lead
Sloan Microelectronics	0.030 × 0.030 × 0.008	±5, ±10	−20 to +300 (50 special)	10 Ω to 510 kΩ	Al
Struthers Electronics	0.025 × 0.075 × 0.010	±5 (0.1 available)	±150	1 to 250 Ω	Pads and tabs
	0.025 × 0.075 × 0.025 to				
	0.100 × 0.100 × 0.010				
TRW Inc: IRC Burlington	0.050 × 0.050 × 0.010	±0.1 to 10	0 to 100 (special ±25, ±50)	20 Ω to 40 kΩ	Cr-Au
	0.010 × 0.006 × 0.0005	±0.1 to 10	0 to 100 (special ±25, ±50)	20 Ω to 40 kΩ	Au beam lead
IRC Philadelphia	0.050 × 0.050 × 0.55	±5 standard	±100	0.025 Ω to 1 kΩ	0.031 OFHC* Cu leads
	0.375 × 0.250 × 0.062	±5 standard ±1 special	±200	0.025 Ω to 1 kΩ	Stud mount
Thick film	0.080 × 0.060 × 0.030	10 to 0.5	±200 (±50 special)	0.1 Ω to 22 MΩ	Wraparound, N-plated with solder tinning, or Au plating
Varadyne	0.050 × 0.050 × 0.025	±20, ±10, ±5,	±200	10 Ω to 330 kΩ	Pd or Ag
	0.075 × 0.050 × 0.025	±2, ±1		390 to 680 kΩ	
	0.015 × 0.050 × 0.025			820 kΩ to 1 MΩ	
	0.150 × 0.050 × 0.025			1.2 to 2 MΩ	
Vishay Resistor	30 mil sq.	to ±0.01	To 1	To 5,000 Ω max	Pads
	0.245 × 0.213	to ±0.001	To 1	To 75 kΩ max	Connection pads, cupron or manganin ribbon

*Oxygen-free high conductivity.

Chip-type resistors offer the advantage of small size in configurations that are compatible with attachment techniques normally employed in thick film hybrid assembly. They can be bridged over conductor runs, thus conserving substrate area. Chip resistors, particularly thick film chip resistors, are also ideal for breadboarding since their performance closely approximates that obtained in the final thick film design.

Encased resistors   Miniature encased resistors with leads can also be utilized in thick film networks to fulfill specific physical or performance needs.   Typical small encased resistors are the $\frac{1}{8}$-W carbon-composition and $\frac{1}{20}$-W fixed-film resistors. The military designation for these resistors is RC05 and RN50, respectively.   The physical outline and dimensions for these resistors are shown in Fig. 10a and b. Table 8 summarizes their performance characteristics.   Precision film resistors smaller

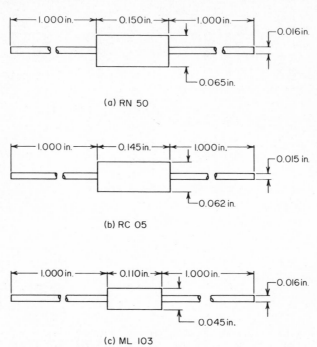

(a) RN 50

(b) RC 05

(c) ML 103

Fig. 10   Outline dimensions of typical miniature fixed resistors.[22-24]

than the RN50 are also available.[22]   Their physical dimensions are shown in Fig. 10c. These resistors are capable of 0.06 W at 125°C and can be obtained in values from 10 Ω to 250 KΩ with tolerances down to ±0.1 percent.   The temperature coefficient is ±50 ppm/°C from −15 to +105°C.

Encased resistors are readily available in a wide assortment of standard values. Table 9 lists several typical suppliers.   Miniature encased resistors can be positioned and attached to thick film circuits in a variety of ways.   They require very little substrate area but do increase the thickness of the overall network appreciably.

Capacitors

Although a wide variety of dielectric materials are presently used in the fabrication of electronic capacitors, only a few are generally accepted for use in thick film hybrid circuits.   Obviously, these are the materials which exhibit the highest capacitance-to-volume ratios.   Figure 11 lists in the order of increasing volumetric efficiency the seven most common capacitor types used in thick film circuits and illustrates the approximate capacitance range of each.   It is interesting to note that despite the small number of dielectric types utilized, the total capacitance range covered is quite respectable.

There are several basic factors which should be considered when selecting and specifying a capacitor for application in thick film circuits. First, the circuit function to be performed and the type and performance parameters of the capacitor required to fulfill the function must be considered. The type of capacitor depends directly on the application. Each capacitor type, because of the inherent properties of its dielec-

TABLE 8   Performance Characteristics of Typical Miniature Fixed Resistors

Performance characteristic	Carbon RC05 type	Metal film RN50 type
Resistance range............................	10 Ω–22 MΩ	10 Ω–0.1 MΩ
Tolerance, %..............................	5, 10	0.1, 1
Power capability, W.......................	$\frac{1}{8}$	1/20
Maximum voltage, V.......................	150	200
Maximum operating temperature, °C........	+70	+125
Temperature coefficient...................	5–13%, depending on value	±25 ppm/°C
Short-term stability, %...................	$\Delta R < 2.5$	$\Delta R < 0.25$
Long-term stability, %....................	$\Delta R < 8$	$\Delta R < 0.5$
Moisture resistance, %....................	$\Delta R < 15$	$\Delta R < 0.5$

SOURCE: Adapted from Refs. 23 and 24.

TABLE 9   Typical Suppliers of Miniature Fixed Resistors

Supplier	Resistor type			
	Carbon composition	Carbon film	Metal film	Other
Airco Speer.....................	x		x	
Allen-Bradley...................	x			
Caddock Electronics.............			x	
Corning Glass...................			x	
CTS Corporation................				Cermet, pellet
Dale Electronics................		x	x	
Electra/Midland................		x	x	
Mepco.........................		x	x	
Ohmite Manufacturing...........	x			
Pyrofilm Resistor...............			x	Pellet, tubular
Stackpole Carbon...............	x			
TRW..........................	x	x	x	

tric material, has certain characteristics which either enhance or restrict its ability to perform certain capacitive functions. Table 10 lists the various capacitive functions and indicates the capacitor types best suited for each. Once the capacitor type has been selected, the applicable performance parameters must be considered and assigned quantitative values. Table 11 lists the performance characteristics of prime interest.

Capacitor configuration and size are the second factors to consider. Film, chip, and encased types are all available for application in thick film circuits. The choice depends on the performance desired, the substrate area available, and the processing and assembly techniques to be used. In thick film microcircuits, capacitor size often becomes a major problem. Sometimes a compromise between size and performance must be made in order to accomplish a critical packaging problem.

Finally, cost and availability must be considered. Most capacitor manufacturers build and stock a variety of standard capacitor values and sizes which are readily

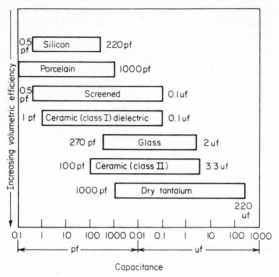

Fig. 11 Approximate value ranges of capacitor types commonly used in thick film hybrid circuits.

TABLE 10   Function Application of Thick Film Hybrid Capacitor Types

Circuit function	Capacitor type						
					Ceramic		
	Silicon	Screened	Glass	Porcelain	Class I	Class II	Tantalum
Bypass.........	x	x	x	x	x	x	x
Blocking.......		x			x	x	x
Coupling.......	x	x	x	x	x	x	x
Tuning........	x		x	x	x		
Timing........			x	x	x	x	x
Filtering.......	x	x	x	x	x	x	x
Energy discharge......						x	
Suppression....						x	x
Temperature compensating..					x		

SOURCE: Adapted from Ref. 115.

available and minimally priced. When capacitor values, tolerances, sizes, or other parameters outside the suppliers' published standards are required, cost and availability can become a critical factor. It is highly recommended that thick film designers contact capacitor suppliers early in the design cycle when requirements tend

to conflict with known practical standards. A list of typical manufacturers of miniature capacitor types is given in Table 12.

In the paragraphs that follow the seven capacitor types commonly used in thick
film hybrid circuits are discussed in detail. Information relative to the performance,
configuration, and size is provided with emphasis on the factors which will assist the
designer in proper selection and application of capacitors to thick film hybrid circuits.

Silicon capacitors[25] Silicon capacitors are a natural extension of the technology
utilized in the fabrication of silicon semiconductor devices. They are made by oxidizing the surface of a slice of silicon, producing a thin layer of silicon dioxide, a
dielectric. The plates of the capacitor are the low-resistivity silicon material, which
is usually gold-backed, and the aluminum pads diffused on the surface of the silicon
dioxide. The value of the capacitor is determined by the size of the chip, the thickness of the dielectric layer, and the area of the aluminum pad. Silicon capacitors are
readily available in chip form with values and performance characteristics as shown
in Table 13. Size and geometry details for several typical single and dual silicon

**TABLE 11   Checklist
of Critical Capacitor
Performance
Characteristics**

Value
Tolerance
Temperature coefficient
Dissipation factor
Dc working voltage
Dc voltage coefficient
Ac voltage coefficient
Frequency response
Insulation resistance
Series resistance
Dielectric absorption
Aging
Q factor

capacitor chips are shown in Fig. 12. Beam-lead versions of these capacitors for
microwave applications are also available.

Silicon capacitors offer the advantage of smaller physical size, better temperature
stability, and lower leakage currents than their ceramic equivalents. Disadvantages
include relatively high series resistance, especially at high frequencies, and relatively
broad tolerances in capacitance value. Silicon capacitors have widest application in
bypass, coupling, tuning, and filtering applications.

Screen-printed capacitors Screen-printed thick film capacitors are formed by
silk-screening a dielectric paste over a base electrode pattern and then screening and
cofiring a second electrode pattern on top of the dielectric. A top and cross-sectional
view of the typical construction technique is shown in Fig. 13. Usually the dielectric
is double-printed in order to reduce the chances of pinholes and short circuits. The
size of a screen-printed capacitor is determined from the standard equation for
parallel-plate capacitors

$$C = \frac{0.225KA}{t} \tag{1}$$

where C = capacitance value, pF
 K = relative dielectric constant of material
 A = opposed plate area, in.^{2}
 t = thickness of dielectric, in.

Although the screen printing process is an economical and practical method of obtaining capacitors in thick film circuits, there are certain limitations and problems. First,

it is extremely difficult to fabricate capacitors to close tolerances. The capacitance value shown in Eq. (1) is a function of dielectric thickness and dielectric constant. Both these factors are difficult to control. Unless extreme care is taken during screen fabrication and the screening process itself, the thickness of the dielectric can vary appreciably, particularly when the dielectric is double-screened. The dielectric constant of available materials varies with formulation and firing profile, time, and tem-

TABLE 12 Typical Suppliers of Chip and Miniature Capacitors

Supplier	Capacitor type						
	Silicon	Porcelain	Materials for screened caps	Class I ceramic	Glass	Class II ceramic	Tantalum
Aerovox....................				x		x	
Alloys Unlimited..............			x				
American Components.........				x		x	
American Technical Ceramics...		x		x		x	
CAL-R.....................				x		x	
Components, Inc..............							x
Corning Glass...............				x	x	x	
Dionics, Inc..................	x						
Du Pont....................			x				
Electro Materials Corp........			x	x		x	
Mallory.....................							x
Monolithic Dielectrics........				x		x	
Motorola....................	x						
San Fernando Electric........				x		x	
Sprague....................				x		x	x
Transitron Electronics........							x
Union Carbide (Kemet Div.)...				x		x	x
USCC-Centralab.............				x		x	
Varadyne...................				x		x	
Vitramon...................		x		x		x	

TABLE 13 Typical Performance Characteristics of Silicon Capacitors[25]

Capacitance range, pF.................	0.5 to 220
Tolerance, %........................	±10
Working voltage, V....................	25 to 75
Temperature range, °C.................	−65 to +200
Dissipation factor, %..................	0.01 to 0.025
Temperature coefficient, ppm/°C........	+35 ± 15
Size, mils............................	20 × 20 to 45 × 45

perature. Figure 14 illustrates typical effects of peak firing time and temperature on the dielectric constant of a commercially available K1200 dielectric composition. As a result of these variables, capacitors with tolerances less than ±20 percent are difficult to achieve with high yields. The capacitors can be adjusted to tighter tolerance limits by air abrasion or similar means, but this technique is costly and potentially detrimental to the performance of the capacitor under environmental conditions.

The second major limitation of screen-printed capacitors is size. Except for capacitor values of several thousand picofarads or lower, screen-printed capacitors consume considerable substrate area. For example, K1200 compositions require 1 in.^{2} of substrate area to obtain capacitors in the 0.05- to 0.1-μF range. With present dielectric compositions a practical capacitance-value limit from a size standpoint is approximately 5,000 pF.

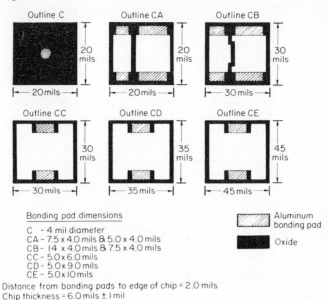

Bonding pad dimensions

C - 4 mil diameter
CA - 7.5 x 4.0 mils & 5.0 x 4.0 mils
CB - 14 x 4.0 mils & 7.5 x 4.0 mils
CC - 5.0 x 6.0 mils
CD - 5.0 x 9.0 mils
CE - 5.0 x 10 mils

Distance from bonding pads to edge of chip = 2.0 mils
Chip thickness = 6.0 mils ± 1 mil

Fig. 12 Configuration and size of typical silicon chip capacitors.[25]

Despite these problems, screen-printed capacitors have definite practical advantages in bypass, coupling, and other applications where low-value wide-tolerance capacitors are acceptable.

Also, since any quantity of screen-printed capacitors can be applied to a substrate simultaneously, they offer a definite economic advantage when several low-value capacitors are required on a single substrate.

A general indication of the range of performance characteristics obtained with screen-printed capacitors is given in Table 14. Since these characteristics are so highly dependent on the dielectric material and process techniques used, it is difficult to be more specific. An example of the type of temperature-coefficient characteristic attainable with the higher-dielectric-constant materials is shown in Fig. 15.

Porcelain capacitors[27] Porcelain capacitors are excellent for high-frequency high-power applications through the UHF range. Fabricated from extremely dense, low-loss porcelain material, these capacitors exhibit higher Q's and lower series resistance than any other equivalent type of capacitor. Low-loss procelain capacitors also have low equivalent noise resistance and high insulation resistance. They are available in chip form with a capacitance range of 0.1 to 1,000 pF. A listing of their typical performance characteristics is provided in Table 15.

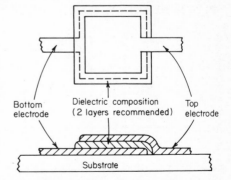

Fig. 13 Typical construction of a screen-printed capacitor.

Because the low-loss porcelain capacitor chips are self-encapsulated by their own highly dense porcelain dielectric molecularly fused into a homogeneous, rugged block, they are impervious to moisture, solvents, shock, vibration, and high altitude. Their low series resistance, high Q, high stability, and high insulation resistance make them ideally suited for tuning, timing, bypassing, and coupling applications when extremely low inductance or high stability is required and for coupling, bypassing, and rf tank circuits in low- and high-power UHF applications.

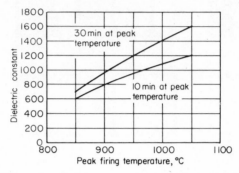

Fig. 14 Typical effects of peak firing temperature and time at peak temperature on the dielectric constant of screen-printed capacitors (Du Pont composition Dp-8289).[26]

TABLE 14   Typical Performance Ranges of Screen-printed Capacitors

Dielectric	Capacitance, pF/in.²	Tolerance, %	Working voltage, V	Dissipation factor, %	Insulation resistance, Ω	Temperature coefficient
Low K.....	2,000–10,000	3 trimmed 20 untrimmed	50–100	0.5–1.5	>10¹⁰	30–180 ppm/°C
High K.....	25,000–160,000	3 trimmed 20 untrimmed	50–100	1–8	>10⁹	−4% +8%

Ceramic capacitors   Ceramic capacitors, because of their ruggedness, wide capacitance range, variety of performance characteristics, high volumetric efficiencies, and comparatively low cost, are used more extensively in thick film hybrid circuits than any of the other capacitor types. Ceramic capacitors are divided into two general dielectric classes, each having certain distinct characteristics. For optimum selection and application, it is advantageous for the design engineer to know and understand the properties and characteristics of these two classes.

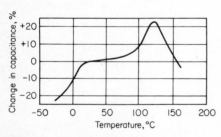

Fig. 15 Typical change in capacitance versus temperature of screen-printed capacitors (Du Pont composition Dp-8289).[26]

Class I dielectrics offer superior stability. Their general characteristics include relatively low dielectric constants (5 to 800), predictable and relatively linear capacitance changes with temperature, low dissipation factor, stability with applied ac and dc voltages, high dielectric strength, high isolation resistance, and no measurable aging effects. Class I materials are identified by an alphameric code, which denotes the direction and slope of the capacitor change with increasing temperature. For example, an N100 material has a negative slope of 100 ppm/°C. Similarly a P150 material has a positive slope of 150 ppm/°C. A material designated as NP0 has a nominal slope of zero, with a guaranteed temperature coefficient of 0 ± 30 ppm/°C. Capacitors made from class I dielectric find wide application in pulse, timing, and tuned circuits, where stability is a prime requirement. Also because of the nearly linear and predictable temperature characteristics they are often used for temperature compensation. The primary disadvantage of this dielectric material is the low range of dielectric constant, which prohibits the fabrication of large-value capacitors within practical size and cost limits.

Class II dielectrics are the general-purpose type which have high dielectric constants (500 to 10,000) and exhibit ferroelectric properties in varying degrees. This latter property means that the class II materials do not vary linearly with temperature, are affected by applied ac and dc voltage, have higher dissipation factors than class I materials, and exhibit measurable aging rates. Class II materials are identified by their dielectric constant, e.g., K1200, K2000, K5000, etc. The significant advantage of class II materials is their high volumetric efficiency and economy, providing high-value capacitors of small size and low cost.

TABLE 15   Typical Performance Characteristics of Porcelain Chip Capacitors[27]

Capacitance range, pF	0.1 to 1,000
Tolerance, %	1 to 20
Working voltage, V	50 to 1,000
Temperature range, °C	−55 to +125
Dissipation factor, %	0.01
Temperature coefficient ppm/°C	+90 ± 20
Size, mils	50 × 50 and 110 × 110

In ceramic capacitors the change in performance parameters with temperature is a result of a change in the crystalline structure of the ceramic. The point at which the ceramic exhibits the highest dielectric constant is known as the *Curie point.* Below this point the crystalline structure is tetragonal and has ferroelectric properties. Above the Curie point a cubic structure occurs, and the ceramic exhibits nonferroelectric properties. The Curie point for pure barium titanate ceramic is about 125°C; however, most commercial dielectric ceramics contain modifiers which tend to shift the Curie point to a lower temperature. Class I materials also have Curie points, but the chemical composition of the dielectric material is such that this point is outside the temperature range of interest. The effect of this phenomenon on the temperature characteristics of several common ceramic dielectric types is illustrated in Figs. 16 and 17. In general, class I dielectrics are quite stable and exhibit only small, linear capacitance-value changes with temperature. Class II dielectrics on the other hand have unstable, nonlinear temperature characteristics which tend to become more pronounced with increasing dielectric constant. Often the temperature characteristics of ceramic capacitors will be stated in terms of standard EIA or military designators. The significance of these standard designators is reviewed in Tables 16 and 17.

The insulation resistance and dissipation factor of ceramic capacitors also change appreciably with temperature. Figures 18 to 20 illustrate typical changes for several common dielectric types. In general the insulation resistance and dissipation factor decrease with increasing temperature. Capacitors made with very thin dielectrics, such as monolithic chip capacitors, exhibit greater changes in insulation resistance with temperature than illustrated in the typical curve of Fig. 18.

Aging is an important factor to consider when utilizing class II high-*K* ceramic

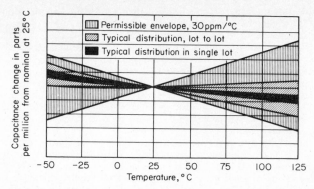

Fig. 16 Typical capacitance change versus temperature for NPO ceramic dielectric capacitors.[28]

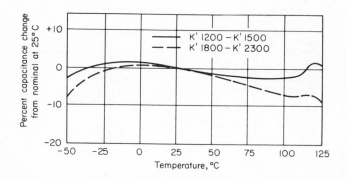

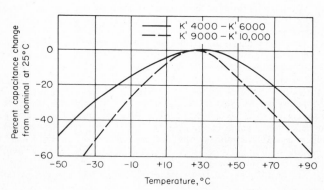

Fig. 17 Typical capacitance change versus temperature for various class II ceramic dielectrics.[28]

capacitors. Once the Curie point is reached, aging begins to take place. This phenomenon is associated with a delay in the change from a cubic structure back to the tetragonal form. The result is a net increase in capacitance value of about 5 percent after the capacitor returns to room temperature. From this point, the capacitor returns to its stable value at a rate of 1 to 2 percent per hour-decade; i.e., after 1 h, the capacitor value will drop 1 to 2 percent. In an additional 10 h, it will drop

TABLE 16   EIA Standard Designators for Ceramic-Capacitor Temperature Characteristics*, [29]

Temperature range for characteristic determination, °C	Letter symbol	Maximum capacitance change over temperature range, %	Letter symbol
+10 to +85	Z5	±1.0	A
−30 to +85	Y5	±1.5	B
−55 to +85	X5	±2.2	C
		±3.3	D
		±4.7	E
		±7.5	F
		±10.0	P
		±15.0	R
		±22.0	S
		+22, −33	T
		+22, −56	U
		+22, −82	V

* EXAMPLE: Capacitor with X5U characteristics has +22%, −56% capacitance change in temperature range of −55 to +85°C.

TABLE 17   Military Standard Designators for Ceramic-Capacitor Temperature and Voltage-Temperature Characteristics*, [30]

Operating-temperature range		Capacitance change with reference to 25°C		
Range, °C	Designator	No voltage applied, %	Dc voltage applied, %	Designator
−55 to +85	A	+15, −15	+15, −40	R
−55 to +125	B	+22, −56	+22, −66	W
−55 to +150	C	+15, −15	+15, −25	X
		+30, −70	+30, −80	Y
		+20, −20	+20, −30	Z

* EXAMPLE: Capacitor with BX characteristic has maximum of ±15% change in capacitance over temperature range of −55 to +125°C.

another 1 to 2 percent, and so on until the decades become so long as to render the capacitance change negligible. Consequently it is not advisable to measure capacitors of this type for at least a day after exposure to high temperature such as those encountered in soldering.

Perhaps the most notable effect resulting from the ferroelectric properties of class II materials is the marked change in capacitance with applied dc and ac voltages. For dc voltages the amount of capacitance change is a function of dielectric thickness and applied voltage in volts per mil. Typical curves of this characteristic for common dielectrics are shown in Fig. 21, from which it is apparent that NP0 material does not change and class II type dielectrics tend to have a positive change at low dc voltages and a negative change as the dc stress is increased. Figure 22 depicts typical capac-

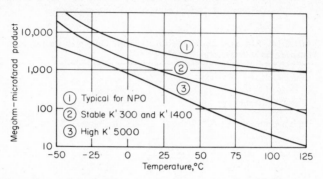

Fig. 18 Typical insulation resistance versus temperature for various ceramic dielectrics.[28]

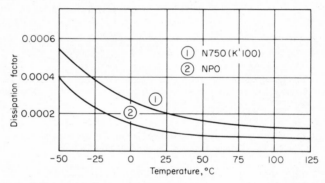

Fig. 19 Typical dissipation factor versus temperature for class I ceramic dielectrics.[28]

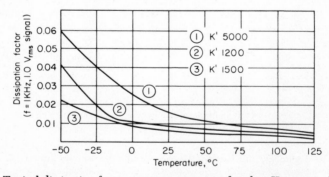

Fig. 20 Typical dissipation factor versus temperature for class II ceramic dielectrics.[28]

itance changes with ac voltage applied. The capacitance changes with applied ac voltage are much more pronounced than those obtained with dc bias and in the opposite direction. Again the NP0 material exhibits very little change.

Ceramic capacitors also exhibit value changes with frequency. Typically, the capacitance will decrease with increasing frequency. The greatest changes are observed with the higher dielectric materials. Figure 23 illustrates the capacitance change with frequency for several common dielectric types.

It should be noted that the curves presented here represent typical performance

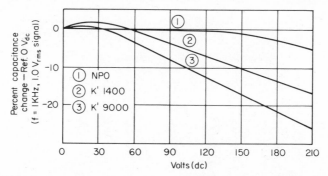

Fig. 21 Typical capacitance change versus applied dc voltage for various ceramic dielectrics.[28]

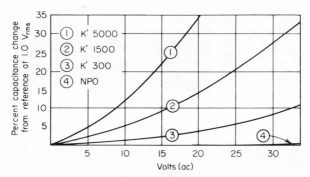

Fig. 22 Typical capacitance change versus applied ac voltage for various ceramic dielectrics.[28]

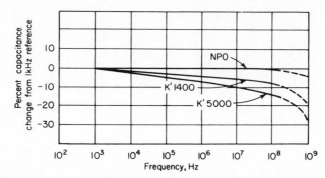

Fig. 23 Typical capacitance change versus frequency for various ceramic dielectrics.[28]

data and are intended only to illustrate the relative degree of change in the various dielectric types and the relative relationship between them. Since each capacitor vendor has different dielectric formulations, these characteristics can differ appreciably from vendor to vendor. The designer should consult the vendor specification sheets to obtain more specific data on these or other characteristics, such as changes in dissipation factor with frequency and applied dc and ac voltages.

Ceramic capacitors can be obtained in ranges from 1 pF to 3.3 μF. Class I materials are available in the 1-pF to 0.1-μF range and class II in the range between 100 pF and 3.3 μF. Tolerance values of $\pm$5 to $\pm$20 percent are common, and values

Fig. 24 Thick film network with through-plate capacitors. (*Centralab Division, Globe-Union Inc.*)

Fig. 25 Thick film networks with encapsulated and unencapsulated ceramic disk capacitors. (*Centralab Division, Globe-Union Inc.*)

down to $\pm$1 percent are possible with some capacitor types and values. Ceramic capacitors are available in a wide variety of package types and configurations. The most basic form of ceramic capacitor in thick film networks is shown in Fig. 24. Here a complete *RC* network is fabricated on a substrate made from high-*K* ceramic material. The substrate thus serves as the dielectric, and the capacitors are formed by screening conductor patterns of the proper size on either side. Resistors are then screen-printed on the same substrate to complete the network. Encapsulated and unencapsulated disk capacitors (Fig. 25) have also been used. This type of capacitor is still widely used in fabricating low-cost thick film networks for the consumer market. However, they require a considerable amount of substrate area and, where possible, are being replaced by the more efficient monolithic-type ceramic capacitor.

Monolithic or multilayer ceramic capacitors, thanks to their high volumetric efficiency and resultant small size, are the most popular form of ceramic capacitor for application in thick film networks. They are constructed by stacking and cofiring thin sheets (down to 1 mil) of metallized green ceramic dielectric. The metallization patterns are offset on the ceramic and are alternately exposed at each end of the ceramic chip, resulting in the construction shown in Fig. 26. The ends are coated with a conductor material which ties the individual layers together and forms a group of parallel connected capacitors within the chip. This accounts for the high volumetric efficiency. Monolithic ceramic capacitors can be obtained in chip form suitable for direct application to thick film conductor patterns. They are available in a wide variety of values, tolerances, sizes, and configurations.

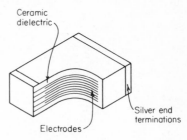

Fig. 26 Pictorial representation of typical monolithic ceramic chip-capacitor construction.

End terminations can be obtained in silver, gold, silver-palladium, gold-palladium, or gold-platinum. Figure 27 shows the relative sizes and shapes of typical ceramic chip capacitors. At present there are no uniform standard sizes established for chip capacitors. Each manufacturer supplies chip capacitors to his own internal size standards.

Fig. 27 Ceramic chip capacitors. (*USCC-Centralab.*)

Thick film designers contemplating the use of chip capacitors from several sources should carefully study the size differences and plan their layout accordingly. Figures 28 and 29 illustrate the typical sizes and capacitance ranges available for chip capacitors constructed from NP0 and K1200 dielectric materials.

Many miniature encased ceramic capacitors are also available for attachment to thick film circuits. Figure 30 shows some popular styles and sizes. Dimensional data and approximate capacitance ranges per package style are provided in Table 18.

Ceramic capacitors of one type or another can be used to fulfill every capacitive circuit function. In addition to bypassing, coupling, and filtering, the stable NP0 capacitors are ideal for critical tuning and timing applications. Other class I dielectrics with predictable linear positive or negative temperature coefficients are excellent temperature-compensating elements. Class II dielectric capacitors can be used for bypassing, blocking, coupling, filtering, energy storage or discharge, suppression, and noncritical time-constant applications.

Glass capacitors[32, 33]   Miniature glass capacitors for hybrid-circuit applications are fabricated from special high-dielectric-constant glass compositions, which provide the same high reliability and consistent performance as the older glass dielectrics but with size and performance characteristics equal to or exceeding those of equivalent ceramic capacitors. High-K glass capacitors can be obtained in several different types of performance characteristics. Typical plots of the performance characteristics of major concern for three types are shown in Figs. 31 to 36. It can be noted that the magnitude and slope of these plots are quite similar to those of the class II ceramic capacitors shown in Figs. 17 to 23. However, since glass does not exhibit ferroelectric properties, as class II ceramic dielectrics do, glass capacitors have minimal dc and ac voltage coefficients and no equivalent aging characteristics.

High-K glass capacitors are made by fusing together layers of ribbon glass and plate-electrode material to form a multilayer, monolithic structure similar to that of a ceramic capacitor. High temperature is then applied to produce microcrystals interspersed throughout the glass, giving the glass its high dielectric constant and resulting high volumetric efficiency. The ends are then metallized, and the resulting capacitor can be left in chip form or provided with leads and encapsulated. A summary of the available values, sizes, and performance characteristics of chip-type miniature glass capacitors is provided in Table 19. Glass capacitors with similar values and performance characteristics can be obtained in encased axial-lead forms ranging in size from 0.090 in. diam. and 0.160 in. long to 0.140 in. diam. and 0.250 in. long.

Glass capacitors are well suited for military and commercial applications, where high reliability and stability are requirements. The larger-value capacitors are ideal for filtering, bypassing, and coupling. The smaller, more stable capacitors can be used in the more critical functional areas, such as timing and tuning. In general, glass capacitors can provide performance equivalent to or better than ceramic capacitors, but they are not as cost-competitive.

Solid-tantalum capacitors   Capacitors made from sintered tantalum and a solid electrolyte system have very high volumetric efficiencies and therefore provide high capacitance values in sizes compatible with the miniaturization achieved with thick film technology. In fact, they are the only capacitor type at present that can physically and economically provide miniature capacitors over 1 μF in value. Because of their method of construction and the fact that they are an electrolyte type of capacitor, solid-tantalum capacitors have certain characteristics different from those of the previously described types. A fuller appreciation of these characteristics can best be obtained by first considering the internal structure of these capacitors.

Solid-tantalum capacitors are made by pressing high-purity tantalum powders into a small slug or pellet, which eventually becomes the cathode of the capacitor. A small tantalum wire is pressed into the pellet to form the anode termination. The slug is then sintered in a vacuum at high temperatures, yielding a firm yet very porous structure. It is this porosity that provides the high cathode plate area and accounts for the high volumetric efficiency. The dielectric is formed by anodizing the slug in an acid bath, forming tantalum pentoxide in the pores as well as on the surface. The thickness of the dielectric is a function of the voltage applied during

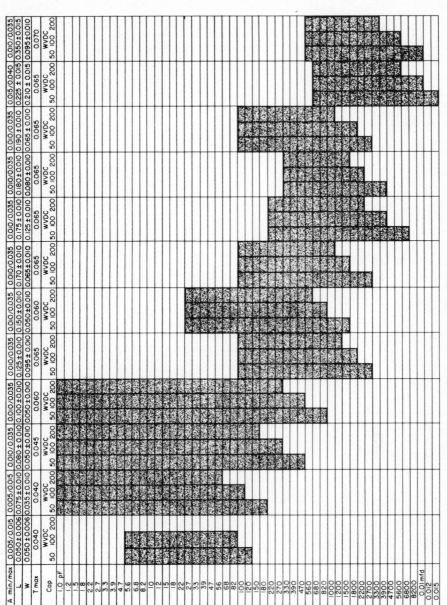

Fig. 28 Typical size and value ranges of NP0 monolithic ceramic chip capacitors. [21]

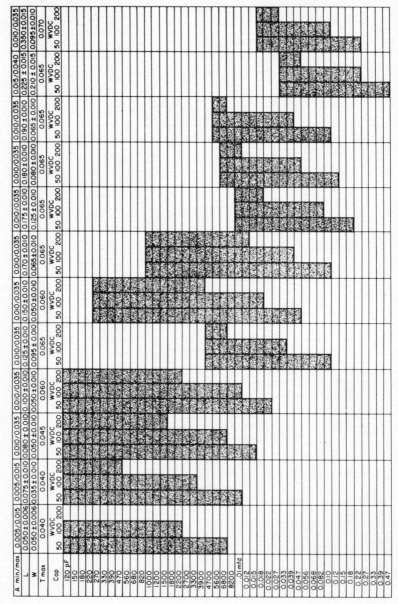

Fig. 29 Typical size and value ranges of K1200 monolithic ceramic chip capacitors.[31]

anodizing and can be accurately controlled. A typical solid-tantalum capacitor has a dielectric thickness of 8 μin. The approximate dielectric properties of the tantalum pentoxide layer are a dielectric strength of 10,000 V/mil and a dielectric constant of 25.

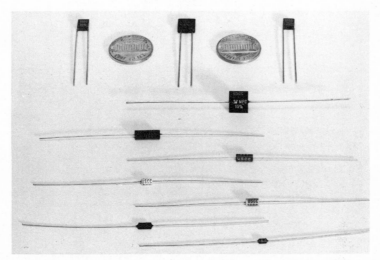

Fig. 30   Miniature encased ceramic capacitors.   (*USCC-Centralab.*)

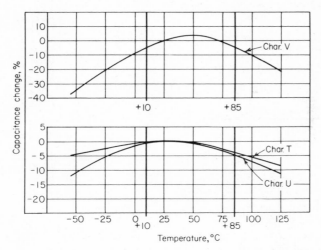

Fig. 31   Typical capacitance change versus temperature for high-K glass capacitors.[32]

Next the slug is dipped into manganous nitrate solution and pyrolytically converted to form manganese dioxide, which serves as the electrolyte and cathode electrode. In addition, the manganese dioxide is a source of oxygen for oxidizing and healing faults which occasionally occur in the dielectric film. Finally the cathode termination is completed by applying carbon and silver paint over the surface of the com-

TABLE 18   Characteristics of Typical Miniature Encased Ceramic Capacitors[31]

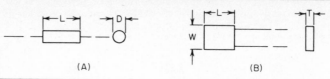

(A)   (B)

Case style	Size, in.				Electrical characteristics			
					NP0		K1200	
	D	L	W	T	50 V	100 V	50 V	100 V
A	0.100	0.250	...	...	0.5–820 pF	0.5–470 pF		
	0.140	0.400	...	...	1,000–3,900 pF	560–2,200 pF		
	0.210	0.515	...	...	4,700 pF–0.018 μF	2,700 pF–0.010 μF		
B		0.200	0.200	0.100	5–1,500 pF	5–1,000 pF	1 pF–0.10 μF	1 pF–0.010 μF
		0.300	0.300	0.100	1,800–5,100 pF	1,200–3,300 pF	0.12–0.50 μF	0.012–0.15 μF
		0.300	0.300	0.150	5,600 pF–0.018 μF	3,900 pF–0.015 μF	0.56–1.0 μF	0.18–0.27 μF
A	0.080	0.150	...	...	...	...	1.0 pF–0.012 μF	
	0.095	0.185	...	...	...	...	0.015–0.027 μF	1.0 pF–0.012 μF
	0.095	0.250	...	...	...	...	0.033–0.056 μF	0.015–0.022 μF
	0.125	0.250	...	...	...	...	0.068–0.12 μF	0.027–0.056 μF
	0.200	0.500	...	...	...	...	0.15–0.25 μF	

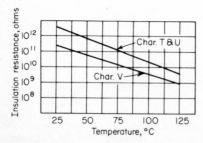

Fig. 32 Typical insulation resistance versus temperature for high-K glass capacitors.[32]

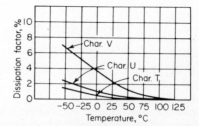

Fig. 33 Typical dissipation factor versus temperature for high-K glass capacitors.[32]

pleted slug.   Figure 37 shows an enlarged cross section of a typical solid-tantalum capacitor slug.

The electrical properties of significant interest in solid-tantalum capacitors are dc leakage current, equivalent series resistance, and maximum working voltage.   Dc leakage is defined as the current flowing through a capacitor after a specified time with a dc voltage continuously applied.   Due to impurities in the tantalum and

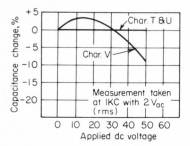

Fig. 34 Typical capacitance change-versus applied dc voltage for high-K glass capacitors.[32]

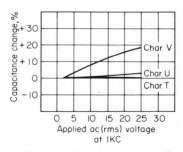

Fig. 35 Typical capacitance change versus applied ac voltage for high-K glass capacitors.[32]

slight imperfections in the extremely thin and irregular dielectric film, solid-tantalum capacitors have leakage paths and consequently exhibit dc leakage current.   At 25°C and with rated voltage applied, the leakage current is quite small.   Typically it is in the range of 1 to 10 μA, depending on capacitor size and value.   However, it is both temperature- and voltage-dependent, increasing with an increase in either temperature or applied voltage.   Figures 38 and 39 show typical leakage current at various temperatures and voltages normalized to the actual value at 25°C and rated dc working voltage.

The minute leakage paths which are the source of the dc leakage current are also believed to be the cause of another phenomenon occasionally encountered in solid-tantalum capacitors, namely, the catastrophic failure of the capacitor while operating in a low-impedance circuit.   It is conjectured that localized heating takes place at the points of high concentration of dielectric imperfections when the capacitor is highly stressed, causing further destruction of the dielectric film.   An avalanche condition results and completely destroys the capacitor.   The self-healing ability of the manganese dioxide electrolyte helps reduce the chances of this type of failure.   Also the failure does not occur when the capacitors are used in a high-impedance circuit, where the current is limited.

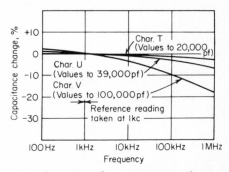

Fig. 36 Typical capacitance change versus frequency for high-K glass capacitors.[32]

The equivalent series resistance (ESR) of the solid-tantalum capacitors is a lumped total of several internal resistances, shown in the equivalent circuit of Fig. 40.   The lead, contact, and electrolyte resistances are independent of frequency, while the dielectric resistance R_2 is dependent on frequency and inversely proportional to it.   Although the equivalent series resistance is a basic parameter, it is not usually measured directly.   Generally the dissipation factor, which is the ratio of the equivalent series resistance to the capacitive reactance at a specific frequency, is used to express this parameter.   Values of 3 to 10 percent are typical for solid-tantalum

TABLE 19 Typical Characteristics of Miniature Glass-Chip Capacitors[33]

Working voltage, V	Size, mils	Capacitance range, pF	Tolerance, %	Temperature coefficient, %	Temperature voltage coefficient, %	Dissipation factor, %	Temperature range, °C
100	95 × 55 × 60 185 × 60 × 60 305 × 85 × 60	10 to 4,700 5,600 to 10,000 12,000 to 51,000	±10, ±20	±15	+15, −25	2.5	−55 to +125
	305 × 85 × 90	51,000 to 100,000	±10, ±20	±15	+15, −40	2.5	−55 to +125
50	95 × 55 × 60 185 × 60 × 60 305 × 85 × 90	5,600 to 10,000 12,000 to 22,000 51,000 to 100,000	±10, ±20	±15	+15, −25	2.5	−55 to +125
	185 × 60 × 60 185 × 80 × 85	270 to 10,000 12,000 to 20,000	±5, ±10, ±20	+2, −10	+2, −10	1.0	−55 to +125
	185 × 60 × 60 185 × 80 × 85	12,000 to 20,000 22,000 to 39,000	±10, ±20	+2, −15	+2, −15	1.5	−55 to +125
	185 × 60 × 60 185 × 80 × 85	22,000 to 51,000 56,000 to 100,000	±10, ±20	+20, −45	+20, −50	3.0	−55 to +125
25	95 × 55 × 60 185 × 60 × 60 185 × 80 × 85 305 × 85 × 90	10 to 27,000 8,200 to 100,000 51,000 × 180,000 100,000 to 330,000	±10, ±20 +80, −20	+22, −66	+22, −66	3.0	−55 to +125

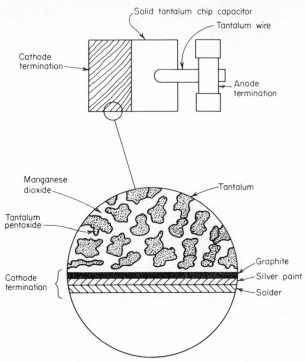

Fig. 37 Typical configuration and structure of a solid-tantalum chip capacitor.

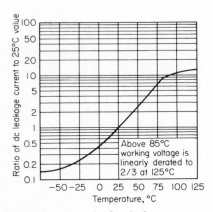

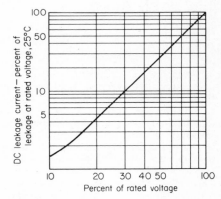

Fig. 38 Typical dc leakage current versus temperature for solid-tantalum capacitors.[34]

Fig. 39 Typical dc leakage current versus applied voltage for solid-tantalum capacitors.[34]

capacitors measured at 120 Hz. The dissipation factor increases with frequency, as shown in Fig. 41.

The dc working voltage (WVDC) is the maximum voltage that can be continuously applied to a capacitor. Because the dielectric in tantalum capacitor is quite thin, excess voltage levels will greatly affect the life of the capacitor. When ac ripple

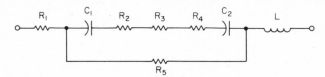

R_1 = Resistance of capacitor terminals and electrodes
R_2 = Represents the ac losses in the dielectric film
R_3 = Resistance of the electrolyte or electrolyte–spacer combination
R_4 = Contact resistance between the metal cathode and electrolyte
R_5 = Represents the dc leakage current
C_1 = Capacitance at the anode observed across the dielectric film
C_2 = Capacitive effect observed at the electrolyte–metal cathode interface
L = Magnetic inductance of terminals, electrodes, and capacitor geometry

Fig. 40 Equivalent circuit of an electrolytic capacitor.[35]

is present, the sum of the dc bias and peak ac ripple should not exceed the WVDC. The maximum dc working voltage of a capacitor is highly dependent on capacitor value. For large-value capacitors the dielectric is maintained quite thin, and consequently the breakdown voltage is reduced. For miniature solid-tantalum capacitors, the WVDC ranges from 50 V for 1,000-pF capacitors to 3 V for the large-value capacitors.

With respect to the other common performance characteristics, such as temperature coefficient and frequency response, solid-tantalum capacitors are comparable with K1200 ceramic capacitors.

Miniature solid-tantalum capacitors are available in a range from 1,000 pF to over 200 μF in both polar and nonpolar form. A polar type is more desirable since for a given capacitance and voltage rating it will be approximately one-half the size, weight, and cost of a nonpolar type. Nonpolar types are required only when the voltage across the capacitor will be both positive and negative. Solid-tantalum capacitors can be obtained in chip form, as shown in Fig. 42, or in a wide variety of encased package configurations. Approximate size, value, and performance re-

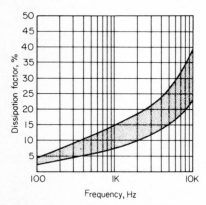

Fig. 41 Typical dissipation-factor range versus frequency for solid-tantalum capacitors.[34]

lationships of typical encased solid-tantalum capacitors are shown in Table 20.

Because of the fragile nature of the wire anode terminator and the difficulties encountered in bonding the silver cathode terminator to a metallized substrate, solid-tantalum chip capacitors have not been widely used in thick film hybrid circuits. Specifically, the major problem involves the deterioration of the silver cathode electrode during reflow soldering of the chip to the substrate. Poor bond strength, increased contact resistance, or even complete loss of the electrode are typical failure modes.

Recent developments, in which copper is flame-sprayed onto the surface of the tantalum slug, forming the cathode electrode, have provided a tantalum chip which can withstand the most rigorous of solder-reflow methods.[37]   In fact, these chips are capable of continuous operation at 175°C, compared to 125°C for capacitors of typical construction.   Regardless of the type used, solid-tantalum capacitors, because of their limited temperature range, should be thoroughly evaluated before they are specified and utilized in an actual product application.   This evaluation should include the subjection of the parts to all the process steps and temperatures that will be encountered during normal circuit fabrication.

The broad range and high capacitance values of solid-tantalum capacitors suggest many possible applications.   They can be used for bypassing, coupling, filtering, and

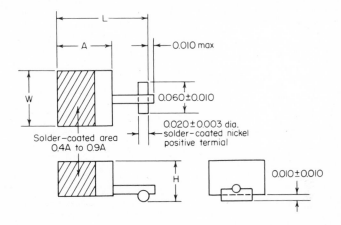

Case size	Capacitance value Range, μF	Dimensions, in.			
		W	L	H max	A
A	0.10 – 4.7	0.060±0.010	0.155±0.010	0.060	0.090±0.010
B	0.33 – 10	0.105±0.010	0.180±0.010	0.060	0.095±0.010
C	0.68 – 33	0.145±0.015	0.210±0.015	0.080	0.140±0.015
D	2.2 – 68	0.150±0.015	0.265±0.015	0.110	0.190±0.015
E	4.7 – 100	0.150±0.015	0.265±0.015	0.150	0.195±0.015

Fig. 42 Outline drawing and capacitance range of typical solid-tantalum chip capacitors.[36]

timing.   Power-supply filtering of rectified sine waves from 5 Hz to 1 kHz is a common application.   Solid-tantalum capacitors can be used in nonprecision time-constant circuits, but circuit design should take into consideration the dc leakage current and equivalent series resistance.   They are not recommended for long time intervals or high-level ac applications, where excessive internal heating is likely to occur.   Care should also be taken to assure that the applied voltage including transients does not exceed the maximum voltage ratings.   As with all components, the manufacturers' data sheets should be consulted for details concerning specific maximum ratings and proper derating information.

Inductive Components

In the past, inductive components were excluded from hybrid circuits because their physical sizes were not compatible with the miniaturization achieved with thick film techniques.   This is no longer necessary, however.   Improvements in magnetic-core materials and wire-winding techniques have made possible a significant reduction in the size of inductive components. Miniature inductors and transformers smaller than some chip-type capacitors are now available in a variety of physical configurations and value ranges. Even tiny tunable inductors can be obtained. These components are ideal for thick film hybrid applications.

For high-frequency circuits, where very small value inductors are required, the inductive function can also be achieved with thick film conductive material in the form of flat spiral patterns. This technique has been used for years in the printed-circuit industry and can be applied to thick film designs with equal success.

TABLE 20   Characteristics of Typical Miniature Encased Solid-Tantalum Capacitors

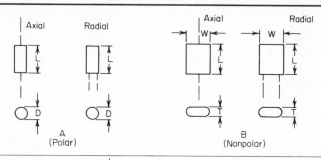

Case style	Size, in.				Electrical characteristics				
	L	D	W	T	Capacitance range, μF	Tolerance, %	Working voltage, V	DC leakage current, μA	Maximum dissipation factor, %
A	0.125	0.070	...	...	0.0010–2.2	+40, −20	20–2	0.5–0.5	6–10
	0.160	0.070	...	...	0.0010–4.7	+40, −20	50–2	0.5–0.5	6–10
	0.200	0.080	...	...	0.33–10	+40, −20	35–2	0.5–0.5	6–10
	0.225	0.100	...	...	0.68–22	+40, −20	35–2	1.0–1.0	6–10
B	0.175* 0.135†	...	0.120	0.070	0.0010–1.0	+40, −20	20–2	0.5–0.5	6–10
	0.210* 0.160†	...	0.120	0.070	0.0010–2.2	+40, −20	50–2	0.5–0.5	6–10
	0.240* 0.200†	...	0.140	0.080	0.15–4.7	+40, −20	35–2	0.5–0.5	6–10
	0.275* 0.235†	...	0.190	0.105	0.33–10	+40, −20	35–2	0.5–0.5	6–10

* Axial.
† Radial.

The selection of inductive components for application in thick film hybrid circuits depends upon several key factors. Obviously, the electrical characteristics such as inductance value, tolerance, quality factor, etc., must be selected to comply with the functional requirements of the circuit. The physical requirements are usually dictated by the size and configuration of the final module package, the processing and assembly methods available, and the maximum temperatures the module experiences during normal fabrication. The last is very important and is often overlooked by thick film designers. Failure to select components with maximum storage temperatures exceeding normal processing temperatures can result in low circuit yields, poor stability, and greatly reduced reliability. A checklist of the critical factors requiring consideration in the selection of inductive components for use in thick film hybrid circuits is provided in Table 21.

In the paragraphs to follow, specific information on the performance, configuration, and application of the various inductive-component types is presented.   Since it is not possible to include complete information on all available types or comment on the many variations possible through custom designs, the designer should consult with inductive-component manufacturers on specific needs.   Table 22 lists typical manufacturers of microminiature inductive components.

TABLE 21   Checklist of Critical Inductive-Component Performance Characteristics

Inductors	Transformers
Inductance value	Number of windings
Tolerance	Turns ratio
Quality factor	Primary impedance
Tuning range (if variable)	Secondary impedance
Resonant frequency	Primary dc resistance
Dc resistance	Secondary dc resistance
Maximum dc current	Primary dc current
Temperature coefficient of inductance	Secondary dc current
Operating-temperature range	Frequency response
Size	Power capability
Configuration	Size
Maximum storage temperature	Configuration
	Maximum storage temperature

TABLE 22   Typical Suppliers of Miniature Inductive Components

	Inductors		Transformers		
Supplier	Fixed	Variable	Audio	Inter-mediate and radio fre-quency	Pulse
Cambridge Thermionic Corporation (Cambion)	x	x			
Consolidated Transformers Unlimited	x		x		x
Delevan Electronics	x	x		x	
Microtran Company	x		x		
Motorola	x*				
Nytronics	x				
Piconics	x	x		x	
Pulse Engineering Inc					x
United Transformer Company	x		x		

* Spiral.

Flat-spiral inductors   Thick-film flat-spiral inductors are made by screen-printing and firing conductor patterns of the appropriate spiral design directly onto the circuit substrate.   The spiral can be either square or circular, as shown in Fig. 43.   The square spiral is easier to design and fabricate and offers higher inductance for a given inside and outside dimension.   The circular spiral of equivalent design tends to exhibit a somewhat higher Q, or quality, factor.   Various equations have been devised for the design of spiral inductors. Two equations found to provide close agreement between calculated and actual values are as follows:

Circular spiral inductors:[39]
$$L = \frac{0.8a^2n^2}{6a + 10c} \quad \text{nH} \tag{2}$$

where $a = (d_o + d_i)/4$
$c = (d_o - d_i)/2$
d_o = outside diameter, mils
d_i = inside diameter, mils
n = number of turns

Square spiral inductors:[40]
$$L = 0.0216S^{1/2}N^{5/3} \quad \text{nH} \tag{3}$$

where S = surface area of coil, mils^2
N = number of turns

For either type of spiral the achievable inductance values are quite low, typically in the 0.05- to 2.0-μH range. The limiting factors are size and acceptable Q factors.

(a) Square spiral   (b) Round spiral

Fig. 43 Typical configurations of thick film spiral inductors.

Equations (2) and (3) indicate that the inductance is directly proportional to the number of turns and the mean radius of the spiral, increasing with an increase in either factor. Since substrate area is usually at a premium in thick film microcircuits, the physical size of the spiral is quite limited. The number of turns also has practical limits. Within a given area the maximum number of turns that can be achieved depends upon the minimum line width and spacing that can be obtained with the screen printing process. Even with 10-mil lines and 10-mil spacings an area of approximately ½ in. square is required to produce a 10-turn spiral. The inductance of this spiral will be less than 0.5 μH. Figure 44 illustrates this inductance, size, and turns relationship for circular spirals with 10- and 20-mil line widths and spacings. These curves tend to indicate that appreciable inductance values could be obtained if sufficient substrate area were available. This is misleading, however, since other factors such as the Q and self-resonance of the inductor actually establish a practical limit on the maximum number of turns. The equivalent circuit of a spiral inductor shown in Fig. 45 will help illustrate this point. In the schematic L_s is the low-frequency inductance, R_s is the series resistance of the spiral, and C_s is the distributed capacitance. Because of skin effect, R_s is frequency-dependent and consequently is shown as a variable resistor in the equivalent circuit. The self-resonant frequency f_r of the spiral inductor is given by [20]

$$f_r = \frac{1}{2\pi\sqrt{L_sC_s}} \tag{4}$$

Consequently the larger the inductance value and/or the higher the distributed capacity, the lower the resonant frequencies. Depending on the dielectric constant of the substrate, the number of turns, and the size of the spiral, distributed-capacitance values from 0.2- to 2.0-pF and self-resonant frequencies from 150 to 1,500 MHz are typical.

The Q, or quality, factor of a thick film spiral inductor is a function of the relationship[20]

$$Q = \frac{\omega L_s[1 - (f/f_r)^2]}{R_s} \tag{5}$$

where $\omega = 2\pi f$
f = frequency of interest
f_r = self-resonant frequency

From the equation and other known effects the following observations can be made concerning the Q of a spiral inductor:

1. The series resistance of the spiral must be kept as low as possible.
2. The Q is frequency-variable. It is low at low frequencies, increases to a maximum, and then decreases to zero at the self-resonant frequency. The factors affecting

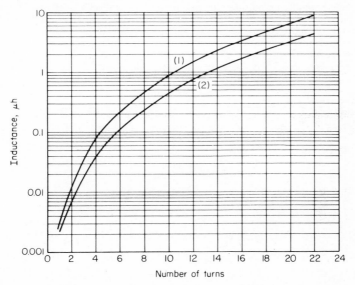

(1) 0.020-in. conductor width and spacing and 0.050-in. inner radius

(2) 0.010-in. conductor width and spacing and 0.025-in. inner radius

Fig. 44 Theoretical inductance versus number of turns for two circular-spiral inductors of typical size.

the Q at high frequencies are the increase in resistance due to skin effect and the distributed capacitance.

3. Since the mean path length (and consequently the resistance of the spiral) increases significantly with each added turn, it is difficult to achieve high-Q spiral inductors. For a Q of 20 or greater it is necessary to limit a spiral inductor to about 10 turns.

Many techniques can be employed to improve the Q and self-resonant frequency of a thick film spiral inductor. Increasing the conductivity of the spiral by using low-resistivity paints, double-screening the pattern, and solder-coating the pattern provides greatest quality-factor improvement. Figure 46 illustrates this point. Improvements in self-resonant frequency will be obtained when substrates with dielectric constants lower than alumina are used. Steatite ceramic is an example. The application of a low-dielectric-constant glaze between the substrate and the spiral is another possibility.

In general, thick film spiral inductors have greatest application at high frequencies when few turns are required and the tolerance of the inductance value is not critical. Typically, spiral inductors can

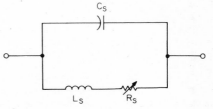

Fig. 45 Equivalent circuit of a spiral inductor.[20]

be screened to within 5 to 10 percent of their calculated values. The major advantage of spiral inductors is the ease of fabrication and the very low cost compared to other inductor types.

Miniature fixed inductors   Most miniature inductors are formed by winding fine wire on a tiny cylindrical magnetic core. The core material can be either ferrite or powdered iron. Typically, ferrites offer higher permeability than powdered iron and have higher operating and storage-temperature capabilities. Ferrite-core inductors can usually be operated to 155°C and can be stored at temperatures around 210°C. A maximum operating temperature of 85°C and a storage temperature of 150°C are typical for powdered-iron-core inductors. Powdered-iron cores in general exhibit much lower temperature coefficients than ferrites.

Miniature fixed inductors are available in a wide range of values and a variety of package configurations.   Outline drawings, package dimensions, and electrical-

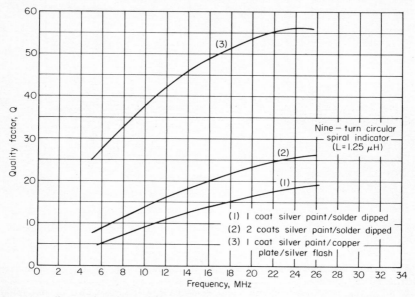

Fig. 46 Effects of various fabrication techniques on quality factor of a spiral inductor.

performance ranges for several available types are provided in Table 23. The first type, style A, is a tiny epoxy-molded axial-lead package. It is capable of meeting the environmental specifications of MIL-C-15305B and has an operating-temperature range of −55 to +125°C. The second type, style B, is also epoxy-molded but is smaller and features a rectangular cross section and radial ribbon leads. Either silver-plated copper or solid-gold ribbon leads are available. The inductor designated as style C in Table 23 is supplied on a ceramic pad which has heavy gold plating on the underside to facilitate bonding to the circuit substrate. The inductor leads are connected to two gold-plated contact areas on the top side. Interconnection between these contact areas to the substrate can be accomplished by any one of a variety of methods, including the wire-bonding techniques commonly used for attaching semiconductor chips. This package can also be furnished with special pattern geometrics to allow direct bonding to the circuit substrate by flip-chip or reflow-soldering techniques. A similar package, shown as style D, is designed for reflow soldering to the substrate in the same manner of attachment as chip-type capacitors. A low-cost miniature inductor, style E, is designed for commercial applications and has an operating-temperature range of −25 to +85°C. The coil and leads are a single piece of self-stripping wire. A thin conformal coat protects the wires and provides a moisture barrier. The last style in Table 23 is a tiny 0.1 by 0.1 by 0.65 in. inductor. It is the smallest of three sizes offered in this configuration and is supplied with plated Be-Cu leads for high thermal conduction. In addition the lead size and spacing are in accordance with the EIA flat-pack standards.

It should be noted that the information in Table 23 is based on manufacturers' standard stock items. Most manufacturers will supply miniature inductors to special requirements or will design and fabricate custom types. For instance, most stock inductors are in the ±10 percent tolerance range, but tolerance values down to ±1 percent or even ±0.5 percent can be obtained upon request. Variations in packaging and lead materials are also possible.

Miniature tunable inductors Several of the fixed inductors described in the preceding paragraphs are also available in variable or tunable form. The inductor listed as style C in Table 23 is available with a core that can be slide-tuned and fixed in place with epoxy without changing the set inductance value.[47] Also available in vari-

Fig. 47 Miniature tunable inductor with side adjustment screw. (*Piconics, Inc.*)

Fig. 48 Miniature tunable inductor with top adjustment screw. (*Piconics, Inc.*)

able form are styles D and F.[48,49] Two devices with self-locking nylon tuning screws are shown in Figs. 47 and 48.[50] Their sizes are illustrated in the dimensioned outline drawings of Fig. 49. They are identical electrically but have different form factors. The part shown in Fig. 49*a* is a side-tuned unit which is used when adjustment is to be accomplished from the end of the substrate. The other form is designed for use in compact designs where edge adjustment cannot be achieved.

Miniature tunable inductors are available with inductance values in the range from 0.023 to 5,800 μH. The maximum tuning range is a function of the inductor size, value, and type. For low-value inductors, a tuning range of approximately 2:1 is typical. A 5:1 tuning span is common in the high-value inductors. Some inductors, such as style C of Table 23, have adjustment capabilities approaching a 10:1 range for the higher inductance values. In general, all other electrical parameters are relatively similar to those specified for fixed inductors of comparable size and form factor.

Tunable inductors find application in resonant circuits or other networks where variation in associated circuit parameters do not permit the use of fixed inductors. The typical useful frequency range of these components is 100 kHz to 500 MHz.

Miniature transformers Transformers are the most difficult inductive component to miniaturize. Typically, they consist of several separate windings of wire wrapped around a core of magnetic material. To reduce the size of a transformer it is necessary to reduce the size of the core and the wire. In order to maintain suitable electrical performance, the miniature cores must be made of materials exhibiting high permeability. Although excellent high-permeability core materials such as silicon,

TABLE 23 Characteristics of Typical Microminiature Fixed Inductors[41-46]

Style	Mfr*	Outline	Major dimensions, in.					Electrical characteristics					
			A	B	C	D	E	L, μH	Q	Frequency, MHz	Temperature coefficient, ppm/°C	DC resistance, Ω	DC inductance, mA
A	1		0.250	0.100	1.25			0.10 to 1,000	40 to 28	690 to 5.2		0.07 to 102	2,000 to 40
B	2		0.180	0.065	0.060 to 0.080	0.060	0.500	0.10 to 4,700	40 to 10	650 to 0.5		0.12 to 600	100 to 5
C	2		0.142	0.065 to 0.095	0.057	0.030	0.102	0.025 to 5,800	60 to 20	1,400 to 0.75	+700 to +65	0.007 to 160	250 to 6.5

Row	Key	Diagram											
D	3		0.160	0.125	0.090 to 0.125	0.031		0.1 to 1,000	40 to 20	25 to 0.79		0.04 to 100	
E	2		0.155	0.075 diam	0.033 diam			0.007 to 5,800	90 to 10	3,000 to 0.6	+300	0.001 to 800	800 to 50
F	4		0.100	0.100	0.065	0.100	0.350	0.015 to 10.0	40 to 40	250 to 40	+200	0.065 to 5.6	735 to 80
			0.150	0.150	0.065	0.100	0.400	12.0 to 100	45 to 45	32 to 10	+200	3.5 to 23.0	110 to 45
			0.250	0.250	0.065	0.200	0.500	120 to 1,000	45 to 35	6.0 to 1.8	+400	10 to 54	85 to 36

* KEY: 1 = Nytronics, Inc.
2 = Piconics, Inc.
3 = Cambion
4 = Delevan Electronics

TABLE 24 Characteristics of Typical Miniature Transformers[51-55]

Style	Mfr*	Outline	Major dimensions in.,					Characteristics					
			A	B	C	D	E	Application	Maximum no. of leads	Maximum no. windings	Typical turns ratio	Power rating mW	Frequency range
1	1		0.100 0.150 0.250 0.250	0.065 0.065 0.065 0.125	0.100 0.100 0.100 0.100	0.012 0.012 0.012 0.012	0.250 0.250 0.250 0.250	Intermediate and radio frequency	4 6 6 6	2 3 3 3	10:1 10:1 10:1 10:1		1–50 mHz 0.5–50 mHz 0.5–50 mHz 0.5–50 mHz 1 Hz–200 mHz (ferrite core)
2	2		0.250	0.250	0.050	0.016	0.375	Output, matching, interstage, isolation, driver	7 standard (10 or more special)	3 center tap primary	10:1:1	80	300 Hz–250 kHz

3	0.260	0.125	0.050	0.015	0.125	Pulse	10 (6 normally used)	3	4:1:2	250		3
4	0.421	0.260	0.025				7	3	22.4:1		600 Hz–2 mHz	4
5	0.410	0.465	0.310	0.100	0.300	Audio, input, output, interstage, isolation, matching	6	3	10:1	50–10	300 Hz–100 kHz	5

* KEY: 1 = Delevan Electronics Corp.
2 = United Transformer Co.
3 = Pulse Engineering Inc.
4 = Consolidated Transformers Unlimited, Inc.
5 = Microtran Company, Inc.

iron, Permalloy, and ferrites exist, the minimum size that can be achieved is still physically limited because sufficient space must be provided for the coils of wire. The degree of reduction in wire size is also limited. Small-diameter wires are extremely difficult to handle and wind into the desired coil configuration. Furthermore they offer high dc resistance and low current-handling capability.

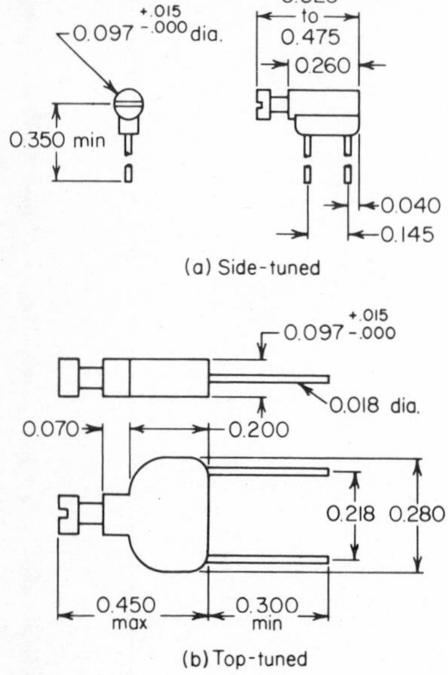

Despite the above restrictions, several types of miniature transformers have been developed and are readily available. Some may be considered quite large compared to other miniature component types, but when space permits, they can be employed to good advantage in thick film hybrid microcircuits. Outline drawings and general information on a cross section of available types are provided in Table 24. These transformers represent a variety of operational modes and a fairly broad frequency spectrum. Information relative to the specific electrical characteristics of these transformer types is not included since there are so many possible variations. Most manufacturers list a number of standard types in their catalogs. Again the designer is urged to consult with the manufacturers for detailed information and for assistance with specific requirements and custom designs.

Fixed transformers are not the only type available in miniature form. Figure 50 pictures a miniature double-tuned transformer,[56] a device designed for interstage and discriminator applications in the 1- to 100-MHz frequency range. An outline drawing and typical available winding configurations are shown in Fig. 51.

Fig. 49 Outline drawings of typical miniature tunable inductors.[50]

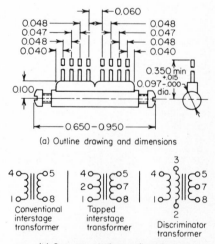

(a) Outline drawing and dimensions

(b) Common winding configurations

Fig. 51 Physical characteristics of miniature double-tuned rf transformers.[56]

Fig. 50 Miniature double-tuned rf transformer. (*Piconics, Inc.*)

Semiconductor Components

Semiconductor components are the heart of thick film hybrid technology. They provide the various active electronic functions which cannot be obtained with any known thick film techniques. Consequently, when coupled with the passive thick film elements, they enable complete functional circuits and even entire subsystems to be fabricated on a single ceramic substrate. Of equal significance is their small physical size. Compared to other active elements, such as vacuum tubes, semiconductor components are extremely small. As a result, full advantage can be taken of the miniaturization achieved with thick film techniques, and high packaging densities can be obtained, particularly when chip-type semiconductors are utilized.

There is very little restriction in the selection of semiconductor components for application in thick film hybrid modules. Basically, if the device meets the electrical-performance requirements and is compatible with the final module configuration, both physically and processwise, it can be used successfully. Nearly all devices listed in manufacturers' catalogs are potential candidates. Possible exceptions are germanium devices because their low maximum junction temperature and lack of a protective passivation layer make them undesirable for thick film hybrid applications. Typical examples of the device types that can be employed successfully are silicon bipolar *npn* and *pnp* transistors, field-effect transistors, silicon diodes of all types, silicon-controlled rectifiers, MOS devices, and monolithic integrated circuits. Furthermore, any combination of these various device types and functions can be integrated into a single thick film network, which is one of the major advantages of the hybrid-circuit approach over monolithic integrated circuits.

Because of the variety of device types that can be utilized and the broad range of performance parameters associated with each, a discussion of the electrical characteristics of semiconductors and the factors associated with the selection and application of these devices to meet circuit functional requirements will not be attempted here. This information can readily be obtained from the many textbooks and manufacturers' application notes available on the subject. This section is devoted instead to a subject more closely associated with thick film applications, namely, the characteristics, merits, and disadvantages of the physical forms and package configurations of the various semiconductor devices.

Semiconductor devices for applications in thick film hybrid circuits can be obtained in either packaged or chip form. Packaged devices can be classified into three categories: standard package form, miniature package form, and chip-carrier form. Chip devices can be classified as faceup-bonded, facedown-bonded, or beam-lead types. In the following paragraphs all these physical forms will be discussed in detail. For quick reference a comparative summary of the essential characteristics of these various package forms is given in Table 25.

In the selection of semiconductor devices for utilization in a specific application, the characteristics of each of the various forms must be evaluated with respect to the requirements of the application in order to ascertain the optimum approach. Care must be taken to assure that all factors are properly weighed. One of the most misleading factors is device cost. To compare cost properly, the basic device cost, the cost of attachment, and the anticipated yields must all be considered. Device availability also deserves some comment. At present a broad assortment of semiconductor types is not available in some of the package forms. Many semiconductor suppliers are willing to package the devices in the desired form if possible. However, the cost may be high unless a high-volume use is realized. In cases of this type it is advisable to get in touch with the various suppliers for specific details. A list of typical suppliers of the various semiconductor device forms is given in Table 26.

Standard semiconductor package forms   When space and application permit, semiconductor devices in standard EIA registered case configurations can be attached directly to a thick film substrate, as illustrated in Fig. 52.

Of the many standard case configurations available, several, because of their size, cost, availability, or shape, have become more widely used in thick film hybrid circuits than others. Outline drawings of the most popular diode, transistor, and monolithic integrated-circuit standard package forms are provided in Figs. 53 to 55. Tran-

TABLE 25 Comparison of Semiconductor-Device Forms

Parameter	Conventional package	Plastic package	Microtab	Miniature plastic package	Ceramic carrier	Plastic carrier	Leadless inverted device	Conventional chip	Flip chip	Beam lead
Relative size	Large	Large	Medium	Medium	Medium	Medium	Medium	Small	Small	Small
Availability	Excellent	Excellent	Fair	Good	Fair	Fair	Good	Excellent	Fair	Good
Relative cost	Medium	Low	Medium	Medium	Medium	Medium	High	Low	High	Medium
Ease of handling	Very good	Very good	Good	Good	Fair	Good	Good	Fair	Good	Fair
Ease of attachment	Good	Good	Good	Good	Fair	Good	Good	Poor	Good	Good
Relative cost of attachment	Low	Low	Low	Low	Medium	Low	Low	High	Low	Low
Potential for automated attachment	Good	Good	Good	Good	Fair	Very good	Very good	Fair	Very good	Very good
Repairability	Good	Good	Good	Good	Fair	Fair	Good	Poor	Fair	Good
Ease of testing	Good	Good	Fair	Fair	Fair	Good	Fair	Poor	Fair	Good
Relative reliability	Good	Good	Good	Good	Fair	Very good	Very good	Good	Very good	Very good
Thermal capability	Fair	Fair	Good	Fair	Good	Fair	Good	Excellent	Fair	Good
High-frequency performance	Fair	Fair	Fair	Fair	Good	Fair	Good	Good	Excellent	Excellent

TABLE 26 Typical Suppliers of Semiconductor Components

Semiconductor package form

Manufacturer	Conventional package	Plastic package	Microtab	Miniature plastic	Ceramic carrier	Leadless inverted device	Plastic carrier	Conventional chips	Flip chip	Beam lead
Amperex	x					x				
Centralab Semiconductor	x				x	x		x		
Crystalonics	x							x		
Dickson					x	x		x		
Dionics				x				x		
Fairchild Semiconductor	x	x						x	x	
General Electric	x	x	x					x		
General Instrument		x								x
Hughes	x								x	
Intel	x							x	x	
Intersil	x							x	x	
ITT	x	x		x				x		
Microsemiconductor Corp.	x							x		x
Motorola Semiconductor	x	x			x	x		x		
National Semiconductor	x	x	x					x		
Pirgo								x		x
Raytheon	x							x		
RCA	x							x		
Signetics	x	x						x		
Siliconix	x	x					x	x	x	
Sprague	x							x		
Stewart Warner	x	x		x		x		x		
Texas Instruments	x	x						x	x	x
Transitron								x		
Unitrode	x									

sistors and diodes in the plastic packages are very popular for consumer and industrial thick film circuit applications. Here cost and not size is critical, and hermeticity is not a requirement. For military applications semiconductors in hermetic packages must be used if the overall thick film module is not sealed. The TO-5, TO-18, and TO-46 metal-can transistor packages and the DO-7, DO-34, and DO-35 glass-sealed-diode forms are most frequently used for this application. The TO-84 and TO-86 flat packs, with their low profile and compact lead configuration, are the most popular integrated-circuit package types. The dual-in-line packages shown in Fig. 56, although popular for printed-circuit-board mounting, have not found wide application

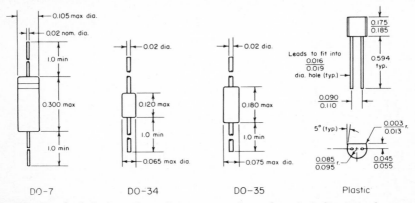

Fig. 52 Examples of thick film networks with attached semiconductors in standard package forms. (*Centralab Division, Globe-Union Inc.*)

DO-7   DO-34   DO-35   Plastic

Fig. 53 Standard diode package forms commonly used in thick film hybrid circuits.

in thick film circuits because their size and particularly their lead configuration are not compatible with normal thick film attachment methods.

Figure 52 also illustrates the most common methods of attaching standard semiconductor packages to thick film substrates. These methods are the edge-, through-hole, and surface-mounting methods. The edge-mounting method is easy and economical and particularly advantageous when thick film circuitry must be placed on both sides of a substrate and interconnections made between the two sides. With proper layout this can be accomplished utilizing the semiconductor as the means of interconnection. The through-hole mounting method offers a neat, compact layout and requires a minimum amount of lead forming, but it requires holes in the substrate, which has a notable effect on substrate cost. The most universal mounting method is the surface-mounting method. Here the devices are laid flat on the substrate and their leads, properly formed, are bonded to the appropriate thick film conductor patterns. Any device, from a diode with two leads to a fourteen-lead integrated-circuit package, can be accommodated with this method. In all the above methods solder applied by a hand, reflow, or dip technique is the normal attachment medium.

The electrical and thermal characteristics of semiconductor devices in standard package forms are well established. This information for each specific device type can be found in the various manufacturers' specification sheets and application notes. Further information on the thermal characteristics of these package types and meth-

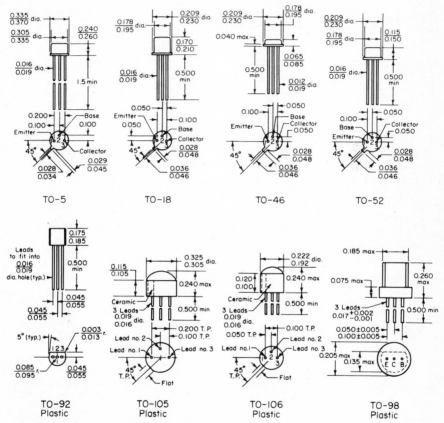

Fig. 54 Standard transistor package forms commonly used in thick film hybrid circuits.

ods of estimating their thermal performance in thick film circuits is provided later in this chapter.

The selection of standard packaged semiconductors for application in thick film hybrid circuits is no different from that for conventional packaging methods. Care must be taken, however, to assure that the maximum junction temperature of the devices is higher than the process temperatures anticipated during final assembly and packaging of the thick film circuit. This is one reason why germanium devices are not recommended. Their low maximum junction temperature, 100°C, is near or below the final cure temperature of many thick film encapsulating materials.

Needless to say, availability and cost of standard packaged semiconductors are excellent. Diodes, transistors, integrated circuits, and most of the other semiconductor active functions are available in one or the other of the standard package forms. Most devices are in high-volume production; consequently, costs are quite low. Low cost is particularly associated with the plastic package types.

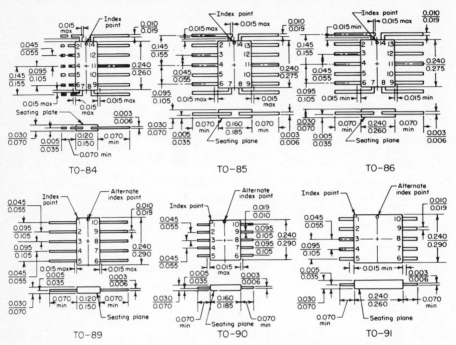

Fig. 55 Standard package forms for monolithic integrated circuits.

A summary of the important advantages and disadvantages of the standard package forms is provided in Table 27. The major disadvantage of these package types is their comparatively large size. They cannot be used when true microminiaturization is desired in thick film form.

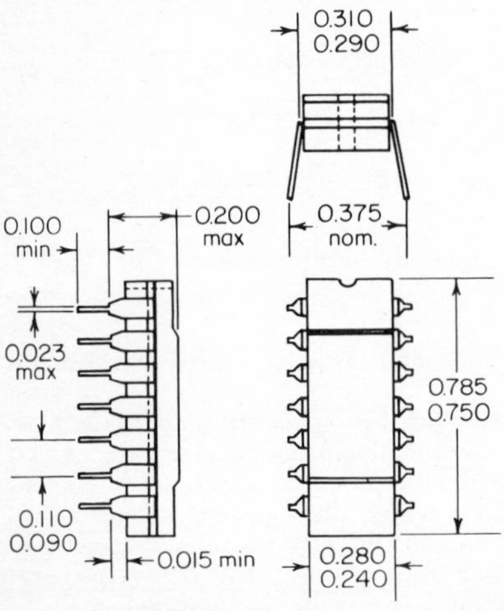

Miniature package forms Miniature package forms are defined here as encased, leaded package forms which are smaller in size than standard package forms. Several versions of this package form exist and are described below. Most were specifically designed for application in miniature hybrid circuits. A summary of the major advantages and disadvantages of these package forms is given in Table 28.

Microtabs. There are several versions of the microtab package, but all basically consist of a ceramic channel with deposited-metal surfaces, to which the semiconductor chip is die- and wire-bonded. Small-diameter wire leads are bonded to the metal pads on the channel, and the chip is covered with a small amount of epoxy for physical protection. This package form is illustrated in Fig. 57. A second version with ribbon

Fig. 56 Outline drawing of typical dual-in-line package.

leads in a T configuration is shown in Fig. 58. An outline drawing of another version carrying an EIA registration number of TO-50 is shown in Fig. 59. Attachment of these device types to the thick film conductor pads is normally accomplished with standard solder or welding techniques.

The electrical performance of microtab devices is similar to that experienced with standard packaged devices. Complete electrical testing can be accomplished on the devices prior to attachment. Their power dissipation is typically in the 150- to 250-mW range in free air. However, this can be greatly increased by bonding the ceramic back directly to the circuit substrate with a good thermal-conductive epoxy.

TABLE 27   Advantages and Disadvantages of Semiconductors in Standard Package Forms

Advantages	Disadvantages
Low cost	Large size
Excellent device availability	Fair heat transfer
Easy to handle	Fair high-frequency performance
Low attachment cost	
Multiple attachment possible	
Easy to test	
Electrical parameters assured	
Good repairability	
Good reliability	
High yields	
Bonds easy to inspect	

TABLE 28   Advantages and Disadvantages of Semiconductors in Miniature Package Forms

Advantages	Disadvantages
Relatively small size	Relatively high cost
Easy to handle	Limited device availability
Low attachment cost	Fair heat transfer
Multiple attachment possible	Fair high-frequency performance
Relatively easy to test	
Electrical parameters assured	
Good repairability	
Good reliability	
High yields	
Bonds easy to inspect	

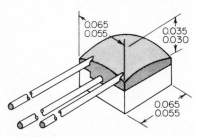

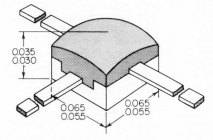

Fig. 57 Microtab package form with wire leads.[57]

Fig. 58 Microtab package form with ribbon leads in T configuration.[57]

Microtab packages, because of their small size and three-lead capacity, are essentially limited to basic semiconductor forms such as diodes and transistors. Typically, most devices packaged in the standard TO-18 case can be placed in this package form. However, at present, few manufacturers offer these forms. Consequently, off-the-shelf availability is quite poor. The cost of microtab packaged devices is generally higher than for equivalent devices in the standard package forms.

Miniature Plastic Forms. The miniature plastic package forms are basically an extension of the T configuration microtab packages described above. The major difference is that a plastic pellet is molded around the semiconductor chip to encapsulate and protect it. Figure 60 illustrates several versions of this package form. The ribbon leads are usually gold-plated Kovar,* permitting either solder or weld assembly. Because of the T-shaped lead configuration, alignment of the package to the substrate is quite simple. Care must be taken in selection and application, however, as the emitter, base, and collector lead breakout can vary from supplier to supplier or from device type to device type from the same supplier.

Fig. 59 Outline drawing of TO-50 microtab package.

Electrical performance of miniature plastic devices is similar to standard packaged devices. The thermal capabilities equal the microtab units described above. Their small physical size suggests one further advantage: good packaging density.

Device availability in this package form is quite good. Several manufacturers offer this package containing *dice*† from most generic families supplied in the standard

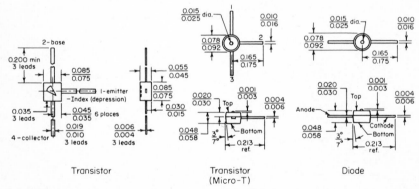

Transistor

Transistor
(Micro-T)

Diode

Fig. 60 Outline drawings of typical miniature plastic package forms.[58, 20] (Micro-T is a trademark of Motorola Semiconductor Products, Inc.)

TO-18 package. Other chips can be placed in this plastic form upon request if the volume is sufficient to justify the cost of conversion. Device functions currently available in this package form include small-signal diodes, dual diodes, *pnp* and *npn* bipolar transistors, and field-effect transistors. Typically, this package type is more costly than the TO-92 plastic package, due primarily to the lower production volumes encountered to date. As volumes increase, the cost should become quite comparable.

Chip-carrier package forms The chip-carrier form of packaging consists of a semiconductor chip bonded to a tiny ceramic, plastic, or other form of header or carrier.

---

* Trademark of Westinghouse Electric Corporation.

† A *die* is the term commonly used to describe a basic semiconductor chip; plural *dice*.

In some types the wire bonds to the semiconductor chip are completed, and the chip is encapsulated with a protective epoxy coating. This package form varies from the miniature package forms in that external wire or ribbon leads are not normally provided. There are several varieties of chip-carrier package forms. Some are specifically designed for multiple or automatic insertion. All are intended for application in miniature hybrid modules. The following paragraphs describe some of the common types.

Ceramic-Channel Carrier. The ceramic-channel carrier is similar to the microtab package previously described except it does not have leads and the chip is usually unencapsulated. Typically, this carrier form is composed of a semiconductor die thermocompression-bonded to the bottom of a U-shaped ceramic channel. The bottom and upper edges of the channel are metallized to facilitate bonding and to provide the necessary pads for interconnection. Fly-wire leads are bonded from the chip pads to the upper edges of the channel. This construction is shown in Fig. 61.

The ceramic-channel carriers are normally attached to the circuit substrate with a good thermally conductive epoxy, which not only retains the chip carrier but also serves as an excellent thermal-conductive medium between the carrier and the substrate. Thermocompression or ultrasonic wire-bond methods are normally used to interconnect the emitter, base, and collector pads of the ceramic channel to the thick film conductor patterns. Although quite small, ceramic-channel-carrier forms are much easier to handle than the basic semiconductor chip. Furthermore, because of the electrical isolation provided by the ceramic channel, these devices can be bonded directly over the substrate conductor patterns, thus conserving substrate area.

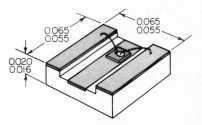

Fig. 61   Semiconductor chip mounted in ceramic-channel carrier.[57]

Since the basic die and wire bonding operations are completed within the channel, no degradation in electrical-performance parameters will be experienced during attachment of the ceramic-channel-carrier devices to a thick film substrate. Consequently, complete testing and device characterization can be accomplished prior to attachment. In general, the electrical performance of a specific device type in this package form will be the same as its standard packaged counterpart. The thermal characteristics will be superior, however, particularly when the carrier is bonded to the substrate with thermally conductive epoxy.

Several manufacturers offer this package form as a standard product. Device availability and variety are only fair at present, due primarily to the introduction of the leadless-inverted-device (LID)° package, a more advanced form discussed later in this section.

The basic cost of a semiconductor in the ceramic-channel carrier is similar to that of an equivalent device in a TO-18 package. Although attachment of the ceramic carrier to the substrate can be economically accomplished, the overall attachment cost is relatively high due to the fly-wire bonds needed to make the electrical connections.

A summary of the major advantages and disadvantages of the ceramic-channel-carrier package is provided in Table 29.

Leadless-Inverted-Device Forms. Leadless-inverted-device (LID) package forms evolved from the ceramic-channel carrier and are specifically designed to eliminate the costly wire bonds and handling problems encountered in attaching the channel-type package. By deepening the walls of the basic ceramic-channel carrier, adding a step-type intermediate level, and making a crosscut across the channel for isolation, the LID carrier is obtained, and all attachment problems associated with the channel carrier are virtually eliminated. Semiconductors are attached to the LID carrier by bonding the basic die to a metallized contact surface at the base of the crosscut channel. This contact surface extends up to the top of two of the four posts pro-

° Trademark of Amperex Electronic Corp.

TABLE 29 Advantages and Disadvantages of Semiconductors in Ceramic-Channel-Carrier Forms

Advantages	Disadvantages
Relatively small size	Relatively high cost
Good heat transfer	Poor device availability
Relatively easy to test	High attachment cost
Electrical parameters assured	Require lead bonds
High yields	Reliability depends upon lead-bond capability
Bonds easy to inspect	Relatively difficult to handle
Good high-frequency performance	Difficult to repair

vided by this configuration. These form the collector-termination pads. The base and emitter of the semiconductor are connected to the metallized steps of the two

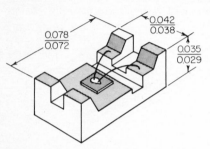

remaining posts by normal bonding techniques. The chip is then covered with an epoxy encapsulation. A sketch of this package form is shown in Fig. 62. This form carries a TO-122 EIA registration. Several variations of this package form exist, some of which are patented. Figure 63 illustrates two other versions. The three-post circular form, or Versa-Pak,* is lower in cost and easier to align than the rectangular form. The 14-post LID is intended for monolithic integrated circuits.

Fig. 62 Semiconductor chip mounted in LID package form.[57]

The LID packages are mounted to a thick film substrate by inverting and attaching the posts or legs to the appropriate conductor patterns. The attachment is usually accomplished by solder-reflow or paste-solder techniques. Pretinning both the mounting pads and the legs of the LID enhances the quality of the bond. Exact positioning is not critical as the

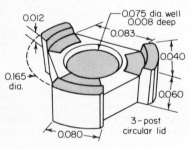

(a) Versa-pak

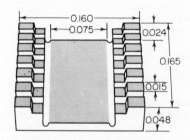

(b) Fourteen-post LID for monolithic integrated circuits

Fig. 63 Alternate LID package forms.[59]

surface-tension forces of the molten solder will pull the LID package into correct alignment, easing the requirement for high position control and permitting simultaneous attachment of many LIDs on a thick film substrate. Utilization of a locating

* Trademark of Frenchtown/CFI, Inc.

jig or template, as shown in Fig. 64, is one common method of accomplishing multiple attachment.

Semiconductor devices in LID form can be purchased with guaranteed electrical performance similar to that provided for standard packaged devices. Because of the low temperature encountered in attachment, no degradation in performance will result. Thermal performance of the LID package is very good and is superior in most respects to standard metal-can packages of the TO-5 and TO-18 variety.

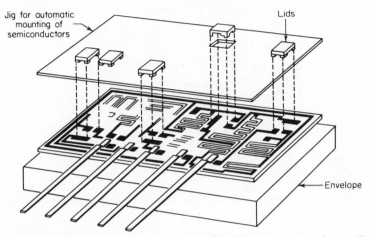

Fig. 64 Template method for multiple attachment of LID devices.[59]

TABLE 30   Advantages and Disadvantages of Semiconductors in LID Package Forms

Advantages	Disadvantages
Relatively small size	Relatively high cost
Relatively good device availability	Bonds difficult to inspect
Easy to handle	
Low attachment cost	
Multiple attachment possible	
Relatively easy to test	
Electrical parameters assured	
Good repairability	
Good reliability	
High yields	
Relatively good heat transfer	
Relatively good high-frequency performance	

Semiconductors in the LID package form are standard products of several manufacturers. Device availability with respect to both type and function is quite good. Bipolar and field-effect transistors, dual diodes, small-signal diodes, zener diodes, and even voltage-variable diodes are available in this package form.

The cost of LID devices is somewhat higher than TO-18 package devices, but volume use should enable the basic device cost to compete with the standard package forms. The low cost of attachment, high yields, and ease of repair of this device will compensate for the basic device cost in many applications.

Table 30 summarizes the advantages and disadvantages of the LID package form.
Plastic Device Carriers. Plastic-carrier forms are a recent addition to the semi-

conductor packaging field. A typical example is the miniMod° package shown in Fig. 65. Primarily intended for monolithic integrated circuits, this packaging concept uses a 35-mm filmstrip transport medium onto which integrated-circuit chips are bonded and encapsulated.

The filmstrip is a polyimide film which is inherently flat and withstands wide tem-

Fig. 65   A miniMod plastic filmstrip carrier.   (*Texas Instruments, Inc.*)

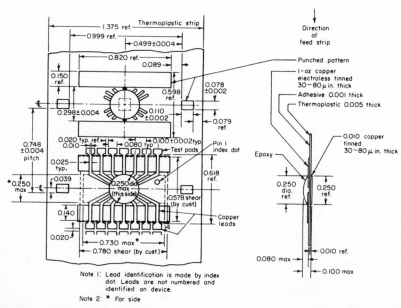

Fig. 66   Outline drawing of the miniMod plastic filmstrip carrier.   (*Texas Instruments, Inc.*)

perature excursions without degradation. The strip is perforated with indexing holes for ease in mechanized processing and testing. Other holes are provided to accommodate the integrated-circuit chips and the lead frame. Detailed package and strip dimensions are shown in Fig. 66.

° Trademark of Texas Instruments, Inc.

To the perforated filmstrip, a 1.4-mil-thick copper ribbon is laminated. Photo-lithographic techniques are used to etch the copper to form the lead frame. In the area where the integrated-circuit chip is to be bonded the leads are etched down to fingers 4 mils wide. Next, the lead frames are tinned in preparation for chip attachment.

Integrated-circuit chips or pellets are specially processed for use with the miniMod lead frame. They are made by fabricating a standard monolithic silicon integrated circuit complete with its aluminum interconnect patterns, depositing a glass overcoat over the entire circuit except for the bonding-pad areas, and depositing on the exposed bonding pads gold bumps which serve as the interconnect between the integrated-circuit chip and the copper fingers of miniMod filmstrip.

The integrated-circuit pellets are aligned under the fingers of the lead frame and assembled to the frame with a gang bonding tool that applies heat and pressure to all fingers simultaneously. This forms an Au-Sn eutectic bond which can withstand temperatures in excess of 280°C. Direct connection of the copper leads to the lead frame eliminates the fine gold wires used in standard chip and wire bonding. This reduces the number of parts and connections needed to package the chip, thus improving reliability and reducing bonding costs.

The miniMod package is completed by surrounding the pellet and interconnect area with an opaque epoxy-type coating, which protects the pellet from physical damages, enhances thermal conductivity, and shields the chip from the effects of light.

The miniMod integrated circuit is removed from the transport film by shearing off the test pads along the lines suggested in Fig. 66. This can be done automatically using the perforated holes for indexing, or for small quantities hand scissors can be used. The miniMod package can be attached to a thick film network using hand soldering, reflow soldering, ultrasonic bonding, conductive epoxy, or any other acceptable bonding technique.

The major advantages of the miniMod package are economy, reliability, and small size. The continuous strip or reel offers ease of handling and automation. With the appropriate equipment a strip of devices can be automatically tested, sheared, and bonded to a thick film network. Also the small size and flexibility of the miniMod scheme offers further cost reductions through higher circuit densities.

Although the miniMod package is presently being used primarily on custom circuits, some standard products are expected to appear in this package form in the future.

Semiconductor chips When true microminiaturization is desired, semiconductor devices in basic chip form can be bonded directly to the thick film substrate. Because the trend in the electronics field is toward microminiaturization, this chip hybrid type of construction is rapidly increasing in popularity and application even though it is generally more costly than thick film circuits containing the more bulky packaged semiconductor devices.

Semiconductor chip devices for application in thick film hybrid circuits are available in several basic forms. These can be classified as faceup-bonded, facedown-bonded, or beam-lead forms. Disregarding cost and availability, the choice of chip form for employment in a specific application is influenced by several key factors. These include the environmental and reliability requirements, the circuit complexity and desired yields, the handling and mounting equipment available, the volume of production anticipated, and the compatibility between the semiconductor chips and the thick film materials to which they are bonded. In the paragraphs to follow the various chip forms are discussed in detail, emphasizing the relative merits of each form with respect to the above factors.

Faceup-bonded Chips. The term faceup-bonded chip is simply another way of describing the conventional semiconductor chip form shown in Fig. 67. These chips are obtained by scribing and breaking a wafer of silicon material that has undergone all passivating, etching, and diffusion processes necessary to produce the desired functional device. Typically, several thousand semiconductor devices are simultaneously fabricated on a single wafer. The process normally used to produce the devices is called the Planar° process. Only silicon devices made by this process are recom-

° Patented process of Fairchild Semiconductor Corp.

mended for application in thick film hybrid circuits. Germanium devices, silicon devices made by other than the Planar process, mesa devices, and some MOS type devices are not recommended unless special provisions are available and extraordinary precautions are taken in the handling and bonding of these chips.

Faceup-bonded devices are usually attached to the thick film conductor pattern using the thermocompression bonding technique. In this method the mounting surface is preheated to a relatively high temperature and the chip is "scrubbed" on the mounting surface either manually or ultrasonically. The scrubbing action breaks down any oxides that may be present on the metallic contact surfaces. The result is a mechanically and electrically reliable silicon-gold eutectic bond. A good electrical bond is essential, as the bottom surface of most transistor and diode chips is a functional electrical contact. For transistors the bottom surface is the collector contact. A poor bond here could greatly affect the collector-emitter saturation voltage. For best bonds it is advisable that the thick film pad and the backing on the semiconductor chip be of a compatible gold system.

Fig. 67 Basic semiconductor chips. (*Texas Instruments, Inc.*)

The use of a gold or gold-eutectic solder preform during bonding aids in reducing the contact resistance. Epoxy bonding of chips is also possible. For diodes and transistors a good electrically conductive epoxy must be used. Since no electric contact is required between the bottom surface of a monolithic integrated circuit and the mounting pad, an integrated circuit can be bonded with any good-quality high-temperature epoxy.

Lead bonding is usually accomplished with thermocompression or ultrasonic techniques using fine gold or aluminum wires. The size of the wire depends on the current-handling requirements, the bonding techniques used, and the size of the bonding pads on the chips. An enlarged photograph of conventional semiconductor chips faceup-bonded on a thick film circuit is shown in Fig. 68. Further information on die and wire bonding of faceup-bonded chip devices is provided in later chapters of this handbook.

One of the major problems in bonding the conventional semiconductor chips directly to a thick film hybrid substrate is yields. To achieve high yields, the chip must be electrically good before bonding, the physical bonds must be made successfully, and the device must not suffer any performance degradation during bonding. It is, in general, quite difficult to ensure that all three of these conditions will be successfully maintained for each chip bonded. First, the electrical performance of a semiconductor device in basic chip form cannot be fully tested and guaranteed (this important factor is discussed in more detail later in this section). Second, the quality of the bond is directly related to the capability, repeatability, and reliability of the bonding equipment and the skill of the operator. Finally, excessive exposure to high temperatures as encountered during normal thermocompression bonding of gold wire to the chips can seriously degrade the electrical performance of the chips. This degradation is a result of an intermetallic action that takes place between the gold wire and the the aluminum pad, a condition commonly known as the *purple plaque*. As a result of these factors, there is a practical limit on the number of faceup-bonded chips that can be attached to a single substrate. An exact number is difficult to state because of the many variables involved, but a reasonable figure for normal thermocompression bonding is around eight to ten devices. This number can be increased if other special bonding techniques are utilized. To illustrate the effect of individual chip-bonding yields on total circuit yields consider the following example. For a circuit containing eight chips the final circuit yield will be 66 percent for an individual chip yield of 95 percent. For higher or lower chip-bonding yields the circuit yields will be as shown in Table 31.

If the die and wire bonding are properly accomplished, the electrical performance of the semiconductor chip should be equal to that of the same die in a standard package form. In fact, the high-frequency performance will often be superior due

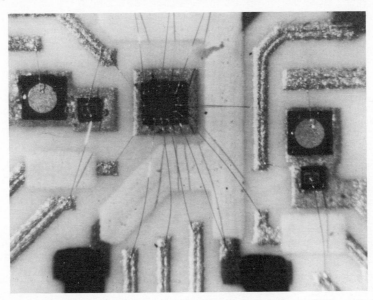

Fig. 68 Semiconductor chips faceup-bonded to thick film network. (*Centralab Division, Globe-Union Inc.*)

TABLE 31 Relationship between Device and Circuit Yield for an Eight-Transistor Circuit

Anticipated yield per device, %	Resulting circuit yield, %
99	92
95	66
90	43
85	27
80	17
75	10

to lower lead inductance and the absence of header capacitance. The major problem is to assure that the device is acceptable before bonding. Semiconductor devices in this chip form are difficult to probe and can be tested only to certain dc performance characteristics. For example, transistors in chip form can be probed for the following parameters only:[60]

I_{CBO} = collector-to-base leakage current
BV_{CBO} = collector-to-base breakdown voltage
BV_{CEO} = collector-to-emitter breakdown voltage
BV_{EBO} = emitter-to-base breakdown voltage
h_{FE} = dc current gain

It is not possible to make small-signal ac or high-current dc measurements on transistor chips either as individual dice or in wafer form. These parameters can be guaranteed with a high degree of probability, however, on the basis of past experience, the dc probe results, and the geometry of the diffusion patterns. Often, to assure acceptance, a quantity of chips can be taken from a lot, bonded in standard packages, and tested to the desired performance characteristics. Passage of these tests assures high yield of the lot to the desired requirements. Similar conditions apply to integrated-circuit, diode, and other semiconductor chip devices. For integrated circuits the full operating temperature range of −55 to +125°C cannot be guaranteed in chip form. However, by tightening probe limits and relying on past experience, a high degree of probability can be assured.

The thermal capabilities of faceup-bonded chips on thick film substrates are excellent. With eutectic bonds the thermal resistance between the chip and the substrate is less than 1°C/W. This is significantly lower than that obtained with any other chip or packaged semiconductor forms. Die bonds made with epoxy, glass frit, or other materials tend to degrade this capability somewhat. More detailed information on this aspect of semiconductor chips and on the thermal characteristics of hybrid circuits in general is provided later in this chapter.

The selection of a conventional faceup-bonded chip involves matching the device parameters after bonding to the performance parameters of the circuit. To do this properly, device specifications should be prepared and adequate visual-inspection criteria and electrical-test provisions stipulated to ensure that high yields to the desired performance characteristics will be obtained after bonding. It is also advisable that the device geometry be specified, because it can vary drastically between suppliers, as illustrated in Fig. 69 for a 2N2222 transistor. Variations in geometry can affect final electrical performance and can also be a problem in bonding because some patterns have larger bonding pads than others. The backing of the chip device should also be specified with regard to both material type and thickness. Variations here can greatly affect the mechanical and electrical integrity of the bond.

Conventional faceup-bonded chips can be procured in any one of a number of different ways. The chips can be obtained in wafer form with the dc electrical parameters sample-tested only and a device yield guaranteed. This is the most economical form from a purchased-price standpoint. The wafer can also be obtained with parameters 100 percent tested and all bad units marked with an ink. In either of these cases the user must be able to scribe and break the wafers into individual chips. In the first case it is also advisable that the user have automatic probing and testing facilities if high yields are desired. This has an added advantage in that the user can sort the family of characteristics on the wafer to the performance ranges of his choosing, perhaps to meet a specific circuit requirement. This assures a lower cost than specifying a single device type when a broader performance range can be utilized.

The second approach is to procure the devices already tested, scribed, and broken into individual chips. Again the devices can be sample-tested or 100 percent tested. The user can obtain all the chips within the wafer or have them sorted to the specific parameters required. The latter method increases the cost of the chips but requires little handling or testing on the part of the user. When working on custom designs where many different chip types of fairly low quantities are required, this is usually the most practical method of procurement.

Faceup-bonded chips are presently the most common type of chip device used in thick film hybrid circuits because of their excellent availability. Also chips of this type are, in general, lower in cost than any other semiconductor device. Nearly all devices available in packaged form can also be obtained in their basic chip form, and most manufacturers of semiconductors are willing to supply them. Many manufacturers in fact have specification sheets, brochures, and catalogs devoted specifically to semiconductor chip devices.

For ease of comparison with other device forms the major advantages and disadvantages of faceup-bonded chips are summarized in Table 32.

Facedown-bonded Chips.   Perhaps the most familiar facedown-bonded chip form

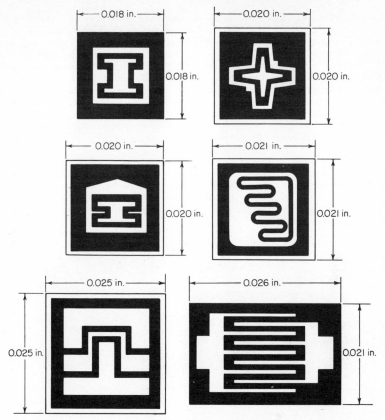

Fig. 69 Typical sizes and geometries of a 2N2222 transistor chip.

TABLE 32 Advantages and Disadvantages of Conventional Semiconductor Chips

Advantages	Disadvantages
Excellent device availability	Difficult to handle
Low cost	High attachment cost
Small size	Difficult to test
Good heat transfer	Electrical performance cannot be assured
Good high-frequency performance	Poor repairability
Bonds easy to inspect	Reliability dependent on bonding capability
	Low yields

is the Flip Chip* device shown in Fig. 70. In this type of device all the electric-contact areas are brought to the top surface of the chip and solder bumps or tiny balls, approximately 0.005 in. in diameter, are attached to the contact areas. The balls are usually made of gold, copper, or aluminum. The chip is then glass-passivated for physical and environmental protection. As their name implies, flip

* Trademark of Digital Equipment Corp.

chips are bonded to a thick film circuit by inverting or flipping the device over and directly mating the projecting bumps to the appropriate thick film pads. This operation is illustrated in Fig. 71. Reflow-soldering or ultrasonic-bonding techniques are normally used to complete the bond. Multiple attachment is possible with the reflow-solder technique, much as described earlier for LID devices. Repair and replacement of flip chips can be accomplished by removing the defective chip and bonding a new chip to the same pads. Because of this ability and the low temperatures experienced during device bonding, it is possible to place a high number of chips on a single substrate and obtain high circuit yields.

Flip chips were designed to improve bond reliability and to reduce the high attachment costs associated with conventional die- and wire-bonding techniques. The improvements obtained with this type of chip form are apparent when one considers that a monolithic integrated circuit in flip-chip form can be attached with one operation. With conventional chips, 1 die bond and 28 wire bonds are necessary to accomplish the same interconnection. In thick film hybrid circuits some of these advantages are offset by certain disadvantages, however. First, with conventional silk-screen techniques it is difficult to obtain uniform thickness of the conductor pat-

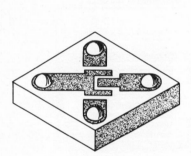

Fig. 70 Flip-chip transistor.

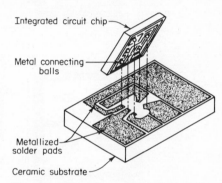

Integrated circuit chip

Metal connecting balls

Metallized solder pads

Ceramic substrate

Fig. 71 Typical method of mounting flip-chip devices.[61]

tern across an entire substrate surface. Consequently, mounting-pad height will vary, and when flip-chip devices with more than three bumps are used, adequate bonds between all device contacts and their corresponding mounting pads cannot be assured. Furthermore, because of the inverted nature of these chips, the bonds are hidden and difficult to inspect. Finally, the close spacing between bumps on a chip (generally in the neighborhood of 10 to 20 mils center to center) requires small pad sizes with very close spacing. This is difficult to achieve with normal silk-screen techniques.

Complete electrical tests and good device characterization and sorting can be accomplished on flip chips prior to mounting. No degradation in performance will result during bonding because of the low bonding temperatures required. Thanks to their very short interconnect paths, flip chips in general exhibit excellent high-frequency and high-speed performance characteristics.

With regard to thermal performance, flip chips are quite poor in comparison with other chip forms. This is due to the fact that the only conductive heat paths between the device junction and the substrate are the tiny balls, or bumps. Thermal performance can be improved by placing a good thermal-conductive epoxy over the mounted chips. Specific thermal-resistance data on flip-chip devices are provided later in this chapter.

Until recently the availability of flip-chip devices was quite poor. Now several manufacturers are making and supplying *npn* and *pnp* bipolar transistors, field-effect transistors, diodes and diode arrays, and even a few integrated circuits in flip-chip form. Device cost is still comparatively high.

A summary of the major advantages and disadvantages of flip-chip devices is provided in Table 33.

Beam-Lead Devices. Beam-lead devices are essentially a member of the facedown-bonded chip family, but their unique physical construction and characteristics suggest that they be treated as a separate chip form. The term *beam lead* refers to the heavy metallic leads that extend in cantilever fashion from the body of the silicon chip, as shown in Figs. 72 and 73. These leads are formed by evaporation processes directly on the individual chips when they are still in wafer form. Gold and aluminum are the most common beam-lead materials. Individual beam-lead chips are separated from the parent wafer by an etching process rather than the scribe-and-break technique used with most other chip forms. One advantage of the etch process is high device yields to physical-inspection criteria after the separation is accomplished. Cracks, chipped edges, and other physical defects experienced with the scribe-and-break technique are avoided with this separation process.

Beam-lead devices possess one further physical advantage, a silicon nitride coating. This is a high-density coating that offers extreme resistance to ionic contaminants and ion migration. It essentially hermetically seals the chip itself, thus eliminating the

TABLE 33   Advantages and Disadvantages of Flip-Chip Semiconductors

Advantages	Disadvantages
Small size	Limited device availability
Relatively easy to handle	Relatively high cost
Low attachment cost	Relatively poor heat transfer
Multiple attachment possible	Bonds difficult to inspect
Relatively easy to test	Fine bonding-pad definition required
Electrical parameters assured	
Relatively good repairability	
Very good reliability	
High yields	
Good high-frequency performance	

need for hermetic packages. It should be noted, however, that the military has not yet accepted this capability for high-reliability applications and still requires that they be contained in a hermetic package.

The major disadvantage of the beam-lead construction is the fragile nature of the tiny beam projections. Special packaging, handling, and bonding techniques are necessary to ensure minimum damage to the chips. Once bonded, however, beam-lead devices are more reliable physically than any of the other chip forms and, in fact, many of the packaged forms.

Beam-lead devices are normally attached to thick film hybrid circuits by inverting the devices and bonding the beam leads to the thick film conductor pads with any one of a number of thermocompression or ultrasonic bonding techniques. Because the size of the beam leads and the spacing between them is so minute (see Fig. 74 for outline drawings of typical beam-lead chips) good thick film fine-line screening definition is essential. Normally a 5-mil line and 5-mil spacing capability is required. By providing depressions in the substrate and screening the appropriate termination patterns around the depression, beam-lead devices can also be bonded faceup. This method enhances the thermal capabilities of the devices.

With regard to application in thick film hybrid circuits, beam-lead devices offer many significant advantages over the other chip forms. Their small physical size permits high circuit density. A beam-lead chip can be attached to a thick film circuit in one basic operation. The high attachment costs associated with the lead-wire technique required for faceup-bonded chips are thus eliminated. Also, because of the low temperature required to bond beam-lead devices, the number of devices that can be placed on a single substrate is not limited. The ease with which defective chips can be removed and replaced further enhances this capability. Tests have shown that with a shear removal process over 10 beam-lead chips can be bonded, removed, and replaced on a single set of thick film mounting pads. Consequently high yields of very complex circuits are possible.

Beam-lead devices offer significant reliability improvement over other chip forms. All junctions are hermetically sealed with the silicon nitride coating. The beam-

Fig. 72   Beam-lead semiconductor chips bonded to a thick film network. (*Centralab Division, Globe-Union Inc.*)

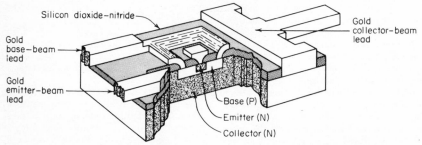

Fig. 73   Typical structure of beam-lead semiconductor chip.[57]

type leads provide improved bond strength and eliminate the many potential failure mechanisms associated with wire leads. As environmental stresses (shock, vibration, acceleration) become more severe, the advantages of beam-lead construction become more apparent. Furthermore, the flexible nature of the beam leads helps to absorb the damaging mechanical stresses that occur during wide thermal variations. Since the beam leads and their associated bonds are visually exposed, the inspection of bond integrity is readily achieved.

Due to their relatively poor thermal characteristics, beam-lead devices are restricted to small-signal applications. Their electrical performance equals or exceeds that experienced with equivalent standard packaged devices. The areas of notable improved performance are in high-frequency and high-speed applications, due primarily to the short, stubby leads. Beam-lead devices are relatively easy to test in chip form and can be tested to parameters other than the basic low-voltage dc characteristics. In fact, beam-lead monolithic integrated-circuit chips can be tested and guaranteed to specific performance-temperature ranges.

The thermal performance of beam-lead devices, as with flip-chip types, is relatively poor in comparison with other chip forms and more closely approximates that obtained with packaged units. The only thermal-conductive path from junction to substrate is through the beam leads. A thermally conductive epoxy placed over the chip will improve the thermal characteristics.

At present beam-lead devices are primarily being utilized in high-reliability, high-frequency, and high-complexity applications. However, as availability improves and costs are reduced, it is quite possible that beam leads will be the most widely used chip form. Several manufacturers are now supplying beam-lead devices, and availability is gradually increasing. Many types of diodes, including microwave diodes, bipolar transistors, and popular DTL and TTL digital monolithic integrated circuits, are available in beam-lead form. The present cost of beam-lead chips is quite high compared to that of the basic faceup-bonded chip but is competitive with several of the standard and miniature packaged forms. Since beam-lead devices consume little more space on a silicon wafer than regular semiconductor chips and high yields are experienced during device separation, it is feasible that prices will drop significantly as the volume used increases. In general, devices with aluminum beam leads tend to be less costly than equivalent devices with gold beam leads. However, availability of devices with aluminum leads is poor at present.

Table 34 summarizes the relative merits and disadvantages of the beam-lead form.

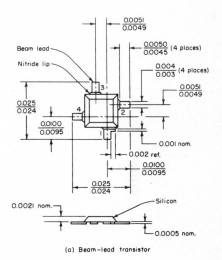

(a) Beam-lead transistor

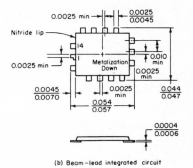

(b) Beam-lead integrated circuit

Fig. 74 Outline drawings and dimensions of typical beam-lead devices.[62, 63]

ENVIRONMENTAL PERFORMANCE GUIDELINES

General

Throughout the design cycle, the designer must maintain a continual awareness of the environmental and physical stresses that the thick film network can be expected to experience during typical in-service conditions. Each circuit component should be selected only after the designer has evaluated its parametric drift during exposure to the anticipated environmental and physical stresses. Component selection should not be based solely on conformance to the electrical requirements at sea level, low relative humidity, and room temperature. In addition to the actual service conditions of the completed product, the stresses encountered in both the thick film processing subsequent to component fabrication or attachment and the additional system assembly operations which the completed thick film module will encounter should be considered. This is particularly important when selecting semiconductors and unpackaged (chip) circuit components.

TABLE 34   Advantages and Disadvantages of Beam-Lead Semiconductor Chips

Advantages	Disadvantages
Small size	Difficult to handle
Relatively good device availability	Fine bonding-pad definition required
Relatively low cost	
Low attachment cost	
Relatively easy to test	
Electrical performance assured	
Good repairability	
Good reliability	
High yields	
Good high-frequency performance	
Relatively good thermal capabilities	
Bonds easy to inspect	

The designer should consider each of the environmental or physical conditions listed below and determine the level of exposure appropriate for the application, as well as considering any stress unique to the application. As an example, if the module is to be subjected to a hydrogen atmosphere, palladium cermet resistor inks should be avoided.

1. Operating-temperature range
2. Storage-temperature range
3. Exposure to sand and dust
4. Immersion in water, salt water, or other liquids
5. Exposure to a saltwater atmosphere
6. Vibration frequency and duration
7. Shock level and duration
8. Acceleration
9. Altitude or pressure variations
10. Rapid temperature variations
11. Humidity or moisture exposure
12. Tensile, bending, and torsion stress on leads
13. Exposure to abnormal atmospheres
14. Exposure to nuclear radiation

When the physical variables have been established, the designer can assess the mechanical suitability of various packaging techniques and compare the parametric drift of potential circuit elements under each condition with the limits established during the circuit-design stage. The package and components best suited for the application can then be selected.

Hermetic versus Nonhermetic

The decision to require hermetic sealing is usually reached after a reliability evaluation shows that it is necessary to meet the reliability goal. This will be covered in detail in the Reliability section of this chapter, but it is important to mention it now because it can have considerable effect on the package configurations, substrate sizes, etc., available to the designer. This will in turn affect circuit partitioning. Consequently this decision must be reached early in the design cycle and prior to any detail-layout work.

Where hermetic sealing is required, it can be attained by attaching hermetically sealed semiconductors to the substrate, by hermetically sealing the area of the substrate which contains the semiconductor chips, or by placing the substrate in a hermetically sealed package. The technique to be employed will depend on the production capabilities available to the designer and the system size, weight, and cost objectives.

Applicable Documents

In the past thick film circuits frequently had to meet the requirements of the military or industrial specification for component parts and were to be tested in accordance with the appropriate military standard or industrial test procedure for that component part. For example, a temperature-compensating ceramic dielectric fixed capacitor within a thick film network might have been specified to conform to the requirements of Military Specification MIL-C-20 and to be tested in accordance with Military Standard MIL-STD-202; or a semiconductor device might have been specified to conform to the requirements of Military Specification MIL-S-19500 and to be tested in accordance with Military Standard MIL-STD-750. This approach to specifying environmental requirements and conducting evaluation tests for thick film circuits gave rise to several difficulties. Various detail requirements, such as physical dimensions, package configuration, terminals, insulation and others did not apply. Unpackaged (chip) and screened thick film components obviously could not be mounted for environmental tests employing the methods specified for packaged components. Finally, the true measure of a component's environmental performance can be assessed only by conducting evaluation tests on the completed thick film circuit in its final mechanical configuration and after the components have been subjected to all the manufacturing processes.

The specifications and standards listed below have been developed or are being developed to provide guidelines for the specification and testing of microelectronic devices including thick film circuits, thus alleviating the problem.

Industrial (Electronic Industries Association, EIA):
EIA/MED-2	Standard on Film Networks
EIA/MED-50	Industrial Specification, Microelectronic Devices, General Specification for

Military:
MIL-STD-883	Test Methods and Procedures for Microelectronics
MIL-M-38510	General Specification for Microcircuits

Other applicable specifications and standards are listed below.

Specifications, federal:
QQ-S-571	Solder; Tin Alloy; Lead-Tin Alloy; and Lead Alloy

Military:
MIL-F-14256	Flux, Soldering, Liquid (Resin Base)
MIL-M-55565	Microcircuits, Packaging of
MIL-C-45662	Calibration System Requirements
MIL-S-19500	Semiconductor Devices, General Specification for

Standards, military:
MIL-STD-100	Engineering Drawing Practices
MIL-STD-129	Marking for Shipment and Storage
MIL-STD-280	Definitions of Terms for Equipment Divisions
MIL-STD-781	Reliability Tests, Exponential Distribution
MIL-STD-806	Graphic Symbols for Logic Diagrams
MIL-STD-1276	Leads, Weldable, for Electronic Component Parts
MIL-STD-1313	Microelectronic Terms and Definitions
MIL-STD-1331	Parameters to be Controlled for the Specification of Microcircuits

Handbook:
MIL-Hdbk-H53	Guide for Sampling Inspection

Publications, federal
Cataloging Handbook H4-1	Federal Supply Code for Manufacturers

Military:
NAVASHIPS 0967-190-4010	Manufacturers' Designating Symbols
NAVORD OD 40778	Procedure for Reliability Prediction of Standard Hardware Program (SHP) Modules

NASA:
   NHB 5300.4 (3C)............ Line Certification Requirements for Microcircuits
EIA:
   JEDEC *Eng. Bull.* 1-B....... Glossary of Microelectronic Terms, Definitions and
                                Symbols

PACKAGE SELECTION

Before a thick film layout can be generated, the end-product package configuration and size must be established. The normal procedure for accomplishing this phase of the design cycle is illustrated in Fig. 75. The first step is to review the cost, size, environmental, and other basic system requirements and to select the fabrication technique and general package style which best conforms to the specified objectives. Next the type and approximate size of all included components must be established. Once these are known, the maximum substrate area needed to fabricate the circuit in thick film form can be calculated. Finally, based on this calculated area, the number of leads, the desired lead configuration, and the circuit power dissipation, a specific package type and size can be determined.

This procedure is based on the assumption that the package can be chosen to meet the physical requirements of the circuit and its included components. This is not always possible, however. Often the package size and shape are predetermined. For instance, a thick film circuit may be desired in a plastic molded dual-in-line package in order to conform to a general system packaging scheme; or a small irregular-shaped package may be specified for system in which high packaging density is required. In such instances it is necessary to deviate from the stated procedure and work backward so to speak. With the package size and configuration known, the problem becomes one of determining whether the circuit can be designed and fabricated in the allotted space. Several attempts at an optimum layout and a careful consideration of component sizes are often necessary when limitations of this nature are imposed.

Although the procedure outlined in Fig. 75 appears quite simple, it is in reality rather complex. This is primarily due to the many variables involved and the compound interrelationships between them. Because of these interrelationships, no simple optimization rules or guidelines can be established. Package selection therefore becomes a trial-and-error process in which the optimum configuration is found by comparing the module requirements with the various available fabrication and packaging alternatives. This is usually a lengthy process relying heavily on input from many phases of the design cycle and on the knowledge and discretion of the designer. Many methods for aiding the selection of an optimum package form have been attempted in the past, such as applying weighted values to the various factors involved. Although these methods help in the final decision process, they do little to reduce the total effort.

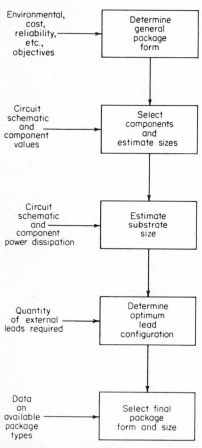

Fig. 75 Typical procedure for selecting final thick film package form.

Little can be done here either to help minimize the overall task except perhaps to indicate in more detail the inputs required and the key factors influencing the final package choice. This information for the major steps outlined in Fig. 75 is provided in the following paragraphs.

General Fabrication and Packaging Considerations

The first step in establishing a final thick film package configuration is to select the appropriate fabrication technique and to determine the general package style required. The information for this step can be obtained from the module or system specifications or from a knowledge of the application. The specific factors are:

1. Cost
2. Size
3. Environment
4. Reliability
5. Electrical performance
6. Power dissipation
7. Application

The application and/or the environmental and reliability requirements will usually dictate whether or not a hermetic package is required. If it is, an appropriate packaging type must be selected. Size and cost goals will usually establish the fabrication techniques and the type of attached components to be utilized. This will further influence the final package form. For instance, if size is the primary goal, a small package type is needed and components in chip or miniature carrier form are in order. If cost is of prime importance, then encased components on a large substrate with a conformal encapsulation will be the most likely approach. In this latter example, semiconductors in individually hermetically sealed packages such as TO-5 cans will be necessary if hermeticity is required. Plastic-covered devices will usually suffice for most other applications.

The electrical-performance requirements and circuit power dissipation also influence the overall fabrication and packaging methods. When a circuit with high power dissipation is encountered, care must be taken to provide sufficient substrate area and a package type capable of dissipating the heat generated. Other factors such as circuit sensitivity and high-frequency emission place further restrictions on the final package form. Here electrostatic shielding, such as a grounded-metal enclosure, may be required.

In brief, the important goal to achieve in this first step is determining the appropriate hybrid fabrication technique and gaining insight into the general end-product package configuration.

Attached-Component Selection

Once the hybrid technique is established and the general physical form of the attached components is known, i.e., encased, chip carrier, chip, etc., the individual components can be selected. At this point it is not necessary to dwell on electrical details, establishing instead the approximate physical size of the components. The dimensions of some typical chip devices are given in Table 35. Additional size information on chip and other component forms can be found in the Component Selection section of this chapter and in the data sheets published by component suppliers. Approximate data for all attached components should be obtained and tabulated. This information is needed for estimating substrate area, the next important step in package selection.

Estimating Substrate Area

There are several methods of estimating the amount of substrate area required for the fabrication of an electronic circuit in thick film form.* Most thick film designers have adopted some special technique of their own, but all are based on the same general concept. One simple, yet fairly accurate method is as follows:[64]

1. List all the resistors, capacitors, and other devices to be included on the thick film substrate and record the maximum power dissipation anticipated for each.

* See also page 4-32.

2. Calculate and list the area required for each circuit element. The values given in Table 35 can be used as typical guidelines. (Note that the resistor-area equation is based on a power density of 25 W/in.^{2}. This equation should be modified if materials with higher or lower power density are used.)

3. Total the individual component areas and multiply the resulting sum by 5. This figure accounts for the area of the lead pads and all interconnections.

The net result of this calculation is the total area required for the thick film circuit. This figure is essential in establishing the final package size or in determining whether a circuit will physically fit within a given package size. An example of this area-estimating technique is provided in Table 36 for the circuit shown in Fig. 76.

TABLE 35   Approximate Sizes of Commonly Used Thick Film Hybrid Components[64]

Resistors (thick film)

$$\text{Area} = \frac{0.04 \text{ in.}^2}{\text{watts}} \times \text{power dissipation (watts)}$$

NOTE 1: All resistors of less than 25 mW should be listed as 25 mW

NOTE 2: Area should be doubled for resistors less than 10 Ω or greater than 50 kΩ.

Active devices (chips)

Type	Size, in.^{2}
Small-signal silicon transistors	0.0015
Signal diodes	0.0010
Integrated circuits	0.0040
Rectifier diodes	0.0040
Power transistors	0.0040

Capacitors (chip, 50-V)

Size, in.^{2}	NP0	K1200	K7000
0.005	2–500 pF	120 pF–0.02 μF	560 pF–0.056 μF
0.008	5–1,000 pF	180 pF–0.04 μF	560 pF–0.1 μF

Although this example depicts a network with chip-type components, the estimating method is equally applicable to thick film networks employing encased components.

External-Lead Considerations

The quantity and spacing of leads required for interconnecting the thick film network to the next higher packaging level will often influence the final package size and shape. As a result, due consideration should be given to this question before selecting the final package form.

Lead quantity can affect both the size and shape of the final package. The size is affected by the simple fact that leads require termination pads and termination pads require substrate area. As the lead quantity is increased, the substrate and package size must be increased accordingly to accommodate them. Often it is this factor rather than the total area required for the thick film network that dictates the final package size. It is difficult to place an exact upper limit on the number of leads a thick film module can have, as lead quantity is highly dependent upon application, circuit complexity, fabrication capabilities, and package-cost goals. A good design practice is to limit the quantity of leads per module to a number commensurate with available, economic, standard packaging methods.

The effect of lead quantity on package shape is illustrated in Fig. 77 for a network requiring 16 leads and a total substrate area of 0.240 in.^{2}. The configuration of

TABLE 36 Example of Estimating Size of Thick Film Hybrid Substrate[64] for Circuit Shown in Fig. 76

Component	Value	Power, mW	Estimated size, in.^{2}
R_1	$2,000\ \Omega$	10	0.001
R_2	$120\ \Omega$	4	0.001
R_3	$390\ \Omega$	50	0.002
R_4	$120\ \Omega$	4	0.001
R_5	$220\ \Omega$	73	0.00292
R_6	$330\ \Omega$	85	0.00340
R_7	$330\ \Omega$	20	0.001
R_8	$330\ \Omega$	85	0.00340
R_9	$330\ \Omega$	20	0.001
R_{10}	$2,000\ \Omega$	10	0.001
R_{11}	$2,000\ \Omega$	10	0.001
R_{12}	$150\ \Omega$	129	0.00516
R_{13}	$220\ \Omega$	81	0.00324
R_{14}	$510\ \Omega$	25	0.001
R_{15}	$150\ \Omega$	8	0.001
R_{16}	$50\ \Omega$	0.1	0.001
Q_1	2N2894	19	0.0015
Q_2	2N2894	23	0.0015
Q_3	2N709	26	0.0015
Q_4	2N709	26	0.0015
Q_5	2N2894	67	0.0015
Q_6	2N2894	67	0.0015
Q_7	2N2894	65	0.0015
Q_8	2N709	22	0.0015
CR_1	FD600	10	0.0010
CR_2	FD600	10	0.0010
CR_3	FD600	10	0.0010
CR_4	FD600	10	0.0010
C_1	1,000 pF	—	0.0050
C_2	0.1 μF	—	0.0080
C_3	0.1 μF	—	0.0080
			0.06712
Multiplier for interconnections......			$\times$ 5
Total estimated substrate area......			0.3356 in.^{2}

Fig. 77b is a simple single-in-line style. Although this configuration is a very popular thick film package form, it is not recommended here because of the impractical length-to-width ratio. The configuration of Fig. 77a with two rows of leads and better size proportionment is more appropriate for this application. Typical package styles exhibiting this configuration are the TO-116 type dual-in-line package and the various flat-pack forms. If the same network required only 12 leads, either configuration would be equally applicable.

Lead spacing primarily affects package size. Typically, a spacing of 0.100 in. center to center or greater is preferred, but spacings as low as 0.050 in. are possible. In general, lead spacing should be selected to fit the application; however, spacings less than 0.100 in. should not be used unless size is truly critical. Most of the newer standard package forms have 0.100-in. lead spacings.

Final Package Selection

The final phase of the package-selection process consists of three basic steps:
1. Selection of the final package form
2. Determination of final package size
3. Determination of the final substrate dimensions

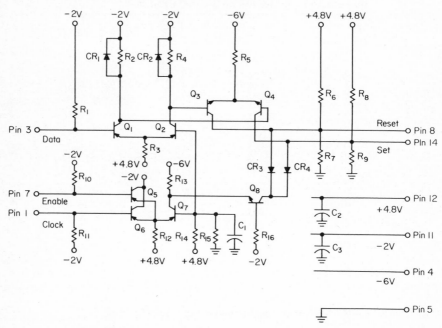

Fig. 76 Circuit utilized in package-selection example.[64]

Once the fabrication techniques, general package form, substrate area, and lead quantity are known, the final package type can be selected. This is accomplished by comparing these known requirements with the characteristics and capabilities of the various available package types and selecting the one which best fits the application in all respects. As stated previously, the designer's knowledge of packaging methods and available package types is a key factor in this final decision. Figure 78 is provided to illustrate the wide variety of thick film package types available. More specific information on thick film packaging is given in Chap. 9 and therefore will not be discussed here. Thick film designers are urged to familiarize themselves with the contents of that chapter and any other literature describing thick film package forms.

As a final aid, a summary of the key factors which influence the final package

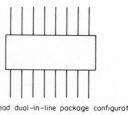

(a) Sixteen-lead dual-in-line package configuration

(b) Sixteen-lead single-in-line package configuration

Fig. 77 Effect of lead quantity and configuration on final package size and shape.

selection and several suggestions which will help narrow the field of choice are provided below in a procedural format.

1. Review the application and the environmental requirements to ascertain whether hermeticity is required.

2. Review the final size requirements (if specified), the estimated substrate area, and the component forms required. These items are very closely interrelated. For example, if the final package size is stipulated, the component types and substrate area must be selected to conform. However, if the final package size is not specified, then everything, including the size and shape of the final package, is left to the discretion of the designer. In this later case, the final determination will most likely be influenced by other factors such as cost, environmental requirements, power dissipation, etc. Also at this point it may be possible to narrow the field of choice appreciably. For instance, if hermeticity is required and a chip-hybrid construction is

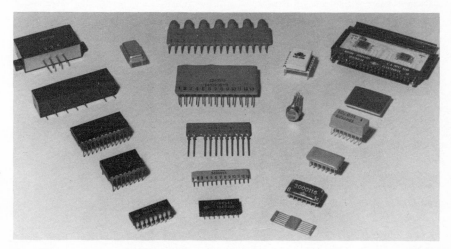

Fig. 78 Common thick film package configurations.   (*Centralab Division, Globe-Union Inc.*)

needed because of size restrictions, then a miniature hermetic package must be employed. Only a few basic package styles are available in this category.

3. Review the number of external leads required. As indicated earlier, lead quantity can greatly influence package size and configuration. This factor alone can preclude the use of some package forms.

4. Review the electrical and physical aspects of the end-product application. Sometimes the electrical requirements will dictate special package considerations such as electrostatic shielding. Physical requirements such as seated height, module-attachment methods, and compatibility of package type with other components in the same assembly will often pinpoint a package form.

5. Review the availability of the various package forms. If conformal coated package forms cannot be used, then EIA registered or other published package forms are recommended since they are readily available and competitively priced. Avoid tight tolerance and irregular-shaped packages, which are difficult to achieve and usually expensive.

6. Review the packaging capabilities and processes available. This is of particular importance if fabrication of the thick film module is to be accomplished in-house. For instance, it would not be judicious for a low-volume run to select a plastic package form which must be injection-molded around the substrate if the tooling and capability do not already exist in-house. In this instance it would be more advantageous to procure an available plastic header and shell and attach the thick film network to it.

7. Determine the most economical approach. Economics is an important factor in any thick film design. In some instances, e.g., consumer-product applications, it becomes a critical factor and greatly limits the choice of package form. In other instances cost can be used as the deciding factor when many alternatives are possible.

8. Select the final package type. After reviewing all the above factors and all the applicable package forms, select the one which best fits the application in all respects.

After the selection of the final package type (or in conjunction with it) a determination must be made to assure that the selected type can be obtained in a size compatible with the power dissipation of the thick film network it is to contain. To accomplish this, determine the total power dissipation of the circuit and compare this figure with the dissipation capabilities of the selected package. Table 37 summarizes typical power-handling capabilities of some of the popular standard package forms. If the power dissipation of the circuit exceeds the capabilities of the selected package, a larger package must be used. Other possibilities include the selection of

TABLE 37   Size and Power-Handling Capability of Common Thick Film Hybrid Package Types[64]

Package	Size, in.	P_d at 25°C	Linear derating, per °C	Area available, in.²
TO-5	0.180*	500 mW	3.3 mW	0.04
	0.250*	500 mW	3.3 mW	0.08
TO-8	0.150*	1.25 W	0.008 W	0.09
	0.185*	1.25 W	0.008 W	0.18
	0.275*	1.25 W	0.008 W	0.27
TO-3		4 W	0.02 W	0.25
	$\frac{1}{4} \times \frac{1}{4}$	500 mW	3.3 mW	0.020
	$\frac{1}{4} \times \frac{3}{8}$	500 mW	3.3 mW	0.045
	$\frac{3}{8} \times \frac{3}{8}$	500 mW	3.3 mW	0.055
	$\frac{1}{2} \times \frac{1}{2}$	750 mW	5 mW	0.15
	$\frac{5}{8} \times \frac{5}{8}$	1 W	0.007 W	0.23
	$\frac{3}{4} \times \frac{3}{4}$	1.25 W	0.008 W	0.29
	1×1	1.5 W	0.01 W	0.64

* Maximum height.

a different package type with greater thermal capabilities or the utilization of special heat-sinking methods.

Finally, after the package size is established, the final substrate dimensions can be exactly determined and the thick film layout started. One word of caution at this point relative to substrate dimensions: because of the shrinkage which occurs during kiln firing of ceramic, ceramic substrates cannot normally be obtained with tight dimensional tolerances, and consequently, thick film designers should become cognizant of the standard acceptable tolerance limits and plan their designs accordingly. This means selecting a substrate size which will fit within the desired package type at its maximum tolerance limit and planning the layout to fit on the substrate when it is at its minimum tolerance limit. Further information on the dimensional properties of ceramic substrates is provided in Chap. 4.

THERMAL DESIGN AND ANALYSIS

All materials and components are limited by inherent maximum operating temperatures, above which their performance may degrade seriously or they may fail catastrophically. In semiconductors, this maximum temperature is usually specified in terms of maximum junction temperature. For other components, maximum body temperatures and hot-spot temperature are commonly used as the figure of merit. In any case, the temperature of an electronic material or component is a direct result of the electric power dissipated in the element and the temperature and heat-transfer characteristics of the environment associated with the component. All components, especially resistors and active devices such as semiconductors, will dissipate heat in the performance of their normal electrical function. The problem is to remove the heat from the power sources as quickly as possible to prevent temperature rises within the heat-generating elements. In the past, this problem was easily solved by widely spacing or heat-sinking the high-heat-generating components or by providing sufficient cooling air to the components in question. In thick film microcircuits, however, the problem is quite different and the solution more complex. Because of their small size and high component densities, thick film microcircuits are subject to high power densities and poor heat-transfer capabilities. As a result, thermal power density rather than component density may be the limiting factor in certain thick film designs. Therefore, thermal-design considerations and thermal-analysis techniques are very important steps in the design cycle. It is quite important that the thick film design engineer become familiar and knowledgeable with regard to thermal-design guidelines, thermal properties of materials, thermal-analysis techniques, and thermal evaluation of thick film circuits.

This section is intended to provide insight into the thermal characteristics of thick film circuits and to present procedures for the thermal design, analysis, and evaluation of practical circuits, particularly with regard to the transfer of heat from the components to the package case. The removal of heat from the case to the ambient air or other cooling medium is quite complex and beyond the scope and intent of this section. The reader is directed to the many text and reference books on heat transfer for more detailed information on this aspect of thermal design and analysis.

Heat-Transfer Fundamentals

Before proceeding into the thermal problems and characteristics of thick film circuits, it may be worthwhile to review some fundamental heat-transfer concepts and indicate their relative importance to the problem at hand.

Heat, or thermal energy, is transferred from one body to another by virtue of a temperature differential. Two important axioms relative to heat transfer are (1) that heat flows only from a high-temperature region to one of a lower temperature and (2) that heat emitted by the high-temperature region must be exactly equal to that absorbed by the lower-temperature region.

When heat is transferred at a steady rate and the temperature at a given point is constant, a steady-state condition exists. Conversely, when heat flow is variable or a function of time, an unsteady or transient state results. Due to the complexity of transient heat-flow analysis, all information and methods of analysis presented in this section will be restricted to steady-state heat flow.

In general, there are three methods, or modes, of heat transfer; conduction, convection, and radiation. They may occur singly or simultaneously.

Conduction is usually defined as the transfer of heat through a solid when a temperature difference exists across the solid. It is considered to be caused by molecular oscillations within the solid.

Convection is the transfer of heat from the surface of a solid to a fluid or within a fluid itself. Natural, or free, convection occurs when a warm object or surface is placed in a still fluid such as air. As the temperature of the fluid next to the heated object increases, the fluid in this region becomes less dense and begins to cause a circulatory type of motion within the fluid. When motion or circulation of the fluid is caused by mechanical or other external means, the heat-transfer method is referred to as *forced convection*.

Radiation is the transfer of thermal energy by electromagnetic radiation. The wavelengths of the radiated waves can range from the long infrared to the short ultraviolet. Radiation is the only form of heat transfer that can occur in a vacuum. If all the incident radiation to a body in space receiving radiant energy is absorbed, zero energy being emitted or reflected, it is a perfect absorber, or *blackbody*. There are no perfect blackbodies in nature, although some surfaces approach blackbody characteristics. Also, in the absence of conduction and convection, a body at thermal equilibrium must emit energy equal to the amount received. Hence, a body which is a good receiver or absorber is also a good radiator or emitter.

The primary thermal consideration in the design of thick film circuits is the transfer of heat from the individual elements or components to the package case. This heat transfer is accomplished almost entirely by conduction. The transfer of heat from the package to the ambient environment, in the absence of conductive heat sinks, is accomplished by convection and radiation. This type of heat transfer is a secondary thermal consideration since methods external to the package (such as heat fins or forced air) can be employed to maintain design requirements. The remainder of this section will be concerned for the most part with the primary problem, namely, the transfer of heat from component to case by means of conduction.

Conductive Heat Transfer

The steady-state transfer of heat through a solid by conduction can be described by the one-dimensional Fourier equation

$$Q = \frac{kA}{L}(T_1 - T_2) = \frac{kA}{L}\Delta T \qquad (6)$$

where Q = heat flow per unit time
k = thermal conductivity of material
A = cross-sectional area perpendicular to direction of heat flow
L = length of the heat-flow path
T = temperature difference across length L

A geometric representation of this equation is shown in Fig. 79.

The thermal conductivity k is of particular importance since it represents the heat-transfer capability of a material. It is a measure of the quantity of heat which flows across a unit area in unit time when the length of the heat path is unity and the temperature gradient across the path is unity. Each material has a different thermal-handling capability, metals being the best conductors and such materials as plastics, glass, and foam being poor conductors, or thermal insulators. Since conduction is the primary mode of heat transfer within a thick film microcircuit, it is important to know and use the best conducting materials in order to obtain optimum heat transfer.

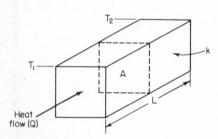

Fig. 79 Geometric representation of thermal conduction.

Table 38 lists the thermal conductivities of the materials commonly used in thick film applications. It should be noted that the values are given in two systems of units: $(W)(in.)/(in.^2)(°C)$, also stated as $W/(in.)(°C)$, and $(Btu)(ft)/(h)(ft^2)(°F)$, or $Btu/(h)(ft)(°F)$. In some literature the thermal conductivity will also be found expressed in units of $(Cal)(cm)/(s)(cm^2)(°C)$ and $(W)(cm)/(cm^2)(°C)$. To translate between the various systems of units, a list of conversion factors is given in Table 39. In this section, the unit $W/(in.)(°C)$ will be used.

Finally, it should be noted that the thermal conductivity is slightly dependent on the temperature of the material. Table 38 specifies values at $20°C$, which for all practical purposes is satisfactory for preliminary analysis.

Electrothermal Analog

Heat flow by conduction is analogous to current flow and bears a similarity to Ohm's law. When Eq. (6) is rewritten as

$$Q = \frac{\Delta T}{L/kA} \qquad (7)$$

it can be seen that Q is analogous to the electric current I; T to the electric potential V; and L/kA to the electric resistance R. The expression L/kA is commonly referred to as the *thermal resistance* θ. The direct analog of this expression is the expression

TABLE 38   Thermal Conductivity of Common Microelectronic Materials at 20°C

Material	Thermal conductivity	
	W/(in.)(°C)	Btu/(h)(ft)(°F)
Silver...........................	10.6	241
Copper...........................	9.6	220
Eutectic bond....................	7.50	171.23
Gold.............................	7.5	171
Aluminum........................	5.5	125
Beryllia 95%.....................	3.9	90.0
Molybdenum.....................	3.7	84
Cadmium.........................	2.3	53
Nickel...........................	2.29	52.02
Silicon...........................	2.13	48.55
Palladium........................	1.79	40.46
Platinum.........................	1.75	39.88
Chromium........................	1.75	39.88
Tin..............................	1.63	36.99
Steel............................	1.22	27.85
Solder (60-40)...................	0.91	20.78
Lead.............................	0.83	18.9
Alumina 95%.....................	0.66	15.0
Kovar............................	0.49	11.1
Epoxy resin, BeO-filled...........	0.088	2.00
Silicon RTV, BeO-filled...........	0.066	1.5
Quartz...........................	0.05	1.41
Silicon dioxide...................	0.035	0.799
Borosilicate glass................	0.026	0.59
Glass frit........................	0.024	0.569
Conductive epoxy.................	0.020	0.457
Sylgard resin....................	0.009	0.21
Epoxy glass laminate..............	0.007	0.17
Doryl cement....................	0.007	0.17
Epoxy resin, unfilled..............	0.004	0.10
Silicon RTV, unfilled..............	0.004	0.10

TABLE 39   Thermal-Conductivity Conversion Factors

From	To			
	$\dfrac{(cal)(cm)}{(s)(cm^2)(°C)}$	$\dfrac{(W)(cm)}{(cm^2)(°C)}$	$\dfrac{(W)(in.)}{(in.^2)(°C)}$	$\dfrac{(Btu)(ft)}{(h)(ft^2)(°F)}$
$\dfrac{(cal)(cm)}{(s)(cm^2)(°C)}$	1	4.18	10.62	241.9
$\dfrac{(W)(cm)}{(cm^2)(°C)}$	2.39×10^{-1}	1	2.54	57.8
$\dfrac{(W)(in.)}{(in.^2)(°C)}$	9.43×10^{-2}	3.93×10^{-1}	1	22.83
$\dfrac{(Btu)(ft)}{(h)(ft^2)(°F)}$	4.13×10^{-3}	1.73×10^{-2}	4.38×10^{-2}	1

for electric resistance $\rho L/A$, where ρ is the electric resistivity of a material. Consequently, the reciprocal of the thermal conductivity is the thermal resistivity. If the symbol ρ_t is used for thermal resistivity, the thermal-resistance equation becomes

$$\theta = \frac{\rho_t L}{A} \tag{8}$$

The similarity between this expression and the expression for electric resistance is now obvious. Employing the thermal-resistance concept, we can now rewrite Eq. (7) as

$$Q = \frac{\Delta T}{\theta} \tag{9}$$

Other thermal properties have similar electrical analogs, given in Table 40.

To illustrate how the analogy can be employed in a typical thermal problem, consider the following example.

A 0.1 by 0.1 in. heat source is eutectically bonded to a 1 by 1 by 0.025 in. thick

TABLE 40   Electrothermal Analogs

Electrical quantity			Thermal equivalent		
Parameter	Symbol	Unit	Parameter	Symbol	Unit
Current...........	I	A	Heat flow............	Q	W/s
Potential..........	E	V	Temperature difference.	T	°C
Resistance........	R	Ω	Resistance...........	θ	°C/W
Capacitance.......	C	F	Capacitance.........	C_t	W-s/°C
Conductance......	G	mho	Conductance.........	k	W/°C
Time constant =			Time constant =		
$R \times C$.........	...	s	$\theta \times C_t$.............	...	s

95 percent alumina-ceramic substrate. The substrate is bonded with conductive epoxy to an aluminum heat sink, which is maintained at 60°C. The eutectic bond is 1 mil thick, and the epoxy is 3 mils thick. The problem is to determine the temperature of the heat source when it is dissipating 100 W. Figure 80a is a pictorial representation of this assembly. The equivalent electric circuit is shown in Fig. 80b. The thermal resistance of the various elements can be determined from Eq. (8) as follows:

$$\theta_{\text{eutectic bond}} = \frac{(1/1.5)\,(0.001)}{(0.1)\,(0.1)} = 0.013°C/W$$

$$\theta_{\text{substrate}} = \frac{(1/0.53)\,(0.025)}{(1)\,(1)} = 0.047°C/W$$

$$\theta_{\text{conductive epoxy}} = \frac{(1/2.07 \times 10^{-2})\,(0.003)}{(1)\,(1)} = 0.145°C/W$$

Since the resistors are in series, the total thermal resistance will be

$$\theta_{\text{total}} = \theta_{EB} + \theta_S + \theta_{CE} = 0.205°C/W$$

The temperature of the heat source can be obtained by

$$Q = \frac{\Delta T}{\theta} = \frac{T_1 - T_2}{\theta}$$

$$100 = \frac{T_1 - 60}{0.205}$$

$$T_1 = 80.5°C$$

This problem can also be solved using a different approach. Suppose it is desirable to ascertain the power dissipated in the heat source when it is operating at a temperature of $100\,^\circ$C. In this case

$$Q = \frac{100 - 60}{0.205} = \frac{40}{0.205} = 195.1 \text{ W}$$

Figure 81 illustrates an assembly in which the thermal-conduction paths are in parallel rather than series. The solution to this problem would be similar except that

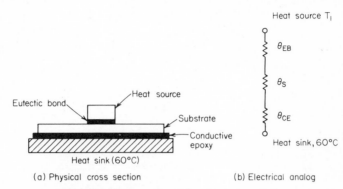

(a) Physical cross section   (b) Electrical analog

Fig. 80   Series-path conductive heat transfer.

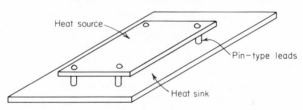

(a) Pictorial representation of example

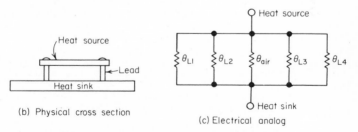

(b) Physical cross section

(c) Electrical analog

Fig. 81   Parallel-path conductive heat transfer.

the equivalent thermal resistance of the parallel combination must be found from the following familiar equation:

$$\frac{1}{Q_T} = \frac{1}{\theta_1} + \frac{1}{\theta_2} + \frac{1}{\theta_3} + \ldots + \frac{1}{\theta_n} \tag{10}$$

Although there are many methods of thermally analyzing and evaluating thick film circuits, the electrothermal analogy has many practical advantages and is widely used for preliminary analysis, especially when the mode of heat transfer is by conduction. It is reasonably accurate, easy to perform, and readily understandable. Also, the electrothermal analogy can be employed early in the design cycle. As a result, it gives the designer a quick method of determining thermal problem areas and a visual aid in evaluating design alternates. Most of the information and data presented in the remainder of this section will be based on this concept.

Thermal Characteristics of Thick Film Elements

Thick film circuits can comprise a wide variety of component types, including film elements, chip devices, and complete encased self-contained components. Since the maximum operating temperatures and thermal characteristics of these components are as varied as the components themselves, a knowledge and understanding of their thermal properties is a basic requirement for good thermal design. Of specific concern are the heat-dissipating elements such as resistor and semiconductor devices. Capacitors are usually not considered in the thermal evaluation since they dissipate very little heat. However, care should be taken to assure that capacitors such as tantalum devices are not thermally overstressed from neighboring high-power-dissipating elements.

Thick film resistors   Thick film resistors, like other electric components, are temperature-limited. A specific maximum operating temperature is difficult to state since it depends upon the particular resistor formulation used. A safe temperature limit for most cermet pastes is between 125 and 150°C. Manufacturerers' data sheets or direct evaluation can be used to obtain more specific operating limits. Even when loaded well above ratings, thick film resistors are not likely to fail catastrophically. In general, the failure mode is exemplified by drift in the resistor values or changes in the temperature coefficient.

Resistor formulations are commonly rated in terms of the power-density capability of the final fabricated resistors. Values between 20 and 50 W/in.^{2} are typical. Although this parameter is quite useful in calculating resistor sizes, it is not all-inclusive and can be misleading if not properly understood. At present, there is no standard method for determining power density, and material suppliers are free to evaluate this parameter in any manner they wish. Typically, it is accomplished by measuring the temperature of small resistors on large substrates when loaded to specified power levels. For instance, a 0.1-inch-square resistor on a 1-in.-square substrate may be found capable of dissipating ¼ W without exceeding a specified temperature limit or experiencing performance degradation. Through normalization to a unit square inch, the material is said to be capable of 25 W/in.^{2}. However, a 1-in.-square resistor of the same material on the same size substrate will be found capable of dissipating only 2 to 3 W in free air before experiencing excessive temperatures, the limiting factor being the heat-transfer capability of the substrate. Other factors such as ambient temperature at which evaluation was made, accuracy of measuring instruments, and mode of cooling (still air, forced air, etc.) also affect the validity of this parameter. However, no matter how a resistor size or power rating is determined, the limiting factor is still the maximum allowable operating temperature. The purpose of a thermal evaluation is to assure that this temperature will not be exceeded.

The thermal performance of thick film resistor networks depends on so many factors, e.g., substrate area and thickness, resistor material utilized, package configuration, lead quantity and configuration, that it is impossible to present equations or design criteria applicable to all design possibilities. There are, however, some general performance characteristics typical of all thick film resistor networks, which can be studied on a specific design and the results applied, at least on a relative basis, to most other thick film configurations. The thick film resistor networks of Fig. 82 will be used to illustrate these general performance characteristics.[65]

The seven networks shown in Fig. 82 were all screened with material of the same resistivity on the same size substrate, and tinned-copper leads were soldered to each of four teminating pads. Modules 1, 2, and 3 contain a single resistor with an inverse aspect ratio of 1:2, but the resistor areas vary in a 4:2:1 ratio, respectively. Networks 4 to 7 have three resistors each. The total resistor area of each of these modules is the same and is equal to the area of the single resistor of module 3. The resistor geometrics of these modules differ however with modules 4, 5, and 6 having

½-, 1-, and 2-square pattern sizes, respectively, and module 7 having a serpentine design.

The power-handling capability of various size resistors of similar design (modules 1 to 3) and R_2 of module 4 on the same size substrate is illustrated in Fig. 83. These curves show that at a given power level the temperature rise is inversely proportional to the resistor area. Thus, resistors should be designed as large as possible in order to obtain minimum operating temperatures. Consequently, if manufacturers' power-density values are used in making size determinations, adequate derating factors must be applied to account for the area reduction anticipated during resistor adjustment and for operation in high-temperature environments.

The concept of power density is illustrated in Fig. 84, where the power levels at the various temperatures depicted in Fig. 83 are divided by their corresponding resistor areas to obtain the power density in watts per square inch. This set of curves reveals that for a given power density the resistor-temperature rise is directly proportional to the resistor-to-substrate area ratio. As previously stated, a given size substrate is capable of dissipating a limited amount of power in still air. For a 1-in.-square high-alumina substrate, this is typically around 2 to 3 W. For best thermal efficiency and minimum resistor-temperature rise, the resistor-to-substrate area ratio should be as small as possible.

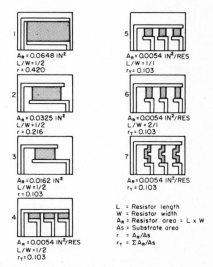

Fig. 82 Thick film resistor designs utilized in thermal study.[65]

A third characteristic is illustrated in Fig. 85, which shows the results of applying equal power levels to modules 3 and 4. Although these two modules have the same resistor aspect ratios and equal resistor areas, the resistors of module 4 experienced a 15 percent lower temperature than module 3. From this result it is apparent that best thermal efficiency can be obtained by providing maximum separation between high-dissipating resistors.

The resistor geometry also affects operating temperature, as can be seen in Fig. 86.

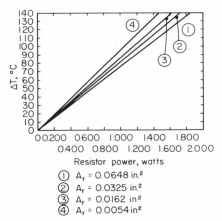

Fig. 83 Effect of resistor area on power capability.[65]

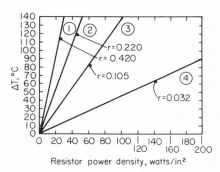

Fig. 84 Effect of resistor-to-substrate area ratio on power density.[65]

Although all these modules have three resistors of equal area, module 4, with an inverse aspect ratio of 1:2, was found to operate cooler than the other configurations. Modules 5 and 6 with 1- and 2-square geometries, respectively, exhibit a temperature rise approximately 5 percent higher than module 4. This may be attributable to the smaller contact area between the resistor material and its terminating conduc-

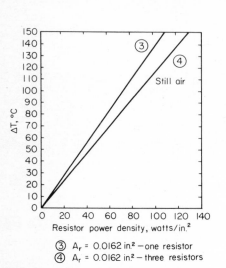

③ A_r = 0.0162 in.² — one resistor
④ A_r = 0.0162 in.² — three resistors

Fig. 85 Effect of distributing heat-generating resistors.[65]

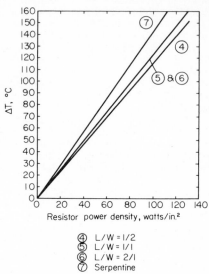

④ L/W = 1/2
⑤ L/W = 1/1
⑥ L/W = 2/1
⑦ Serpentine

Fig. 86 Effect of geometry on thermal characteristics of resistors.[65]

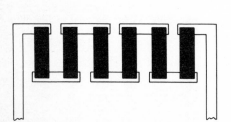

Fig. 87 Recommended serpentine resistor design for high-power applications.

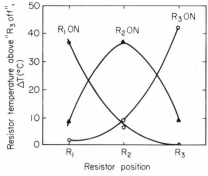

Fig. 88 Effect of resistor position on thermal profile of substrate.[65]

tor electrodes. It is important that sufficient overlap onto the terminating conductors be provided to minimize this potential source of temperature hot spots. The serpentine resistors of module 7 can be seen to operate at temperatures 20 percent higher than module 4. From a thermal standpoint, this type of resistor configuration is not recommended since each bend or corner represents an area of high current density and a potential source of thermal hot spots. If long resistors are required in high-current applications, the design technique of Fig. 87 is recommended.

When one resistor at a time on module 4 is energized to a power level of 1.2 W, the thermal profile of Fig. 88 results. The fact that R_3 experiences a higher operating temperature than the other resistors illustrates the need for sufficient substrate area in the vicinity of a resistor to obtain best heat transfer. High-power-dissipating resistors should be located away from the edges of the substrate unless heat sinks, leads, or other thermal conductors are available in the immediate area.

Thick film resistors properly applied and fired to a ceramic substrate are essentially isothermal, and the thermal resistance in the bond between the resistor film and the substrate surface is negligible. Thermal hot spots can occur in thick film resistors, but with proper design and fabrication they can be minimized and the isothermal condition assumed. Consequently, thermal analysis of a thick film resistor in applications where heat transfer is primarily by conduction is a simple matter of determining the thermal resistance of the substrate and the material bonding the substrate to the heat sink. Three common methods of determining thermal resistance through materials of dissimilar cross sections are illustrated in Fig. 89.

Method A assumes that the junction between any two dissimilar materials is isothermal. This assumption is often used in conductive-heat studies, particularly when the junction is between two excellent thermal conductors such as metals. In the analysis of thick film networks this assumption or technique can be used if the heat-dissipating element covers the majority of the substrate or if many elements of nearly equal dissipation cover the substrate surface area. For this case, the thermal

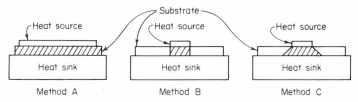

Fig. 89 Methods of determining thermal resistance.

resistance between the source and sink is a simple resistance and can be computed from Eq. (8), where A is the total substrate area.

Method B is referred to as the *shadow method* because thermal conduction is assumed to take place only in the area directly below, i.e., in the shadow of, the heat-dissipating element. Equation (8) is also used to determine the thermal resistance for this method, but area A is the area of the dissipating element only. This method provides a conservative figure and is often used when a few elements on a substrate are dissipating far more power than the other remaining elements.

Method C is based on the assumption that heat flows from source to sink at a spreading angle of $45°$. For this method, the thermal resistance, commonly called *spreading resistance*, is obtained from the relationship[66]

$$\theta_R = \frac{1}{2k(a - b)} \ln \frac{a}{b}\left(\frac{2L + b}{2L + a}\right) \tag{11}$$

where θ_R = thermal resistance
 a = length of dissipating elements
 b = width of dissipating elements
 L = thickness of conducting material
 k = thermal conductivity of material

Although this method may be the most precise, methods A and B are much easier to use and provide results which are satisfactory for initial evaluation.

Encased semiconductors A wide variety of encased semiconductors, including diodes, transistors, silicon-controlled rectifiers, monolithic integrated circuits, etc., in a broad assortment of package types can be utilized in thick film hybrid circuits. Since nothing can be done to enhance the internal thermal characteristics of these devices, the thick film designer must understand their specified thermal characteristics and design accordingly.

The thermal characteristics of encased components, particularly diodes and transistors, can be found on manufacturers' data sheets in the form of thermal-resistance values from junction to case θ_{JC} or junction to ambient θ_{JA}. The junction temperature T_J for a given power dissipation P_p and ambient temperature T_A can be calculated from

$$T_J = P_p\theta_{JA} + T_A \qquad (12)$$

Thermal-resistance values are not usually given for monolithic integrated circuits. Digital circuits are usually specified in terms of maximum power dissipation and maximum ambient temperature. Linear integrated circuits are also specified in terms of power dissipation, and a derating factor in milliwatts per degree Centigrade is usually included. One factor not found on specification sheets is the path of maximum heat transfer from the inside to the outside of the component case. Steps to assure optimum thermal performance within a given thick film design can be taken if these paths are known. Table 41 provides thermal data on many of the common package types used in thick film applications, including information on the best conductive heat paths.

TABLE 41   Thermal Resistance of Common Semiconductor Packages

Package type	Thermal resistance, °C/W			Location of best conductive heat path
	Junction to case	Case to ambient	Junction to ambient	
TO-5	35–100	115–200	150–250	On cap as close to flange as possible
TO-18	80–150	270–350	350–500	On cap as close to flange as possible
TO-46	80–150	220–300	300–450	On cap as close to flange as possible
TO-92 plastic	100–200	150–300	250–400	Collector lead
Miniature plastic	—	—	450–500	Collector lead
TO-116 plastic dual-in-line	100–150	50–150	200–300	One of the package leads (usually the ground lead, pin 14)
TO-86 ceramic flat pack	35–100	150–250	250–350	Bottom ceramic surface

Faceup-mounted semiconductor chips The sources of heat in semiconductor devices such as diodes, transistors, and monolithic integrated circuits are the junction areas and the diffused resistors. These elements are located in the top 1-mil surface of the semiconductor chip. The primary heat-flow path from this surface to the substrate is through the silicon body, normally around 0.007 in. thick, and the chip bonding material. Heat transfer through the gold or aluminum wires which form the electrical connection to the substrate is very small and can be neglected. Depending on the device geometry, the source of heat can be very small with respect to the surface area or can essentially cover the entire chip surface area. An estimate of the size of the heat-dissipating area can usually be made by inspecting the chip geometry visually. The approximate thermal resistances of various sizes of square silicon chips 7 mils thick with various heat-dissipating areas is given in Fig. 90.

The chip-to-substrate bond can be a source of high thermal resistance and should be given prime consideration. Common methods of bonding chips to a metallized pattern on a thick film substrate include eutectic bonding, adhesives, glass frit, and solder. The thermal resistance of these materials for various size bonding areas is

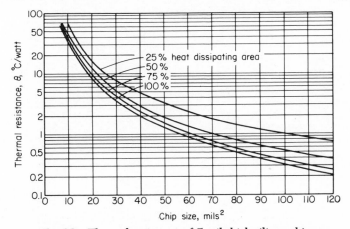

Fig. 90 Thermal resistance of 7-mil-thick silicon chip.

shown in Fig. 91. The eutectic and solder joining materials offer the least thermal resistance. However, these are not as compliant as the adhesives, and irregularities in the chip or substrate surfaces can cause voids in the joining material. As a result, the bonding contact area is somewhat less than 100 percent. For design purpose, a figure of 80 percent can normally be assumed. For best heat transfer, the bonding material should be a good thermal conductor, should be thinly applied, and should contact as much of the bonding area as possible, preferably 100 percent.

Facedown-bonded chips In facedown-bonded chips, e.g., flip-chip and beam-lead devices, the heat flow is through the parallel paths offered by the bumps or beam-type leads and the air gap between the chip and the substrate to which it is bonded. The electrothermal analogy for these devices is similar to that shown in Fig. 81c. For all practical purposes, the heat flow through the air gap can be ignored, and a uniform distribution of heat on the base of the chip can be assumed. An equivalent amount of heat will then flow through each of the interconnecting elements. Based on these assumptions, the approximate thermal resistance of flip-chip devices with various pad sizes, quantities, and materials is as shown in Fig. 92. Figure 93 illustrates the approximate thermal resistance of beam-lead devices. Here the thermal path is assumed to be perpendicular to the thin cross section of the gold beam and to have an effective length equal to one-half the total beam length, or about 0.004 in. In general, facedown-bonded chips exhibit appreciably higher thermal-resistance characteristics than faceup-mounted devices. It is recommended that a good thermal evaluation be performed when these devices are utilized.

Thermal resistance of final package Thick film circuits can be enclosed in a wide variety of package configurations. Many of these package types are described in Chap. 9. The exact thermal resistance of each package type is difficult to state since there are so many variations in size, shape, material composition, and application. However, if the configuration and physical characteristics of a specific package are known, an approximate evaluation can be realized by using the electrothermal analogy. Four of the most common package types utilized in thick film modules and their simplified substrate-to-ambient thermal-resistance analogs are illustrated

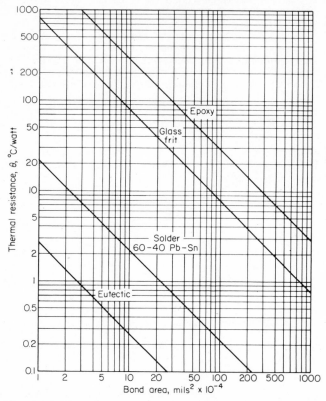

Fig. 91 Thermal resistance of common semiconductor-chip bonding materials (material thickness 2 mils).

in Fig. 94. Note that the thermal resistance for each package type is composed of both internal and external elements. When the analysis is concerned with only the primary thermal consideration, namely, the transfer of heat from source to case by conduction, only the internal resistance need be considered. These resistances can be found by calculation from known package dimensions and materials. The external thermal resistance is quite complex, involves all three modes of heat transfer, and is highly dependent upon such factors as mounting methods and materials, heat-sink size, material, and efficiency, cooling-air temperature and velocity, proximity to other heat-generating elements, and so forth. When analyzing the transfer of heat from source to case in thick film modules, the external thermal problem can be bypassed by assuming that the package-to-ambient heat-transfer interface is maintained at a specified maximum operating temperature. This, of course, assumes that sufficient cooling methods will be provided in the final system to prevent this inter-

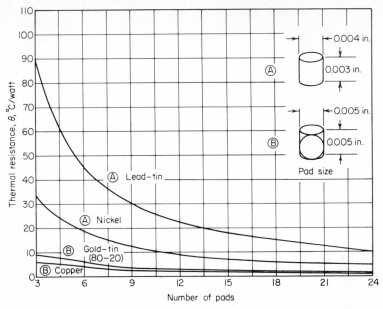

Fig. 92 Approximate junction-to-substrate thermal resistance of flip-chip semiconductors.[67]

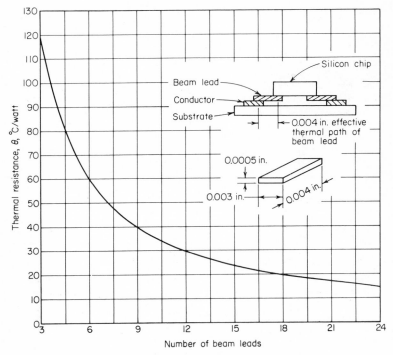

Fig. 93 Approximate thermal resistance of semiconductors with gold beam leads.

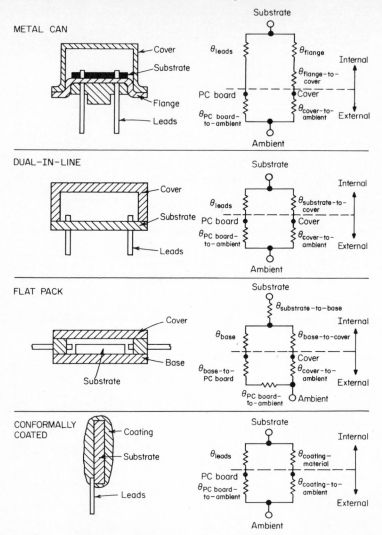

Fig. 94   Thermal equivalents of common hybrid-circuit package types.

face temperature from being exceeded. This is a reasonable assumption if sufficient thermal data are provided on the thick film product.

Thermal Design and Evaluation Example

The following example illustrates how the information in the foregoing paragraphs can be used to assist in the preliminary thermal evaluation of thick film. microcircuits.

A thick film package with four resistors and two transistors is shown in Fig. 95. Circuit details are as follows:

Resistors:
Size........................ 0.040 by 0.040 in.
Dissipation................ 100 mW max

Transistor, Q_1:
- Size.................... 0.025 by 0.025 in.
- Dissipation............. 250 mW in 50% of surface area

Q_2:
- Size.................... 0.100 by 0.100 in.
- Dissipation............. 1 W in 50% of surface area

Substrate................... 1 by 1 by 0.030 in., 95% alumina
Leads...................... 14 copper pins, 0.0005 in.^{2} in cross-sectional area

All transistors are eutectically bonded to the substrate metallization. The package is mounted 0.1 in. above a printed-circuit board, which is maintained at an ambient temperature of 60°C. The problem is to assure that the junction temperatures of Q_1 and Q_2 will not exceed 150°C.

Visual inspection shows that transistors Q_1 and Q_2 are operating at the highest power levels and are therefore most critical thermally. Q_2 will be analyzed first. The analysis can be simplified by assuming that all heat from the transistor is conducted

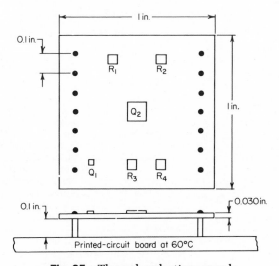

Fig. 95 Thermal-evaluation example.

away from the package by only six center pins. This worst-case assumption and its equivalent electrical analogy are shown in Fig. 96. The analysis begins with the determination of each of the thermal-resistance elements and the summation of these elements from source to sink:

$$\theta_{\text{silicon}} = 0.55°\text{C/W} \qquad \text{(Fig. 89)}$$

$$\theta_{\text{eutectic bond}} = \text{negligible} \qquad \text{(Fig. 90)}$$

$$\theta_{\text{substrate}} = \frac{0.5}{(0.66)(0.030)(0.3)} = 84.2°\text{C/W}$$

This is the thermal resistance of the substrate from the center of the transistor chip to one set of package pins. Since there are two such paths and they are in parallel, the net thermal resistance of the substrate is 42.1°C/W.

$$\theta_{\text{pins}} = \frac{0.1}{(9.6)(0.0005)} = 20.8°\text{C/W}$$

Since there are six pins offering parallel thermal-resistance paths,

$$\theta_{\text{pins}} = \frac{20.8}{6} = 3.5°\text{C/W}$$

$$\theta_{\text{total}} = 0.55 + 42.1 + 3.5 = 46.2°\text{C/W}$$

The junction temperature of Q_2 can now be found, using Eq. (9),

$$1 = \frac{T_2 - 60}{46.2}$$

$$T_2 = 46.2 + 60 = 106.2°C$$

This is below the specified maximum of 150°C.

Now for Q_1, assume that the heat is removed only by the two pins directly adjacent to the transistor chip. Then

$$\theta_{silicon} = 7.6°C/W \quad \text{and} \quad \theta_{eutectic\ bond} = 0.48°C/W$$

$$\theta_{substrate} = \frac{0.2}{(0.66)\,(0.2)\,(0.030)} = 50.5°C/W$$

$$\theta_{pins} = \frac{20.8}{2} = 10.4°C/W$$

$$\theta_{total} = 7.6 + 0.48 + 50.5 + 10.4 = 68.9°C/W$$

The resulting junction temperature T_2 can be calculated as before and will be found to be 77.2°C, which is well below the 150°C maximum.

It is interesting to note that if epoxy rather than the eutectic were used to bond Q_1, θ_{total} would be about 518°C/W. The junction temperature would then be 189.5°C, which is above the allowable maximum.

Thermal-Design Guidelines[67]

Several basic thermal-design rules and guidelines can be derived from the foregoing discussion. Utilization of these guidelines will help assure optimum thermal performance of a thick film circuit and will minimize costly redesign late in the design cycle.

1. Select a substrate material with good thermal conductivity, such as beryllia or alumina.

2. Use a thick substrate for best heat transfer and distribution unless it is to be bonded to a heat sink, in which case the thinnest possible substrate should be used.

3. Distribute power sources as evenly as possible over substrate surface to reduce hot spots and maintain a uniform thermal profile.

4. Locate highest power sources over or near heat sinks.

5. Do not locate high power sources near substrate corners or edges unless heat sinking is provided in that vicinity.

6. Design power resistors as large as possible and assure that sufficient area will be maintained after adjustment.

7. High-heat-dissipating components should be attached to the substrate in such a manner as to provide the lowest possible thermal resistance between component and substrate.

8. Components should not be attached over other heat-dissipating elements, i.e., chip capacitors over thick film resistors.

9. For best thermal performance, make the junction between a back-down-mounted semiconductor chip and the substrate with material of low thermal resistance, as thin as possible, and providing a uniform contact area, free from voids.

Assumed heat-conduction area

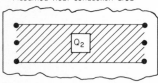

(a) Pictorial representation

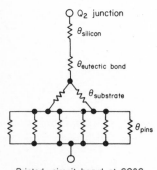

Printed–circuit board at 60°C

(b) Electrothermal analog

Fig. 96 Assumed conductive heat paths of evaluation example.

10. When wire leads or pins are attached to a substrate:
    a. Make leads or pins as short as possible.
    b. Select pin material of high thermal conductivity and maximum possible cross-sectional area.
    c. Use as many leads as possible.
    d. Locate heat sources as near the pins as possible.
11. In final packaging:
    a. Select a package configuration which will provide best thermal conductivity for the specific application.
    b. Provide a good thermal-conduction path between substrate and package case.
    c. Avoid air spaces within package if possible. Plastics with high thermal conductivity can be used to fill voids and improve heat transfer.
    d. Heat sinking a package can improve the power-handling capability by at least one order of magnitude.

THICK FILM LAYOUT GENERATION

Preliminary Physical Layout[1]

During the early phases of the design cycle the designer will normally require an initial sizing estimate or a preliminary layout to verify that the thick film form of the circuit function will meet the system size, weight, and volume requirements. At a minimum this consists of placing all components in the relative positions they are ultimately expected to occupy, arranging lead pads at the appropriate edges of the substrate, and estimating the total area required for all elements and interconnections. Frequently it will be determined at this point that one or more of the design goals cannot be met. When this occurs, the designer must review the design alternatives, select one or more, and repeat this preliminary step until all the design goals and functional requirements are satisfied or an acceptable compromise has been reached. Listed below are several design alternatives that can be considered.

1. The package style and/or package size may be changed.
2. Alternate circuit partitioning may be possible.
3. Large-value film resistors or capacitors may be replaced with miniature discrete components either within the module package or as discrete components external to the thick film module.
4. Alternate lead arrangements may be considered.
5. Packaged components may be replaced with unpackaged (chip) components.
6. Both sides of a substrate may be utilized for screening thick film elements and interconnections or mounting discrete components.
7. Multiple substrate or multilayer thick film screening techniques may be employed.

The thermal, environmental, and reliability requirements should also be given consideration at this point.

A typical procedure for making an initial size estimate and a preliminary layout is described in the following paragraphs.

Analyze existing data The following information and parameters, which should have been generated in preceding stages of the design cycle, are required to make a preliminary sizing estimate or layout:

1. Preliminary circuit schematic
2. Specified physical and functional requirements
3. Selection of basic technology to be employed for each component (discrete, thick film, chip, etc.)
4. Physical and electrical characteristics of each component
5. Preferred package style

These data should be carefully reviewed prior to any preliminary layout effort.

Functional schematic orientation Analyze and redraw the preliminary schematic to eliminate or minimize the number of crossovers, to position the external leads at the substrate edges to conform to the desired package configuration, and to place all components in the relative positions they are to occupy. All power components should be evenly distributed, and inputs and outputs should be well separated. An

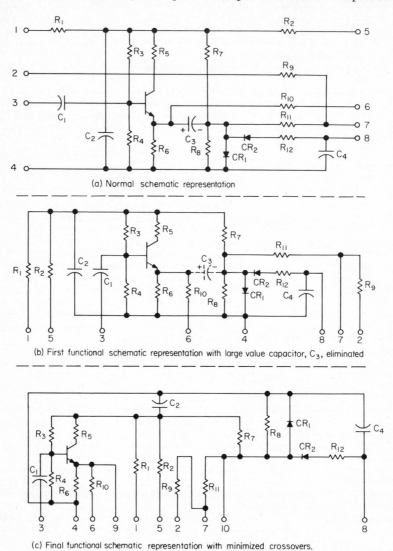

(a) Normal schematic representation

(b) First functional schematic representation with large value capacitor, C_3, eliminated

(c) Final functional schematic representation with minimized crossovers.
C_3 to be connected externally between pins 9 and 10

Fig. 97 Conversion of normal to functional schematic.

example of the development of a schematic representation of this type is shown in Fig. 97.

Particular consideration should be given to the elimination of crossovers. Since thick film resistors and inductors are constructed on a single plane, an isolation barrier is required at each crossover point. This is usually accomplished by applying one or two coats of a dielectric glaze, which adds fabrication processes to the thick film network, increasing the end-product cost and introducing a potential failure mechanism.

Estimating substrate size A procedure for estimating substrate size is presented in the section on Package Selection. Using the principles outlined in that section, a suitable initial estimate of the required substrate size can be made. From this estimate the overall package size and configuration can be established and a preliminary determination of whether the specified size goals will be achieved can be made.

Preliminary layout If the calculated area is very close to the available area, a detailed preliminary physical layout is recommended. In some instances, e.g., high-frequency applications, prototype thick film circuits are required for design evaluation, and a detailed preliminary layout is automatically required. If situations of this type exist, proceed with the thick film layout following the final physical-layout procedures which follow. A preliminary layout will enable the designer to determine more precisely the required substrate area and will aid him in selecting design alternatives when the initial design does not meet the specified dimensional goals.

Final Physical Layout Considerations[1]

This section establishes the procedures and guidelines for the final circuit layout. This final layout procedure consists of positioning the various thick film and add-on circuit elements, the circuit-element interconnections, and the terminal pads on an enlarged outline of the substrate. The most common scale is 10:1. A scale of 20:1 is sometimes used with designs in which thick components and conductor sizes are less than the minimums shown in this section. A separate piece of artwork is generally required for each material, process step, or thick film layer required. Preparation of this artwork is discussed in Chap. 2. The remainder of this section describes only those steps necessary to produce the final composite thick film layout.

Review existing data Prior to the final circuit layout the designer will have the final circuit schematic and parts list, a preliminary physical layout or an estimate of the substrate size, the preferred package and lead breakout, the final technology selection for each circuit element (screened thick film, encapsulated add-on, chip add-on, etc.), and confirmation from the worst-case design and preliminary thermal and reliability analyses that the projected design will meet the functional requirements. This information should be reviewed before the final layout is started and should be available for ready reference as the layout progresses.

Functional schematic orientation If neither the circuit schematic nor the package style has been changed subsequent to the preliminary physical layout, the functional schematic orientation will remain the same. If changes have been made, redraw the final schematic following the procedures outlined for the preliminary physical layout.

Substrate size determination The preliminary physical layout established a substrate size compatible with the circuit-element density and the preferred package. Subsequent preliminary thermal and reliability analyses should have confirmed the suitability of the selected substrate size. If any changes have been made subsequent to these steps, they should be repeated and an appropriate substrate size selected.

Key-element positioning During the preliminary physical layout the elements requiring unique placement were identified. The circuit and worst-case analyses, preliminary thermal and reliability analyses, or the breadboarding steps may have identified additional elements which require unique placement. Review these steps and list all key elements.

Physical layout The general layout procedure is outlined below:
1. Determine the size and configuration of all thick film circuit elements.
2. Locate all key elements.
3. Locate external-lead bonding pads.
4. Locate other circuit elements and draw interconnect patterns.
5. Orient all chip elements to minimize wire-bond and interconnect lengths.
6. Locate and indicate crossover patterns and interconnections.
7. Identify all circuit elements, circuit-element characteristics, trimming directions, and materials and processes necessary to fabricate the design.

Final Physical Layout Guidelines

The appropriate guidelines for implementing this procedure are given in the following paragraphs. Both minimum and preferred design dimensions are provided. The preferred design dimensions are recommended and are representative of existing production capabilities. Designing to minimum dimensions should be avoided. However, as the state of the art advances and improved manufacturing capabilities are implemented, high-production yields to the minimum design dimensions will also be practical. Before proceeding into the final thick film layout, there is one last important factor to consider. The layout should be prepared so that it will fit on any substrate which complies with the dimensional tolerances specified for the selected

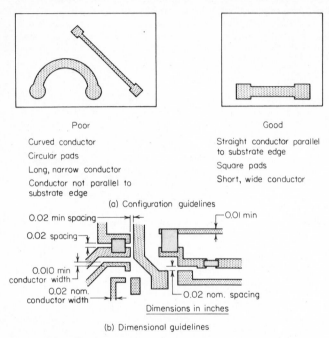

Poor

Curved conductor

Circular pads

Long, narrow conductor

Conductor not parallel to substrate edge

Good

Straight conductor parallel to substrate edge

Square pads

Short, wide conductor

(a) Configuration guidelines

Dimensions in inches

(b) Dimensional guidelines

Fig. 98   Thick film conductor-design guidelines.[68]

substrate. Therefore, the minimum substrate dimensions must be used to draw the substrate outline which will be used for the final layout.

Conductor guidelines   In the design and layout of thick film conductors and interconnects, the following guidelines are recommended:

1. Straight-line patterns parallel to the edges of the substrate and square pads should be used wherever possible to conform to the screen mesh and facilitate screening (see Fig. 98a).

2. Conductors should be kept as short and wide as possible to minimize added circuit resistance, stray capacitance, and increases in thermal coefficient of resistance (TCR), particularly when terminating low-value resistors and for ground or transistor collector paths (see Fig. 98a).

3. The absolute minimum conductor width is 0.010 in. A width of 0.020 in. is the preferred minimum (see Fig. 98b).

4. The spacing between conductors and other conductor lines and pads must be at least 0.010 in. A spacing of 0.020 in. is the preferred minimum. A spacing of 0.020 in. minimum is recommended between adjacent elements that are parallel for more than 0.200 in. (see Fig. 98b).

5. If possible, only one side of the substrate should be used. Screen printing and

firing on both sides complicates processing and reduces yield.   If conductors are printed on both sides of the substrate, the same reference should be used for layout and screening of each side.

6. Conductors should not be run around the edge of the substrate.   Edge printing requires special tooling, reduces yield, and introduces a potential failure mechanism (see Fig. 99).

7. The border between a conductor and the edge of the substrate should be a minimum of 0.025 in.   A border of 0.030 in. is the preferred minimum.   Conductors may become damaged during the assembly operations or provide a leakage or shorting path to the case in certain package types if they are placed too close to the edge

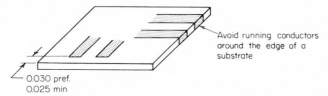

Fig. 99   Design guidelines for the location of thick film conductors near the edge of a substrate.   (Dimensions in inches.)

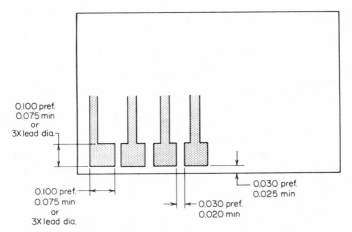

Fig. 100   Guidelines for the design of external-lead pads. (All dimensions in inches.)

of the substrate.   Proper spacing will also reduce screen-alignment problems (see Fig. 99).

External-lead pads   Pads for attaching external leads should be as large as possible to provide adequate bond strength.   Pad width should be at least 2 to 3 times the lead diameter or 0.075 in., whichever is greater.   The preferred minimum width is **0.100 in.**   Pad length should be at least 0.075 in.   The preferred minimum pad length is 0.100 in.   The minimum spacing between the edges of adjacent lead pads should be at least 0.020 in.   The preferred minimum spacing is 0.030 in. (see Fig. 100).

Crossovers   Avoid crossovers whenever possible.   They add to the manufacturing cost, reduce yield, and introduce another potential failure mechanism.   When crossovers are required, the following techniques are available.

Screened Conductor Crossovers.   Screened conductor crossovers require a minimum of two extra screening operations (one for the crossover dielectric and a second one for the conductor) and one or two extra firings.   The crossover-dielectric pad should be a minimum of 0.030 in. larger than the conductors it is insulating.   The preferred minimum crossover-dielectric pad is 0.040 in. larger than the conductors which it is insulating (see Fig. 101).

Crossover dielectrics present a pinhole hazard which could result in conductor-to-conductor shorts or leakage paths. Consequently the recommended practice is to screen two dielectric layers. This adds additional manufacturing cost. When two dielectric layers are used, the recommended size for the first layer is 0.040 in. larger than the conductor it is insulating, and the recommended size for the second layer is 0.020 in. larger than the conductor it is insulating. The second metallization layer (the crossover metallization) should overlap the first metallization layer by at least 0.020 in. Figure 102 shows the step-by-step construction for this type of crossover.

Wire-Conductor Crossovers. When wire-conductor crossovers are used, a dielectric screening operation is required to insulate the portion of the circuit under the crossover wire from the crossover wire. Ultrasonic or thermocompression bonding of

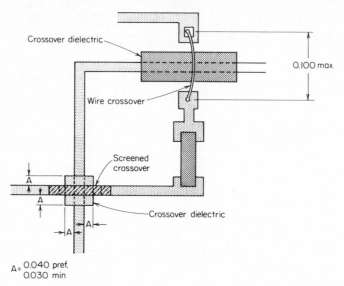

$$A = \begin{matrix} 0.040 \text{ pref.} \\ 0.030 \text{ min} \end{matrix}$$

Fig. 101 Crossover techniques used in thick film networks. (All dimensions in inches.)

fine-gage aluminum or gold bonding wire may be used as a crossover medium. To prevent shorting to adjacent components or conductors, the maximum length of wire used as a crossover should be 0.100 in. For longer runs an intermediate bonding pad or an alternate crossover method should be used (see Fig. 101).

Small-gage copper wire or ribbon may also be used to form the crossover. Using a discrete-component lead as a crossover wire is not recommended because it is very difficult to achieve good strong solder joints consistently at more than one position on a component lead.

Pads for holes and slots Holes and slots in the ceramic substrate should be avoided or minimized because they are costly to tool and difficult to produce to tight dimensional tolerances. If holes or slots are used, Chap. 4 of this handbook should be consulted before hole diameters, positions, and tolerances are selected.

Pads for Feedthrough Connections. An eyelet or small-gage wire bonded to a pad on each side of the substrate is a common method of providing an electrical feedthrough or side-to-side interconnect function. Through-hole screening, a technique in which a vacuum is employed to draw the conductor paste through the holes in the substrate during the screening process, is another method of obtaining the interconnect function. However, unless special precautions are taken, this latter process is quite susceptible to metallization smears, high-resistance connections, burnoff, and other problems associated with high-production operations.

For either of the feed-through interconnect methods described above, the pads surrounding the substrate hole may be round or square. Square pads are recommended to facilitate screen fabrication and the actual screening process. The pad length and width (or diameter) should be 3 times the diameter of the wire or eyelet which will

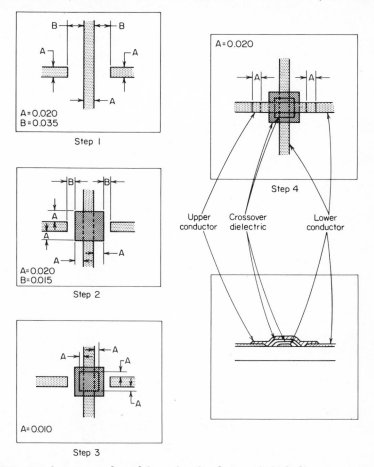

Fig. 102 Development and guidelines for the design of thick film crossovers. (All dimensions in inches.)

be inserted into the hole but in no case less than 0.060 by 0.060 in. The center of the pad, the area centered above the hole, may be blocked out to prevent the conductor paste from flowing into the hole during screening. If this technique is employed, the minimum diameter of the blocked-out area should be 0.010 in. greater than the maximum hole diameter (see Fig. 103).

Pads for Components and Header Pins. A pad is required on only one side of the substrate if its function is to provide a mounting pad for a discrete-component lead or header pin inserted through a hole in the substrate. The guidelines given above for feedthrough interconnect pads also apply to the design of pads for component leads and header pins when holes are utilized. If slots are used, a rectangular pad with a width 4 times the slot width and a length 4 times the slot length is common practice. In no case should the pad length or width be less than 0.060 in. As with holes, the area centered above the slot may be blocked out. The width of the blocked-out area should be 0.010 in. greater than the maximum slot width and the length 0.010 in. greater than the maximum slot length.

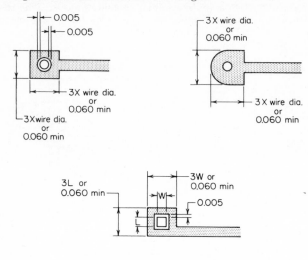

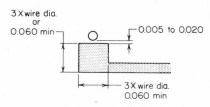

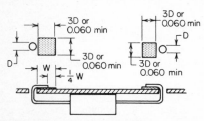

Fig. 103 Guidelines for the design of terminating pads around holes and slots in the substrate. (All dimensions in inches.)

Pads should completely surround the hole or slot. If sufficient substrate area is not available to accomplish this, the pad may be reduced. Reduced pads should surround the hole by at least 180°. The length and width of this reduced pad should be at least 3 times the lead diameter or 0.060 in., whichever is greater. The pad should be positioned a minimum of 0.005 in. and a maximum of 0.020 in. from the hole or slot. If the pad size is reduced in this manner, the lead must be formed to match the pad (see Fig. 104).

Fig. 104 Design guidelines for encased-component terminating pads using through-hole mounting techniques. (All dimensions in inches.)

Pads for surface mounting of leaded discrete components The width of pads used for the attachment of leaded devices should be 3 times the lead wire diameter or 0.060 in., whichever is larger. The pad length is determined by the type of component to be attached. Figures 105 and 106 show pad dimensions for typical component types. Transistors may also be mounted on the edge of the substrate. Figure 107 shows the recommended pad dimensions for this technique.

Pads for attachment of chip capacitors and chip resistors.[69] All chip components should be aligned in the same direction on the substrate, if at all possible, to facilitate assembly. If the attachment pad is connected to conductors wider than 0.030 in., the use of a solder dam (a 0.020-in.-wide band of glaze, resistor material, etc.) across the conductor is recommended to prevent a buildup of solder at the capacitor termination. Figure 108 shows the recommended pad dimensions for attachment of chip capacitors, with standard EIA dimensional tolerances, when a

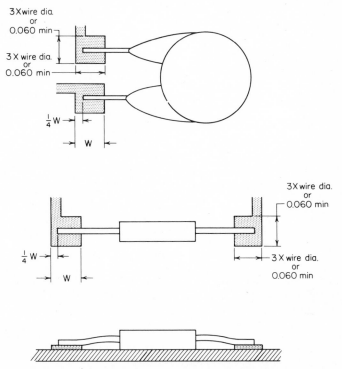

Fig. 105 Design guidelines for encased-component terminating pads using surface-mounting techniques. (All dimensions in inches.)

solder-reflow attachment technique is employed. Figure 109 shows the recommended pad dimensions for attachment of chip capacitors, with standard EIA dimensional tolerances, when an epoxy attachment technique is employed. Figure 110 contains a general procedure for determining pad dimensions for other chip devices.

Thick film resistor guidelines

Calculations for Thick Film Resistors. The basic equation used to calculate resistance is

$$R = \rho \frac{L}{A} \tag{13}$$

where R = resistance
ρ = bulk resistivity (resistance per unit volume)
L = length of resistor
A = cross-sectional area of resistor

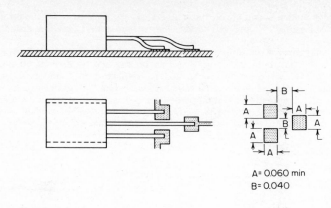

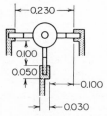

A= 0.060 min
B= 0.040

Fig. 106 Guidelines for the design of terminating pads for surface-mounted transistors. (All dimensions in inches.)

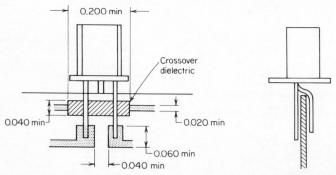

Fig. 107 Guidelines for the design of terminating pads for edge-mounted transistors. (All dimensions in inches.)

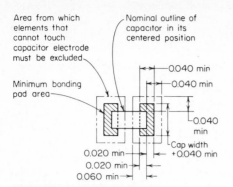

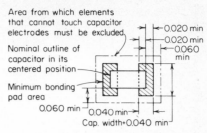

Notes: 1. For epoxy bonding, the minimum allowable capacitor length of about 0.120 in. is recommended.

2. No bare conductor should be allowed under any capacitor.

Fig. 108 Design guidelines for capacitor attachment with a solder-reflow process.[69] (All dimensions in inches.)

Fig. 109 Design guidelines for conductive-epoxy capacitor attachment.[69] (All dimensions in inches.)

The cross-sectional area A of the resistor is equal to the product of the film thickness t and the width of the resistor W:

$$A = tW \tag{14}$$

Substituting (14) in (13) yields

$$R = \rho \frac{L}{tW} \tag{15}$$

$$R = \frac{\rho}{t} \frac{L}{W} \tag{16}$$

The thickness of screened thick film resistors is relatively constant. Variation in resistance values is primarily associated with changes in the length or width of the applied material. Consequently, thick film resistor inks are calibrated in terms of unit length and width only. This procedure yields a design resistivity termed *sheet resistivity* ρ_s, which is the bulk resistivity ρ divided by the film thickness t:

$$\rho_s = \frac{\rho}{t} \tag{17}$$

Sheet resistivity is expressed in ohms/per square (also written ohms/□ or $\Omega/\square$). Substituting (17) in (16) yields

$$R = \rho_s \frac{L}{W} \tag{18}$$

For example, the resistance of a resistor screened with an ink with a sheet resistivity equal to 100 Ω/sq which was 0.120 in. long and 0.040 in. wide would be determined using Eq. (18):

$$R = \rho_s \frac{L}{W} = 100 \frac{\Omega}{\text{sq}} \frac{0.120 \text{ in.}}{0.040 \text{ in.}} = 300 \ \Omega$$

The sheet resistivity of a resistor ink is determined by screening and measuring a resistor of unit length and width. When

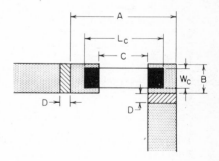

L_C = Length of chip device
W_C = Width of chip device
A = L_C maximum + 0.040
B = W_C maximum + 0.020
C = L_C minimum − 0.040
D = Width of solder dam, 0.020 in. min

Fig. 110 Guidelines for the design of terminating pads for chip devices. (All dimensions in inches.)

the length and width of a film resistor are equal, the resistor is termed a 1-square resistor, and its resistance is equal to the sheet resistivity of the ink.

Aspect Ratio. The ratio of a resistor's length to its width is also defined as its *aspect ratio.* The aspect ratio of a thick film resistor is equal to the number of squares in the resistor.

The aspect ratio of a thick film resistor should not be greater than 10:1 or less than 1:3 (see Fig. 111). An aspect ratio less than 1:1 (the resistor's width is greater than its length) is called an *inverse* aspect ratio.

Selection of Resistor Inks. Since a separate piece of artwork, a separate screen, and a separate screening operation are required for each resistor ink, the number of resistor inks required for a given design should be kept to a minimum and should not exceed three. All resistors should be positioned on the same side of the substrate if possible; if this is not possible, select the combination of resistivities for each side of the substrate which result in the fewest resistor inks per side.

Rectangular-Resistor Design. The variability in the as-fired value of a thick film resistor is affected by the absolute size of the resistor as well as by the screen-printing-process variables, e.g., substrate nonuniformities; screen mesh, emulsion, and tension variations; screening machine drift; and position of the resistor on the substrate. In

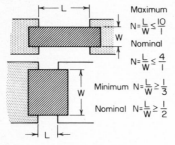

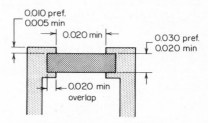

Fig. 111 Resistor aspect-ratio design guidelines.[68]

Fig. 112 Dimensional guidelines for the design of rectangular resistors. (All dimensions in inches.)

general, this variability can be reduced by designing resistors as large as possible. The maximum resistor size is limited by available substrate area, substrate characteristics, and material cost. As the length of a resistor approaches 0.200 in., substrate camber and flatness become progressively more important. Beyond this point, substrate surface grinding may be necessary. The minimum resistor size is limited by the required power dissipation and production capabilities. The minimum practical size is 0.020 by 0.020 in. The preferred minimum size is 0.030 by 0.030 in. (see Fig. 112).

Resistor patterns should be rectangular or hat-shaped. Curved, zigzag, or other unusual resistor patterns should be avoided because they are difficult to screen and adjust (see Fig. 113).

The recommended procedure for the design of rectangular thick film resistors is outlined below. An example is included to illustrate the procedure.

Assume that a 2,000-Ω ±1 percent ½-W resistor is required for a system with an operating temperature range of 0 to 70°C.

1. Select a resistor ink. If the designer had resistor inks with sheet resistivities of 500, 1,000, and 3,000-Ω/sq available, he could obtain a 2,000-Ω resistor by designing a 4-square resistor (aspect ratio of 4:1) with the 500 Ω/sq ink, a 2-square resistor (aspect ratio of 2:1) with the 1,000 Ω/sq ink, or a ⅔-square resistor (aspect ratio of ⅔:1) with the 3,000 Ω/sq paint. Each alternative is acceptable. The ink which will result in the minimum number of screening operations will normally be selected. When power levels are low and size reduction is of primary importance, the ink which will yield the smallest resistor is selected. In our example let us assume that the ink with a resistivity of 1,000 Ω/sq is chosen.

2. Determine appropriate film power density. Compare the maximum operating temperature of the system with the power-density data for the selected resistor ink and derate the film power density, if necessary. In our example assume that the resistor ink is rated for 40 W/in.² at 70°C. The maximum operating temperature is 70°C; therefore, no derating is necessary. Since the resistors will be trimmed by reducing the width of the screened resistor, all resistors should be designed to ensure that the maximum film density (derated if necessary) is not exceeded in case of maximum resistor-width reduction during trimming. In our example, assume that the maximum allowable resistor-width reduction during trimming is 50 percent. The film power density used for design of untrimmed resistors would be 20 W/in.² (40 −20 W/in.²).

3. Determine untrimmed resistance value. The distribution of untrimmed resistance values is approximately ±15 to ±20 percent. Untrimmed resistors must be designed so that they do not exceed the upper tolerance limit for the required resistor. The

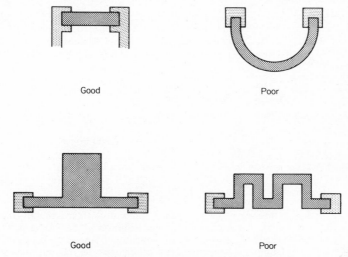

Good   Poor

Good   Poor

Fig. 113   Resistor-configuration design guidelines.

resistance value to be used for the untrimmed resistor is determined by the resistor tolerances and production capability. If no production data are available, design untrimmed resistors to 70 percent of the desired nominal resistance. In our example, assume that we know from production experience that the 1,000 Ω/sq resistivity ink in production yields untrimmed resistance values which vary approximately ±20 percent from the designed value. Then we design our untrimmed resistor to be 80 percent of the desired nominal resistance, or 1,600 Ω (80 percent of 2,000 Ω). This provides an additional safety factor.

4. Determine untrimmed resistor length or width. For resistors with inverse aspect ratios (number of squares less than 1), the length is calculated first using Eq. (19). If the calculated length is less than the 0.020 in. minimum, the length is increased to at least 0.020 in. and preferably to 0.030 in.

$$L = \sqrt{\frac{PR_u}{P_d\rho_s}} \tag{19}$$

where L = untrimmed resistor length
P = resistor power dissipation at maximum ambient temperature
R_u = desired untrimmed resistor value
P_d = film power density at maximum ambient temperature (derated if necessary)
ρ_s = sheet resistivity of selected ink

Once the length is determined, the untrimmed resistor width W is calculated as follows:

$$W = \frac{L\rho_s}{R_u} \tag{20}$$

For resistors with aspect ratios greater than 1:1 (number of squares greater than 1), the width is calculated first from Eq. (21). If the calculated width is less than the 0.020 in., the width is increased to at least 0.020 in. and preferably to 0.030 in.

$$W = \sqrt{\frac{P\rho_s}{P_d R_u}} \tag{21}$$

Then the length is calculated:

$$L = \frac{W R_u}{\rho_s} \tag{22}$$

In our example, we require an untrimmed resistance value of 1,600 Ω, and we are using an ink with a resistivity of 1,000 Ω/sq. Therefore the aspect ratio will be greater than 1:1 (1:1.6 in this case), and the number of squares will be greater than 1 (1.6 squares in this case); so we will use Eq. (21) to calculate the width of the untrimmed resistor first:

$$W = \sqrt{\frac{P\rho_s}{P_d R_u}} = \sqrt{\frac{(\frac{1}{2})(1,000)}{(20)(1,600)}} = \sqrt{0.0156} = 0.125 \text{ in.}$$

This is an acceptable resistor width, and so the length is calculated using Eq. (22):

$$L = \frac{W R_u}{\rho_s} = \frac{(0.125)(1,600)}{1,000} = 0.200 \text{ in.}$$

This is an acceptable resistor length. Check using Eq. (18):

$$R_u = \rho_s \frac{L}{W} = 1,000 \frac{0.200}{0.125} = 1,600 \ \Omega$$

This checks with the desired value of 1,600 Ω, and if the expected distribution of ± 20 percent is achieved, the highest resistance values will be less than the upper tolerance limit of the required resistor and the lowest resistance values can be brought within tolerance by trimming away less than 50 percent of the untrimmed resistor width, thus fulfilling all physical requirements. The power-dissipation rating of the resistor under these conditions is

$$\begin{aligned} P &= P_d \times A_{t,\,min} \\ &= P_d \times L \times W_{t,\,min} \\ &= P_d \times L \times \tfrac{1}{2} W_u \end{aligned} \tag{23}$$

where P = power rating of the resistor
P_d = film power density at maximum ambient temperature
A_t = minimum resistor area after trimming
L = resistor length
$W_{t,\,min}$ = minimum resistor width after trimming = $\frac{1}{2} W_u$ in our example
W_u = untrimmed resistor width

$$P = 40 \,(0.200)\,(\tfrac{1}{2})\,(0.125) = 0.500 \text{ W}$$

Therefore, the worst-case resistor (the resistor requiring maximum trimming) will dissipate the required $\frac{1}{2}$ W at 70°C.

Hat Resistor Design. The hat resistor configuration should be considered whenever aspect ratios greater than 6:1 are required (more than 6 squares are required). Resistors with aspect ratios greater than 10:1, the maximum recommended aspect ratio for rectangular resistors, are readily attained with hat resistors. The use of hat resistors can significantly reduce the number of required resistor inks.

There are many variations of hat resistors, but the two basic configurations are shown in Fig. 114. The broken lines show the portion of the resistor removed in trimming. The configuration with straight edges facilitates layout generation, screen fabrication, and screening. This is the preferred configuration.

The general guidelines presented previously for rectangular resistors apply to hat resistors as well, but since hat resistors are generally designed so that they can be adjusted over a large range, precise calculations are not required.

Preferred configuration

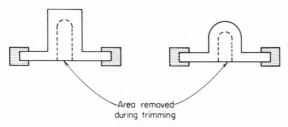

Fig. 114   Hat-resistor configurations.

The recommended procedure for the design of a hat resistor is outlined below.

1. Select resistor ink. The ink resulting in the minimum number of screening operations and an aspect ratio of 6:1 or greater will normally be selected. When power levels are low and size reduction is of primary importance, the ink which will yield the smallest resistor is selected.

2. Determine appropriate film power density. Compare the maximum operating temperature of the system with the power-density data for the selected resistor ink and derate the film power density if necessary. The resistor width will remain constant during trimming; therefore no additional power derating of the film is required.

3. Determine untrimmed resistance value. Resistors should be designed so that the untrimmed resistance value is below the upper tolerance limit. The resistance value to be used for the untrimmed resistor is determined by the resistor tolerance and the production capability. If no production data are available, design untrimmed resistors to 70 percent of the desired nominal resistance.

4. Determine resistor width using Eq. (21):

$$W = \sqrt{\frac{P \rho_s}{P_d R_u}}$$

For the resistor width W in inches see Fig. 115. If the calculated width is less than 0.020 in., the width is increased to at least 0.020 in. and preferably to 0.030 in.

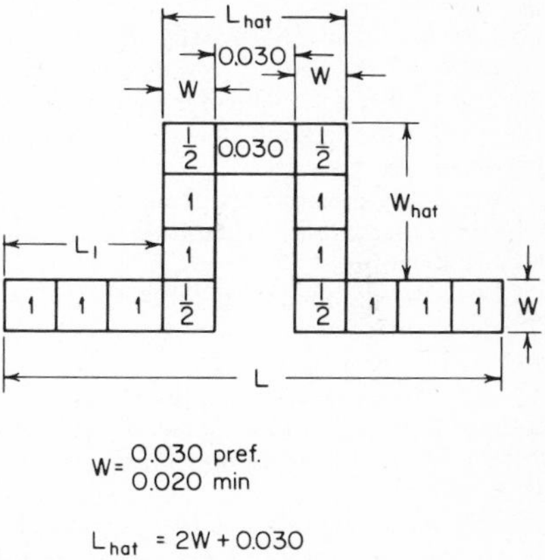

5. Determine untrimmed resistor length L. The hat dimensions (W_{hat}, L_{hat}) are not considered when calculating the untrimmed resistor length. The resistor length is calculated using Eq. (22):

$$L = \frac{W R_u}{\rho_s}$$

For the untrimmed resistor length L in inches see Fig. 115. If the calculated length is less than $5W$, a rectangular resistor should be considered because the hat length (L_{hat}

$$W = \frac{0.030 \text{ pref.}}{0.020 \text{ min}}$$

$$L_{hat} = 2W + 0.030$$

Fig. 115 Dimensional guidelines for the design of hat resistors. (All dimensions in inches.)

in Fig. 115) should be $2W + 0.030$ in. to allow for trimming and hat leg lengths (L_1 in Fig. 115) of at least 1 square are recommended. The actual untrimmed resistor values will be less than the calculated untrimmed resistor values due to the non-uniform voltage gradient in the hat region. The resistance of this region will be approximately 0.5 to 0.7 ($L_{hat}/W)\rho_s$. The designer compensates for this inaccuracy by designing the hat to allow sufficient trimming to attain the desired resistance.

6. Determine hat length. As discussed above, the minimum hat length allows 0.030 in. for trimming (see Fig. 115). Thus

$$L_{hat} = 2W + 0.030 \text{ in.} \tag{24}$$

7. Determine hat width. The hat width (W_{hat} in Fig. 115) is selected to provide the desired final resistance value. The resistivity of corner squares is approximately one-half the resistivity of a normal square, as shown in Fig. 115. Thus the resistance of a fully trimmed hat resistor as shown in Fig. 115 is

$$R = \frac{(L - L_{hat})\rho_s}{W} + 4(\tfrac{1}{2})\rho_s + \frac{0.030}{W}\rho_s + 2\frac{(W_{hat} - W)\rho_s}{W} \tag{25}$$

To compensate for previous approximations and to ensure that the desired final resistance value is attained, W_{hat} will be selected to yield a final calculated resistance value 1.5 times the desired resistance value. (More precise corrections can be made if production data are available on the selected ink.) Equation (25) thus becomes

$$1.5R = \frac{(L - L_{hat})\rho_s}{W} + 2\rho_s + \frac{0.030\rho_s}{W} + 2\frac{(W_{hat} - W)\rho_s}{W} \tag{26}$$

Simplifying and solving for W_{hat} yields

$$W_{hat} = W\left(0.75\frac{R}{\rho_s} + 1\right) - 0.5L \tag{27}$$

8. Modification. Normally there are many hat configurations which can be used to achieve the desired resistance. The procedure presented above will yield one acceptable design. The designer may wish to alter this configuration to optimize the layout. If the original design is modified, the following guideline applies. When the hat width is increased X in., the untrimmed resistor length L must be decreased $2X$ in. Several examples of modifications are shown in Fig. 116.

Resistor Location. The following guidelines should be followed with respect to resistor location:

1. Resistors should be located as far from the edge of the substrate as possible. The clearance between any resistor and the edge of the substrate should be 0.030 in. minimum (see Fig. 117).

2. Resistors with high power dissipation should be evenly distributed on the substrate and should not be placed near the substrate edge. If resistors are screened on both sides of the substrate, high-wattage resistors should not be placed opposite each other.

3. Resistors should not be positioned underneath any unencapsulated components.

Resistor Orientation. For ease in processing and to obtain maximum yields in volume production, the following resistor-orientation guidelines are recommended:

1. Orient all resistors in either the x or y direction (parallel to the substrate edges) to facilitate trimming and to reduce the variability introduced during the screening operation (see Fig. 118).

2. Attempt to orient all resistors in the direction of the travel of the screening-machine squeegee (usually parallel to the substrate's major axis). If this is not possible, try to orient the long resistors in this direction.

3. If multiple-head trimming equipment is used, attempt to design resistors in line and maintain a spacing between resistors to be trimmed simultaneously which is consistent with the capability of the trimmer.

Closed Loops. Closed resistor loops should be avoided if possible so that resistors can be measured and trimmed individually. Measurements in closed loops are slow and costly. If a resistor loop is necessary, the loop should be left open until all the resistors in the loop have been trimmed. It may then be closed with a wire bond or a small-gauge wire or ribbon. The metallized pattern on each side of the gap must be designed to provide an acceptable pad for whichever bridging technique is selected (see Fig. 119).

Resistor-Conductor Interface. The resistor should overlap the conductor pad by a minimum of 0.020 in. on each end in the direction of current flow. The width of the conductor pad should exceed the width of the resistor by 0.020 in. The metallization should extend a minimum of 0.005 in. beyond the resistor termination (perpendicular to the direction of current flow) (see Fig. 112).

Resistor Location with Respect to Other Components. The following guidelines apply to the location of thick film resistors with respect to other components, lead pads, etc.:

1. The side of the resistor which will not be trimmed should be located a minimum of 0.020 in. from any conductor pattern, pad, resistor, or any other screened area. The preferred minimum spacing is 0.030 in.

2. The side of the resistor which will be trimmed should be located a minimum

Fig. 116 Equivalent hat-resistor designs.

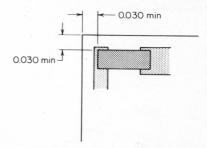

Fig. 117 Resistor-location design guidelines. (All dimensions in inches.)

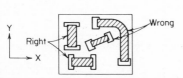

Fig. 118 Resistor-orientation guidelines.[70]

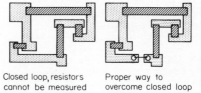

Closed loop, resistors cannot be measured

Proper way to overcome closed loop

Fig. 119 Design guidelines for closed-loop resistors.[68]

of 0.040 in. from all other screened patterns. The preferred minimum spacing is 0.050 in. (see Fig. 120).

Resistor Termination Pad. The following guidelines are recommended for resistor termination pads:

1. A contact pad must be provided at both ends of each resistor to accommodate the probes used to monitor resistance during the trimming operation. Frequently no terminal pad, component pad, or other suitable area is available; in such cases a pad 0.040 by 0.040 in. or larger must be provided.

2. When resistor inks with different sheet resistivities terminate on a common pad, they should be separated by at least 0.010 in. (see Fig. 121).

3. When a resistor termination pad also serves as a contact pad for a terminal or an add-on component, the pad must be designed so that bleed-out from the resistor does not affect that portion of the pad which will be used to bond the terminal or add-on component. Several methods of doing this are shown in Fig. 122.

General Comments. Resistors designed utilizing the preferred guidelines established above and fabricated from compatible, commercially available materials under controlled conditions will vary from the design value approximately ±15 to ±20

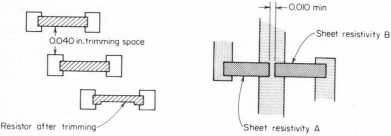

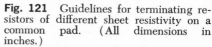

Fig. 120 Recommended spacing between resistors to accommodate trimming.[70]

Fig. 121 Guidelines for terminating resistors of different sheet resistivity on a common pad. (All dimensions in inches.)

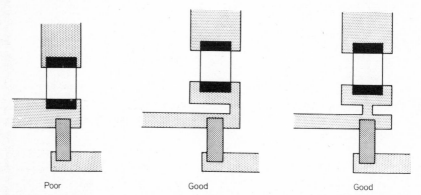

Poor   Good   Good

Fig. 122 Guidelines for terminating thick film resistors and add-on components on a common pad.

percent. However, when resistors are designed to the minimum lengths and widths or other design minimums are exceeded, the designer must consider termination effects and other geometry-related effects. Because these effects are material-dependent, they have not been included, but several references are cited.[71-80]

Metallization Resistors. Conductor materials are occasionally used to form low-value wide-tolerance resistors. The maximum practical resistance value for a metallization resistor is approximately 5 Ω due to the low sheet resistivity of conductor materials. Resistors normally consist of a series of shorting bars between long, narrow parallel conductors. The minimum conductor width should be 0.020 in., and 0.030 in. is the preferred minimum. A metallization resistor is adjusted by removing the shorting bars with standard resistor-trimming equipment. Consequently, the minimum length of the shorting bars (spacing between parallel conductors) should be 0.030 in. to allow for adjustment. The metallization resistor should be coated with a layer of glaze (it must be glazed if the substrate will be immersed in solder). A metallization resistor is normally trimmed after the adjacent conductors are soldered. Location of the test probes used to monitor resistance in the trimming operation is very important with any low-value resistor. The designer should indicate the probe positions required to obtain the desired resistance value (see Fig. 123).

Thick film-capacitor guidelines[81]

Calculations for Thick Film Capacitors. The basic capacitance equation was presented in the section on Component Selection [Eq. (1)]:

$$C = \frac{0.225KA}{t}$$

where C = capacitance value, pF
K = dielectric constant of material
A = opposed plate area, in.[2]
t = thickness of the dielectric, in.

This basic formula applies to thick film capacitors which have screened dielectrics and which are formed by using the ceramic substrate (normally a high-K ceramic body) as the dielectric material.

Design parameters for thick film capacitors can be determined rapidly from the nomograph in Fig. 124, which consists of six vertical calibrated scales and a blank turning line. The basic equation is solved with

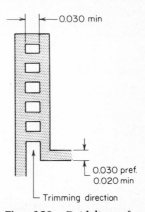

Fig. 123 Guidelines for designing low-value resistors using thick film conductive materials. (All dimensions in inches.)

the five scales on the left (K, t, blank, C, and A); the two additional scales on the right help to determine trade-offs between length and width.

1. K scale. Although early dielectric constants used in thick film work were in the 1 to 300 range, recently announced paste developments are rated at 1,300 K. To make full use of the state of the art, the K scale covers 2 to 2,000.

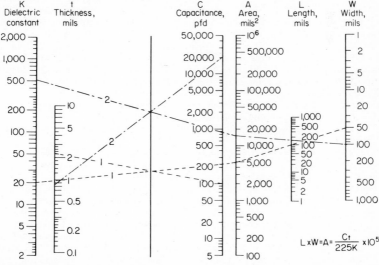

Fig. 124 Thick film capacitor-design nomograph. The six scales and the blank turning line cover most of the practical thick film capacitor sizes. See text for further discussion.[81]

2. *t* scale. Thick films are usually 0.5 mil and up. Normal dielectric thicknesses range from about 0.8 to 2.5 mils, depending upon whether a single or double printing is used. Larger values can be used where the substrate is the dielectric material. The particular construction of the nomograph allows extending the range of the scale in both directions to cover any unusual techniques of capacitor construction.

3. *C* scale. For low values of dielectric constant, industry sources indicate they are able to print capacitors in the 5- to 500-pF range. For material of high *K*, capacitors of from 500 to 50,000 pF have been printed, 20,000 pF being typical.

4. *A*, *L*, and *W* scales. Although areas as high as 1 in.² are unusual, they are not considered impossible. Typical values of *A*, *L*, and *W* would fall near the center of these scales.

Examples of Thick Film Capacitor Calculations Using Nomograph

1. If we want to design a 100-pF capacitor having an area of 5,000 mils² using a material of dielectric constant 20, we can find the dielectric thickness by following this procedure (lines identified with 1 in Fig. 124 apply to this example). Draw a straight line between 20 on the *K* scale and 5,000 mils² on the *A* scale. Note the point where the line crosses the center blank turning scale. Draw a second straight line between the 100-pF value on the *C* scale and the point on the turning scale and extend it to the left where it intersects the *t* scale at 2.25 mils. (Note that the two scales *K* and *A* are used together and likewise the two inside scales *t* and *C*.) Various values of *L* and *W* can be determined by joining the 5,000 mils² on the *A* scale with values on the *L* and *W* scales. With this example, *W* = 50 mils and *L* = 100 mils.

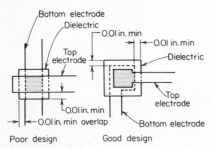

Dielectric ink should overlap bottom electrode by 0.010 in. on all edges. Top electrode should fall completely within bottom electrode area.

Fig. 125 Thick film capacitor-design guidelines.[68]

2. If we are confronted with the need for finding the capacitor area for a dielectric constant of 500, a dielectric thickness of 0.8 mil, and a capacitance of 20,000 pF, we proceed as follows (lines identified with 2 in Fig. 124 apply to this example.) Draw a line connecting the 0.8-mil value on the *t* scale and the 20,000-pF value on the *C* scale. Note the turning-scale crossing point. Connect the 500 value on the *K* scale with this point and continue to the intersection with the *A* scale at 14,000 mils² (remembering that the *K* and *A* scales are joined and that the *t* and *C* scales are used together). Drawing a line from the *A* value to a 100-mil width gives the length at 142 mils.

Capacitors with Screened Dielectrics. The following guidelines apply to typical thick film capacitors with screened dielectrics (see Fig. 125):

1. The screened dielectric should overlap the lower electrode (plate) a minimum of 0.010 in. on each side.

2. The top capacitor electrode (plate) must fall completely within the bottom electrode area or cross over the bottom electrode in an insulated region.

Alternate thick film capacitor configurations have been suggested to protect the dielectric from degradation due to moisture, to lower the electrode resistance, to accomplish trimming while exposing a minimum of electrode area, and to build up multilayer screen-printed capacitors.[82, 83] In the discussion of conductor guidelines, the designer was cautioned about the normally undesirable stray capacitance introduced in long parallel closely spaced conductors. Occasionally, long parallel closed spaced conductors are employed to achieve very low value capacitors. The capacitance resulting from this technique can be increased substantially by screening a high-*K* dielectric material over the two parallel lines.[82] Since these techniques are not widely used, few design data are available and designs employing these techniques should be evaluated thoroughly.

Double screening the dielectric is recommended to minimize shorts and dielectric breakdown failures. The main disadvantages of screening double dielectric layers are the significant reduction in capacitance and the cost of an additional manufacturing operation.

It is very difficult to hold dielectric thickness (particularly if double-screened), peak firing temperature, and peak firing time constant throughout a production run. Therefore, it is difficult to produce thick film capacitors to close tolerances with acceptable yields. Trimming techniques (air abrasion, laser, etc.) are costly, and very few data are available on the long-term effects of these techniques on capacitor parameters.

Capacitors Employing High-K Substrates. When a high-*K* ceramic substrate is used for the capacitor dielectric, the screened plates (electrodes) should be positioned as far as possible from the edge of the high-*K* substrate to reduce edge effects. If several capacitors are formed on the same substrate, the sets of screened plates should be positioned as far from each other as possible.

Thick film spiral inductors The design of circular and square spiral inductors was presented in the Component Selection section. Conductor line widths and spacings may have to be reduced to 0.010 in. or less to achieve the required inductance. Layout generation, screen fabrication, and the screening operations are less complex for square-spiral inductors. This becomes particularly important when normal design guidelines are exceeded.

Glazes

For Conductors. The use of an insulation glaze in conjunction with screened crossovers and wire crossovers has been covered in other sections. However, other applications exist.

1. If a substrate is to be immersed in solder, a layer of glaze can be screened over conductors, such as gold conductors, which should not come in contact with the solder and over other areas which are not intended to receive a coating of solder, e.g., a conductor terminating a very low value resistor with a tight tolerance which was trimmed prior to the soldering operation. The glaze should overlap the area to be protected by at least 0.010 in. on each side.

2. If a glaze screening operation is required, adding glaze strips to form solder dams at terminal and add-on component termination pads, as shown in Fig. 110, will facilitate assembly by improving the uniformity of solder coatings on these pads. This technique can also be used to control the flow of adhesives used in the assembly operation.

For Resistors. A layer of glaze is sometimes screened over resistors to protect them during trimming or to protect them from atmospheres which degrade the resistor. If protection from detrimental atmospheres is desired, the glaze must be applied after the resistors have been trimmed. The effect of the glaze firing on the resistor should be studied carefully, and the long-term effect of any glaze-resistor interactions should be determined before this technique is selected.

Bonding-pad layout guidelines for semiconductor chips The following guidelines apply for the design of bonding pads for semiconductor chips:

1. Bonding pads for a semiconductor die should be a minimum of 0.005 in. greater than the die in each direction. A minimum of 0.010 in. greater than the die in each direction or 0.040 by 0.040 in., whichever is greater, is the preferred minimum pad size. Larger bonding pads will facilitate assembly. If possible, the pad size should be increased to allow the bonding of a second die should the initial die not function properly. When a bonding pad is designed to accommodate multiple dice, identification of the separate die-pad areas, by necking or any other technique, may facilitate bonding operations (see Fig. 126).

2. The base of some diode chips is the cathode, while the base of other diode chips is the anode. Similarly, some transistor chips have a collector contact on the surface of the chip, and others have the collector contact at the base of the chip. This information should be obtained before preparing the detail layout and consistency between suppliers should be ensured if there are multiple suppliers for any of the dice.

3. Pads for wedge, stitch, or ultrasonic bonds should be a minimum of 0.015 by 0.015 in.

4. When using bonding wire for interconnections, the maximum length of the wire should be 0.100 in.   A wire length of less than 0.050 in. is preferred.

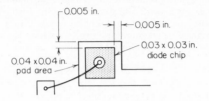

Chip-to-pad clearance should be at least
0.005 in. to allow for adequate contact area.

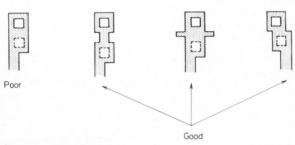

Fig. 126   Design guidelines for semiconductor die bonding pads.[68]

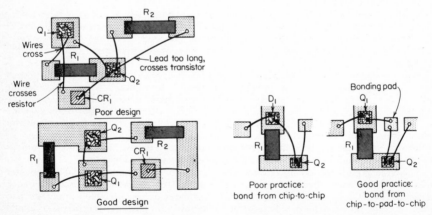

Fig. 127 Guidelines for the placement of faceup-bonded semiconductor wire bonds.[68]

Fig. 128 Guidelines for interconnecting faceup-bonded semiconductor dice.[68]

5. Pads for die and wire bonds should be a minimum of 0.050 in. from the edge of the substrate.

6. Wires should not cross over semiconductor dice or another wire.   When a wire crosses over a conductor, the portion of the conductor under the wire should be covered with an insulating layer (see Figs. 101 and 127).

7. Do not bond directly from die to die but use an intermediate bonding pad (see Fig. 128).

8. Pad patterns for flip-chip, beam-lead, LID, and other semiconductor chips are designed to conform to the contact points for the particular device. The pad sizes and spacings are a function of the tolerances of the chip bonding points and screening capabilities. Pad sizes should be designed as large as possible to facilitate chip bonding.

9. All chip-bonding pads should be aligned in a row (parallel to a substrate axis) if at all possible, to facilitate die and wire bonding and reduce the time during which the device is exposed to die and wire bonding temperatures.

HIGH-FREQUENCY CONSIDERATIONS[40]

Several unique advantages of thick film techniques enable this technology to be utilized in the fabrication of high-frequency circuits. Conductor patterns are generally quite short, thus minimizing path resistance and parasitic inductance. Also, stray capacitance can be kept quite low because of the low dielectric constant of the substrate materials, the ability to place a ground plane on the reverse side of a substrate, and the low profile and short leads of most attached components. Circuits in the VHF, UHF, and (with strip-line techniques) SHF frequency ranges have been fabricated using thick film techniques.

TABLE 42   Electrical Properties of Common Ceramic Substrates

Substrate type	Dielectric constant at 1 mHz	Dissipation factor at 1 mHz
Alumina............	8.5–10	0.0001–0.0008
Steatite............	5.5–6.5	0.004 –0.022
Beryllia............	6.0–6.5	0.0002–0.0005

The conversion of a high-frequency circuit to its thick film counterpart is usually more difficult than for low-frequency circuits. Often it may be necessary to lay out or rework thick film patterns several times in order to achieve the desired performance or to reduce intercoupling effects. Nevertheless, thick film designers with a basic understanding of high-frequency fundamentals and a general knowledge of the behavior of thick film materials at high frequencies should be able to obtain good designs with minimum difficulty through the VHF range. Above this frequency range, thick film design becomes considerably more complex and specialized.

This section provides general design guidelines and materials information with respect to designing thick film circuits up to approximately 250 MHz. The reader should consult other sources for design in the UHF and SHF frequency range, as this technology is beyond the scope of this section.

Substrates

Since substrates are the base elements upon which thick film circuits are fabricated, they must exhibit a very low dielectric constant if stray capacity and its associated intercoupling effects are to be minimized. Three of the most common ceramic substrate materials used for high-frequency applications are steatite, high alumina, and beryllia. A comparison of their electrical properties is provided in Table 42.

Steatite   Of the three ceramic materials listed above, steatite has the lowest dielectric constant and consequently is widely used in the fabrication of coil forms, tube sockets, and other products associated with high frequency. As a substrate material, however, steatite lacks the physical strength and the good thermal properties of alumina. Also it is not compatible with many of the noble-metal conductor and cermet resistor systems used in fabricating thick film networks for industrial and military applications. Steatite substrates are most frequently used in the consumer-product area, where the silver and carbon-resistor thick film systems are employed.

Alumina   Although high-alumina ceramic does not have the low dielectric constant of steatite, its excellent physical strength, relatively low cost, and good thermal characteristics make it highly desirable as a thick film hybrid substrate.   Despite its higher dielectric constant, high alumina can provide excellent performance beyond 250 MHz.   Further improvements can be made by applying a low-dielectric-constant glaze onto the substrate and fabricating the circuit on the glaze or by screening a ground plane on the back of the substrate.   This will help reduce the stray capacitance between the circuit components.

Beryllia   Beryllia, with a dielectric constant less than that of alumina and a thermal conductivity far exceeding it, is an excellent substrate for high-frequency high-power applications.   However, it is more costly and lacks the high physical strength of alumina.   Also beryllia in fine-particle form is highly toxic.

Conductors

In designing high-frequency thick film circuits, it is important to minimize all capacitive and inductive parasitic effects.   This can be accomplished by keeping conductor lengths as short as possible and conductor widths and spacings as broad as possible.   However, when substrate area is at a premium it may not be possible to obtain the conductor widths and spacings desired, as they are directly related.   In such cases it will be necessary to determine which of the two factors most seriously affects circuit performance and compromise accordingly.

Thick film crossovers should be avoided.   At high frequencies the few picofarads of capacitance between the crossover patterns can produce detrimental effects on circuit performance.   Care must also be taken with regard to the layout in general. The signal path should be kept very short, and proper isolation should be provided between stages and between inputs and outputs of each individual stage.   This is necessary in order to prevent undesirable feedback effects such as oscillations.   Thick film conductor materials offering highest conductivity such as silver and gold should be employed if possible.

Resistors

Thick film resistors exhibit significant changes in effective resistance value at frequencies above several megahertz.   These changes are attributed to residual inductive and capacitive effects within the resistive material and its associated termination points.   The direction and degree of the value change are highly dependent upon the particular material used.   In general, resistors manufactured with cermet compositions will experience variations as shown in Fig. 129.   Resistors made from high-resistivity materials will tend to decrease in value with increased frequency, and resistors made with low-resistivity material will tend to increase in value.   Resistor geometry also has an effect on the change in resistor value with frequency.   A square resistor offers the best high-frequency performance, and a resistor with an inverse aspect ratio exhibits the worst performance.   This latter effect is attributed to the closely spaced resistor termination pads; they provide a significant shunt capacitance across the resistor that becomes predominant as frequency is increased.   Before attempting any high-frequency design the thick film designer should have available high-frequency performance data on the specific materials to be utilized.

Capacitors

At high frequencies capacitor values tend to be quite small.   Excellent advantage can be taken of thick film techniques at these frequencies by utilizing screened-on capacitors or small chip capacitors.   Although screened-on capacitors cannot be fabricated to close tolerances, they can provide an economic means of performing noncritical coupling or bypass functions when values under 1,000 pF are desired. However, if substrate area is critical, if larger values are needed, or if better performance is desired, chip-type capacitors are recommended.   In general chip capacitors soldered directly to thick film conductor patterns provide superior performance at high frequencies over their encased counterparts.   This is due to the elimination of lead inductance and resistance.

NP0 ceramic capacitors with temperature coefficients of ± 30 ppm/$°$C, tolerances to ±1 percent and values from 1 pF to 0.1 μF can be obtained in chip form. They are available in a wide variety of sizes and end-termination materials. Because of their stability, value range, and small size, they are widely used in high-frequency applications.

Porcelain capacitors provide the ultimate in high-frequency performance. They equal or exceed NP0 stability, and, most important, they exhibit low dissipation factors even in the high microwave frequency range. Values from 0.1 to 1,000 pF are available in chip form. Further information on these and other capacitors is given in the Component Selection section of this chapter.

Inductors

Utilization of inductors in thick film circuits also becomes possible and practical at high frequencies. Screened-on inductors in spiral or other form can be employed

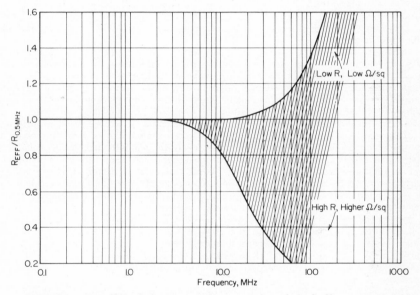

Fig. 129 Typical high-frequency characteristics of thick film resistors.[40]

above 100 MHz or so. Tiny rf coils, chokes, and transformers are available in a wide variety of values, sizes, and package configurations. The Component Selection section of this chapter provides additional information concerning the capabilities and application of the various available inductive component types.

Semiconductors

High-frequency semiconductors can be included in thick film circuits the same as for low-frequency applications, but it is usually desirable to have measured performance data on the devices before attaching to the thick film circuitry. This necessitates encased components, as high-frequency measurements on semiconductor dice are not possible. As a result the popular package types for high-frequency applications are the small metal-can headers, such as the TO-46, the microtab type package, and the ceramic carrier or LID. Devices such as UHF transistors and pin diodes are now becoming available in beam-lead form. Because of their small size, short flat leads, and ease of assembly, the beam-lead devices are very likely to become the prime device configuration for high-frequency application in the future.

RELIABILITY OF THICK FILM MODULES

Inherent Reliability

The use of thick film technology brought with it a promise of an inherent reliability improvement. The major cause of breakdowns in electronic systems has been failures in component interconnections. The number of manually made component interconnections can generally be reduced 30 to 60 percent by employing thick film technology. These interconnections are replaced by chemically bonded material interfaces on the substrate, which reduces a module's susceptibility to wiring errors and damage due to dynamic environments such as shock, vibration, and acceleration. Localized heating and hot spots within resistive elements are minimized, thanks to the direct bond between the thick film resistors and a substrate with high thermal conductivity. The use of thick film technology reduces the mass of a device, making it less susceptible to damage due to shock, vibration, and acceleration. A reliability improvement of from 10:1 to 20:1 over a discrete version of a circuit function can be anticipated when thick film technology is employed. An early prediction of the failure-rate range for typical hybrids which have been carefully designed and fabricated when employed in moderate environments was from 0.05 to 0.005 percent per thousand hours.[84]

Reliability data can be generated by accurate recording of field experience, long-term life testing at design electrical and environmental stresses, or carefully designed accelerated life-test experiments where modules are subjected to severe electrical and environmental stresses to accelerate failure mechanisms. Field experience has substantiated the reliability improvements which have been predicted, but the technology has been somewhat constrained by the lack of published reliability data. Much of the field experience is either not documented or is considered proprietary. Failure rates are so low that units must be tested for millions of total test hours at design electrical and environmental stress levels to generate meaningful reliability data or to verify predicted failure rates at meaningful confidence levels. This is generally prohibitive from a cost standpoint. Great care must be exercised when conducting accelerated life tests because detailed knowledge of thick film failure mechanisms is required to ensure that new failure mechanisms are not introduced at the higher stress levels. This information is not readily available.

Substantial reliability data have been published by IBM on the solid-logic technology thick film circuits used in the IBM 360 computer systems. Module failure rates for this system have followed a normal learning curve and are currently reported to be approximately 0.001 percent per thousand hours.[85] No thick film resistor failures have been experienced in this program, with over 90 billion hours of operation.[86]

The Space and Tactical System Corporation has collected detailed reliability data on hybrid modules. Over 3,800,000 ground-based module hours and 4,200,000 module hours in orbit have been recorded without any failures.[87]

Reliability Predictions

A preliminary reliability evaluation is normally performed early in the design phase of the design cycle to determine if the anticipated design is capable of meeting the reliability goals. This is normally performed after the preliminary schematic has been established, circuit elements have been defined, a preliminary physical layout or size estimate has been prepared, the preliminary package style has been selected, and the preliminary thermal and worst-case analyses have been completed.

The final reliability evaluation is performed after all the steps in the circuit-design, topology-design, packaging-design, and prototype-development phases of the design cycle have been completed. At this point the design is frozen and ready for production.

As discussed above, testing thick film hybrid circuits to establish or verify failure rates is very difficult and costly because of the low failure rates and the lack of detailed knowledge concerning the relationship between the performance of thick film devices under normal conditions and their performance under conditions of accelerated electrical and environmental stresses. Therefore there is a need for an accurate reliability prediction model to demonstrate conformance to the reliability allotment, reliability requirement, or reliability goal set forth in the general system requirements. Thick film technology is very dynamic, with frequent materials and process changes and new devices, any one of which can introduce new failure modes and/or failure mechanisms. Consequently care must be exercised when using previous experience, and the development of reliability prediction models has been retarded and is still in the infant stage.

The large variety of custom designs, unique processes, and singular configurations employed in the manufacture of thick film circuits precludes the use of generalized prediction techniques and requires a model which details each circuit element. Two

TABLE 43   Comparative Failure Rates for Various Bonding Techniques in Percent per Thousand Hours

Case	Interconnection	One lead	14-lead device	150-lead device
1a	Thermocompression wire bonds	0.00013	0.0018	0.02
1b	Ultrasonic wire bonds	0.00007	0.001	0.014
2a	Face bond	0.00001	0.00014	0.0015
2b	Beam lead	0.00001+	0.00014+	0.0015+

TABLE 44   Solder-Terminal Reliability Compared with Best-Quality Wire-Bonded Connections[89]

Source	Interconnection reliability comparison	
	Type of interconnection	Fails per connection per 1,000 h, %
RADC.............	Gold wire with 2 thermocompression bonds	0.00012
RADC.............	Gold wire with 2 ultrasonic bonds	0.00006
IBM Fed. Syst......	Gold wire to aluminum land with 2 thermo-compression bonds plus flight level screening	0.00005
IBM Computer Div..	SLT copper ball connection (1966)	0.00004
IBM Computer Div..	SLT copper ball connection (1968–1969)	0.00001–0.00002*
IBM Computer Div..	Controlled-collapse connection	0.00001*

* Best estimate.

reliability-prediction models are outlined below. The piece-part failure rates for the discrete add-on devices employed in the thick film module can be estimated using the procedures outlined in MIL-STD-756A and Mil Handbook 217; Ref. 93, vol. 2; or other acceptable sources.

Additional published data on the failure rates of microcircuit interconnections are given in Tables 43 and 44 and Figs. 130 and 131.

In Fig. 130*a*, the interconnections marked 1 and 2 are thermocompression bonds of gold wires. The listed failure rate of 0.00006 percent per thousand hours is based on observed failures in a number of Air Force systems. In Fig. 130*b* interconnections 1 and 2 are ultrasonic bonds of aluminum wire. The connection marked 3 in Fig. 130 may be either a welded or soldered joint connecting a Kovar lead to a printed-circuit board. The failure rate of 0.000012 percent per thousand hours at a 90 percent confidence level, for both soldered and welded connections, is based on the results of a completed Air Force contract sponsored by RADC.

Face bonding is generally defined as any technique by which chips are attached directly to a conductor pattern on a substrate of some type. It includes ultrasonic, solder-reflow, and solder-ball techniques. A typical face bond is shown in Fig. 131. The silicon chip is bonded directly to the conductor pattern on a glass or ceramic substrate or in some cases directly to a printed-circuit board by solder or ultrasonic techniques. The results of an Air Force program indicate that with very close process controls, a failure rate may be attained that under normal conditions will equal, or be less than, the failure rate of a conventional soldered or welded joint.

Beam leads are generally defined to include all techniques by which lead frames are attached to silicon chips, i.e., vacuum and chemical deposition, diffusion, thermocompression techniques, welds, etc. The beam-lead approach is illustrated on the right side of Fig. 131. The beam-leaded device shown is attached directly to a pad on the conductor-supporting substrate. It is estimated that the reliability of the junction between the chip and the lead, point 1, is at least an order of magnitude better than a soldered or welded joint, giving the estimated failure rate of 0.000001 percent per thousand hours. The reliability of the bond between the beam and the conductor pad, point 2, is estimated to be at least as reliable as a welded or soldered joint, resulting in a failure rate of 0.00001 percent per thousand hours.

Interconnection	Failure rates a	Failure rates b
1	0.00006	0.00003(est)
2	0.00006	0.00003
3	0.00001	0.00001
Totals	0.00013%/1000hr	0.00007%/1000hr

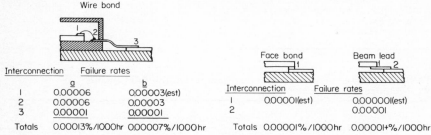

Fig. 130 Failure-rate summary for wire-bond interconnection systems.[88]

Interconnection	Failure rates (Face bond)	Failure rates (Beam lead)
1	0.00001(est)	0.000001(est)
2		0.00001
Totals	0.00001%/1000hr	0.00001+%/1000hr

Fig. 131 Failure-rate summary for face-bond and beam-lead interconnection systems.[88]

The Department of Defense (DOD) Reliability Analysis Center (RAC) is expanding its operation to include hybrid microcircuitry. In this data-collection system, hybrid data are broken down into their constituent elements, the package, the substrates (with their deposited circuit components), and the attached active and passive chip elements. When sufficient data are published, reliability-prediction models can be updated to reflect these data and therefore increase their accuracy.[90]

Preliminary reliability analysis[1] This model is intended to provide a quick and easy preliminary approximation to the thick film hybrid microcircuit failure rate:

1. For each active and passive element, calculate or make best estimate of the junction or hot-spot temperature.
2. Determine average substrate temperature.
3. Using data from steps 1 and 2 and the failure-rate data from Table 45, determine the base failure rate for each active or passive element, wire bond, and crossover (use average substrate temperature for temperature adjustment of wire-bond and crossover failure rates, hot-spot or junction temperature for active and passive elements).
4. Calculate the base failure rate for the hybrid microcircuit from

$$\lambda_b = \lambda_{b1} + \lambda_{b2} + \lambda_{b3} + \cdots + \lambda_{bn} \tag{28}$$

where λ_{b1}, . . . are the individual failure rates, at operating temperature, for each circuit element, wire bond, and crossover. When more precise values are required, reference should be made to Ref. 93, vol. 2, or similar source.

5. Calculate the substrate complexity factor π_c.
   a. For each substrate, determine the ratio of the actual film area to the total substrate area.

TABLE 45   Hybrid-Circuit Element Base-Failure Rates in Percent per Thousand Hours[1]

Element	Temperature, °C				
	25	50	75	100	125
Thick film resistor..............	0.0005	0.0010	0.0015	0.002	0.0025
Chip capacitor.................	0.001	0.0015	0.0025	0.006	0.025
Wire bonds:					
Au-Al ball.................	0.000005	0.00002	0.0001	0.001	0.006
Al-Au.....................	0.00001	0.00001	0.00001	0.00001	0.00005
Al-Al.....................	0.00001	0.00001	0.00001	0.00001	0.00001
Au-Au....................	0.000004	0.000004	0.000004	0.000004	0.000004
Crossovers...................	0.000005	0.000005	0.000006	0.000008	0.00001
Transistor chips:					
Low power.................	0.0001	0.0003	0.0009	0.0027	0.007
Power.....................	0.005	0.010	0.03	0.09	0.27
Diode chips..................	0.0001	0.0003	0.0009	0.0027	0.007
Microcircuits:					
Quad gate or equivalent........	0.002	0.0036	0.018	0.082	0.24
Dual F/f or op. amp equivalent.	0.004	0.0072	0.036	0.164	0.48
SSI (equivalent of 25 gates).....	0.0125	0.0225	0.1125	0.512	1.5
MSI (equivalent of 50 gates)....	0.025	0.0459	0.225	1.02	3.0
LSI (equivalent of 100 gates)...	0.050	0.09	0.45	2.04	6.0

b. Add 1 to the value found in part a to obtain the substrate complexity factor C_s.

c. Repeat steps 1 and 2 as required for additional substrates.

d. Calculate the total complexity factor from the relationship

$$\pi_C = C_{S1} + C_{S2} + \cdots + C_{SN} \tag{29}$$

where

$$C_{S1} = 1 + \frac{\text{total film area of substrate 1}}{\text{total area of substrate 1}}$$

$$C_{S2} = 1 + \frac{\text{total film area of substrate 2}}{\text{total area of substrate 2}}$$

6. Determine quality factor π_Q, which depends upon the level of process control, screening, and testing required on production devices. These factors correspond to the three quality levels defined in MIL-STD-883:

Level	π_Q
1	1
2	10
3	30

The quality level π_Q is assumed to apply at all levels of element procurement, hybrid assembly, and testing. For example, the chip lead bonds would be more tightly controlled by sample testing, etc., and there would be better control of hybrid-element procurement on level 1 devices compared to levels 2 and 3.

7. Determine the environmental factor π_E. Choose the appropriate factor from Table 46. For commercial aircraft the factor is 5.0. The environmental factors π_E have been excerpted from Ref. 93, vol. 2.

8. Determine the package factor π_p using the equation

$$\pi_p = 2 + (N - 14)(0.04) \tag{30}$$

where N is the number of external package pins. The package factor π_p is based on the RADC-TR-69-350 model for large microcircuit packages. Package reliability relates to the total seal area and number of leads which must be sealed.

9. Calculate the hybrid microcircuit failure rate from the equation

$$\lambda_M = \lambda_b \pi_C \pi_Q \pi_E \pi_P \tag{31}$$

using the values found in steps 1 through 8.

10. The complete hybrid microcircuit failure-rate prediction model is

$$\lambda_M = \lambda_b \pi_C \pi_Q \pi_E \pi_P$$

where λ_M = hybrid microcircuit failure rate, % per 1,000 h
  λ_b = base failure rate for all circuit elements and lead bonds combined
  π_C = hybrid substrate complexity adjustment factor
  π_Q = quality factor
  π_E = environmental adjustment factor
  π_P = package factor

Failure-rate evaluation procedures for thick film hybrid microcircuits[91]

1. In the case of hybrid microcircuits as applied herein, passive components, such as resistors and capacitors together with their interconnections, are produced on

TABLE 46   Environmental Adjustment Factor π_E For Hybrid Microcircuits[91]

Environment	Symbol	π_E
Laboratory	L_O	1.0
Ground, fixed	G_F	2.0
Portable	G_P	6.0
Mobile	G_M	7.0
Satellite, orbit	S_O	1.5
Launch	S_L	8.0
Airborne, inhabited	A_I	5.0
Uninhabited	A_U	7.0
Missile	M	10.0

passive substrates using thick film techniques. Active microdiscrete components, such as diodes and transistors, are then added separately. Consequently, the failure rates of such hybrid microcircuits are calculated by summing the failure contributions of their microdiscrete components, substrates, passive networks, and overall package factors and then applying the environmental adjustment factor. The utilization of this approach of combining all essential segments in the formulation and in the calculation of failure rates of hybrid microcircuits is illustrated in progressively expanded detail.

2. The general mathematical model for hybrid microcircuit failure rate λH_μ is

$$\lambda H_\mu = \lambda_b \pi_E \tag{32}$$

where λ_b = base failure rate
  π_E = environmental adjustment factor (Table 46)

3. The base-failure-rate λ_b model for hybrid microcircuits is

$$\lambda_b = \Sigma \lambda_{DD} + \lambda_{sub} + \lambda_{HK} + \lambda_{HP} \tag{33}$$

where $\Sigma \lambda_{DD}$ = sum of discrete-device failure rates, computed as shown in Eq. (34)
  λ_{sub} = hybrid-circuit substrate failure rate, computed as 0.00012% per 1,000 h multiplied by area of substrate in square inches (if actual area of the substrate is unknown, use 80% of the external package area)
  λ_{HK} = thick film network substrate pattern failure rate, computed as shown in Eq. (35)
  λ_{HP} = hybrid package failure rate, computed as shown in Eq. (36)

4. The failure rate for discrete devices in hybrid microcircuits $\Sigma \lambda_{DD}$ is the sum of the failure rates of all the discrete devices used in a hybrid microcircuit and forms a portion of the total base failure rate as shown in the prediction procedure for general hybrid microcircuits. The failure rate is computed from

$$\Sigma \lambda_{DD} = N_{CC}\lambda_{CC} + \Sigma \lambda_{CD} + \Sigma (\lambda_{DS}\pi_T) \tag{34}$$

TABLE 47 Failure Rate λ_{DS} for Discrete Semiconductor Devices (Chips) in Hybrid Microcircuits[91]

Semiconductor (chip) description	λ_{DS} failure rate, % per 1,000 h	
	Bonded wire lead (faceup) devices	Flip chips, bumped beam-lead, or tab-lead devices
Diodes:		
Si low-power switching.........	0.00015	0.00020
Si general purpose.............	0.00025	0.00030
Si rectifiers...................	0.00045	0.00050
Ge switching..................	0.0010	0.0010
Ge rectifiers..................	0.0020	0.0020
Zener........................	0.00025	0.00030
Varactor......................	0.0020	0.0020
Tunnel.......................	0.0020	0.0020
Transistors:		
Si low-power switching.........	0.00034	0.00040
Si general purpose.............	0.00055	0.0006
Si power > 1.0 W..............	0.0033	0.0033
Ge low-power switching.........	0.0007	0.0008
Ge general purpose............	0.0013	0.0013
Ge power > 1.0 W.............	0.004	0.004
FET.........................	0.0007	0.0008
Unijunction...................	0.0007	0.0008
Silicon controlled rectifiers (SCR):		
Low power $\leq$ 1.0 A...........	0.0008	0.0008
Power > 1.0 A................	0.004	0.004

Monolithic integrated circuits

Obtain failure rate of monolithic integrated circuit, excluding π_E (environmental factor), and multiply by 0.5 for use in hybrid microcircuits

where N_{CC} = number of chip capacitors

$\quad\lambda_{CC}$ = failure rate for chip capacitors = 0.0005% per 1,000 h

$\quad\Sigma\lambda_{CD}$ = sum of failure rates of conventionally packaged devices used in construction of the hybrid microcircuit (glass-packaged diodes, molded resistors, or capacitors), obtained from specific part type by computing failure rate in the normal manner excluding environmental factors

$\Sigma (\lambda_{DS}\pi_T)$ = sum of failure rates of all discrete semiconductor flip chips (bumped beam or tab-lead), and faceup wire-lead-connected chips

$\quad\lambda_{DS}$ = individual discrete semiconductor chip failure rate, found in Table 47

$\quad\pi_T$ = failure-rate adjustment factor as function of hybrid substrate operating temperature, found in Fig. 132

5. The failure rate for thick film networks in hybrid microcircuits λ_{HK} is based on the pattern complexity, the number of screen and fire operations, the number and

tolerance requirements of the resistor elements, and the temperature of operation of the network substrate. The failure rate is computed from

$$\lambda_{HK} = [A_S \lambda_{CK} + N_{SF} \lambda_{SF} + \Sigma (N_{TR} \lambda_{TR})] \pi_T \tag{35}$$

where A_S = area of the substrate, in.^{2}

λ_{CK} = failure-rate contribution due to the network complexity: values found in Fig. 133 as a function of number of separate conductive areas (resistors and conductive paths) per square inch of substrate*

N_{SF} = number of screen and fire cycles required to form thick film pattern on substrate

λ_{SF} = 0.00004% per 1,000 h = additive-failure-rate term for each screen and fire operation

$\Sigma (N_{TR} \lambda_{TR})$ = sum of additive failure rates for each resistor as a function of required resistance tolerance

N_{TR} = number of thick film resistors of given tolerance

λ_{TR} = failure rate to be used for each resistor of given tolerance as specified in Table 48

π_T = failure-rate adjustment factor for thick film networks as function of substrate operating temperature (values of this π_T factor are shown in Fig. 132)

The resulting failure rate for the thick film network λ_{HK} is entered into the general hybrid-microcircuit failure-rate formula in Eq. (33) to obtain the overall hybrid-microcircuit base failure rate.

6. The hybrid-microcircuit package failure rate λ_{HP} is a function of the package style or configuration and the materials used in its construction. The equation is

$$\lambda_{HP} = \lambda_{PF} \pi_{PF} \tag{36}$$

where λ_{PF} = 0.0002% per 1,000 h = normalized value of failure rate for all hybrid microcircuit packages

π_{PF} = adjustment factor which modifies λ_{PF} as a function of package style and materials used in its construction, tabulated in Table 49 for various combinations of style and material

The hybrid-microcircuit-package failure rate λ_{HP} obtained from the calculation in step 6 is used in the general hybrid-microcircuit failure-rate formula in Eq. (33) to obtain the overall hybrid-microcircuit base failure rate.

Failure Modes and Mechanisms

Detailed knowledge of thick film hybrid microcircuit failure modes and failure mechanisms is the key to producing reliable devices. A thorough understanding of the causes of failure allows meaningful improvements in the areas which will yield reliability improvements:

1. Design standards
2. Component and material selection
3. Process controls
4. Design of acceptance tests
5. Design of screening procedures
6. Design of step-stress and accelerated life tests

The thick film industry is just beginning to develop a comprehensive fund of knowledge of failure modes and failure mechanisms. These data are difficult to obtain for the following reasons:

1. The industry is dynamic, and frequent material, design, and process changes have made some of the previous data obsolete and have introduced additional failure modes and failure mechanisms.

2. Typical thick film hybrid failure rates are so low that to obtain failure data by

---

* In some instances the λ_{CK} factor calculated using this technique will not lie within the graph, and any extrapolation would yield very high λ_{CK} factors. A substrate size of approximately ½ by ½ in. may be the practical minimum when employing this model.

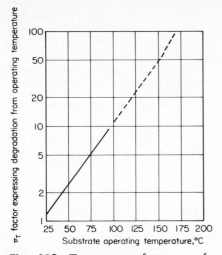

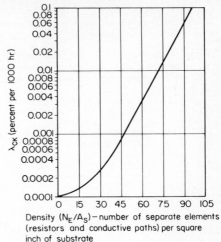

Fig. 132 Temperature factor π_T for hybrid microcircuits.[91]

Fig. 133 Failure rate, λ_{CK} for hybrid networks.[91]

TABLE 48   Failure Rate λ_{TR} for Hybrid Network Thick Film Resistors in Percent per Thousand Hours[91]

Resistor tolerance, ±%	λ_{TR}
0.1–<1.0	0.00005
1.0–<2.0	0.000025
2.0–<5.0	0.000020
5.0–10.0	0.000015
>10.0	0.000005

TABLE 49   Adjustment Factor π_{PF} for Hybrid-Microcircuit-Package Failure Rate[91]

Type of package	Substrate bonding	π_{PF}
Flatpack:		
Kovar, solid..........	Eutectic solder	1.0
Other, metal..........	Eutectic solder	2.0
Alumina..............	Eutectic solder	2.0
Glass................	Glass frit	2.5
TO-5:		
Kovar header.........	Eutectic solder	1.5
Glass header..........	Glass frit	2.0
Axial lead:		
Metal................	Eutectic solder	1.0

conducting laboratory tests at design stress levels requires millions of device test hours, which is not only costly but also introduces considerable time delays.

3. Many of the data which have been collected have not been published.

The main steps in a failure analysis procedure are to determine the following:[92] °

1. Effect, e.g., an electrical short or open or a deviation beyond specifications limits

2. Failure mode, e.g., broken wire, separated bond, incomplete aluminum intraconnection, crack

3. Characterization of abnormalities, i.e., scientific characterization of abnormalities associated with failure mode in terms of size, location, chemical composition, physical structure, and physical properties

4. Failure-mechanism hypothesis, i.e., relation of available data on environmental or operational stresses, namely, thermal, electrical, chemical, and mechanical, to probable reactions resulting in the abnormalities with the chemical, structural, and physical characteristics determined in the preceding step

5. Verification, i.e., experimental verification of probable mechanism or mechanisms; if not verified, repeat previous two steps and this one until verification is obtained

6. Corrective action, i.e., apply action such as material, process, design, or use-environment change based on the understanding of the failure mechanism

The foregoing procedure will provide information which ranges between two conditions: (1) complete quantification and understanding in simple cases, e.g., the action of mechanical stress resulting in a tensile failure of burnout by a known electrical overload of a resistor; or (2) a conceptual but nonquantitative understanding which, however, permits good corrective action.   An illustration of this condition is the conceptual, nonquantitative understanding of the Au-Al bond failure mechanism developed for the Minuteman II program and used to implement corrective actions.

A failure-analysis program must, of course, be tailored to the specific product.   A general failure-analysis procedure for microcircuits is given in method 5003 of MIL-STD-883.   This procedure is more applicable to monolithic devices, but it can be used as a guideline to establish a failure-analysis procedure for thick film devices.

Since considerable failure data are available for add-on discrete active and passive components, this section will discuss only failure modes and failure mechanisms which are unique to thick film circuit elements, chip (unpackaged) components, and thick film devices as a unit.   Data regarding failure modes (the electrical property of the device in which the failure is detected)[93] are more prevalent than data regarding failure mechanisms (a basic physical process or change, at the atomic or molecular level, which is responsible for the observed failure).[93]   The information in this section is intended to advise the designer of certain problem areas.   The data have been derived from both laboratory tests and field experience from various sources with units employing various design standards, topology, loading levels, materials, process techniques, resistor-adjustment methods, termination techniques, packages, acceptance testing and screening procedures, and exposed to a variety of environments.   Additionally, both catastrophic failures and failures due to parametric drift are reported. Catastrophic failures are not very common with thick film devices.   Consequently most reported failures are due to parametric drift, and the definition of allowable parametric drift is a function of the design and the application.   Great care must be exercised by the designer in evaluating these data.   References should be thoroughly studied to interpret them properly.

The most common thick film module failures are due to defective bonds and faulty hermetic seals.[92, 94]   Frequent bond failures are the result of an increased quantity of semiconductor bonds, one of the most common failures of semiconductor devices. The number of hermetic-seal failures is a function of the package size.   Thick film devices frequently employ large packages, which are more susceptible to leak failures.

Substrate.   The only known failure mode related solely to the substrate is catastrophic failure due to cracking.   Substrate cracking may result from the thermal shock incurred during soldering operations, improper bonding to the package, a mismatch in the thermal coefficient of expansion between the substrate and the package materials and adhesives or the encapsulant. When all organic materials are not removed from the substrate surface during cleaning operations, poor adhesion of thick film materials to the substrate may result. This may yield weak bonding pads and produce failures due to open bonds where add-on components are joined, leads are attached, or the substrate is bonded to the header.

Metallization.[96-100] Poor adhesion (bond strength) of thick film metallizing may result from improper firing cycles or the presence of organic materials on the substrate surface due to improper substrate cleaning prior to conductor screening.

Bond strength can be seriously reduced due to the dissolution of the thick film conductor material in solder (leaching) and/or the formation of gold-lead-tin intermetallic compounds during soldering operations.[98, 100] Bond failures can result at the time add-on components, leads, or headers are attached or later in service. When exposed to high temperatures (150°C) solder-coated thick film conductors experience a loss of adhesion, a reduction of conductivity, and a reduction of current-carrying capability with time which is a function of the conductor material, flux, solder composition, solder temperature, immersion time, amount of agitation, and soldering technique (immersion in solder pot, wave soldering, hand soldering with iron, etc.), ambient temperature, and time.

It has been noted that the higher the tin concentration of the solder, the more interactions between metals occur and the lower the strength tends to be both before and after stressing. It has been shown that when a Pd-Au conductor soldered with

TABLE 50   Composition of Silver-Saturated Solder Baths[96]

Low-temperature solders.......	62Sn-36Pb-2Ag
	10Sn-86Pb-4Ag
High-temperature solders......	97.5Pb-2.5Ag
	97.5Pb-1.5Ag-1Sn

60-40 lead-tin solder is subjected to heat aging, extensive tin penetration occurs, right down to the glaze-conductor interface, leaving a lead-rich tin-depleted region at the solder-conductor interface. The high tin content of the intermetallic gives rise to a threefold expansion in the conductor-layer thickness. Under these conditions, the adhesion progressively degrades. This mechanism has been confirmed by the use of tin-free solders, which give better adhesion.[96] When conductor materials incorporating gold or components with gold-plated leads are employed, gold-lead-tin intermetallic compounds are formed in soldered connections, producing brittle connections with reduced bond strength.

These problems can be minimized by controlling solder operations to keep bonding times to a minimum; by minimizing the exposure of the unit to high temperatures (150°C) subsequent to the solder operations; by pretinning components with gold-plated leads prior to attachment to the thick film conductor (the gold content of solder pots used for this purpose should be monitored and controlled);[100] or by saturating the solder with silver if silver thick film conductors are employed. Several typical silver-loaded solder compositions are listed in Table 50.

Poor bonds may also be caused by poor solderability of the thick film conductor, which may be the result of conductor oxidation, improper firing, glass buildup due to firing, or degradation due to firings subsequent to the metallization firing.

Silver-bearing thick film conductors are subject to silver migration. In the presence of moisture and an applied field, silver will migrate through a moisture layer, causing an intermittent short circuit.[96]

Poor wire and die bonds may be due to oxidation of the bonding pads.

Capacitors.[1, 96, 101] Thick film capacitors are subject to opens (due to cracked dielectric), shorts or degrading insulation resistance (due to dielectric pinholes, breakdown or surface peaks in electrode material), and excessive drift (due to air-abrasive adjustment, oxidation of the dielectric, crazing of the dielectric reaction of dielectric with the electrode material or exposure to hydrogen). Table 51 shows the effect of hydrogen gas on thick film capacitors of several different compositions.

Unencapsulated multilayer monolithic capacitors are susceptible to the termination problems (loss of adhesion or bond strength, formation of undesirable intermetallic compounds in solder connections, etc.) discussed under metallization. They may also experience loss in capacitance due to delamination of the layers or interelectrode shorts. Decreased insulation resistance and/or interelectrode shorting may result if inner electrodes are exposed. Monolithic capacitors may fracture or delaminate when

TABLE 51   Effect of Hydrogen Gas on Thick Film Capacitors and Resistors[96]

Description		Initial measurement	24 h after exposing to H_2	48 h after exposing to H_2	24 h after exposing to ambient (after testing to H_2)
			Capacitors		
E.S.L.:					
EX4510 unglazed electrode	A.	3,560 pF	4,400 pF	4,050 pF	4,400 pF
5800 B (Pt-Au)	B.	3,750 pF	4,550 pF	4,400 pF	4,500 pF
EX4510 unglazed electrode	A.	4,500 pF	4,600 pF	4,500 pF	4,460 pF
8831 (Au)	B.	4,500 pF	4,700 pF	4,500 pF	4,500 pF
Dupont 8289 glazed electrode	A.	0.0105 μF	0.0105 μF	0.0104 μF	0.0098 μF
ESL 9600 (Pd-Ag)	B.	0.0102 μF	0.0102 μF	0.0100 μF	0.0095 μF
	C.	0.079 μF	0.0820 μF	0.0770 μF	0.0740 μF
O.I. 062755 unglazed electrode	A.	355 pF	380 pF	380 pF	367 pF
061405 (Pd-Au)	B.	349 pF	370 pF	370 pF	360 pF
	C.	762 pF	819 pF	819 pF	790 pF
	D.	760 pF	805 pF	805 pF	780 pF
	E.	3,300 pF	3,500 pF	3,500 pF	3,400 pF
	F.	355 pF	378 pF	378 pF	368 pF
	G.	349 pF	370 pF	370 pF	360 pF
	H.	762 pF	820 pF	819 pF	790 pF
	I.	762 pF	820 pF	820 pF	800 pF
	J.	3,325 pF	3,525 pF	3,525 pF	3,420 pF
Chip cap....................	A.	0.097 μF	0.096 μF	0.096 μF	0.096 μF
			Resistors		
Du Pont Birox, unglazed,	R_1	34 kΩ	34 kΩ	34 kΩ	34 kΩ
trimmed	R_2	130 kΩ	130 kΩ	130 kΩ	130 kΩ
	R_3	17 kΩ	17 kΩ	17 kΩ	17 kΩ
	R_4	449 kΩ	450 kΩ	450 kΩ	450 kΩ
	R_5	8 kΩ	8 kΩ	8 kΩ	8 kΩ
E.S.L. 7000 series, unglazed,	R_1	11 kΩ	4.2 kΩ	4.3 kΩ	440 kΩ
trimmed	R_2	4 kΩ	1.6 kΩ	1.6 kΩ	175 kΩ
	R_3	550 kΩ	>20 MΩ	>20 MΩ	>20 MΩ

exposed to temperature variations if excess solder is built up on the termination during the attachment operation.

Resistors.[1, 84] Catastrophic failures in thick film resistors are very rare. Most failures are the result of excessive parametric drift or parametric instability. Cited as potential explanations for these failures are chemical reactions due to soldering fluxes,[97] absorbed gases,[96] adhesives, solvents, and encapsulants; agglomeration or segregation of constituents; internal stress relief; high field pulses;[102] ion migration; stress-relief cracking at the interface of the resistor film and its protective glaze; overspray of resistor surface with an abrasive trimmer, hot spots created by misalignment of resistor and conductor, or excessive adjustment, diffusion of the conductor into the resistor;[103] and mechanical stresses generated by thermal expansion and contraction of casting resins during hardening.[104] The material systems which have been studied and the comprehensiveness of the investigations vary considerably. The designer should review the references thoroughly to assess their applicability.

The instability of Ag-Pd resistors when subjected to hydrogen has received the most publicity. When subjected to a reducing atmosphere, the resistance of Ag-Pd resistors generally drops considerably; however, if excessive amounts of hydrogen are present, behavior may be erratic and resistance may even increase due to the formation of palladium hydride.[105-107] Hydrogen gas may be encountered in many ways; semiconductors may be bonded under forming gas; the unit may be encapsulated with

TABLE 52   Effect of 100 Percent Hydrogen on Pd-PdO Resistors (Not Adjusted) [107]

Nominal kΩ/sq	Polyurethane coating	Resistance change after 75 min, %
0.5	No	−57.7
3.0	No	−14.1
8.0	No	+13.5
20.0	No	+37.9
0.5	Yes	−13.6
3.0	Yes	−9.3
8.0	Yes	−3.6
20.0	Yes	+1.0

TABLE 53   Effect of 50,000 and 5,000 ppm of Hydrogen on Pd-PdO Resistors[*, 107]

Nominal kΩ/sq	H_2, ppm	Abrasion adjusted	Time, h	Change, %
3.0	50,000	No	17.5	−4.35
8.0		No	17.5	−1.05
3.0	50,000	Yes	17.5	−71.00
8.0		Yes	17.5	−31.50
0.5	5,000	No	40.0	−1.93
3.0		No	40.0	−3.00
8.0		No	40.0	−0.11
0.5	5,000	Yes	40.0	−43.20
3.0		Yes	40.0	−62.00
8.0		Yes	40.0	−29.00

* All resistors uncoated and powered at 10 W/in.²

an organic coating where hydrogen is involved;[108-110] the resistor may be subjected to outgas products in a hermetic package evolving hydrogen or amines (epoxies, adhesives, solvents, solder fluxes,[111] etc.); [112] or the resistor may be employed in a closed system where hydrogen is generated by dry cells, the reaction of moisture with metals, coolants,[105] seepage from wet slug capacitors, or the pyrolysis of overheated plastics.[107] The degree of degradation is influenced by length of exposure, hydrogen concentration, the method and degree of adjustment, the use of protective coatings, and the Pd-PdO–glass compositions.[107] The effect of these variables on typical Pd-PdO thick film resistor is shown in Tables 52 and 53 and Figs. 134 to 139. This problem has been overcome by the use of other resistor systems, encapsulation materials which use anhydrides as a catalyst, or by placing a protective glass coating over the resistor.[113]

Semiconductors. The semiconductor failure mechanisms associated with semiconductor-die and with die and wire bonding, which have been experienced by the semiconductor industry, apply to thick film hybrids. Many failure-mechanism studies have been published. Many old problems are accentuated, and several new problems

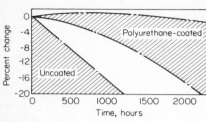

Fig. 134 Effect of 1,000 ppm of hydrogen on typical Pd-PdO resistors under power.[107]

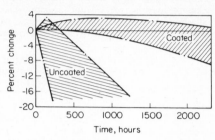

Fig. 135 Effect of 500 ppm of hydrogen on typical Pd-PdO resistors under power.[107]

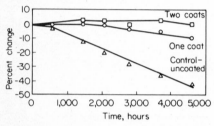

Fig. 136 Effect of polyurethane coating on unadjusted 500 Ω/sq resistors under power in atmosphere consisting of nitrogen and 500 ppm of hydrogen.[107]

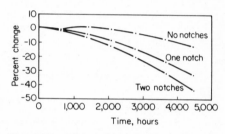

Fig. 137 Effect of notch adjustment on resistors exposed to 500 ppm of hydrogen in nitrogen (coated and powered).[107]

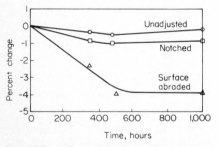

Fig. 138 Effect of method of adjustment on resistors exposed to 500 ppm of hydrogen in nitrogen.[107]

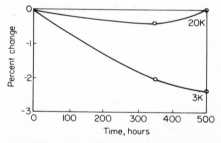

Fig. 139 Effect of 5,000 ppm of hydrogen on 3- and 20-kΩ resistors with glass encapsulant.[107]

are introduced with thick film hybrids. The use of multiple dice increases the time that dice are exposed to die and wire bonding temperatures. Dice from several production runs and from multiple manufacturers may be mounted on the same substrate. This may present bonding problems with fixed bonding conditions due to variations in metallization thickness, surface characteristics, etc.[84] Lead-wire lengths tend to be greater, which increases the susceptibility to bonding-wire shorts and other wire-bond defects due to shock, vibration, and acceleration. Oxidation of the conductor may result in marginal die and wire bonds. Poor bonds may result from placement of wire bonds on areas where the Au-Si eutectic, formed during bonding, has spread. Significant changes in the resistivity of conductive expoxies used for die bonding have been found during temperature cycling.[96] The effects of the various materials and processes used to hermetically seal or to coat conventional discrete packaged semiconductors are well known. The new materials and new or modified process techniques used to seal or coat semiconductor dice on thick modules must be thoroughly evaluated to assess their effect on the die and the die and wire bonds.

Reliability Improvement and Screening Procedures[114]

The reliability of thick film hybrid microcircuits has been increasing thanks to greater knowledge and standardization of design techniques, significant improvements in thick film materials, introduction of automated processes and fabrication techniques which afford increased process controls and reduction of process variations, use of automated and more sophisticated acceptance-testing equipment, development of effective screening procedures, employment of failure-analysis programs which eliminate identifiable failure mechanisms, and reliability-assurance programs.

Screening procedures for thick film hybrids have been included in MIL-STD-883 (method 5004) and are being developed by the EIA Med 50 committee. These screening procedures list several assurance levels. There are no failure-rate guarantees associated with the various screening levels, but a lesser risk can be assumed when devices are subjected to higher-level (more severe) screens. Units are tested to assure conformance to various visual, electrical, mechanical, environmental, and life tests. Defective and substandard or marginal devices will not be completely eliminated due to 100 percent screening, but there are indications that the screens are effective and that there is a relationship between severity and effectiveness.[94]

Screening procedures should be included in all high-reliability programs. They minimize failure rates, reduce subsequent assembly losses (which allows the system manufacturer to maintain a reduced inventory), reduce system troubleshooting time, and lower the user's spare-parts inventory. However, 100 percent screening is expensive to conduct, and rework of rejected devices is generally not feasible, which means that completed units must be scrapped, entailing further expense.

Screening tests must be designed to reject defective and substandard parts, due to excessive parametric drift or catastrophic failure, but not to degrade good parts. Therefore, the effectiveness of any screening procedure depends on the knowledge of device failure modes and failure mechanisms. As more detailed knowledge of thick film failure modes and mechanisms is obtained, screening procedures will become more effective and less costly.

REFERENCES

1. Thornell, J. W., et al.: Hybrid Microcircuit Design and Procurement Guide, Boeing Company, document available from National Technical Information Service, Springfield, Va., *Doc.* AD 705 974.
2. Scheffler, H. S., and F. R. Terry: Description and Comparison of Computer Methods of Circuit Analysis, Autonetics Div., North American Aviation, Inc., *Rep.* EM-6839, June 30, 1961.
3. Staff Report: Electronics Guide to CAD Programs, *Electronics*, Apr. 13, 1970.
4. Zobrist, G. W.: Thinking of Getting into CAD?, *Electronics*, Mar. 30, 1970.
5. Cordwell, W. A.: Transistor and Diode Model Handbook, *IBM Fed. Sys. Div. Tech. Rep.* AFWL-TR-69-44, October 1969, Air Force Weapons Lab., New Mexico.
6. Glass, C. M., and G. K. Pritchard: Model Conversion for CIRCUS, TRAC, SCEPTRE, and NET-1, *Air Force Weapons Lab. Tech. Rep.* AFWL-TR-69-37, July 1969.
7. *Electronics*, Sept. 19, 1966–July 10, 1967, series of articles on computer-aided design, pts. 1–10.
8. Root, C. D.: CIRCUS Means Versatility as a CAD Program, *Electronics*, Feb. 2, 1970.

9. IBM 1620 Electronic Circuit Analysis Program (ECAP), IBM Corp., Data Processing Division, White Plains, N.Y.; Application Program 1620-EE-02X, 1965.
10. Shirley, F. R.: You Don't Have to Be a Programmer to Use a Time-shared Computer to Solve Design Problems, *Electron. Des.*, Apr. 12, 1967.
11. Decker, F. R., and A. J. Welling: Need a Circuit Design Program? Why Not Custom-tailor One to Fit Your Needs? *Electron. Des.*, Oct. 11, 1966.
12. Solomon, Bert: Program to Design a Bridged-T Attenuator Is BASIC, *Electron. Des.*, Jan. 7, 1971.
13. Spitalny, A., and M. J. Goldberg: On Line Graphics Applied to Layout Design of Integrated Circuits, *Proc. IEEE*, November 1967.
14. Quick, R.: A Simple Computer Aided System for the Production of Hybrid Interconnection Artwork and Integrated Circuit Masks, *Proc. 1968 Hybrid Microelectronics Symp.*
15. Chitayat, A. K., and J. Lauria: Efficiency and Programming of Automatic Artwork Generators, *Solid State Technol.*, November 1970.
16. Integrated Circuit Engineering Corporation: ICEMAP: A Tool for the Integrated Circuit Design Engineer.
17. Topfer, Morton L.: Uses and Advantages of Chip Resistors, *Solid State Technol.*, November 1970.
18. CTS Microelectronics, Inc.: Ceradot Pellet Resistors.
19. American Components, Inc.: Miniature Metal Film Resistors for Hybrid and Integrated Circuit Applications.
20. Motorola Semiconductor Products, Inc.: "The Semiconductor Data Book," 5th Ed.
21. Flynn, George: Survey of Chip Resistors, *Electron. Prod.*, Jan. 18, 1971.
22. Caddock Electronics: Microminiature Precision Resistor Model ML 103.
23. MIL-R-10509, Resistors, Fixed, Film (High Stability), General Specification for.
24. MIL-R-11, Resistors, Fixed, Composition (Insulated), General Specification for.
25. Dionics, Inc.: MOS Capacitor Chips for Hybrid Circuits.
26. E. I. du Pont de Nemours & Co.: K1200 Capacitor Dielectric Composition DP-8289.
27. American Technical Ceramics: ATC UHF/Microwave Chip Capacitors.
28. San Fernando Electric Mfg. Co., West-Cap Division: Ceramic Chip Capacitor Handbook.
29. Electronic Industries Association: Ceramic Dielectric Capacitors, Classes 1 and 2, Standard RS-198, November 1957.
30. MIL-C-11015D, Capacitors, Fixed, Ceramic Dielectric (General Purpose), General Specification for.
31. USCC-Centralab: Miniature Ceramic Capacitors.
32. Corning Electronics: Corning CYK Capacitors.
33. Corning Electronics: Corning Glass–Ceramic Chip Capacitors.
34. Components, Inc.: Tantalum Capacitors Performance Characteristics.
35. Transistor Electronics, Inc.: Parameters of Tantalum Capacitors.
36. Union Carbide Corporation: Tantalum Chip Capacitors.
37. Union Carbide Corporation: Tantalum Chip Capacitors Pack High Value into Hybrid Circuits.
38. Components, Inc.: MINITAN Microminiature Solid Tantalum Capacitors.
39. Burkett, Frank S., Jr.: Improved Designs for Thin Film Inductors, *Proc. 21st Electron. Components Conf., 1971.*
40. Hoft, Don: Design Considerations for Building High Frequency Hybrid IC's, *EEE*, May 1969.
41. Nytronics, Inc.: Standard Component Catalog 9-SCG-1.
42. Piconics, Inc.: D Series Molded Inductor.
43. Piconics, Inc.: C Series and P Series (miniature inductors).
44. Cambridge Thermionic Corporation: Coil, Micro-miniature (Fixed), (Range .1 H to 1 mH).
45. Piconics, Inc.: JC Series: Low Cost Microelectronic Inductor.
46. Delevan Electronics Corporation: Delevan Micro-i™ Series Coils.
47. Piconics, Inc.: G Series and PV Series (miniature tunable inductors).
48. Cambridge Thermionic Corporation: Micro-Min Coil (Semi-fixed), (Range .06 H to 2 mH).
49. Delevan Electronics Corporation: Micro-i™ Variable, Series PV150 and PV250.
50. Piconics, Inc.: B Series and TT Series Tunable Inductors.
51. Delevan Electronics Corporation: Micro-i™ Transformers, Series Design.

52. United Transformer Company: Product Catalog, 1970.
53. Pulse Engineering, Inc.: Instrumentation Applications: Pulse Transformers (components catalog).
54. Consolidated Transformers Unlimited, Inc.: Technical data sheet on micro-mini audio and pulse transformers.
55. Microtran Company, Inc.: Microtran Transformers, Cat. 701.
56. Piconics, Inc.: Double Tuned Transformers.
57. Bond, M. V.: Comparison Simplifies Micro-packaging Selection, *EDN*, October 1968.
58. Texas Instruments, Inc.: Types A3T2221A, AST2222A; NPN Epitaxial Planar Silicon Transistors, *Bull.* DL-S6810872.
59. EDN Staff: Anatomy of the LID, *EDN*, February 1968.
60. Transitron Electronic Corporation: Semiconductor Slice and Dice Catalog.
61. Stern, Lothar, and Irwin Carroll: The Why's and How's of Semiconductor Chips, *Solid State Technol.*, November 1970.
62. Texas Instruments, Inc.: BLT2369, NPN Epitaxial Planar Silicon Transistor.
63. Motorola Semiconductor Products, Inc.: Motorola Microcircuit Components.
64. Bristol, R. G.: Design Guide to Hybrid Package Size, *Electron. Eng.*, September 1969.
65. Hatzipanages, D., and J. H. Powers: Thermal Characteristics of Film Modules, *Proc. Electron. Components Conf. 1968.*
66. Balents, L., R. D. Gold, A. W. Kaiser, and W. R. Peterson: Design Considerations for Power Hybrid Circuits, *Proc. 1969 Hybrid Microelectron. Symp.*
67. Mandel, A. P.: Designing Hybrid Modules Using Basic Thermal Guidelines, *Electron. Packag. Prod.*, July 1969.
68. Keister, F., and D. Auda: Thick Film Hybrid Design: The Right Way, *EDN*, June 10, 1968.
69. Kobs, D.: Attachment Techniques for Ceramic Chip Capacitors, *Proc. 1971 Hybrid Microelectron. Symp.*
70. Some Design Layout Rules for Thick Film Circuits, *Circuits Manuf.*, September 1970.
71. Kuo, C. Y., and H. G. Blank: The Effects of Resistor Geometry on Current Noise in Thick-Film Resistors, *Proc. 1968 Hybrid Microelectron. Symp.*
72. Kuo, C. Y.: The Contact Resistance in Thick-Film Resistors, *Proc. 1969 Hybrid Microelectron. Symp.*
73. Loughran, J. A., and R. A. Sigsbee: Termination Anomalies in Thick Film Resistors, *Proc. 1969 Hybrid Microelectron. Symp.*
74. Herbst, D. L., and M. Greenfield: Voltage Sensitivity versus Geometry of Thick Film Resistors, *Proc. 1969 Hybrid Microelectron. Symp.*
75. Kuo, C. Y.: Termination Resistance in Thick-Film Resistors, *EDN*, Mar. 15, 1970.
76. Jefferson, C. F.: Geometry Dependence of Thick Film Resistors, *Proc. 1970 Electron. Components Conf.*
77. Peckinpaugh, C. J., and W. G. Proffitt: Termination Interface Reaction with Non-Palladium Resistors and Its Effect on Apparent Sheet Resistivity, *Proc. 1970 Electron. Components Conf.*
78. Riemer, D. E.: The Effect of Geometry on the Characteristics of Thick Film Resistors, *Proc. 1970 Electron. Components Conf.*
79. Garvin, J. B., and S. J. Stein: The Influence of Geometry and Conductive Terminations on Thick Film Resistors, *Proc. 1970 Electron. Components Conf.*
80. Bristow, C. W. H., W. L. Clough, and P. L. Kirby: The Current Noise and Non-Linearity of Thick Film Resistors, *Proc. 1970 Hybrid Microelectron. Symp.*
81. Young, C. W.: A Thick Film Capacitor Design Nomograph, *Electron. Packag. Prod.*, July 1971.
82. A Technique for Producing Thick Film Capacitors, "DuPont Thick Film Handbook," Thick Film Technology, ser. A-76961 no. 2.2.3.3, 9/71.
83. Thick Film Dielectrics, "DuPont Thick Film Handbook," Thick Film Technology, ser. A-76958 no. 2.2.3, 9/71.
84. Minuteman Microelectronics: Application Guide Revision D, June 30, 1968.
85. Platz, E. F.: Solid Logic Technology Computer Circuits Billion Hour Reliability Data, *1968 Proc. Annu. Symp. Reliab.*
86. Platz, E. F.: Reliability of Hybrid Microelectronics, *1968 West. Electron. Show Conv.* (WESCON).
87. Frissora, J. R.: Packaging of Space Systems Utilizing Hybrid Microelectronic Techniques, *1968 West. Electron. Show Conv.* (WESCON).

88. McCormick, J. E.: On the Reliability of Microconnections, *Electron. Packag. Prod.*, June 1968.
89. Totta, P. A.: Flip Chip Solder Terminals, *Proc. 1971 Electron. Components Conf.*
90. Huenemann, R. G.: Reliability Analysis Center Hybrid Reliability Data Classification System, *Proc. 1971 Hybrid Microelectron. Symp.*
91. Procedure for Reliability Prediction of Standard Hardware Program (SHP) Modules Code Ident. 10001 NAVORD OD40778.
92. Browning, G. V.: Survey of Failure Mechanisms for Hybrid Microcircuits, *Proc. 1969 Natl. Electron. Packag. Prod. Conf.*
93. Vaccaro, J., and H. C. Groton (eds.): "RADC Reliability Physics Notebook," AD-624769.
94. Straub, R. J., and J. P. Farrell: The Effectivity of Screening Hybrid Microcircuits per MIL-STD-883, *Proc. 1971 Electron. Components Conf.*
95. Hays, C. W.: Are Hybrid Circuits Reliable? *1968 Hybrid Microcircuits Appl. Conf. Dig.*
96. Gundoters, V. K.: Failure Modes in Thick Film Hybrids, *Proc. 1971 Electron. Components Conf.*
97. Tsunashima, E.: The Reducing and Oxidizing Phenomenon of Soldering Flux to the Surface of Thick Film Materials, *Proc. 1971 Hybrid Microelectron. Symp.*
98. Anjard, R. P.: Leaching During Solder Immersion: Thick Film Conductors, *Proc. 1970 Natl. Electron. Packag. Prod. Conf.*
99. Crossland, W. A., and L. Hailes: Thick Film Conductor Adhesion Reliability, *Proc. 1970 Hybrid Microelectron. Symp.*
100. Wild, R. N.: Effects of Gold on Solder's Properties, *Proc. 1968 Natl. Electron. Packag. Prod. Conf.*
101. Holden, J. P.: The Development of Glaze Capacitors for Thick Film Circuits, *Radio Electron. Eng.*, December 1968.
102. Pakulski, F. J., and T. R. Touw: Electric Discharge Trimming of Glaze Resistors, *Proc. 1968 Hybrid Microelectron. Symp.*
103. Skulnik, D.: Metal Diffusion between Conductors and Ruthenium Oxide Resistors, *Proc. 1970 Hybrid Microelectron. Symp.*
104. Tsunashima, E., and K. Terasaka: The Solution of Parasitic Effects on Thick Film Hybrid Circuits Resulting from Encapsulation Process, *Proc. 1969 Hybrid Microelectron. Symp.*
105. Solomon, J., and R. N. Wild: Hydrogen Degradation of Silver-Palladium Oxide Resistor Materials, *1968 Hybrid Microcircuits Appl. Conf. Dig.*
106. Robinson, W. L., J. Boros, and H. Goldfarb: The Effect of Atmospheres Containing Hydrogen on Pd/PdO Type Thick Film Resistors, *Am. Ceramic Soc. 21 Pac. Coast Reg. Meet., Pasadena, Calif.*, Oct. 23–25, 1968.
107. Fanelli, L. H., W. L. Robinson, and V. J. Leggis: Factors Affecting Solder Wetting and Resistance Drift in Thick Film Circuits, *Proc. 1969 Electron. Components Conf.*
108. Asama, K., Y. Nishimura, and H. Sasaki: Study on the Thick Film Resistance Abrupt Change by Resin Packaging, *Proc. 1969 Hybrid Microelectron. Symp.*
109. O'Connell, J. A., et al.: Bidirectional Electrochemical Trimming of Thick Film Resistors, *Proc. 1967 Electron. Components Conf.*
110. Carey, J. P.: Encapsulation of Thick Film Substrates, *Proc. 1967 Natl. Electron. Packag. Prod. Conf.*
111. Tsunashima, E.: Effect of Soldering Flux on the Reliability of Thick Film Hybrid Circuits Resulting from Encapsulation Process, *Proc. 1969 Hybrid Microelectron. Symp.*
112. Himmel, R. P.: PdAg Thick Film Resistor Stability in Hermetic Package, *Proc. 1970 Hybrid Microelectron. Symp.*
113. Hirsch, H., and F. Koved: Problem of Degradation of Encapsulated Thick-Film Resistors Solved, *Insulation*, June 1966.
114. Cole, E. M., and D. L. Fan: A Reliability Improvement Program for Hybrid Circuits, *Proc. 1969 Hybrid Microelectron. Symp.*
115. MIL-STD-198, Capacitors, Selection and Use of.

Chapter **2**

Photofabrication Operations, Screens, and Masks*

STEWART G. STALNECKER

Microcircuit Engineering Corporation,
Medford, New Jersey

INTRODUCTION

The selective deposition of thick film materials on a suitable substrate base involves the use of a stencil screen or mask. The screen or mask in turn must be made by some photofabrication method involving specialized materials and processes.

The various uses of photofabrication are too numerous to define. The contents

* The author wishes to express his appreciation to the many people who assisted in supplying information for this chapter: Richard Volk, Lewis H. Cronis, Gerard Daniel, John Busher, Richard Zeien, Paul Miller, Ron Headly, and Philip Ferinde. In addition he wishes to thank Joan Stalnecker for her efforts throughout the preparation of the chapter and especially during the completion of the final manuscript.

of this chapter are limited to the application of photofabrication to the preparation of thick film screens and masks. Additional information on photofabrication methods may be found in Ref. 1.

GLOSSARY

Some terms commonly used in photofabrication and for defining screens and masks are listed in the glossary of terms.

PHOTOFABRICATION PROCESS STEPS

The basic steps for preparing stencil screens or masks are outlined in Table 1.

Circuit layout drawings for master artwork Before preparation of the artwork, a circuit-layout drawing of the desired pattern geometry is required.

TABLE 1   Process Steps for Screen or Mask Preparation

1. Circuit-layout drawing
2. Artwork generation
3. Camera reduction
4. Screen- or mask-frame preparation
5. Image transfer or mounting

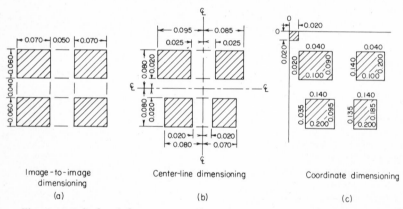

Fig. 1 Methods of dimensioning drawings prior to artwork preparation.

Dimensioning. This layout may be in the form of a dimensioned sketch or scaled drawing on a paper or Mylar° grid sheet. Figure 1 illustrates three methods of dimensioning a circuit pattern, i.e., a group of four resistor elements.

For relatively simple patterns, such as those illustrated in Fig. 1, all three of the methods shown are easy to interpret. However, with more complex arrays the technique illustrated in Fig 1a may be troublesome for the technician preparing the artwork. For example, if there were 25 resistor elements in a row and each element were spaced differently, each dimension would have to be added to the preceding dimension to locate the next position. This type of dimensioning would therefore necessitate numerous computations in order to generate the artwork pattern. Any computational error would result in inaccurate artwork, which could easily pass undetected until the circuit patterns were actually printed. This requirement for additional computation would not be necessary if the designer or draftsman used centerline or coordinate dimensioning, as illustrated in Fig. 1b and c respectively.

° Registered trademark of E. I. du Pont de Nemours & Co.

Centerline dimensioning is useful if the image layout is symmetrical about both axes. However, coordinate dimensioning is the favored method of dimensioning drawings because (1) dimension lines are not required, (2) more dimensions can be placed on the drawing without interference, and (3) the dimensions are most practical for artwork preparation by means of a coordinatograph. When using a coordinate drawing the pattern dimensions may be placed on the image lines or in close proximity, as illustrated in Fig. 2.

Layout drawings prepared on grid sheets are very convenient since the grid lines can serve as a basis for referencing the dimensions. For dimensional stability it is recommended that a Mylar grid be used in place of paper. Mylar grids up to 36 by 48 in. size are available with tolerances of ±0.002 in. over the entire gridded area.[2] With Mylar or an accurate paper grid sheet it is possible to hand cut an artwork by tracing the pattern off the grid sheet directly. This method is both rapid and reasonably accurate, especially if the artwork scale is 20× and a drawing board with Metal T square and triangles are available. How-

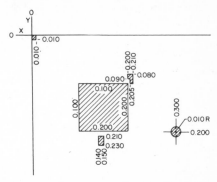

Fig. 2 Placing coordinate dimensions on the circuit-layout drawing.

ever, when tracing an artwork, the original paper or Mylar grid sheet should be used if possible. If copies of paper or Mylar grid drawings are made, the copy should first be checked dimensionally. Errors of 0.1 in. along a 24-in. dimension can result from stretching in the copy machine. Where possible, artwork drawings should be made up so that patterns fall on the grid lines or halfway in

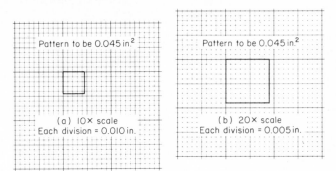

Fig. 3 Identical pattern layouts on grid sheets.

between at 10× or on the grid lines at 20× if the drawings are not to be dimensioned (see Fig. 3). Thus it is best if all patterns can be designed with line widths, lengths, and spacings of 0.005 in. or multiples thereof.

Artwork Scale Factors and Dimension Lines. All artworks should be labeled either with a scale factor such as 5×, 10×, 20× or a dimension line. The dimension line must be accurately prepared, and for this reason it is not recommended for determining the scale of the artwork. The problem arises in the photographic operation, where the cameraman must determine the ratio to be used. For example, an inked artwork has a dimension line of 10.100 in. drawn between two dimension lines and is labeled "Reduce to 1.000 ±0.001 in." A second inked artwork has the same labeled dimension; however, the inked dimension measures out to 10.050 in. To hold the tolerance specified of 1.000 ±0.001 in. the two artworks would have to be reduced by different factors; that is, 10.100 in./10.1× = 1.000 in., whereas 10.050 in./10.05× = 1.000 in. This would mean two different camera settings. In this situation both artwork patterns may have been prepared accurately with only the dimension lines themselves in error. If artworks are prepared accurately, only the scale factor should be specified, such as 10× or 20×. If it is necessary to compensate the artwork reduction to allow for a shrinkage factor, the scale factor could simply be changed to read 16× or "Reduce 20 in. to 1.250 ±0.001- in."

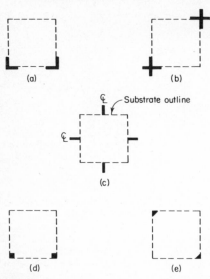

Fig. 4 Substrate registration marks used as positioning aids.

Positioning Aids on Circuit Patterns. During the printing of thick film patterns it is often desirable to know where the pattern should be located on the substrate. Therefore it is helpful to incorporate registration marks on the artwork which serve as positioning aids. Figure 4 illustrates some typical arrangements.

The registration marks shown in Fig. 4a to c would be located off the substrate for applications where there would not be room enough to print them on the substrate. It would also be desirable to block out these areas on the screen or mask so that ink would not pass through and smear the edge of the substrates. The centerlines shown in Fig. 4c have an added advantage which is utilized during the screen- or mask-making process. The centerlines themselves serve as a positioning aid for locating the images in the center of screen or mask frame. Therefore, during multiple printings, only minor machine adjustments are necessary to position the subsequent patterns on the substrates. The positioning aids illustrated in Fig. 4d and e are so placed that if

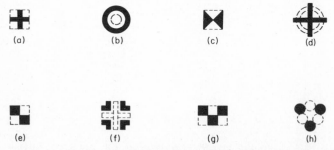

Fig. 5 Circuit registration marks used as positioning aids.

desired they can be printed on the substrate provided space is available. Other positioning aids or registration marks are illustrated in Fig. 5.

The dotted lines represent *companion patterns* which would be printed with a series of screens. As an example, all the squares in Fig. 5g might be printed on the substrate one at a time using a total of six different screens.

Artwork Generation

The master artwork is an enlarged accurately scaled reproduction of the circuitry patterns. The degree of enlargement will be determined by (1) the method of artwork preparation, (2) the required accuracy of the final size imagery, and (3) the maximum size the camera copyboard will accept for photoreduction.

In most applications, artworks for thick film microcircuitry are prepared at 10× or 20× scale. The dimensional requirements and tolerances for artwork and screens are outlined in Table 2.

Positive and negative artworks The preparation of the master artwork may be in one of two forms, either positive or negative. The following examples differentiate between a positive and negative artwork. Figure 6 illustrates a positive artwork and subsequent processing steps for preparing a stencil screen.

TABLE 2   Dimensional Guidelines and Tolerances for Thick Film Circuits

	Screen, mils	Artwork size, mils	
		10×	20×
Conductor widths, minimum..............	5 ± ½	50 ± 5	100 ± 10
Resistor widths, minimum................	20 ± ½	200 ± 5	400 ± 10
Conductor and resistor spacings..........	10 ± 1	100 ± 10	200 ± 20

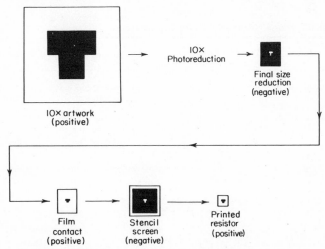

Fig. 6 Positive artwork and subsequent processing steps.

Working backward, if a black resistor on the substrate is considered a positive image, then the stencil screen has the reverse pattern, or negative image. The screen in turn was produced from a positive image, consisting of black emulsion on a clear sheet of film. The positive image was made by making a contact print from the final size negative, which in turn was produced by making a 10× camera reduction of a 10× positive artwork. If the original artwork is prepared as a negative, the sequence of preparing the printed substrate is as shown in Fig. 7.

From Figs. 6 and 7 it is clear that one process step can be eliminated if the original artwork is prepared as a negative instead of a positive. Table 3 is a comparison of positive and negative artworks.

Methods and materials for artwork preparation The master artwork may be prepared by any number of methods, such as applying opaque tapes to white paper or film, inking on a matte-surface film, or using a cut-and-peel two-layer Mylar-backed film. Table 4 compares the stability of materials used as bases for artwork.

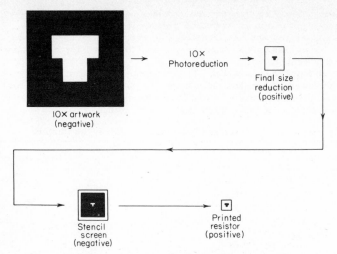

Fig. 7 Negative artwork and subsequent processing steps.

TABLE 3 Positive- and Negative-Artwork Comparison

	Positive artwork	Negative artwork
Normal preparation method....................	Inking or taping	Strippable film
Orientation in camera copyboard...............	Right side up	Upside down
Masking around artwork required..............	No	Yes
Camera reduction (one step)..................	Negative produced	Positive produced
Pattern and emulsion orientation...............	Right-reading emulsion up	Right-reading emulsion down
Contact print required for making stencil screen..	Yes	No

TABLE 4 Stability of Materials Used as Bases for Artwork[1]

Material	Temperature		Relative humidity, ppm/%	Aging in 5 years, ppm	
	ppm/°C	ppm/°F		78°F, 60% RH	90°F, 90% RH
Triacetate film..............	63	35	60	−300	−4700
Vinyl film..................	54	....	?		
Acrylic plate...............	70	40	80		
Estar* polyester film, 0.004 in..	27	15	21	−250	
Estar polyester film, 0.007 in..	27	15	16	−250	−100 +200
Polyester film, drafting......	27	15	13		
Photographic glass..........	4.5	2.5	Nil		
Fused quartz...............	0.5	0.3	Nil		
Celanar† polyester film.......	20	11	11		

* Trademark of Eastman Kodak Co.
† Trademark of Celanese Corp.

Table 5 is a comparison of the three most frequently used methods for artwork preparation.

Although artworks for thick film patterns are seldom prepared by taping, taping is often useful for making corrections to existing artworks for decreasing line widths or lengths. For this application, red cellophane or so-called *lithographers' tape* is employed. Inking of artworks on a matte-surface film is also used infrequently. Of all the methods for preparing artwork, inking is the most time-consuming, and the end results are often unacceptable. The problem of opacity of the finished inked areas may result in difficulty during the camera reduction. If inking is used for the preparation of artworks, a light table should be used to check the opacity of the pattern areas.

The strippable-film method has been the favored means of artwork production. A typical strippable film is constructed using a clear Mylar base, to which a red gelatin or plastic overcoat is adhered. The clear base is available in thicknesses of 0.003, 0.005, and 0.0075 in.[3] For most applications the 0.005-in. or "thick," material is satisfactory. The material may be cut with an X-acto knife, scalpel, or even a razor

TABLE 5   Comparison of Artwork-Preparation Methods

Type of artwork	Features	Advantages	Disadvantages
Taping on Mylar or acetate	Similar to printed-circuit-board artwork	Relatively simple, fast, low cost, good contrast	Accuracy of line widths, not all tape widths available, poor stability
Inking on matte Mylar	Similar to mechanical drawings	Any line width can be produced, good stability	Very slow, opacity not consistent, corrections difficult to make
Strippable film	Two-layer, Mylar-backed film	Relatively simple to use, good stability, excellent contrast	Some technique required for cutting, light table required

blade with good results. The artwork is prepared by cutting through the gelatin layer only, following which the material is lifted and peeled off the base film (Fig. 8). If the cuts are made with a sharp tool, the edges of the patterns will be sharp and well defined. The strippable-film materials are also very opaque photographically and provide excellent contrast and resolution during the camera reduction.

Hand-Cut Artworks. A typical setup for hand cutting a strippable-film artwork at 10× would include the items listed in Table 6. A light table is useful for backlighting the strippable film to illuminate the cut marks, which should be just deep enough to cut through the red or amber film layer. Metal triangles and T squares are preferred over acrylic tools for stability and ease of handling. Plastic drafting tools are susceptible to cut marks when sharp cutting blades are used.

Manual Coordinatograph. Figure 9 illustrates a manual coordinatograph which is an extremely useful tool for preparing master artworks. Direct dial readings to 0.001 in. are possible over the entire work surface.[4] Scale factors of 10:1, 20:1, and 50:1 are usually available so that artworks can be prepared at 2×, 5×, 10×, 20×, or 50× without having to convert the drawing dimensions to the artwork scale. In addition to making cuts at right angles, a polar-model coordinatograph is available for cutting angles by rotating the entire cutting surface (see Fig. 10). Small circle cutters and beam compasses provide means for making circular cuts usually to diameter tolerances of ±0.002 in. Digital readout attachments may be substituted for the dial scales to increase speed and efficiency up to 30 percent.

Table 7 gives a comparison between three methods for generating artwork using strippable film.

Machine-generated artwork has been prepared for thick film circuit patterns, but this method is seldom used often enough to justify the high cost of the equipment.

Table 8 summarizes the accuracy and precision of artwork prepared by the three methods outlined in Table 7.

Step-and-repeat artwork If an array of equally spaced, identical patterns is re-

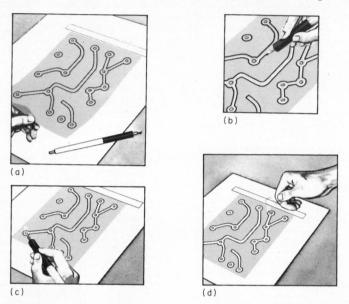

Fig. 8 Cut-and-peel artwork preparation. (*a*) Cut a piece of Ulano Rubylith large enough to cover the area to be masked. Tape it down firmly at the top with the dull side up. (*b*) With the Ulano swivel or other cutting tool, outline the areas to be masked. Do not cut through the polyester backing sheet. (*c*) Using the special Ulano peeler, lift up a corner of the film, thus separating it from the backing sheet. (*d*) Now carefully peel off the film as outlined, leaving a completed mask, positive or negative, that corresponds exactly to the desired pattern. Ulano and Rubylith are registered trademarks of the Ulano Companies. (*Ulano Companies*)

TABLE 6 Hand-Cut Artwork-Preparation Setup

1. Light table with metal edges
2. Mylar grid with 10 squares per lineal inch
3. Steel scale with graduations of 0.01 in.
4. Stainless-steel T square and triangles
5. Swivel knife, scalpel, or single-edged razor blades
6. Compass with scalpel blade
7. Loupe (5× to 10×)
8. Strippable film
9. Needle-point pricker for lifting film

quired, then a *step-and-repeat* pattern must be prepared. The step-and-repeat matrix may be accomplished in several ways. In some cases the circuit patterns are relatively simple, i.e., squares and rectangles, and can be reproduced on the original artwork. For more complex images, a single artwork is prepared and then repeated on a layout, as outlined in Table 9. This procedure is particularly useful where step-and-repeat cameras are not available (see below).

Figure 11 illustrates a step-and-repeat pattern which also includes several types of positioning aids.

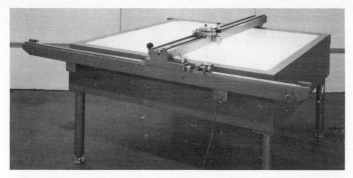

Fig. 9 Manual coordinatograph, rectangular model. (*Consul-Mutoh Ltd.*)

Fig. 10 Manual coordinatograph, polar model. (*Consul-Mutoh Ltd.*)

TABLE 7 Methods of Generating Artwork Using Strippable Film

Cutting method	Advantages	Disadvantages	Comments
Hand cut............	Inexpensive tools used, little space required	Slow, poor accuracy	10× or 20× scale usually used, depending upon skill of operator
Manual coordinatograph	Rapid, accurate, good precision possible	Equipment expensive, considerable space required	Operator must be reasonably skilled, 10× scale most often used
Machine-drawn (automatic)	Rapid, accurate, high productivity	Very expensive, corrections difficult to make, program required	Good for large volume of work, usually not justified due to cost

TABLE 8   Typical Accuracy and Precision of Artwork[1]

Drawing means	Accuracy, in.	Pre-cision,* ppm	Finest line, in.
Manual............	0.010	±250	0.010
Machine-aided (manual coordinatograph)...............	0.001	±25	0.004
Machine-drawn:			
Automatic scribe or cut plotter.....................	0.001	±25	0.002
Automatic photoplotter........	0.0005	±12	0.002

* Based on artwork 1 m square (40 by 40 in.).

TABLE 9   Procedure for Preparing a Step-and-Repeat Artwork with Complex Imagery

1. Prepare one master negative artwork at 10✕ or 20✕
2. Cut centerlines on the artworks within the pattern area
3. Cut centerline widths to 0.002 in. at step-and-repeat size
4. Reduce master negative artwork to 5✕ or 10✕, making positive images
5. Scribe a 5✕ or 10✕ step-and-repeat grid on a sheet of clear Mylar with a coordinatograph if possible
6. Use the 0.002-in. centerlines to align the individual images to the scribed Mylar grid
7. Attach the 5✕ or 10✕ positive images to the step-and-repeat grid with clear tape or spray adhesive
8. Identify the right-reading-side of the array and label
9. Reduce the step-and-repeat matrix to the final size (5✕ or 10✕)

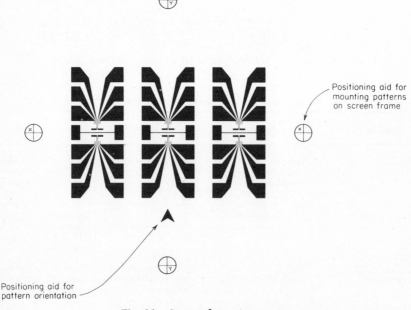

Positioning aid for mounting patterns on screen frame

Positioning aid for pattern orientation

Fig. 11   Step-and-repeat pattern.

All artworks should be carefully checked for dimensional accuracy and errors in preparation before proceeding further. Table 10 is an artwork-preparation checklist.

Photographic Operations

When the master artwork of the circuit patterns is complete, a camera reduction is required to obtain the final-size imagery. The camera, therefore, is an important tool which is used to provide an exact reproduction of the circuit artwork at a reduced size.

TABLE 10   Artwork-Preparation Checklist

1. Artwork scale, 10×, 20×, etc.
2. Artwork will fit into camera copyboard
3. Positive or negative artwork
4. Artwork or drawing number, including revisions
5. Date of artwork preparation
6. Positioning aids
7. Step-and-repeat centerlines
8. Right-reading side
9. X- or Y-axis designation

TABLE 11   Camera-Selection Factors

Item	Camera consideration	Recommendation
Artwork size.....	Copyboard size	50 by 50 in. to handle 20× artwork for 2 by 2 in. substrate size
Reduction ratio...	Lenses	6-in. focal length for 10× and 20×; 14- or 24-in. focal length for 5×
Film or plate size.	Film or plate holder	Up to 20 by 24 in. for doing step-and-repeat work at 5× or 10×
Camera type.....	Low-bed or overhead type	Low bed for limited space; overhead for high-volume operation
Camera precision	High accuracy ±0.001 in. repeatability or better	For metal-mask fabrication and multilayer patterns; fine line work
	Medium accuracy, ±0.002–±0.003 in.	For resistor, conductor, and capacitor networks of nominal complexity

Reduction cameras   There are many cameras available on the market which are suitable for making photoreductions at the 10× or 20× reduction ratio. Camera reductions of artwork may be done within a company already equipped with a photographic facility or by an outside laboratory, which should be capable of meeting all the specifications. However, if a camera is to be purchased for producing screens or masks, several factors should be taken into consideration, as outlined in Table 11.

All cameras operate on the same basic principles and have three main features in common: (1) a film back or plate holder, (2) a copyboard, and (3) a lens (see Fig. 12). The film or plate holder and the copyboard of the camera must be parallel and at right angles to the optical axis of the lens. Therefore, the three main parts of any camera are usually rigidly supported and mounted on a level bed or track. Two of the three main parts of the camera must be movable in order to obtain different reduction ratios, as illustrated in Fig. 13.

The cameras illustrated in Fig. 14*a* and *b* are so designed that the lens and the copyboard are the two movable elements. The film or plate holder is fixed in position and protrudes into the darkroom. This type of camera is known as a *dark room camera*. The camera illustrated in Fig. 15 is so designed that the movable parts are the lens and the film or plate holder.

Artwork Illumination. To perform a photoreduction with a camera, the artwork must be mounted in the copyboard and illuminated. Front lighting is seldom used for illuminating artworks unless the circuitry has been laid out on a base material which is not sufficiently transparent. Contrast ratios of only 40:1 or 50:1 are realized with front illumination, compared to ratios of up to 1,000:1 for rear-illuminated translucent artworks.[5] As mentioned previously, the strippable-film artworks have excellent contrast and should be used with rear illumination. Figure 16 illustrates the basic arrangement for illuminating the artwork with a backlighting system.

Lenses. The function of the camera lens is to "gather," transmit, and project the

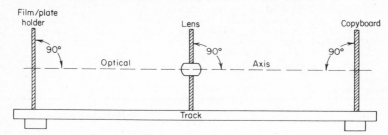

Fig. 12   Main features of a process camera.

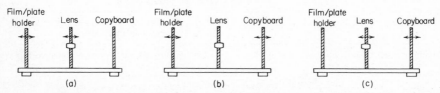

Fig. 13   Movable parts for three different types of process cameras.

light rays traveling from the artwork to the focal plane. Thus the lens will allow an exact likeness of the image to be reproduced on the film or glass plates. For thick film screen and mask making, *process lenses*, or lenses used for copying, are normally used. Process lenses are specifically designed for photographic flat-surface copy (flat field lens) and have no appreciable depth of field. The focal length of a process lens normally falls in the 6 to 24-in. range and is an important factor in achieving the correct enlargement or reduction ratio. Table 12 illustrates the reduction ratios for different focal-length lenses. The maximum reduction with a lens of a given focal length can be determined by the relationship[5]

$$R = \frac{u - f}{f}$$

where u = subject distance
f = focal length

The choice of the lens for a camera can be established by determining which lens focal length would cover the diagonal dimension of the camera's maximum film-size rating at a 1:1 or 100 percent size.[6] For example the diagonal of 20 by 24 in. film is 31.3 in. A 19-in. process lens at f/11 covers a diagonal of 32.2 in. Therefore, the standard lens choice for a 20 by 24 in. process camera would be a 19-in. focal-length lens. Process lenses with long focal lengths generally have small apertures and are therefore incapable of resolving fine patterns.

For thick film screens or masks, the line width and spacing requirements are not as critical as for semiconductor work. If the resolution of a process lens is 1,000 lines per inch, then line widths and spacings of 0.001 in. are theoretically possible.

(a)

(b)

Fig. 14 Process camera. (*a*) overhead model. (*Robertson Photo-Mechanix, Inc.*) (*b*) Low-bed model. (*R. W. Borrowdale Co.*)

Most thick film screens and masks are currently being produced with minimum line widths and spacings down to 0.003 in.

If 10× or 20× artwork is to be reduced, the reduction should be made in one step since each successive reduction causes a slight loss in resolution.[7] If 50× or 100× reductions must be made in two steps, the final step should be made at the

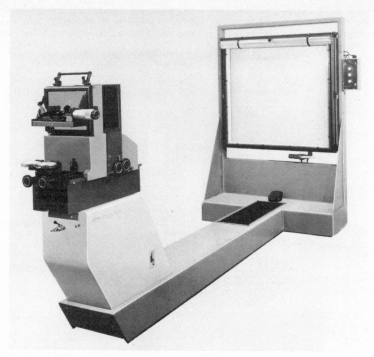

Fig. 15 Microphotography camera system for 10× and 20× reduction ratio. (*HLC Engineering Co.*)

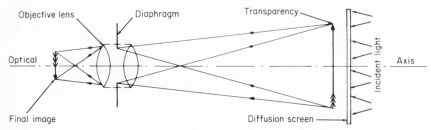

Fig. 16 Basic arrangement of elements for diffuse illumination of transparent originals.[5]

TABLE 12 **Enlargement and Reduction Ratios for Different Focal-Length Lenses**

Lens focal length, in.	Enlargement or reduction range
24	5× reduction–3× enlargement
14	10× reduction–6× enlargement
6	10× reduction–25× reduction

highest practical reduction. Resolution of the image is also affected by several other factors, namely, the type of sensitized material used, the film-processing conditions, and the field of view of the lens employed. For testing the quality of microimages the test pattern shown in Fig. 17 can be used. The purpose of this pattern is to determine maximum useful resolution with good image quality, not limiting resolution.

Camera Operation. When making photoreductions, the objective of the cameraman should be to reproduce the artwork pattern at the correct size with the best possible image definition. To achieve the proper image definition, the setup and

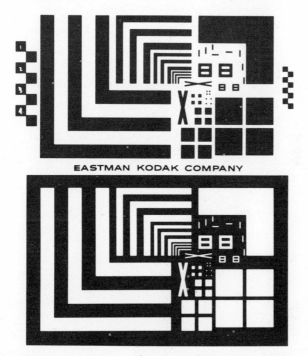

EASTMAN KODAK COMPANY

Fig. 17 A test pattern to aid in the evaluation of microimages.[5]

operation of the camera should take into consideration a number of important factors, as outlined in Table 13.

When making photoreductions of artworks, the image size will be as important as obtaining the sharpest focus of the image, if not more important. Resistor-element line widths in particular must be accurate, in addition to the spacing between terminations, which determine the effective length of the resistor. On the other hand, slightly rounded corners on images do not have any appreciable effect on the electrical performance of the printed patterns. It is advisable to check a known dimension on the reduced image against the same dimension of the artwork. For a reduction-ratio size check the longest accurately determined dimension on the artwork would be preferred. An alternative approach to determining the reduction ratio would involve the use of a *test strip* prepared on a stable base material such as strippable film. Figure 18 illustrates a 20-in. test strip and a traveling-stage microscope for measuring images up to 4 in. long. With the setup shown in Fig. 18 size checks for reductions of 5× to 20× are made to tolerances of 0.0001 in. The correct image size can be accomplished by changing the object-to-lens distance and

TABLE 13   Factors Affecting Image Definition

Factor	Requirement	Recommendation
Artwork.............	Sharp images	Cut artwork using strippable film
Artwork illumination..	High contrast	Backlight copyboard.
Camera vibration.....	Vibration-free	Shock mounting
Focus...............	Determine point of best focus	Compare series of exposures while making small lens changes
Camera alignment....	Focal plane, lens, and object plane parallel	Make periodic alignment checks on camera
Camera lens.........	Good resolving power, adequate field coverage, minimum chromatic aberration	Run practical photographic tests to compare theoretical and actual lens performance
Sensitized material....	Fine granularity, high contrast, high acuteness, high resolution, relatively low photographic speed	Compare characteristics of sensitized films and plates to achieve correct results

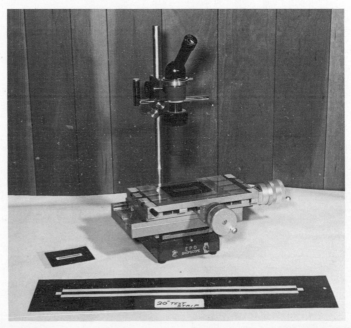

Fig. 18 Setup for measuring the camera reduction ratio. (*Microcircuit Engineering Corporation.*)

the lens-to-image distance on the camera. The following series of equations provides a system for calculating the required changes.[8]

$$\Delta u = -RF \frac{\Delta X}{X} \qquad \Delta v = \frac{F \Delta X}{RX}$$

where Δu = required change in object-to-lens distance
Δv = required change in lens-to-image distance
R = reduction ratio, defined so that when image is smaller than object, $R > 1$
F = focal length of lens
ΔX = change in image size required to go from the measured size to desired size
X = desired image size

$$\frac{1}{F} = \frac{1}{u} + \frac{1}{v}$$

Reduction: $R = \frac{u-F}{F} = \frac{F}{v-F} = \frac{u}{v} = \frac{1}{m} = \frac{h}{h'}$

Lens to image: $v = \frac{Fu}{u-F} = mu = \frac{u}{R} = F\left(\frac{1}{R} + 1\right)$

Subject to image: $u + v = \frac{(m+1)^2F}{m} = \left(\frac{1}{R}+R+2\right) F$

Lens to subject: $u = \frac{v}{m} = vR = \frac{Fv}{v-F} = (R + 1) F$

Fig. 19   Useful optical formulas.[5]

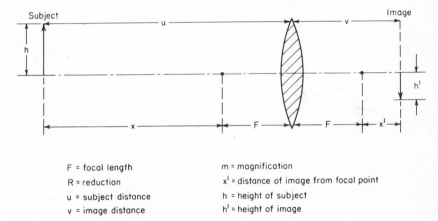

F = focal length

R = reduction

u = subject distance

v = image distance

m = magnification

x¹ = distance of image from focal point

h = height of subject

h¹ = height of image

Fig. 20   Subject-lens-image positions.[5]

The minus sign in the first equation indicates that as the object-to-lens distance increases, the image size X decreases. Thus if the image X is too large, ΔX is negative and the object-to-lens distance Δu is positive or must be increased.

For multilayer thick film circuitry requiring very close registry, it is advisable that all layers be photographed in succession with the artwork oriented in the same field of the lens. If at a later date one layer is redesigned, it may be advisable to rephotograph the entire set of artworks to assure optimum registration between layers. Figure 19 lists some useful optical formulas and Fig. 20 illustrates the subject-lens-image positions.

Photosensitive Materials. The preparation of stencil screens or metal masks is accomplished by the transfer of an image from a film or glass plate to an emulsion or photoresist material. The film or glass plates in turn are prepared by a camera reduction of the original artwork or by preparing a contact positive or negative from the camera reduction. Some different types of films and plates for use in preparing thick film screens or masks are listed in Table 14.

The support base for the materials listed in Table 14 is an important characteristic of these films. Both Estar° and Cronar‡ are thermoplastic materials, consisting of polyethylene terephthalate, a high polymer material belonging to the class called *polyesters*.[9] These polyester films exhibit very good dimensional stability; however, the films do change size during processing, with aging, and with changes in humidity and temperature. Even glass plates will show size changes under certain conditions. Table 15 lists some of the major factors affecting film size. Figures 21 to 24 correspond to the stability factors presented in Table 15.

During the actual film processing, the different factors affecting the dimensional stability may occur simultaneously. Therefore, depending upon the laboratory conditions, these size changes may be additive or tend to partially cancel each other. As an example, an increase in air temperature would cause an increase in size, which would often be offset by a lower relative humidity, normally a cause of decrease in size. For achieving the best dimensional control the following practical considerations should be noted:

1. The artworks or film transparencies and unprocessed film should be stored in the same area where they will be in equilibrium with the temperature and relative humidity. The polyester-base films will return to equilibrium in a matter of minutes after a change in temperature. However, if changes in relative humidity occur, the time to attain equilibrium is considerably longer, as shown in Table 16.

2. The areas where film is stored, exposed, and processed should be maintained at a temperature of 70°F and 50 percent relative humidity. The degree of environmental control necessary will depend upon the importance of the dimensional stability required. Table 17 lists some guidelines for polyester-base films.

3. The film being used should be uniformly processed by following the manufacturer's recommendations. The overall net time should be as short as possible.

4. Film drying should be done at room temperature. If forced drying is utilized, the relative humidity of the air entering the drier and the drier temperature may be controlled for optimum size holding, as illustrated in Table 18.

If the size changes of the polyester-base films are still a serious problem, the camera reduction and contact printing work should be done on glass plates. A comparison chart for polyester base films versus glass plates is shown in Table 19.

Film transparencies or glass plates must be inspected for pinholes, scratches, and other defects in the image area. Artworks and the camera copyboard must be clean to avoid flaws such as broken runs, short circuits between conductors, and missing pattern areas. If opaquing is required, the material buildup on the emulsion side should be kept to a minimum, otherwise poor contact may result during the subsequent image-transfer operation. With film transparencies, the opaquing may be done on the *base* (opposite side from the emulsion). Following the inspection and retouching steps the films or glass plates should be inserted and stored in negative preservers or glassine envelopes. Storing the film and plates in this manner should help eliminate some of the scratching and fingerprints caused by additional handling. In addition, duplicate or master patterns should be filed in case of loss or damage to the film or plates in current use.

Step-and-repeat cameras   If two or more identical images are to be printed on a substrate, a step-and-repeat operation will be required to produce the desired array. Step and repeat may be done on the artwork, as described earlier. However, if a considerable amount of work is contemplated and a high degree of accuracy is required, a step-and-repeat system is often necessary. Some photoreduction cameras can be fitted with step-and-repeat attachments, which are mounted as backs in place of the film or plate holders. Figure 25 illustrates a step-and-repeat back for the camera shown in Fig. 15. Step-and-repeat cameras incorporate index controls for positioning each exposure as the stepping and repeating is accomplished. The

° Registered trademark of Eastman Kodak.
‡ Registered trademark of E. I. du Pont de Nemours & Co.

TABLE 14 Photosensitive Materials for Preparing Photoreductions

Name	Manu- facturer	Support base	Contrast	Resolving power	Comment
Kodalith Ortho film 4556 thick base type 3	Kodak	Polyester	Extremely high	Extremely high	Excellent for intermediate reductions and intermediate step-and-repeat work
Cronar† Ortho S Litho Cos-7	Du Pont	Polyester	Very good	High	Excellent for intermediate reductions and intermediate step-and-repeat work
Kodagraph* Superneg* film 4922 (thick base)	Kodak	Polyester	High	Very high	Excellent for final reductions, free from pinholes, exceptionally clear in transparent areas
Kodak* high-speed duplicating film 4575 (thick base)	Kodak	Polyester	Moderately high	Very high	Intermediate reductions, positive-working
Kodak* high-resolution film SO-343 (thick base)	Kodak	Polyester	Extremely high	Ultra high	Best resolution for a film material, slow speed
Kodak* high-resolution plate.	Kodak	Glass	Extremely high	Ultra high	Very good stability, best resolution, slow speed

* Registered trademark of Eastman Kodak Company.
† Registered trademark of E. I du Pont de Nemours & Co.

stepping movement on some machines is manual, whereas on some automatic models the matrix is programmed on cards by the operator.

If break-apart ceramics are to be printed, the step-and-repeat pattern should match the center-to-center spacing of the substrates. Not only should the substrate

TABLE 15 Factors Affecting Dimensional Stability of Polyester-Base Films[9]

Factor affecting size change	Effect on film	Amount of size change	Graphic illustration	Comment
Humidity.....	Humidity increases, film size increases	0.002% per 1% RH (0.007 in. film base)	Fig. 21	Change in film dimension reversible
Temperature..	Temperature increases, film size increases	0.001%/°F	Fig. 22	Reversible and nonpermanent change
Processing....	Swelling of emulsion during wet processing, shrinkage of emulsion during drying	Varies	Fig. 23, Table 18	Film does not always dry back to original dimensions
Aging........	Shrinkage due to emulsion's exerting a compressive force on film base	0.02% after 5 years	Fig. 24	Insignificant size change for film used within several weeks after procdssing

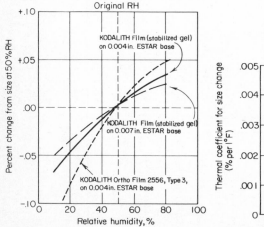

Fig. 21 Film-size changes due to relative-humidity changes.

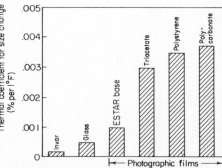

Fig. 22 Thermal coefficients for size change of Estar-base films compared with those of other materials.

drawing be used to specify the centerline spacings, but the ceramic substrates should be checked as well. If the ceramic shrinkage has not been carefully controlled, it is possible that the center-to-center dimensions of the individual substrates may not match the drawing dimensions. However, laser scribing of substrates after firing now allows the ceramics manufacturers to step and repeat the break lines on the substrates with a high degree of accuracy. For laser scribing the theoretical center-to-center spacing on the substrate drawing should closely match the true centerlines of the actual substrates. Figure 26 illustrates a fully automatic step-and-repeat machine.

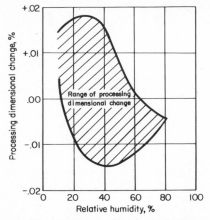

Fig. 23 Processing dimensional change of film as a function of relative humidity.

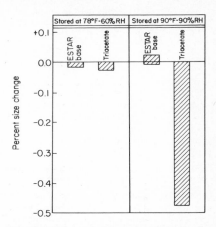

Fig. 24 Typical film-size changes due to aging.

TABLE 16 Approximate Time to Reach Equilibrium after a Change in Relative Humidity[10]

Estar film base	Equilibrium, %			
	40	50	80	100
4 mils................	10 min	30 min	80 min	4 h
7 mils................	1 h	2 h	4 h	8 h

TABLE 17 Environmental Control and Dimensional Stability for Polyester-Base Films[10]

Allowable dimensional variation	Control required
3 mils (0.003 in.) or more in 10 in...	Control required for comfort is adequate
Between 1 mil (0.001 in.) and 3 mils (0.003 in.) in 10 in.	Temperature and relative-humidity control beyond that required for comfort
1 mil (0.001 in.) or less in 10 in....	Very close temperature and relative-humidity control and choice of films with the best dimensional characteristics

SCREENS AND MASKS

The screens and masks are the end products produced from the artwork and photoreduction operations. However, screens and masks are only intermediate parts, or tooling, which must in turn be used on a machine to selectively deposit the

TABLE 18   Guide for Drying Conditions for Optimum Size Holding*,[10]

Relative humidity of surrounding air, %	Nominal range of drier temperature for minimum processing size change, °F
30	70–75
40	75–85
50	85–100
60	100–110

* The values in this table are for drying in a cabinet and as such are not fully representative for drying in a processor such as the Kodak Supermatic Processor, model 242, since the drying time and therefore the amount of overdrying are not as great.

TABLE 19   Polyester-Base Films versus Glass Plates

Characteristic	Polyester-base film (0.007 in. thick)	Glass plate (high-resolution)	Comment
Thermal coefficient* of expansion	0.001%/°F	0.00045%/°F	Glass superior to film
Humidity coefficient* of expansion	0.0015%/1% RH change	0	Glass far superior to film
Contrast classification†.....	High	Extremely high	Glass plate has finer granularity, less emulsion thickness
Resolving-power class......	Extremely high	Ultra high, 50,000 lines per inch	Very-high-quality optical system required to take advantage of superior resolving power of glass plates
Emulsion speed (ASA) exposure index (based on 10-s exposure time with tungsten)	6	0.025	Film 240 times faster than high-resolution plates; high-aperture lenses required for glass plates
Ease of handling during processing	Vacuum film holder required, unbreakable	Glass plate holder required, breakable	Film easier to handle from the standpoint of setup and breakage
Ease of handling during screen or mask making	Very easy to handle for screens and masks	Difficult handling for screens, few problems for masks	Film preferred for both screens and masks
Cost....................	4 × 5 in. size 12–15 cents per sheet	2 × 2 in. plate 65–70 cents each	Film cost considerably lower than glass

* Techniques of Microphotography, Kodak *Publ.* P-52, p. 48.
† Kodak *Publ.* P-1AA; *Pam.* P-47.

thick film paste materials onto the desired substrates. In addition to allowing paste materials to be deposited in certain areas on the substrate, the screens and masks can be used to control the thickness of the deposits as well.

In order to explore the merits of both screens and masks more fully, each of these tools will be examined separately. The materials of construction, the various applications, and the manufacturing techniques for both screens and masks will be presented with comparisons between them where helpful.

Screens versus masks To differentiate between screens and masks the term *screen* will apply to woven mesh materials with plastic or gelatin emulsion coatings called *stencils,* whereas *masks* will be defined as having an all-metal construction. Thus the masks will often be referred to as metal masks.

Fig. 25 Manual step-and-repeat back. (*HLC Engineering Co.*)

Stencil Screens

The screen frame Screen frames are used for supporting the woven mesh and are produced from wood, plastics or phenolics, and metal. Table 20 compares the merits of the three most commonly used materials for making frames.

As indicated in Table 20, wooden frames are unsatisfactory for use in preparing screens for thick film printing applications. Most screen frames are made from cast aluminum, primarily because of light weight, strength, stability, and reusability. However, aluminum frames are often warped or racked as a result of the casting process. To eliminate these defects, the cast frame should be machined to a specific thickness with the bottom surface flat and parallel to the top surface. Machining also eliminates uneven stress points, such as burrs and projections on the mounting surfaces for the screen mesh. In addition, machined frames with similar patterns can readily be interchanged on the screen-printing machine while maintaining the same screen-to-substrate separation or breakaway distance.

All-metal frames also lend themselves to numerous methods of mesh attachment, such as (1) metal strips crimped to the mesh and screw-mounted onto the frame

Fig. 26 Automatic step-and-repeat machine. (*Royal Zenith Corp.*)

TABLE 20   Comparison of Materials for Stencil-Screen Frames

Characteristic	Frame material		
	Wood	Plastic or phenolic	Metal
Dimensional stability..	Poor	Good	Excellent
Torsional stability.....	Poor	Good	Good to excellent
Flatness..............	Poor	Good to excellent (machined)	Fair (aluminum casting); excellent (machined)
Mesh mounting.......	Staples, glue, cord	Epoxy adhesive, crimped tubing	Epoxy adhesive, crimped tubing, mesh straps with screens, floating frame or bars
Solvent resistance.....	Fair to good	Good	Excellent
Relative cost.........	Very low	Moderate	Moderate to high
Reuse...............	Impractical	Possible	Generally reusable
Designed to fit thick film screen printing machines	No	Yes	Yes

(Fig. 27); (2) mesh crimped in slots with bars or tubing (Fig. 28); (3) beaded filler clamped with frame bar (Fig. 29); (4) epoxy adhesive (Fig. 30); and (5) channel frames with floating bars (Fig. 31). Frames with the mesh attached by screw mounting, like those illustrated in Figs. 27, 29, and 31, are so constructed that the mesh can be retensioned during use. These same frames are relatively easy to reprocess by removing the old mesh material and attaching a new insert or filler.

Screen frames are available in a variety of sizes and are usually specified by the inside dimensions. The range of screen-frame sizes for thick film printing is 3 by

Fig. 27 Screw mounting of mesh with strips to metal frame. (*Graining Equipment Co.*)

Fig. 28 Bar-in-slot mounting of mesh on metal frame. (*Industrial Reproductions, Inc.*)

3 to 12 by 12 in. However, two popular sizes in widespread use are the 5 by 5 and 8 by 10 in. frames. Selection of the proper screen-frame size depends upon (1) the size of the substrate being printed and (2) the design of the screen-printing machine. During printing, the frame should be large enough to prevent the squeegee from traveling too close to the edge of the frame, thereby creating too high a strain on the mesh, as illustrated in Fig. 32.

Several rules of thumb have been applied in selection of the proper frame size by considering the substrate size or the squeegee size. One recommendation is that the frame should be 2 to 4 in. larger around all four sides of the substrate.[12] Another suggestion is that the inside area of the frame be approximately 4 times the length and width of the substrate.[13] In other words, for a 1 by 1 in. substrate the minimum inside dimensions of the screen frame should be 4 in. on each side. A third rule of thumb states that the area traversed by the squeegee should be a minimum of 1 in. less than the inside screen-frame dimensions on all four sides.[11] Additional recommendations may be found in Ref. 11.

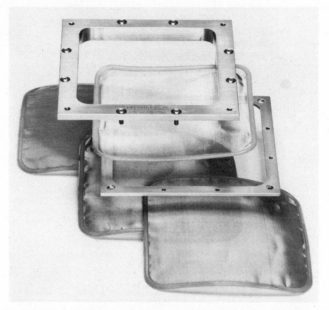

Fig. 29 Beaded-filler mounting of mesh in metal frame. (*Engineered Technical Products.*)

Fig. 30 Epoxy-adhesive method for mounting mesh on metal frame.

The second factor influencing the choice of the screen-frame size is the model or design of the screen-printing machine. Some machines are designed to handle only one size frame, and others can be modified to hold various sizes. A printer designed to hold an 8 by 10 in. screen frame might be fitted with an adapter plate to handle smaller frames, especially if substrates 1 by 1 in. or smaller are to be printed. Smaller frames are more economical and easier to store and handle.

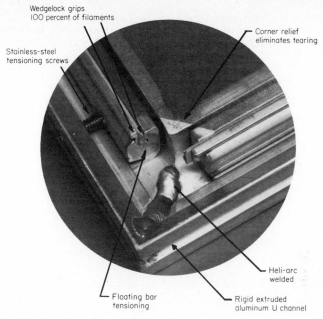

Wedgelock grips
100 percent of filaments

Corner relief
eliminates tearing

Stainless-steel
tensioning screws

Heli-arc
welded

Floating bar
tensioning

Rigid extruded
aluminum U channel

Fig. 31 Dia-Print screen chase. (*Lycoming Screen Printing Co.*)

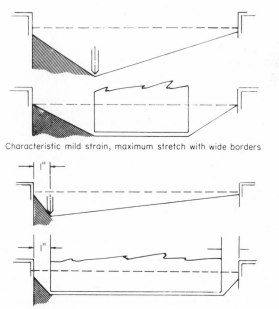

Characteristic mild strain, maximum stretch with wide borders

1"

1"

Characteristic sharp strain, minimum stretch with narrow border

Fig. 32 Effect of border width on screen stretch and strain during printing.[11]

For maximum flexibility in the screen-printing operation, a square frame is recommended in preference to a rectangular one. Frames with tooling holes evenly spaced on the four corners can be mounted any one of four ways in the printing machine so that screening can be performed in any one of four directions. If rectangular frames are used, it is important that the stencil images be placed with the correct orientation

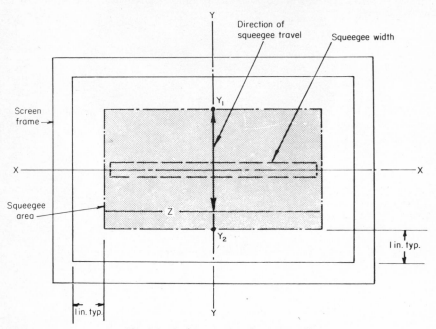

Fig. 33 Reference axes for screens.[11]

to the frame and in the proper position on the mesh. To assist in identifying the reference axis of screens the direction of squeegee travel is along the Y axis of the screen, as illustrated in Fig. 33. Therefore if a series of patterns is to be printed using rectangular screen frames, the Y-Y axis of each pattern should be specified, preferably on both the master artwork and on the film positive. In some applications where rectangular substrates or step-and-repeat patterns are being employed, it is possible that neither the screen nor the substrate-holding fixture can be oriented to accommodate a screen image which is 90° out of position on the mesh. In this situation the screens cannot be used and must be reprocessed with the images oriented in the proper direction.

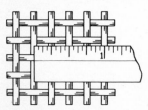

Fig. 34 A section of 4-mesh wire cloth.[15]

The screen mesh The screen mesh is the material mounted on the screen frame which serves as a support for the stencil. The term *mesh* refers to the number of openings and fractional parts of an opening, per lineal inch, counting from the center of any wire to a point exactly 1 in. away.[14] The mesh *count* therefore is the number of open spaces per lineal inch. Figure 34 illustrates a section of a 4-mesh fabric, which means the material has four openings per lineal inch, each measuring ¼ in. from center to center of two adjacent wires. This mesh is also a square mesh, meaning that the mesh count is the same in both directions of the weave. For very fine meshes, the exact mesh count should be gaged by using a microscope. However, in normal practice this method is tedious and time-consuming. For an approximate mesh count, a lunometer (Fig. 35) can be used by placing the scale over the mesh and determining where the axis center of the interference lines falls on the scale.

The mesh *opening* refers to the space between two adjacent parallel wires, as illustrated in Fig. 36, where the mesh opening is ½ in. Finer meshes should also be checked under a microscope or with an optical comparator to determine exact mesh openings.

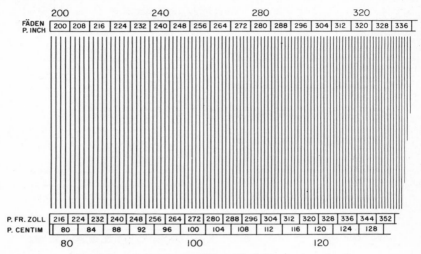

200		240		280		320	

FÄDEN P. INCH: | 200 | 208 | 216 | 224 | 232 | 240 | 248 | 256 | 264 | 272 | 280 | 288 | 296 | 304 | 312 | 320 | 328 | 336 |

P. FR. ZOLL: | 216 | 224 | 232 | 240 | 248 | 256 | 264 | 272 | 280 | 288 | 296 | 304 | 312 | 320 | 328 | 336 | 344 | 352 |

P. CENTIM: | 80 | 84 | 88 | 92 | 96 | 100 | 104 | 108 | 112 | 116 | 120 | 124 | 128 |

| 80 | | 100 | | 120 | |

Fig. 35 A lunometer. (*Wire Cloth Enterprises, Inc.*)

To calculate the width of the opening for any mesh the following formula is applied:[15]

$$\text{Width of opening} = \frac{1 - DM}{M}$$

where D = thread diameter
M = mesh count

Two other specifications for screen mesh are the thread diameter and the percentage of open area. Thread diameter must be specified to differentiate between materials with the same mesh count. For example, 120-mesh stainless steel is available with four different diameters, 0.0025, 0.0026, 0.0027, and 0.0037 in.[16] Monofilament nylon mesh is available in three thread diameters for the same mesh count, and this holds true for almost the entire range of meshes in nylon. The percentage of open area for any mesh is determined by using the formula[15]

$$\frac{(OM)^2}{1}(100) = (OM)^2(100)$$

where O = size of opening
M = mesh count

Another important characteristic of the mesh materials used for screen printing is the weave. The two most popular types of weave used for stencil screens are the plain and the twill weave, illustrated in Fig. 37. The twill weave is used so seldom for making thick film printing screens that it must be clearly specified, otherwise plain-weave

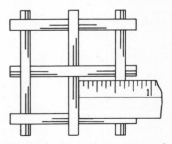

Fig. 36 A mesh opening of ½ in.[15]

material will be used. Twill weave is used for making stainless-steel mesh; in particular 270, 325, 400, and finer meshes. Nylon 280 mesh and finer are sometimes twill, as well as polyester mesh of 230 mesh and finer.[17]

All fabrics are woven into rolls or bolts from 12 to 87 in. wide and 180 ft long. Since the filaments running the length of the roll, or warp threads, are not woven as tightly as the weft threads, which run the width of the fabric, it is possible to obtain different deflection characteristics during printing depending upon which way the squeegee travels relative to the warp or weft.

Screen Fabrics. The three most commonly used materials for screen mesh are monofilament nylon, monofilament polyester, and stainless steel. Silk is not used in

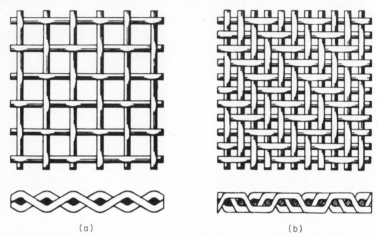

(a)   (b)

Fig. 37 Part of the illustration is an end view, the other part is a top view. End views of weaves: (*a*) plain, (*b*) twilled.[14]

TABLE 21   Disadvantages of Natural Silk for Use as a Thick Film Screen Fabric

1. Silk cannot be woven in meshes finer than 230
2. Percentage of open area is too small for adequate paste transfer
3. Abrasion resistance is poor
4. Thread diameter is not controllable
5. Silk cannot be reprocessed with sodium hypochlorite
6. Meshes are multifilament, further restricting paste transfer

the preparation of screens for thick film printing for a number of reasons, outlined in Table 21. The terms *silk screen* and *silk screening* as applied to thick film printing have been supplanted by *stencil screen* and *screen printing*, respectively. The nylon and polyester or Dacron° are also called synthetic meshes and will often be referred to as such. Table 22 gives a general comparison of the three fabrics.

In addition to the characteristics of the mesh material itself, it is important to consider the end use of the finished screen. Table 23 offers some guidelines to selecting the best mesh material for handling a particular screen-printing application. The specifications for monofilament nylon are shown in Table 24. Full rolls or bolts of monofilament nylon mesh are normally 60 yd long by 40 to 87 in. wide. Usually the material is priced by the running yard. Most nylon-mesh material is woven outside of the United States, primarily in Switzerland, France, Germany, Italy, and Japan. Monofilament nylon is available in a wide range of mesh count. However,

° Registered trademark of E. I. du Pont de Nemours & Co.

TABLE 22 General Comparison of Screen Mesh Materials

Mesh characteristic	Nylon	Polyester	Stainless steel	Comment
Tensile strength	Very good	Very good	Superior	Nylon strength reduced 10–15% when wet
Chemical resistance	Very good	Very good	Excellent	Mesh usually not affected by chemicals; resistant to solvents
Resilience	Superior	Excellent	Fair	All have high immediate elasticity for off-contact printing; screen-to-substrate separation less for wire
Abrasion resistance of mesh	Very good	Good	Excellent	Synthetic meshes tend to be abraded more by paste materials and substrates than stainless-steel mesh
Stability	Good	Excellent	Superior	Stainless steel best for close registration
Resistance to denting and coining	Excellent	Excellent	Poor	Synthetics do not coin readily; wire very susceptible
Availability of fine thread diameters	Good	Fair	Excellent	Wire mesh has the finest wire diameter and highest mesh count
Adhesion of stencil emulsion	Very good	Very good	Good	Synthetics allow emulsion to be exposed on inside (squeegee side) of mesh for greater adhesion
Mesh mounting to high tension	Fair	Good	Very good	Nylon difficult to tension to high values
Relative cost (materials)	Low	Moderate	Most expensive	Low material cost for nylon may be offset by higher mounting cost; stainless steel easier to stretch and mount
Ability to be decoated and reprocessed	Good	Good	Not recommended	Stainless-steel screens coin and are difficult to decoat with some emulsions
Abrasion effects on printer squeegee	Low	Low	High	Stainless steel more detrimental than synthetics, more rapid squeegee wear; urethane squeegees recommended
Moisture absorption	~4%	~0.4%	None	Stainless superior to synthetics
Sensitivity to heat	High, degrades over 250°F	High, degrades over 250°F	Very low	Nylon should be processed carefully if high temperatures (epoxy curing) experienced

TABLE 23 Guidelines for Selecting Screen Mesh Material for Various Screen-Printing Applications

Screen-printing application	Nylon	Polyester	Stainless steel	Comment
Printing on flat substrates, holding very close registration	Not recommended	Good	Excellent	Nylon stretches too much
Printing on bowed substrates, to registration of ±0.003 to ±0.005 in.	Very good	Very good	Poor	Stainless steel does not conform as well to substrate contours; depends upon squeegee head as well as mesh material
Printing multilayer substrates without complimentary patterns	Good	Good	Fair	Stainless steel applicable for limited number of layers or if patterns are compensated
Printing into recesses on ceramic packages	Very good	Poor	Not recommended	Stainless steel coins too easily
Printing over chips or other devices previously attached to substrate	Very possible	Possible but difficult	Not recommended	Nylon conforms better than other materials
Printing glaze compositions with sharp particles	Not recommended	Not recommended	Good	Glass particles will eventually cut through synthetic meshes
Printing very fine lines in the 0.003 in. range	Not recommended	Not recommended	Very good	Stainless mesh is available in very fine meshes with high percentages of open area
Printing solder cream with very heavy deposits	Very good	Very good	Good	Heavy emulsion coatings are easier to apply to the synthetics for printing heavier ink deposits

TABLE 24   Specifications for Monofilament Nylon Fabric (Nitex*, [18])

Mesh no.	Thread diameter, in.	Mesh opening, in.	Open area, %
16-T	0.0138	0.0465	59
20-T	0.0118	0.0374	57½
25-T	0.0110	0.0295	54
33-T	0.0087	0.0209	50
40-T	0.0079	0.0167	47
50-T	0.0068	0.0140	46½
63-S	0.0047	0.0108	48½
63-T	0.0055	0.0104	43
70-S	0.0039	0.0098	51
70-T	0.0047	0.0094	44½
70-HD	0.0055	0.0088	38
77-S	0.0039	0.0087	47
77-T	0.0047	0.0083	41
74-HD	0.0055	0.0079	34½
83-S	0.0035	0.0081	45½
83-T	0.0039	0.0077	41
83-HD	0.0047	0.0075	38
90-S	0.0028	0.0085	57
90-T	0.0035	0.0069	41
90-HD	0.0039	0.0071	41½
103-S	0.0024	0.0075	58½
103-T	0.0031	0.0064	43
103-HD	0.0035	0.0059	35
109-S	0.0024	0.0067	55
108-T	0.0031	0.0059	43
109-HD	0.0035	0.0057	38
114-S	0.0024	0.0063	53½
114-T	0.0028	0.0061	47½
114-HD	0.0031	0.0056	41
125-S	0.0024	0.0059	52
120-T	0.0028	0.0055	44½
123-HD	0.0031	0.0050	37½
130-S	0.0024	0.0053	49
132-T	0.0028	0.0047	40
130-HD	0.0031	0.0045	35
137-S	0.0019	0.0054	55
138-T	0.0024	0.0049	46½
139-HD	0.0028	0.0044	38
152-S	0.0019	0.0048	52
149-T	0.0024	0.0044	43
149-HD	0.0028	0.0040	34½
157-S	0.0019	0.0044	49½
157-T	0.0024	0.0041	40½
157-HD	0.0028	0.0036	32
166-T	0.0024	0.0037	37½
185-S	0.0017	0.0036	46½
185-T	0.0019	0.0035	43
185-HD	0.0024	0.0030	31
196-T	0.0017	0.0033	44
198-HD	0.0019	0.0031	38
206-S	0.0015	0.0034	49
206-T	0.0017	0.0031	42½
206-HD	0.0019	0.0029	35
230-S	0.0014	0.0030	46½
230-T	0.0015	0.0028	42
225-T	0.0017	0.0027	38½
230-HD	0.0019	0.0025	32½

TABLE 24   Specifications for Monofilament Nylon Fabric
(Nitex*, [18]) (Continued)

Mesh no.	Thread diameter, in.	Mesh opening, in.	Open area, %
240-S	0.0014	0.0027	44
240-T	0.0015	0.0026	39
242-T	0.0017	0.0024	35
240-HD	0.0019	0.0022	28½
260-S	0.0014	0.0024	40½
260-T	0.0015	0.0023	36
260-HD	0.0017	0.0021	31
283-S	0.0012	0.0023	45½
283-T	0.0014	0.0022	37
283-HD	0.0015	0.0020	32
306-S	0.0012	0.0021	41
306-T	0.0014	0.0019	33½
306-HD	0.0015	0.0018	29
330-S	0.0012	0.0018	37
330-T	0.0014	00.017	30
350-S	0.0012	0.0016	34
350-T	0.0014	0.0015	26
380-S	0.0012	0.0014	30
380-T	0.0014	0.0012	21½
465-S	0.0012	0.0010	21

* Registered trademark of Zurich Bolting Cloth Mfg. Co., Ltd.

with the 350 to 465 meshes the percentage of open area is so low that it becomes extremely difficult to print thick film materials with such screens. Conversely the very coarse meshes not only make it difficult to resolve and print fine-line patterns but also result in extra heavy deposits during printing. When ordering Nitex† monofilament-nylon mesh, the thread diameter or the suffix S, T, or HD should be specified to avoid confusion over which thread diameter is required. The letters S, T, and HD indicate light, regular, and heavy, respectively, as applied to the thread-diameter classification. If nylon mesh is to be used for making a screen, the material must be stretched properly. Nylon should be stretched between 3 and 6 percent in both directions, as indicated in Table 25. A procedure for mounting nylon mesh is outlined in the following steps:[12]

1. Pretension the mesh by taking up the initial slackness and creases.
2. Mark two parallel lines in both directions (warp and weft) using a known dimension.
3. Moisten the mesh with water using a wet sponge or cloth.
4. Tension the screen slowly by applying equal amounts of force to all four sides.
5. Repeat step 4 at 15- to 30-min intervals while keeping the nylon wet until full tension is applied.
6. Check percentage of stretch using line markings from step 2.

Insufficient tension of nylon-mesh screen may cause numerous problems during printing such as:[12]

1. Poor registration
2. Loss of definition, smeared and distorted patterns
3. Poor snap-off
4. Shorter screen life due to extended breakaway distance
5. Slower squeegee speed
6. Loss of open mesh area; insufficient paste deposit

The principal drawback in using nylon mesh is the tendency of the material to stretch, or "walk," during the screening process. As the squeegee travels across the

† Registered trademark of Zurich Bolting Cloth Mfg. Co., Ltd.

inside surface of the screen, the mesh will shift or move in the direction of the squeegee stroke. The amount of shift is a function of the variables listed in Table 26.

The tendency of nylon mesh to shift during printing may be used to advantage; e.g., by purposely mounting the mesh with lower tension, it may be possible to print into recesses or over attached components on the substrate. The screen would not coin in either case; however, registration tolerances would have to be relaxed and the breakaway distance increased to provide adequate snap-off. Another distinct advantage of nylon mesh is its ability to transmit light. When the precoated screen is exposed during processing, it is possible to expose the emulsion on the inside of the screen as well as on the contact side. Exposed emulsion on the inside

TABLE 25 Percentage of Stretch for Nylon Meshes during Mounting[19]

Mesh count	Stretch, %
54	3
60–83	3–3.5
92–137	3.5
148–175	4
186–240	4.5
260–305	5.5
330 and finer	6

TABLE 26 Factors Affecting Screen Shift during Printing with Nylon Mesh

Variable affecting nylon-mesh shift	Comment
Screen mesh.........	Use coarser mesh whenever possible for greater stability; tension properly
Screen tension........	Keep screen tension as high as possible
Breakaway distance....	Screen should be as close to substrate as possible with good release and without smearing
Squeegee pressure......	As low as possible
Squeegee speed........	Check for uniformity and effects of low and high speeds
Squeegee length.......	As short as possible and still provide uniform deflection
Squeegee stroke.......	As short as possible to cover substrate and provide adequate flood coating
Squeegee durometer....	Hard enough to eliminate gripping action of soft squeegee on mesh

will result in a better-adhering stencil as well as a protective coating to prolong the sharpness of the squeegee blade. Transmission of light through the nylon threads is not always beneficial, especially if the light scatters and exposes the screen emulsion adjacent to the mesh in the image areas. This phenomenon is similar to halation in films, where light is reflected from the rear surface of the film back into the emulsion. To minimize the amount of light scattering between the mesh filaments, special dyes are used for coating the threads. The common dye colors are orange and yellow, but red and purple have also been used. Since the exposure setup is important for producing fine-line screens, tests should be conducted to determine the advantages of the dyed nylon versus the white material. In general, colored fabrics require longer exposure times than undyed materials. The darker the color, the longer the exposure time. This characteristic could be advantageous, since it increases the range for proper exposure.

Nylon mesh has been used extensively in the decorating industry for printing on irregular-shaped containers and dinnerware. Due to regristration problems, nylon mesh has not been widely accepted for printing thick film circuits. However, as previously mentioned, nylon does have unique characteristics which should be considered for specialized printing applications.

Monofilament polyester falls in a category somewhere between nylon and stainless-steel mesh. As indicated in Tables 22 and 23, polyester combines such characteristics as excellent stability and stencil adhesion with resistance to coining. However, polyester screen mesh is not currently available with fine thread diameters for printing 0.002 to 0.003-in. line widths, as indicated in Table 27.

Polyester materials can be mounted in the same manner as nylon mesh, the main difference being the amount of stretch required for adequate tension. Typically,

TABLE 27   Specifications for Monofilament Polyester Fabrics[20]

Mesh count	Thread, diameter, in.	Mesh opening, in.	Fabric thickness, in.	Open area, %*
74	0.0047	0.0088	0.0080	42.4
86	0.0039	0.0076	0.0080	42.7
109	0.0032	0.0059	0.0053	41.4
124	0.0028	0.0051	0.0047	40.0
140	0.0024	0.0047	0.0045	43.3
157	0.0024	0.0039	0.0045	37.5
170	0.0024	0.0035	0.0043	35.4
178	0.0016	0.0039	0.0031	48.2
198	0.0016	0.0034	0.0031	45.3
232	0.0015	0.0028	0.0025	42.2
244	0.0015	0.0025	0.0024	37.2
262	0.0015	0.0022	0.0025	33.2
284	0.0015	0.0019	0.0027	29.1
305	0.0015	0.0016	0.0028	23.8
380	0.0014	0.0011	0.0024	17.5

* Percentage of open area $= (OM)^2(100)$, where O = mesh opening, M = mesh count.

TABLE 28   Percentage of Stretch for Polyester Meshes during Mounting[19]

Mesh count	Stretch, %
38–123	1
131–206	1.5
230–305	2
330 and finer	2.5

polyester should be stretched between 1 and 2½ percent, as illustrated in Table 28. Tensioning is faster with polyester than with nylon since polyester does not relax and require periodic retensioning during stretching. Also, polyester mesh should be stretched dry. Like nylon mesh, the polyester materials are available with white and dyed threads. Exposure tests should be run on all materials to evaluate the merits of using the antihalation mesh. Standard rolls or bolts of the material are approximately 30 yd long by 40 to 80 in. wide. Normally the material is priced by the running yard. Polyester mesh has not achieved widespread success in the United States but is being used extensively in England and the European countries for thick film printing as well as other applications.

The most popular of all the materials used for making screens is stainless-steel wire mesh, excellent dimensional stability and the availability of mesh with very small wire diameters being the primary reasons. The extremely small threads with wider diameters down to 0.0008 in. allow for a relatively large percentage of open area, especially in the higher mesh count. Table 29 is a specification sheet for stainless-steel wire cloth. Tolerances for wire diameters for some of the mesh listed in Table 29 are as shown in Table 30. For wire diameters below 0.0045 in. the tolerance on the wire diameter will fall within ±0.0001 in.[21] Recently a tolerance of ±4 percent for the stainless steel wire diameter has been specified.[17] Tolerance in mesh count are in accordance with Table 31.

TABLE 29.   Specifications for Stainless-Steel Wire Cloth (Screen-Printing Grade)

Mesh count	Wire diameter, in.	Mesh opening, in.	Open area, %*
80	0.0037	0.0070	31.4
84	0.0039	0.0080	45.2
105	0.0030	0.0065	46.6
120	0.0025	0.0058	48.4
120	0.0026	0.0057	46.8
120	0.0027	0.0056	45.2
120	0.0037	0.0046	30.5
130	0.0026	0.0051	44.0
135	0.0023	0.0051	47.4
145	0.0022	0.0048	48.4
150	0.0026	0.0041	37.8
165	0.0019	0.0041	45.8
165	0.0020	0.0041	45.8
180	0.0018	0.0038	46.8
180	0.0019	0.0037	44.4
180	0.0020	0.0036	42.0
200	0.0016	0.0034	46.2
200†	0.0020	0.0030	36.0
200	0.0020	0.0030	36.0
200	0.0021	0.0029	33.6
230	0.0014	0.0028	41.5
230	0.0015	0.0029	44.5
250	0.0016	0.0024	36.0
250†	0.0014	0.0025	39.1
270	0.0014	0.0022	35.3
270†	0.0016	0.0023	38.6
280	0.0012	0.0024	45.2
280	0.0014	0.0022	37.9
325	0.0011	0.0020	42.3
325	0.0012	0.0019	38.1
325†	0.0014	0.0017	30.5
400	0.0010	0.0015	36.0
400	0.0009	0.0016	41.0
450†	0.00106	0.00116	27.2
508†	0.00108	0.00092	21.8
508†	0.000866	0.0011	31.2

* Percentage of open area $= (OM)^2(100)$, where O = mesh opening, M = mesh count.
† Twilled weave.

As with monofilament nylon, the wire diameter should be specified in addition to the mesh count when ordering stainless-steel material or screens. This is particularly important for 200-mesh cloth, where both 0.0016 and 0.0021-in. wire-diameter mesh are used extensively. The range of mesh count used for making thick film screens or masks runs from 80 to 400 mesh. Coarser-mesh materials are difficult to mount on the screen frames and do not offer enough support for the emulsion stencil, which must bridge the gap between the mesh. For metal masks the support requirement is not as critical unless the pattern is made up of small lines and spacings. The finer-mesh materials, such as 450, 500, and 508, are available with twill weave only and do not offer any advantages for use in printing thick film compositions. The percentage of open area for a mesh count of 450 and finer is less than 32 percent due to the increased number of threads per inch and the smaller mesh openings. As a result, it is difficult to obtain deposition through the screen with good continuity and print thickness.

Of all the stainless-steel compositions available, types 316 and 304 are the most commonly used. Type 304 is preferred over type 316 since the latter is more brittle due to small traces of molybdenum. Full rolls of the material are normally 100

TABLE 30. Tolerances for Wire Diameters[14]

Carbon steel		Stainless steel and all nonferrous	
Wire diameter, in.	Tolerance (plus or minus), in.	Wire diameter, in.	Tolerance (plus or minus), in.
0.500 and coarser..........	0.003	0.500 and coarser.........	0.002
0.4375–0.080..............	0.002	0.4375–0.063.............	0.0015
0.072–0.035...............	0.001	0.054–0.047..............	0.001
0.032–0.028...............	0.0008	0.041...................	0.0008
0.025–0.020...............	0.0006	0.035...................	0.00075
0.018–0.016...............	0.0005	0.032...................	0.0006
0.015–0.011...............	0.0004	0.028–0.020..............	0.0005
0.010–0.006...............	0.0003	0.018–0.012..............	0.0004
0.0055–0.0045.............	0.0002	0.011–0.008..............	0.0003
		0.0075–0.0045...........	0.00025

TABLE 31   Tolerances in Mesh[14]

Mesh size	Tolerance in average mesh count, %	
	Warp wires (plus or minus)	Shoot wires (plus or minus)
30 and coarser..	2	5
Finer than 30 but not finer than 200.....................	2	4
Finer than 200..	3	4

ft long and 12 to 60 in. wide. Usually the material is priced by the square foot. Most stainless-steel fabrics finer than 100 mesh are produced outside of the United States, primarily in Germany, Holland, Switzerland, and Japan.

A specification comparison chart is shown in Table 32 to illustrate the dimensional similarities of the three mesh materials.

If it appears desirable to change the screen mesh material for a particular printing application, an evaluation of the new mesh should be conducted. The type of screen mesh used influences numerous factors in the screen-printing operation, such as:

1. Breakaway distance
2. Deposit thickness
3. Squeegee pressure

TABLE 32 Specification Comparison Chart for Stainless-Steel, Polyester, and Nylon Mesh

Stainless steel				Monofilament polyester				Monofilament nylon			
Mesh count	Wire diameter, in.	Mesh opening, in.	Open area, %*	Mesh count	Thread diameter, in.	Mesh opening, in.	Open area, %*	Mesh count	Thread diameter, in.	Mesh opening, in.	Open area, %*
84	0.0039	0.0080	45.2	86	0.0039	0.0076	42.7	83-S	0.0035	0.0081	45.2
105	0.0030	0.0065	46.6	109	0.0032	0.0059	41.4	103-T	0.0031	0.0064	43.5
120	0.0027	0.0056	45.2	124	0.0028	0.0051	40.0	120-T	0.0028	0.0055	43.6
150	0.0026	0.0041	37.8	157	0.0024	0.0039	37.5	157-T	0.0024	0.0041	41.4
165	0.0020	0.0041	45.8	178	0.0016	0.0039	48.2	185-T	0.0019	0.0035	41.9
200	0.0016	0.0034	46.2	198	0.0016	0.0034	45.3	196-T	0.0017	0.0033	41.8
200	0.0021	0.0029	33.6	200PE	0.0022	0.0029	33.6	206-HD	0.0019	0.0029	35.7
230	0.0014	0.0028	41.5	232	0.0015	0.0028	42.2	230-T	0.0015	0.0028	41.5
250	0.0016	0.0024	36.0	244	0.0015	0.0025	37.2	260-T	0.0015	0.0023	35.8
270	0.0014	0.0022	35.3	262	0.0015	0.0022	33.2	283-T	0.0014	0.0022	38.8
280	0.0012	0.0024	45.2					283-S	0.0012	0.0023	42.4
				305	0.0015	0.0016	23.8	306-HD	0.0015	0.0018	30.3
325	0.0011	0.0020	42.3					306-S	0.0012	0.0021	41.3
325†	0.0014	0.0017	30.5					330-S	0.0014	0.0017	31.5
400	0.0010	0.0015	36.0					350-S	0.0012	0.0016	31.4
400	0.0009	0.0016	41.0	380	0.0014	0.0011	17.5	465-S	0.0012	0.0010	21.6
508†	0.00108	0.00092	21.8								

* Percentage of open area = $(OM)^2(100)$, where O = mesh opening, M = mesh count.
† Twilled weave.

4. Squeegee speed
5. Squeegee angle of attack
6. Screen tension
7. Emulsion thickness on the screen

These variables therefore make it difficult to interchange mesh materials even though they have approximately the same physical dimensions.

All screen fabrics must be cleaned thoroughly before applying the emulsion coating or stencil. Stainless steel should be cleaned particularly well because of the oils and contaminates left on the mesh as a result of the weaving process. Nylon or polyester materials are not as contaminated as wire cloth, but they should also be cleaned well to ensure good adhesion of the stencil for maximum screen life. Degreasing the mesh in a vapor degreaser or by washing with solvents is recommended, followed

TABLE 33   Commercially Available Screen-Cleaning Agents

Supplier	Name	Mesh cleaned	Remarks
Azoplate Corporation[22]	Azokleen	All fabrics	Concentrated liquid, hot-water rinse required
Albert Rose[23]	Pregan A9-HV	All fabrics	Concentrated liquid, cold- or warm-water rinse
Advance[24]	Nylon mesh prep	Nylon	Powder; cleans and roughens threads for better stencil adhesion
	Stainless-steel mesh prep	Stainless steel	Concentrated liquid, rinse with hot water
	All-mesh detergent	All fabrics	Powder form, scrub into mesh, rinse with hot water
Chemical supply house	Caustic soda	Nylon, polyester	20% solution, hazardous to work with, reacts with aluminum, neutralize with 6% acetic acid
	Cresylic acid	Nylon	Roughens threads, will dissolve nylon, must be completely removed

by a detergent, alkaline, or acid cleaning. Table 33 lists some commercially available screen-cleaning agents. Scouring powder, although useful for roughening synthetic fabrics, is not recommended because there is always the possibility that the powder particles will clog the fabric, especially in finer-mesh materials. All cleaning agents should be applied to the mesh and scrubbed into both sides of the weave with a nylon brush. After soaking for 10 to 15 min the scrubbing should be repeated and the screen rinsed thoroughly with a water spray. Flame treatment of stainless steel is definitely not recommended. Burning will make the material brittle and render it more susceptible to damage.

Mesh Tension. Screen mesh tension is a variable which affects the pattern registration, definition, and the print thickness during screening. Screens which are too loose and show ripples or sagging on the frame will not release properly unless the screen-to-substrate separation is greatly increased. Then the mesh will yield even further, until the screen is completely useless. If a series of screens of a particular size and mesh have the same tension, it is possible to maintain the same breakaway distance, squeegee pressure, and other printing parameters between screen changes. Screen tension varies as a result of fabrication, ambient temperature, handling, cleaning, image application, and use.[25]

Mesh tension has been measured by various techniques devised by equipment manufacturers and screen users. Ottaviano has measured screen deflection by applying a force of 1 lb per inch of squeegee blade to the screen and recording the displacement with a fixed height gage.[25] This setup was also utilized to correlate the screen deflection and the screen-to-substrate separation for determining the effects of off-contact spacing on resistor deviation.

Fig. 38 Screen-tension tester. (*Affiliated Manufacturers, Inc.*)

Fig. 39 Screen-tension tester. (*Engineered Technical Products, Inc.*)

Other methods for determining screen tension include *tension testers*, illustrated in Figs. 38 to 40, which are utilized by applying weights to the shaft of the dial indicator and determining the deflection in mils per unit of weight applied. Tables 34 and 35 show recommended screen tensions for 5 by 5 and 8 by 10 in. ID frames.

Mesh Selection. Information concerning the selection of the proper screen mesh material has already been outlined in Table 23; however, once the material is chosen, the correct mesh count and thread or wire diameter must be selected. Table 36 presents some considerations for choosing the proper screen mesh.

As outlined in Table 36, the mesh specification for a particular printing application should take into consideration screen life, pattern definition, pattern line widths and spacings, substrate surface profile and camber, and deposit thickness. In addition, the paste itself must pass through onto the substrate without hanging in the screen. The mesh opening in particular must be large enough to ensure that the larger ink particles do not clog the screen. As a rule of thumb it has been recommended that the mesh opening be approximately 2½ to 5 times larger than the average particle size of the paste material.[28]

Deposit thickness of the paste can be varied to a great extent by using a different screen mesh with a different wire diameter and/or mesh count. Headly has printed the same compositions with 105-, 165-, 200-, and 325-mesh stainless-steel screens

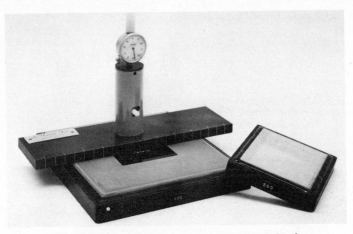

Fig. 40   Screen-tension tester.   (*De Haart, Inc.*)

TABLE 34   Recommended Tensions for Stainless-Steel Screens[26]

Mesh count	Wire diameter, in.	Mesh opening, in.	Deflection, mils*	
			5 by 5 in. frame	8 by 10 in. frame
80	0.0037	0.0070	35	45
105	0.0030	0.0065	35	50
150	0.0026	0.0041	35	55
165	0.0020	0.0042	50	70
200	0.0016	0.0034	55	70
200	0.0021	0.0029	45	55
230	0.0015	0.0029	55	70
250	0.0016	0.0024	50	65
325	0.0011	0.0020	65	85

* Deflection tolerances: ±12 mils/lb for standard frames, ±10 mils/lb for machined frames. Data measured on a Presco STG-3 tension test gage with a 1-in.-diam ball under an applied force of 1 lb.

and then measured the dried and fired film thicknesses, as shown in Table 37. Table 38 is an analysis of the data in Table 37.

Several conclusions can be drawn from the data in Table 38; e.g., the average shrinkage factor or ratio of dried film thickness to fired film thickness is 1.62. Also the dried thickness and fired thickness are both directly proportional to either the wire diameter or the mesh opening. In other words, as the wire diameter or the mesh opening decreases, the fired or dried film thickness decreases in the same proportion. Additional tests were run by Headly to determine the effects of different screen mesh on print thickness (Fig. 41). Riemer has also run experiments on print

TABLE 35   Recommended Tensions for Stainless-Steel Screens with Mesh Mounted at 45° and 90° to Frame[27]

Mesh count	Angle, deg	Wire diameter, in.	Mesh opening, in.	Deflection, mils*	
				5 by 5 in. frame	8 by 10 in. frame
200	45	0.0016	0.0034	40	50
200	90	0.0016	0.0034	40	50
250	45	0.0016	0.0024	45	55
250	90	0.0016	0.0024	45	55
325	45	0.0011	0.0020	50	60
325	90	0.0011	0.0020	50	60

* Deflection tolerance ±8 mils/lb for machined frames.   Data measured on a Presco STG-3 tension test gage with a 1-in.-diam ball under applied force of 1 lb.

TABLE 36   Considerations when Selecting Screen Mesh for Different Printing Applications

Printing application	Consideration	Recommendation
Very fine line printing (0.002–0.004 in.)	Wire interference with pattern, nominal deposit thickness	325- or 400-mesh stainless steel with pattern oriented at 45° to mesh weave
Fine line printing (0.005–0.010 in.), long runs	Screen life, sawtoothing, 0.7- to 1.0-mil deposit thickness	230-, 280-, or 325-mesh stainless steel with pattern oriented orthogonal or at 45° to mesh weave
Long production runs: Resistors...............	Screen life, maintaining good tension, minimum line widths of 0.020 in.	Coarse mesh for strength, 165- or 200-mesh stainless steel
Conductors.............	Screen life, maintaining good tension, minimum line widths of 0.020 in.	Coarse mesh for strength, 200- or 230-mesh stainless steel or a polyester mesh
Short runs, glaze compositions for resistors or capacitors and, long runs, solder or braze compositions	Irregular surface, coining, heavy deposits	Coarse-mesh polyester or nylon (83 to 150 mesh)
Long runs, glaze, solder, or braze compositions on smooth substrates	Screen life, heavy deposits	80- to 150-mesh stainless steel
Printing conductors for crossovers without complementary patterns	Irregular surface, registration of ±0.003 to ±0.005 in., 0.7- to 1.0-mil deposit thickness, sawtoothing	283-S or 306-S nylon with pattern oriented orthogonal or at 45° to mesh weave

TABLE 37   Effect of Mesh Parameters on Film Thickness and Electrical Properties of Various Resistor Compositions[29]

Du Pont composition	Mesh	Film thickness, mils		Resistance, Ω/sq	Hot (125°C) temperature coefficient of resistance
		Dry	Fired		
7800	105	1.23	0.66	0.15	133
	165	1.00	0.64	0.16	421
	200	0.59	0.45	0.21	643
	325	0.32	0.29	0.29	461
7826	105	1.64	0.99	153	662
	165	0.88	0.66	313	646
	200	0.75	0.45	532	594
	325	0.39	0.36	1,034	547
7827	105	1.55	0.94	1,245	404
	165	1.13	0.77	1,480	405
	200	0.82	0.55	2,051	319
	325	0.56	0.28	3,953	229
7828	105	1.48	0.98	2,650	148
	165	1.09	0.73	3,520	124
	200	0.84	0.58	4,720	81
	325	0.57	0.31	7,860	2.8
7832	105	1.59	1.10	5,560	65
	165	1.29	0.75	6,430	58
	200	0.99	0.53	8,230	31
	325	0.73	0.35	13,180	19
7860	105	1.68	0.98	9,830	403
	165	1.27	0.76	14,380	365
	200	0.89	0.50	32,900	262
	325	0.62	0.28	75,500	205

TEST DATA: Four-square test patterns (0.25 by 0.75 in.) printed onto 1 by 1 by 0.025 in. AlSiMag 614 substrates with a Presco 100B printer. Terminations were Du Pont 7553 Pt-Au fired at 1000°C and a 10-min soak. Resistors fired at 760°C peak and 45-min cycle through a continuous-belt furnace. Dry films measured optically with a vernier optical gage. Fired films measured by a Starrett model 652 gage.

thickness as a function of screen mesh and wire diameter (Table 39). Table 39 was prepared by calculating the volume of ink transferred from a screen using the parameters illustrated in Fig. 42. The wet thickness T_W is determined by the equation[31]

$$T_W = D\left(2 - \frac{D\pi}{2ma}\right)$$

where D = fabric wire diameter, mils
$m = 1{,}000/M$, mils
M = mesh count per inch
$a = \cos(\arctan DM)$

The values of film thickness in Table 39 are based on screen mesh with no emulsion buildup on either side of the fabric and on the assumption that all the ink in the screen mesh is transferred to the substrate. In actual practice, some ink residue is left in the screen, and some emulsion buildup is present on both sides of the screen. However, the figures in Table 39 are useful as approximations for determining wet deposit thickness. Riemer defines a shrinkage factor S using the relationship[31]

$$S = \frac{T_W}{T_f}$$

where T_W = wet thickness
T_f = fired thickness

TABLE 38 Effect of Mesh Parameters (Analysis of Data from Table 37)

| Stainless steel mesh count | Wire diameter D, mils | Mesh opening O, mils | Open area, % | Average film thickness, mils | | Shrinkage factor T_D/T_F | Ratio of thickness to wire diameter | | Ratio of thickness to mesh opening | |
				Dry T_D	Fired T_F		T_D/D	T_F/O	T_D/D	T_F/O
105	3.0	6.5	46.6	1.53	0.94	1.63	0.51	0.31	0.24	0.14
165	1.9	4.2	48.0	1.11	0.72	1.54	0.58	0.38	0.26	0.17
200	1.6	3.4	46.2	0.81	0.51	1.59	0.51	0.32	0.24	0.15
325	1.1	2.0	42.3	0.53	0.31	1.71	0.48	0.28	0.27	0.16

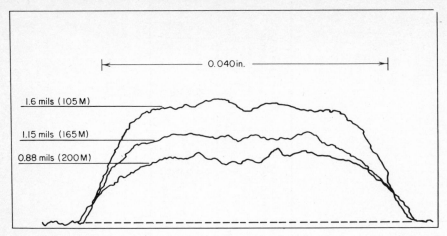

0.040 in.

1.6 mils (105 M)

1.15 mils (165 M)

0.88 mils (200 M)

Fig. 41   Stylus-type graphic measurement of dried print thickness for different mesh screens.[30]

TABLE 39   Wet Deposition Thickness T_W as a Function of Screen Mesh M and Wire Diameter D[31]

Mesh count M	Wire diameter D, mils	Wet thickness T_W, mils
105	3.0	4.44
120	2.6	3.85
145	2.2	3.25
165	2.0	2.92
165	1.9	2.82
180	1.9	2.74
200	2.1	2.70
200	1.6	2.35
230	1.4	2.06
250	1.6	2.12
250	1.4	1.99
270	1.6	2.02
270	1.4	1.91
325	1.4	1.75
325	1.1	1.55
400	1.1	1.37
400	1.0	1.32

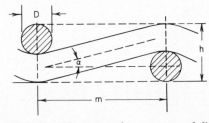

Fig. 42   Geometry of a screen mesh.[31]

The shrinkage factor can be determined for any paste composition by using optical methods for measuring the wet thickness and a mechanical device for checking the fired thickness.

The screen stencil   The function of the screen stencil is that of a mask or mold which determines the geometry and to some extent the thickness of a particular printed pattern.   The stencil on a screen is usually a plastic or gelatin coating; it is referred to as an *emulsion* although metal stencils are also used in the production of metal masks.   To avoid confusion the metal stencils will be referred to as *masks,* as explained earlier.

In the previous discussion of deposit thickness, it was stressed that the screen mesh itself does in fact greatly influence the print thickness during the screening operation.   However, as illustrated in Fig. 43, the emulsion buildup also contributes to the deposit thickness.   If the emulsion buildup is represented by the dimension b, the theoretical wet deposit thickness is $T = T_w + b$.   There are both advantages and disadvantages in increasing the emulsion thickness below the mesh of the screen.   One main advantage in emulsion buildup applies to printing fine-line conductors in the range below 0.010 in., where the patterns are to have sharp edges and adequate thickness.   By using the finer screen meshes, that is, 325 or 400 stainless steel, and applying 0.8 to 1.2 mils of emulsion below the mesh it is possible to print extremely sharp conductor lines with reasonable print thickness.   Figure

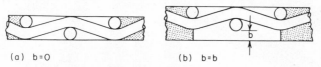

(a) b=0   (b) b=b

Fig. 43   Screen without and with emulsion buildup b.[31]

44 illustrates fine-line patterns printed with screens having different emulsion thicknesses.   The effect shown in Fig. 44a, known as *stairstepping* or *sawtooth edge,* is caused by a number of factors primarily related to the screen.   If there is little or no emulsion buildup below the mesh of the screen, the squeegee will print the mesh pattern on the edges of the stencil, as illustrated in Fig. 44a.   With the emulsion buildup the screen paste should pass through the mesh around the filaments and conform to the cavity, or mold, formed by the stencil below the mesh.   The analogy of a gasket between the mesh and the substrate might better describe this characteristic.   The danger of varying the emulsion buildup to control deposit thickness is discussed by Riemer and is illustrated in Fig. 45.   In this situation, if large open areas are to be printed, the squeegee may depress the mesh into the open cavity, thereby decreasing the actual deposit on the substrate.   The mesh count and wire diameter as well as the squeegee contribute significantly to this phenomenon.   Additional information may be obtained from Ref. 32.

Pattern Alignment.   In addition to the emulsion buildup below the mesh of the screen, the orientation of the pattern to the mesh is also important, especially when printing fine-line patterns.   At one time fine-line patterns were considered to be line widths and spacings in the 0.005- to 0.010-in. range.   However, more recently the term refers to lines and spacings in the 0.002- to 0.005-in. category.   Several approaches to preparing screens with 0.002- to 0.005-in. lines have been used with varying success.   One technique involves designing the open areas of the screen to coincide with the mesh openings in the screen as shown in Fig. 46.   It has been recommended that a line to be printed should have a width of at least two mesh openings.   With one mesh opening the printed line can be broken if the mesh or pattern are not closely aligned and the image "steps" over a wire.[11]   Therefore, for a 400-mesh screen with a mesh opening of 0.0016 in. the smallest practical line width is two mesh openings, or 0.0032 in.   This design guideline works well only under certain circumstances.   If the mesh openings are not constant or the mesh wires are not uniformly parallel and perpendicular, problems may arise.   The finer-mesh stainless-steel materials such as 325 and 400 are not always square mesh and show variations in the mesh openings due to faults in weaving. An alternative solution to aligning the images to the mesh openings is to orient either the image at 45° to the mesh or mount the mesh on the frame at 45° and apply the image in the normal manner. Figure 47 illustrates two methods of aligning the mesh to the

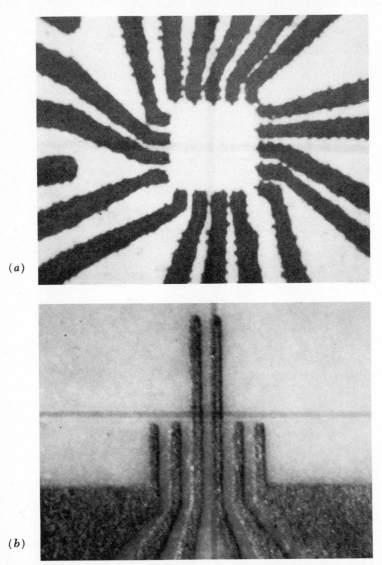

(a)

(b)

Fig. 44 Screen-printed pattern with (a) sawtooth edges and (b) 0.003-in. lines and spacings. (*Microcircuit Engineering Corporation.*)

frame. Figure 48 illustrates a stencil screen with a 0.002-in. line aligned at 45° to the mesh weave. Nesselroth and Zeien report that alignment of the image lines and spacings to the screen mesh is nearly impossible when applying the patterns parallel and at right angles to the screen mesh. Placing the image at a 45° angle to the screen mesh not only eliminated the alignment problem but also improved the uniformity of the line width between the X and Y conductor runs.[93]

Although many types of materials are used for preparing stencils on the various screen mesh, most screens for printing thick film circuits are made from liquid- or film-type emulsions which are processed photographically. The three basic types of

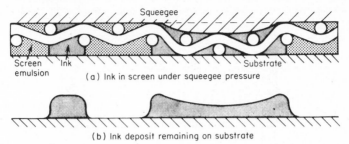

(a) Ink in screen under squeegee pressure

(b) Ink deposit remaining on substrate

Fig. 45 Wet film thickness as function of screen opening size for screen with emulsion buildup.[31]

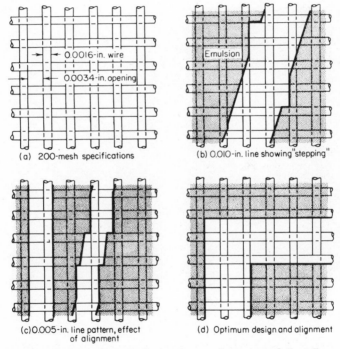

(a) 200-mesh specifications

(b) 0.010-in. line showing "stepping"

(c) 0.005-in. line pattern, effect of alignment

(d) Optimum design and alignment

Fig. 46 Specifications for a 200-mesh stainless-steel screen and the effect of pattern alignment.[1]

stencils are (1) the indirect, (2) direct, and (3) hybrid, or indirect-direct, type, as outlined in Table 40. The metallic stencil or metal mask is included as a basis for comparison.

The indirect emulsion differs from the direct and hybrid emulsions in the method used to apply the stencil to the screen. With the indirect emulsion the pattern is developed in the emulsion film before attaching to the screen rather than after the screen is coated. Most screen emulsions are prepared as water solutions of gelatin, polyvinyl alcohol, or polyvinyl acetate. The sensitizing agents are normally compounds of ammonium, potassium, and sodium dichromate or diazo systems. When sensitized and applied to the screens, the emulsions are sensitive to heat, moisture, and actinic light. The dichromate-sensitized solutions must be mixed just before use because of the limited shelf life of these materials. Oxidation of the emulsified resin occurs in addition to breakdown of the dichromate in acid solutions.[35] The

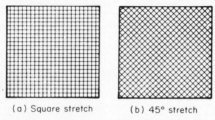

(a) Square stretch   (b) 45° stretch

Fig. 47 Alignment of mesh to frame.[11]

Fig. 48 Stencil screen with a 0.002-in. line opening oriented at 45° to the mesh weave. (*Microcircuit Engineering Corporation.*)

effect of humidity on dichromate-sensitized emulsion sensitivity is shown in Fig. 49. Screens prepared with these emulsions should be processed as soon after coating and drying as possible. Refrigeration may prolong the shelf life of the coated screens up to 3 or 4 days under controlled light and humidity conditions. Additional information on dichromated colloid solutions has been reviewed by Kosar.[35] Unlike the dichromate sensitizers, the diazo compounds provide a more stable emulsion system which has a much longer shelf life. Namely, 2 months when stored at room temperature and 4 months when refrigerated. Coated screens may be subjected to warm air to accelerate drying but should be stored at 70°F and away from light. Emulsions prepared with diazo sensitizers exhibit a lower light sensi-

TABLE 40 Comparison of Screen Stencils for Thick Film Printing[34]

Type of stencil	Features	Advantages	Disadvantages	Comment
Indirect.........	Presensitized emulsion film on backing sheet	Inexpensive, rapid processing, good detail, screen can be reprocessed	Becomes brittle, poor life, one film thickness available	Good for short runs only, life unpredictable
Direct.........	Sensitized liquid emulsion	Inexpensive, durable, easy handling, screen can be reprocessed	Difficult to control emulsion buildup; possibility of sawtooth effect	Excellent durability, excellent detail possible if properly prepared, good solvent resistance
Hybrid (indirect-direct)	Emulsion film applied to screen, then processed	Good for detail; coarser-mesh screens can be used, available in different emulsion-film thicknesses, screen can be reprocessed	More expertise required, handling more difficult, may become brittle	Good durability, most satisfactory on synthetic meshes, excellent detail possible, controllable emulsion thickness
Metallic.........	Electroformed metal, plated mesh, etched solid metal, or etched foil applied to mesh	Sharp detail possible, heavy deposits with heavier metal backing, longer life, controlled deposition unaffected by solvents	Fragile and very susceptible to squeegee or substrate problems, more expensive	Must be handled carefully, with proper treatment should have extra-long life

tivity, which may influence the type of light source used and the exposure time required.

Before outlining the process steps for preparing screens a list of the equipment and services required for the image application on emulsion-type screens is presented

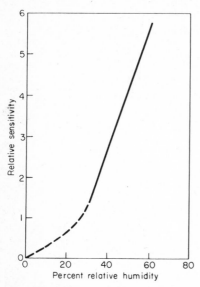

Fig. 49 Effect of relative humidity on dichromate-sensitized emulsion sensitivity.[35]

(Table 41). The vacuum frame is used for holding the film positive in close contact with the stencil emulsion during the exposure process. A deep-well type of vacuum frame can be used for processing any type of emulsion system. Standard contact printers used in photography do not have a blanket flexible enough to conform to the inside of a screen frame. Figure 50 illustrates the function of the flexible blanket used on a deep-well vacuum frame. This same vacuum unit can be used for processing sheets of emulsion film for making indirect-emulsion screens. Fig. 51 illustrates a commercially available deep-well exposure unit complete with vacuum frame and built-in light source. Special light sources are required to expose the emulsion systems properly and thus define the outline of the stencil image. Since the emulsions are negative-working, the exposed areas of emulsion remain on the screen. Polymerization of the exposed material renders the emulsion insoluble in water, allowing the unexposed material to be washed out of the screen. Actinic light must be used for exposing both the dichromated and diazo-sensitized emulsion systems.

The spectral energy distribution of the light emitted from the source is the most important characteristic rather than candlepower or power output. The dichromated and diazo-sensitized materials are most sensitive to light in the 0.3- to 0.5-μm range, as illustrated in Fig. 52. As indicated

TABLE 41   Equipment and Services Required for the Image Application on Emulsion-Type Screens

Equipment	Services
Vacuum frame	Vacuum source
Exposure unit	Compressed air
Light box	Hot and cold water
Refrigerator	Air conditioning
Drying cabinet or fan	110/220 V ac
Sink and trays	Yellow-light darkroom
Spray gun or nozzle	
Microscope or Loupe	

in Fig. 52, the xenon source tested produces most of its energy between 0.8 and 1.0 μm, which is within the infrared range.

Indirect-Emulsion Stencil. The indirect emulsions are prepared as unsupported films or as a supported film with a plastic backing sheet. In addition both presensitized and unsensitized films are commercially available. Since the indirect stencil is processed before application to the screen mesh, it is possible to reproduce very fine detail in the emulsion. The screen mesh does interfere with the development of the pattern although some difficulty may be encountered when transferring the image to the screen. Placing the image at a 45° angle to the weave may alleviate the problem of mesh interference. Indirect stencil screens are prepared by the following process steps:

1. Set up for the screen-making process under yellow room light.

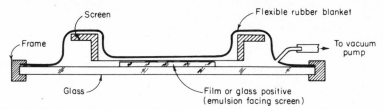

Fig. 50   Typical flexible-rubber vacuum frame.

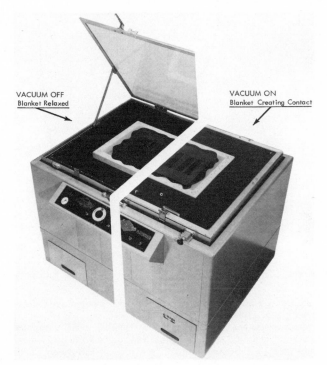

Fig. 51   Commercial deep-well flip-top vacuum frame. (*NuArc Co.*)

2. Cut the indirect film to size and expose it in a contact printer using a film positive with the emulsion side in direct contact with the emulsion film. For this application the photo positive must be right-reading with the film emulsion down.

3. Expose the screen using an actinic light source. A series of step exposures may be required to achieve the correct exposure time depending upon the light source used and the distance from the work surface.

4. Develop the exposed film in the recommended solution for 1 to 6 min.

5. Wash out the image by carefully running warm water (110 to 120°F) over

the image until the unexposed material is completely removed.   Rinse in cold water to set the emulsion.

6. Place the film emulsion side up on a flat support large enough to hold the entire piece of emulsion film but small enough to fit inside the screen frame.   Align the screen mesh to the image and place the contact side of the screen against the emulsion film.

7. Press the mesh down gently to assure complete contact with the stencil. Allow the stencil film to partially encapsulate the mesh before removing the excess water with absorbent paper, such as clean newsprint.

8. Allow the film to dry completely before slowly peeling off the plastic backing

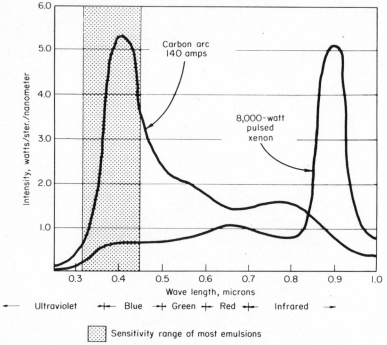

Fig. 52 Spectral energy distribution of carbon arc versus pulsed-xenon lamp.[36]

sheet.   Any adhesive from the backing sheet can be removed from the stencil with a solvent such as naphtha or toluene.

9. Inspect the screen for stencil adhesion to the mesh, pinholes, alignment, image distortion, and tension.   Water-soluble blockouts with good solvent resistance are commercially available for touching up pinholes or unwanted open areas in the stencil.

Figure 53 is a flow chart which summarizes the process steps outlined, and Fig. 54 shows a cross section of an indirect-emulsion screen.   The emulsion thickness below the mesh for an indirect stencil is controllable to a very small degree and depends primarily upon how much of the gelatin-film material is removed during the washout step.   Therefore, the best method for varying the print thickness with such stencils is to use different screen meshes.

Direct-Emulsion Stencil.   The sensitizers for direct emulsions and the characteristics of these systems have been discussed earlier.   Processing screens with direct emulsions involves applying the material in liquid form directly on the mesh prior to exposure, as outlined in the following steps:

1. Set up for the screen-making process under yellow room light.

2. Mix the sensitizer and emulsion together thoroughly according to the manufacturer's directions, using a stainless-steel spatula or a plastic rod. Prepare and store the material in sealed glass or plastic containers.

3. Use a nylon paint brush to apply the emulsion to the mesh, brushing it into the weave from both sides of the screen. Squeegee the excess material from both sides of the mesh and store in a separate container. A stiff piece of stainless steel or an aluminum or plastic squeegee should be used for removing the surplus emulsion.

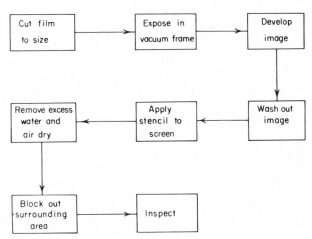

Fig. 53 Flow chart for the preparation of indirect-emulsion screens.

4. Dry the screen at room temperature when using dichromate-sensitized emulsions or at temperatures up to 120°F maximum with the diazo-sensitized materials.

A circulating-air drying cabinet with filters should dry the screens in 15 to 30 min. The screens should be oriented with the outside (printing contact side) down.

5. Recoat the screen on the contact side only to build up the emulsion thickness below the mesh.

6. Align the film positive to the mesh using magnification so that the filaments are parallel and perpendicular to the images if possible. The emulsion of the film positive

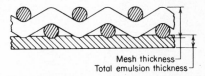

Fig. 54 Cross section of an indirect-emulsion screen.[1]

or glass plate should be in direct contact with the emulsion on the contact side of the screen. Special fixtures may be used at this time to locate the pattern very accurately on the screen frame.

7. Expose the screen in a deep-well vacuum frame. Exposure times vary depending upon the mesh, emulsion type, and coating thickness. For optimum resolution a step exposure should be run to determine the best time and distance settings.

8. Wash out the unexposed-image areas by first soaking the screen in warm water (100 to 120°F) for 1 to 2 min. Then direct a fine spray of warm water at the mesh, concentrating on the contact side of the screen. With stainless-steel screens very little water spray should be directed on the inside of the wire mesh.

9. Use a spray gun with filtered compressed air to dry the screen. This method will dislodge all the water and loose emulsion from the open areas of the screen.

10. Inspect the screen under a microscope for incomplete washout using transmitted yellow light. Rewash the screen if necessary to remove residual emulsion.

Block out pinholes and open areas around the frame with the emulsion stored separately in step 3. Figure 55 is a flow chart outlining these process steps, and Fig. 56 shows a cross section of a direct-emulsion screen.

The emulsion coating below the mesh for direct-emulsion screens can be built up to thicknesses of 0.008 in. by recoating the screen 3 or 4 times. However, exposure of these heavy-emulsion coatings is quite difficult unless a high-intensity light source is available. Nylon or polyester mesh will allow the light to pass through the threads, thereby exposing the emulsion behind the mesh for better adhesion of the stencil.

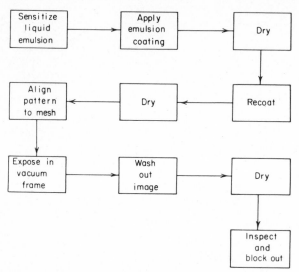

Fig. 55 Flow chart for the preparation of direct-emulsion screens.

In some cases it is possible to expose the screen on the inside by using a positive which is reversed from the original, i.e., original is right-reading emulsion up, and reversal is right-reading emulsion down. Such heavy-emulsion screens are being used for printing solder and braze compositions with good success.

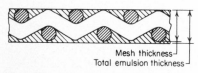

Fig. 56 Cross section of a direct-emulsion screen.[1]

Hybrid, or *Indirect-Direct, Stencil.* In 1967 several new emulsion systems called *hybrid* or *indirect-direct* stencils were developed. The name hybrid was derived from the characteristics and processes for preparing the stencil, which combine the indirect- and the direct-emulsion systems. The main advantage of the hybrid system is the availability of durable emulsion films which can be applied to the mesh like direct-emulsion coatings. Therefore, good detail or resolution of the image is possible, together with good adhesion of the stencil to the screen. The following process steps describe the methods for making a hybrid-stencil screen:

1. Set up for the screen-making process under yellow room light.
2. Sensitize the liquid emulsion according to the manufacturer's directions.
3. Cut a piece of unsensitized emulsion film to size and place the sheet on a clean flat surface with the emulsion side up.
4. Place the screen with the contact side of the mesh resting on the emulsion film. (A flat surface is important for the mesh to make contact over the entire emulsion-film surface.)
5. Pour a bead of sensitized emulsion along one edge of the inside of the screen.

Use a soft rounded squeegee and draw the emulsion across the mesh. Make one or two smooth passes using light pressure. During this process, the emulsion film is both sensitized and adhered to the screen mesh.

6. Dry the screen with the contact side down for 15 to 30 min in a circulating-air drying cabinet.

7. Remove the backing sheet from the emulsion film and align the positive to the mesh using red or yellow transmitted illumination.

8. Expose the screen in a deep-well vacuum frame using the same schedule as for dichromate-sensitized emulsions.

9. Wash out by soaking for 1 to 2 min in warm water (100 to 120°F) and then apply a water spray gently to both sides of the screen. Complete the washout by directing a more forceful spray on the contact side of the screen.

10. Dry the screen with clean compressed air to remove all traces of water and small particles of emulsion.

11. Inspect for complete washout and rewash if necessary. Block out pinholes and open areas around the mesh.

Figure 57 is a flow chart outlining the process steps in the preceding paragraphs.

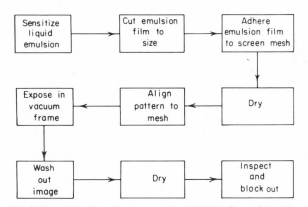

Fig. 57   Flow chart for the preparation of hybrid-emulsion screens.

Hybrid stencils make it possible to prepare screens with controlled emulsion thickness. However, technique must first be developed to assure repeatability in the coating process.

Handling and Storage of Stencil Screens. Common problems in making screens are outlined in Table 42. After preparation of a stencil screen it is important that the screen be used and stored properly to prolong its life. The following list of procedures is recommended for handling and storing screens:

1. Label screens with pattern number, revision number, or date.

2. Store new and used screens in plastic bags to protect from dust and dirt.

3. Store screens on end in racks or with cardboard spacers for protection against dents and punctures.

4. Store screens away from heat and extremes in humidity. Under low humidity screens may dry out and crack. High humidity causes swelling.

5. Clean screens thoroughly after use to remove ink residues in the pattern areas.

6. Set up the screen on the printing machine carefully to avoid tearing or coining of the mesh.

7. When using new solvents to clean the screen, test the emulsion outside the pattern area for compatibility of the solvent and the emulsion.

There is often the question of the life of a screen measured in terms of the number of prints that can be obtained from it. When printing posters, book covers, or printed-circuit boards, a screen may last for 50,000, 100,000, or even more impressions. However, screens used for printing thick film compositions are subject to a number of adverse conditions which have a pronounced affect on the useful life of the screen.   As indicated in Table 43, screen life is related to the chemicals in contact with the screen and the setup of the screen in the printing machine.

Types of Metal Masks

The woven-mesh screens described in the previous section play an important part in the production of thick film circuits.   The preparation of the emulsion stencil has been discussed in detail, i.e., the preparation methods and the end uses for the finished screens.

TABLE 42   Troubleshooting Chart for Making Stencil Screens

Problem	Possible cause	Remedy
Emulsion blisters or falls off screen	Mesh not clean, emulsion underexposed, uneven emulsion thickness, washout water too hot, emulsion not sensitized properly, clear areas of positive tinted	Clean mesh thoroughly according to instructions, increase exposure time or shorten distance to light source, check instructions for preparing emulsion, check clear areas of positive for transparency
Excess fisheyes in emulsion coating	Emulsion contains air bubbles, air entrapped	Do not mix emulsion vigorously, allow air to escape from emulsion before using, apply emulsion to screen mesh carefully
Lint, dirt, or other foreign matter trapped in emulsion coating	Coating environment is dirty, drying cabinet not clean, emulsion contaminated	Coat and dry screens in a dust-free area, wear lint-free clothing when coating screens, filter emulsion, do not reuse emulsion
Heavy deposit of emulsion in center area of screen	Mesh tension too low, screen mesh not supported properly during emulsion application; review technique for using emulsion applicator	Increase mesh tension, check instructions for applying emulsion, apply more pressure to emulsion applicator
Screen difficult to wash out...	Emulsion accidentally exposed to ultraviolet light, coated screen subjected to elevated temperatures, images on positive not completely opaque, emulsion not sensitized properly, shelf life of coated screen expired	Check instructions for sensitizing emulsion, store coated screens away from heat and light, use coated screens as soon after drying as possible
Thin haze of emulsion on edges or in open areas	Screen overexposed, screen has not been washed out long enough, positive positioned backward, emulsion too old, poor contact between positive and emulsion	Wash out screen again using water spray, position positive emulsion to emulsion during exposure, shorten exposure time, use fresh emulsion, check contact in vacuum frame
Excess of pinholes in screen...	Film or glass positive dirty, glass on vacuum frame dirty, emulsion contaminated	Clean positives and vacuum frame before exposures, use filtered or new emulsion for each screen
Poor alignment of image to screen mesh	Mesh not applied straight on frame, positive has shifted prior to exposure, poor artwork	Check alignment of mesh to frame, attach positive securely to emulsion, check original pattern artwork, apply image at 45° to mesh or mesh at 45° to frame

This section is devoted to metal masks, which are also used for selectively depositing different paste materials onto substrates.   Following the discussion of the different types of metal masks, the design criteria and applications of metal masks will be presented.

Metal masks are currently produced by a variety of processes utilizing certain solid metals as well as some combinations of dissimilar metals.   Two basic categories of masks have been outlined, as shown in Table 44.   The terms *indirect* and *direct* metal mask have been adopted primarily as a carryover of the terminology used for classifying emulsion-stencil screens.   Basically the term indirect as applied to metal masks refers to a mask which is produced by using a separate etched or woven-wire mesh onto which a metallic stencil is applied in a separate operation.   On the other hand, the term direct refers to a metal mask which is produced by etching or electro-forming a mesh and stencil pattern in a single operation. Direct metal masks are usually monolithic or bimetal structures, as will be explained later. Not all the masks outlined in Table 44 are in widespread use for printing thick film compositions; the common types are listed in Table 45.

TABLE 43   Factors Affecting the Life of Emulsion Screens

Factor affecting screen	Effect on screen	Comment
Particle composition in screening paste	High-glass compositions cause greater emulsion wear	Particle shape may be improved, screens should be cleaned as carefully and infrequently as possible
Particle size distribution in screening paste	Larger particles may lodge in finer-mesh screens	Use coarser-mesh screens or reduce maximum particle size if possible
Vehicle in screening paste....	May attack emulsion alone or in combination with cleaning solvent	Test the screen emulsion outside the pattern area with a new paste material
Cleaning solvent............	May remove emulsion directly or cause swelling during washup	New solvents should be tested on obsolete screens or outside the image area on new screens; clean screens as infrequently as possible
Screening machine: Squeegee pressure........	Excess pressure may coin or tear wire-mesh screens and stretch or tear synthetic-mesh screens	Increase squeegee pressure slowly and check each setup before printing on a different screen
Excessive screen-to-substrate separation	Causes stretching or tearing of screen	Use a minimum screen-to-substrate separation for each setup; replace screens with too low tension
Storage conditions..........	Patterns are clogged with ink or dirt, emulsion is dried out and cracked	Clean screens thoroughly after use and store in plastic bags between 60 and 75°F and 40 and 60% RH

TABLE 44   Types of Metal Masks

I. Direct metal masks
    A. Combination etched mesh and cavity
        1. Etched solid-metal mask
        2. Electroformed mask
II. Indirect metal masks
    A. Etched or electroformed metallic stencil
        1. Woven-mesh support
            a. Adhesive-bonded
            b. Brazed or welded bond
        2. Etched-mesh support
            a. Adhesive-bonded
            b. Brazed or welded bond
    B. Deposited metallic stencil
        1. Nickel plated on woven mesh
            a. Etched-back stencil pattern

Direct metal masks

Etched Solid-Metal Mask. Etched solid-metal masks, or coetched masks are pro-duced by etching holes in a solid sheet of metal. Most etched solid-metal masks are produced by etching a cavity pattern on one side of the metal and a mesh pattern on the reverse side. However, some masks are being produced with completely open cavities with no mesh support at all in the image areas.[37] The manufacturing

TABLE 45 Comparison of Metal Masks for Thick Film Printing

	Direct metal masks		Indirect metal masks	
	Etched solid metal	Electroformed	Bonded stencil	Deposited stencil
Construction..........	Solid metal foil etched from both sides	Etched core material, electrodeposited mesh and cavity apertures	Etched or electroformed stencil bonded to woven mesh	Metal deposited on woven mesh and etched-back stencil pattern
Preferred materials of construction	Molybdenum	Nickel, beryllium-copper	Molybdenum, stainless steel	Nickel, stainless steel
Registration of cavity to mesh	Random or specific	Random or specific	Random	Random
Amount of artwork required..	Both mesh and cavity layout patterns	Both mesh and cavity layout patterns	Single layout pattern for image apertures	Single layout pattern for image apertures
Artwork compensation required (negative photoresist)	Positive compensation	Negative compensation	Positive compensation for etched stencil and negative compensation for electroformed stencil	Positive compensation
Recommended printing mode.	Contact	Contact	Off-contact	Off-contact
In-house fabrication.........	Possible but difficult	Not recommended	Possible	Possible

process utilized in the preparation of etched solid-metal masks involves the use of photoresist masking for double-sided etching of a metal foil. Figure 58 is a flow chart outlining the process steps for making etched solid-metal masks. The properties of the metals commonly used for producing etched masks are listed in Table 46. Bronze and brass have also been used to produce masks but with only limited success. Molybdenum is the most desirable material because of its high strength and fine grain structure, which result in smoother edges on the etched holes.

The construction of a typical etched solid-metal mask is shown in Fig. 59. As indicated there, the cavity depth is typically 50 percent of the mask thickness. The critical dimensions of any mask are the mesh thickness, the orifices of the mesh pattern, the cavity depth, and the cavity aperture. Table 47 lists some guidelines for the design parameters for etched solid-metal masks and for other metal masks. In order to hold the dimensions of the mesh orifices and the cavity apertures, the etch factor or the amount of undercutting must be determined. The etch factor is defined as the ratio of etch depth to undercut distance, as illustrated in Fig. 60. Etch factors must be determined experimentally and depend upon:

1. Type of metal being etched
2. Type of photoresist
3. Thickness of the metal stock
4. Geometry of the pattern
5. Etched depth of the mesh and cavity patterns
6. Etchant
7. Type of etching equipment
8. Controls on the etching process

Etchants commonly used for preparing metal masks are listed in Table 48. Once the etch factor is determined for a particular material, the cavity aperture or the mesh opening for a particular mask geometry must be compensated; this is done on the original artwork before the photoreduction. To determine the compensation required for a phosphorbronze mask, an etch factor of 2 might be used for the processing conditions outlined in Table 48.

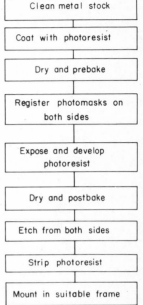

Fig. 58 Flow chart for the preparation of etched solid-metal masks.

TABLE 46 Properties of Metals Commonly Used for Etched Masks[1]

Metal	Tensile strength, psi × 10^{-3}	Yield strength at 0.2% offset, psi × 10^{-3}	Thermal expansion at 25–500°C, μin./(in.)(°C)	Melting point, °C
Copper.................	32	10	18.3	1083
Beryllium-copper.........	60–120*	28–112*	17.8–300°C	
Nickel....................	55	15	15.2	1440
Stainless steel (304).......	105	45 at 0.5% offset	18.3	1475
Molybdenum.............	70	56	5.7	2622
Tungsten................	260	...	4.6	3410
Iron-cobalt-nickel alloys...	90	50	6.2	1450

* Depend upon degree of temper of available grades.

For example, a 0.004-in.-thick phosphor-bronze mask is to be prepared with a cavity depth of 0.002 in. X is the depth of etch, and U is the undercut.

$$\text{Etch factor} = \frac{X}{U} = 2$$

$$\frac{0.002}{U} = 2$$

$$U = 0.001$$

Thus, the undercutting is 0.001 in. at each edge, and the photoresist pattern must be compensated so that the original aperture is undersize by an equivalent amount. If a cavity aperture width of 0.025 in. had to be maintained, the photomask aperture

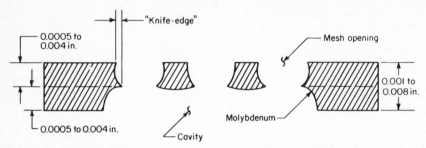

Fig. 59   Construction of a typical etched solid-metal mask.

width would have to be compensated by 0.002 in.   The artwork image, in turn, if prepared at 10×, would indicate an aperture of 0.250 − 0.020 or 0.230 in.

In practice, the etch factor may not be the same for all the apertures on the same mask.   The etch rate for very narrow lines will not be the same as for large open-

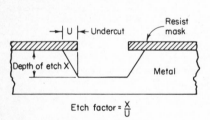

Fig. 60   Etch factor.[1]

ings.   Therefore, the amount of compensation must be different in order to hold the same aperture tolerance for all the openings. Due to the complexity of the design compensation required, it is often more desirable to allow the mask manufacturer to supply the original artwork in addition to the finished mask.

Two additional design considerations of etched metal masks are the mesh-pattern geometry and the mesh-to-cavity registration.   The mesh-pattern geometry is part of the overall grid structure.   The grid porosity and geometry have a direct effect on the metering of the paste into the cavity and the strength of the mask in the printing areas.   The grid or mesh openings should be positioned so that the cavity will be filled but not overfilled during the printing operation.   Unfortunately, there is little information available to establish design criteria for determining where the mesh holes should be located and what size they should be relative to the cavity.   However, for small cavities a good rule of thumb would be to design the mesh with a 50 percent open area based on the size of the cavity.   Layout of the mesh openings relative to the cavity apertures must be done very carefully to eliminate the possibility of leaving voids in the printed patterns. Figure 61 illustrates the importance of good mesh-to-cavity registration.   Figures 62 to 64 illustrate three of the most commonly used grid geometries for metal masks; round, square, and hexagonal.   The registration of the grid to the cavity may be done in three ways; (1) 100 percent random registration, (2) 100 percent specific registration, and (3) a combination of random and specific registration. With 100 percent random registration the grid openings do not follow the aperture

TABLE 47 Guidelines for Design Parameters for Metal Masks

Parameter	Direct metal masks		Indirect metal masks	
	Etched solid metal	Electroformed	Bonded stencil	Deposited stencil
Cavity-aperture tolerance	±0.0003 in.*	±0.0002 in.*	±0.0003 in.	±0.0003 in.
Cavity-depth tolerance	±0.0003 in.*	±0.0001 in.*	±0.0002 in.*	
Mesh-orifice tolerance	±0.0004 in.*	±0.0002 in.*	±4% on mesh count; ±0.0001 in. on wire diam	±4% on mesh count; ±0.0001 in. on wire diam
Mask-thickness range	0.002 to 0.010 in.	0.0015 to 0.030 in.	0.0045 to 0.030 in.	0.003 to 0.008 in.
Percentage of open area of mesh	30 to 60	30 to 60	20 to 50	20 to 50
Smallest spacing between parallel lines	0.002 in.	0.002 in.	0.008 in.	0.002 in.
Mounting-frame flatness and parallelism	Very critical ±0.001 in.	Very critical ±0.001 in.	Noncritical ±0.005 in.	Noncritical ±0.005 in.

* Typical.

of the cavity precisely, as illustrated in Figs. 63 and 64. With 100 percent specific registration the grid openings are aligned to the cavity over the entire image area, as illustrated in Fig. 62. Random and specific registration are used together if certain fine-line areas of the pattern must have specific registration, and other large open areas may be designed with random and specific grid-to-cavity registration as well as round and square grid structures applied to specific areas. Figure 66 illustrates a substrate printed with the mask shown in Fig. 65.

TABLE 48   Etchants Commonly Used in Processing Metal Masks[1]

Material	Etchant	Concentration	Temperature, °F	Etch rate, in./min (fresh solution)	Typical etch factor
Copper and copper alloys	FeCl₃	42°Bé	120	0.002	2.5–3.0:1
	CuCl₂ solutions	2M CuCl₂ in 6 N HCl (typical solution)			
	CuCl₂	35°Bé	130	0.00055	2.5–3.0:1
	Chromic-sulfuric	20–30% H₂SO₄ 10–20% chromate	120	0.0015	2–3:1
	(NH₄)₂S₂O₈	20%	90–120	0.001	2–3:1
	NH₄Cl sat. with NaCl				
Molybdenum	H₂SO₄:HNO₃:H₂O	1:1:1–5	130	0.001 at 130°	
	HNO₃:HCl:H₂O	1:1:1–2			
	K₃Fe(CN)₆ (200 g/l):NaOH (20–25 g/l):Na₂C₂O₄ (3–3.5 g/l)				
	NaOH(10–20%),* Na₂C₂O₄(5%)				
Nickel and nickel-iron, alloys	FeCl₃	42–49°Bé	110–130	0.0005–0.001	1–3:1
Nickel-silver, alloys	FeCl₃	42°Bé	130		
	Chromic-sulfuric		120		
	(NH₄)₂S₂O₈	20%	90–120		
Phosphor bronze	FeCl₃	42°Bé	80	0.0005	2–1
	Chromic-sulfuric		80	0.0005	2–1
	(NH₄)₂S₂O₈	20%	80	0.0003	2–1
	FeCl₃(42°Bé):HCl(2°Bé)	9:1	110–120		

* For electrolytic etching.

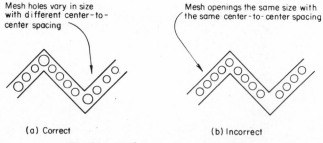

Mesh holes vary in size with different center-to-center spacing

Mesh openings the same size with the same center-to-center spacing

(a) Correct   (b) Incorrect

Fig. 61   Mask-to-cavity registration.

The advantages of etched solid-metal masks are the monolithic construction, the feasibility of designing the mask with 100 percent specific registration of the mesh-to-cavity openings, and the ability to form the mesh and cavity openings in one chemical etching operation. Disadvantages of this same mask include limitation of 0.004 in. in foil thickness for fine lines, aperture tolerances more difficult to control than for electroformed masks, and the need for different etch factors for different apertures (see Table 47). Etched solid-metal masks should be used for contact printing with no more than a 0.005-in. mask-to-substrate separation.

Fig. 62 Circular grid structure.[38]

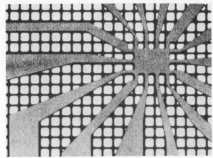

Fig. 63 Square grid structure.[38]

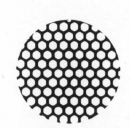

Fig. 64 Hexagonal grid structure.[38]

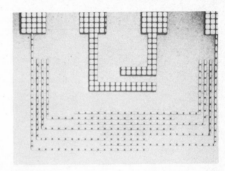

Fig. 65 Direct-metal-mask test pattern, 0.002-in. lines.[38]

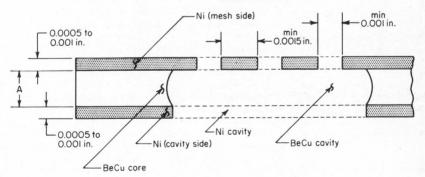

"A" dimension may be varied from 0.0005 to 0.030 in. Average is 0.001 in.

Fig. 66 Construction of a typical electroformed metal mask (Ni–Be-Cu–Ni).

Electroformed Masks. Electroformed metal masks are also prepared from solid-metal stock; however, dissimilar metals are usually used to form a bimetal or trimetal structure. Figure 66 illustrates the cross section of a typical electroformed mask using a nickel–beryllium-copper–nickel construction. Other core materials used for fabricating electroformed masks are copper, bronze, and brass. However, beryllium-copper is preferred because of (1) good strength, (2) good etchability, and (3) fine grain structure. As with the etched solid-metal mask, the electroformed mask is formed with a mesh or grid pattern on one side and a cavity on the reverse side. However, instead of using photoresist and etching to form the mesh openings and cavity directly, the mask is plated with nickel, which in turn becomes the resist for selectively etching the core material. Figure 67 is a flow chart for the preparation of electroformed metal masks. Additional process information for preparing electroformed masks may be found in Refs. 1 and 7. The grid structures and registration arrangements for the etched solid-metal masks also apply to the fabrication of electroformed metal masks. Figure 68 illustrates an electroformed nickel–beryllium-copper-nickel mask with 100 percent specific registration. Artwork compensation is also required when preparing electroformed masks with very close tolerances on the aperture dimensions. If a negative photoresist is used, the patterns must be made smaller to allow for the increase in the lateral dimensions of the image over the resist material, as shown in Fig. 69, where the increase in width of the deposited metal E at the edge of the pattern is approximately equal to the thickness of the deposit T. Therefore, the increase in the overall width of the pattern would be approximately twice the thickness of the deposited nickel. Artwork compensation will be negative, or the pattern should be designed smaller. Electroformed masks must also be used for contact printing with a 0.003-in. maximum mask-to-substrate separation. The relative merits of electroformed nickel–beryllium-copper-nickel masks are outlined in Table 49.

Indirect metal masks   As earlier defined, the indirect metal masks are prepared by utilizing etched or woven wire mesh onto which separate metallic stencils are applied. Most indirect masks are prepared with commercially available stainless-steel wire mesh although some success has been achieved by bonding metallic stencils to sheets of metal having mesh patterns etched into them.

Bonded-Stencil Mask. The bonded-stencil masks fall into two subcategories (Table 44) and differ only in types of mesh used for supporting the stencil. This type of mask was designed to take advantage of the dimensional accuracy, sharp line definition, and durability of metallic stencils which could be used in applications where off-contact printing is required. Most indirect metal masks fall into the woven-mesh category of stencils, also known as *floating* or *suspended* metal masks. This type of mask is prepared by etching or electroforming a metallic stencil, which is then attached to a stainless-wire mesh. The mesh is usually premounted on a metal printing frame. Figure 70 illustrates an indirect metal mask prepared by bonding a metallic stencil to a stainless-steel wire mesh. The metallic stencils are usually 0.001 to 0.003 in. thick although masks have been prepared with stencils up to 0.030 in. thick.[98] Etching or electroforming is accomplished by following the same process steps as outlined in Figs. 58 and 67, respectively. However, instead of fabricating the mesh and grid structure on opposite sides of the foil, the entire cavity pattern is etched through the metal from both sides. The exact methods for bonding the metallic stencils to the wire on the etched mesh are for the most part proprietary; however, adhesives, welding, and brazing techniques are generally used. Registration of the stencil to the mesh is usually random unless the pattern is designed to correspond with the mesh openings in the woven wire mesh. Figure 70 illustrates excellent alignment of the stencil to the mesh which appears to be 100 percent specific registration. If the thickness

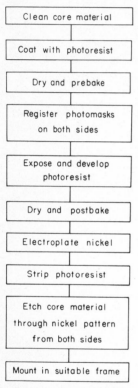

Fig. 67 Flow chart for an electroformed metal mask. (*Microcircuit Engineering Corporation.*)

of the metallic stencil is to be 0.005 in. or greater, the artwork should be compensated to allow for the etch factor of the stencil material and etching process. For stencils less than 0.005 in. it is feasible to etch from one side and then bond the beveled side of the metal to the woven mesh or other support structure. The rela-

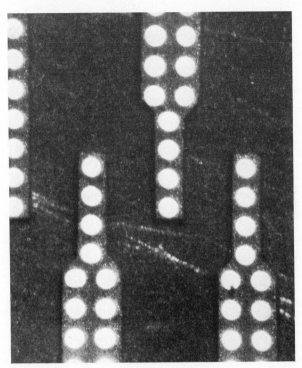

Fig. 68 Electroformed Ni–Be-Cu–Ni metal mask with 100 percent specific registration.

tive merits of bonded-stencil masks are outlined in Table 49. The bonded-stencil masks are recommended for off-contact printing with a 0.005- to 0.020-in. separation.

Deposited-Stencil Mask. The deposited-stencil mask involves the use of a woven wire-mesh material onto which a layer of metal is deposited and then etched back to the desired stencil pattern. The materials of construction for this type of mask are normally stainless-steel wire cloth electroplated with nickel. The main advantage of this type of mask is the excellent adhesion of the stencil to the mesh and the ability to reproduce isolated land patterns in the stencil. Disadvantages include random registration of the stencil cavity to the mesh, difficulty in holding line-width tolerances due to single-sided etching, and the inability to clear out the open areas during etching. The last difficulty is caused by the initial plating process, in which the wires of the mesh are plated on three sides. The end result is a significant loss of open area or weakened mesh caused by overetching (see Table 49). As with the etched solid-metal mask, the pattern design must be compensated for the etch factor of the process and material. Deposited-stencil masks

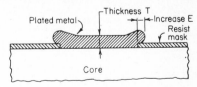

Fig. 69 Profile of an electroformed pattern.[1]

TABLE 49 Comparison of Metal Masks

Type of mask	Advantages	Disadvantages
Etched solid metal (molybdenum)	Monolithic construction, can be designed with 100% specific registration, single chemical-etching step required, has good strength	Available in. thicknesses up to 0.004 in. only, more difficult to control aperture tolerances, must be used for contact printing only
Electroformed (nickel–beryllium–copper–nickel)	Tolerances on mesh and cavity aperture can be controlled very accurately, mask can be made in any thickness, very small line widths possible, can be designed with 100% specific registration	More fabrication steps needed, not practical for printing large open areas, must be used for contact printing only, relatively fragile
Bonded stencil to wire mesh	Wide range of wire mesh available for bonding stencil to, metal foil can be etched from both sides with high accuracy, can be used for both contact and off-contact printing	Difficult to bond patterns with spacings under 0.008 in., isolated land areas cannot be utilized, possibility of delamination of stencil from mesh, random registration of mesh-to-cavity pattern
Deposited stencil on wire mesh	Excellent adhesion of stencil to wire mesh, ability to reproduce isolated land areas in pattern	Difficult to control aperture widths, random registration of mesh-to-cavity pattern

should be used for off-contact printing with a 0.005- to 0.020-in. separation. Figure 71 illustrates a deposited-stencil mask.

Design considerations for metal masks Having discussed the various techniques for fabricating metal masks, it is now appropriate to review the design considerations for them. Basically the mask performs the same function as the woven-mesh emulsion screen, i.e., the selective deposition of a specific volume of paste material onto a substrate. The mask therefore must contact the substrate during the printing cycle, receive a finite quantity of paste as the squeegee traverses the pattern, and release the material from the cavity and mesh as the mask and substrate are separated. The factors listed in Table 50 should be carefully considered when selecting and designing metal masks.

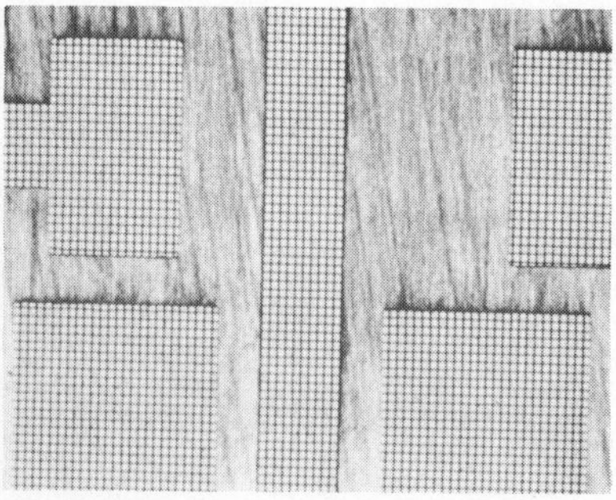

Geometry of the Image. When considering which type of metal mask to use, the geometry of the desired pattern will dictate which type or types of mask are most applicable. For very fine line geometries with fine lines and spacings less than 0.005 in., the electroformed and possibly the etched solid-metal masks would be employed. The mesh-to-cavity registration should be 100 percent for all the fine-line apertures. For printing patterns with large open areas the indirect metal masks would be more appropriate due to their great strength.

Characteristics of the Paste Material. The rheology of the paste to be printed through the mask has a direct bearing on the choice of masks. For highly thixotropic

Fig. 70 Line definition, indirect metal mask.[38]

materials both the electroformed and etched solid-metal masks are recommended. The porosity of the mesh pattern would certainly influence the amount of ink transferred into the cavity. The function of the mesh is to fill but not overfill the cavity during printing. A mask that performed well with one type of paste might not work well with pastes having a different rheology. Compositions with large particles would require masks with large mesh openings, in which case the indirect metal masks might prove more useful.

Thickness of Deposit. The deposit thickness is directly proportional to the cavity depth. For printing 0.002- or 0.003-in. beam-lead bonding pads, an electroformed mask would be the most practical type. Etched solid-metal masks are difficult to

Fig. 71 A deposited-stencil mask. (*Screen Printing Systems, Inc.*)

TABLE 50 Selection Factors for Metal Masks

1. Geometry of the image
2. Characteristics of the paste material
3. Thickness of deposit
4. Substrate characteristics
5. Life expectancy of the mask
6. Printing-machine design
7. Mask cost and preparation time

produce with line widths under 0.004 in., especially if the overall mask thickness must be 0.004 in. or greater. Consider a situation in which a fired film thickness of 0.001 in. is required. If the shrinkage factor of the paste is 2, the cavity depth is 50 percent of the mask thickness, and 100 percent of the ink is transferred from the cavity, then a 0.004-in. mask would be required. In actual practice only 50 to 90 percent of the material is transferred from the cavity, which means that the mask thickness might vary from 0.004 to 0.008 in. in overall thickness. The mask thickness should be determined by taking all the above-mentioned factors into consideration. This same procedure was followed when selecting an emulsion-stencil screen earlier in the chapter. For printing solder and brazing compositions as well as heavy deposits of glass materials, the indirect metal masks are generally used.

Substrate Characteristics. When printing with metal masks the substrates must be reasonably flat. This requirement is most important, especially with contact printing. Contact printing with electroformed and etched solid-metal masks is accomplished by forming a closed cavity consisting of the substrate and the mask stencil. During printing this cavity must be filled and the paste contained. Substrates which are cambered result in bleedout of the ink, with the resultant loss of definition and even shorting between closely spaced conductor runs. For bowed substrates it may be necessary to use indirect metal masks with off-contact printing.

Life Expectancy of the Mask. A metal mask should outlast an emulsion-stencil screen by virtue of the materials of construction. Some printing pastes are extremely abrasive and will shorten the life of an emulsion screen in fewer than 1,000 prints. Metal masks should not wear out as rapidly. On the other hand, metal masks must be treated with a great deal more care than, say, a polyester-mesh emulsion screen. Etched solid-metal masks and electroformed masks in particular are very fragile, depending upon the geometry of the images. Indirect metal masks in general are more durable than direct metal masks and therefore should be considered for greater life expectancy. Table 51 lists additional parameters affecting the life of metal masks.

Printing-Machine Design. The printing machine must be set up properly and maintained with the proper adjustments. The substrate-holding fixture must be

TABLE 51   Printing Parameters Affecting Deposition and Life of Metal Masks

Parameter	Recommendations and comments
Cavity depth.................	Directly proportional to deposit thickness
Mesh porosity (% of open area)..	Determines volume of ink metered into cavity during printing
Grid structure (mesh)..........	Influences passage of ink into cavity and strength of mask
Squeegee pressure..............	Minimum pressure to move paste through mask and provide good mask-to-substrate contact
Squeegee durometer............	Hard squeegee (80 durometer or higher)
Squeegee attack angle.........	Determined by printing samples and evaluating results
Squeegee speed................	Rheology of the paste in conjunction with squeegee speed affects deposit thickness; effect of changing squeegee speed must be determined
Squeegee-mask-platen alignment.	Squeegee, mask, and substrate to be in parallel planes
Mask-to-substrate separation....	In contact or as small a separation as possible
Substrate-holding fixture........	Recessed to allow substrate to lie flush with platen surface
Substrate surface and flatness....	Substrate should be as smooth and as flat as possible
Paste rheology................	Mesh openings and porosity are a function of paste rheology
Paste particle size and shape.....	Large, coarse particles abrade mesh; particles must pass freely through mesh openings
Mask cleaning agents...........	Cleaning agents must be compatible with metal or adhesives; careful physical handling necessary

TABLE 52   Specifications for Metal Masks

1. Type of mask
2. Cavity thickness
3. Cavity-aperture tolerances
4. Mesh thickness
5. Mesh-aperture tolerances
6. Mesh-to-cavity registration (random or specific)
7. Mesh-to-cavity registration tolerance
8. Knife-edge (molybdenum only)
9. Mask surface conditions (size of pits and dents)
10. Mask frame or mounting arrangement
11. Mask tension when mounted (mils per 100 g)
12. True positional accuracy tolerance
13. Squeegee direction during printing

TABLE 53 Recommended Applications for Metal Masks

Application	Direct metal masks		Indirect metal masks	
	Etched solid metal	Electroformed	Bonded stencil	Deposited stencil
Fine-line conductor printing (0.001 or 0.002 in.)	Possible	Very good	Not recommended	Not recommended
Fine-line beam-lead printing (0.003 to 0.006 in.)	Good	Very good	Not recommended	Not recommended
Printing solder pastes	Good	Good	Very good	Fair
High-volume printing of resistor elements:				
Contact printing	Very good	Very good	Possible	Not recommended
Off-contact printing	Not recommended	Not recommended	Very good	Very good
Printing patterns with large open areas (over 0.200 in.)	Not recommended	Not recommended	Very good	Very good
Printing very thixotropic materials	Very good (with specific registration)	Very good (with specific registration)	Not recommended	Not recommended
Recommended printing mode	Contact (or up to 0.005 in.)	Contact (or up to 0.003 in.)	Off-contact (0.005 to 0.020 in.)	Off-contact (0.005 to 0.020 in.)

properly designed to eliminate mask coining, and the mask should be used (either in contact or off-contact) with a minimum of deflection by the squeegee. The setup of the printer must also be such that the squeegee, mask, and substrate platen are all parallel and coplanar. Any misalignment of the mask to the squeegee and the substrate could result in irregular printing and damage to the mask. If a direct metal mask is to be used for contact printing, the printing machine must be designed to operate in this mode. The mask should be designed to be mounted on an acceptable frame for the printing machine. Most masks are mounted on machined-aluminum frames using mechanical rollpins, screw tension devices, or epoxy adhesives.

Mask Cost and Preparation Time. The cost and preparation time for metal masks may have a direct bearing on which type of mask will be used. If random registration of the mesh-to-cavity pattern is acceptable and fine-line printing is not required, an indirect metal mask may be acceptable. Design time for direct metal masks is longer, especially if specific registration of the mesh-to-cavity pattern is required. The cost of 100 percent specific registration can increase the master artwork preparation and tooling cost by as much as a factor of 10. However, if high volume and high yield are the primary considerations in the selection of a metal mask, either of the direct metal masks with 100 percent specific mesh-to-cavity registration may be justified.

Table 52 lists some items which should be included when buying or specifying metal masks.

Applications for metal masks   Table 53 outlines some recommended applications for metal masks. The selection of a metal mask in preference to a woven-mesh emulsion screen can be accomplished only after careful consideration of the intended use of the mask or screen. For prototype work involving fewer than 500 units the emulsion screen is more economical and can be produced in less time. However, if the geometry of the pattern is such that emulsion screens will not produce the desired results, metal masks must be used. For high-volume high-precision printing, metal masks may prove more economical.

REFERENCES

1. Ryan, R. J., E. B. Davidson, and H. O. Hook: chap. 14 in C. A. Harper, "Handbook of Materials and Processes for Electronics," McGraw-Hill, New York, 1970.
2. Bishop Graphics, Inc.: manufacturer's literature.
3. Ulano Company: manufacturer's literature.
4. Consul-Mutoh Ltd.: manufacturer's literature.
5. Techniques of Microphotography, Eastman-Kodak *Tech. Publ.* P-52.
6. Brown Manufacturing Co.: "Cameraman's Handbook."
7. Ingraham, R. C.: Photolithographic Masks for Integrated and Thin Film Circuitry, *Solid State Technol.*, March 1965.
8. Wilson, E. T.: Camera Adjustments for Accurate-sized Images, *J. Photogr. Sci.*, vol. 12 (1964).
9. Physical Properties of Kodak Estar-Base Films for the Graphic Arts, *Eastman Kodak Pam.* Q-34.
10. Dimensional Stability Required for Precision, *Kodak Compass*, P-1-69-1.
11. Hughes, D. C., Jr.: "Screen Printing of Microcircuits," Dan Mar, Somerville, N.J., 1967.
12. Zurich Bolting Cloth Mfg. Co., Ltd., Switzerland: Nitex (instruction manual).
13. Place, T.: Practical Applications for Thick-Film Resistors and Conductors, paper presented at *1st Thick Film Symp., Los Angeles, February 1967.*
14. Industrial Wire Cloth, Commercial Standard CS 232-60, U.S. Government Printing Office.
15. Gerard Daniel and Co.: manufacturer's literature.
16. Wire Cloth Enterprises, Inc.: data sheet.
17. Gerard Daniel (Gerard Daniel and Co.): private discussion.
18. Kressilk Products, Inc.: manufacturer's literature.
19. Swiss Bolting Cloth Mfg. Co., Ltd., Zurich: Selected Fabrics for Screen Printing.
20. Colonial Printing Ink Co., Inc.: data sheet.
21. Miller, Paul (P. Miller & Sons Wire Works, Inc.): private discussions.

22. Azoplate Corporation: data sheet.
23. Albert Ross Chemicals, Ltd.: data sheet.
24. Advance Process Supply Co.: General Catalog.
25. Ottaviano, A. V.: Repeatability in Screen Printing Hybrid Microcircuits, *Proc. Int. Soc. Hybrid Microelectron.*, *1969.*
26. Industrial Reproductions, Inc.: data sheet.
27. Microcircuit Engineering Corporation: product specification.
28. Tarcza, W. (Corning Glass Works): private discussions.
29. E. I. duPont de Nemours & Co.: "Thick Film Handbook," sec. R-2, 1967.
30. Headly, R. C.: Reproducibility of Electrical Properties of Thick Film Resistors, *1st Thick Film Symp., Los Angeles, February 1967.*
31. Riemer, D. F.: The Direct Emulsion Screen as Tool for High Resolution Thick Film Printing, *Proc. Electron. Components Conf., 1971.*
32. Riemer, D. F.: The Effect of Geometry on the Characteristics of Thick Film Resistors, *Proc. Electron. Components Conf., 1970.*
33. Nesselroth, M. D., and R. H. Zeien: LSI Interconnection Techniques: Fabricating a High Resolution Hybrid Array, *Electron. Packag. Prod. Conf., June 1971.*
34. Clough, W. L.: A Compatible Patterning Technique for Custom Built Thin and Thick Film Circuits, *Electron. Equip. News,* June 1967.
35. Kosar, J.: "Light Sensitive Systems," Wiley, New York, 1965.
36. Strong Electric Corp.: product bulletin.
37. Miller, L. F.: Paste Transfer in the Screening Process, *1968 SAE Microelectron. Packag. Interconnection Conf., Palo Alto, Calif., 1968.*
38. Coronis, L. H.: Indirect and Direct Etched Metal Masks for Deposition Control and Fine Line Printing, *Proc. Int. Soc. Hybrid Microelectron. Conv. 1969.*

Chapter 3

Facilities, Equipment, and Manufacturing Operations for Circuit Deposition and Testing

RENE E. COTE

Integrated Circuits, Inc., Bellevue, Washington

INTRODUCTION

Origin of the screen printing process   The screen printing process, one of the oldest existing art forms, dates back as far as 500 B.C.

Silk screening has been used to decorate glassware and chinaware and to print posters, fabrics, and similar graphics.   In the late 1930s, the process was first used in the electronics industry for applying the electrodes or plates to ceramic disk capacitors.   Sometime later, it was found that curable-carbon-composition resistors could be added to the ceramic capacitor bodies to achieve the first *hybrid circuit* resistor-capacitor networks and that passive filters could be produced by screen printing.

In the late 1940s and early 1950s, RCA extended the process with more development work on two projects.   The first was Tinkertoy, where printed and fired ceramic substrates were stacked to the base of vacuum tubes.   The second was Micromodule; contracts were awarded to Electra-Midland and CTS to develop high-stability precision resistor networks on alumina substrates.   Although the overall project was not totally successful, the development work on thick film printed and fired resistor networks proved the feasibility of the process.   This opened the way to a new method of packaging electronic circuitry.

When IBM announced that they had totally committed their System 360 computer to thick film hybrid technology, the process was established as a major electronic packaging technique.   Henceforth more systems manufacturers began to look to thick films as a way of solving some of their packaging problems, and the component houses accelerated their development work to keep pace with the growing need. The material suppliers responded with new and improved materials, and the equipment manufacturers also speeded up their efforts with new and improved equipment, so that today thick films are a convenient, cost-effective, reliable method of packaging electronic subsystems.

Purpose of the chapter   This chapter examines the facilities, processes, and equipment required to screen-print, fire, and adjust thick film components.   The topic is explored in enough detail to allow the reader to understand the process, along with its advantages and disadvantages, in sufficient depth to be able to accomplish his aims.

If he plans to set up a thick film facility, this chapter gives him the information needed to avoid the usual pitfalls.   No attempt is made to recommend or compare the commercially available equipment.   Instead, each area is reviewed in detail so that the reader can then examine his own needs and fit the available equipment to them.

In addition, a method of costing the entire process is presented.   This method has proved to be a convenient and fast way of estimating costs of the process and should be of interest to both the user and the producer.   The user will now find a way of estimating the cost of the thick film approach for his packaging needs.   The producer will find a quick way of estimating the pricing for thick film arrays.

Organization of the chapter   The chapter is organized to correspond with the normal process flow of the product, so that the user can understand and follow through the various steps in the process.   It is further designed to allow easy traceability through the various steps required to produce a thick film array.   In this way all the major points can be covered in chronological order.

The chapter begins by detailing a typical facility layout.   Figure 1 shows a medium-scale facility with built-in room for expansion.   This facility does not include any provisions for circuit assembly but concentrates solely on screening, firing, and adjusting resistor networks.   These networks may be an end product in themselves or provide the substrates for subsequent thick film manufacture.   The facility includes all operations from incoming inspection and material preparation to visual and electrical final inspections.

Consideration is given to substrate, screen, and screening-material preparation.   In all cases an attempt is made to give typical specifications for processing these materials before detailing the screen printing process.

The screen printing section not only considers the process itself but also some of the more important variables which should be taken into account to properly specify the equipment. Both contact and off-contact printing are discussed, and a list of suppliers is given for handy reference. A typical process specification is outlined.

The firing process follows, with details of equipment and manufacturing procedures. The discussion includes process monitoring techniques. Furnace profiling equipment, procedures, and equipment are also included. An attempt is made to relate the equipment and facility to the specific task or level of activity desired from the operation. Again, a list of manufacturers is given for reference.

Air-abrasive and laser resistor trimming are discussed, together with procedures and process monitoring techniques which have been used to control the process more closely, achieving tighter tolerances and improved yield.

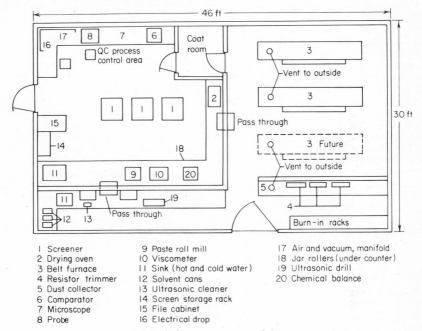

I Screener	9 Paste roll mill	I7 Air and vacuum, manifold
2 Drying oven	IO Viscometer	I8 Jar rollers (under counter)
3 Belt furnace	I I Sink (hot and cold water)	I9 Ultrasonic drill
4 Resistor trimmer	I2 Solvent cans	20 Chemical balance
5 Dust collector	I3 Ultrasonic cleaner	
6 Comparator	I4 Screen storage rack	
7 Microscope	I5 File cabinet	
8 Probe	I6 Electrical drop	

Fig. 1   Thick film screening and firing area.

Finally, a method of costing the entire process is presented. The considerations not only will help price this portion of the process, but properly used can also be used as a monitor to keep the process in control.

OVERALL FACILITY AND LAYOUT

It is not possible to design a single facility which will be all things to all people. With this in mind, the first and most important step to take in designing the facility is to determine into which area of concentration the group will fall. Key considerations in designing the facility are whether the facility will be used solely as a prototype facility or as a production facility or for a combination of the two. Further, it must be decided whether the end products will be simple resistor networks or complex hybrids. Another question to be faced is whether the products will be geared to commercial or military markets.

All this may seem elementary, but unless the definition is made at the outset, the resulting facility can be unsuitable for the job the designer envisioned or later finds necessary. The proper choice of screen printers, furnaces, and trimming equipment is very dependent on the end product to be produced, and equipment manufacturers have designed their products to meet specific applications. Although there is some flexibility, the proper long-range results will be obtained only by precise specification at the outset. For example, a 4-in.-wide belt may be more than suitable for a medium-scale production facility, but it will be totally unsuitable for high-volume resistor networks requiring a constant output. By the same token, a manual resistor trimmer is suitable and economical for a prototype facility but would be a bottleneck for any other type. A high-speed laser trimmer may do an excellent job for repetitive high-volume work but may require excessive setup time for a prototype operation, negating its advantages.

Some general comments can still be made about facilities which will apply in any event. Figure 1 also shows a layout for a screening, firing, and resistor-adjusting area which can be considered as a compact medium-scale facility and in which prototype work and some production work can be accomplished. The facility has a controlled-atmosphere area for material preparation and screen printing with pass boxes for product and material access into and out of the room. The furnaces are located outside the screen printing area to simplify management of the atmosphere-control system, which is more important in the printing operation than in the firing of the printed and dried substrates. All furnaces are vented to the outside to eliminate the organic by-products of the burn-off cycle of the firing process.

The resistor-trimming area differs greatly, depending on whether laser trimming or air-abrasive trimming is used. Each of the different types of trimming (as well as screening, firing, and other details of the thick film process) is covered in more detail later in this chapter.

The services, facilities, and equipment for each of the areas are considered separately and specifically related to each of the steps in the process.

It is difficult to state unequivocally that the thick film facility should be air-conditioned, humidity-controlled, and/or dust controlled. As the technology improves and as line widths narrow, dust control in the screening area becomes a necessity, especially for eliminating large foreign particles. Temperature control has an effect on viscosity. Humidity control has an effect on the rate of solvent evaporation, including emulsion life of the screens themselves. Therefore, control definitely helps, especially in cases where state-of-the-art circuits are being processed.

As a rule of thumb, most thick film manufacturers state that absolute temperature and humidity are not that important and are more dependent on operator comfort. However, variations are to be reduced to a minimum. In other words, the temperature may range from 70 to 80°F and the humidity from 30 to 50 percent, but once the absolute level is set, it should be controlled at that level as uniformly as practical. The temperature should be held $\pm 2°F$ and humidity ± 5 percent or better in the state-of-the-art facilities.

The nebulous but nevertheless real advantage that clean-room facilities offer is the state of mind it produces in the operators. Where sophistication is required, there is no substitute for well-trained and properly motivated personnel. Clean rooms have a way of instilling an atmosphere conducive to precision work.

SUBSTRATE PREPARATION AND INSPECTION

It will be assumed that substrates are purchased or treated as purchased material received by the thick film fabrication area rather than being fabricated totally within the facility. The substrates may be alumina, Al_2O_3; beryllia, BeO; barium titanate, $BaTiO_3$; or steatite. All are used for fabricating thick film circuits, and inspection, sorting, and cleaning are essentially the same for all. Table 1 lists important characteristics of the various substrates.

When small quantities of nonstandard substrates are required, a substrate scriber (Fig. 2) can be used. Accuracies of 0.001 in. can be attained using up to 4 by 4 in. starting material. This can reduce tight prototype schedules by several weeks.

The operation of the scriber consists of inserting the workpiece into a holding chuck and turning the indexing crank until the scribe path is aligned with the diamond scribe point. The toolholder assembly is then slid on the guide rails to the forward limit stop. This point automatically drops the diamond point to the preset scribing level.   The tool is then drawn across the substrate under preset scribing force to the back limit stop, which, when engaged automatically, raises the diamond to the clear position for the next stroke.

The part can then be indexed in 0.005-in. increments.   The procedure is repeated until all lines are scribed in one direction and then repeated again for the lines in the other direction.

The parts can be separated by breaking manually or by using an automatic fracturer, shown in Fig. 3.

Dimensional tolerances, camber, and warpage are the key parameters tested at incoming inspection.   If the substrate is within the proper length and width toler-

TABLE 1   Typical Substrate Properties

	Alumina	Beryllia	Titanates
Surface roughness, μin., CLA	20–40	20–40	30–100
Thermal conductivity, $(cal)(cm)/(s)(cm^2)(°C)$	0.04	0.60	0.02
Thermal expansion coefficient, ppm/°C	6	7	10–16
Dielectric strength (typical), V/mil	300	300	150
Volume resistivity, Ω-cm	10^{16}	10^{16}	$10^{8}–10^{14}$
Dielectric constant	9	7	15–10,000
Dissipation factor	0.0001	0.001	0.0001–0.03
Specific gravity	3.85	2.90	4–6
Flexural strength, psi	50,000	25,000	5,000–20,000

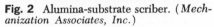

Fig. 2 Alumina-substrate scriber. (*Mechanization Associates, Inc.*)

Fig. 3 Substrate fracturer. (*Mechanization Associates, Inc.*)

ances, especially if the part has holes in it, it is immediately apparent.   However, variations in camber or thickness, which may be noncritical for purely commercial applications, assume quite a different perspective when exact control of the resistor's electrical properties is required.   The screen-to-substrate distance is one of the prime controls in regulating the amount of material deposited on the substrate during the printing process.

Precision line printing and the electrical properties of the resistor are very much controlled by the breakaway or snap-off distance between the screen and the substrate.   A 1-mil change can affect the value of the resistor as much as 1 percent. This is not as critical, however, as the effect this parameter can have on other electrical properties, such as temperature coefficient and long-term stability.   Each resistor system has its own optimum thickness which must not be reduced or exceeded.   Substrate-thickness variation contributes more to the problem than machine variation.

The equipment required to inspect substrates thoroughly includes calipers, micrometers, drop-weight micrometers, low-power optical equipment, and, for sophisticated work, a surface analyzer. All this equipment and the operation itself should be under the supervision of incoming-quality-control inspection. Unless the parts being fabricated are ultraspecial, sampling procedures are used rather than 100 percent inspection. The parts should not be handled without protective coverings, such as lintless gloves or finger cots, since it is almost impossible to remove all traces of fingerprints which may be left on the parts.

In general, substrate manufacturers exercise care in handling the parts, which, due to the very nature of the high-temperature firing and chemical composition, are inherently clean. This allows the parts to be introduced directly into the front end of the process without any elaborate preprocessing. This can be verified at incoming inspection.

If the process has been running consistently and the user has been standardizing on particular substrate vendors, a great deal of care should be exercised when considering a change. Economics might dictate changing to other sources—even foreign sources. Because of the differences in chemical composition of the basic raw materials a complete evaluation of the alternate source is advisable before any radical change is made. The basic reason for this is the nature of the fluxes which will predominate on the substrate surfaces to affect both electrical properties and adhesion.

If it is found that the substrates require precleaning, vapor degreasing and/or ultrasonic cleaning with trichloroethylene or Freon* are advisable. This is a suitable method of removing some grease, oil, or particulate material which may have deposited on the parts. Chlorinated solvent residues must be thoroughly removed. If more drastic methods, such as strong oxidizing or reducing agents, must be used, the procedures must be very carefully controlled since they can affect the "glassy phase" on the surface of the ceramic, which is very important to resulting adhesion. Residues from such cleaning procedures must be totally removed since they can be trapped within microscopic voids on the surface of the substrate and subsequently alter the predicted electrical and/or mechanical properties of the fired films.

After inspection, the substrates should be stored in a clean, dry, dust-free area prior to use.

SCREEN PREPARATION AND INSPECTION

The basic tooling required for the fabrication of a thick film circuit is the screen. The most common screens (Fig. 4) in use today include direct emulsion screens, indirect or transfer emulsion screens, suspended metal masks, and solid etched metal masks, all covered in detail in Chap. 2.

The screens are also an important factor in determining the thickness of the deposition which will be printed during the screen-printed process. Table 2 shows the relation of mesh number to wire diameter and percent open area, and Table 3 shows the effect of mesh size on film thickness and electrical properties of one of the resistor compositions. The behavior observed is even more critical on other resistor systems which have a tendency to peak. Along with mesh size, emulsion thickness must also be controlled and held constant.

Direct emulsion screens are coated using a sensitized emulsion which is squeegeed onto the mesh and then allowed to dry. One supplier, Industrial Reproductions, Inc., Nashua, N.H., offers storable emulsion screens which are unsensitized and can be stored 6 months or more. These screens are presensitized before use and processed like screens coated with presensitized emulsion.

A film positive is then aligned to the substrate side of the screen. A microscope or other optical method is used to align the pattern to the wire mesh. Figure 5 shows a frame suitable for this purpose. The screen and film are placed with the positive side to the glass on a vacuum frame. Exposure time varies with the light source and distance. Pulsed xenon light sources at 45-A line current and at 30 to 36 in. distance require 2½ to 3 min exposure time. A carbon-arc source at the same distance requires 4 to 5 min.

* Trademark of E. I. du Pont de Nemours & Co.

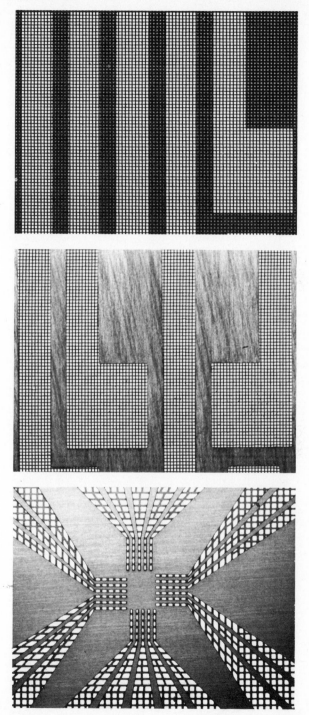

Fig. 4 Emulsion and metal-mask patterns. (*Industrial Reproductions, Inc.*)

TABLE 2   Mesh or Fabric Number

Mesh	Stainless-steel wire	Nitex	Japanese silk	Wire diameter, in.	Mesh opening, in.	Open area, %
165	165			0.0019	0.0042	47.1
180	180			0.0018	0.0042	46.6
185		185		0.0017	0.0037	47.5
186			20×	0.0024	0.0030	31.0
196			25×	0.0027	0.0034	44.0
200	200			0.0016	0.0034	46.2
206		206		0.0015	0.0035	50.0
229		229		0.0014	0.0030	46.5
230	230			0.0014	0.0029	44.0
240		240		0.0014	0.0028	
250	250			0.0018		45.0
260		260		0.0014	0.0024	
270	270			0.0016		
283		283		0.0012	0.0024	44.5
306		306		0.0012	0.0021	41.0
325	325			0.0011	0.0020	42.2
330		330		0.0012	0.0018	37.0
350		350		0.0012	0.0016	34.0
380		380		0.0012	0.0014	30.0
400	400			0.0010	0.0015	36.0
465		465		0.0012	0.00098	21.0

TABLE 3   Effects of Mesh Size on Film Thickness and Electrical Properties of Resistor Compositions

Du Pont no.	Mesh	Film thickness, mils		Resistance	Hot (125°C) temperature coefficient of resistance
		Dry	Fired		
7826	105	1.64	0.99	153 Ω	662
	165	0.88	0.66	313 Ω	646
	200	0.75	0.45	532 Ω	594
	325	0.39	0.36	1,034 Ω	547
7827	105	1.55	0.94	1,245 Ω	404
	165	1.13	0.77	1,480 Ω	405
	200	0.82	0.55	2,051 Ω	319
	325	0.56	0.28	3,953 Ω	229
7828	105	1.48	0.98	2.65 kΩ	148
	165	1.09	0.73	3.52 kΩ	124
	200	0.84	0.58	4.72 kΩ	81
	325	0.57	0.31	7.86 kΩ	2.8
7832	105	1.59	1.10	5.56 kΩ	65
	165	1.29	0.75	6.43 kΩ	58
	200	0.99	0.53	8.23 kΩ	31
	325	0.73	0.35	13.18 kΩ	19

SOURCE: Data from E. I. du Pont de Nemours & Co. data sheet A-53826-3/67.   Test method was to screen-print a pattern of four squares 0.250 by 0.750 in. on a 1 by 1 by 0.025 in. alsimag 614 substrate with a Presco 100B, oil check and rigid squeegee.   Terminations were Du Pont 7553 Pd-Au.   Terminations fired in box furnace, 1000°C, 10-min soak; resistors fired standard curve, 760°C peak in continuous-belt furnace.   Dry film thickness measured by calibrated optical focus.   Fired film measured with mechanical gage.

The washout procedure is accomplished by soaking the exposed screen in water at 90 to 110°F for 30 to 60 s. The screen is then rubbed lightly from the squeegee side of the mesh with a soft cloth, sponge, or polyurethane foam to remove any excess emulsion. It is helpful to agitate the water bath while rubbing. The objective is to remove only the emulsion on the surface of the mesh. An aerated, lightpressure water spray from the substrate side of the screen is used to wash the emulsion from the pattern area. The open area is then cleaned with 40 to 50 psi maximum filtered air spray applied primarily from the substrate side.

The screen is allowed to dry thoroughly. This is very important to screen life. If heat is used to dry the screen, it should not exceed 120°F. After drying, any voids or damage spots can be repaired using blockout material.

The equipment required to do the job consists of a vacuum frame to assure that the film or glass-plate positive will be in intimate contact with the emulsion. This

Fig. 5   Pattern-alignment fixture. (*De Haart, Inc.*)

TABLE 4   Recommended Screen Tensions for 5 by 5 inch Frame

Mesh size	Wire diameter, in.	Mesh opening, in.	Deflection, in.*
105	0.0030	0.0065	0.035
150	0.0026	0.0041	0.035
165	0.0020	0.0042	0.050
200	0.0016	0.0034	0.055
200	0.0021	0.0029	0.045
230	0.0015	0.0029	0.055
250	0.0016	0.0024	0.050
325	0.0011	0.0020	0.065

* Deflection tolerance ±0.012 in.

ensures the best pattern definition. A microscope is required to align the pattern to the mesh, a light source is required to expose the patterns, and a temperature-controlled bath with water-spray provisions is required to soak and wash out the screens. A filtered-dry-air source is also required.

Indirect emulsion masks are fabricated by exposing the presensitized film in a manner similar to that described above and then applying this film to the mesh screen. In general this type of screen is less durable but gives excellent line definition for short runs.

Metal masks are generally purchased except for very sophisticated operations. In general a double etching is used with the pattern etched from the substrate side and the mesh from the squeegee side of the metal foil.

Screen tension must be controlled if fine pattern definition and thickness control are to be maintained. Table 4 gives recommended screen tensions for 5 by 5 in. frames. Figures 6 and 7 show equipment available for measuring screen tension and deflection.   Tension control not only leads to thickness control and line definition but assures high yields in the long run.

Manufacturers of screen process equipment include:

Advance Process Supply Co.
Chicago, Ill. 60639

Affiliated Manufacturers, Inc. (AMI)
Box 248
Whitehouse, N.J. 08888

Aremco Products, Inc.
P.O. Box 145
Briarcliff Manor, N.Y. 10510

De Haart, Inc.
Burlington, Mass. 01803

General Research, Inc.
Sparta, Mich. 49345

Graining Equipment Co.
Nashville, Tenn. 37203

Industrial Reproductions, Inc.
P.O. Box 888
Nashua, N.H. 03060

Microcircuit Engineering
Medford, N.J. 08055

Towne Laboratories, Inc.
Somerville, N.J. 08876

Ulano
New York, N.Y. 10028

Fig. 6   Screen-tension gage. (*De Haart, Inc.*)

Fig. 7   Screen-tension gage. (*Engineered Technical Products.*)

MATERIAL CHECKOUT AND MONITORING

All resistor, conductor, and dielectric thick film compositions are thoroughly dispersed during manufacture. Since most of these are dispersions of metal and glass frits in organic binders, they have a tendency to separate or settle during shipment and/or improper storage.

In order to assure that the proper viscosity is observed at incoming inspection and, more important, that a homogeneous composition is achieved prior to use, the materials must be thoroughly redispersed into their resin systems. Elaborate equipment is not necessary for this task, which can be accomplished easily and effectively by

Fig. 8 Three-roll mill. (*Paul O. Abbe, Inc.*)

simple hand stirring with a stainless-steel spatula until all the lumps are removed and/or no sediment is observed on the bottom of the container.

Some people prefer three-roll mills (Fig. 8), which are commercially available for this purpose and widely used in the ink and paint industries. They have a distinct advantage over hand stirring in accomplishing the task quickly and uniformly.

Certain disadvantages and restrictions must be taken into consideration. These include sample size, since 100 percent recovery is almost impossible. If large batches of materials are being handled, the three-roll mill is obviously preferred, since the task is almost impossible by hand. For small samples, the loss on a mill can be considerable and expensive. It may even be impossible to get the sample onto a mill if small quantities are involved, since the material may simply coat the rollers.

Also to be considered are the length of time and the pressures used to redisperse the materials. Extreme care must be taken not to alter the particle size or the composition of the thick film materials by grinding the frits or metallic particles during the rolling operation. The reduction of particle size due to milling adversely affects both print thickness and electrical characteristic.

The roller speed must also be monitored and controlled. If stirring equipment is used, the speed should be kept as slow as possible to affect uniform dispersion but prevent air from being beaten into the mixture and trapped there. This would affect the apparent viscosity drastically and could cause actions to be taken in error. It should be remembered that this is a temporary situation, not a permanent change in the material, which can be rectified by placing the jar on a roller mill for 12 to 16 h before attempting to measure the viscosity or using the paste.

Viscosity is a very important parameter to control since it affects both print thickness and line definition. Each paste manufacturer recommends the viscosity he feels best for his products. The user, however, must determine the viscosity which will best suit his process. Once this has been determined, control and monitoring are not only advisable but absolutely necessary. A viscometer is used to measure the viscosity of the paste, as discussed in detail below.

The pastes are composed of solids, a resin binder system, and a compatible solvent system, which is generally a high-boiling low-vapor-pressure liquid. Since this solvent has a tendency to evaporate with time, it must be replaced. Again, each manufacturer has a recommended solvent which must be used if the proper composition and balance are to be maintained. Only enough solvent to maintain the proper viscosity should be blended in, since too much will affect film thickness and the electrical properties of the paste.

The most widely used viscometers are the Brookfield series (Fig. 9), available in several models and various spindles. Models are available with different speeds, and some are variable. Viscometers are also discussed in Chap. 5. The flexibility permits checking a variety of composi-tions.

Fig. 9 Viscometer. (*Brookfield Engineering Laboratories.*)

Since most of the materials are thixotropic, monitoring can also be checked as a function of viscosity versus rotation speed and evident changes can help show rheological changes which might not be evident at a single speed. The user should attempt to follow the directions of the paste manufacturer and duplicate the speed and spindle originally used to measure the shipping viscosity.

Brookfield models HAT, HBT, HBF, LVT, and RVT, with several spindles and holders, are the most common viscosimeters. Du Pont recommends the following procedure:

1. Stir the material for test until it is homogeneous. (Note special precautions for vigorously stirred compositions.)

2. The composition volume should be sufficient to provide a minimum depth of 2 in. Special adapters are obtainable for reading smaller samples. (Brookfield has a small-sample cup.)

3. Immerse spindle in sample.

4. Adjust sample and spindle to $25 \pm 1°C$. (Use a water bath if necessary.)

5. Stir sample with spindle to assure homogeneity.

6. Attach spindle to instrument and adjust immersion to groove indicated on spindle shaft.

7. Start instrument; allow indicating printer to stabilize at a fixed dial position for at least three revolutions before taking a reading.

8. Convert the dial reading to centipoises,* using the factor finder provided with the instrument.

Once the materials have been dispersed, blended, and/or adjusted to the proper viscosity, they should be stored on a slowly rotating machine. Rotating the jars constantly prevents settling and caking; the result is more uniform mixtures, which make redispersion much easier or eliminate it completely. Figures 10 and 11 show typical thixotropic mixers used for this purpose.

Manufacturers who supply equipment for this phase of the process include:

Paul O. Abbe, Inc.
Little Falls, N.J. 07424

Brookfield Engineering Laboratories
Stoughton, Mass. 02072

J. H. Day Company
Cincinnati, Ohio 45212

Engineered Technical Products
P.O. Box 1465
Plainfield, N.J. 07061

Ferranti-Packard Electronics
Toronto 15, Ont.

Mechanization Associates, Inc.
Mountain View, Calif. 94040

Norton Company
P.O. Box 350
Akron, Ohio 44309

U.S. Stoneware Company
P.O. Box 350
Akron, Ohio 44309

SCREEN PRINTING

The manufacture of thick film circuitry is basically a screen printing technique. The circuit is fabricated by successively printing each layer until the desired circuitry is achieved. The basic tools for accomplishing this are the squeegee and the screen. The system which incorporates these basic tools and provides adequate means for controlling them precisely to make them useful in the fabrication of electronic circuitry is the *screen printer*.

Since this is the heart of the process, a great deal of thought must go into the selection of the screen printer. At this point, the decision made earlier which type of facility it is to be plays an important part in the selection since the variety of printers is very wide. Only the proper choice of equipment will allow the required precision and accuracy.

Equipment Considerations

The basic functional parts of a hybrid-circuit screen printer can be divided into screen mounting, substrate-holding tool, squeegee, travel, and pressure system, transfer mechanism to move substrate from the load to print position, and X, $Y + Z$, and θ (angular) adjustment of the screen to the substrate position.

Screen mounting The screen mounting must provide firm yet adaptable mounting screws or clamps to suit the particular screen frame and hold it rigidly in place while still allowing for quick changes to other screens.

In addition, the mounting should be substantial enough to remain rigid throughout the travel of the squeegee. Any movement or shifting must be avoided during the printing cycle, especially from print to print. Another important feature is that the mounting must be parallel to the substrate holder and stay parallel. The design should be such that a maximum deviation of 0.002 in. is maintained over any range.

With these features, the mounting plate must also allow easy accessibility to the screen frame for changes and the bottom for easy cleaning during the process. As much open area as possible must be provided for preliminary setup, substrate alignment, and ink or paste addition.

Substrate-holding tool The substrate holding tool must be capable of accurately locating and registering the substrate, providing a method of rigidly fixing or holding the substrate in that position. It must wear well and be easy to replace for changeover to different substrate designs or due to wear. It should also provide completely flat surfaces to keep the screen and the substrate parallel throughout the process.

* In this handbook the ANSI abbreviation for centipoises, cP, is used; in the industry, however, the form cps is common.

The simplest and most generally used method for locating substrates is small-diameter pins or dowels. Because of the abrasive nature of the ceramic, especially alumina, these are generally made of hardened steel, carbide, or (in the case of high-speed high-volume equipment) sapphire.

These pins are arranged according to the shape of the part to be printed. The most common plate-type locators generally have one pin centered on the short side

Fig. 10   Thixotropic storage rack. (*Mechanization Associates, Inc.*)

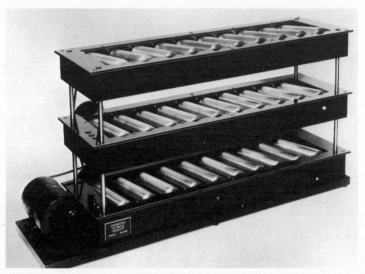

Fig. 11   Thixotropic storage rack. (*Engineered Technical Products.*)

and the remaining two spaced at one-fourth and three-fourths the length of the long side. If the substrate is to be printed on both sides, the pin arrangement should be reversed to assure that the same points of reference are being used in all cases.

Although there are many ways of holding the part in place after it has been registered, the easiest and most convenient method is to use vacuum cavities, ranging from a simple hole to more complex patterns designed to create enough holding power to keep the part located throughout the screening process. The vacuum system should be sufficient to provide uniform vacuum from part to part and machine to machine in the series hookup common in most facilities.

Squeegees, travel, and pressure system   The shape of the squeegee and the material from which it is made have been the subject of several studies.   The material must be compatible with the resins and solvents used in the thick film process.   Since it must not swell, distort, or wear excessively, it is limited to neoprene, polyurethane, and Viton.[*]   The hardness of the material is generally between 50 and 90 durometer.   The choice is rather subjective; one group prefers the softer 50 to 70 durometer, reasoning that a more pliable squeegee will be allowed to conform to the irregular printing surfaces more readily, and the second group reasons that the harder material, 80 to 90 durometer, should be used to maintain a constant attack angle, cut off the ink or paste sharply, clean the screen surface, and provide longer life.

Experiment will show which material is most suited to any particular application; once chosen, it should be maintained for uniformity.   Polyurethane and Viton presently appear to be the most commonly used materials.

The shape of the squeegee also depends to some degree on personal preference,

TABLE 5   Some Significant Properties of Elastomeric Materials

	Practical hardness range or durometer	Abrasion resistance	Solvent resistance					Acid resistance
			Hydrocarbons		Oxygenated solvents, alcohols, ketones	Oil and gasoline		
			Aliphatic	Aromatic				
Polyurethane (Vulcan, Elasta Cast, Vulkolan, Multrathane, Adiprene, Vibrathane)[*]	50–95+	Outstanding	Excellent	Good	Poor	Excellent		Poor
Neoprene	40–95	Very good	Good	Fair	Poor	Good		Good
Nitrile (Buna N, Hycar, Krynac, Paracril, Butaprene)[*]	40–95	Good	Excellent	Good	Poor	Excellent		Fair
Butyl	40–80	Good	Poor	Poor	Very good	Poor		Very good
Silicone	40–85	Poor	Poor	Poor	Fair	Fair		Fair
Fluorocarbon (Fluorel, Viton)[*]	60–90	Good	Excellent	Excellent	Poor	Excellent		Very good

[*] Other names, trade names, or base materials sometimes used interchangeably.

especially when attack angle is considered.   The main purpose of the squeegee is to present an even, sharp edge to the screen, the angle generally being 45 to 60°.   This pushes a roll of ink or paste ahead of the squeegee, creating a pressure which forces the material through the screen mesh.   The sharp edge of the squeegee deflects the screen ahead of it and allows for a sharp breakaway and more printing accuracy in off-contact printing.   Table 5 gives some of the material properties.   Pressure is applied to the squeegee by dead weight, spring loading, or pneumatic cylinders.   All are used effectively, depending on the desired results.   The most common method is pneumatic cylinder.   Figure 12 shows a picture of a Presco head with some of its adjustments.

The pressure applied to the squeegee affects both the definition and the uniformity of the print achieved.   The exact pressure for best results depends on the ink or paste viscosity, the screen tension, and the breakaway distance.   The range of pressure is 1 to 10 lb per linear inch of squeegee and should be determined and set for each facility.

Too much pressure is to be avoided to prevent coining the screen and substantially reducing its life.   In order to prevent overtravel of the squeegee, a mechan-

[*] Trademark of E. I. du Pont de Nemours & Co.

ical stop is also provided in most printers. Figure 13 shows a squeegeeometer which can be used to measure pressure.

Squeegee travel is as important as squeegee pressure. Slower speeds generally require higher pressure. Many methods are used to move the squeegee across the screen, including locomotive linkage, flat disk cams, and air cylinders, which appear to be the most common. The primary objective is smooth, uniform travel. This becomes most important when uniformity is required within a lot and in print thickness, the ultimate consideration when printing resistors on a single board. The travel speed of the squeegee must be controlled. This is commonly done by means of an air-oil check system mounted in-line or parallel to the squeegee travel system.

Fig. 12 Squeegee head and pressure-control system. (*Precision Systems Co., Inc.*)

The ratio of squeegee length to screen size is one of the factors affecting printing. A rule of thumb is that the area swept by the squeegee (length times the stroke) should be less than the screen area by a 1-in. border all around.

Single-direction printing requires some method of returning the ink or paste. This is generally accomplished by means of a flood blade, although dual-headed and reservoir blades have also been employed.

Fig. 13 Squeegeeometer. (*De Haart, Inc.*)

Transfer mechanism to move substrate from load to print position By far the most popular method used to transfer the substrates to the print position is the carriage. A flat-bed carriage is used, generally traveling on parallel rods, to which the substrate-holding fixture is attached. The carriage should be heavy enough to ensure rigidity under print conditions and free from play. The most significant design feature is its ability to be repeatedly positioned under the screen. Positioning accuracies of 0.001 to 0.0005 in. are attainable.   The same mechanisms used to move the squeegee may be used to move the carriage.   The most common method is an air cylinder.

Three automatic feed systems are shown in Figs. 14 to 16.

Fig. 14   Syntron bowl substrate feed mechanism. (*De Haart, Inc.*)

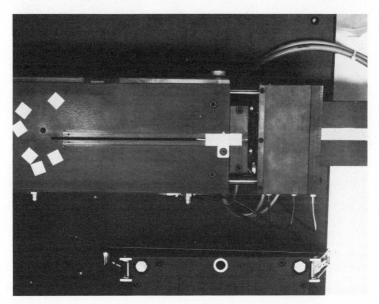

Fig. 15   In-line substrate feed mechanism. (*De Haart, Inc.*)

Adjustments   Adjustments are in *X, Y, Z,* and are required to register the substrate to the screen.   The *X, Y,* and θ (angular) adjustments are required to correct for variations in the screen pattern position with respect to the work holder in the print position.   These should be lineal to provide direct follow-through.

The Z adjustment is required in off-contact printing to set the vertical distance of the screen to the substrate.   This sets the spacing, or breakaway, distance to ensure definition and accuracy of the print. Very simple mechanisms should be used for this purpose. Micrometer dials may add accuracy and speed the setting up.

The worktable upon which the screener is to be mounted should be consistent with operator convenience and handling equipment to be used and at a comfortable height. The air should be filtered to remove dirt and moisture, and where cylinders are used, metered oil mist should be provided. An optical comparator is very useful in aligning the pattern to the substrate and monitoring the process. This should be a reflecting type of instrument.

Summary   The screen printer should be specified using the following considerations.

Design Features

1. A strong, rigid, accurately made overall framework

Fig. 16   Lineal substrate feed mechanism. (*De Haart, Inc.*)

2. A parallel relationship between squeegee guides, screen mounting, and workholder (in print position) accurate to within 0.002 in.

3. The ability to adjust the screen to workholder orientation, X, Y, and θ, plus Z, adjustment between screen and workholder, positively, with ease over a range of at least ½ in. or 6°, with positive locking action

4. A means of moving the substrate from a load position to print position under the screen repeatedly within ±0.001 in.

5. A rigid fixed screen mounting which maintains frame accuracy without stressing it

6. A squeegee design and system which assures that the attack angle of the blade remains constant, that travel across the screen is smooth and even, and that speeds can be controlled and altered to suit conditions, that the pressure is applied positively and is measurable, controllable, and mechanically limited against overtravel

Figures 17 to 25 show some common printers. Tables 6 and 7 may be useful for determining production rates of screen printing.

The two basic printing techniques in use today, *contact* and *off-contact*, are similar and require the same basic tools but differ in the breakaway distance between screen and substrate. In contact printing this distance is only 1 to 2 mils, whereas in off-contact printing it is 20 to 30 mils.

Fig. 17 Forslund printer. (*Forslund Engineering Co.*)

Fig. 18 De Haart model SP-SA-5 printer. (*De Haart, Inc.*)

Fig. 19   Presco model 330 printer. (*Precision Systems Co., Inc.*)

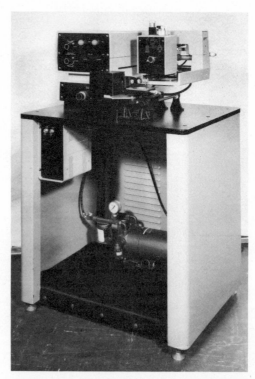

Fig. 20   Sel-Rex printer. (*Sel-Rex Electromaterials Laboratories.*)

Fig. 21   AMI automatic printer. (*Affiliated Manufacturers, Inc.*)

Fig. 22   De Haart model AOL 7HL printer. (*De Haart, Inc.*)

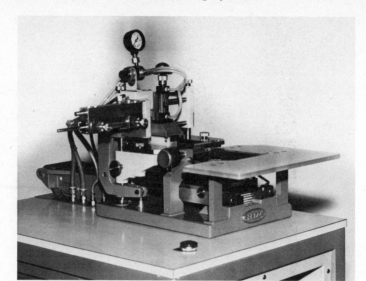

Fig. 23 Presco model 350 printer. (*Precision Systems Co., Inc.*)

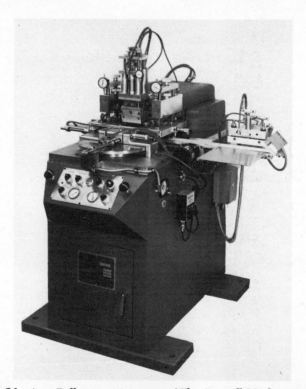

Fig. 24 Auto Roll automatic printer. (*The Autoroll Machine Corp.*)

Fig. 25 De Haart automatic printer. (*De Haart, Inc.*)

TABLE 6 Production Rates for Various Screen Printing Parameters, prints per hour per machine

Type part positioning	Type part loading	Average total machine cycle time, s	Substrate sides printed per machine cycle					
			1 up	2 up	3 up	4 up	10 up	20 up
Manual carriage.....	Hand	10	360	720	1,080	1,440	3,600	7,200
Automatic carriage..	Hand	6	600	1,200	1,800	2,400	6,000	12,000
	Automatic load	5	720	1,440	2,160	2,880	7,200	14,400
		4	900	1,800	2,700	3,600	9,000	18,000
		3	1,200	2,400	3,600	4,800	12,000	24,000
Dial..............	Automatic load	2	1,800	3,600	5,400	7,200	18,000	36,000
		1	3,600	7,200	10,800	14,400	36,000	72,000

SOURCE: Precision Systems Co., Inc., Somerville, N.J.

In contact printing, the screen is in contact with the substrate through the entire travel of the squeegee, and the printer is mechanized to produce a sharp, controlled breakaway. This results in sharper and more uniform depositions. It also allows the use of rigid metal masks, which do not have the ability to flex like the stainless-steel or Nitex° screens used in off-contact printing.

Off-contact printing can be done on a contact printer by setting the gap differently; however, contact printing cannot be done on off-contact printers since (because

TABLE 7   Production Rates for Various Loading Factors

Prints per hour per machine	Prints required per circuit per machine	Maximum production rate, finished circuits per machine per shift			
		Per day (6 h)	Per week (5 days)	Per month (21 days)	Per year (256 days)
720	1	4,320	21,600	90,720	1,105,920
	2	2,160	10,800	45,360	552,960
	3	1,440	7,200	30,240	368,640
	4	1,080	5,400	22,680	276,480
	5	864	4,320	18,144	221,184
	6	720	3,600	15,120	184,320
900	1	5,400	27,000	113,400	1,382,400
	2	2,700	13,500	56,700	691,200
	3	1,800	9,000	37,800	460,800
	4	1,350	6,750	28,350	345,600
	5	1,080	5,400	22,680	276,480
	6	900	4,500	18,900	230,400
1,200	1	7,200	36,000	151,200	1,843,200
	2	3,600	18,000	75,600	921,600
	3	2,400	12,000	50,400	614,400
	4	1,800	9,000	37,800	460,800
	5	1,440	7,200	30,240	368,640
	6	1,200	6,000	25,200	307,200
1,800	1	10,800	54,000	226,800	2,764,800
	2	5,400	27,000	113,400	1,382,400
	3	3,600	18,000	75,600	921,600
	4	2,700	13,500	56,700	691,200
	5	2,160	10,800	45,360	552,960
	6	1,800	9,000	37,800	460,800
3,600	1	21,600	108,000	453,600	5,529,600
	2	10,800	54,000	226,800	2,764,800
	3	7,200	36,000	151,200	1,843,200
	4	5,400	27,000	113,400	1,382,400
	5	4,320	21,600	90,720	1,105,920
	6	3,600	18,000	75,600	921,600

SOURCE: Precision Systems Co., Inc., Somerville, N.J.

they are not mechanized to break away after the printing cycle) smearing occurs as the substrate is removed.

The number of variables which contribute to the overall control of the process can be described by considering the six basic factors which affect printing: the squeegee, the ink or paste, the screen, the substrate, the printer, and the postprint treatment. Broken down into individual variables, these 6 basic factors can be expanded into 50 specific items.

° Trademark of E. I. du Pont de Nemours & Co.

1. The squeegee
   a. Type of material
   b. Hardness
   c. Size and shape of edge
   d. How held
   e. Method of applying pressure
   f. Amount of pressure
   g. Method of travel across the screen
   h. Uniformity of action
   i. Travel speed
   j. Ratio of size and stroke to screen size
   k. Cleanness of breakaway
   l. Orientation of pattern geometry
2. The ink
   a. Viscosity
   b. Homogeneity
   c. Composition
3. The screen
   a. Mesh count
   b. Wire diameter
   c. Material
   d. Type of coating
   e. Thickness of coating
   f. Accuracy of pattern
   g. Alignment of pattern to mesh
   h. Flatness
   i. Tension of mesh
   j. Frame type
   k. Method of mounting mesh to frame
   l. Method of supporting frame for printing

3. The mask (alternate)
   a. Etched mesh count
   b. Mask thickness
   c. Material
   d. Method of support
   e. Elasticity
   f. Flatness
   g. Method of supporting frame for printing

4. The substrate
   a. Material
   b. Surface finish
   c. Flatness or camber
   d. Dimensional tolerance
5. The printer
   a. Breakaway
   b. Rigidity
   c. Workholding
   d. Registration
   e. Reproducibility
   f. Production rate
6. Postprint treatment
   a. Settling
   b. Drying
   c. Firing
   d. Adjusting

The list appears awesome. Some of the variables are easy to control, while others are nebulous. The interrelation cannot be discounted, and even though each variable is not necessarily completely understood, the right type of tooling and screen printer, coupled with adequate process control, allows remarkable results and reproducibility to be achieved.

The manufacturing procedure for screen printing can be outlined as follows:

1. The specified screen is selected and attached to the screen printer.
2. The printer is then set up and aligned as directed.
3. The specified substrate is placed on the substrate holder of the printer.
4. The print cycle is actuated.
5. The printed substrate is removed and placed on a firing boat.
6. The print is dried a minimum of 10 min at room ambient.
7. The printed substrates are then placed in an oven and dried at 125°C for approximately 5 min.
8. The parts are then removed and forwarded to the firing area.

During the setup cycle it is an excellent idea to run a preproduction lot of approximately 10 to 25 pieces to assure that the design and resistivity are correct. All data should be recorded and kept for future use. During the production phase, samples should be tested periodically to assure that there has been no change in the process. This information should also be recorded. These data will result in an excellent monitoring and control of the process and generate very useful design and reliability data.

Manufacturing Procedures

The easiest way to detail the manufacturing process for screen printing is to present a typical manufacturing specification and process flow chart. It should be noted that this is a typical specification, not intended to imply that this is the only process used.

Process Specification

1.0 Title
    1.1   Substrate screening procedure
2.0 Scope
    2.1   The following is the detailed procedure to be followed when screening substrates for use in the fabrication of thick film hybrid circuits. Refer to flow chart FC-XXX.
3.0 Substrate cleaning procedure
    3.1   The substrates shall be placed loosely in a specially designed basket (see drawing #SK-XXX) and lowered *slowly* into the vapor side of the ultrasonic degreaser. They shall be kept in the vapor approximately 1 min and then transferred to the liquid side. The generator shall be turned off and the parts returned to the vapor side and withdrawn slowly.
    3.2   The substrates shall then be soaked in warm denatured alcohol for 15 min. The alcohol shall be poured off and replaced with new material and the process repeated.
    3.3   The substrates shall then be transferred to a beaker and rinsed under a flow of deionized water for 30 to 60 s.
    3.4   The substrates shall then be placed in a suitable container and dried for a minimum of 30 min at 150°C.
4.0 Setup procedure
    4.1   The process lot card must be obtained from the supervisor and must accompany each lot through the process. Each operator shall date and initial the card after accomplishing the designated task. The designated screen shall be obtained from the screen rack, weighed, and recorded on the paste-usage form. The screen shall then be mounted to the screen holder and loosely secured to the screen printer. It shall be aligned roughly by sighting through the screen onto the substrate.
    4.2   The specified paste shall be drawn from the storage cabinet. Jar weight shall be recorded on the paste-usage form. Approximately ½ oz of paste shall be applied to the screen. (*Too little paste will cause irregular printing and result in skips and voids.*)
    4.3   The paste shall be stirred thoroughly with a stainless-steel spatula for approximately 1 min or until a smooth mixture is obtained. There shall be no sediment or deposit on the bottom of the jar. The viscosity shall be checked using the Brookfield viscometer and adjusted, if necessary, to 100,000 cP, using butyl carbitol acetate. Resistor pastes shall be run on the three-roll mill two to three cycles if they have stood unagitated for any appreciable length of time (2 to 3 weeks).
    4.4   The machine shall be cycled and final alignment of pattern accomplished by printing and visually checking the pattern on the substrate with an optical comparator and/or microscope. Small rectangular or TO-5 substrates may best be aligned by measuring and setting the indexes as noted on the detail part drawing.
    4.5   Before proceeding with the lot, the supervisor shall be called to verify the alignment and the uniformity of the deposition.
5.0 Normal operation
    5.1   Screening and firing sequence
        5.1.1   Conductors
            a. Gold side 1 (dry)
            b. Pt-Au side 1 (dry and fire)
            c. Pt-Au side 2 (dry)
            d. Hole-throughs (dry and fire)
        5.1.2   Resistor tests
            5.1.2.1   Resistor tests are used to choose the paste or pastes required for the production lot. They are necessary and should not be bypassed unless specifically allowed.

5.1.2.2   Resistor tests shall consist of screening five parts with the paste nearest to that specified on the lot card, five plates with the value above that specified, and five plates with the value below that specified. If more than one resistor paste is required on the substrate, all are to be screened in the sequence to be followed during production. NOTE: When two-sided screening of resistors is necessary, it will be found that the second firing will affect the resistance and compensation will be necessary.

5.1.2.3   After firing the tests, the resistance measurements shall be read and recorded and the paste which is 70 to 90 percent of design value used. If the pastes checked do not fall within the range, a new test shall be run before attempting production.

5.1.3   Hole-throughs

5.1.3.1   If the part requires hole-throughs, a continuity test will be made and the parts repaired prior to proceeding further.

5.1.4   Resistors

  a. Lowest value side 1 or 2 (dry; fire if only 1)
  b. Resistor 2, side 1 or 2 (dry and fire)
  c. Resistor 3 (dry and fire) NOTE: All resistors on one side will be screened, dried and fired before starting the other side.
  d. A sample of 10 substrates shall be read and recorded from each lot at $\frac{1}{2}$-h intervals through the firing process. NOTE: Backside resistors shall be fired on edge using suitable carriers.

5.1.5   Dielectrics

  a. Side 1 (dry and fire)
  b. Side 2 (dry and fire)

It should be noted that it is impossible to separate the firing operations from the screening operations since the process flow must be continuous and the various materials fire at successively lower temperatures.

At this point the substrates go through a quick visual and electrical inspection before being transferred to the resistor-adjusting area. The substrates should be kept in a convenient storage area in such a way that pattern damage such as scratches and nicks will be prevented.

Manufacturers of thick film printing equipment include:

Affiliated Manufacturers, Inc.
Box 248
Whitehouse, N.J. 08888

Aremco Products, Inc.
P.O. Box 145
Briarcliff Manor, N.Y. 10510

Castle Rubber Co.
Butler, Pa. 16001

De Haart, Inc.
Burlington, Mass. 10803

Engineered Technical Products
Somerville, N.J. 08876

Forslund Engineering Co.
Petaluma, Calif. 94952

Pelmor Laboratories, Inc.
Newtown, Pa. 18940

Joseph E. Podgor Co.
Pennsauken, N.J. 08109

Precision Machine & Development Co.
New Castle, Del. 19720

Precision Systems Co., Inc. (PRESCO)
Box 148
Somerville, N.J. 08876

Wells Electronics, Inc.
Weltek Division
South Bend, Ind. 46613

Western States Autoroll Corp.
Woodland Hills, Calif. 91364

FIRING

The thick film process consists of screening the various elements on a ceramic substrate and then firing these elements at elevated temperatures. The furnace, or kiln, used in the firing operation must be specified with the same care as the screen printer. The firing of thick film circuitry requires careful control over temperature and time. Once the firing profile has been established for a certain material, it must be maintained not only during the specific firing but from day to day.

The firing of thick film circuits is a rather complex process in which the organic binders and solvents burn out in the first phase, the metallic elements are either oxidized or reduced to develop the required characteristics of resistivity and temperature and voltage coefficients, and then a sintering of the colloidal glass materials occurs to anchor the film to the substrate and protect the metallic elements. Controlled cooling to room ambient temperature follows.

After the film has been printed on a substrate, it should be allowed to "settle" since printing through the screen tends to produce mesh lines in the pattern. The settling time varies with each material and depends on the viscosity of the system; however, times vary from about 5 min to approximately 20 min. This should be determined experimentally for each process.

Once a settling time has been established, the films should be dried before firing. This removes the only volatiles, and the binders remain. Desirable drying temperatures depend on the solvents used and range from 100 to 150°C for 5 to 15 min. The resulting films should be tough enough to permit handling and subsequent printings.

Several methods have been used for drying the films, including hot plates, ovens, heat guns, and infrared lamps. Many users feel that infrared energy with a wavelength longer than 3 μm is most suitable. This energy penetrates the films and permits drying without forming a crust over the surface, entrapping the vehicle, which would subsequently blister during firing to form voids, etc.

Major Equipment Considerations

Thick films have been fired in a variety of chambers. However, in order to achieve the control required, an electric continuous-belt furnace is necessary. Many considerations enter into the specifications of such a furnace.

Electric furnaces are built in a variety of ways but share common features. For operating temperatures of 1000 to 1100°C, the heating elements consist of resistance wire or ribbon of Nichrome* or Kanthal.† The heater elements, supported by ceramic tiles, are placed on the sides of the chamber or sometimes completely around the chamber. Since the elements and the tiles emit contaminants in the form of gases and fine dust which could affect thick film circuits, it is important to maintain a clean atmosphere around the substrates.

The most convenient way to accomplish this is to use a muffle, usually of ceramic, metal, or quartz. Ceramic muffles are generally constructed of high alumina, ranging in thickness from ½ to 1 in. depending on size. They have a high thermal lag and may be quite load-dependent if the product moves rapidly through the kiln. The metal muffles are generally ⅛ to ¼ in. thick and are common for reducing or inert atmospheres, but in oxidizing atmospheres they tend to form oxides, which may scale and contaminate the product.

Fused-quartz muffles are used in oxidizing atmospheres below 1200°C. They range in thickness from 3/16 to ¾ in., and since they are "transparent" to heat do not exhibit the thermal lag observed in ceramic muffles.

The kiln should have muffles to provide a clean atmosphere. The muffles must be kept purged of all contaminants which may result from the firing cycle itself. This must be done by continuously flushing the contaminants from the muffles by a counterflow so that clean gas is drawn over the parts in the hotter zone. Two methods are used to flush the kiln: (1) forced air with adjustable baffles and flue vents and (2) an inclined muffle with a tilt of approximately 2° which induces flow due to a chimney effect.

In the first method, care must be taken to produce a fairly laminar nonturbulent flow. The tilted muffle automatically produces this effect. If some turbulence must be present, it can be induced and controlled.

Profiling Procedures and Equipment

The temperature in the furnace is controlled by thermocouples, which must be strategically located. They must respond quickly to load changes and be kept in precise position. The most convenient circuitry utilizes silicon-controlled rectifiers with proportional power controllers. A good system should maintain the same tem-

---

* Trademark of Driver-Harris Co.
† Trademark of Kanthal Corp.

perature profile regardless of load. It is common procedure, however, to send dummy substrates ahead of the parts to be fired. This acts to stabilize the furnace by forcing any conditions within the furnace to react to a load before introducing the actual work.

In order to set and maintain accurate profiles, multizone furnaces are recom-

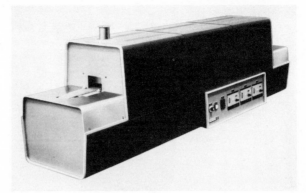

Fig. 26 Lindberg Hevi-Duty furnace. (*Lindberg Hevi-Duty Division, Sola Basic Industries.*)

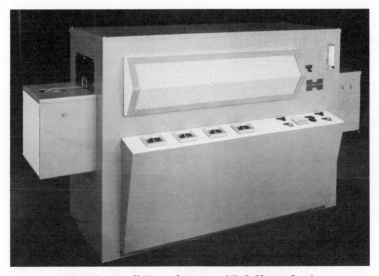

Fig. 27 Small Hayes furnace. (*C. I. Hayes, Inc.*)

mended, each zone with its own controllers. Three- to six-zone furnaces are common in the thick film industry, with more adventurous experimenters increasing this number significantly. Figures 26 to 34 show typical thick film multizoned furnaces. Economics generally dictates the best approach consistent with the size of the kiln itself. Four- and six-zone furnaces are most common.

Temperature should be controlled to better than ±3° across the belt, the material of which should be chosen carefully, depending on the operating temperature

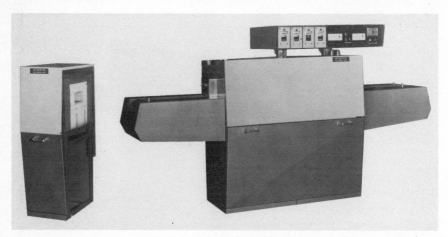

Fig. 28   Small Keith furnace. (*W. P. Keith Co., Inc.*)

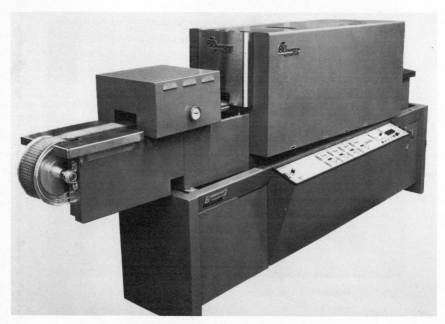

Fig. 29   Medium BTU furnace. (*BTU Engineering Corp.*)

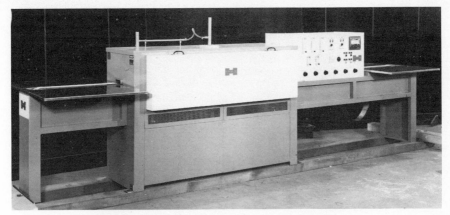

Fig. 30   Medium Hayes furnace. (*C. I. Hayes, Inc.*)

Fig. 31 Large Linberg Hevi–Duty furnace. (*Linberg Hevi-Duty Division, Sola Basic Industries.*)

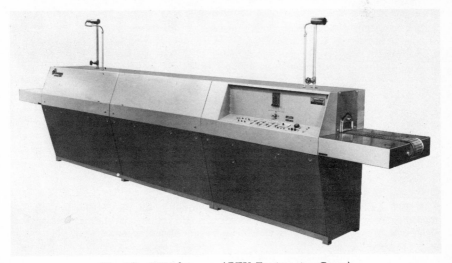

Fig. 32   BTU furnace. (*BTU Engineering Corp.*)

and load. Common materials are Inconel,* Nichrome V,† and stainless steel. The weave of the belt depends on the load and the drive mechanism.

The belt-drive mechanism should be such that the speed is adequately and positively maintained. Once the firing cycle has been set, it must be repeated accurately for reproducibility from batch to batch.

Fig. 33 Keith furnace. (*W. P. Keith Co., Inc.*)

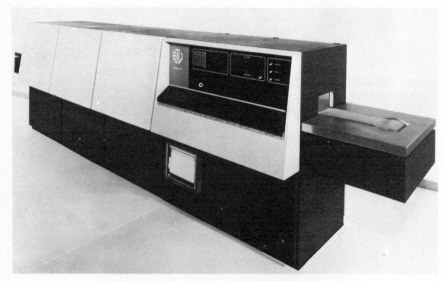

Fig. 34 BTU furnace. (*BTU Engineering Corp.*)

A good set of profiling thermocouples is required to set the profiles initially and then monitor them periodically. Although a single couple may be used, it is more significant to use three couples attached to a multipoint recorder so as to measure

* Trademark of International Nickel Co.
† Trademark of Driver-Harris Co.

the center and two sides of the belt. In this way not only the shape of the profile but the variation across the belt can be measured.

Since heat is generated from the kilns, it is generally desirable to have them in a separate room with window pass-throughs for substrates before and after firing.

When it comes to the atmosphere in which thick films should be fired, there are probably as many answers as there are systems to be fired. No doubt some materials are more sensitive than others. Air is considered to be a good atmosphere for firing thick film circuits. For most systems it is more than adequate. For other systems, oxygen-rich atmospheres with controlled dew points are required.

Reducing atmosphere should never be used because the films will be destroyed, since all are basically oxides, and reducing atmospheres will result in powdery metallic films.

The shape of the profile is very dependent on the process and the materials being used. There is no universal profile to be used. Some common points are that the burnout phase should be at a rate not to exceed $200°C/min$, to avoid trapping carbon particles, reduce bubbling, and reduce cracking of the films. Once an optimum temperature has been established, it should be reproduced. Very significant changes can be observed in sheet resistivity of certain resistor materials with a change in temperature. Once all the reactions have taken place, the substrate can be cooled quite rapidly, consistent with properly stress-relieving the ceramic to avoid thermal fracturing.

Major manufacturers of firing equipment include:

BTU Engineering Corp.
Waltham, Mass. 02154

C. I. Hayes, Inc.
Cranston, R.I. 02910

W. P. Keith Co., Inc.
Pico Rivera, Calif. 90660

Lindberg Hevi-Duty Division
Sola Basic Industries
Chicago, Ill. 60612

Thermco Products Corp.
Orange, Calif. 92668

Trent, Inc.
Philadelphia, Pa. 19127

Watkins Johnson Co.
Stewart Division
Scotts Valley
Santa Cruz, Calif. 95060

RESISTOR ADJUSTING AND TESTING

Once all the elements have been screen-printed onto the substrate to achieve the desired configuration, methods of adjusting the resistors to closer tolerance and monitoring the entire process must be provided.

Since it is not often economical to control all the variables in the screening and firing process to maintain resistor tolerances, an adequate method of tailoring the resistor elements must be available. The method must be fast, convenient, and economical. It should be capable of achieving tolerances of $±0.1$ percent of reproducibility.

There are several methods of adjusting resistors. Some manufacturers refire, based on the initial results, to modify the observed values. Resistor values have been altered using spark-gap trimming, high-power pulsing, electron beam, rf pulsing, laser trimming, hot-gas oxidation, electrochemical oxidation, and several other techniques.

The most common method remains the air-abrasive method (see Figs. 35 to 39), in which the length-to-width ratio is altered by removing material to reduce the width of the element. This method is positive and easily controlled, and since it removes material, it does not change the bulk characteristics of the resistor. There is no heat involved to stress the ceramic thermally, and the part can be cleaned after trimming to remove any contaminants.

Air-abrasive trimming is a high-accuracy method of resistor adjustment because a system can be built which incorporates the decision-making function coupled with a specially designed reservoir and feed mechanism to minimize the processing variation. The decision-making function monitors the resistor continuously from the start of the trim cycle to shutoff and then automatically tests the finished part and accepts or rejects it. The reservoir and feed system maintains the particle content of the air-abrasive stream constant.

Laser techniques (Figs. 40 and 41) require higher production rates to justify their higher cost. These techniques lend themselves to greater automation since

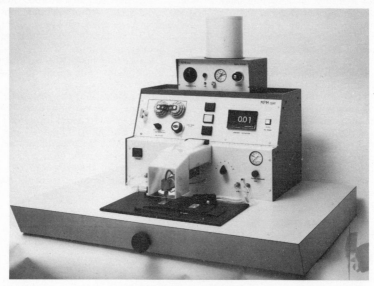

Fig. 35 Trimmer. (*MPM Corp.*)

Fig. 36 Trimmer. (*Comco Supply, Inc.*)

they can be coupled with automatic feed systems and computer-controlled. The process tends to heat the part, and care must be taken that the temperature coefficients do not affect the final acceptance check values. Another prime advantage of the laser is in functional or active trimming (see Fig. 42). Since the method consists of burning material away, circuits with active components can be adjusted without having to protect the semiconductors with gel coatings.

Air-abrasive trimming has two major limitations: (1) cuts cannot be less than 0.005 to 0.010 in., and (2) the alumina powder must be removed during the trimming process by means of a dust collector.

A schematic of the basic air-abrasive unit is shown in Fig. 43. The basic unit

Fig. 37   Trimmer. (*De Haart, Inc.*)

Fig. 38   Trimmer. (*S. S. White Co., Division of Pennwalt Corp.*)

consists of an air supply, a powder reservoir with a vibrator, and a mixing chamber and holder fitted with an appropriate nozzle. When the air-in solenoid is energized, the system becomes pressurized. Next, a single electric switch opens the pinch valve ahead of the nozzle and at the same time causes the vibrator to pulse at 3,600 c/m.

This vibration trickles powder into the airstream. The amount depends on the vibration amplitude. A rheostat regulates the amplitude and thereby the intensity of the cutting action.

The resistor trimming system is achieved by adding a monitoring and measuring system to the basic unit. A controlled feed mechanism and a quick-acting stop

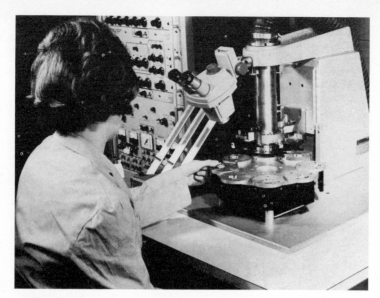

Fig. 39 Trimmer. (*S. S. White Co., Division of Pennwalt Corp.*)

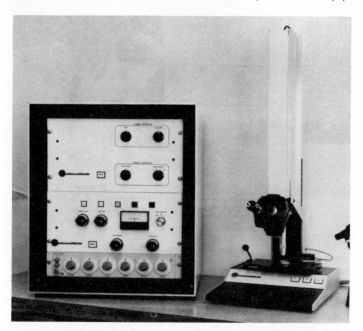

Fig. 40 Laser trimmer (*Apollo Lasers, Inc.*)

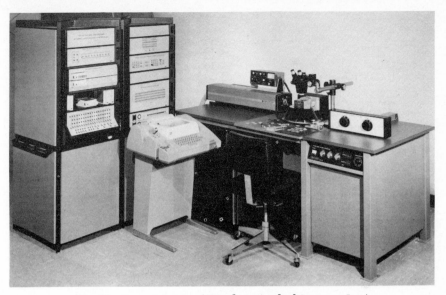

Fig. 41   Laser trimmer. (*Teradyne Applied Systems, Inc.*)

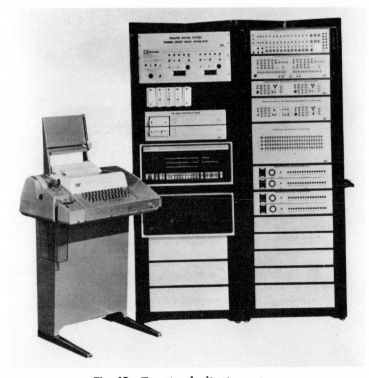

Fig. 42   Functional adjusting system.

action for the nozzles completes the system. The operator places the part to be adjusted on the workholder and actuates the cycle, and the machine takes over.

The laser system replaces the air-abrasive unit with a laser micromachining tool. Both Q-switched yttrium-aluminum-garnet (YAG) and continuous carbon dioxide, CO_2, laser beads are used. The YAG laser has certain advantages, but the CO_2 system is cheaper and has proved satisfactory in many cases.

The advantages of the YAG laser include a higher-speed solid-state Q switch which allows faster trimming. The Q-switched YAG pulse has single pulse control versus the two- to three-pulse overshoot of the CO_2 laser. This allows more accuracy at higher speeds. The YAG allows better material-removal rate and is capable of a 0.001-in. spot size. The laser allows extremely rapid rates (4 in./min).

Laser manufacturers suggest that the air-abrasive method has the following disadvantages:

1. Abrasive trimming leaves an exposed edge on the resistor, making it more susceptible to environmental contamination than the original outer surface.

2. The method has a relatively slow material-removal rate which limits trimming speed.

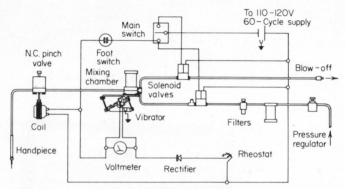

Fig. 43 Schematic of air-abrasive system. (*S. S. White Co., Division of Pennwalt Corp.*)

3. The use of abrasive particles in an airstream requires a mechanical shutoff system which inherently has a slow actuation time. This long shutoff time, coupled with bounce, precludes accurate trims at high material-removal rates because of overshoot.

4. Abrasive trimming using commercially available hardware is relatively expensive due to down time for probe and nozzle setup, nozzle replacement, and the multiple loading and unloading of multiresistor patterns.

5. Air-abrasive trimming is not a clean process either from an environmental or a circuit standpoint.

In contrast, they offer the following advantages for laser trimming:

1. Laser trimming cuts a path through the resistor by vaporizing the material under the focused laser beam. Material melts only at the focused spot and reseals the trimmed resistor as it solidifies.

2. The high-power high-repetition-rate YAG and CO_2 lasers offer high material-removal rates, making the speed of the XY table and the bridge response time the limiting factors.

3. The use of laser trimming provides direct electronic control of the beam turnoff, giving microsecond rather than millisecond response times. This allows maximum use of the trimming speed without losing final-tolerance accuracy.

4. Laser trimming is actually less expensive since the operator spends virtually

all his time actually trimming resistors rather than in nontrimming maintenance and setup.

5. Laser trimming is an extremely clean process. It allows trimming to be performed immediately adjacent to active devices without any damaging effects.

With all the apparent advantages, the fact remains that the number of laser systems in use is a distinct minority. Where they do exist they are generally used primarily for dynamic trimming. Laser manufacturers feel that this condition is due to lack of planning of the total system prior to installation.

The initial cost of a laser trimming system is approximately an order of magnitude greater than the equivalent air-abrasive system. This fact certainly makes it unattractive to the small manufacturer or the prototype facility.

There is no doubt that the laser system is essential for certain applications, including the fabrication of extremely close tolerance networks and the production of functionally adjusted hybrid circuits. In these cases the advantages are immediate, and the equipment has a very fast payoff time.

Factors that affect trimming in the air-abrasive unit include the type of abrasive power, gas pressure, abrasive-particle flow, nozzle speed, nozzle configuration, and hardness and thickness of the resistor material.

Most of these can be made constant, leaving only the nozzle speed as a variable. Resistors ranging in value from 10 Ω to over 1 MΩ have been successfully trimmed to ± 0.1 percent tolerance. This has been accomplished using the following operating conditions:

1. No. 1 powder (27-μm aluminum oxide powder)
2. A filtered plant air supply at about 85 psi
3. An abrasive flow of 3 to 5 g/min
4. A trimming speed of $\frac{1}{2}$ to $1\frac{1}{2}$ in./min

Settings of 40 to 80 psi have been found most suitable in removing material without attacking the underlying substrate. The trimming speed is the most critical parameter and must be controlled to minimize overshoot.

Accuracy is traded for faster trimming speeds. At a nozzle speed of 1 in./min accuracies of ± 0.1 percent can be maintained. At nozzle speeds of 6 in./min, one is lucky to maintain ± 0.5 percent. For looser tolerances, this can be used to advantage.

Material-removal rates are highest when the nozzle width is almost the same as the length of the resistor path. Generally, a narrow cut using a round nozzle produces the most stable resistor with the lowest noise. It requires the most time. Wide cuts approaching the full length of the resistor introduce noise which can be reduced by reencapsulation.

Overshoot has been mentioned several times. Usually it is constant for a particular batch of parts and can be anticipated, and therefore allowances can be made for it.

All these facts make the air-abrasive method the most widely used method of adjusting resistors, and the economics dictates this approach for most users.

Systems have been assembled using multiple heads and multiple electronics as well as automatic feeding equipment for volume production.

Capacitors can be trimmed to a lower value using air-abrasive techniques. This is accomplished by removing electrode area. The air-abrasive tool is used to cut through the dielectric material and through the capacitor electrodes.

Once thick film elements have been trimmed and the substrates are ready for further processing, process monitoring and control should be considered. Every facility must have resistance- and capacitance-measuring equipment, which can be used for monitoring as well as final inspection.

A good digital voltmeter with an ohm scale is an invaluable tool. Some capacitance meters have voltage outputs which can be fed into the digital voltmeter to produce a digital capacitance bridge at reasonable cost.

Automatic resistor and resistor-capacitor testers can be coupled to probe stations to test several or all of the thick film elements at one time. This eliminates the need to probe each element separately and speeds the process up considerably.

Manufacturing Procedures

As in the substrate-screening procedures, the easiest way to detail the resistor-adjusting process is to present a typical manufacturing specification, subject to the same qualifications stated for the screening process.

1.0   Title
    1.1   Resistor testing and adjusting procedure
2.0   Scope
    2.1   The following is a detailed procedure to be followed when testing and adjusting resistors on substrates for use in the fabrication of thick film hybrid circuits.
3.0   Adjusting of resistors
    3.1   Testing and adjusting of thick film resistors is done to correct the resistance of the fired resistor to closer tolerance. This procedure consists of abrading the resistor to change its length-to-width ratio and thereby increase the value of the resistor to the specified value and tolerance. The procedure consists of the following steps.
        3.1.1   Setup (mechanical)
            3.1.1.1   The proper index plate shall be affixed to the trimmer and a sample substrate placed in proper position.
            3.1.1.2   Using the XY mechanism, the substrate shall be positioned under the nozzle.
            3.1.1.3   The nozzle which best conforms to the geometry of the resistor shall be selected. The rule shall be to make the widest cut possible.
            3.1.1.4   The resistor element to be trimmed shall be positioned so that the cut is started 5 to 10 mils ahead of it.
            3.1.1.5   The measurement electrodes shall then be positioned to make contact in the proper specified locations. NOTE: Care shall be exercised to avoid touching the nozzle with the electrodes before, during, or after the trimming cycle.
        3.1.2   Setup (electrical)
            3.1.2.1   The final value shall be dialed into the top row of decade switches marked *final value.*
            3.1.2.2   A percentage of the final value shall be dialed into the bottom row of decade switches marked *minimum start value.* This is the value below which trimming is not desirable and is 50 percent for resistors rated above 500 mW, 60 percent for those rated 100 to 500 mW, and 75 percent for values rated below 100 mW.
            3.1.2.3   The *multiplier* which gives the best sensitivity shall then be dialed in.
            3.1.2.4   The *stop value* shall then be set so that the resistor consistently falls between ± 2 percent for 10 pieces in a row or the resistor tolerance for resistors ± 1 percent below.
        3.1.3   Setup (speed)
            3.1.3.1   The speed of nozzle travel shall be set to the fastest speed necessary to obtain the most consistent cut.
4.0   Cleaning
    4.1   After resistor adjusting, the parts shall be cleaned by a suitable method, approved by Engineering, to remove any excess material or abrasive powder remaining on the substrate.

The reader will notice that the above "specification" is based upon an air-abrasive method of resistor adjustment. This is not meant to imply that laser trimming cannot be used. The discussion and presentation is only to show a typical operation, and methods can be substituted without changing the general procedure.

It might be useful at this point to mention a few techniques which have been shown to be effective in achieving tighter tolerances. The considerations noted here will be limited to the trimming process rather than design, which is covered elsewhere.

One method of achieving a closer tolerance is to make multiple passes at the resistor. The first cut is made to a nominal 2 to 5 percent, and the subsequent cut, which has less effect, can be controlled more closely to achieve a tight tolerance.

A second method is to use an L-shaped cut. The forward cut is made until approximately 2 to 5 percent tolerance is reached and then the cut is angled to run lengthwise with the resistor, again affecting the rate at which the resistor value is altered and allowing much tighter tolerances to be achieved.

Major manufacturers of resistor adjusting equipment include:

Apollo Lasers, Inc.
Los Angeles, Calif. 90045

Arvin Systems, Inc.
Dayton, Ohio 45404

Comco Supply, Inc.
Burbank, Calif. 91506

De Haart, Inc.
Burlington, Mass. 01803

Electro Scientific Industries (ESI)
Portland, Ore. 97229

Hughes Aircraft Co.
Torrance, Calif. 90509

Korad
Subsidiary of Union Carbide Corp.
Santa Monica, Calif. 90406

Micronetics, Inc.
Watertown, Mass. 02172

MPM Corp.
Cambridge, Mass. 02140

Spacerays, Inc.
Northwest Industrial Park
Burlington, Mass. 01803

Teradyne Applied Systems, Inc.
Chicago, Ill. 60634

TRW/Instruments
El Segundo, Calif. 90245

S. S. White Co.
Division of Pennwalt Corp.
New York, 10017

INSPECTION OF COMPLETED SUBSTRATES

Final electrical and visual inspection is a precaution to ensure that rejectable material does not filter into the assembly area inadvertently. This inspection can be accomplished satisfactorily using available equipment.

The following outline of the workmanship and quality standards generally used for a thick film printed and adjusted substrate is in specification form.

1.0 Title
    1.1 Inspection criteria for printed and adjusted thick film substrates
2.0 Scope
    2.1 The following is a detailed procedure including acceptance and rejection criteria for use while inspecting completed thick film substrates.
3.0 Visual inspection procedures and criteria
    3.1 Each lot of substrates shall be visually inspected using a microscope at 30X magnification. Both front and back illumination shall be used as required. Refer to visual standards prepared by Engineering.
    3.2 The distance from the edge of the substrate and any circuit element (conductor or resistor) shall be 0.005 in. minimum unless otherwise noted on the detail part specification.
    3.3 The alignment of one pattern to another pattern, such as conductor overlap, shall be at least 50 percent of design width.
    3.4 The resistor-conductor overlap must be continuous and at least 0.002 in. minimum.
    3.5 The average width of conductive paths shall be 80 percent minimum and 150 percent maximum of the design dimensions. In no case shall the element be reduced less than 50 percent of design or a minimum of 0.005 in.
    3.6 Inclusions or trapped foreign material shall be cause for rejection if they cause the element not to conform to paragraph 3.5.
    3.7 The cut end of the resistor shall be clean and not show signs of feathering or shadowing exceeding 10 percent of the trim depth or 0.010 in. maximum.
    3.8 Damage to any adjacent element due to trimming shall be cause for rejection if it fails paragraph 3.5.
    3.9 The resistor cut shall be started on one end, and in no case shall the element be less than 0.008 in. This criterion shall also apply if the resistor is inadvertently cut inside either edge.
    3.10 The maximum cut allowed through resistors shall be:
        3.10.1 75 percent of design width but not less than 0.008 in. for resistors rated less than 100 mW.
        3.10.2 60 percent of design width for 100 to 500 mW.
        3.10.3 50 percent of design width for resistors rated over 500 mW.
    3.11 Scratches and/or voids shall be cause for rejection if they exceed 50 percent of design width. In all cases a minimum of 0.005 in. of unexposed ceramic shall be required. There shall be no excessive buildup of material at the end of the scratch.
    3.12 All die-attaching and wire-bonding areas shall be continuous and thick enough to allow proper attaching and bonding.

3.13 There shall be no evidence of chips or cracks in the ceramic substrate.

3.14 The minimum spacing between circuit elements shall be 50 percent of design dimension. In no case shall the spacing be less than 0.002 in.

3.15 Any hole in the substrate which accommodates a lead wire shall be free of foreign material and/or excess conductor material.

3.16 When required, the dielectric covercoat of resistor or conductor shall be continuous and meet all dimensional requirements outlined above.

4.0 Electrical inspection procedures and criteria

    4.1 The substrates shall be checked electrically according to the following sampling plan:

        4.1.1 For resistor tolerances greater than ± 3 percent a lot tolerance percent defective (LTPD) of 10 percent with a minimum acceptance number of 2 shall be used.

    4.2 Temperature coefficient of resistance (TCR) shall be checked on each lot as follows:

        4.2.1 A sample of three pieces shall be taken from the lot, and resistor values shall be measured and recorded at room temperature.

        4.2.2 The parts shall be placed on a hot plate set at $125 \pm 5°C$ and allowed to stabilize.

        4.2.3 The resistor value shall be read and recorded.

        4.2.4 The TCR shall be calculated using the following equation:

$$\text{TCR} = \frac{10^6 (R_t - R)}{R(T - T_{rt})} \quad \text{ppm/}°C$$

where R_t = resistance at $125°C$, Ω

      R = resistance at room temperature, Ω

      T = temperature $(125°C)$

      T_{rt} = room temperature

        4.2.5 The TCR shall be within specified tolerance.

    4.3 Any evidence of peeling, blistering, or poor adhesion shall be cause for rejection.

5.0 Transfer to production

    5.1 After the substrates have been inspected and the lot traveler signed and stamped, the lot shall be transferred to production or storage as designated.

All the above can be condensed to suit less elaborate needs into the following statements:

A. Electrical. All resistors shall be measured on a go no-go basis. Low resistors shall be returned for rework. The TCR shall be checked on a sample basis using a hot plate at $125°C$.

B. Visual. All substrates shall be inspected at $30\times$ using visual standards as a guide for acceptance or rejection. Areas of concern are:

1. Major
      a. Indexing and pattern definition
      b. Skips and voids exceeding 25 percent of the element width
      c. Inclusions exceeding 10 percent of the element width
      d. Shadowing of resistor cut exceeding 10 mils or 10 percent, whichever is smaller
      e. Resistors overabraded (maximum cut allowed):
         (1) Less than 100 mW: 75 percent of design width, not less than 8 mils
         (2) 100 to 500 mW: 60 percent of design width
         (3) Over 500 mW: 50 percent of design width
      f. Damage to any other circuit element
      g. Spacing between circuit elements less than 3 mils
2. Minor
      a. Bubbles in the glass
      b. Excess material
      c. Unabraded resistor (check value)
      d. Scratches in the glass
      e. Electrode material spilling over edge
      f. "Dirty" substrates
         Major defects are cause for rejection. Minor defects should be reported to the production foreman for corrective action.

It can be seen that this is a much more concise statement of criteria. It is very effective in any case.

COST CONSIDERATIONS AND METHODS

Once the facility has been properly specified and the equipment and the process are operating successfully, the next most important area of concern is cost. The costing of the operation is important from two main viewpoints. First it is important to know costs so as to maintain a production operation in control. Second, it is just as important to have adequate cost procedures in order for the sales and marketing people to use in pricing both standard and custom products.

A method which has been used quite successfully considers the key elements of the various steps of the process as follows:

Operation	Side	Rate/h	Hours per 1,000 resistors	Yield, %
Clean substrate.................	—	—	0.1	100
Screen conductor 1............	1	300	3.3	99
Screen conductor 2............	1	300	3.3	99
Fire conductor.................	1	*	—	100
Screen conductor 3.............	2	300	3.3	99
Fire conductor.................	2	*	—	100
Screen resistor 1...............	1	300	3.3	95
Screen resistor 2...............	1	300	3.3	90
Screen resistor 3...............	1	300	3.3	80
Fire resistors..................	1	*	—	100
Screen resistor 1...............	2	300	3.3	95
Screen resistor 2...............	2	300	3.3	90
Screen resistor 3...............	2	300	3.3	80
Fire resistors..................	2	*	—	100
Screen glass...................	1	*	—	98
Fire glass.....................	1	*	—	100
Screen glass...................	2	300	3.3	98
Fire glass.....................	2	*	—	100

* Depends on size of parts and furnace calculated on belt speed and part size to equal loading and unloading time only.

Using the above information, charts can be prepared, classified into different categories such as:

CASE 1: Single-sided screening without protective glass
      a. One resistor screening
      b. Two resistor screenings
      c. Three resistor screenings
CASE 2: Single-sided screening with protective glass
CASE 3: Double-sided screening with or without glass

Also, using this table, the data can be amplified to include multilayer work or any other variation. Once the charts are established and modified to accommodate the specific manufacturer's labor rates, the data might take the form:

CASE 1: Single-sided Screening without Glass

	No. of resistor screenings		
	1	2	3
Yield, %................................	90	81	65
Screen rate, h per 1,000 resistors............	11	14	18

CASE 2: Single-sided Screening with Protective Glass

	No. of resistor screenings		
	1	2	3
Yield, %................................	88	79	64
Screen rate, h per 1,000 resistors............	15	18	21

Pricing can be accomplished relatively easily using the following formula.

> *Substrate Screening and Firing*
> Direct labor
>   Screening_____ hours at $_____/hour_____
>   Firing _____ hours at $_____/hour_____
> Total direct labor
> Overhead at __%   _____
> Raw materials:
>   Conductor and resistor inks _____
>   Substrate ____ × ____ in. ____
> Substrate manufacturing cost
> Substrate cost/overall yield __% × __%

The same treatment can be made for the trimming process. This operation is not as complex as the screening since it is a single process operation. Assuming 250 passes per hour, 4 h per 1,000 resistors is necessary. For a yield estimate assume 0.5 percent per resistor less for 5 percent resistors, 1 percent per resistor for 2 percent and 2 percent per resistor for 1 percent resistors.

Developing a table similar to that for screen printing yields the following:

Number of resistors	1	2	3	4	5	6	7	8	9	10	11	12	13	14	15	16	17	18	19	20
Yield at 5/10%	98	98	98	98	97	97	96	96	95	95	94	94	93	93	92	92	91	91	90	90
At 2%	97	97	97	96	95	94	93	92	91	90	89	88	87	86	85	84	84	83	82	81
At 1%	94	94	94	92	90	88	86	85	83	81	80	78	77	75	74	72	71	69	68	67
Hours per 1,000 resistors	4	8	12	16	20	24	28	32	36	40	44	48	52	56	60	64	68	72	76	80

Pricing can then be done using the following formula:

> *Resistor Adjusting*
> Direct labor
>   Adjusting ____ hours at $_____/hour _____
>   Overhead at __%
>   Total labor
>
>   Yield __%

When this yielded and burdened cost of manufacture is added to the burdened yielded substrate cost, the result is a quick and easy total cost of the entire end operation.

This treatment can be extended to the entire hybrid-circuit fabrication process. Figures 44 and 45 are a sample quote form for the entire circuit. The first page is used to identify the customer and his circuit. Engineering follows through, identifying the procedures to be used. The detail bill of materials is listed to be costed at various quantity levels.

The resulting information is then translated to cost using charts and tables developed as previously described. The labor costs are burdened using the overhead rate peculiar to the direct manufacturing operation. These numbers are then transferred to the summary and multiplied by another number X, which is derived relating the variable cost of manufacture to the fixed costs including general and administrative cost; profit and sales costs yields the selling price.

The selling price is adjusted for quantity using standard learning-curve factors. The table shown assumes an 80 percent learning curve for quantities below 1,000 pieces and 90 percent for quantities above 1,000 pieces.

The system has been used effectively many times and has proved to be a quick and reasonably thorough review of a hybrid-circuit application.

QUOTE NO. _____

HYBRID CIRCUIT COST ESTIMATE

DATE: _____

CUSTOMER: _____

ADDRESS: _____

CUSTOMER PART NO. _____ MFG. PART NO. _____

DEVICE: _____

MANUFACTURING NOTES:

SCREENING: _____

ADJUSTING: _____

DIE ATTACHING: _____

COMPONENT ATTACHING: _____

WIRE BONDING: _____

PACKAGING: _____

TESTING: _____

SPECIAL PROCESSING: _____

BILL OF MATERIALS: (PURCHASED PARTS)

ITEM		QTY./ CKT.	UNIT COST	$/CKT.	UNIT COST	$/CKT.	UNIT COST	$/CKT.	UNIT COST	$/CKT.
DESC.	PART NO.									
B of M X FACTOR										

START UP COSTS:

ENGINEERING & TOOLING: _____

_____

SHEET 1 of 2

Fig. 44 Cost sheet, front.

REFERENCES

1. E. I. du Pont de Nemours & Co., Electrochemicals Dept.: "The Thick Film Handbook," Wilmington, Del.
2. Kulischenko, W.: "Thick Film Resistor Trimming by AJM," S. S. White Publication.
3. Cote, R. E.: "Problems in Screen Printing and Firing," NEPCON Central, 1971.
4. Burns, F.: A Compact Low Cost Thick and Thin Film Laser Resistor System, *ISHM Proc.* 1970.
5. Cote, R. E.: "Thick Film Design, Fabrication and Packaging," Unitek Corporation/Weldmatic Division Workshop, January 1970.

CUSTOMER _____ QUOTE NO. _____
PART NO. _____ DATE _____

A. <u>SUBSTRATE MFG.</u>
   DIRECT LABOR:
      SCREENING _____ HRS. @ RATE/HR. _____
      ADJUSTING _____ HRS. @ RATE/HR. _____

      TOTAL DIRECT LABOR   _____
      OVERHEAD @ XXX%   _____

   MATERIALS:
      SUBSTRATE _____ + INKS _____   _____

   SUBTOTAL: _____
      SUBSTRATE YIELDED COST _____ % X _____ %   1. _____

B. <u>ASSEMBLY + TEST</u>
   DIRECT LABOR: NO. $/EA.
      DIE ATTACH   _____ @ _____
      WIRE BOND   _____ @ _____
      PACKAGING   _____ @ _____
      TESTING   _____ @ _____
      SPECIAL   _____ @ _____

      TOTAL DIRECT LABOR   _____ X 1000 = _____
      OVERHEAD @ XXX%   _____

   TOTAL ASSEMBLY + TEST COST   _____   2. _____

C. <u>TOTAL COST + SELLING PRICE</u>
   1. SUBSTRATE COST: _____
   2. CIRCUIT ASSY. + TEST: _____
                  SUB-TOTAL: _____
   3. YIELDED MFG. COST: _____ %   _____

   4. YIELDED COST X "F" = BASIC COST: _____   _____
      "F" = FACTOR FOR OVERHEAD,
         G & A, PROFIT, ETC.

D. <u>SELLING PRICE (BASIC COST X FACTOR)</u>

QUANTITY	1-99	100-249	250-499	500-999	1K	2.5K	5K	10K
MFG. COST								
MATERIALS								
SELLING PRICE								

QUANTITY	1-99	100-249	250-499	500-999	1K	2.5K	5K	10K
FACTOR	4.0	2.75	2.0	1.5	1.25	1.0	.95	.90

SHEET 2 of 2

Fig. 45   Cost sheet, back.

6. Thompson, G. W., III: Air Abrasive Resistor Trimming, *Solid State Technol.,* April 1970, p. 69.
7. Van Hise, J. A.: Controlling the Variables in Thick Film Resistor Fabrication, *Elec. Packag. Prod.,* April 1970, pp. 48–58.
8. Waters, R. L., and M. J. Weiner: Resistor Trimming and Other Micromachining with a YAG Laser, *Penn. State Univ. Eng. Sem. New Ind. Technol.,* July 1969; *Solid State Technol.,* April 1970, pp. 43–49.
9. Austin, B. M.: Thick Film Screen Printing, *Solid State Technol.,* June 1969, pp. 53–58.
10. Beck, J. H.: The Anatomy of a Furnace, NEPCON, 1969.
11. Early, R. C.: Thick Films: Setting up a Prototype Facility, *Electron. Pack. Prod.,* April 1969.
12. Hughes, D. D., Jr.: Recent Developments in Screen Printing Technology, NEPCON, 1969.
13. Ruth, S. B.: Hybrids . . . Thick and Thin, *Electron. Eng.,* October 1969.
14. Stone, G. B.: Automatic Laser Resistor Trimming: A Systems Approach to the Design of an Industrial Laser System, *1969 Hybrid Microelectron. Symp., Dallas, October 1969.*
15. Auda, D., and M. Schneider: Thick Film Materials on a Thin Budget, *Electron. Pack. Prod.,* August 1968.
16. Beck, J. H.: Firing Thick Film Integrated Circuits, *ISHM Proc.* 1967, pp. 114–119.
17. Hughes, D. C., Jr.: "Screen Printing of Microcircuits," Dan Mar, Somerville, N.J., 1967.
18. Marcott, C.: Prototyping Microcircuits, *Electron. Prod.,* December 1967.
19. Spigarelli, D. S.: Selection of Furnace Equipment for Thick Film Firing, *ISHM Proc. 1967,* pp. 37–43.
20. Minard, R. A.: Viscosity as a Process Variable, *Instrum. Autom.,* July, 1958.

Chapter **4**

Substrates for Thick Film Circuits

Charles A. Harper
Westinghouse Electric Corporation, Baltimore, Maryland

INTRODUCTION

Although substrates provide the base onto which all thick film circuits are fabricated, little attention is usually given to substrates, compared with other components of the overall thick film circuit device. While this is understandable, generally speaking, due to the more sensitive and critical nature of other components of the system, a better understanding and application of substrate properties can frequently lead to improvements in design, fabrication, and reliability of the device. It is the purpose of this chapter to present data and guidelines toward these objectives. It should be mentioned at this point that while there are many substrate materials and substrate forms, this chapter will focus primarily on those of potential use in the field of thick film technology. However, all ceramic materials of potential use for thick film substrates will be discussed, along with the primary thick film substrate, which is alumina. Also, some comparisons of other substrate types will be made to ceramic thick film substrate materials, for a broader understanding of the points which are important to thick film technology.

BASIC CERAMIC AND THICK FILM SUBSTRATE MATERIALS

As was mentioned above, ceramics are most widely used as thick film substrates. Ceramic materials found early use in substrate applications due primarily to their high mechanical strength, their high electrical resistivity over broad temperature ranges, and their chemical inertness relative to the variety of process conditions to which they are subjected in the course of generating the circuit placed on them. The earliest materials used were the common electrical porcelains and steatite and related compositions, both types of materials being processed from naturally occurring raw materials.

Ceramic materials offer mechanical and electrical properties that make them especially suitable for the electrical and electronics industries. The bulk of electrical ceramics are used either as insulators or as dielectrics. Electrical-grade ceramics have more exacting property requirements than the ceramics used for refractory or structural purposes. Properties such as dielectric strength, dielectric constant, dissipation factor, and thermal and electric conductivity are closely related to microstructure as well as to composition and processing.

In common with other insulating materials, ceramic insulators must have low dielectric constants to avoid capacitance effects, adequate dielectric strength to withstand the applied voltage without breakdown, low dissipation to avoid excessive electrical losses, and mechanical strength sufficient to withstand service conditions.

Oxide ceramics are characterized by their chemical inertness, oxidation resistance, moderately high refractoriness and resistivity, and by their low thermal expansions, thermal conductivities, and densities. Both the electrical resistivity and the thermal conductivity of oxide ceramics decrease with increasing temperature.

Basic Substrate Requirements

One reason for the popular acceptance of ceramics as thick film substrates is the ability of the ceramic substrate to withstand temperatures far in excess of 1000°C. This is important since thick film materials are fired at temperatures of about 1000°C and lower. Aside from the higher softening temperatures available in ceramic materials, their principal advantage over glass is their high-thermal conductivity. This is fully realized only in high-purity alumina and beryllia since intergranular heat barriers can drastically reduce the thermal conductivity of a ceramic. For example, the addition of 5 percent aluminum oxide to pure beryllium oxide decreases its thermal conductivity to two-thirds of its original value.

There are many factors to be considered in selecting a substrate for a given thick film system. These include the following but are not necessarily limited to them:

Material	Permeability
Surface finish	Flexural strength
Flatness	Compressive strength

Camber

Thermal coefficient of expansion

Thermal conductivity

Dielectric constant

Water absorption

Specific gravity

Volume resistivity

Dielectric strength

Loss tangent

Loss factor

Operating-temperature capabilities

Compatibility to thick film materials

Considering the number of variables this list could present to the design engineer, it is fortunate that the thick film industry has pretty much standardized on 96 percent alumina, Al_2O_3. Although other ceramics, such as forsterite, beryllia, titanium dioxide, and steatite, can be used and are used, the bulk of applications are accomplished with 96 percent alumina. The primary reason is that this material is both reasonably priced and contains the necessary properties, both physical and electrical, that make it essentially compatible with the resistor, conductor, and insulating materials commonly used for the fabrication of hybrid thick film microcircuits.

The temperatures utilized in the manufacture of the ceramic substrates are approximately 1500 to 1900°C, and this high-temperature fabrication makes them naturally immune to the lower temperatures (1000°C and less) of thick film processing.

Basic Ceramic Compositions

As was mentioned above, ceramics of potential use as substrates cover many compositions. The primary characteristics of these ceramic compositions, or bodies, are summarized below. Typical physical properties of these ceramics are given in Table 1.

Electrical Porcelain A typical electrical porcelain body consists of approximately 50 percent clay, $Al_2Si_2O_5(OH)_4$, and 25 percent each of flint, SiO_2, and of feldspar, $KAlSi_3O_8$. The high clay content gives the green body plasticity, which facilitates easy fabrication. Feldspar reacts with the clay at high temperatures (1200 to 1300°C) to give mullite and a viscous liquid phase. Solution of flint (1300 to 1400°C) increases the viscosity of the liquid phase and helps to maintain the shape of the body during firing. These compensating changes give porcelain an unusually long firing range and a great tolerance for compositional variations.

The high loss factor of porcelain is due to the large glass content and to the high mobility of the alkali ions. These ions also cause porcelain to have a relatively low resistivity. The electrical properties can be improved by replacing the alkali ions with larger and less mobile alkaline-earth ions (Ca^{2+}, Mg^{2+}, Ba^{2+}) and by lowering the glass content. High-alumina porcelains have a dielectric constant nearly constant through the temperature range of most interest in electronic-device application (-50 to $+250$°C).

Steatite Steatite porcelains are low-loss materials that are more commonly used as components for variable capacitors, coil forms, electron-tube sockets, and general structural insulation (bushings, spacers, support bars, etc.) but which offer potential for some substrate applications. These bodies can be manufactured to close dimensional tolerances using automatic dry pressing and extrusion methods. Unlike clay-flint-feldspar porcelains, steatite bodies require close control of the firing temperature, since the firing range to obtain vitrified materials is short. Commercial steatite compositions are based on 90 percent talc, $Mg_3Si_4O_{10}(OH)_2$, plus 10 percent clay. Feldspar additions greatly extend the firing range, but they degrade the electrical properties because of the introduction of alkali ions. Low-loss steatite compositions use additional magnesia to combine with the excess silica; barium oxide is used as the fluxing agent. The fired body consists of enstatite, $MgSiO_3$, crystals bonded together by a glassy matrix.[1]

Steatite is characterized by a dielectric constant and a dissipation factor that increase with temperature at low frequencies, but these are relatively independent of temperature at microwave frequencies (see Fig. 1). The physical properties of steatite are listed in Table 1.

Cordierite The low thermal-expansion coefficient, and consequently the high thermal-shock resistance, makes cordierite, $Mg_2Al_4Si_5O_{18}$, bodies useful for high-tem-

TABLE 1 Typical Physical Properties of Ceramics[29]

	Vitrified products						
	1	2	3	4	5	6	7
Material	High-voltage porcelain	Alumina porcelain	Steatite	Forsterite	Zircon porcelain	Lithia porcelain	Titania, titanate ceramics
Typical applications	Power-line insulation	Spark-plug cores, thermocouple insulation, protection tubes	High-frequency insulation, electrical appliance insulation	High-frequency insulation, ceramic-to-metal seals	Spark-plug cores, high-voltage high-temperature insulation	Temperature-stable inductances, heat-resistant insulation	Ceramic capacitors, piezoelectric ceramics
Specific gravity, g/cm³	2.3-2.5	3.1-3.9	2.5-2.7	2.7-2.9	3.5-3.8	2.34	3.5-5.5
Water absorption, %	0.0	0.0	0.0	0.0	0.0	0.0	0.0
Coefficient of linear thermal expansion, at 20-700°C, 10^{-6} in./(in.)(°C)	5.0-6.8	5.5-8.1	8.6-10.5	11	3.5-5.5	1	7.0-10.0
Safe operating temperature, °C	1,000	1,350-1,500	1,000-1,100	1,000-1,100	1,000-1,200	1,000	
Thermal conductivity, (cal/cm²)/(cm)(sec)(°C)	0.002-0.005	0.007-0.05	0.005-0.006	0.005-0.010	0.010-0.015		0.008-0.01
Tensile strength (psi)	3,000-8,000	8,000-30,000	8,000-10,000	8,000-10,000	10,000-15,000		4,000-10,000
Compressive strength, psi	25,000-50,000	80,000-250,000	65,000-130,000	60,000-100,000	80,000-150,000	60,000	40,000-120,000
Flexural strength, psi	9,000-15,000	20,000-45,000	16,000-24,000	18,000-20,000	20,000-35,000	8,000	10,000-22,000
Impact strength (½-in. rod), ft-lb	0.2-0.3	0.5-0.7	0.3-0.4	0.03-0.04	0.4-0.5	0.3	0.3-0.5
Modulus of elasticity, psi $\times 10^{-6}$	7-14	15-52	13-15	13-15	20-30	10-15	10-15
Thermal shock resistance	Moderately good	Excellent	Moderate	Poor	Good	Excellent	Poor
Dielectric strength (¼-in.-thick specimen), volts/mil	250-400	250-400	200-350	200-300	250-350	200-300	50-300
Resistivity at room temperature, ohm/cm³	10^{12}-10^{14}	10^{14}-10^{15}	10^{13}-10^{15}	10^{13}-10^{15}	10^{13}-10^{15}		10^{8}-10^{15}
Te value, °C	200-500	500-800	450-1,000	above 1,000	700-900		200-400
Power factor at 1 MHz	0.006-0.010	0.001-0.002	0.0008-0.0035	0.0003	0.0006-0.0020	0.05	0.0002-0.050
Dielectric constant	6.0-7.0	8-9	5.5-7.5	6.2	8.0-9.0	5.6	15-10,000
L grade (JAN Spec. T-10)	L-2	L-2-L-5	L-3-L-5	L-6	L-4	L-3	

* Reprinted from A. R. Von Hippel, "Dielectric Materials and Applications," by permission of the MIT Press, Cambridge, Mass.

Semivitreous and refractory products

Material	8	9	10	11
	Low-voltage porcelain	Cordierite refractories	Alumina, aluminum silicate refractories	Massive fired talc, pyrophyllite
Typical applications	Switch bases, low-voltage wire holders, Light receptacles	Resistor supports, burner tips, heat insulation, arc chambers	Vacuum spacers, high-temperature insulation	High-frequency insulation, vacuum-tube spacers, ceramic models
Specific gravity, g/cm³	2.2-2.4	1.6-2.1	2.2-2.4	2.3-2.8
Water absorption (%)	0.5-2.0	5.0-15.0	10.0-20.0	1.0-3.0
Coefficient of linear thermal expansion, at 20-700°C, 10^{-6} in./(in.)(°C)	5.0-6.5	2.5-3.0	5.0-7.0	11.5
Safe operating temperature, (°C)	900	1,250	1,300-1,700	1,200
Thermal conductivity, (cal/cm²)(sec)(°C)	0.004-0.005	0.003-0.004	0.004-0.005	0.003-0.005
Tensile strength, psi	1,500-2,500	1,000-3,500	700-3,000	2,500
Compressive strength (psi)	25,000-50,000	20,000-45,000	15,000-60,000	20,000-30,000
Flexural strength, psi	3,500-6,000	1,500-7,000	1,500-6,000	7,000-9,000
Impact strength (½-in. rod), ft-lb	0.2-0.3	0.2-0.25	0.17-0.25	0.2-0.3
Modulus of elasticity, psi × 10^{-6}	7-10	2-5	2-5	4-5
Thermal shock resistance	Moderate	Excellent	Excellent	Good
Dielectric strength (¼-in.-thick specimen), volts/mil	40-100	40-100	40-100	80-100
Resistivity at room temperature, ohm/cm³	10^{12}-10^{14}	10^{12}-10^{14}	10^{12}-10^{14}	10^{12}-10^{15}
Te value, °C	300-400	400-700	400-700	600-900
Power factor at 1 MHz	0.010-0.020	0.004-0.010	0.0002-0.010	0.0008-0.016
Dielectric constant	6.0-7.0	4.5-5.5	4.5-6.5	5.0-6.0
L grade (JAN Spec. T-10)				

perature applications. Like steatite, vitrified bodies are difficult to make because of the short firing range. When the intended use is for other than electrical applications, feldspar is added as the fluxing agent in order to increase the firing range. Typical physical properties are listed in Table 1.

Forsterite. In contrast to both steatite and cordierite, forsterite, Mg_2SiO_4, bodies present few firing problems. The absence of alkali ions in the vitreous phase gives forsterite insulators a higher resistivity and a lower electric loss with increasing temperature than steatite bodies. Since these low-loss dielectric properties persist at high frequencies, forsterite is used for small microwave tubes such as nuvistors.

The high thermal-expansion coefficient makes forsterite suitable for ceramic-to-metal seals, but it also causes the material to have poor thermal-shock resistance. Table 1 lists the physical properties of forsterite.

Alumina As we mentioned above, the desirable electrical and mechanical properties possessed by alumina, Al_2O_3, make it the most widely used material for thick film substrates.

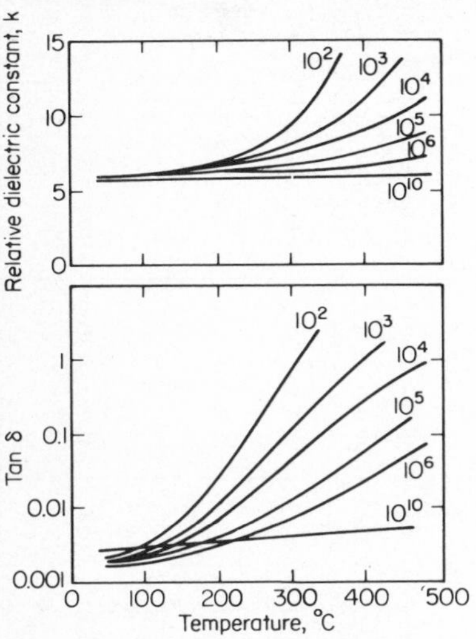

The dielectric constant and the loss tangent of alumina are affected by impurities, e.g., Si, Ti, Mg, and Ca.[2] Substitution of either Si^{4+} or Ti^{4+} for Al^{3+} in alumina creates donor levels at the impurity sites and acceptor levels at the compensating cation vacancies. Coversely, the substitution of either Mg^{2+} or Ca^{2+} for Al^{3+} creates acceptor levels at the impurity sites and donor levels at the compensating interstitials. These donors and acceptors contribute charge carriers which affect the dielectric and loss characteristics of alumina. Since the solubility of Si^{4+} in Al_2O_3 is limited, excess SiO_2 leads to the formation of a glassy phase at the grain boundaries and to ionic conduction.

Ultra-low-loss alumina can be made by eliminating the glassy phase. This is accomplished by sintering pure, fine-grained powder at high temperatures (1800 to 1900°C) to produce a low-porosity body having very small grain size. The problem of residual porosity in Al_2O_3 ceramics affects both the thermal and the optical properties. Below a red heat, thermal conductivity decreases with increasing porosity, since the pores act as a thermal impedance (see Fig. 2). The effect of even a small amount of porosity on the optical transmission of Al_2O_3 is drastic (see Fig. 3) and emphasizes the difficulty involved in making a translucent ceramic. High-density polycrystalline alumina bodies have been made with very high translucency.

Alumina ceramics have high elastic moduli ($\sim50 \times 10^6$ psi) and high strengths (bend strength 20 to 40×10^3 psi), giving them the highest fracture strength among the refractory oxides. As in other oxide ceramics, the strength begins to decrease rapidly above 1000°C.

Alumina is stable in air, vacuum, water vapor, hydrogen, carbon monoxide, nitrogen, and argon at temperatures up to 1700°C. Hydrogen fluoride will react with alumina. At high temperatures ($\sim1700°C$) alumina will vaporize as Al_2O in the presence of water vapor or reducing atmospheres. A compilation of the electrical and physical properties of alumina is listed in Table 2.

Beryllia BeO offers the unusual combination of high thermal and low electric conductivity. At room temperature, the thermal conductivity is about one-half that of copper, but with increasing temperature the conductivity of beryllia decreases much more rapidly than that of copper. The high strength and the high thermal conductivity give BeO good thermal shock resistance.

Fig. 1 Dielectric constant and tan δ for a steatite ceramic over a range of temperature and frequencies.[1]

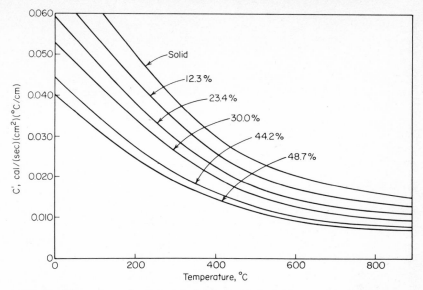

Fig. 2 Thermal conductivity of alumina with various amounts of porosity.[3]

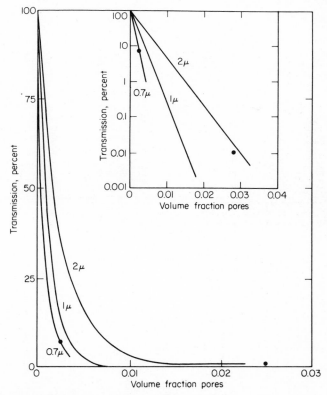

Fig. 3 Transmission of polycrystalline alumina containing small amounts of residual porosity.[1]

TABLE 2 Mechanical, Thermal, and Electrical Properties of Coors Alumina and Beryllia Ceramics

Property*	Test	(1) AD-85	(2) AD-90	(3) AD-94	(4) AD-96
Specific gravity:					
Typical	} ASTM C 20-46	3.42	3.58	3.62	3.72
Minimum		3.37	3.53	3.57	3.67
Hardness (typical), Rockwell 45N	ASTM E 1867	75	79	78	78
Surface finish, μin. (arithmetic avg.):					
Typical, as fired	} Profilometer (0.030-in. cutoff)	65	65	65	65
Typical, ground		45	40	50	50
Ultimate, lapped		13	3	10	10
Crystal size, microns:					
Range		2–12	2–10	2–25	2–20
Average	ASTM C 373-56	7	4	12	11
Water absorption		None	None	None	None
Gas permeability†		None	None	None	None
Color		White	White	White	White
Compressive strength (typical), psi × 10⁻³:					
At 25°C	} ASTM C 528-63T	280	360	305	300
At 1000°C		—	75	50	—
Flexural strength, psi × 10⁻³:					
At 25°C:					
Typical	ASTM C 369-56 (½-in.-diam. rods)	43	49	51	52
Minimum†		39	44	46	47
At 1000°C:					
Typical		25	—	20	25
Minimum†		20	—	17	20
Tensile strength (typical), psi × 10⁻³:					
At 25°C	} Brazil test	22	32	28	28
At 1000°C			15	15	14
Modulus of elasticity, psi × 10⁻⁶	} Sonic method	33	39	41	44
Shear modulus, psi × 10⁻⁶		13	16	17	18
Bulk modulus, psi × 10⁻⁶		20	23	24	25
Sonic velocity, 10³ m/sec		8.2	8.7	8.9	9.1
Poisson's ratio		0.22	0.22	0.21	0.21
Maximum use temperature, °C(°F)		1400 (2550)	1500 (2725)	1700 (3100)	1700 (3100)

Thermal coefficient of linear expansion, 10⁻⁶ cm/(cm)(°C):		°C	°F	°C	°F	°C	°F	°C	°F
−200 to 25°C	} ASTM C 372-56	3.4	1.3	3.4	1.9	3.4	1.9	3.4	1.9
25 to 200°C		5.3	3.0	6.1	3.4	6.3	3.5	6.0	3.3
25 to 500°C		6.2	3.5	7.0	3.9	7.1	3.9	7.4	4.1
25 to 800°C		6.9	3.8	7.7	4.3	7.6	4.2	8.0	4.4
25 to 1000°C		7.2	4.0	8.1	4.5	7.9	4.4	8.2	4.6
25 to 1200°C		7.5	4.2	8.4	4.7	8.1	4.5	8.4	4.7

Thermal and electrical properties table (values by specimen condition and frequency):

Property (ASTM method)	Mat 1 AC 25°C	Mat 1 DC 500°C	Mat 2 AC 25°C	Mat 2 DC 25°C	Mat 3 AC 25°C	Mat 3 AC 500°C	Mat 3 AC/DC 800°C	Mat 3 DC	Mat 4 AC 25°C	Mat 4 AC 500°C	Mat 4 AC/DC 800°C	Mat 4 DC
Thermal conductivity, cal/(sec)(cm²)(°C/cm): (ASTM C 408-58)												
20°C	0.035		0.040		0.043				0.043			
100°C	0.029		0.032		0.035				0.035			
400°C	0.016		0.017		0.017				0.017			
800°C	0.010		0.010		0.010				0.010			
Specific heat at 100°C, cal/(g)(°C) (ASTM C 351-61)	0.22		0.22		0.21				0.21			
Dielectric strength, volts/mil (avg. rms values): Specimen thickness, in.: (ASTM D 116-69)												
0.250	240	—	235	—	220			—	210			—
0.125	340	940	320	920	300			1050	275			740
0.050	440	1250	450	1100	425			1100	370			840
0.025	550	1550	580	1300	550			1140	450			900
0.010	720	1750	760	1600	720			1150	580			980
Dielectric constant: (ASTM D 150-65T)												
1 kHz	8.2	13.9	8.8		8.9	11.8			9.0	10.8		
1 MHz	8.2	8.9	8.8		8.9	9.7			9.0	9.6		
100 MHz	8.2	—	8.8		8.9	—			9.0	—		
1 GHz	8.2	8.3	—		8.9	9.1			8.9	9.4		
10 GHz	—		8.7		8.7		9.4		8.7		9.9	
50 GHz												
Dissipation factor: (ASTM D 150-65T)												
1 kHz	0.0014	0.580	0.0006		0.0002	0.215			0.0011	0.200		
1 MHz	0.0009	0.024	0.0004		0.0001	0.008			0.0001	0.0039		
100 MHz	0.0009	—	0.0004		0.0005	—			0.0002	—		
1 GHz	0.0014	0.003	—		0.0010	0.002		0.004	0.0006	0.0009		0.0028
10 GHz	0.0019		0.0009		0.0021				0.0068			
50 GHz												
Loss index: (ASTM #??)												
1 kHz	0.011	8.06	0.005		0.002	2.54			0.010	2.16		
1 MHz	0.007	0.214	0.004		0.001	0.078			0.001	0.037		
100 MHz	0.007	—	0.004		0.004	—			0.002	—		
1 GHz	0.011	0.025	—		0.007	0.018		0.038	0.001	0.008		0.028
10 GHz	0.016		0.008		0.018				0.005			
50 GHz									0.059			
Volume resistivity, ohm/cm²/cm: (ASTM D 1829-66)												
25°C	>10¹⁴		>10¹⁴		>10¹⁴				>10¹⁴			
300°C	4.6×10¹⁰		1.4×10¹¹		9.0×10¹¹				3.1×10¹¹			
500°C	4.0×10⁸		2.8×10⁸		2.5×10⁷				4.0×10⁸			
700°C	7.0×10⁶		7.0×10⁶		5.0×10⁵				1.0×10⁸			
1000°C	—		8.6×10⁶		5.0×10⁵				1.0×10⁶			
Te value, °C	850		960		950				1000			

Note: Volume resistivity values use LaTeX scientific notation: $>10^{14}$, 4.6×10^{10}, 4.0×10^{8}, 7.0×10^{6}, 1.4×10^{11}, 2.8×10^{8}, 8.6×10^{6}, 9.0×10^{11}, 2.5×10^{7}, 5.0×10^{5}, 3.1×10^{11}, 1.0×10^{8}, 1.0×10^{6}.

* Footnotes are on p. 4-14.

4-9

TABLE 2 Mechanical, Thermal, and Electrical Properties of Coors Alumina and Beryllia Ceramics* (Continued)

Property*	Test	(5) AD-99	(6) AD-995	(7) AD-998
Specific gravity:				
Typical.	ASTM C 20-46	3.83	3.84	3.82
Minimum.		3.78	3.80	3.78
Hardness, typical Rockwell 45N.	ASTM E 1867	80	81	79
Surface finish, μin. (arithmetic avg.):				
Typical, as fired.	Profilometer (0.030-in. cutoff)	55	55	—
Typical, ground.		35	35	—
Ultimate, lapped.		3	3	—
Crystal size, microns:				
Range.		5-50	10-50	10-35
Average.	ASTM C 373-56	22	20	12
Water absorption.		None	None	None
Gas permeability†.		None	None	None
Color.		White	Pink	Ivory
Compressive strength (typical), psi $\times 10^{-3}$:				
At 25°C.	ASTM C 528-63T	345	330	320
At 1000°C.		130	140	—
Flexural strength, psi $\times 10^{-3}$:				
At 25°C:				
Typical.	ASTM C 369-56 (½-in.-diam. rods)	48	45	48
Minimum‡.		45	40	43
At 1000°C:				
Typical.		30	33	28
Minimum‡.		25	28	—
Tensile strength (typical), psi $\times 10^{-3}$:				
At 25°C.	Brazil test	31	28	—
At 1000°C.		12	15	—
Modulus of elasticity, psi $\times 10^{-6}$.	Sonic method	50	52	50
Shear modulus, psi $\times 10^{-6}$.		21	22	21
Bulk modulus, psi $\times 10^{-6}$.		29	30	—
Sonic velocity, 10^3 m/sec.		9.4	9.7	9.4
Poisson's ratio.		0.21	0.21	0.21
Maximum use temperature, °C(°F).		1725 (3140)	1750 (3180)	1950 (3540)

Thermal coefficient of linear expansion, 10^{-6} cm/(cm)(°C):

Property	Test	AD-99 °C	AD-99 °F	AD-995 °C	AD-995 °F	AD-998 °C	AD-998 °F
−200 to 25°C.	ASTM C 372-56	3.4	1.9	3.4	1.9	3.4	1.9
25 to 200°C.		6.3	3.5	6.3	3.5	6.7	3.7
25 to 500°C.		7.3	4.0	7.3	4.1	7.3	4.1
25 to 800°C.		7.9	4.4	7.8	4.4	7.8	4.4
25 to 1000°C.		8.2	4.6	8.1	4.5	8.0	4.5
25 to 1200°C.		8.4	4.7	8.3	4.6	8.3	4.6

Property	ASTM	Material A	Material B	Material C
Thermal conductivity, cal/(sec)(cm²)(°C/cm):	ASTM C 408-58			
20°C		0.070	0.075	0.070
100°C		0.055	0.065	0.055
400°C		0.015	0.028	0.025
800°C		—	0.017	0.015
Specific heat at 100°C, cal/(g)(°C)	ASTM C 351-61	0.21	0.21	0.21

Dielectric strength, volts/mil (avg. rms values): (ASTM D 116-69)

Specimen thickness, in.:	Material A — AC	Material A — DC	Material B — AC	Material B — DC
0.250	215	800	225	—
0.125	290	900	310	840
0.050	390	—	450	1050
0.025	480	980	550	1150
0.010	600	1100	625	1300

Dielectric constant: (ASTM D 150-65T)

	Material A AC 25°C	Material A AC 500°C	Material A DC 800°C	Material B AC 25°C	Material B AC 500°C	Material B DC 800°C
1 kHz	9.4	11.3	—	9.4	—	—
1 MHz	9.4	10.0	—	9.4	10.3	—
100 MHz	—	—	—	—	—	—
1 GHz	9.4	10.0	10.5	9.4	10.4	11.0
10 GHz	9.4	10.0	10.4	9.4	9.8	10.2
50 GHz	—	—	—	—	—	—

Dissipation factor: (ASTM D 150-65T)

	Material A AC 25°C	Material A AC 500°C	Material A DC 800°C	Material B AC 25°C	Material B AC 500°C	Material B DC 800°C
1 kHz	0.0042	0.1500	—	0.0004	—	—
1 MHz	0.0002	0.0047	—	0.0001	0.0023	—
100 MHz	—	—	—	—	—	—
1 GHz	0.0002	0.0003	0.0005	0.0001	0.0002	0.0003
10 GHz	0.0002	0.0002	0.0006	0.0001	0.0003	0.0006
50 GHz	—	—	—	—	—	—

Loss index: (ASTM # ??)

	Material A AC 25°C	Material A AC 500°C	Material A DC 800°C	Material B AC 25°C	Material B AC 500°C	Material B DC 800°C
1 kHz	0.040	1.70	—	0.004	—	—
1 MHz	0.002	0.047	—	0.001	0.024	—
100 MHz	—	—	—	—	—	—
1 GHz	0.002	0.003	0.005	0.001	0.002	0.003
10 GHz	0.002	0.002	0.006	0.001	0.003	0.006
50 GHz	—	—	—	—	—	—

Volume resistivity, ohm/cm²/cm: (ASTM D 1829-66)

	Material A	Material B	Material C
25°C	$>10^{14}$	$>10^{14}$	$>10^{14}$
300°C	1.0×10^{13}	1.5×10^{11}	$>10^{13}$
500°C	6.3×10^{10}	1.4×10^{9}	6.3×10^{11}
700°C	5.0×10^{8}	4.0×10^{7}	3.4×10^{9}
1000°C	2.0×10^{6}	8.0×10^{5}	7.8×10^{6}
Te value, °C	1050	980	1140

* Footnotes are on p. 4-14.

TABLE 2 Mechanical, Thermal, and Electrical Properties of Coors Alumina and Beryllia Ceramics* (Continued)

Property*	Test	(8) AD-999	(9) Vistal	(10) BD-995-2
Specific gravity:				
Typical	} ASTM C 20-46	3.96	3.99	2.90
Minimum		3.94	3.98	2.86
Hardness (typical), Rockwell 45N	ASTM E 1867	90	85	67
Surface finish (typical), μin. (arithmetic avg.):				
Typical, as fired	} Profilometer (0.030-in. cutoff)	20	25	22
Typical, ground		35	35	20
Ultimate, lapped		<1	<1	—
Crystal size, microns:				
Range	} ASTM C 373-56	1–6	50–45	10–40
Average		3	20	24
Water absorption		None	None	None
Gas permeability†		None	None	None
Color		Ivory	Translucent white	White
Compressive strength (typical), psi × 10^{-3}:				
At 25°C	} ASTM C 528-63T	550	370	310
At 1000°C		280	70	40
Flexural strength, psi × 10^{-3}:				
At 25°C:				
Typical	} ASTM C 369-56 (½-in.-diam. rods)	95	41	40
Minimum‡		89	—	35
At 1000°C:				
Typical		70	25	—
Minimum‡		65	—	—
Tensile strength (typical), psi × 10^{-3}:				
At 25°C	} Brazil test	48	30	20
At 1000°C		32	15	5
Modulus of elasticity, psi × 10^{-6}		56	57	51
Shear modulus, psi × 10^{-6}		23	23.5	20
Bulk modulus, psi × 10^{-6}	} Sonic method	—	—	—
Sonic velocity, 10^3 m/sec		9.9	9.9	11.1
Poisson's ratio		0.22	0.22	0.26
Maximum use temperature, °C(°F)		1900 (3450)	1900 (3450)	1850 (3360)

Thermal coefficient of linear expansion, 10^{-6} cm/(cm)(°C):

Property*	Test	AD-999 °C	AD-999 °F	Vistal °C	Vistal °F	BD-995-2 °C	BD-995-2 °F
−200 to 25°C	} ASTM C 372-56	3.6	2.0	3.4	1.9	2.4	1.3
25 to 200°C		6.5	3.6	6.5	3.6	6.4	3.6
25 to 500°C		7.4	4.1	7.4	4.1	7.7	4.3
25 to 800°C		7.8	4.3	7.8	4.3	8.5	4.7
25 to 1000°C		8.0	4.5	8.0	4.5	8.9	4.9
25 to 1200°C		8.3	4.6	8.3	4.6	9.4	5.2

Property	ASTM	Material 1 (AC / 25°C)	Material 1 (DC / 500°C)	Material 1 (800°C)	Material 2 (AC / 25°C)	Material 3 (AC / 25°C)	Material 3 (DC)
Thermal conductivity, cal/(sec)(cm²)(°C/cm):	ASTM C 408-58						
20°C		0.074			0.095	0.67	
100°C		0.055			0.070	0.48	
400°C		0.031			0.030	0.20	
800°C						0.07	
Specific heat at 100°C, cal/(g)(°C)	ASTM C 351-61	0.21			0.21	0.31	
Dielectric strength, volts/mil (avg. rms values):	ASTM D 116-69						
Specimen thickness, in.:		AC	DC		AC	AC	DC
0.250		240	—		230	260	—
0.125		325	920		340	340	830
0.050		460	1050		510	490	—
0.025		590	1200		650	610	—
0.010		800	1450		—	800	—
Dielectric constant:	ASTM D 150-65T	25°C	500°C	800°C	25°C	25°C	
1 kHz		9.9	—	—	10.1	6.8	
1 MHz		9.8	—	—	10.1	6.8	
100 MHz		—	—	—	10.1	6.8	
1 GHz		9.8	10.3	10.7	—	6.8	
10 GHz		—	—	—	10.1	6.7	
50 GHz		—	—	—	—	—	
Dissipation factor:	ASTM D 150-65T						
1 kHz		0.002	—	—	0.0005	0.001	
1 MHz		0.0002	—	—	0.00004	0.0003	
100 MHz		—	—	—	0.00006	0.0006	
1 GHz		0.00006	0.0003	0.0008	—	0.0006	
10 GHz		—	—	—	0.00009	0.0003	
50 GHz		—	—	—	—	—	
Loss index:	ASTM #??						
1 kHz		0.020	—	—	0.005	0.007	
1 MHz		0.002	—	—	0.0004	0.002	
100 MHz		—	—	—	0.0006	0.004	
1 GHz		0.0006	0.003	0.009	—	0.004	
10 GHz		—	—	—	0.001	0.002	
50 GHz		—	—	—	—	—	
Volume resistivity, ohm/cm²/cm:	ASTM D 1829-66						
25°C		$>10^{15}$			—	$>10^{17}$	
300°C		1.0×10^{15}			—	$>10^{15}$	
500°C		3.3×10^{12}			—	5.0×10^{13}	
700°C		9.0×10^{9}			—	1.5×10^{10}	
1000°C		1.1×10^{7}			—	7.0×10^{7}	
Te value, °C		1170			—	1240	

* Footnotes are on p. 4-14.

Footnotes to Table 2

SOURCE: Coors Porcelain Company.

* Ceramic property values vary somewhat according to method of manufacture, size, and shape of part. Closer control of values is possible.

† No helium leak through a 1-in.-diameter plate, 0.001 in. thick, measured at 3×10^{-7} torr vacuum versus approximately 1 atm helium pressure for 15 sec at room temperature.

‡ Minimum flexural strength is a minimum mean for a sample of 10 specimens.

(1) AD-85: Nominally 85 % Al_2O_3. A good all-around high-alumina ceramic for both electrical and mechanical applications.

(2) AD-90: Nominally 90 % Al_2O_3. A tough, fine-grained alumina ceramic especially well suited for demanding mechanical applications.

(3) AD-94: Nominally 94 % Al_2O_3. A very good alumina ceramic for metallizing; ideal for all but the most critical electrical and mechanical applications.

(4) AD-96: Nominally 96 % Al_2O_3. An excellent alumina ceramic for special electronic applications and many mechanical applications.

(5) AD-99: Nominally 99 % Al_2O_3. A very strong, impervious alumina ceramic designed for virtually all kinds of critical electrical and mechanical applications.

(6) AD-995: Nominally 99.5 % Al_2O_3. An extremely low-loss alumina ceramic used widely in many electronic applications and some mechanical applications.

(7) AD-999: Nominally 99.9 % Al_2O_3. The hardest, strongest, purest alumina ceramic available; recommended for use in ultrasevere mechanical applications and/or highly hostile environments.

(8) Vistal: Registered trademark. Nominally 99.9 % Al_2O_3. A translucent, high-purity alumina ceramic for highly critical electrical and electronic applications; strong, excellent resistance to chemical attack.

(9) BD-995-2: Nominally 99.5 % BeO. A much-improved beryllia ceramic possessing low dielectric loss, high electrical resistivity, good dielectric strength; meant for use where high thermal conductivity is required.

GENERAL NOTES:

All measurements are typical for the materials shown. For some specific applications it may be necessary to measure values for use in design formulas. Dielectric constant specifically can be controlled.

Dashes (—) and blank spaces indicate values not measured at this time. All values shown are measured at room temperature unless otherwise specified. All data are typical unless otherwise specified.

Composition Control: Alumina and beryllia contents of Coors ceramics are controlled using chemical, spectrographic, and x-ray fluorescent methods for quantitative determination of minor ingredients.

Chemical Resistance: Coors sintered alumina and beryllia ceramics are highly resistant to chemical attack and corrosion. For optimum material selection, it is recommended that specific data on chemical resistance be obtained for particular applications.

Beryllia is stable in air, vacuum, hydrogen, carbon monoxide, argon, and nitrogen at temperatures up to 1700°C.[5] Beryllia will react with water vapor above 1650°C to form volatile $Be(OH)_2$, and it will decompose in atmospheres containing halogens or sulfur.

BeO has a lower dielectric constant than alumina but one with about the same temperature dependence. The temperature dependence of the resistivity of BeO is again similar to that of Al_2O_3.

In massive form, BeO is not dangerous; however, in the form of powder or dust, beryllia may be hazardous to health. BeO suppliers can give reliable safety advice.

Magnesia MgO is a better electrical insulator than Al_2O_3, particularly at high temperatures. Because of its high thermal expansion and low strength, it has poor thermal-shock resistance. Although MgO shows little tendency to hydrate in large masses, it will hydrate in powdered form. The combination of high thermal and low electric conductivity makes MgO suitable for insulating thermocouple leads and for heating-core elements.

Zirconia Dense, crack-free zirconia, ZrO_2, ceramics are difficult to produce because of the disruptive volume change ($\sim$10 percent) that occurs during transformation from the tetragonal to the monoclinic phase at 1000°C. This disruptive phase change is eliminated by stabilizing the cubic form of zirconia with solid solutions of CaO, Y_2O_3, Yb_2O_3, Nd_2O_3, or Sc_2O_3. Stabilized ZrO_2 has relatively poor thermal resistance because of its high thermal-expansion coefficient [$\sim$11 × 10^{-6} cm/(cm) (°C), about 1½ times that of alumina] and its relatively low thermal conductivity [0.004 cal/(s)(cm²)(°C/cm), about one-fourth that of alumina].

THERMAL, ELECTRICAL, AND MECHANICAL PROPERTIES OF SUBSTRATES

Aside from fabrication and surface properties of substrates, discussed in following sections of this chapter, the most important properties to be considered in design use of thick film substrates are their thermal, electrical, and mechanical properties, reviewed in detail at this point. Wherever possible, an attempt has been made to present important data over the useful range of important variables instead of merely presenting point data. This should aid designers in predicting performance trends, which is frequently a very useful design consideration.

Thermal Properties

Thermal expansion Figure 4 shows the thermal expansion of various substrate materials in comparison with the thermal expansion of other materials frequently used in the construction of thick film circuit devices. These data will be useful in analyzing the system for thermal-expansion differentials, which are usually the most important consideration in thermal-force problems. It should be noted that all these values are appreciably lower than the thermal-expansion values for adhesives, encapsulants, and other plastic materials sometimes used in the construction of thick film circuit devices. Since the variations are so great for the large volume of filled and unfilled plastic formulations, data sheets of the individual supplier should be consulted for specific plastic formulations.

Thermal conductivity Thermal conductivity is one of the most used design characteristics of substrates, due to the frequent design requirement for maximum heat dissipation from the thick film circuit device. Figure 5 shows the thermal conductivity of various high-density substrate materials, and Table 3 shows the relative thermal conductivities of several groups of materials related to copper as a standard. As would be expected, porosity, or reduced density, will lower the thermal conductivity for a substrate material. This was shown earlier in Fig. 2 as a function of porosity and is shown also in Table 4 as a function of the alumina content of an alumina substrate material.

Beryllia is known for its high thermal conductivity, compared with other substrate materials. This is clearly shown in Fig. 5 and Table 3. As was shown earlier for alumina, the density, purity, and porosity of the beryllia substrate can have a significant effect on the thermal conductivity of the substrate. This is shown both in Fig. 6 and in Table 5.

Although glazed substrates are not common in thick film substrates, as they are in thin film substrates, due to the differences in nature of the deposited circuitry elements, it should be noted that if used, a glazed surface increases the thermal impedance of the substrate. For example, the thermal resistance of 1 mil of glaze is equivalent to that of 30 mils of alumina or of 190 mils of beryllia. Glazing beryllia, therefore, almost entirely negates any thermal advantage which this ceramic might otherwise offer. Typical data for a glaze material are included in Table 6.

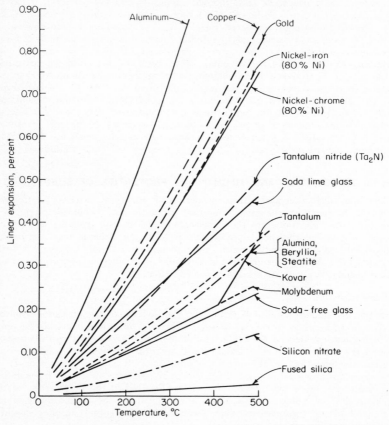

Fig. 4 Linear thermal expansion of various substrates and associated materials as a function of temperature.

Specific heat The specific-heat data for alumina and beryllia materials are compared with values for other associated materials in Table 7.

Electrical Properties

Resistivity properties All ceramic materials used for substrates are electrical-grade ceramics possessing very high electrical resistivity values at temperatures in which they are commonly used in the electronics industry. The comparison of volume resistivity for various ceramics as a function of temperature is shown in Fig. 7. Note that all resistivity values decrease sharply as a function of temperature. This

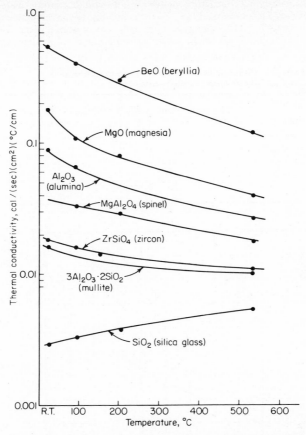

Fig. 5   Thermal conductivity of various high-density substrate materials.[6]

TABLE 3   Relative Thermal Conductivities of Several Important Material Categories[7]

Material	Percentage of thermal con- ductivity of copper
Silver...........................	105
Copper...........................	100
High-purity beryllia, BeO...........	62
Aluminum........................	55
Beryllium........................	39
Molybdenum......................	39
Steel............................	9.1
High-purity alumina, Al_2O_3.........	7.7
Steatite.........................	0.9
Mica............................	0.18
Phenolics, epoxies.................	0.13
Fluorocarbons....................	0.05

is common in nearly all insulating materials. Plastic insulating materials exhibit this same functional characteristic but at a considerably lower temperature, of course.

Dielectric strength   The dielectric strength of ceramics, like the dielectric strength of most insulating materials, varies considerably as a function of temperature,

TABLE 4   Effect of Alumina Content on the Thermal Conductivity of Alumina Substrates

Alumina, %	Thermal conductivity, cal/(s)(cm²)(°C/cm)	Change in thermal conductivity, %
99	0.070	
98	0.061	−13
96	0.043	−39
85	0.035	−50

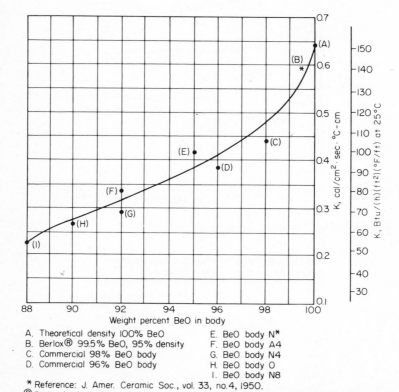

A. Theoretical density 100% BeO
B. Berlox® 99.5% BeO, 95% density
C. Commercial 98% BeO body
D. Commercial 96% BeO body
E. BeO body N*
F. BeO body A4
G. BeO body N4
H. BeO body O
I. BeO body N8

* Reference: J. Amer. Ceramic Soc., vol. 33, no. 4, 1950.
® Berlox is the registered trademark of National Beryllia Corp. (beryllium oxide).

Fig. 6   Thermal conductivity of beryllia as a function of BeO content at 25°C.[8]

frequency, material thickness, density, porosity, purity, and other variables. The first three are perhaps the most important, due to their more common application occurrence. Table 8 compares the room-temperature dielectric strength of various other insulating materials with that of glass at seven frequencies ranging from 60

Hz to 100 MHz. Table 9 shows the effect of temperature on dielectric strength at 60 Hz for alumina. The dielectric-strength values of alumina, beryllia, and steatite as a function of test-piece thickness are shown in Fig. 8.

Dielectric constant Another important electrical property of insulating materials is dielectric constant. The key variables of application importance are temperature and frequency. While dielectric constant increases with temperature, the change is not drastic in most cases over the normal temperature range for electronics applications. Composition and purity are the factors of greatest influence. Figure 9 shows

TABLE 5 Effect of Porosity on Thermal Conductivity of Beryllia[7]

	Percent of thermal conductivity of 100% pure, 100% dense BeO		
BeO, %	BeO+ porosity	BeO+ alumina	BeO+ silica
100*	100	100	100
99	98.8	92.1	85.2
98	96.8	85.2	78.8
97	94.8	78.8	62.0
96	93.2	74.8	58.6
95	90.6	66.0	48.0

* No porosity, no alumina, no silica.

TABLE 6 Properties of Ceramic Substrates and Glazes[9]

				Dielectric properties at 1 MHz and 25°C		
Material	Tensile strength, psi	Expansion coefficient, μin./ (in.)(°C)	Coefficient of heat transfer, (W)(in.)/ (in.^{2})(°C)	Relative dielectric constant	Dissipation factor, %	Volume resistivity at 150°C, pΩ-cm
Alumina.............	25,000	6.4	~0.89	9.2	0.03	>100
Beryllia..............	15,000	6.0	5.8	6.4	0.01	>100
Corning 7059 glass......	~10,000	4.6	~0.03	5.8	0.1	>100
Modified BaTiO₃.......	4,000	9.1	0.007	6,500	1.8	0.2
Modified TiO₂.........	7,500	8.3	0.017	80	0.03	0.5
Glaze for alumina:						
2.5% sodium oxide....	~10,000	5.5	~0.03	6.3	0.16	>100
Alkali-metal-free......	~10,000	5.3	~0.03	7	0.2	>100

the dielectric constant of several ceramic materials compared to fused silica. Note that the range of values for various purity levels of alumina is greater than the range of values for a given composition up to 200°C, which is well beyond the operating temperature for most electronic industry applications.

Dielectric constant as a function of frequency for several ceramics is shown in Table 10.

Dissipation factor This electrical property is a measure of the electrical-loss characteristics of an insulating material and is usually of greatest importance at higher frequencies. As with dielectric constant, the most important variables of this characteristic are temperature and frequency. Ceramics are relatively stable in this respect compared to most plastic materials. The dissipation-factor data for several ceramics, including three alumina compositions, as a function of frequency and temperature are shown in Table 11.

Mechanical Properties

The specific mechanical data for ceramic materials which are used as substrates are given in Tables 1 and 2. Basically, ceramics are brittle materials, giving rise

TABLE 7   Specific-Heat Values for Substrate and Associated Materials[10]

Temperature, K	Specific heat, cal/(g)(°C)				
	Beryllia, BeO	Alumina, Al₂O₃	Silicon	Molybdenum	Nickel
100	. . .	0.03	. . .	0.06	
200	0.15	0.12	. . .	. . .	0.09
300	0.26	0.20	0.17	0.06	0.11
400	0.31	0.23	0.19	0.06+	0.12
600	0.40	0.27	0.21	0.06+	0.16*
800	0.45	0.28	0.22	0.06+	0.13
1000	0.48	0.29	0.23	0.07	0.13
1200	. . .	0.30	0.23	0.07	0.14
1400	. . .	0.31	0.24		

* Magnetic transformation of nickel.

NOTE: Magnesia is very similar to alumina. Aluminum exhibits an almost linear increase in specific heat from 0.20 to 0.30 over the temperature range 200 to 900 K. Copper undergoes an almost linear increase from 0.09 to 0.12 cal/(g)(K) over the temperature range 200 to 900 K. The specific heat of silver is 0.055 to 0.065 from room temperature to its melting point. Titanium exhibits a gradual increase in specific heat from 0.12 to 0.17 as the temperature increases from room temperature to 1140 K. At this temperature a polymorphic change from alpha to beta form results in a drop in the specific heat to 0.15. At temperatures above 1400 K its rate increases until the specific heat is 0.2 at 1800 K.

to breakage problems associated with handling and nonuniform stresses. Substrate failure can be thermally or mechanically induced. Both ceramics and glasses are brittle at normal application temperatures in the electronics industry, following Hooke's law to the point of fracture, which is different from fracture in metals because it occurs without plastic deformation. The brittle nature of ceramics means that strength properties are not best measured by tensile tests. Transverse-bending or modulus-of-rupture tests are recommended. In these tests, the specimen is positioned on two supports with increasing loads applied at the center of the span until fracture occurs. Needless to say, standardization of tests is mandatory, and tests are quite sensitive to specimen uniformity and defects such as microcracks, voids, etc. For reader guidance, the test methods used to gather the data in Table 2 are recorded in the column adjacent to the test description. Some further mechanical information is given below.[4,13]

Compressive strength   Ceramics generally exceed metals in compressive yield strength. When used in a design that capitalizes on compressive strength, alumina ceramics offer the design engineer one of the strongest materials available—at a cost considerably below that of exotic materials.

Eighty-five percent alumina has a typical compressive strength of 290,000 psi at room temperature. As the alumina content of dense ceramic increases, so does the

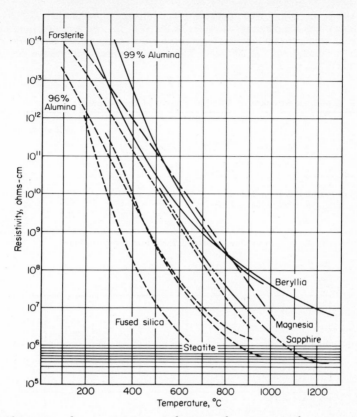

Fig. 7 Change in volume resistivity as a function of temperature for various ceramic materials.[10]

TABLE 8   Dielectric Strengths of Various Insulating Materials at Several Frequencies[11]

Material	Thick-ness, mils	Dielectric strength, rms V/mil						
		60 Hz	1 kHz	38 kHz	180 kHz	2 MHz	18 MHz	100 MHz
Polystyrene (unpigmented)	30	3,174	2,400	1,250	977	725	335	220
Polyethylene (unpigmented)	30	1,091	965	500	460	343	180	132
Polytetrafluoroethylene (Teflon*)	30	850	808	540	500	375	210	143
Monochlorotrifluoroethylene (Kel-F)†	20	2,007	1,478	1,054	600	354	129	29‡
Glass-bonded mica	32	712	643	—	360	207	121	76
Soda-lime glass	32	1,532	1,158	—	230	90	55	20‡
Dry-process porcelain	32	232	226	—	90	83	71	60‡
Steatite	32	523	427	—	300	80	58	56‡
Forsterite, (AlSiMag-243)	65	499	461	455	365	210	112	74
Alumina, 85% (AlSiMag-576§)	55	298	298	253	253	178	112	69

* Trademark of E. I. du Pont de Nemours & Co., Inc., Wilmington, Del.
† Trademark of Minnesota Mining and Manufacturing Co., St. Paul, Minn.
‡ Puncture with attendant volume heating effect.
§ Trademark of American Lava Corp., Ridgefield, N.J.

compressive strength. For example, 99 percent alumina has a compressive strength, at room temperature, in excess of 380,000 psi. As a comparison, hardened, tempered, high-strength alloy steels have room-temperature compressive yield strengths ranging from 275,000 to 319,000 psi. Typical compressive strengths are shown in Fig. 10.

TABLE 9   Dielectric Strength of Alumina at Several Temperatures[10]

Ceramic	V/mil at 60 Hz			
	25°C	300°C	800°C	1000°C
Alumina porcelain (0.125 in. thick)................	400	200	50	30–40
Alumina, 99% (electrode spacing 0.038 in.)..........	450	180	55	35
	250*	70*	35*	17*

* Safe maximum values; sample withstood 2,000 h at these voltages.

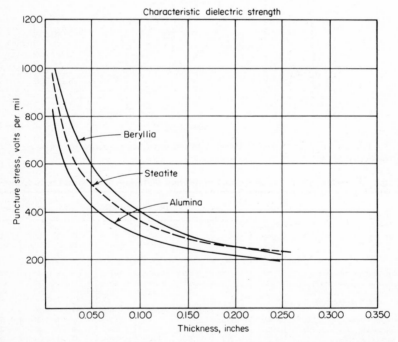

Fig. 8   Dielectric strength of three ceramics as a function of specimen thickness.[12]

At high temperatures, alumina ceramics maintain a much greater percentage of their room-temperature compressive strengths than metals do. Alumina ceramics retain their shapes at temperatures above 3000°F, temperatures far above the melting points of most metals.

Alumina ceramics show no loss of compressive strengths at cryogenic temperatures.

In fact, ceramics exhibit compressive strengths slightly higher at temperatures approaching that of liquid nitrogen ($-320°F$).

Tensile strength   Ceramics have 2 to 5 times the tensile strength of electrical porcelains and far greater tensile strength than such materials as glass, plastic, and steatite. Tensile strengths of ceramics range from 17,000 psi to over 35,000 psi, somewhat less than that of metals.

To take advantage of the unique physical and electrical properties of ceramics, the tensile strength can be accommodated by utilizing a thicker cross section of the ceramic. If the thicker cross section is not practical, it may be possible to revise the design to change the stress from tensile to compressive. The compressive strengths of ceramics are approximately 10 times greater than their tensile strengths.

Modulus of elasticity   High modulus of elasticity in alumina ceramics assures the design engineer of minimal distortion under conditions of high loading. Typical Young's modulus values, at room temperature, are 33×10^6 psi for 85 percent alumina and 52×10^6 psi for 99.5 percent alumina. These values define alumina ceramics as exceptionally rigid materials. Unlike metals, alumina ceramics show only slight elastic deformation under high loads (see Fig. 11). When the load is removed, the near-perfect elasticity of these materials causes the ceramic parts to return to their original dimensions with no plastic deformation.

Dimensional stability   The properties of high hardness, rigidity, and minimum thermal expansion of ceramics combine to provide a material that is highly stable dimensionally. Metals warp or distort during machining or when heated. Plastics tend to flow and to change dimensions under pressure, or when heated, or upon aging. Within the limits of tensile and compressive strengths, alumina ceramics are completely rigid and maintain machined tolerances under conditions of high loads and high temperatures.

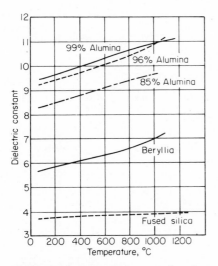

Fig. 9   Dielectric constant as a function of temperature measured at a test frequency of 4 GHz.[10]

Alumina ceramics have no plastic deformation and only slight elastic deformation under high loads. Even when minute flexing occurs, these materials always return to their exact previous shape because they are almost perfectly elastic. They expand

TABLE 10   Dielectric Constant for Several Ceramics as a Function of Frequency[10]

Ceramic	1 MHz		1 GHz		10 GHz		25 GHz	
	25°C	500°C	25°C	500°C	25°C	500°C	25°C	500°C
Fused silica...	3.78	3.78	3.78	3.78	3.78	3.78	3.78	3.78
Steatite......	5.7	6.7	5.5	6.5	5.2	6.0	5.2	
Forsterite.....	6.2	...	5.9	...	5.8	6.3	5.8	
Beryllia, 99%.	6.4	6.9	...	...	6.1	6.3	6.0	6.3
Alumina, 96%.	9.0	10.8	...	9.5	8.9	9.4	8.7	9.0
99%.......	9.2	11.1	9.1	9.88	9.0	9.86	8.9	9.85

upon heating according to low, uniform, and predictable thermal-expansion characteristics.

Mechanical failure in thermal shock   The ability of substrates to withstand rapid temperature changes without damage is generally referred to as their thermal-shock resistance.   This complex quantity is related to the thermal conductivity, the specific heat, and the density of the material since these properties together determine the temperature of the material during the change.[13]   It is also dependent on the coefficient of thermal expansion and the elastic modulus of the substrate, because these are the

TABLE 11   Dissipation Factor of Several Ceramics as a Function of Temperature and Frequency[10]

Tem-pera-ture, °C	Alumina			Beryllia, 99%	Forsterite	Steatite	Fused silica
	85%	96%	99%				
1 MHz							
25	0.0004	0.0003	0.0002	0.0001	0.0004	0.002	0.0002
300	0.002	0.003	0.0006	0.0001	. . .	0.006	
500	0.009	0.013	0.002	0.0004	. . .	0.06	
800	0.06	0.09	0.005	0.003			
1 GHz							
25	0.001	0.0003	0.0002	0.0002	0.0005	0.0015	0.0001
300	0.002	0.0007	0.0003	0.0003	. . .	0.004	
500	0.004	0.0015	0.0015	0.0006	. . .	0.015	
4 GHz							
25	0.00.3	0.0007	0.0002	0.0003	. . .	. . .	0.0004
200	0.0013	0.0008	0.0003	0.0004	. . .	. . .	0.0002
400	0.0018	0.0009	0.0003	0.0004	. . .	. . .	0.0003
600	0.0033	0.002	0.0004	0.001	. . .	. . .	0.0005
800	0.005	0.003	0.0008	0.002	. . .	. . .	0.0009
1000	0.009	0.007	0.002	0.003	. . .	. . .	0.002
1100	. . .	. . .	0.006	0.004	. . .	. . .	0.004
10 GHz							
25	0.0015	0.0006	0.0001	0.0001	0.0009	0.003	0.0001
300	0.002	0.001	0.0001	0.0001	0.001	0.004	0.00008
500	0.003	0.002	0.0002	0.0001	0.0013	0.005	0.00009
800	. . .	0.006	0.0005	0.0005			
25 GHz							
25	. . .	0.0007	0.0003	0.004			
300	. . .	0.0009	0.0003	0.004			
500	. . .	0.002	0.0003	0.004			
800	. . .	0.005	0.0004	0.006			

factors which govern the resulting stress. Thermal-shock resistance is often used empirically by referring it to the behavior of different materials in a particular thermal cycle. To obtain a more quantitative and universal figure of merit, a coefficient of thermal endurance F has been defined by Winklemann and Schott:[18]

$$F = \frac{P}{\alpha E}\sqrt{\frac{k}{\rho c}}$$

where P = tensile strength
α = linear coefficient of thermal expansion
E = Young's modulus
k = thermal conductivity
ρ = density
c = specific heat

Although the equation yields only qualitative agreement with experimental observations, it is of value in comparing different substrate materials with each other.

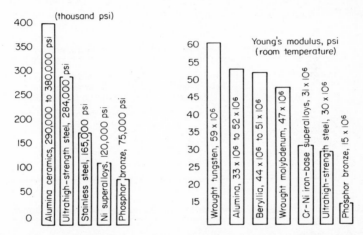

Fig. 10 Compressive yield strengths of typical metals compared to alumina ceramics.[4]

Fig. 11 Young's modulus comparison data for ceramic and metal materials.[4]

Table 12 shows the relative shock resistances of four materials together with their expansion coefficients, since the latter exert the greatest influence on the substrate's ability to withstand shock. It is of interest to note that fused silica has the highest shock resistance while other glasses are at the bottom of the list.

The resistance to thermal stress differs from the shock resistance insofar as it refers to the ability of a body to withstand nonuniform stress. Thermal stress is generated if the expansion of one part of a substrate is impeded by adjacent material which does not expand. This condition may arise from a number of causes.[1] Examples are anisotropic polycrystalline bodies and two-phase materials such as glass ceramics, glazed ceramics, or glazed metals. Thermal stresses are also produced in homogeneous bodies if temperature gradients are present. During heating and cooling, for instance, the substrate surface responds more rapidly to the induced change than the interior; hence, a temperature gradient exists perpendicular to the surface. Transverse gradients are caused by electrically loading thick film components which cover only a part of the substrate.

It is difficult to assign quantitative values to the stress resistance of substrates. Qualitatively it is probably related to the thermal shock resistance, and the same order of merit prevails as in Table 12. Failures due to excessive stress alone are probably very rare. Substrate cracking or chipping are more likely to be due to a combination of factors including mechanical and thermal shock as well as thermal stress.

SUBSTRATE SURFACE FACTORS

The substrate surface requirements for thick film deposition ($\sim 1 \times 10^{-3}$ in.) are not as stringent as those for thin film deposition (~ 100 Å or 4×10^{-7} in.). Consequently, either glass or a glazed ceramic can be used for thin film deposition whereas lower cost, as-fired ceramic substrates can be used for thick films. The rougher, as-fired surface of ceramics promotes adhesion of thick film inks, which is a desirable advantage. The surface finish of a drawn glass is typically smoother than a polycrystalline ceramic. The surface finish of a ceramic depends upon its microstructure and density. A high-density ceramic composed of small-sized grains will form a smoother substrate than a ceramic composed of larger grains. As-fired alumina substrates with a surface finish of under 5 μ in. are commercially available, and a typical specification is shown in Table 13. While such fine surfaces are not predominantly required in thick film technology, they could offer advantages for micro-

TABLE 12   Relative Shock Resistance of Substrate Materials[13]

Material	Thermal-endurance factor F	Thermal coefficient of linear expansion, ppm/°C, for 0–300°C
Silica..............	13.0	0.56
Alumina..........	3.7	6.0
Beryllia..........	3.0	6.1
Glass.............	0.9	9

TABLE 13   Technical Specifications for Fine-Surface, As-fired Alumina Substrates[14]

Physical properties:
Composition.......... 99.5% Al_2O_3
Surface.............. As-fired, glaze-free finish characterized by a 3- to 4-μin. centerline average surface smoothness; extremely low incidence of macro-defects such as burrs, blisters, etc.
Density.............. 3.91–3.94 g/cm³; substrates impervious as measured by dye-penetration tests
Thickness............ Current standard thickness 0.026 in.
Thickness control...... ±0.0015 in.
Maximum camber..... 0.003 in./in.
Average camber....... 0.002 in./in.
Grain size............ 1 μm
Flexural strength...... 70,000 psi
Miscellaneous......... Substrates can be supplied in a prescored condition; fired substrates may be diamond-scribed for controlled separation

Electrical properties:
Dielectric constant..... 9.6 at 1 MHz
Dissipation factor...... 0.00025 at 1 MHz
Thermal properties:
Thermal conductivity.. 0.088–0.090 (cal)(cm)/(s)(°C/cm²)
Thermal expansion:
At 25–300°C........ 63×10^{-7} in./(in.)(°C)
At 25–600°C........ 71×10^{-7} in./(in.)(°C)
At 25–800°C........ 73×10^{-7} in./(in.)(°C)

wave applications and, of course, for general thin film work. Development of the surface characteristics of alumina ceramics is further discussed in the section on Substrate Fabrication and Design Factors.

Surface Characteristics and Measurements

The primary surface characteristics may be considered as roughness (or smoothness), waviness, and flatness. Surface measurements for identifying these characteristics and some pertinent comments about these characteristics are presented below.

Measurement of surface characteristics   One of the most practical methods for measuring surface characteristics is profilometry, which has been used for surfaces where resolution greater than 1 or 2 μ in. is not necessary. Briefly, this method involves the use of a fine stylus which is attached to a tracer arm. As the stylus is made to traverse the surface, a magnified profile of the surface is drawn on a chart. This method suffers from a lack of two-dimensional resolution. Linear resolution is also limited by the size of the stylus tip. As a result, this method can reproduce

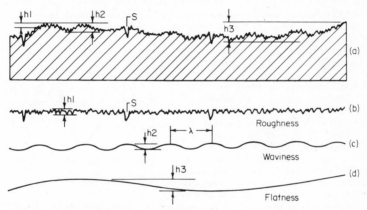

Fig. 12   (a) Stylus trace of a surface with several features; (b) roughness $h1$; (c) waviness $h2$; and (d) flatness deviation, $h3$.[13]

only scratches, high-defect clusters, and macroscopic imperfections. However, the resolution obtainable with this technique is quite satisfactory for usual thick film work.

A practical advantage of the technique of stylus-profilometry surface measurement is the fact that it yields a quantitative measure of surface roughness. The statistical significance of the various figures of merit derived from stylus instruments is given below.[13]

The Root-Mean-Square (rms) Roughness.   A typical stylus trace of a substrate surface is shown in Fig. 12a. The irregularities of the surface may be thought of as consisting of three components of different periodicities, referred to as roughness, waviness, and flatness in Fig. 12b, c, and d. Roughness is the property which stylus instruments measure; it can be characterized numerically by the average deviation of the trace from an arbitrary mean. Referring to the trace in Fig. 13, the rms value is derived by dividing the peaks and valleys into narrow segments of height y, summing overall y^2, and subsequent averaging:

$$\text{rms} = \sqrt{\frac{y_1^2 + y_2^2 + y_3^2 + \cdots + y_n^2}{n}}$$

As this illustration shows, the rms value is an integral quantity which is useful in assessing the total of a quantity fluctuating over a period of time. An example is the power conveyed by an alternating current. However, the rms value does not constitute a direct measure of the surface roughness, although it is occasionally used for that purpose.

The Arithmetic Average (AA or CLA) Roughness. The present standard of expressing surface roughness is called the *arithmetic average* (AA) in the United States (ASA B 46.1-1962) or the *centerline average* (CLA) in England. To derive this value, the abscissa of a trace as shown in Fig. 13 must be drawn over an assigned length of surface so that the areas under the curve above and below the centerline are equal. The mathematical definition of this quantity is

$$\text{AA} = \frac{a + b + c + d + \cdots}{ML}$$

where $a, b, c, d, \ldots$ = areas under peaks or above valleys of trace (see Fig. 13)
L = assigned length of stylus travel
M = vertical magnification used

Typically, the AA value is 10 to 30 percent lower than the rms value for the same trace. It has the advantage of an unambiguous mathematical definition, it can readily be measured from a stylus trace by means of a planimeter, or it can be derived electronically if an integrating instrument is available. Most stylus instruments are equipped for this option and provide the AA value directly.

In using statistical averages to characterize substrate surfaces, it should be clear that a single number can give only a very incomplete description of the surface texture. Low values may mask the presence of deep scratches (feature s in Fig. 12) which may cause discontinuities in deposited thin films but are not as critical in thick films. Similarly, surface waviness and flatness are not reflected in rms or AA

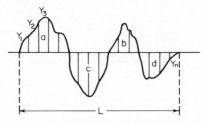

Fig. 13   Schematic of a surface trace.[13]

values, although both properties are important in photoresist work. Finally, there is some ambiguity in both methods because surface profiles having the same amplitudes but different periodicities yield identical average values. Other factors to consider when stylus traces are interpreted are the tip radius (typically 0.05 to 0.1 mil), which limits the fine resolution in the direction of travel, and the different magnifications utilized in the horizontal and vertical axes of the profile. Three typical profilometer measurements are shown in Fig. 14.

Surface flatness   Flatness or warpage constitutes one of the greatest problems associated with thick film substrates. A particularly important potential problem is stability of deposited resistors. Printing resistors onto a warped substrate, followed by flattening the substrate in a subsequent sealing operation, represents a practical case of potential problems. The sealing process flattens the substrate, which stresses not only the substrate itself but the ink on it, as well. This stressing affects the resistance. The longer the resistor or the bigger the substrate the worse the effect. Substrate flatness would play a significant part in the manufacture of ladder networks (very sensitive resistor-conductor networks), for instance. Using a thicker substrate (50 mils or more) is one way to overcome the stressing problem. Basically, however, substrate suppliers have standardized on a 25-mil substrate, for cost-versus-flatness factors. For very tight-tolerance work, low thermal coefficient of resistance (TCR), or electrical stability, for instance, consideration should be given to use of thicker substrates.

Some screen-printing equipment comes with a torsion-bar system which accommodates for camber, up to a point, thereby permitting more uniform ink deposits, even on a nonflat substrate. However, in substrates with compound camber (nonflatness in the length and width) even these improved screen printers are not sufficient to overcome the flatness problem.

SUBSTRATE FABRICATION AND DESIGN FACTORS

In applying substrates to thick film applications, some knowledge of substrate-fabrication techniques, as well as dimensional and design factors, will provide the designer with a better understanding of the materials with which he is working, and hence the design limitations imposed by the nature of ceramic-fabrication technology.

The Basics of Ceramic Processing

A brief explanation of the nature of ceramics will help in understanding these materials.[4]   Alumina ceramics, as stated earlier, are fired ceramic compositions in which the major crystal phase is alpha-alumina, or corundum (sapphire).   The aluminum oxide content must be 80 percent or higher for the material to be considered a "high alumina" ceramic.   Common aluminas range from 85 percent alumina content (by weight) to 99.5 percent, according to the alumina content.

The alumina powder (in a finely ground state having a particle size of 5 μm or smaller) accompanied by fluxing agents such as silica, magnesia, or calcia, are compounded together at high pressures.   They are then sintered to form a high-strength, dense, nonporous ceramic material.   During the sintering or firing process, tempera-

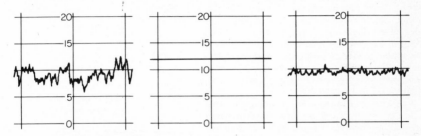

Fig. 14   Talysurf surface-profile measurement of a 23-μin. CLA as fired alumina surface (*top*), a 0.3-μin. CLA glazed surface (*center*) and a 7-μin. CLA as fired surface (*bottom*). AlSiMag is a trademark of American Lava.[12]

ture, atmosphere, and time are carefully controlled to produce the desired polycrystalline structure of the particular ceramic part.   After firing, the ceramic is composed mostly of the pure alumina crystals, carefully controlled for size.   The bond is achieved by the fluxing-compound matrix (a tough glassy phase) and the interlocking crystals.

Thus, the apparent surface hardness is that of the pure alumina crystals, which compose the largest percentage of the surface.   However, the small interstices between the crystals are filled with the relatively softer glassy-phase matrix and are therefore subject to erosion by impingement by ultrasmall particles.   When a surface is polished, the crystals may be perfectly flat, but the spaces in between may be lower, because the softer material is more quickly worn away, presenting a profile of plateaus and valleys reflecting the crystal size of the alumina crystals.   The alumina crystals can be polished to smooth, flat surfaces of extreme hardness, but designers must consider the small valleys possible in all ceramics.   Designers can turn these small valleys into advantages, e.g., for improving and controlling adhesion of thick film inks.

Ceramics are subject to attack by the fluoride ion because of the minute amount of exposed glassy-phase matrix at the crystal interstices.   However, the alumina crystal remains completely inert.

The phenomenal compressive strength of alumina ceramics is due to the crystal formation.   With most of the crystals lying against each other, the compressive strength of the ceramic approaches the strength of the pure crystal.   Consequently, the compressive strength increases as the amount of the alumina content of the ceramic increases.   The tensile strength of the ceramic is lower than the compressive strength because it is derived from the combined strength of the glassy-phase matrix and the mechanical strength of the interlocking crystals.

Temperature stability, even under repeated cycling, is achieved because the ceramic structure does not change until temperatures reach the softening point of the ceramic.

Substrate Fabrication Methods

Ceramic substrates can be made by several different forming or fabrication methods, in which the substrate part is formed and then fired to the desired production. Finishing operations such as grinding or glazing may also be required. A flow chart indicating three alternate forming methods (pressing, extrusion, and casting) is shown in Fig. 15. The forming methods are discussed in more detail below.[13]

Powder pressing In powder pressing, dry or slightly dampened powder is packed into an abrasion-resistant die under a sufficiently high pressure (8,000 to 20,000 psi) to form a dense body. This process allows rapid or automatic production of parts with reasonably controlled tolerances since the shrinkage during the sintering process is slight. There are limitations, however. Holes cannot be located too close to an outside edge. Pressure variations from uneven filling of long or complex dies lead to defects like inhomogeneous properties and excessive warp. As a rule, powder pressing is not recommended for parts larger than 6 in. square.

Isostatic pressing In contrast to dry pressing, this method applies uniform pressure to the powders. Here, dry powders are enclosed in an elastic container and inserted into a cavity which is first evacuated and then filled with a liquid like water or glycerin. Pressures ranging from 5,000 to 10,000 psi are applied through the liquid and yield a uniformly compacted piece. Production of 1,000 to 1,500 pieces per hour is feasible. An important advantage of isostatic pressing is that it permits fabrication of pieces with relatively large length-to-width ratios. For example cylinders of 14-in. diameter and 24-in.

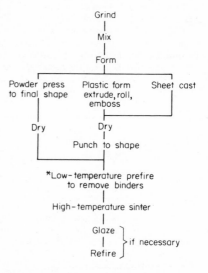

*May be separate heat treatment or combined with high-temperature sintering step.

Fig. 15   Ceramic-substrate-forming processes.[13]

length can be made. However, the surfaces which contact the elastic container must be machined since they are not smooth enough for substrate use. This extra step adds to the cost of isostatic pressing and makes it uneconomical for the production of most substrates.

Extrusion If oxide powders are mixed with certain organic materials, the mixture becomes plastic enough to be forced through a die. Extrusion is particularly useful in forming long pieces of uniform cross section such as cylindrical resistor cores. Extrusion is a fast process, it can reproduce fine detail, and it is economical. However, considerable shrinking occurs during drying and firing so that close final tolerances are hard to obtain. Thin sheets up to 0.1 in. thick can be prepared by passing the extrusion through successive preset rollers until the proper prefiring thickness is obtained.

Sheet casting (green tape) This is a low-pressure process of considerable versatility. The oxides are prepared as a slurry by adding liquid or organic binders, plasticizers, and solvents. The slurry is spread onto a carrier film of Mylar or cellulose acetate which moves at a constant speed under a metal knife blade positioned a short distance above the film. Such a mechanism is shown in Fig. 16. As the film and slurry move under the blade, a thin sheet of wet ceramic forms. The thickness of this sheet is controlled by adjusting the height of the blade over the carrier film. The resultant ceramic sheet, usually referred to as *green sheet* or in the *green state*, is air-dried to remove the solvents. Typically, a 30 to 50 percent alignment edges and orientation marks are subsequently punched out of the flexible green sheet. The sheets may then be used either individually or as laminates.[15]

Sintering of oxides After the substrate has been formed by one of the above processes, it is often necessary to prefire at 300 to 600°C to remove organic binders, lubricants, and plasticizers used as forming aids. Further heating at higher temperatures to densify the aggregate of small particles is called *sintering*. This results in a ceramic body of considerable mechanical strength. Sintering of solids consolidates the fine particles by recrystallization, a process whereby the larger grains grow at the expense of the smaller ones. Excessive grain growth is responsible for the undesirable roughness exhibited by many as-fired ceramics. Recently, it has become possible to inhibit grain growth during firing and to produce high-purity

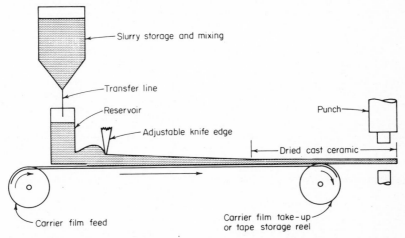

Fig. 16 Schematic of sheet-casting method for forming thin ceramic substrates.[13]

dense alumina with markedly improved surfaces and improved mechanical and electrical properties, as noted in Table 13.

During sintering, ceramics shrink as much as 18 to 25 percent in their linear dimensions or up to 50 percent in volume because of loss of binder and particle coalescence. Numerical examples of the nominal shrinkage accompanying various sintering processes are given in Table 14, along with the tolerances that can be obtained commercially.

Glazing Undesirable as-fired surfaces can be improved by coating one or both sides of the substrate with thin glassy layers called *glazes*. These consist of low-melting silicate glasses which are squeegeed, dipped, or sprayed on. Another firing cycle melts and fuses the glaze to the body. Glazes may also be applied by the so-called *transfer-tape process*.[16] This consists of mixing the glaze in the form of a fine powder with suitable organic binders and plasticizers. The resulting slurry is cast onto a carrier sheet such as Mylar or cellulose acetate. The glaze layer may be coated with an organic adhesive if a sticky surface is desired. It is then pressed over the substrate surface to be covered, the carrier layer is stripped, and the glaze is fired as described above. This method permits selective application of the glaze at thicknesses between 15 and 250 μm (0.0006 to 0.010 in.) to virtually any material. While adherence of glazes to substrates is not a problem, the coefficients of expansion of the body and glaze materials should be close to prevent excessive strain (see the earlier section on Mechanical Failure in Thermal Shock).

Hot pressing   In this method, powders are densified by the simultaneous application of high temperatures and pressures. It is used to form materials which normally sinter with difficulty or which require very high sintering temperatures. In contrast to normal sintering, densification occurs without grain growth, but the process is slow. Since it requires long heating and cooling times for the sample as well as for the dies and associated equipment, hot pressing is a relatively expensive method of substrate production.

Substrate Dimensional and Design Factors

Some typical dimensional design criteria and other design considerations for use with ceramic materials are presented at this point. As mentioned in the previous section and detailed in Table 14, shrinkage of ceramic substrates is quite high in the forming operations. This general knowledge and supplier data for a particular

TABLE 14   Substrate Shrinkage and Tolerances for Various Ceramic-forming Processes[13]

Forming process	Material	Firing shrinkage, %	Commercial tolerances
Powder pressing.............	Alumina, 94–99% Steatite Beryllia, 94–98% Barium titanate	16 10 16 14	±1% but not less than ±0.005 in. in any dimension
Extrusion..................	Same as for powder pressing	13–14	±½% but not less than ±0.003 in. in any dimension
Isostatic pressing............	Alumina, 94–99% Steatite Beryllia, 94–98%	16–18 10–12 16–18	Same as for powder pressing
Sheet casting...............	Alumina, 94–99% Steatite Beryllia, 94–98% Barium titanate	18–22 18–22 18–22 18–22	Length and width: ±½%; thicknesses up to 0.040 in.: ±10%

product will be required by the product designer. Also, as general guidance, dimensional design criteria for a commercial 95 percent alumina and a 99.6 percent alumina are shown in Table 15.

Estimating Substrate Area and Density for Hybrid Microcircuits

When converting discrete circuits to hybrid microcircuits, it is necessary to know the substrate-area requirements and the details of the component density. One useful technique which has been developed is that of the unit system.[18] It consists of assigning a number of units of area to each circuit component and then adding up the total number of units to determine the required area. One unit is defined as the substrate area required for one general-purpose thick-film resistor, rated at 100 mW after trimming to ±2 percent tolerance. Table 16 shows the number of units normally allocated for other types of conventional circuit elements. In each case, the number of units specified includes whatever additional area is required for terminations, resistor trimming, wire-bonding pads, and spacing between adjacent components.

If the circuit has been defined but the package has not been chosen, the substrate area can be estimated. Start with an optimum one-unit resistor area of 0.015 ins. Assign the proper number of units to each component in the circuit, according

TABLE 15 Dimensional Criteria for a 95% and a 99.6% Alumina[17], *

Thickness		Hole size		Length and width	
Thickness range, in.	Tolerance, in.	Size, in.	Tolerance, in.	Dimensions, in.	Tolerance,† in.
As-fired:		0.014–0.059	±0.002	0.050–0.199	±0.002
0.008–0.014	±0.001	0.060–0.187	±0.003	0.200–0.299	±0.003
0.015–0.029	±0.002	...	...	0.300–0.399	±0.004
0.030–0.050	±0.003	...	...	0.400 and up	±1%
Ground:					
0.010–0.050	±0.001				

Surface finish, Cleveland measuring equipment, 0.050 cutoff, 0.0005 radius stylus

	Alumina	
	95%	99.6%
As-fired ceramic...	40 μin.	10–20 μin.
Diamond-polished ceramic..................................	10 μin.	As low as 2 μin.
Minimum fired thickness....................................	0.008 in.	0.010 in.
Camber: 0.004 in./in. standard; 0.003 in./in. available on request; NLT 0.002 in. overall		

* Data are for both types except where noted.
† Outside-dimension tolerances may be held less than indicated.

TABLE 16 Typical Unit-System Component Areas for Estimating Substrate Area and Density[18], *

Component type	Per component
Resistors (cermet, thick film):	
General-purpose (up to 100 mW).........................	1.0 unit
Precision, ratio tracking, aspect ratio ≤ 4:1, ≤ ±1%.......	2.0 units
Capacitors:	
Screened (cermet)......................................	270 pF/unit
Chip capacitors 0.1 by 0.1 in...........................	2.0 units
Diodes, passivated chip:	
Signal/switching.......................................	0.5 unit
Zener/reference..	0.5 unit
Schottky/hot carrier...................................	0.5 unit
Transistors, passivated chip:	
Bipolar small signal...................................	0.5 unit
Bipolar low or medium power...........................	1.0 unit
JFET...	0.5 unit
Integrated circuits, passivated:	
Linear (741, 710, 107, etc.)............................	2.0 units
Digital (935, 946, 7400, etc.)..........................	4.0 units
MOS arrays (3101, etc.)................................	0.5 unit/lead
MSI devices (74145, etc.)..............................	0.5 unit/lead

* For metric-system applications, the unit system can be used as shown, except that the one-unit resistor area becomes 9,677 mm².

to the table, and total the number of units for the circuit. Determine the substrate area required as follows:

$$A_s = (0.015 \text{ in.}^2)U_T$$

where A_s = required substrate area
U_T = total number of units

A package choice based on available substrate area can then be made from the many different styles available.

When the circuit and package designs are both firm, the component density in square inches per unit can be estimated. Add up the total number of units for the circuit, according to the table. Determine the available substrate area in the package size you intend to use. Calculate the component density as follows:

$$D = \frac{A_s}{U_T}$$

A component density of 0.015 in.^{2}/unit is considered a moderate density level if the supplier of your final package has the freedom to make pin assignments. Using more specialized fine-line screening techniques, densities to 0.006 in.^{2}/unit have been achieved. Note that component density is appraised in area per unit, instead of units per area. Area per unit is more directly correlated to layout dimensioning.

Some components require special consideration because of their large size or an unusually large number of leads. Since the unit system is strictly a shortcut method, a considerable amount of good judgment and common sense must be applied. The following general guidelines should be considered:

• It may be easier to deduct the area required for large capacitor chips before applying the unit system.

• High component densities are easier to achieve in smaller packages because of the proximity to pins.

• Some circuits naturally flow from input to output without complex feedback interconnections.

• Packaging efficiency is usually greater in larger hybrid packages, but yield and testing help set the limit.

SUBSTRATES FOR MICROWAVE APPLICATIONS

While the use of thick film technology does not predominate in the area of microwave applications, the economic and manufacturing efficiencies of thick film technology do lead to frequent microwave applications. There are, of course, certain uniquely important parameters for the use of substrates for microwave applications. These parameters primarily have to do with minimizing losses and maintaining the greatest possible uniformity of substrate properties. Hence, control of electrical properties such as dielectric constant and dissipation factor are very important, as are control of substrate purity, density, and surface properties. Some useful data and information on these topics will be presented in this section.

Substrate Parameters

Chemical purity, substrate density, and surface finish The circuit loss will depend directly upon the physical properties of the substrates.[20] Chemical purity will obviously affect not only the absolute value of the dielectric constant and loss tangent but also their variation over the substrate. Different substrate manufacturers add their own grain-growth inhibitors and sintering aids; there are differences in the impurities in the starting alumina powder and any impurities added unintentionally during processing. For example, ball milling may introduce foreign matter from both the balls and the mill, and pressing can introduce material from the dies. It has been shown that Na_2O is the worst offender from this point of view and that even a nominal 99 percent alumina may have a very high loss tangent due to this source. However, the loss tangent and dielectric constant are generally more closely related to the density of the material than to its chemical impurities. A good alumina substrate should have a dielectric constant approaching 10 and a tan δ of less than 1×10^{-3}. These properties must be uniform over the whole substrate area if losses are to be minimized. Density variations due to pores, excessive grain growth, or agglomerations of the glass phase will seriously influence these properties. With high-alumina bodies, the desired electrical properties can be achieved.

Substrate surface finish influences not only the conductor definition but the resistive loss. It has been shown that the loss approaches a plateau of the order of 0.2dB/in. at about 10-μin. CLA when the tests are done on wide conductor lines, and little apparent improvement was obtained by using 2-μin. CLA substrates.[21] This improvement of a factor of 5 in surface finish incurs an increase in substrate cost of a similar factor. If fine lines are required, particularly at high frequencies, however, there is little alternative, since even 10-μin. CLA substrates have large voids on the surface.

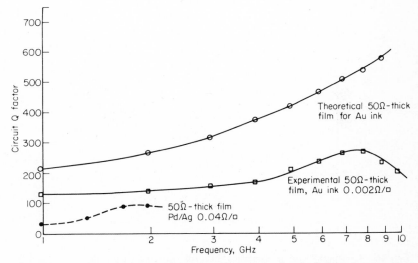

Fig. 17   Circuit Q factor as a function of frequency for thick film circuitry.[20]

Performance data for thick film circuits   Optimization and performance data have been developed by several investigators. [20,22] Generally, 325-mesh stainless-steel screens were found to give the best line definition.[20] A major disadvantage is that it is very difficult to print smaller than 0.004-in. lines on a 0.004-in. space. Even for wider lines, the edge definition is not as good as etched thin film.

Reasonable resistivity inks are Pt, Pd, Au, or Ag alloys, which are in the range of 0.02 to 0.1 Ω/sq, compared with pure Au inks in the range 0.002 to 0.02 Ω/sq. About 0.0005 in. of material is deposited onto the substrate. Bonding to the ceramic is achieved via the glass frit in the ink, which reacts with the glassy phase in the alumina. Obviously, therefore, a compromise has to be reached in terms of dielectric loss on a debased alumina with some glassy phase to give a good bond. Surface finish also affects adhesion. The most satisfactory results have been obtained from a 97.5 percent alumina with 8-μin. CLA surface finish when all these points were taken into account.[20]

As can be seen in Fig. 17, the results obtained on a simple resonant circuit do not approach that predicted theoretically for the nominal resistivity of the ink. This is because of the raggedness of the sides of the conductors on the one hand and the effect of the glass phase at the substrate-conductor interface on the other. This latter could account for anomalous variations of the Q of the circuit at higher frequencies. The Au ink contained very little glass, whereas the Pd-Ag ink had substantially more glassy phase. The loss per unit length in the case of the Au ink is a factor of 2 worse than a thin film plated up at about 2 GHz, which means that thick film could find a place in the manufacture of L-band radar and communications circuits.

In addition to the data given above, Fig. 18 shows a comparison of propagation loss for a thick film and a thin film microwave circuit.

Based on the investigation by Bosnell,[20] the following conclusions were drawn relative to substrates for microwave applications:

• Density or grain size is the major factor in determining dielectric loss in alumina substrates, with the proviso that Na_2O can sometimes cause trouble.

• Variation of electrical parameters over substrate area and batch to batch is less for 99 percent alumina bodies than for 95 to 96 percent alumina substrates.

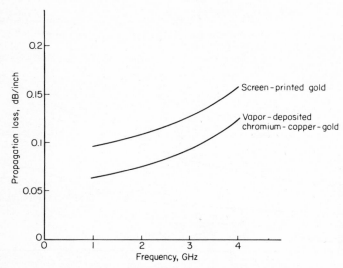

Fig. 18   Comparison of propagation loss for thick film and thin film circuitry.[22]

• Problems can occur due to devitrification of the glassy phase in the substrate under certain high-temperature manufacturing processes, particularly if the major glass constituents are CaO and SiO_2 only.

• Very little difference was found between the Mo or Cr seed-layer systems in the thin film process for microstrip manufacture or between Au or Cu in terms of loss per inch in strip line.

• For substrates with a surface finish of less than approximately 10-μin. CLA little or no change in loss was observed. Bow and camber must be a minimum.

• The Q of a circuit manufactured by thick film techniques was found to be about half that of the same circuit in thin film, up to about 8 GHz, at which point a severe drop occurred. Obviously, therefore, if short lines are required at lower frequencies, thick film is a useful manufacturing technique for microstrip.

• It was found that the best substrate compromise for thick film was a 97.4 percent alumina, 10-μin. CLA material. A 325-mesh screen using an indirect emulsion was found to give optimum results.

SUBSTRATE CLEANING

Since presence of contamination on the surface of substrates can produce adverse effects on both the adhesion of deposited thick film materials and electrical performance of those materials in their circuitry functions, substrate cleaning before further processing is recommended. While much debate is possible on selection of cleaning materials and methods, some basic guidelines can be followed.   Adaptation of these basics to the individual situation is preferable to specifying some cleaning technique without applying basic logic.   There are, however, some practices which are generally satisfactory for common type thick film applications.

Basic Substrate Cleaning Considerations

Common soils may be classified a number of ways. They may be water-soluble or not. They may form reaction products with acids or alkalies. They may be thermally attacked or not. Most often, since the type of soil on the surface of substrates is not accurately known, basing the cleaning procedure on chemical principles is not possible. However, the most common soils are oil-based soils of various types. Hence, vapor degreasing with trichloroethylene or ultrasonic cleaning with trichloroethylene or Freon are good general-purpose substrate-cleaning techniques for thick film substrates. Fortunately, substrate cleaning is not usually as critical in thick film technology as it is in thin film technology. This is because of the normally rougher surface of thick film ceramic substrates, which greatly improves the adhesion of deposited thick film materials. Because of the adhesion advantage of ceramic substrates, compared to glass or glazed ceramic thin film substrates, vapor or ultrasonic degreasing usually suffices as a cleaning method. If other specific soils are known or suspected, the correct chemical attack should be determined. Where water-soluble soils are suspected, deionized water or high-purity alcohol should be used. Regarding vapor or ultrasonic degreasing, the hot vapors of vapor degreasing give better solubility of oils and organic contaminants, while ultrasonic cleaning, with the associated agitating actions of this cleaning method, is better for dislodging particles from the interstices of rough-surfaced substrates.

In addition to degreasing, oil or alcohol cleaning, or chemical cleaning, heat cleaning should be considered. Many cleaning operations leave films or residues which also have an effect on subsequent operations. Such films or residues increase as cleaning solution deteriorates or is expended with continued use. Strict control of cleaning-solution cleanliness is most important. In any event, due to the high-temperature stability of unglazed ceramics, film residues or uncleaned surfaces can often be cleaned by firing at high temperatures (1000°C or so). Open flame may even be used, but thermal differentials and soot formation can be problems.

Surface-Cleanliness Tests

Once a substrate has been cleaned, it is frequently desirable to test the effectiveness of the cleaned surface.[13] Several tests which are useful to assess the cleanliness of substrate surfaces are listed in Table 17. The first three methods are based on the wettability of surfaces and may help in recognizing the presence of hydrophobic contaminants on normally hydrophilic materials or vice versa. Impurities having the same affinity to water as the substrate material are not detected. Although the tests are quite sensitive, the interpretation of the observed surface condition is somewhat subjective. Furthermore, the rougher the surface the more difficult the characteristic wetting patterns are to recognize. The contact-angle method, too, relies on surface wettability but yields somewhat more quantitative information.

The coefficient-of-friction method is perhaps more valuable for special thin film work, since it is one of the few techniques which can be applied within vacuum systems as well as outside. Although it yields numerical values, a correlation between the measured coefficient and the concentration of contaminants on the surface is not easily established. In addition, the test may be destructive.

Adhesion tests in one form or another are invariably destructive insofar as one cannot assume that the original surface condition persists throughout the test. Therefore, this method requires test samples representing the same cleaning technique as applied to the substrates. Also, adhesion tests are strictly functional and merely indicate whether a particular cleaning method yields the desired result or not. Consequently, the usual practice is to establish one test empirically as being indicative of a surface condition which will yield the desired film property. Some form of adhesion test on a sampling basis may be the most practical for thick film work.

TABLE 17   Methods of Testing Surface Cleanliness[13]

Method	Description of test	Sensitivity, monolayer	References
Black-breath figures	Surfaces which are free of hydrophobic contamination allow condensation of a smooth specular water film of low reflectance; the smooth water film has a lower refractive index than the glass and forms an antireflection coating; as it evaporates, the film thickness decreases, and interference occurs; contact angles of droplets in the black-breath figure approach zero, while on less clean surfaces, the contact angles are appreciably larger; such water films are called *gray-breath figures*; clean glass showing black-breath figures has an abnormally high coefficient of friction ($\mu_s = 1$, see below)	. . .	Holland[23]
Atomizer test	The dry substrate surface is exposed to a fine water spray; on a contaminated surface, the sprayed droplets coalesce into larger ones, whereas on a clean surface, the fine mist remains	$\frac{1}{10}$	ASTM F21–62T
Water-break test	If a clean substrate is slowly withdrawn from a container filled with pure water, a continuous film of water remains on the surface	1	ASTM F22–62T
Contact angle	The contact angle between a water droplet and a surface is a quantitative measure of the surface wettability; on completely wettable surfaces, the contact angle is 0°; on nonwettable surfaces, the droplets assume spherical shapes with contact angles approaching 180°	. . .	Longman and Palmer[24]
Coefficient of friction μ_s	The resistance to sliding of a metal or glass object on a substrate surface depends on the cleanliness of the surface; the closer μ_s approaches 1, the cleaner the surface	$\frac{1}{2}$	Holland[23]
Coefficient of indium adhesion σ	The surface is tested by measuring the coefficient of adhesion σ between it and a clean piece of indium; σ is defined as the ratio of tensile force required to produce adhesion failure to the joining force; it ranges from 0 dirty surfaces to as high as 2 for clean ones. The method works equally well for hydrophobic and hydrophilic surface contaminants; freshly fractured silicon of high purity and rigorously cleaned glass yield coefficients of 2	$\frac{1}{2}$	Krieger and Wilson[25]
Film adhesion	The degree of adhesion of metal films on substrate surfaces also indicates surface cleanliness; however, lack of adhesion is not always caused by surface contamination		
Fluorescent dyes	Certain surface contaminants such as oils absorb fluorescent dyes, or dyes may be added to such materials as resists; residues of this type are readily detected when illuminated with ultraviolet light	1	Missel et al.[26]

SUBSTRATE SOURCES AND COSTS

There are many sources of substrates available for almost any type of substrate desired. A listing of a large number of substrate manufacturers and processors, indicating the type of products and/or services of each, is shown in Table 18. This listing covers not only the United States but Europe and Japan as well. While this listing may not be all-inclusive, it will provide the user or purchaser with an excellent set of suppliers from which to choose.

Regarding costs of substrates, these data can be accurate only for specific cases, of course. Nevertheless, some guiding approximations are possible. Such a set of cost approximations is shown in Table 19, using soda-lime glass as the base of ap-

TABLE 19. Cost of Selected Flat Substrates Relative to Soda-Lime Glass[13]

Material	Cost, arbitrary units based on soda-lime glass = 1
Soda-lime glass (drawn)............................	1
Alkali-free glass (drawn)...........................	3
Fused silica (polished).............................	6
Glass-mica (polished).............................	5
Glass-ceramic (polished)...........................	15
Alumina, 94–96%, as-fired........................	5
Glazed.....................................	7–8
Polished....................................	15
99.5%, as-fired...............................	7
Polished....................................	21
99.9%, as-fired...............................	200
Polished....................................	300
Sapphire, 0.040-in., polished, one side............	400
Both sides.................................	450
0.20-in., polished, one side...................	350
Both sides.................................	380
Beryllia, 99%, as-fired...........................	25
Glazed.....................................	30
Polished....................................	125
Steatite (as fired)...............................	5
Forsterite (as fired).............................	5
Barium titanate, as-fired..........................	12
Polished....................................	35

proximation. Even though soda-lime glass is not a usual substrate for thick film circuitry, it does provide a convenient base, owing to its generally low cost compared to other substrate materials. Needless to say, costs are not always straightforward, and some key considerations should be taken into account in dealing with substrate costs:

• The volume of substrates to be purchased, along with amortization costs for any special tooling, can be a major factor in determining cost per substrate. Long-term open orders may offer advantage.

• Special operations such as inspection, cleaning, machining, surface treating, etc., will influence cost. Grinding or polishing may double or triple the basic substrate cost. As mentioned earlier, as-fired fine-surface ceramics are available, and these may be more economical than polished substrates.

• Designing configurations that are difficult to produce will reduce yields and increase costs. It is suggested that designs be reviewed with suppliers to establish the most simple combination of design and fabrication method. (See earlier section of this chapter on fabrication techniques.)

TABLE 18 Substrate Manufacturers and Processors[13],*

United States

		Glasses					Glass ceramic	Glass mica	Glazed ceramic	Alumina		Beryllia	Steatite	Forsterite	Alkaline-earth porcelain	Titanate	Single-crystal materials	
M = manufacturer, P = processor	Soda-lime	Alkali boro-silicate	Alkali-free	Fused silica	Special	Glass ceramic	Glass mica	Glazed ceramic	≤96%	>96%	Beryllia	Steatite	Forsterite	Alkaline-earth porcelain	Titanate	Sapphire	Silicon
American Feldmuehle, Glenbrook, Conn.	M										x							
American Lava, Chattanooga, Tenn.	M,P								x	x	x	x	x	x	x	x		
Amersil, Hillside, N.J.	M				x													
Basic Ceramics, Hawthorne, N.J.	M,P								x	x	x	x	x	x				
Brush Beryllium, Elmore, Ohio	M,P											x						
Carborundum, Latrobe, Pa.	M,P								x	x	x		x					
Centralab, Milwaukee, Wis.	M,P								x	x	x		x			x		
Cermetron, San Diego, Calif.	M,P									x	x		x			x		
Coors Porcelain, Golden, Colo.	M,P								x	x	x	x						
Corning Glass, Corning, N.Y.	M,P		x	x	x	x	x		x	x								
Crownover Manufacturing, La Jolla, Calif.	M								x	x								
Degussa, New York, N.Y.	M,P									x	x							
Diamonite, Shreve, Ohio	M,P									x	x							
Dow Chemical, Midland, Mich.	M									x								x
Electro-Ceramics, Salt Lake City, Utah	M,P								x	x						x		
Electra Manufacturing, Independence, Kans.	M,P														x			
Erie Technological, State College, Pa.	M,P															x		

Company	Type																				
Frenchtown Porcelain (CFI), Frenchtown, N.J.	M,P								x												
General Electric, Schenectady, N.Y.	M	x		x						x											
Haveg Industries, Taunton, Mass.	M,P					x															
INSACO, Quakertown, Pa.	P			x	x			x	x	x				x							
Intellux, Goleta, Calif.	M		x																		
J. M. Freed, Perkasie, Pa.	P	x	x	x													x				
Lapp Industries, Leroy, N.Y.	M,P													x							
Linde, Union, N.J.	M																		x		
Meller, Providence, R.I.	P						x		x	x									x		
Molecular Dielectrics, Clifton, N.J.	M,P						x														x
Monsanto Chemicals, St. Louis, Mo.	M							x													
Mycalex, Clifton, N.J.	M,P																				
National Beryllia, Haskell, N.J.	M,P									x											
NGK Spark Plugs, Los Angeles, Calif.	M,P							x	x	x						x					
Owens-Illinois, Toledo, Ohio	M,P	x	x	x																	
Rosenthal China, New York, N.Y.	M,P							x	x	x	x	x	x								
Royal Worcester, Charlotte, N.C.	M,P							x	x	x			x								
Semi-Metals, Westbury, L.I., N.Y.	M																			x	
Texas Instruments, Dallas, Tex.	M																			x	
Trans-Tech, Gaithersburg, Md.	M,P												x								
United Mineral and Chemical, New York, N.Y.	M,P							x	x	x	x	x	x		x						
Ventron, Bradford, Pa.	M																				x
Vitta, Wilton, Conn.	M			x				x													
Western Gold and Platinum, Belmont, Calif.	M,P							x	x	x	x										

4-41

TABLE 18 Substrate Manufacturers and Processors[13].* (Continued)

M = manufacturer, P = processor		Glasses								Polycrystalline materials							Single-crystal materials	
		Soda-lime	Alkali-boro-silicate	Alkali-free	Fused silica	Special	Glass ceramic	Glass mica	Glazed ceramic	Alumina ≤96%	Alumina >96%	Beryl-lia	Stea-tite	Forster-ite	Alkaline-earth porcelain	Titan-ate	Sap-phire	Silicon
Europe																		
Andermann and Ryder, London, England	M,P	x								x	x							
Englass, Leicester, England	M,P		x						x	x	x							
Hoboken Chemical, Hoboken, Belgium	M,P									x	x							x
Plessey, Chessington, Surrey, England	M,P								x	x			x			x		
Rosenthal Technische Werke, Frankfurt, Germany	M,P									x	x	x	x	x	x			
Royal Worcester, Tonyrefail, Glamorgan, England	M,P									x	x				x			
Smiths Industries, Rugby, Warwickshire, England	M,P								x	x								
Steatite and Porcelain, Stourport-on-Severn, Worcestershire, England	M,P									x	x		x	x	x	x		
Wacker, Munich, Germany	M,P																	x
Japan																		
Chisso, Chiyoga-ku, Tokyo	M,P																	x
Komatsu Electronic, Hiratsuka-shi, Kanagawa-ken	M,P																	x
Kyocera, Nakagyo-ku, Kyoto	M,P									x	x	x	x	x	x			
NGK Spark Plugs, Mizuho-ku, Nagoya	M,P									x	x					x		

• For optimum yields and costs, keep length-to-width ratios as close to 1 as possible, keep area as small as practical, and do not design overly thin substrates, preferably not under 0.025 in. These guides will minimize warpage and nonuniform shrinkage, which is a problem with ceramics which have a high firing shrinkage, as noted in Table 14. Length-to-width ratios higher than 4 should be avoided.

• Many suppliers stock certain standard substrate sizes and configurations. A listing of these should be maintained, as their use will often result in lower costs and much faster deliveries. Delivery schedules on custom configurations are frequently a major problem.

Break-apart or Postage-Stamp Substrates

Numerous suppliers of thick film substrates offer substrates which are delivered in a large sheet but which are prescored, allowing gang printing of deposited circuitry, followed by breaking apart after all deposition operations. This, of course, offers substantial economy in screening or other deposition operations. While the basic cost of the substrate may be slightly higher, the cost of the substrate with deposited circuitry may be substantially lower. Some trade names and suppliers of this type substrate are Strate-Breaks (Coors), Snap-Strates (American Lava), and Snap-A-Part (Cermetron).

REFERENCES

1. Kingery, W. D.: "Introduction to Ceramics," pp. 420–426, Wiley, New York.
2. Atlas, N. M., and H. H. NaKamura: Control of Dielectric Constant and Loss in Alumina Ceramics, *J. Am. Ceram. Soc.*, vol. 45, no. 10, pp. 467–471, October 1962.
3. France, J., and W. D. Kingery: Thermal Conductivity, IX: Experimental Investigation of Effect of Porosity on Thermal Conductivity, *J. Am. Ceram. Soc.*, vol. 37, no. 2, pp. 99–107, February 1954.
4. Coors Porcelain Company: "Coors Technical Data Book" on Ceramics Properties.
5. Smith, R., and J. P. Harve: Beryllium Oxide, *Proc. 1st Int. Conf. BeO*, North-Holland Amsterdam, 1964.
6. Lynch, J. F., et al.: Engineering Properties of Ceramics, *U.S. Dept. Comm. Bull.* AD803-765.
7. Hessinger, P. S.: How Good Are Beryllia Ceramics?, *Electronics*, Oct. 18, 1963.
8. National Beryllia Corporation: National Beryllia Technical Bulletin on Berlox BeO.
9. Martin, J. H.: The Manufacture of Ceramic-Based Microcircuits, *Sprague Tech. Pap.* TP-66-10, Sprague Electric Company.
10. Comeforo, J. E.: Properties of Ceramics for Electronic Applications, *Electron. Eng.*, April 1967.
11. Frisco, L. J.: Frequency Dependence of Electric Strength, *Electro-Technol.*, vol. 68, pp. 110–116, August 1961.
12. American Lava Corporation: *Alsimag Tech. Bull.* 691 and 652.
13. Brown, Richard: Thin Film Substrates, in Leon I. Maissel and Reinhard Glang (eds.), "Handbook of Thin Film Technology," McGraw-Hill, New York.
14. Materials Research Corporation: MRC Technical Bulletin on High Alumina Substrates.
15. Stetson, H., and B. Schwartz: *Abstr. Am. Ceram. Soc. Bull.*, vol. 40, p. 584 1961.
16. Ettre, K., H. D. Doolittle, P. F. Varadi and R. F. Spurck: *Proc. Electron. Components Conf., Washington, D.C., May 1963.*
17. Centralab Division, Globe-Union, Inc.: Centralab Technical Bulletin on Technical Ceramics.
18. Pittroff, L. F.: Estimating Substrate Area and Density for Hybrid Microcircuits, *Electronics*, Dec. 18, 1972.
19. Winklemann, A., and O. Schott: *Ann. Phys. Chem.*, vol. 51, p. 730, 1894.
20. Bosnell, J. R., and K. H. Lloyd: Substrate Composition and High Microwave Circuit Performance, *Electron. Packag. Prod.*, November 1972.
21. Bosnell, J. R.: The Effect of the Physical Properties of Substrates on Their Use in Microwave Hybrid Circuits, *Microelectronics*, vol. 3, no. 10, Spring 1971.
22. Foster, T. M.: Thick Film Microwave Integrated Circuits, *Electronics*, July 15, 1973.

23. Holland, L.: "The Properties of Glass Surfaces," Chapman & Hall, London, 1964.
24. Longman, G. W., and R. P. Palmer: *J. Colloid Interface Sci.*, vol. 24, p. 185, 1967.
25. Krieger, G. L., and G. J. Wilson: *Mater. Res. Stud.*, July 1965.
26. Missel, L., D. R. Forgeson, and H. M. Wagner: *Electron. Packag. Prod.*, May 1966.
27. Harper, C. A.: "Handbook of Materials and Processes for Electronics," McGraw-Hill, New York, 1970.
28. Harper, C. A.: "Handbook of Electronic Packaging," McGraw-Hill, New York, 1969.
29. Von Hippel, A. R.: "Dielectric Materials and Applications," MIT Press, Cambridge, Mass.

Chapter **5**

Conductor Materials, Processing, and Controls

WILLIAM T. HICKS

Electronic Products Division, Photo Products Dept.,
E. I. du Pont de Nemours & Co., Niagara Falls, New York

INTRODUCTION

Conductors play a major role in thick film circuits. They constitute the largest quantity of material screened on a typical thick film circuit, and very complex properties are demanded of both the fluid composition and final fired conductor product. In some electronic devices, conductors provide the only thick film contribution, e.g., the interconnection of integrated-circuit devices to phenolic boards. Conductors must be processed under a considerable range of firing temperatures when other compositions, such as resistors and dielectrics, are used with them. Yet they must retain excellent conductivity, solderability, and bonding capability since they must finally provide the ultimate connection between the circuit and the outside world.

The purpose of this chapter is to:
1. List the widely varied functions of thick film conductors
2. Describe how thick film conductors are prepared
3. Instruct how to test thick film conductors
4. Explain the variations in properties of conductors with composition
5. List the typical properties of available commercial compositions
6. Describe how best to process thick film conductors

The reader is warned that thick film conductor technology is in a rapid state of change due to the vigorous competition between manufacturers. Thus, many compositions with improved properties over those shown in this chapter may be available by the time this handbook appears in print. The consumer of thick film conductor compositions must keep in close contact with the manufacturer to assure that he is using the optimum material for his particular application.

For the purpose of this chapter "thick film conductors" will be limited to those which can be fired in air and which are therefore largely composed of precious metals. Refractory-metal compositions must be fired in controlled atmospheres, and this important technology will not be discussed here.

THE FUNCTIONS OF THICK FILM CONDUCTORS

The functions of a thick film conductor are so varied that no one composition can fulfill them all. This explains why there is such a multitude of compositions with different properties. Most of the functions are listed in the following sections.

Fig. 1   Strap leads soldered to conductor pads and extending from side of circuit. Telemeter oscillator or rescue set. (*Telefunken.*)

Fig. 2   Pins swaged through substrate and soldered to conductor. (*Semiconductor Electronic Memories, Inc.*)

Conductor Interconnections

The most obvious application of a conductor is to transmit a signal from one portion of a thick film circuit to another. Thus high conductivity is the most important property here. Solder dipping a conductor may sometimes substantially increase its conductivity.

Soldered-Lead and Device Attachment

Eventually, thick film circuits must be connected to the rest of the electronic device or instrument. This is frequently carried out by soldering leads from the side of the circuit substrate or by soldering the conductors to pins swaged into holes in the substrate (see Figs. 1 and 2). Often discrete resistors, capacitors, or solid-state devices are connected to thick film conductors by soldering when that device is not economically available in a thick film version. Here, the important properties are ease of solder wetting, resistance to solder leaching, and adhesion of the pad to the substrate. Fine definition is often required, e.g., when flip chips are soldered to conductors.

Thick Film Resistor Terminations

Thick film conductors are used to terminate and define the length of resistors. Here the most important property is low contact resistance, which results from smooth, trouble-free overlaps. To ensure a low-noise-level resistor, there must be good adhesion between the conductor and fired resistor, and the connection must be stable and strain-free. A very small diffusion zone of conductive components into the resistor is desirable to yield a resistor which scales in almost linearly. Obviously, the conductivity of the conductor must be high compared with that of the resistor. In some cases, catastrophic interaction can take place between fired conductors and

Fig. 3 A ruthenium-based resistor cofired with a Pd-Ag conductor composition.

resistors, e.g., when Pd-Ag resistors are fired over silver conductors (see the section Conductor-Resistor Overlap Cosmetics). Generally, conductors are prefired before the resistors are printed and fired; but cofiring is possible with certain combinations and is frequently a desirable economy. Figure 3 illustrates a smooth cofired termination.

Crossover Connections

Complex modern circuits frequently require that conductor runs must cross over each other without leaking a signal from one to the other. It is important here that the conductor not diffuse into the dielectric, and the conductor must have good adhesion to the dielectric after firing. It is also important that the top conductor should not delaminate from the dielectric during cofiring. In some circuits, the top conductor must be solderable to maintain high conductivity. Figure 4 shows a conductor crossover configuration.

Capacitor Electrodes

Capacitors with a wide range in values may be produced by screen printing a conductor electrode and firing and then printing a dielectric composition as a second layer and a top conductor electrode as a third layer. It is important here, again, that the top conductor have good adhesion to the dielectric and that little diffusion of the conductor occur into the dielectric during firing. It is also important that the conductors not reduce the dielectric constant of the insulating layer. It simplifies processing if the top conductor can be cofired with the dielectric. Figure 5 pictures a screen-printed capacitor.

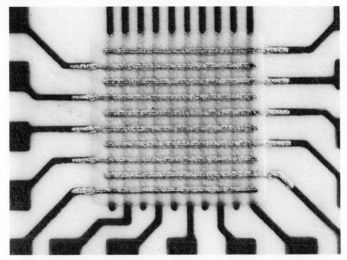

Fig. 4 A conductor crossover configuration.

Fig. 5 A screen-printed capacitor.

Chip and Die Bonding

Active devices, e.g., transistors, diodes, or integrated circuits, are frequently bonded to thick film conductors fired on substrates, as shown by Fig. 6. Generally Au-Si eutectic bonding is used here. Thus, the conductor must have a moderately high gold content. The conductor must also have good adhesion to the device and to the substrate. Good conductivity is required if the back of the chip is used to carry conductive signals.

Face bonding of active devices is often accomplished through the use of beam leads or soldered flip chips. Gold thermocompression bonding is frequently used for beam-lead attachment so that conductors of many compositions can be used. Thus either a capability for gold thermocompression bonding or good solderability is required when beam-lead or flip-chip devices are attached to thick film conductors. Figure 7 illustrates the attachment of flip chips to a soldered thick film conductor.

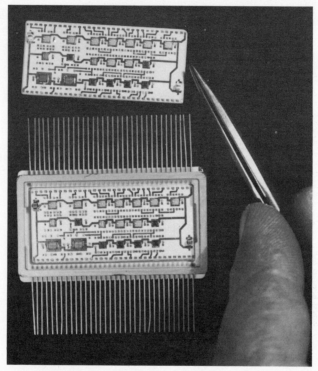

Fig. 6 Integrated circuits die-bonded to gold conductor pads. Note thermocompression gold wire bonds to face of integrated circuits. Computor memory module. (*General Electric Co.*)

Wire Bonding

Frequently, face connections are made to die-bonded active devices using gold thermocompression or aluminum ultrasonic wire bonding. In either case, good adhesion is required between the wires and the conductor under the conditions available in the bonding equipment. Again, good conductivity and fine-line definition are required in these applications. Figure 6 shows a gold wire thermocompression-bonded to gold conductor leads and integrated circuits.

Low-Value Resistors

At times a labyrinth conductor pattern is used as a low-value resistor (0 to 10 Ω) (see Fig. 8). Here the main requirement is that the conductor have a reproducible resistance from lot to lot.

Packaging of Thick Film Circuits

Thick film conductors are sometimes used to form part of a seal to protect that circuit from the atmosphere. Sometimes metal caps or lids are soldered or brazed to thick film conductors to form such a seal. In this case, good brazing capability and good adhesion to the substrate are required of the conductor. Frequently, compatibility with sealing glass is required in these applications.

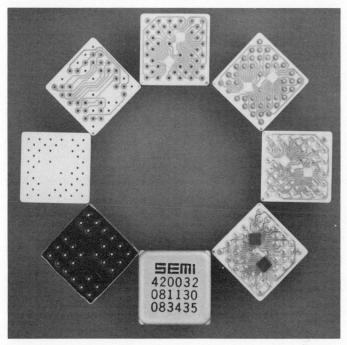

Fig. 7   Steps involved in the production of a semiconductor electronic memory device incorporating flip-chip integrated circuits soldered to a fine-line Pd-Au conductor array.   (*Semiconductor Electronic Memories, Inc.*)

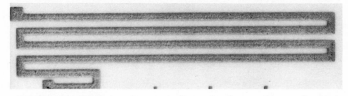

Fig. 8   Labyrinth conductor pattern which might be used as a low-value resistor.

PRODUCTION OF THICK FILM CONDUCTORS

Constituents of a Thick Film Conductor

Thick film conductor compositions consist of three main ingredients:

1. Finely divided metal of carefully controlled size and composition, or alternately in whole or in part, organometallic compounds which decompose on firing

2. Finely divided vitreous and oxide phases, also of controlled size, shape, and composition

3. An organic medium which suspends the inorganic constituents until the paste is fired

Fired conductor composite During firing of the screened and dried conductor, all the organic matter is volatilized or oxidized, leaving behind a composite made of a mixture of metal and glass. At the common firing temperatures (700 to 1000°C) the metal particles sinter and alloy while the vitreous binder softens and wets the metal and substrate. The conductivity of this system will be a function of the conductivity of the metal alloy used and the relative quantities of metal and glass in the final system. The frits serve to bond the metal particles to the substrate through their reaction with the glassy phase of the alumina substrates. The glass may also serve to protect the metal to some degree from leaching in the solder bath. After firing, the glass is not uniformly distributed through the composite but generally tends to concentrate at the substrate surface. If this did not happen and all the metal particles were uniformly wetted by the glass, it would be impossible to solder the fired conductor and the composite would have no electric conductivity since there would be no continuous conducting path. Since the glass viscosity and surface tension are temperature-dependent, firing temperature plays an important role in determining many characteristics of a particular conductor, e.g., solderability, solder-leach resistance, and adhesion.

Thus although the composition of the conductor fluid is important, the final qualities of the fired-on conductors are highly dependent on the processes used by both the material supplier and the circuit manufacturer.

Organic medium The organic medium in combination with the particular controlled-size metal and frit must give the proper rheological characteristics to the conductor for the particular method chosen for its application to the substrate. Frequently, a conductor with a particular metal and solids composition is available in several different organic vehicles for different types of application, e.g., mask printing or screen printing.

Manufacturing Processes

The thick film conductor manufacturer obtains or produces metal and frit of controlled sizes and shapes and combines them with organic media in combination with extraneous agents, such as dispersing or flow agents. The ingredients are mixed, and the solid components are carefully dispersed in the organic medium for screen-printable conductors. To ensure good quality control and reproducible characteristics from lot to lot of conductors the manufacturer must impose stringent quality-control tests at various stages of production of the thick film conductor. These are classified in the following section.

Manufacturers' Quality-Control Tests

Tests of initial ingredients Generally the metals or alloys used are tested for particle size through measurement of surface area and for purity by emission spectrographic analysis. Frequently solids or chemical-analysis tests must be included. Glass frits are generally tested through chemical analysis and softening-point measurements. Here too impurities are generally monitored through emission spectrographic analysis. The organic media used must also be tested for chemical purity and viscosity.

Tests on conductive paste The degree of dispersion of the solid in the organic medium is generally tested through use of a standard grind gage, where the fluid is drawn down in a groove which becomes progressively shallower. The larger particles are detected through their appearance as streaks or specks when proper lighting is used on the surface of the wet film. Generally if the paste is to be used for mask printing, more careful dispersion and finer particle sizes must be obtained. The total solids in the fluid are tested through a combustion method, and the viscosity is tested using equipment discussed in a later section. Sometimes the viscosities are specified for at least two different shear rates. Frequently at this point additional tests are made, such as chemical analysis or emission spectrographic testing for impurities in the conductor fluid.

Tests on the fired product Samples of all lots of conductor are printed, dried, and fired according to a prescribed schedule. The resulting fired composite is then tested for electric resistance, solder-wetting capability, and soldered adhesion if it is a solderable conductor.

Since conductor manufacturers generally produce a large number of products with subtle differences in properties in small batches, carefully testing each lot of these many products often means considerable expense. Thus, even base-metal compositions must be priced high enough to cover the careful quality control and technical service which must be provided to the consumer of these products.

MEASURING THICK FILM CONDUCTOR PROPERTIES

Careful testing of thick film conductor properties is important to both the producer and the consumer of these materials. The producer needs these tests to

develop and choose the best metal, glass, and organic vehicle compositions to give the optimum properties for a particular application, while the consumer must choose among the various products offered by each of the thick film conductor producers for his particular application. Both the producer and the consumer must make continuous quality-control tests to assure that the product remains as closely as possible the same as that found initially to meet the circuit requirements. The last section described briefly what tests manufacturers frequently use to control production. In this section, these test methods will be given in detail for the use of both producers and consumers along with more extensive tests used in developing conductor products. In a later section conductor tests which might best be used by the consumer as a routine incoming quality-control check will be listed.

Fig. 9 Viscometer with spindle attached suitable for control-test measurements on fluid in stock container. (*Brookfield Engineering Laboratories, Inc.*)

Viscosity Measurement

Brookfield viscometers The Brookfield Engineering Laboratories equipment is probably the most convenient and economical to use for routine quality-control testing of thick film conductor paste.[*] In using this equipment, the conductor fluid must be first brought to a specified temperature, generally $25 \pm 0.2°C$, and then a suitable spindle is immersed to a mark on the spindle shaft. The spindle is held by a spring and rotated in the fluid at prescribed rates by an electric motor. This spring is distorted by the stress made on the spindle by the fluid, and this distortion is registered on a scale at the top of the instrument, which rotates with the spindle. A number of instruments are available from the Brookfield Engineering Laboratories which incorporate springs of different constants and allow spindle rotation at various speeds. Figure 9 shows a Brookfield viscometer with spindle attached. For most thick film conductive fluids, which have a relatively high viscosity, the HBT instrument is used. A number of spindles are available for this instrument, ranging from a straight shaft to a fairly wide disk on a shaft. For quality-control purposes, it is convenient to use an appropriate one of these spindles with the fluid in an open jar. For precious-metal conductors, frequently the small-sample adapter (SSA 14/6) is used with the HBT instrument. The temperature of the sample chamber in this case is controlled by a water bath so that the fluid is at the desired temperature.

[*] Brookfield Engineering Laboratories, Stoughton, Mass.

The disadvantage of the conventional Brookfield equipment is that the shear rate of the fluid is poorly defined since it varies considerably from the region directly adjacent to the spindle to the outside surface of the jar or cup. The Brookfield equipment is calibrated using Newtonian fluids, and the Brookfield Engineering Laboratory supplies calibration constants for each combination of spindle and instrument.

Most conductor fluids are at least to some degree pseudoplastic; i.e., their true viscosity tends to decrease as the shear rate is increased. Thus, the viscosity values determined with Brookfield equipment must be considered apparent viscosities, which allow consistent quality control but are not the true viscosities when defined as the rate of increase in stress divided by the rate of increase in shear. Because of the variation in shear rate through the fluid, the measured apparent viscosities may vary considerably from one type of equipment and spindle to another, but this is unimportant for quality-control purposes as long as the same spindle and instrument are always used for the same composition. The shear rate is better defined when the small-sample adapter is used since the space between the spindle and cup is small compared to the diameter of the spindle or the bore of the cup.

Brookfield has recently developed a cone and plate viscometer called the Wells-Brookfield Micro-Viscometer. The advantages of this arrangement are described in the following section.

Rotovisko viscometer The Rotovisko* viscometer is another popular piece of equipment for measuring the viscosities of thick film conductor fluids. Generally, the cone and plate assembly, which comes with this equipment, is used to measure thick film conductor viscosities. Since with this assembly the distance between the moving surface and the stationary surface increases directly in proportion to the difference in speed between the two surfaces, the shear rate can be calculated and is constant from the center of the cone to the outside. Here, the true viscosity of the fluid can be determined if a plot is made of variation in stress as a function of the shear rate. Such a plot would be a straight line for a Newtonian fluid, but for most conductor fluids the rate of change of stress decreases as the shear rate is increased. Generally the apparent viscosity, i.e., the total increase in stress divided by the total increase in shear rate, is the measurement used.

The Rotovisko viscometer also has the advantage of using only a small sample of conductor paste. Temperature is controlled by water circulating through the plate and cone assembly.

On the other hand, the Rotovisko equipment is more cumbersome to use than the Brookfield viscometers and requires very careful adjustment of the cone-to-plate distance. Generally, it is necessary to calibrate the Rotovisko each time it is used by making comparative measurements with Newtonian standards.

Haake is now introducing a viscometer that can linearly vary shear rate with time according to a prearranged program.

Comparison of measurements made with Brookfield and Rotovisko viscometers. The apparent viscosity values measured using the Rotovisko equipment are generally much lower than the apparent viscosities measured with Brookfield equipment since much higher shear rates are used with the Rotovisko viscometer. As mentioned previously, all conductor fluids tend to be pseudoplastic; i.e., their viscosity decreases with increasing shear rate. For comparison, the following measurements were made on a Pd-Au conductor fluid prepared for mask printing in a highly pseudoplastic vehicle. The apparent viscosity measured with the Brookfield HBT small-sample adaptor SSA 14/6 decreased from 32,000 P (poises, p)† at a shear rate of 0.20 s^{-1} to a value of 3,550 P at a shear rate of 4 s^{-1}. On the other hand, with the Rotovisko viscometer, the apparent viscosity of the same lot of the same conductor composition decreased from a value of 410 P at a shear rate of 98 s^{-1} to a value of 260 P at a shear rate of 295 s^{-1}. These measurements made by both instruments are consistent

---

* Trademark of Haake Instruments, Inc.

† In the industry these values would be written, for example, as 3,200 Mcps, where the M stands for thousand and cps for centipoises; the P used in the text is the ANSI abbreviation for poise.

when the large difference in shear rate and the highly pseudoplastic nature of this composition are considered.

Ferranti-Shirley viscometer The Ferranti-Shirley° viscometer, which uses a plate and cone assembly similar to the Rotovisko with the same constant shear rate throughout the sample, is designed to give a continuously and linearly varying rate of shear with time, and the stress and shear rate are recorded. Figure 10 illustrates this viscometer complete with recorder, water bath, and associated control equipment.

A measurement made on this instrument is shown in Fig. 11 for a typical Pd-Ag conductor composition. Here the shear rate, the independent variable, is plotted as ordinate while the dependent variable, stress, is plotted as abscissa according to the style of most textbooks on viscosity. This equipment is useful for making sophisticated rheological measurements on conductor fluids and comparing them with printing behavior.

Fig. 10 Cone and plate viscometer with programmed linearily variable shear rate. The viscometer is shown complete with recorder, water bath, and associated control equipment. (*Ferranti Electric, Inc.*)

Solids Content

A solids determination provides a relatively convenient and rapid measurement of the relative amounts of solid materials, such as metal and frit, compared with the quantity of organic vehicle in the conductor fluid. This analysis is used by the manufacturer to detect gross weighing errors in the production of the thick film conductor. The solids content of a particular composition is directly related to the fired-film thickness obtained after firing. Solids-content measurement may also be used as a means of detecting loss of solvent during storage or use of a conductor fluid. If the solids content increases, the appropriate solvent may be added to decrease it to its original value.

A solids analysis is carried out as follows. Approximately 2 g of the fluid conductor is weighed into a previously fired and tared porcelain crucible using an analytical balance. The conductor is then heated at 100 to 200°C until all or most of the solvent has evaporated. For a gold or silver conductor, the resulting composite is then fired for at least 20 min at 750°C in a muffle furnace. For the less volatile mixed-metal conductors, e.g., Pd-Ag, Pd-Au, or Pt-Au, the composite is fired at 1050°C for 20 min to prevent oxidation of palladium where present. The fired mate-

° Ferranti Electric, Inc., Plainview, N.Y.

rial is then cooled to room temperature and reweighed in the analytical balance. The solids content is then calculated as the weight of the final fired product divided by the original weight of the fluid conductor. Solids contents for most conductors vary from 60 to 90 weight percent.

Film Thickness

Measurement of fired-film thickness is a good monitor of printing quality and may give an explanation for variations in conductor resistance and solder-leaching resistance. Rough measurements can be provided by a microscope with a calibrated focusing knob. More accurate measurements, as well as a monitor of surface roughness, are provided by such instruments as the Clevite Brush Surfanalyzer 150.° A typical Surfanalyzer tracing of three 80-mil-square fired conductor pads is shown in Fig. 12. Note that the height of the pads and therefore the roughness of the con-

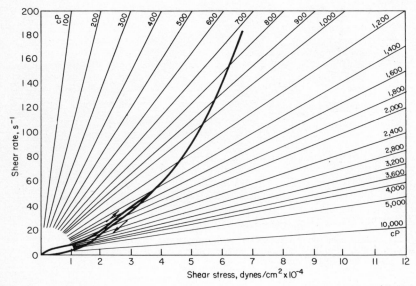

Fig. 11   Shear rate versus shear stress curve for a Pd-Ag conductor composition produced by a Ferranti-Shirley viscometer system.

ductor and the warp of the substrate are greatly amplified in the recording of this instrument.

Conductivity

As previously mentioned, conductivity is the most important property of a fired thick film conductor. Conductivities vary widely from very high values for gold conductors to much lower values, about a fiftyfold decrease, for Pt-Au conductor compositions. Often the conductivity is a good monitor of printing quality and film thickness for a particular run of conductor composition.

Generally, the resistance of the fired thick film conductor is measured using a four-point probe ohmmeter, which requires very low current flow for operation. To increase the actual resistance measured, special labyrinth patterns of fired conductor are usually used, varying from 200 to 400 squares in length. The resistance values for conductors are generally calculated in milliohms per square.

° Brush Instrument Division, Clevite Corp., Cleveland, Ohio.

If all the conductor runs in the circuits are solder-coated after firing, the resistance of the thick film conductor by itself is not as important, since solder dipping usually increases the conductivity severalfold.

Wettability

There is a wide variation in how rapidly different fired-film conductors are wetted by the solder. Most solderability tests involve dipping a fired metallized substrate for a given time in a particular solder with a specified flux and temperature. After the fired and soldered pattern is washed off with solvent, the part is examined under a microscope and the solderability of the conductor is rated according to the number and size of voids in the solder film over the fired conductor. This provides a qualitative rating useful for most purposes, but often incapable of showing subtle differences in solder wettability of conductors.

A more quantitative solder-wetting test involves measuring the degree of spreading of a measured quantity of solder over a fluxed, fired conductor when the conductor has been held for a certain time at a specified temperature. The following test was developed by the North American Rockwell Electronics Group:[1]

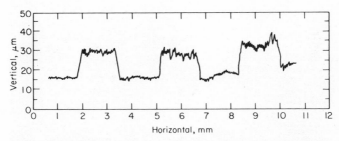

Fig. 12 Brush Surfanalyzer tracing of thick film conductor fired on alumina.

1. Print a ½-in.-diam conductor pattern on an alumina substrate and dry and fire the pattern at the normal temperatures. Multiple refires might be employed if this is important in the final circuit where the conductor is to be used.
2. Place the fired substrate on a hot plate at a specified temperature high enough to cause the solder to flow. This would amount to 220°C for a 63-37 tin-lead solder.
3. Place 3 drops of the desired fluxing agent on the fired conductor pattern with an eyedropper and allow to stand 1 min.
4. Drop a preweighed piece of solid solder wire of the desired composition approximately 35 mils in diameter by 3 in. long directly onto the fluxed area using a glass tube.
5. Allow the solder to stand on the fluxed, heated conductor for 2 min.
6. Remove the substrate and allow it to cool in ambient air in a horizontal position.
7. Remove the flux with an appropriate solvent.
8. Measure the height H of the solder above the fired conductor surface.
9. Calculate the spreading factor SF as follows:

$$D = 1.2407 \sqrt[3]{\frac{W}{\rho}}$$

$$SF = \frac{100\,(D - H)}{D}$$

where W = weight of solder
ρ = density of solder

By this procedure, a highly solderable Pt-Au conductor composition O was found to have a spreading factor of 74.3 percent (see Table 7).

Solder-Leach Resistance

When a fired-film conductor pattern is immersed in a pot of molten solder, the metal gradually dissolves in the molten solder until the point is reached where the solder dewets from the substrate because there is not enough metal left in the conductor pattern. This dewetting is often confused with the nonwetting that occurs when a conductor has a poor solder-wetting capability. This may usually be cleared up by microscopic examination. If the nonwetted area still shows the original color of the conductor and the conductor has its original thickness, then poor solder wetting may be assumed. Backlighting the substrate helps in this type of examination. If practically all the metal is gone except for a few black specks and most of the substrate shows through, solder leaching has occurred.

If the dewetting is caused by solder leaching, the dewetted areas should get larger with additional exposure to molten solder. However, if the voids are caused by poor wetting, the voids should get smaller with additional solder dipping.

Many tests have been used for solder-leach resistance where a fired conductor is immersed in solder for long periods of time and the time is measured until obvious

Fig. 13   End point of multiple-dip test for solder-leach resistance on a 20-mil line pattern.

leaching occurs, either as detected by eye or under a microscope. The weakness of this test lies in the fact that the solder surrounding the immersed metal area becomes saturated with the particular metal being leached, and therefore the test is sometimes insensitive. Here again the North American Rockwell Electronics Group has suggested a test which provides a more quantitative measure of solder-leach resistance, especially where a consumer wishes to solder discrete parts to a thick film conductor and subsequently replace defective parts so that a particular area of thick film conductor may have to be resoldered many times:[1]

1. Print and fire a conductor pattern containing 20-mil lines.

2. Heat the solder desired to its normal operating temperature or a higher temperature if one wishes a faster test.

3. Dip the fired conductor pattern in flux, and then dip the pattern in solder for 10 s.

4. Withdraw the pattern and hold it horizontal for 3 s.

5. Dip the part into trichloroethylene to quench the substrate and dissolve the flux.

This whole cycle is counted as one dip. After each dip, the 20-mil lines are examined under a microscope using backlight. The dips are repeated for a particular conductor until the 20-mil line is leached through as detected by a discontinuity in the solder coverage at some location on the line. Fig. 13 illustrates such a failure.

A number of different conductor compositions can be run simultaneously in this test to get a good comparison of their relative solder-leach resistance. The different conductors should be dipped in rotation to assure as uniform conditions for the different conductors as possible.

Fairly drastic conditions must be used with this test to rate Pt-Au conductors. Here 63-37 tin-lead solder heated to 250°C is used with an active flux. A quality Pt-Au conductor, like composition P in Table 7, may survive up to 30 dips, according to this test, before leach-through of a 20-mil line. On the other hand, a silver conductor will probably not last through even one dip with such severe conditions. To rank such conductors, milder conditions should be used, such as a 62-36-2 Sn-Pb-Ag solder at 215°C with a mildly active flux.

Adhesion

The adhesion of a fired conductor to a substrate is one of the most difficult properties to measure reproducibly. Most methods used with solderable thick film conductors involve pulling a soldered lead from the substrate, but there is a wide variation in the configuration of this soldered lead, and these variations lead to different modes of conductor or solder failure.

Normal tension measurement   The following steps describe methods to test a conductor in tension normal to the substrate:

1. Small areas of conductor (possibly 50 by 50 mils) are printed and fired according to the specified schedule.

2. The fired parts are dipped into the specified solder for a given time and temperature and pulled out slowly at a 45° angle.

3. A solder-coated copper wire with one end deformed to make a head approximately 40 mils in diameter is attached to the solder-coated pad using a reflow technique with a small soldering iron or a resistance heating device. Sometimes large rivets are used in this test, and the solder is reflowed by passing the parts through a belt furnace at the desired solder-melting temperature.

4. The headed wires, or rivets, are pulled normal to the substrate using an Instron* testing instrument or other equipment which will pull the wires at a specified speed, for example, 0.2 to 0.5 in./min. An instrument which provides a chart recording is convenient, but a device which shows the maximum pull experienced before failure is sufficient.

Generally, in this test, abnormally thick substrates must be used because with many conductors the substrates will break before the conductive pad is pulled from the substrate.

Peel-type adhesion tests   In actual practice, most leads to a thick film circuit fail as a result of peeling or twisting of the lead against the soldered conductor. Even where protection is provided against such peeling by swaging pins into holes in the substrate, shear-type failures have been noted. Peel failure is especially evident where leads are directly outboarded from the side of a thick film circuit and where the subsequent packaging does not protect the leads adequately.

Peel or twist failures result at much lower forces than normal tension failures because of the brittle nature of the conductor-to-substrate bond. For this reason, the adhesion properties of the conductors in this chapter are described in terms of peel-type tests. A specific test of this type is described as follows:[2]

1. A conductor pattern is printed which involves rows of 80-mil squares (see Fig. 14).

2. The conductor is fired according to the specified process.

3. A solder-coated 20-gage copper wire is cleaned with solvent, unrolled to a length of 48 in., and stretched 2 in. more to yield a straight length of wire.

4. The wires are cut into 4-in. lengths, and one end of each wire is bent into a type of shepherd's crook, as shown in Fig. 15.

5. The wires are fitted snugly over a row of conductor squares so that a straight length of wire lies flat against the rows of conductor squares directly in contact with each square. The wires are carefully aligned using index marks so that they are centered with the row of 80-mil squares. Three wires are placed on each substrate,

* Instron Engineering Co., Quincy, Mass.

and generally at least four substrates are prepared in this manner for each conductor to be tested.

6. With the wires clamped on with a pair of tweezers at the other end from the shepherd's crook, the substrates are dipped in a specified flux followed by the appropriate solder at the desired temperature. For example, use a 10-s dip in 62-36-2

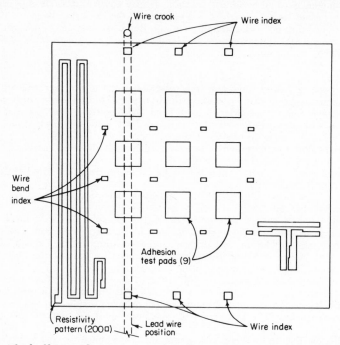

Fig. 14 Thick film conductor test pattern for peel-type adhesion, resistance, and printing-definition tests.

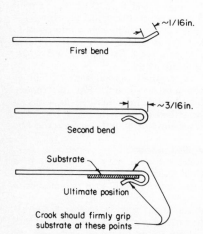

Fig. 15 Wire-lead-forming steps for peel-type adhesion test on thick film conductor compositions.

Sn-Pb-Ag solder at 220°C for Pd-Ag conductors. The solder temperature and time should be optimized for each conductor composition. In some cases, active fluxes decrease the adhesion value, while with certain conductors active flux is needed to get good solder wetting. The solder conductor pad should be slowly withdrawn from the solder bath while held at a 45° angle to form a continuous fillet between the conductor pad and the wire.

7. The substrate with the soldered wire and conductor pads should be held level while the substrate cools and the solder solidifies.

8. The flux should be washed off with an appropriate solvent, e.g., trichloroethylene.

9. The soldered parts should be aged at room temperature for a specified period, such as 16 h, before being subjected to the pulling test. Figure 16 shows how the adhesion value measured on a Pd-Au conductor composition varied with aging time at room temperature. After an overnight

period of 16 h, little further change occurs.   The increase in adhesion values is thought to be due to an annealing process in the solder which tends to make it more ductile, thus allowing it to spread the load more uniformly to the fired conductor.

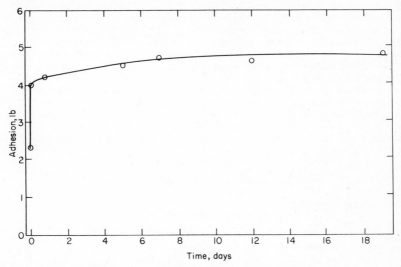

Fig. 16   Peel-type adhesion values of Pd-Au conductor composition K after soldering in 63-37 tin-lead solder as a function of room temperature aging.

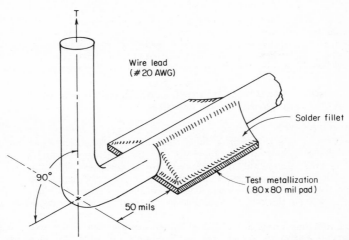

Fig. 17   Bending wire lead for peel-type adhesion test of thick film conductor compositions.

10. After the aging period, the wires are bent sharply perpendicular to the substrate at an index mark 50 mils from the edge of the conductor pad (see Fig. 17).   It is very important that this distance from the 90° bend to the edge of the fired conductor pad be kept constant.   A metal plate is held against the wires while this bend is made manually to ensure as sharp a bend as possible.   If the distance from the

bend to the edge of the conductor pad is increased, less force will be required to pull the conductor from the substrate. If the distance from the edge of the conductor to the bend is decreased, a higher force will be required to pull the conductor pad from the substrate.

11. The substrate and wires are clamped in an Instron tensile tester, and the wire is pulled from the substrate at a speed of 0.5 in./min. The maximum force required to pull the conductor pad from the substrate is recorded. Generally, only the first pad in each row of three is used for this test.

Comparison of tensile and peel-type test results   Table 1 illustrates the values obtained using the two different types of adhesion test described in the preceding sections. Both conductors were fired at 850°C in a belt furnace, and 63-37 tin-lead solder was used throughout. A Pt-Au composition O gave a tensile test value of 5 lb, or 2,040 psi, when pulled from a 50-mil square. The same conductor fired in the same manner gave an adhesion value of only 2 lb when a peel-type test was used with a larger 80-mil-square pad. A high adhesion-type Pd-Au conductor M gave a much higher value in tension, 9 lb, or greater than 3,600 psi, at which point the substrate failed; i.e., the conductor pad was never pulled from the substrate. How-

TABLE 1   Tensile versus Peel Tests for Pd-Au and Pt-Au Conductors

Code	Tensile 50 mils		Peel, 80 mils, lb
	lb	psi	
O	5	2,040	2
M	>9*	>3,600*	5

* All substrates broke.

ever, in a peel-type test with a larger 80-mil-square pad, the conductor pad was peeled from the substrate at a value of 5 lb. Thus, it is apparent that either of these conductors is more likely to fail in a peel-type mode when actually used in a practical electronic circuit. In this case the conductors have the same approximate relative rankings regardless of which tests are used, but this may not always be true.

EXPLANATION OF VARIATION IN CONDUCTOR PROPERTIES WITH COMPOSITION AND FIRING TEMPERATURES

Electrical conductivity

The electric conductivity of a fired thick-film conductor depends on (1) the resistivity of the bulk-metal alloy involved, (2) the relative quantities of nonconductive glass or other compounds incorporated into the composite structure, and (3) the density of the resulting composite, i.e., the number of voids remaining in the structure after the organic vehicle has been fired out. Figure 18 shows the typical structure of a fired thick film conductor as revealed by a scanning electron microscope.

Table 2 presents for single- and double-metal conductors of different compositions the calculated resistance values for the bulk-metal alloy of the same composition spread in a continuous sheet over the area covered by the same weight of metal in a thick film conductor composition. It is assumed that the thick film conductor composition is printed with a 200-mesh screen giving a 30-μm wet-film thickness. The coverage of the thick film conductor compositions was calculated from the relative volumes of metal, frit, and vehicle in the compositions. The resistivities of the bulk alloys were taken from International Critical Tables.[3] For comparison, Table 2 also shows the resistance values actually measured for the same thick film conductor compositions printed with a 200-mesh screen.

The resistance value calculated for the same weight of gold conductor is approximately the same as the measured value. The measured value for the silver conductor is approximately 50 percent higher than the value calculated for this composition. A recently developed series of Pd-Ag conductor compositions shows measured resistance values close to that calculated for the same weight of bulk alloy over that area. However, some of the older series of Pd-Ag compositions have approximately twice these resistance values. The Pd-Au and Pt-Au conductor compositions show meas-

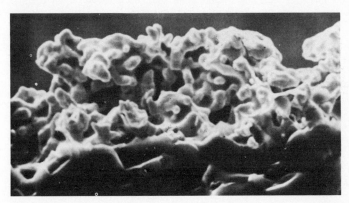

Fig. 18 Scanning electron micrograph of a fired Pt-Au conductor composition in cross section (2,000×).

TABLE 2   Measured and Calculated Resistance Values for Various Thick Film Conductors

Code	Metal A	Metal B	Wt. A / Wt. B	Resistance, mΩ/sq		Fired-film thickness, μm	
				Calc.	Meas.	Calc.	Meas.
A	Ag	...	...	5.0	7.7	4.2	6
	Ag	Pd	3.5	18.8	20	9.2	16
I	Ag	Pd	2.5	23.3	25	9.1	16
J	Ag	Pd	2	31.2	35	9.2	16
R	Au	...	...	2.6	3	12.7	
K	Au	Pd	2.5	38.0	104	7.0	14
L	Au	Pd	2.5	38.0	73	7.0	11
O	Au	Pt	3.5	41.4	150	6.3	12

ured resistance values 2 to 3 times higher than the values calculated for the same weight of bulk-metal alloy spread over the same area. The higher resistance values measured for the Pd-Au and Pt-Au conductor compositions are probably a result of the combination of the effect of the higher melting points of these alloys and the higher atomic weights of gold and platinum compared to palladium and silver, which allow less diffusion and a lower degree of sintering when fired at the same temperature.

Naturally, the thickness of the fired thick film conductor is much greater than when the same weight of bulk-metal alloy is spread over the same area. The last two columns of Table 2 compare the calculated thicknesses for the fired thick film compositions knowing the volume content of the components and assuming no voids as compared with the actual measured thickness of the final fired product. Generally, the actual fired thicknesses are 50 to 100 percent greater than those calculated for void-free composites. Thus, it is apparent that fired thick-film conductors have an electric conductivity 2 to 6 times lower than the same volume of bulk-metal alloy of the same composition. The fact that the measured resistance values come as close as they do to calculated resistance values on the basis of the weight of the metal on a unit area implies that there is relatively little isolation of metal particles by the glass frit.

The two Pd-Au conductor compositions shown in Table 2 have quite different measured resistance values despite the fact that they incorporate the same metal alloy composition and frit content. The fact that the second composition has a lower measured resistance value than the first is the result of processing of intermediates, which results in a higher fired-film density, as shown by the measured fired-film thicknesses.

Effect of firing temperatures As firing temperatures are increased, the degree of compaction increases, possibly glass isolation of metal particles decreases, and this results in lower resistance values. For example, a typical Pd-Ag conductor D has a resistance value which decreases from 52 mΩ/sq when fired at 700°C to 29 mΩ/sq when fired at 1050°C.

**TABLE 3   Solubility of
Precious Metals in
Molten Tin at 250°C**

Metal	Solubility, wt. %
Au	15.0
Ag	6.0
Pd	~0.5
Pt	1.0

Solder-Wetting Capability and Solder-Leach Resistance

Both the wettability and leach resistance of a fired-film conductor are tied to the degree of interaction between the metal in the conductor and the solder. Solderability, or solder-wetting capability, is defined as how readily a fired thick film conductor is wet by molten solder. This encompasses both how rapidly the conductor is wet by the solder and how completely; i.e., to what degree small voids are left in the solder film. Mildly active solder fluxes are generally used to remove dirt and surface oxides from the fired-film conductors.

Solder leaching results when the metal in the fired-film conductor dissolves into a large mass of molten solder. This process proceeds to the point where sufficient metal is no longer left in the composite matrix to maintain good solder coverage, and at this point dewetting of the solder takes place. The relative tendency for different thick film conductor alloy systems to dissolve in high-tin solders may be surmised by examining the binary phase diagrams between each of the metals involved and tin.

Table 3 shows to what extent each of the four most common metals used in conductors dissolve in molten tin at 250°C before a solid phase precipitates out.[4,5] Gold shows the highest solubility, with silver next, while palladium and platinum show considerably less solubility. Thus, gold conductors are very difficult to solder because they leach rapidly into a tin-lead solder, such as 63-37 tin-lead. Silver conductors can be soldered in eutectic tin-lead solders which have been saturated with silver, such as 62-36-2 Sn-Pb-Ag, but leach much more rapidly in simple eutectic solder. By itself, palladium does not wet readily because it usually has an oxide film on its surface. By itself, platinum may be soldered but usually shows poor wetting because of its slow solution. The best soldering is obtained with alloys of palladium or platinum with gold or silver, since these alloys give the proper compromise between no solubility and too rapid solution into the molten-solder pot.

As the relative quantity of palladium or platinum is increased in a Pd-Au, Pd-Ag, or Pt-Au conductor composition, solder leaching decreases. Pt-Au compositions are considerably more leach-resistant than Pd-Au compositions, apparently due to the high atomic weight of platinum, which would give slower diffusion rates compared to palladium, as well as to the low solubility of platinum in molten tin.

The solder-leach resistance of Pd-Ag compositions is also enhanced by using a solder which already contains appreciable silver, for example, 62-36-2 Sn-Pb-Ag. Pd-Ag conductors containing amounts of palladium large enough to give a silver-to-palladium ratio of 2:1 or less show poor solderability when fired at a high temperature, such as 850°C followed by a refire at 500°C as occurs during encapsulation of a resistor composition. This is related to the formation of a stable oxide by palladium and palladium alloys at intermediate temperatures. This particular type of poor refire solderability may be solved by using Pd-Ag conductors with lower palladium content.

Effect of firing temperature   Most conductor compositions show an optimum firing temperature range for best solderability and solder-leach resistance. Generally, both solderability and solder-leach resistance improve as firing temperature is increased to the point where good alloying takes place. Leach resistance becomes poorer when still higher firing temperatures are used, possibly due to the diffusion of the protective frit to the substrate surface.

Conductor-Resistor Overlap Cosmetics

Thick film conductor compositions must make smooth overlaps when used as terminations for thick film resistor compositions. Figure 3 shows an example of such a smooth bubble-free termination obtained when a Pd-Ag conductor was cofired with a thick film resistor composition. When a poor combination of thick film conductor and resistor is used, even when separately fired, considerable bubbling can sometimes occur in the overlap region, occasionally resulting in actual splitting of the fired composite between the conductor and resistor with an open circuit. Other times the split is not as obvious but may show up upon temperature cycling of the thick film circuit. In still other cases, the overlaps may appear all right to the naked eye but result in high-noise-level resistors. Figures 19 and 20 show examples of poor overlap terminations. Frequently, however, a certain amount of bubbling may be present in the overlap termination, and yet the terminations are still continuous and remain trouble-free.

Palladium-silver resistor termination   Many early thick film resistor compositions consisted of palladium oxide, silver, and frit suspended in an organic vehicle. This type of resistor shows disastrous bubbling when printed over a silver conductor and fired even when the silver conductor is fired before the resistor composition is printed. This bubbling probably is the result of the reaction

$$PdO + Ag \rightarrow Ag\text{-}Pd \text{ (alloy)} + 1/2O_2(g)$$

At the 700 to 800°C firing temperature of the resistor composition, oxygen bubbles are created through this reaction which then bubble through the molten frit. Since the terminations are cooled down before the bubbling has stopped, bubbles are left solidified in the frit of the conductor-resistor system. Sometimes the gas escapes between layers of resistor and conductor, causing splitting at these boundaries.

These reactions can be prevented by previously alloying the silver in the conductor with palladium before it comes in contact with the palladium oxide of the resistor system. Thus, Pd-Ag conductors with an Ag-Pd ratio of 3:1 or less make smooth terminations when prefired and then overprinted with Pd-Ag resistor systems.

Sometimes for economy or simplicity of operation, thick film consumers prefer to cofire the resistor composition with the conductor composition. In this process, the conductor composition is generally printed first and dried; then the resistor composition is printed over the conductor at the terminations and dried. Finally, the combination is cofired at the recommended firing temperature for the resistor. Pd-Ag conductors can be successfully cofired with Pd-Ag resistor compositions providing there is sufficient palladium in the conductor. This usually requires an Ag-Pd ratio of 2:1 or less.

High-precision resistor systems   A number of manufacturers have brought out a line of ruthenium-based resistor systems which give more reproducible resistance values, lower temperature coefficients of resistance, and less drift in resistance than the older Pd-Ag resistor systems.   Many of these resistor compositions may be used with conductors which have low palladium content, since silver does not react with the components of these resistor compositions.   Cofiring these resistor systems with conductors with high palladium content usually does cause some bubbling, however.

Fig. 19   Disruption occurring when a Pd-Ag resistor is fired over a Pd-Ag conductor with a high Ag-Pd ratio (incident light, 100-mil-wide resistor).

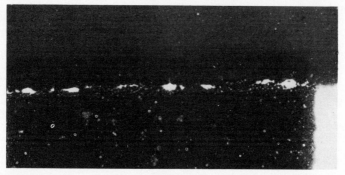

Fig. 20   Disruption occurring when a Pd-Ag resistor is fired over a Pd-Ag conductor with a high Ag-Pd ratio (transmitted light, 100-mil-wide resistor).

This probably occurs through the following combinations of reactions.   Unalloyed palladium in the unfired conductor composition oxidizes during the firing process to give palladium oxide.   At temperatures above 700°C, the silver in the conductor begins to alloy with the palladium in the palladium oxide, causing the liberation of oxygen as the alloying proceeds.   Above 700°C, the glass in the resistor systems is soft enough to trap this gas and cause bubbling.

Figure 21 shows the weight changes which occur when a Pd-Ag mixture is heated through this firing-temperature range.   Considerable weight loss still occurs above 700°C, and this is the temperature region where the bubbling appears to occur in the resistor-conductor terminations.   Conductors containing small quantities of palladium may be successfully cofired with these ruthenium-based resistor systems, as illustrated in Fig. 3.

Loss of Soldered-Lead Adhesion on Thermal Aging

Frequently, thick film conductor compositions are tested to see how much loss in adhesion occurs when a conductor with a lead soldered on is aged for a period at an elevated temperature, for example, 48 h at 150°C. Consumers do not generally expect their thick film circuits ever to see these temperatures in operation, but this test is carried out to show what loss of adhesion might be expected through long exposures to lower-temperature environments, e.g., several years of operation at

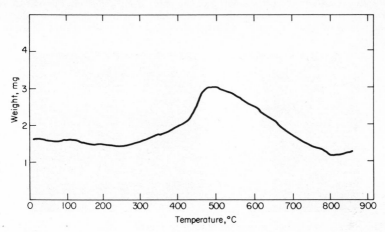

Fig. 21 Weight of a 1:2 Pd-Ag powder mixture when heated from 25 to 850°C in a 30-min period.

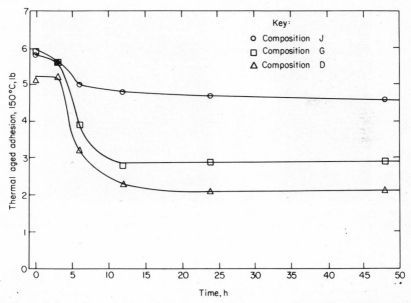

Fig. 22 Adhesion of Pd-Ag conductor compositions as a function of aging time at 150°C.

60°C. At present, greater losses are generally experienced in adhesion of conductors containing silver than those containing gold, e.g., Pd-Au, or Pt-Au conductors.

Figure 22 shows the adhesion behavior as measured by a peel test when three Pd-Ag conductors having the same palladium-to-silver ratio are aged with the leads soldered on for extended periods at 150°C. Note that most of the loss occurs in the first 12 h and the loss after 48 h is only slightly greater than after 12 h.   Other experiments show that even after several hundred hours not much further loss in adhesion occurs.   Thus, for rapid comparison of a number of compositions, a 16-h exposure test at 150°C should be adequate for good aged-adhesion comparisons.

Electron-microprobe studies show that the loss in adhesion on thermal aging is closely related to the diffusion of tin from a solder into the thick film conductor all the way down to the substrate.   Figures 23 and 24 show the relative positions of

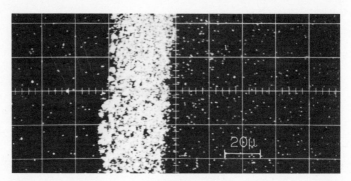

Fig. 23 Silver distribution in a fired Pd-Ag conductor soldered with eutectic solder and aged 48 h at 150°C as revealed by electron-beam microprobe x-ray fluorescence analysis.

Fig. 24 Tin distribution in a fired Pd-Ag conductor soldered with eutectic solder and aged 48 h at 150°C as revealed by electron-beam microprobe x-ray fluorescence analysis.

tin relative to the silver of a Pd-Ag conductor after exposures to a temperature of 150°C for 48 h.   Since the loss in adhesion seems to be related to diffusion of tin into the conductor, loss of adhesion on thermal aging may be greatly reduced by using a solder containing only a small amount of tin, for example, 10-90 tin-lead. Figure 25 summarizes results showing how the use of a low-tin solder greatly reduces the loss of adhesion on thermal aging of a Pd-Ag conductor.

Crossland and Haile[6] have shown that the loss of adhesion on thermal aging is related to the diffusion of tin into the palladium-bearing conductor to form $PdSn_3$, an intermetallic compound.   Their curves show a somewhat different time-temperature dependence of adhesion, but this may be because they used tensile measurements.

PROPERTIES OF THICK FILM CONDUCTORS
AS A FUNCTION OF COMPOSITION

There are a large number of thick film conductor manufacturers producing many products in each composition class to be described in this section. Table 4 presents a partial list of these manufacturers. To measure the properties important to thick film conductors, such as conductivity, solderability, solder-leach resistance, adhesion, resistor compatibility, wire-bonding capability, and die-bonding capability, using

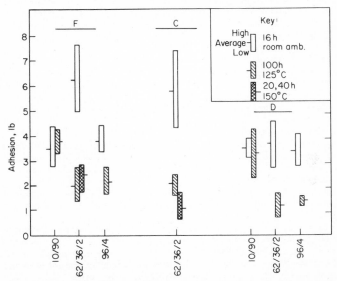

Fig. 25 Initial and thermal-aged adhesion of Pd-Ag conductor compositions *F*, *C*, and *D* soldered with 10-90 Sn-Pb, 62-36-2 Sn-Pb-Ag, and 96-4 Sn-Ag solders.[2]

TABLE 4   Principal United States Manufacturers of Thick Film Conductors

Airco Speer, Division of Air Reduction Co., Inc., St. Marys, Pa. 15857
Alloys Unlimited Inc., Electronic Materials Division, Mellville, N.Y. 11747
Cermalloy, Cermet Division, Bala Electronics Corp., Conshohocken, Pa. 19428
E. I. du Pont de Nemours & Co., Electronic Products Division, Photo Products Dept., Wilmington, Del. 19898
Electro Materials Corp. of America, Mamaroneck, N.Y. 10543
Electro-Science Labs, Inc., Pennsauken, N.J. 08110
Engelhard Industries, Inc., Hanovia Liquid Gold Division, East Newark, N.J. 07029
Sel-Rex Electromaterials Laboratories, Santa Ana, Calif. 92705
Owens-Illinois, Electronic Materials, Toledo, Ohio 43601

standard tests for all the products made by all manufacturers in each class would be an overwhelming task. Literature can be obtained from each manufacturer, but generally the tests and qualitative descriptions rating conductors vary greatly from one manufacturer to another.

For the purposes of this section, typical properties are given for a number of products in each class of conductor compositions in terms of Du Pont products. In most cases, other manufacturers will have products with similar properties, but the consumer must take it upon himself to use the tests described in this chapter to determine subtler differences in properties represented by products made by different manufacturers of the same type.

It will be seen that the products vary in their properties with composition in the general manner outlined in the preceding section.

Palladium-Silver Conductor Compositions

As a class Pd-Ag conductor compositions are the most economical of the thick film conductor compositions and therefore probably enjoy the most widespread use. These conductors show the best initial adhesion of any of the conductor classes. They show good solderability and moderate solder-leach resistance. Pd-Ag conductors with sufficiently high palladium content have the best termination cosmetics when cofired with Pd-Ag resistors. Low-palladium-content silver conductors generally show the best compatibility when cofired with ruthenium-base resistor compositions. Most Pd-Ag conductor compositions give good terminations when separately fired with most common resistor systems. Pd-Ag conductors generally have good aluminum and gold thermocompression wire-bonding capability. Die bonding is not possible with Pd-Ag conductors since it requires the formation of an Au-Si eutectic. The soldered adhesion of these conductors drops off more rapidly with thermal aging than for conductors containing gold. Pd-Ag conductors, as well as the other class of conductor compositions, are generally available in a number of vehicle formulations presenting a variety of fine-line definition capabilities depending on how much care the consumer wishes to take in printing.

Table 5 presents measured properties of a number of typical Pd-Ag conductor compositions fired in a belt furnace with a peak of 7 min at 850°C and a total cycle time of approximately 1 h. The first column presents the weight ratio of silver to palladium in the metal portion of the conductor. The second column gives fired-film thicknesses of 80-mil square pads as determined by a Brush Surfanalyzer. The resistance values were measured on a 200-square labyrinth pattern and are given in units of milliohms per square.

Initial adhesion values were measured by the peel test described previously using 20-gage wire peeled from 80-mil squares. The wires and pads were dipped in 62-36-2 Sn-Pb-Ag solder for 10 s at 220°C after being fluxed in Dutch Boy 115° mildly active flux. The soldered samples were held overnight at room temperature before the wires were pulled on Instron test equipment. The aged adhesion values were measured with the same type of test except that the soldered samples were aged for 48 h at 150°C before the wires were pulled off the substrates.

Leach resistance was also measured according to the test described in a previous section. A fired conductor pattern with 20-mil lines was dipped with the lines perpendicular to the surface of the solder for 10 s in 62-36-2 Sn-Pb-Ag solder at 230°C. The samples were fluxed each time in Dutch Boy 115 mildly active flux, held level for 3 s after the solder dip, and then quenched while still warm in trichloroethylene. After each dip, the samples were examined with a microscope at approximately 50 power with transmitted and incident light. The dips were repeated until leach-through was observed at some point along the length of the 20-mil lines in an area of the pattern which was at least ¼ in. below the surface of the solder during the dipping operation.

The resistor compatibility of a conductor is rated according to both the appearance of the termination cosmetics and the coefficient of variation for both prefired and cofired conductor compositions. With either PdO-Ag or ruthenium-based resistor compositions, if a composition gives smooth terminations when cofired with a resistor system, it is marked CF. All compositions which showed good compatibility when cofired with a resistor system, also showed good compatibility when prefired. Compositions recommended as being compatible with a particular resistor system on prefiring but not on cofiring are designated PF. Compositions not recommended as being compatible with a particular resistor system on prefiring with a particular resistor system are designated NR; this means that the combinations are not recommended because of termination cosmetics, room-temperature drift, or high contact resistance between this particular conductor and the resistor system.

In this table, conductor compositions are grouped according to their frit system.

° Trademark of National Lead Co.

As Table 5 shows, the fired-film thicknesses vary according to the relative quantities of frit and the total solids content of a particular group of conductors. The electric resistance of Pd-Ag conductors increases with the palladium content in each system. This is directly related to the change in resistance of the bulk alloys, as shown by Table 2. The last two conductors, *I* and *J*, show the lowest electric resistance, apparently because of the frit system used for these conductors.

These conductors also show the highest initial adhesion and the smallest loss of adhesion on thermal aging. Generally, adhesion values are lower if these conductors are fired at either higher or lower temperatures.

Also, as previously mentioned, solder-leach resistance increases with the palladium content. Again the last two conductors, *I* and *J*, appear to have the highest leach resistance at equivalent palladium content because of the frit system involved.

For reasons described in the last section, the resistor compatibility increases with

TABLE 5   Typical Properties of Pd-Ag Compositions Fired in Belt Furnace at 850°C

Code	$\dfrac{\text{Wt. Ag}}{\text{Wt. Pd}}$	Fired-film thickness, μm	Electric resistance, mΩ/sq	Adhesion, lb		Leach‡ resistance, dips	Resistor compatibility§	
				Initial*	Aged†		Ag-PdO	Other¶
A	∞	6	8	5	0.6	1	NR	NR
B	3.5	12	28	4	1	3	PF	PF
C	2.5	12	37	4	1–2	3	PF	PF
D	2	12	52	4	1–2	4	CF	PF
E	4	11	25	5	1	4	PF	PF
F	3	11	29	5	1–2	5	PF	PF
G	2	15	68	5	1–2	5	CF	PF
H	12	13	10	5–6	2	2	NR	CF
I	2.5	16	25	5–6	4	5	PF	PF
J	2	16	35	5–6	4	8	CF	PF

* Peel; 80-mil-square pads. Held overnight at room temperature after dipping 10 s in 62-36-2 solder at 220°C (DB 115 flux)·
† As above but held 48 h at 150°C after soldering.
‡ Number of dips required for leach-through of 20-mil line. For each dip sample was dipped in mild flux (DB 115) then held 10 s at 230°C in 62-36-2 solder and quenched in trichloroethylene.
§ See text for explanation of coding.
¶ Ruthenium-based resistor systems.

palladium content when these conductors are used to terminate Pd-Ag resistor compositions. Compatibility with ruthenium-based compositions becomes worse with increase in palladium content. However, when the conductors are prefired, good compatibility is obtained in all cases.

All these conductors show good solderability in eutectic 62-36-2 Sn-Pb-Ag solders. The conductors with high palladium content, *D*, *G*, and *J*, show poor solderability if initially fired at a high temperature, such as 760 or 850°C, and then refired at 500°C, as for encapsulating resistors. The last two conductors, *I* and *J*, show poor solderability when used with 10-90 tin-lead solder. Conductors *B*, *C*, and *D* show the best solder wetting in 10-90 tin-lead solder.

All these compositions show good compatibility with dielectric compositions used to make conductor crossovers. A special Pd-Ag conductor must be used, however, if good solderability is desired over the crossover.

Sometimes, thick film capacitors are made by printing and firing a layer of conductor and a special high-*K* dielectric composition and finally a top electrode conductor composition. Frequently, the two top layers are cofired. Compositions *B*, *C*, *E*, *G*, *H*, *I*, and *J* are recommended as having good compatibility with high-*K* dielectric compositions; but peeling may be experienced if composition *D* is used as a top electrode. A special Pd-Ag conductor is available for giving the highest dielectric constants.

Most of these compositions show good aluminum wire-bonding capability, but optimization of aluminum wire-bonding variables has been studied most extensively with composition *F* (see subsequent section on this subject).

One further word of caution must be given to users of Pd-Ag conductor compositions: silver conductors containing little or no palladium are susceptible to *silver migration;* i.e., fired, closely spaced silver conductor leads sometimes short out by forming dendritic silver growths from one lead to another when subjected to high humidity in combination with a high voltage gradient.[7] Compositions *A* and *H* would be particularly subject to this type of failure. When such compositions are used, the circuits must be carefully encapsulated to prevent contamination by moisture. The other compositions shown in Table 5 have enough palladium in them to ensure that they are not subject to this type of failure. With this high a palladium content, migration failure will occur from solder components before it will from migration of silver.

Thus this family of Pd-Ag conductors presents the thick film circuit designer with a wide variety of properties which should meet the needs of most thick film circuits. The high-palladium-content conductors provide the best solder-leach resistance

TABLE 6   Typical Properties of Pd-Au Compositions Fired in Belt Furnace at 850°C

| Code | Wt. Au / Wt. Pd | Fired-film thickness, μm | Electric resistance, mΩ/sq | Adhesion, lb | | Leach‡ resistance, dips | Resistor compatibility§ | |
				Initial*	Aged†		Ag-PdO	Other¶
K	2.5	14	100	4	3	5	PF	PF
L	2.5	11	73	5	4–5	4	PF	PF
M	3	10	83	5	4–5	7	PF	PF
N	2.5	13	56	4.5	4	5	PF	PF

* Peel test. 80-mil-square pads. Held overnight at room temperature after soldering 10 s in 63-67 (DB 115 flux) at 240°C.
† As above but held 48 h at 150°C after soldering.
‡ Number of dips required for leach through of 20-mil lines. For each dip sample was dipped in active flux (Kester 1544) then held 10 s at 250°C in 63-37 solder and quenched in trichloroethylene.
§ PF = prefire.
¶ Ruthenium-based resistor systems.

and best compatibility for the Pd-Ag resistor compositions, while the low-palladium compositions provide better electric conductivity, ease of soldering, and compatibility, with high-precision resistor systems. The latter compositions also provide the most economy in use.

Palladium-Gold Conductor Compositions

As a whole, this class of compositions provides greater ease of soldering in low-tin solders and lower loss of adhesion on thermal aging with high-tin solders than Pd-Ag conductor compositions. In addition, these compositions give better solder-leach resistance when used in eutectic solder containing no silver. Furthermore, some military specifications exclude the use of silver in thick film circuits because of previous experience with silver-migration failure. In this case, Pd-Au conductor compositions provide the most economical solution.

Typical properties of this class of conductors are summarized by Table 6. Again the conductor compositions were fired in an 850°C belt-furnace cycle. The properties were measured in the same manner as the Pd-Ag conductors with the following exceptions. In the adhesion measurements, the conductors were dipped in 63-37 tin-lead solder for 10 s at 240°C, since this combination of conditions provides the best solder adhesion. The samples were held overnight at room temperature before being pulled on the Instron tester. Samples were held for 48 h at 150°C for the aged adhesion test.

In the solder-leach-resistance experiments, an active flux, Kester 1544° was used,

° Kester Solder Co., Chicago, Ill.

and the samples were dipped in 63-37 tin-lead solder for 10 s at 250°C before quenching. These more stringent leaching conditions provide a better comparison of these conductors with the Pt-Au conductor compositions in the next section.

The fired-film thicknesses were lower for conductors L through N than for conductor K because different intermediates used in their manufacture provide a higher fired-film density.

As a further result of the use of these intermediates, the electric resistances of conductors L through N are lower than that for conductor K because of their higher fired-film density, which, in turn, provides higher values of initial adhesion and less loss in thermal aging. Conductor L shows the best solder-leach resistance of these Pd-Au conductors.

It is recommended that all these conductors be prefired before they are overprinted with resistor compositions.

Solderability of all these conductors is excellent, but K, L, and M provide the best solder-wetting capability. Composition N has a higher solids content than the similar composition L and incorporates a special vehicle system optimized for mask printing: 5-mil lines with 5-mil spaces can be printed using this conductor compo-

TABLE 7   Typical Properties of Pt-Au Compositions Fired in Belt Furnace at 850°C

Code	$\dfrac{\text{Wt. Au}}{\text{Wt. Pt}}$	Fired-film thickness, μm	Electric resistance, mΩ/sq	Adhesion, lb		Leach‡ resistance, dips	Resistor compatibility§	
				Initial*	Aged†		Ag-PdO	Other¶
O	3.5	13	150	2	2	11	PF	PF
P	3.5	17	90	5	4–5	30	PF	PF

* Peel test. 80-mil-square pads. Held overnight at room temperature after soldering 10 s in 63-37 (DB 115 flux) at 240°C.
† As above but held 48 h at 150°C after soldering.
‡ Number of dips required for leach through of 20-mil lines. For each dip sample was dipped in active flux (Kester 1544) then held 10 s at 250°C in 63-37 solder and quenched in trichloroethylene.
§ PF = Prefire.
¶ Ruthenium-based resistor systems.

sition. This composition provides excellent soldering with 10-90 tin-lead solder; and this, together with its good line definition, makes it an excellent candidate for use in flip-chip soldered circuits.

These conductors cannot generally be readily soldered when used as the top conductor over a crossover dielectric composition.

Pd-Au compositions, as a class, show good gold thermocompression, bonding capability and can be die-bonded under suitable conditions.

Platinum-Gold Conductor Compositions

This class of thick film conductors is the most expensive per unit area printed but gives the best leach resistance for thick film conductor compositions. These conductors are recommended especially in circuits where discrete components are attached with solder and where certain components must be replaced frequently because of the complexity of the circuit.

Table 7 gives the typical properties of two of these compositions when fired at 850°C in a belt furnace. The properties were measured in the same way as the Pd-Au conductor compositions.

The more recent composition P is obviously superior to the older composition O. P has a higher fired-film thickness, a 40 percent lower electric resistance, and 2 to 3 times the initial and aged adhesion of conductor O. In addition, conductor P has 3 times the solder-leach resistance of conductor O. Both these compositions have higher leach resistance than any of the Pd-Au conductors, however. It is recommended that these Pt-Au compositions be prefired when they are used with resistor compositions.

The older composition, O, has the best solderability and can be soldered when fired over a crossover dielectric composition. Compositions O and P are both suitable for aluminum wire bonding, gold thermocompression bonding, or die bonding. However, in die bonding, less flow-out of the Au-Si eutectic is experienced than with a gold composition.

Die and Wire Bonding—Various Compositions

Silicon-die bonding Silicon-die bonding is frequently used to fasten silicon dies to thick film circuits where connections are subsequently made to the face of the silicon chip using aluminum wire bonding or gold thermocompression bonding. Since the Au-Si eutectic is used to form the adhesion bond, only conductors containing appreciable gold can be used for this purpose.

A number of the above compositions containing gold, along with gold compositions designed only for die and wire bonding, were tested for die-bonding capability by Headley.[2] Included in this test were some new compositions containing small quantities of platinum to give a solder-bonding capability. Headley used 30 by 30 by 5 mil silicon bareback dice with a Kulicke and Soffa model 642 bonder.* He used a

TABLE 8 Comparative Die-Bonding Performance of Gold-Bearing Compositions

Code	Type	Firing temperature, °C	Shear force,* lb	Failure mode†
Q	Pt-Au	850	19	D
		1000	19	D
P	Pt-Au	850	19	S
M	Pd-Au	850	11	S
K	Pd-Au	850	10	S
O	Pt-Au	850	8	M
R	Au	850	6	M

* Die size 30 by 30 by 5 mils. Average of nine samples. Standard deviation 2 to 4 lb.
† D = die fracture; S = failure of metallizing; M = mixture of D and S.

column temperature of 380°C with a bonding time of 3 s and a clamping force of 90 g. The substrates were kept at a temperature of 380°C from 4 to 6 min. Table 8 summarizes silicon-die bonding data measured by Headley and ranks the conductors in order of decreasing bond adhesion. The bond strengths listed are measured according to the force required to dislodge the die from the fired conductor composition and not from the amount of flow-out of the Si-Au eutectic, which often is only a measure of the gold content of the conductor. The highest force was measured for the low Pt-Au conductor composition Q. The next best bond strength was shown by Pt-Au composition P. The table lists the type of die-to-conductor adhesion failure found in each case. The fact that gold composition R ranks last even though there was some evidence of die failure here implies a weakening of the die due to the Au-Si eutectic formation. Generally, gold compositions have been preferred in die bonding since a more profuse flow-out of the eutectic is obtained with them, and this allows convenient, nondestructive inspection.

Gold thermocompression bonding Gold thermocompression bonding is generally used to fasten gold wires between the face of a die that has been back-bonded to a thick film circuit and conductor fingers, which approach the bonded die on the thick film circuit. Most of the compositions shown in the previous tables show good capability for gold thermocompression wire bonding.

Aluminum wire ultrasonic bonding Aluminum wire ultrasonic bonding provides an alternative way of making connections between the face of a silicon die and thick

* Kulicke and Soffa Manufacturing Co., Fort Washington, Pa.

film circuit conductors. Aluminum wire bonds may be made with Pd-Ag conductors as well as gold conductors.

Figure 26 shows a histogram summarizing a study of ultrasonic aluminum wire bonding with gold composition R. The conductor was fired at 850°C in a belt furnace; 2-mil aluminum wire was used containing 1 percent silicon and elongated 1 percent. A Kulicke and Soffa model 484 aluminum wire-bonder was used with a clamping force of 24 g. A power setting of 78 mW for 80 ms was used for these experiments.

Figures 27 and 28 summarize data obtained by Headley for aluminum wire bonding on Pd-Ag compositions E and G.[2] Here he used 2-mil aluminum wire containing 1 percent silicon hard-as-drawn. The power setting was ⅓ to ½ W for 100 to 250 ms, and a clamping force of 40 g was used. Thus good bonding was achieved through the use of more energy than for the gold composition described above.

The Royal Aircraft Establishment of the United Kingdom has carried out sophisticated aluminum-wire-bonding studies on composition F[8] using 1-mil aluminum wire containing 1 percent silicon, elongated 2 percent. They fired the conductor at 925°C. peak belt-furnace cycle of 8 min at over 900°C. They used a 30-g clamping force and found the optimum power setting to be 1.75 W at 0.225 s. Fig. 29 shows a

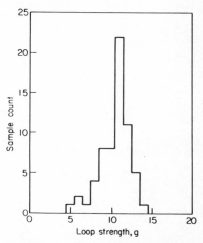

Fig. 26 Loop strength of 1-mil aluminum wires ultrasonically bonded to gold conductor composition R fired to an 850°C peak in a ½-h belt-furnace cycle.[2]

histogram of their experiments. The average of 12 g shows an excellent bonding capability with the proper settings. Figure 30 shows the optimization of the time and power settings for this conductor.

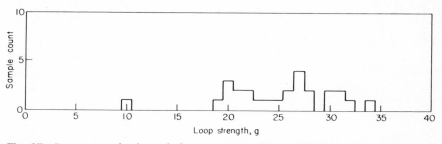

Fig. 27 Loop strength of 2-mil aluminum wire ultrasonically bonded to Pd-Ag conductor composition E fired to an 850°C peak in a 1-h belt-furnace cycle.

PROCESSING THICK FILM CONDUCTOR COMPOSITIONS

The following sections are intended to give the neophyte user of thick film conductors practical advice on processing these materials.

Storage

Most of the thick film conductor compositions described above cause no problems as far as shelf life is concerned. The viscosities of more recent compositions are high enough to ensure that little settling occurs, and roller storage is not recommended.

Some of the high-palladium Pd-Ag compositions occasionally show rheopectic behavior when stored on slowly moving rollers; i.e., their viscosities increase drastically with prolonged or high-level shear stress.

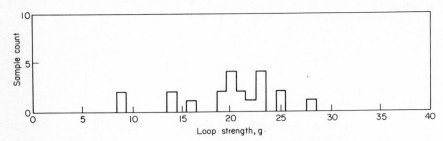

Fig. 28 Loop strength of 2-mil aluminum wire ultrasonically bonded to Pd-Ag conductor composition G fired to an 850°C peak in a 1-h belt-furnace cycle.

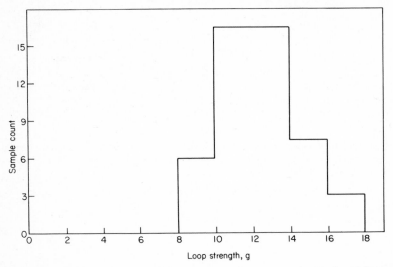

Fig. 29 Loop strength of 1-mil aluminum wire ultrasonically bonded to Pd-Ag conductor composition F under optimum conditions.

Printing

Many conductor compositions come in several vehicle variations, the same solid content yielding the same final fired properties but allowing different types of printing.

Mask printing Mask printing allows very high levels of definition. Highly pseudoplastic or shear-thinning vehicles have been created to give the optimum mask printing. Figure 31 shows the variation in viscosity with shear rate of such a composition compared to the normal screen printing compositions. Most thick film conductor compositions are at least somewhat pseudoplastic; i.e., the apparent viscosity tends to decrease to some degree with increasing shear rate. The highly pseudoplastic compositions, however, which show a large decrease of viscosity with increasing shear rate, allow the conductor to be readily pushed through the mask by the squeegee and yet set up quickly on the substrate, allowing retention of good definition. Frequently, this type of composition appears gelled when a jar of it is first opened, but with stirring the decreasing viscosity of the composition is apparent.

A bifurcated squeegee, in which the conductor fluid is pumped through the center of the squeegee, allows these compositions to be printed without manual redistribution of the paste upon the mask.

Screen printing Here less pseudoplastic conductors are desired since high production rates are generally used, and the conductor fluid or ink must flow sufficiently to ensure that when the squeegee retraces its path, sufficient paste is in front of the squeegee to give good printing. Also the paste must flow enough to level and not

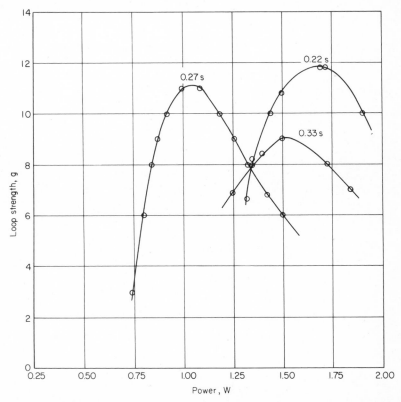

Fig. 30 Loop strength of 1-mil aluminum wire ultrasonically bonded to Pd-Ag conductor composition F as a function of time and power settings.

leave obvious screen marks. Even here for best resolution, 325-mesh screens must be used and relatively high-viscosity paste is specified. For high-production work, where less definition is required, lower-viscosity material is generally used.

Kobs and Voigt[9] have given a good discussion of the optimization of screen printing variables. They recommend the use of a high-level-hardness squeegee (85 durometer) since this allows good printing over a wider range of conditions. Proper adjustment of squeegee speed and pressure also helps to achieve a good definition with ease of printing. Larry[10] describes techniques for improving resolution by using special coatings on screens and substrates.

Leveling When conductors are screen-printed, it is good practice to allow the conductor to level in ambient air for about 15 min. Blowing air over the conductor during this period helps improve definition.

Drying

Most conductor compositions should be dried for at least 15 min at 150°C using a muffle furnace or belt furnace. In some cases, infrared lamps are used for this purpose, but care must be taken not to heat the freshly printed conductor too rapidly, for fast evaporation of solvents could cause surface roughness or poor definition. Some of the latest highly pseudoplastic compositions require a drying temperature of 80°C or less for best definition.

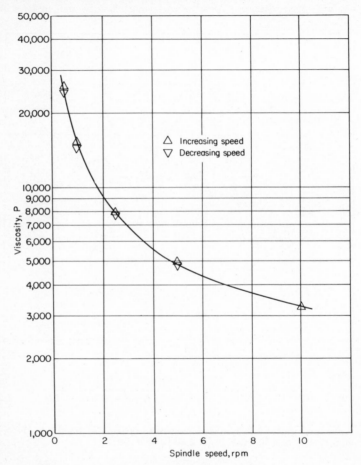

Fig. 31 Viscosity versus spindle speed for highly pseudoplastic gold composition R measured at 25°C.

Firing

Use the profile recommended by the conductor manufacturer. In all cases, a good draft is a prime requirement. Fresh, clean air should be introduced at the downstream end of the furnace, and this draft should pass through the high-temperature zone and exit at a point well before the parts reach the maximum temperature to draw off the combustion products and prevent carbon deposition on the conductive surfaces. Poor ventilation frequently causes solderability problems, especially when conductors are refired many times. Very slow cooling can sometimes cause soldering problems with very-high palladium Pd-Ag conductors, since PdO may form on the surface at intermediate temperatures.

Soldering Fluxes

Generally, it is best to use the mildest flux that will give an adequate solder coverage. Active fluxes may sometimes give lower adhesion values with conductor compositions designed to give excellent solder wetting with mild fluxes. In the tables shown in previous sections, Pd-Ag compositions A through G, all the Pd-Au compositions, and Pt-Au composition O should be soldered using only mildly active solder flux, meeting military specification MIL-F-14256 Type A, such as Dutch Boy 115, Alpha 611, or Kestor 196. The soldering flux should always be rinsed off with a solvent, such as trichloroethylene. More active fluxes such as Dutch Boy 120 or Kestor 1571 may be used with compositions I, J, and P.

Trying New Compositions

New conductor compositions should always be tested using the full firing schedule contemplated in the final-circuit design. For example, this might include an 850°C conductor fire, an 850°C dielectric fire, an 850°C top-conductor fire, and a 760°C resistor fire. When putting a new conductor composition into production, it is good practice to order several small lots produced at different times by the manufacturer and try out the final firing schedule needed for your circuit. Then, larger quantities of material, 50 to 100 troy ounces, should be ordered and evaluated before full production is scheduled. Frequently, a different firing schedule may give improved adhesion or soldering properties to the final conductor.

Customer Specifications

Once a thick film consumer has found a composition suitable for his needs and thoroughly tested it in its final application, he usually imposes some specifications on incoming new lots of material to ensure that the properties he measured will remain the same. For many applications, specifying the solids content and viscosity range will prove sufficient. Generally, it is best to select the same viscosity instrument and range specified by the manufacturer in the production of the product. Occasionally, the customer may prefer an end-use test, e.g., solderability and adhesion measurements, to further control the materials submitted to him. Probably a single-dip solderability test with a specified solder, time, and temperature would be best for these purposes. A visual examination may be used to eliminate conductor lots which give too many voids. Tests requiring actual chemical analysis of the conductor compositions are possible but much more time-consuming. Generally, gross errors in manufacture of conductor compositions show up clearly in the solids analysis or the end-use tests. It is not good practice to restrict the manufacturer too closely with a very narrow solids or viscosity range or by specifying, for example, resin-to-solvent ratios, as this may so limit the manufacturer that he is never able to produce a material which meets specifications. Too tight specifications, i.e., specifications which are more restrictive than necessary for good end use of the product, will probably result in a product which is more expensive to produce than is necessary for the application.

Troubleshooting Thick Film Conductor Compositions

Table 9 summarizes the problems that thick film conductor users frequently encounter and attempts to show the normal causes for certain symptoms and means of remedying these situations.

REFERENCES

1. Prather, J., and R. Slowinski: private communication.
2. Headley, R. C.: private communication.
3. Washburn, E. W.: "International Critical Tables," McGraw-Hill, New York, 1926.
4. Hansen, M.: "Constitution of Binary Alloys," McGraw-Hill, New York, 1958.

5. Elliott, R. P.: "Constitution of Binary Alloys, First Supplement," McGraw-Hill, New York, 1965.
6. Crossland, W. A., and L. Haile: Thick Film Conductor Adhesion Reliability, *Solid State Technol.*, February 1971.
7. Short, O. A.: Silver Migration in Electric Circuits, *Tele-Tech Electron. Ind.*, February 1956.
8. Fisher, H. D.: private communication.
9. Kobs, D. R., and D. R. Voigt: Parametric Dependencies in Thick Film Screening, *Solid State Technol.*, February 1971.
10. Larry, J.: The Influence of Surface Energies on Line Resolution in Screen Printing, *IHSM 1971 Int. Microelectron. Symp.*

TABLE 9   Troubleshooting Conductor Problems

Difficulty	Remedy
Printing	
Will not print:	
Squeegee stopped too high	Lower stop
Squeegee pressure too low	Increase pressure
Screen clogged	Clean screen
Fluid too thick	Thin with appropriate solvent if viscosity and solids higher than specified
Conductor fluid lifts off substrate with screen	Increase distance from screen to substrate to 40 mils for 5×5 in. screen; increase pressure on squeegee; use harder squeegee; use slower squeegee speed; screen too old: replace; dirty substrates; fluid too thick: thin as above
Poor resolution as printed	Squeegee pressure too great: decrease; fluid or room too warm: cool; dirty substrates; fluid viscosity too low: consult manufacturer
Drying	
Does not dry	Dry longer or at higher temperature
Loss of definition on drying	Dry at lower temperature or heat more slowly
Rough surface after drying	Allow to stand longer before heating
Firing	
Rough or bubbly prints	Fire at lower temperatures in recommended range; use more compatible materials: consult manufacturer
Loss of definition on firing	Fire at lower temperature or heat more slowly
Soldering	
Poor solder wetting	Increase ventilation during firing; fire at higher temperature or for longer period; fire at lower temperature; increase solder temperature; use more active flux; dip for longer period in solder with agitation
Solder leaching	Lower solder temperature; shorten exposure to solder; for Ag-bearing conductors use solder containing Ag; use composition with more Pd or Pt; use lower firing temperature

Chapter **6**

Resistor Materials, Processing, and Controls

GILBERT C. WAITE
Information Systems Department
Honeywell, Inc.

INTRODUCTION

Before the emergence of multilayer ceramics as a viable packaging technique, the thick film resistor was a fundamental reason for the existence of many hybrid-circuit manufacturers. Only in the manufacture of stable, economic, and physically small resistors did the thick film firm excel. Thin film houses with the line-definition capabilities of photolithography could make smaller, closer-spaced conductors; active-device manufacturers excelled in semiconductor chip mounting and bonding; the integrated circuit provided higher component density; and welded or cordwood modules had the economic advantage. Even in the area of resistor manufacturing thick films came off second best to thin films for precise-value or low-temperature-coefficient requirements. Resistor materials have improved hand in hand with thick film processing, until today they find application in every phase of the electronics industry from home entertainment to space exploration.

In preparing this chapter on the development of materials and processes an attempt has been made to refer to techniques and materials covering all phases of industry development. However, with 17 currently identified suppliers of resistor materials, several of whom offer more than one product line, it was impossible to include all the data available. The aim was then redefined to present selected data to cover the great majority of applications. During the writing one vendor or another dropped or added a product line, and as a result no final version could be written.

Specific equipment references have been kept to a minimum since the intent has been to stress principles and trends rather than conditions. Processing conditions, where stated, are primarily for reference purposes or are typical rather than optimum. Where optimum conditions or characteristics can be clearly established on the basis of scientific data, they are identified as such.

Historical Background

The genealogy of some thick film materials can be traced back more than 100 years to a class of organometallics involving gold chlorides and rosin. These rosinates, as they were called, were screened and fired for decorative purposes on china and glass tableware and were formulated not unlike the resinates of today. The very first resistor materials, which appeared in the 1940s, consisted of powdered carbon in phenolic materials. These systems were used in home-entertainment applications until early in the 1960s and are finding renewed interest for direct application to printed circuit boards. In 1950 a patent was issued to J. D'Andrea, of E. I. du Pont de Nemours & Co., for a system of electrodes formulated with palladium powders in glass frit. Work with these conductors led to the discovery that when metals of the platinum group are powdered and fired with frit, they exhibit a more gradual change in resistivity with increasing metal content than other noble metals. Figure 1 gives relative curves of resistivity vs. metal content for some of the metals and the Pd-Ag combination. Palladium-silver was chosen for the first particulate cermet resistor system because of the temperature-coefficient-of-resistance characteristics that could be achieved with the bulk metals.[1] Figure 2 shows the resistivity and TCR for bulk-metal alloys and cermet materials of palladium and silver. In 1958 a patent was awarded to J. D'Andrea for a particulate resistor system of Pd-Ag in frit, and the current era of thick film materials began.

Other systems of resistive glazes were developed during this period, but since they were for in-house use by the respective developers, they had no impact on the commercial market.

MATERIALS

Materials for the manufacture of thick film resistors can be most generally classified into organometallic, or resinate, systems and particulate metal in glass frit, or cermet systems. The system of carbon powder in phenolic is beyond the scope of this chapter since current use is restricted to proprietary programs.

Resinates

Originally, precious-metal resinates were simply solutions of the metal chlorides in organic solvents. These solutions, however, proved to be corrosive and unstable. The newer noble-metal preparations are actually organometallic compounds in which the metal atom is attached to a sulfur or oxygen atom linked to a carbon atom.[2] When these compounds are thermally decomposed, a film is deposited on the substrate. The process requires that this decomposition take place in an oxidizing, or air, atmosphere. Decomposition takes place at temperatures from 500 to 700°F. Under these conditions, noble-metal compounds deposit metallic films, and other materials deposit metal-oxide films. Single-application thicknesses of such films reach 10 μin. (2,500 Å) maximum. For this reason their application as resistive elements in thick film circuitry is limited to substrates with highly polished or preglazed surfaces. The relatively low deposition temperatures, however, permit the use of glass substrates, on which most fritted materials could not be fired. The films produced by resinate system have some porosity but are continuous; therefore the mechanism of conduction and the resistivities of the resulting films are like those of the bulk material sources.

The resinate process can be used to produce a series of cermet resistor materials. By using an organic solution containing gold, palladium, rhodium, and several base metals and performing the deposition on glass particles rather than on a surface a particulate system of resistor materials has been produced.[3]

Cermets

The name cermet is usually applied to materials which result in a fused structure of conductive or resistive materials in a

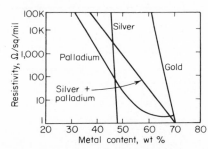

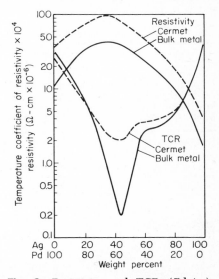

Fig. 1 Resistivity vs. concentration of metal powders in PbO-B₂O₃-SiO₂ frit. (*L. C. Hoffman, Precision Glaze Resistors, Amer. Ceramic Soc. Bull., vol. 42, no. 9, pp. 490–493, September 1963.*)

Fig. 2 Resistivity and TCR (Pd-Ag). (*D. L. Herbst and M. Greenfield, Theory of Conduction in Thick Film Conductors, INTERNEPCON 1972 Proc., pp. 106–114.*)

vitreous, nonconductive binder. These resistive elements may include such materials as oxides or oxide compounds of indium, thallium, ruthenium, palladium, tungsten, osmium, iridium, and rhodium as well as gold, silver, platinum, and palladium, either in elemental form or in compounds. Some compositions contain modifiers for TCR such as zinc, bismuth, manganese, or nickel oxide. In finely divided powder form, these elements or their compounds become the conducting particles of the cermet resistor. Particle size of the conducting phase generally ranges from 0.2 to 2.0 μm.

In the formulation stage, the vitreous nonconducting binder of the fired cermet is also a finely divided material. Average particle sizes of the frit range from 0.5 to 10 μm. Composition of the glass varies with the manufacturer's intended firing temperature and may range from a mixture of lead bismuth borosilicate and zirconium oxide for an intended firing temperature of 1000°C to a mixture of lead bismuth and lead oxide for temperatures of 750°C.

Some currently used frit formulations are listed in Table 1 with the recommended firing temperature. The frit and the metals and/or oxides constitute the fired film, but in paste form they represent approximately two-thirds of the composition by weight. The remainder consists of such resins as ethyl cellulose and such solvents as terpineol or butyl Carbitol° acetate with traces of surfactants and wetting agents to give the desired printing characteristics. Table 2 indicates the composition of a typical cermet resistor paste. Since the percentages shown are according to weight, use of alternate materials having different density for the resin or solvent function could markedly change the composition. Variations would also be detectable according to the intended resistivity for certain classes of conducting-phase materials (see Fig. 1). In the resulting fired film the dependence of resistivity on gold or silver concentrations is very steep. Silver powder, for example, has a resistivity of less than 1 $\Omega/(sq)(mil)$ at 48 percent and 100kΩ at 56 percent. Palladium, on the other hand, has useful resistivities over a range of 33 to 70 percent. For a palladium-powder system, therefore, the percent by weight of metal would vary from high resistivity to low, while for a silver-powder system the change would be relatively small.

TABLE 1 Frit Formulations and Recommended Peak Firing Temperatures

Name	Formulation	Temperature, °C
Lead bismuth borosilicate and zirconium oxide*.....	PbBiBSiO$_2$ + ZrO$_2$	980–1050
Lead bismuth borosilicate*.........................	PbBiBSiO$_2$	850
Lead bismuth and lead oxide†.....................	PbBi + PbO	775
Lead zirconate and silica.........................	PbZrO$_2$ + SiO$_2$	760

SOURCE: Data from Ref. 17.   * High-melting.   † Low-melting.

TABLE 2 Typical Composition of Resistor Paste

Material	Function	Total weight, %
Metals or oxides.............................	Conduction	27
Glass frit....................................	Flux (binder)	40
Resin..	Rheology (binder)	6
Solvent......................................	Dilution	26
Surfactant...................................	Dispersion	1

Mechanism of conduction   The primary mechanism of conduction in cermet resistors is point-to-point contact. There was some doubt about the conductive mechanism for Pd-Ag systems because of the large number of trace elements found in the fired films of some early formulations.[4] Efforts to determine the mechanism through thermodynamic analysis were inconclusive. It has been shown, however, that for both the ruthenate and Pd-Ag systems the particle-to-particle contact mechanism satisfies all observed reactions.[5] Factors such as particle size[6] and firing conditions[7] can affect resistivity, TCR, and chemical composition of the conducting particles, but conduction still occurs through a chainlike structure of conducting particles. It is apparent that alteration in the frit material can affect the microstructure of the fired film. Dispersion of conducting particles in fired films of pastes using similar conducting materials are reported to be alternately in a thin layer close to the substrate-resistor interface[5] and uniformly dispersed throughout the glass phase.[5,8] Resistor materials may also react with materials in the interfacing conductor or electrode system to give resistance values other than expected,[9] but these effects alter only the density of the particle-to-particle contacts or the bulk resistivity of the conducting particles themselves; the conducting mechanism is not changed.

Types of formulations   Although the conducting phase or material of a resistor system may consist of any of several elements, their oxides, or alloys, and compounds

° Registered trademark of Union Carbide Corp.

of the elements and oxides, the more generally used systems consist of palladium, ruthenium, and iridium in compound or oxidized form. Selection of these elements is based both on availability and applicability to required processing conditions. Glass-processing temperatures are involved; materials that form stable conducting or semiconducting compounds after processing at such temperatures are gold, silver, and the platinum-metals group i.e., platinum, ruthenium, palladium, iridium, rhodium, and osmium. Of these, gold, silver, and platinum are precluded in elemental form since their bulk resistivities would yield films of less than 2 Ω/sq for film thicknesses greater than 0.3 mils.[10] Iridium and rhodium are very expensive, and osmium is not only expensive but rare. Most resistor paste now used is formulated around the other elements, namely palladium and ruthenium.

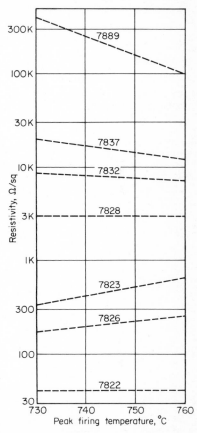

Palladium-based Resistor Systems. There are two primary formulations within the category of palladium-based systems. In one the conducting medium is made up of palladium oxide and silver, and in the other of palladium and silver. Pastes formulated from these specific materials, together with modifications and other possibilities, are described in detail in the patent literature.[11–14] A material of the first classification is the 7800 Certifired* series manufactured by Du Pont. This material, the first of the generally marketed thick film resistor material, was designed to be fired at temperatures between 730 and 760°C. Figure 3 indicates sheet resistivity vs. peak firing temperature for the members of this system. This system is typical of *static* types, i.e., those without a chemical reaction during the firing cycle.[7] The relatively large gaps in available sheet resistivity are narrowed by paste blending. The typical furnace profile for firing the 7800 series (Fig. 4) is bell-shaped. This particular example indicates a time above 700°C of slightly more than 12 min. The TCR for the various members for excursions of temperature from 25 to 1250°C and from +25 to −55°C are indicated in Fig. 5. This range of −50 to +750 ppm/°C and the deviation noticed for relatively small changes in firing profile were prime reasons for the development of other systems.[14] A series of similar formulations produced by Electro Science Laboratories is called series 7000. Resistivities of the series members range from 1 Ω/sq to 5 MΩ/sq. Recom-

Fig. 3 Resistivity vs. firing temperature for 7800 series. (*Du Pont, The Thick Film Handbook, Bull. A71753, August 1970.*)

mended firing temperature is 780°C peak in a bell-shaped profile with approximately 17 min above 700°C. Figure 6 indicates the relationship of the TCR to the fired resistance value for temperature excursions from 25 to 125°C. The curve indicates both the target values for paste manufacture and the generalized curve that includes values to be expected with normal variations in process conditions.

In the second type of formulation with the palladium-based group the palladium is in metallic form. This Pd-Ag–frit system has been classified as dynamic since determination of resistivity is governed by the oxidation of palladium in the presence

---

* Registered trademark of E. I. du Pont de Nemours & Co.

of silver at temperatures above 400°C. The 8000 series produced by Du Pont and the 6900 series by Electro Science Laboratories are typical of this type of formulation. Du Pont recommends a slightly lower firing temperature for the 8000 series than for the 7800. Figure 7 shows a typical profile for 8000 series with a peak

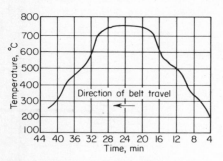

Fig. 4 Firing profile for 7800 series resistor compositions. (*Du Pont, The Thick Film Handbook, Prod. Bull. A80631, May 1973.*)

Fig. 5 TCR vs. resistivity for 7800 series resistor compositions with 8151 Pd-Ag terminations.[4]

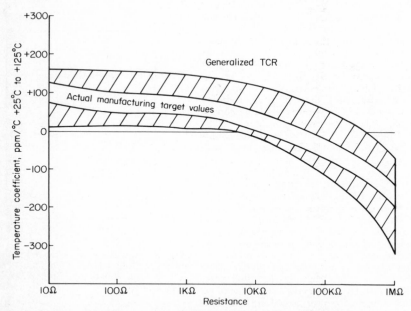

Fig. 6 TCR vs. resistivity for ESL 7000 series (25 to 125°C fired at 780°C). (*S. J. Stein and L. Ugol, Effect of Firing Conditions on Stability and Properties of Glaze Resistors, Electro Science Labs., Inc., Pennsauken, N.J., September 1968.*)

temperature of about 730°C and a dwell above 700°C of approximately 10 min. Figure 8 indicates resistivity vs. belt temperature for selected members of each of Du Pont's 7800 and 8000 series. Figure 9 indicates the 8000 series resistivities vs. firing temperature in degrees Celsius. Note that Fig. 9 does not indicate the curve for composition 8020, which is intended only as a blending number.

Ruthenium-based Resistor Systems. Ruthenium can exist as a ruthenate with several of the heavy metals or rare earths or as a rutile structure in ruthenium di-oxides. Its oxides are semiconducting, but the TCRs are positive, indicating metallic conduction.[15] Ruthenium is now the most common material in the constituents of thick film resistor systems. Of the 14 manufacturers listed or mentioned here at least 9 offer a resistor-paste system based upon one of the oxides of ruthenium. One of the early systems containing this material was the Firon* series. The material was formulated with a frit system requiring higher-temperature firing than the prevalent Pd-Ag systems. Figure 10 gives sheet resistivity vs. firing temperature for the members of this system. Recommended peak furnace temperature for this series is 980°C.

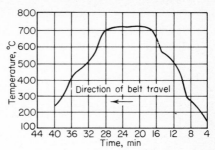

Fig. 7 Firing profile for 8000 series resistor compositions. (*Du Pont, The Thick Film Handbook, Prod. Bull. A80631, May 1973.*)

Figure 11 shows the effects of time at temperature $t = 980°C$. Recommended time at peak temperature is 15 min.[16] The TCR as a function of furnace tempera-

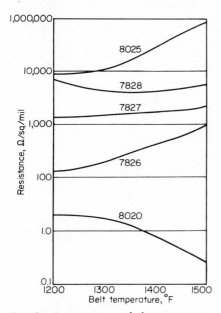

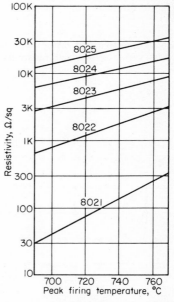

Fig. 8 Resistivity vs. belt temperature for some members of Du Pont's 7800 and 8000 series.[4]

Fig. 9 Resistivity vs. firing temperature for 8000 series resistor compositions. (*Du Pont, The Thick Film Handbook, Prod. Bull. A78372, November 1971.*)

ture is shown in Fig. 12. The combination of the peak temperature and time at temperature recommended for these materials generates a trapezoidal furnace profile. A system based on a ruthenate, marketed by Du Pont under the name Birox,† is

* Registered trade name of Electro Materials Corporation of America.
† Registered trade name of E. I. du Pont de Nemours & Co.

formulated with a frit that produces best results at furnace temperatures between 800 and 900°C. The manufacturer's recommendation for firing is 850°C; data on

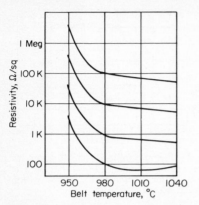

resistivity vs. temperature are shown in Fig. 13. These compositions are also relatively insensitive to time at temperature, as indicated by Fig. 14. The TCR, however, does vary with firing temperature, as indicated in Fig. 15 for hot TCR (25 to 125°C) and Fig. 16 for cold TCR (+25 to −55°C). The manufacturer's suggested profile is shown in Fig. 17. Peak temperature indicated is 850°C, with 9 to 10 min at peak in a total cycle of 60 min. Like several of the other manufacturers, Du Pont makes Birox formulations available in various TCR ranges for economic advantage to both manufacturer and user as well as various frit compositions to accommodate such applications as potentiometer wiping surfaces.

Fig. 10 Resistor vs. temperature for EMCA Firon series. (*G. Lane, Designing with High Reliability Thick Film Resistors, Proc. 2d Symp. on Hybrid Microelectronics, October 1967.*)

Another system based on ruthenium is series G by the Electronic Materials Division of Plessey. Resistivity vs. peak firing temperature for this series is shown in Fig. 18 for basic resistivities up to 1 kΩ/sq and Fig. 19 for resistivities up to 1 MΩ/sq. The higher-resistivity compositions indicate a slight increase in sensitivity to firing conditions as evidenced by Fig. 19. Figures 20 and 21 show variation of sheet resistivity with time at temperature, and again a slight

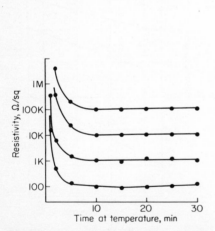

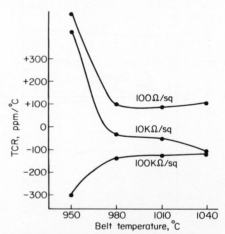

Fig. 11 Resistivity vs. time at temperature $t = 980°C$ for EMCA Firon series. (*G. Lane, Designing with High Reliability Thick Film Resistors, Proc. 2d Symp. on Hybrid Microelectronics, October 1967.*)

Fig. 12 TCR vs. firing temperature for EMCA Firon series. (*G. Lane, Designing with High Reliability Thick Film Resistors, Proc. 2d Symp on Hybrid Microelectronics, October 1967.*)

increase in sensitivity is indicated in the high-resistivity members over that indicated for resistivities up to 1 kΩ/sq. Figures 22 and 23 give the TCR for 25 to 125°C on the high-temperature characteristic and +25 to −30°C for the low-temperature side.

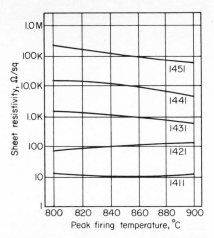

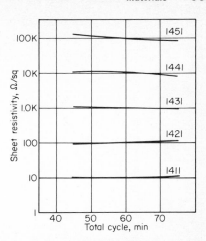

Fig. 13 Typical effect of peak firing temperature on sheet resistivity of Birox 1400 series resistors (60-min total cycle, Pd-Au 8651 terminations). (*Du Pont, The Thick Film Handbook, Suppl. Data 1400-3, serial no. A83591, October 1972.*)

Fig. 14 Typical effect of length of firing cycle on sheet resistivity of Birox 1400 series resistors (850°C peak temperature, Pd-Au 8651 terminations). (*Du Pont, The Thick Film Handbook Suppl. Data 1400-3, serial no. A83591, October 1972.*)

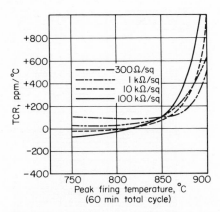

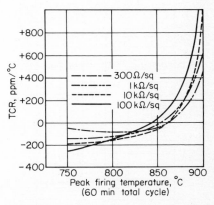

Fig. 15 Typical effect of peak firing temperature on hot TCR (25 to 125°C). (*Du Pont, The Thick Film Handbook, Prod. Bull. A69108, January 1970.*)

Fig. 16 Typical effect of peak firing temperature on cold TCR (−55 to +25°C). (*Du Pont, The Thick Film Handbook, Prod. Bull. A69108, January 1970.*)

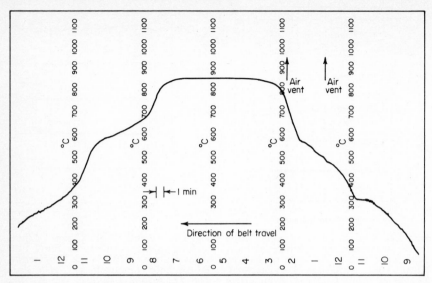

Fig. 17 Manufacturer's suggested furnace profile for Birox 1400 series. (*Du Pont, The Thick Film Handbook, Prod. Bull. A69108, January 1970.*)

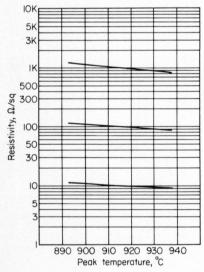

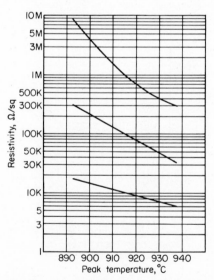

Fig. 18 Resistivity vs. peak firing temperature for Plessey EMD series G resistors (up to 1 kΩ/sq).[20]

Fig. 19 Resistivity vs. peak firing temperature for Plessey EMD series G resistors (up to 1 MΩ/sq).[20]

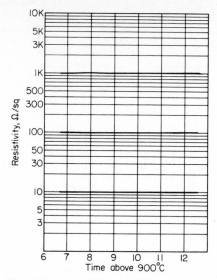

Fig. 20 Resistivity vs. firing time for series G resistors (up to 1 kΩ/sq).[20]

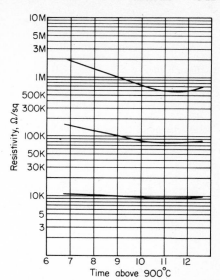

Fig. 21 Resistivity vs. firing time for series G resistors (up to 1 MΩ/sq).[20]

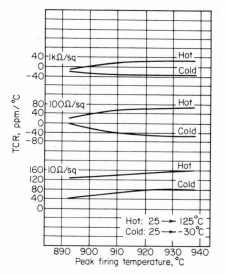

Fig. 22 TCR vs. typical peak firing temperature for series G resistors for 25 to 125°C.[20]

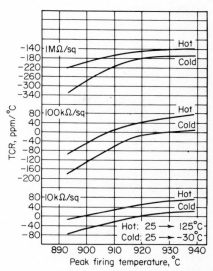

Fig. 23 TCR vs. typical peak firing temperature for series G resistors for +25 to −30°C.[20]

Another ruthenium-based system has been introduced as Powerohm* 850 series. A companion system, Powerohm 780, complements the 850 series by accommodating a lower firing temperature. Resistivity vs. peak firing temperature for the 850 series is shown in Fig. 24, and typical firing-cycle effects on resistor formula resistivity are indicated in Fig. 25. The TCR vs. peak firing temperature for several of the system members is given in Fig. 26, and 27 indicates the effect of increased firing time on TCR. The TCR for high temperatures is indicated to 150°C for this system; cold TCRs are taken from +25 to −55°C.

Electro Science Laboratories produces a ruthenium-based system in their 2800 series designed to be fired with the same profile specified for the 7000 series. Re-

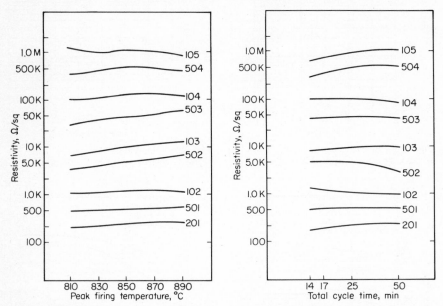

Fig. 24 Typical firing-temperature effects on resistor-formula resistivity for Powerohm 850 series (25-min firing cycle; conductor: Conductrox Pd-Ag 3412; screen: 175-mesh polyester). (*Thick Film Systems, Inc., Bull. R-850-6.*)

Fig. 25 Typical firing-cycle effects on resistor-formula resistivity for Powerohm 850 series (850°C peak firing temperature; conductor: Conductrox Pd-Ag 3412; screen: 175-mesh polyester). (*Thick Film Systems, Inc., Bull. R-850-6.*)

sistivity with respect to peak firing temperature for this series is shown in Fig. 28. The TCR for this series is plotted with respect to resistivity rather than firing time or temperature to show the more negative TCR values with higher resistivities (Fig. 29). Values represent temperature excursions of from 25 to 125°C for hot TCR and from +25 to −55°C for cold TCR.

Platinum-, Gold-, and Iridium-based Systems. One commercially available system differs from the platinum-group systems mentioned in that it contains no ruthenium. It is also produced by Electro Science Laboratories (the 3800 series).[17] This system is based on iridium and is designed to fire at 980°C. Resistivity plots for the various members for firing temperatures from 930 to 1000°C are given in Fig. 30.

Resistivity for time at peak firing temperature is shown in Fig. 31. A trapezoidal profile is recommended for this system with a 10- to 15-min dwell at peak temperature. The TCR is plotted vs. peak firing temperature in Fig. 32, and data on TCR

* Registered trade name of Thick Film Systems, Inc.

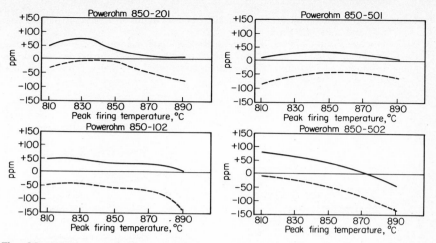

Fig. 26 TCR vs. peak firing temperature (Powerohm 850 series). (*Thick Film Systems, Inc., Bull. R-850-6.*)

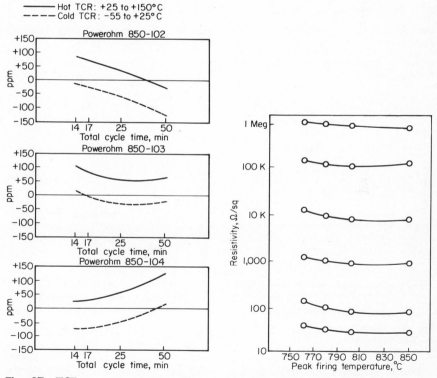

Fig. 27 TCR vs. time at temperature (Powerohm 850 series; 850°C peak firing temperature; conductor: Conductrox Pd-Ag 3412; screen: 175-mesh polyester). (*Thick Film Systems, Inc., Bull. R-850-6.*)

Fig. 28 Firing-temperature effects on resistivity for ESL 2800 series. (*C. Huang and S. J. Stein, Thick Film Resistors for High Yield Processing, Electro Science Labs., Inc., Pennsauken, N.J.*)

vs. sheet resistivity are indicated in Fig. 33. Hot TCR for this series is plotted from 25 to 125°C and cold TCR from +25 to −55°C.

Miscellaneous Formulations. For many resistor materials reported in the literature little information exists,[4] e.g., indium oxide, tungsten-tungsten carbide and molybdenum disilicide. A thallium oxide system is reported by Speer Carbon Co., and some details can be found in the literature.[18]

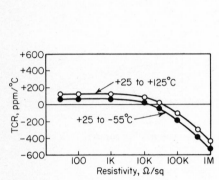

Fig. 29 TCR vs. resistivity for ESL 2800 series. (*C. Huang and S. J. Stein, Thick Film Resistors for High Yield Processing, Electro Science Labs., Inc., Pennsauken, N.J.*)

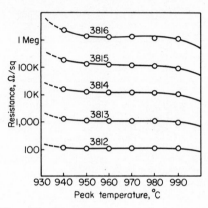

Fig. 30 Resistance vs. peak temperature for ESL 3800 series.[22]

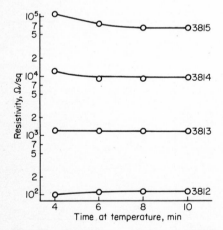

Fig. 31 Resistivity vs. time at 980°C for ESL 3800 series.[22]

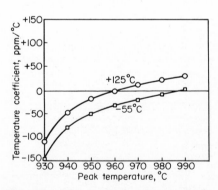

Fig. 32 TCR vs. peak temperature for ESL 3800 series.[22]

Vehicle systems   The actual constituents of the vehicles used in commercially available resistor systems are not disclosed. General characteristics and requirements are described by Herbst[10] without describing formulations. Miller[19] describes the functions and formulas of useful vehicle systems.

The Vehicle Formula. The liquid binder which carries the particles of metal and glass and forms a pastelike structure commonly has at least three ingredients: a resinous or polymeric binder to provide the basic rheological properties; a solvent, to dilute the resin, which can be evaporated later to dry the image; and sometimes a surface-active agent to permit the solid particles to be wetted by the vehicle and properly dispersed in it.

The volatility of the solvent system is one of the most important factors for ease of screening. Excessively volatile solvents tend to dry out during use; this causes fluctuations in the obtained resistance values since the amount deposited changes as the solvent leaves the paste. A solvent of very low volatility is therefore desirable, but this puts extra burdens on the formulator to control the flow of the print on the substrate and to provide good wetting of the pigment (metal and glass) particles; low-volatility liquids tend to have lower dissolving and wetting powers than smaller, more volatile molecules.

Common resistor vehicles contain such resins as ethyl cellulose or methacrylates diluted with such solvents as terpineol, Carbitol,* or Cellosolve* and their relatives and

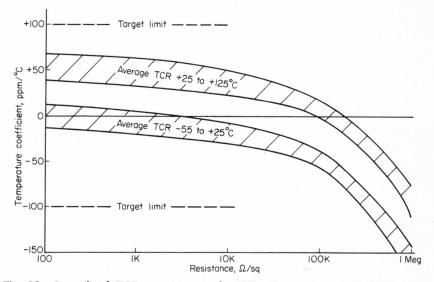

Fig. 33 Generalized TCR vs. resistivity for ESL 3800 series. (*S. J. Stein, J. B. Garvin, and M. Vail, Thick Film Pastes for High Performance Use, Hybrid Microelectronics Symp. ISHM, 1969.*)

derivatives. Some manufacturers recommend that if further dilution is required a less volatile solvent such as butyl Carbitol acetate be used. Figures 34 and 35 give typical volatilization characteristics for several resins and solvents useful for the purposes described.

Rheology. The most important rheological characteristics are the viscosity at the shearing rate used during the screening operation, the viscosity at very low rates of shear (gravity) which are experienced by the print after screening, and the time rate of the thixotropic buildup, i.e., the ability to reform a gel structure and thus become reimmobilized. Many screening systems use a gelatin agent to provide such a gel structure.

Material Selection

The proper choice of resistor compositions is essential to the successful manufacture of thick film hybrid microcircuitry. The resistive elements of thick film networks require the greatest precision and stability of electrical properties, while their properties are often the most sensitive to variations in processing and use. In selecting resistor compositions, there are several points to consider:

Fired resistor properties

Design flexibility

* Registered trademarks of Union Carbide Corp.

Sensitivity to processing variables

Effects of subsequent processing steps

These points must be considered together to determine which resistor compositions offer the best combination of fired properties and process characteristics. Factors

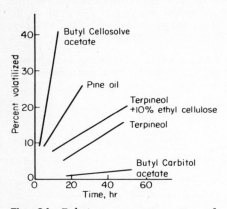

Fig. 34 Relative evaporation rates of solvents at room temperature in flowing air.[19]

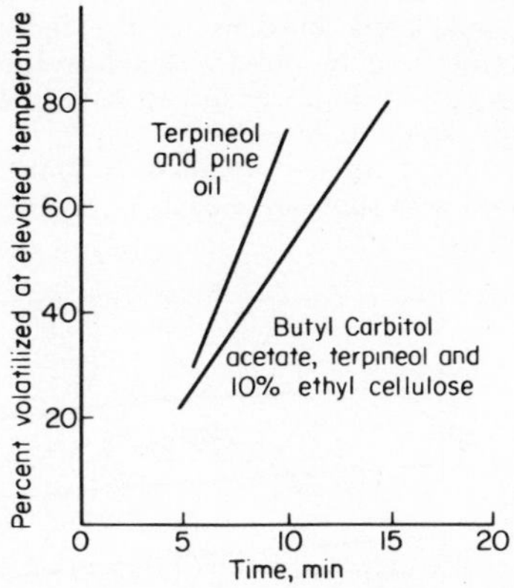

Fig. 35 Relative evaporation rates under radiant heat (~85°C).[19]

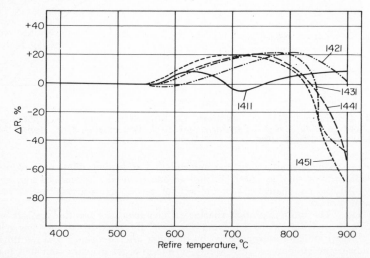

Fig. 36 Resistance change on refiring for Birox 1400 series resistors (10-min refire cycle; 400 to 900°C; Pd-Au 8651 terminations). (*Du Pont, The Thick Film Handbook, Suppl. Data 1400-17, serial no. A83590, October 1972.*)

such as resistor noise and termination compatibility and interaction will be discussed in the section on design parameters.

Occasionally fired resistor materials must be subjected to high temperatures during subsequent processing steps. Under certain conditions and with some series of resistor pastes, quite large changes in resistance values can result.

Figures 36 and 37 show the effect of various refiring times and temperatures on Du Pont's Birox 1400 series and Thick Film Systems' Powerohm 850 series, respectively.   Reactions of this nature are generally quite repeatable and predictable, but individual data must be generated for each process condition.

Table 3 lists several suppliers of thick film pastes with partial characteristics of their formulations.

Paste Storage and Preparation

Storage   Thick film materials vary with respect to proper storage methods.   It is of course necessary to prevent evaporation of the solvents.   In some vehicle systems the metal particles can agglomerate if the paste is not kept on a jar roller.   Rollers for this purpose (available commercially) roll a cylindrical jar roll about its axis at very slow rates (1 r/min or less).

Other gelled systems do not require rolling and in fact may change characteristics if subjected to such storage for as long as 2 weeks.   Refrigeration has been sug-

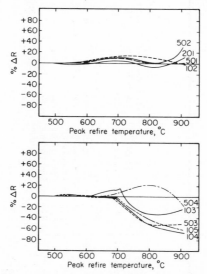

Fig. 37   Typical resistance changes on refiring for Powerohm 850 series (14-min refire cycle; same chip fired at 400, 500, 600, 700, 800, and 900°C).   (*Thick Film Systems, Inc., Bull. 850-1.*)

gested by several suppliers to help prolong shelf life.   In any event specific recommendations should be sought from the manufacturer.

Blending   Usually the particular resistivity desired for a given design cannot be achieved from a jar of paste labeled with that resistivity by the manufacturer because the supplier and user do not have identical processing conditions and cannot achieve identical results.   In such cases resistor materials can be blended to achieve the desired resistivity.

Most resistor-paste suppliers provide curves of blending parameters.   Examples of two such curves are shown in Figs. 38 and 39.   Figure 38 indicates the resistivity of different blends of end members whose base resistivity is exactly one decade apart, that is 10 kΩ/sq and 1 kΩ/sq.   In this curve the high-value member is plotted on the left margin, and the resulting resistivity declines as higher proportions of the lower-value member are added according to percentages indicated to the right of the left margin.   Figure 39 shows resistivities of blends between members separated by more than three decades.   This curve is taken from early literature, and while blending such values is undoubtedly possible, it is not recommended.   As these figures indicate, the resulting resistivity points do not necessarily fall on a straight line in a semilog plot.   It follows then that predictability of final resistivity will be better if the end-member values are closer to each other.

TABLE 3 Selected Resistor Pastes and Characteristics[a]

Manufacturer	Trade name	Series	Firing Time, min	Firing Temp,[b] °C	Firing Profile	TCR range, ppm/°C	Noise range,[c] dB	Resistor size, in.	Glaze required?
Airco Speer		TGA	9	570	Bell	-400 +50	-30 -2	....	Yes
American Components Inc.		2000	7.5	960	Trapezoidal	-150 +150	-35 -5	....	No
		3000	7.5	960	Trapezoidal	-250 +250	-30 0	....	No
Cermalloy		1000	9	875	Trapezoidal	-100 +100	....	....	No
		2000	9	875	Trapezoidal	-200 +200	....	....	No
		500	9	875	Trapezoidal	-50 +50	....	....	No
		7000NPS	8	925	Trapezoidal[d]	-500 +250	0 +25	0.250 × 0.050	No
Conshohocken Chemicals		CC100	[e]	[e]	....	-150 +150	....	....	Yes
		CC200	10	610	Trapezoidal	-100 +100	....	....	Yes
Du Pont	Certifired	7800	7	760	Bell	-300 +300	-18 +22	0.200 × 0.100	Yes
	Certifired	8000	7	730	Bell	-250 +250	-18 +4	0.200 × 0.100	Yes
	Birox	1300	10	850	Trapezoidal	-250 +250	-25 0	0.200 × 0.100	No
	Birox	1400	10	850	Trapezoidal	-100 +100	-30 -6	0.200 × 0.100	No
	Firon	5000	15	985	Trapezoidal	-140 +140	-25 +5	0.050 × 0.050	No
Electro Materials Corp.		5000	10	800	Trapezoidal	-200 +200	-35 0	0.100 × 0.100	No
		6800	7	760	Bell	-300 +300	-15 +10	0.125 × 0.125	Yes
Electro Science Laboratories		7000	10	780	Bell	-300 +150	-15 +8	0.125 × 0.125	Yes
		3800	15	980	Trapezoidal	-700 +700	-35 0	0.125 × 0.125	No
		2800	10	780	Trapezoidal	-250 +100	-30 ...	0.125 × 0.125	No
		2800 C	25	600	Trapezoidal	-200 +200	-22 +3	0.080 × 0.080	No

Manufacturer	Composition	Time,[a] min	Peak temp,[b] °C	Profile					Resistor size, in	
Englehard	Resinate RR	10	600	Trapezoidal	+100	+200			....	Yes
Matthey Bishop Inc.	Blend-Ohm	5	760	Bell	-125	+50	-33	-0.5	0.200 × 0.100	No
Methode Development Co.	44R	7	810	Bell	-100	+100	-27	+15	0.320 × 0.080	No
	75P	10	775	Trapezoidal	-100	+100				
	LTR-X2000	180	163	Batch Oven	-500	+500				
Plessey Electronics Materials Div.	A	8.5	760	Bell	-150	+230	-35	+10	0.200 × 0.100	No
	B	12.5	845	Bell	-100	+100	-35	+5	0.200 × 0.100	No
	G	9	915	Bell	-75	+75	-35	+12	0.200 × 0.100	No
Sel Rex	Ohm Path	10	950	Trapezoidal	-125	+50	-35	0	0.125 × 0.125	No
Thick Film Systems, Inc.	Power-ohm 780	11	780	Trapezoidal	-200	+200	-35	0	0.080 × 0.040	No
	Power-ohm 850	11	850	Trapezoidal	-100	+100	-30	+10	0.080 × 0.040	No
Transcene	Ohm Resist 150	10	775	Trapezoidal	-600	+350	-20	+25	....	Yes

[a] Characteristics taken from product data published by respective manufacturers. Used by permission.

[b] Temperatures are *peak*. In trapezoidal profiles the given times are time at peak temperature. For bell-shaped profiles time represents the period of the profile above an arbitrary temperature usually 10 to 30°C below peak temperature; Du Pont Certified 8000 series 7 min above 700°C.

[c] Observed values for noise index depend upon resistor area; larger sizes indicate lower noise levels. Where known, the resistor dimensions used to develop the above data are given.

[d] These resistors must be fired in an atmosphere of 50 to 75 ft³/h of N₂.

[e] Value-dependent.

Paste suppliers generally provide recommendations and procedures for blending pastes in their literature. Some general considerations and guidelines are applicable, however, to all thick film paste blending.

End-Member Determination. The specific characteristics of each of the members or constituents of the proposed blend must be determined. Presumably some of this information was available when it was realized that a blend is required; i.e., the resistivity of one member is indicated. These data can be erroneous, however, if the precise conditions under which the data were obtained are not duplicated for the other end member. If, for example, the resistivity of the member to be added was determined previously on patterns of different resistor size, different conductor interface, or different firing conditions, a source of possible error has been introduced. One method of reducing this error is to characterize each individual jar of paste as

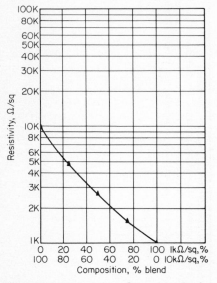

Fig. 38 Typical blending curve[20] for Plessey EMD series G resistors; 1 and 10 kΩ/sq.[20]

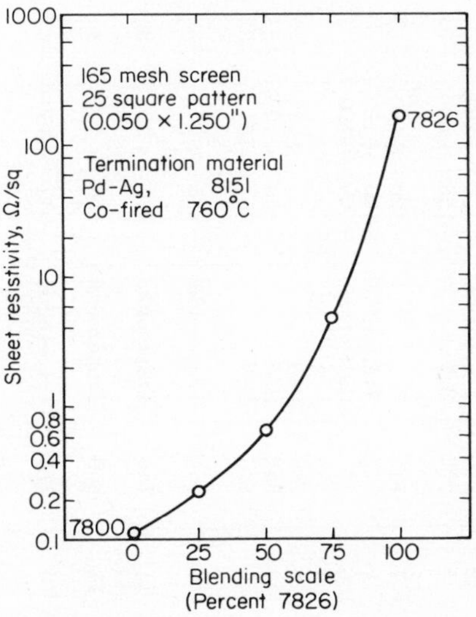

Fig. 39 Blending curve for 7800 and 7826 resistor compositions. (*Du Pont, The Thick Film Handbook, Prod. Bull. A71753.*)

it is received by firing a sample on a standard pattern with standard conductor interface in a furnace operating to a standard profile. Even then, the thickness of the print must be monitored since different screen-printer settings will cause different print thicknesses resulting in different resistivities. Because of solids content and viscosity variations between the blend members, both the end members and the resulting blend may have different print thicknesses, but if these thicknesses are known, the blender can better judge the necessary modifications to the blend to achieve the desired resistivity.

End-member determination must include (1) standard or common processing, including screening and firing parameters, (2) pattern and conductor thickness monitoring of both end members and trial blend (3) resistivity determined on a large enough sample of substrates to ensure statistical realism.

Physical Blending When preparing blends of pastes, each jar of composition should be stirred thoroughly before the required amount is removed. When a small amount of one component is to be blended with a large amount of another, a comparable volume of the larger component should be added to the smaller one, followed by good agitation. Subsequently, the larger component should be added in small increments and stirred to uniformity after each addition until the blending is complete.

It is desirable to blend adjacent members of the series rather than compositions covering a wide range of resistivity values.

Percentages by weight of the respective end members should be measured into a preweighed container on an analytical balance according to the desired weight of the final blend. Mixing should be thorough but not necessarily rapid since agglomeration of particles and polarization or polymerization of organics can occur under high shear conditions. A ground-glass plate and a muller or spatula are useful for final blending. Loss of solvent during this operation will affect viscosity and the resulting screen print deposition. For this reason the operation must be performed with minimum exposure, or solvents must be replaced. Recommended thinners or solvents are listed by most manufacturers of thick film pastes.

PROCESS

The thick film process as it affects the manufacture of resistors encompasses design, screen printing, drying and setting, furnace firing, and trimming. Some of the descriptions that follow may be supplemented or even duplicated in other chapters because similar functions, such as screening, must be seen from different viewpoints, for example, line definition for conductors vs. controlled volume deposit for resistors.

Design

In addition to the obvious requirement for topology to achieve the desired interconnection of prescribed resistors, other aspects of thick film design that require consideration include the effect of termination interaction with conductors, noise factors, high-frequency response, and thermal characteristics. The final design is the result of several compromises determined on the basis of available space, proposed trimming method, ranges from highest to lowest resistor values, the factors of application, and a host of considerations covered in other chapters with respect to packaging, component type, and attachment. Design factors in this chapter will be limited to those affecting the selection and processing of resistor materials as if resistors were the only component on the substrate involved. For this reason, recommendations should be considered as guidelines rather than hard-and-fast rules.

Resistor calculations   The basic equation for resistance is

$$R = \frac{\rho L}{A}$$

where
R = resistance
ρ = resistivity in bulk form
L = length
A = area

This equation can be further simplified to

$$R = \frac{\rho L}{A} = \frac{\rho L}{tw}$$

where t = thickness of screened film
w = width of screened resistor

$$\rho_s = \rho \frac{(bulk)}{t}$$

where ρ_s = sheet resistivity, Ω/sq

$$R(\Omega) = \rho_s \frac{L}{W} \quad \text{since} \quad \frac{L}{W} = \frac{length}{width}$$

$$= \text{number of squares}$$

From the above, the resistivity of the fired paste can be determined for a resistor of predetermined size. Conversely, the ratio of length to width, or aspect ratio, of a resistor of desired value can be calculated for a predetermined paste. An option between design technique and processing method is presented here. If it is assumed that the process of material deposition has inherent characteristics to prevent screening resistors to within 1 percent, then in order to produce resistors with 1 percent tolerance, trimming is necessary. The highest value in the as-fired distribution of resistors must be below the upper acceptable limit if upward trimming is to be used (see the section on Trimming).

The mean of the distribution must be some value below nominal, usually 15 to 30 percent. This can be accomplished by either (1) assuming that the value for the as-fired sheet resistivity is as measured and adding to the width (decreasing the aspect ratio) of the resistor or (2) calculating the sheet resistivity according to the formula and blending paste to fire at a fraction of calculated value; i.e., a 1,000-Ω/sq paste fires at 850 Ω/sq. With the second approach, the paste is referred to for design purposes as 1,000 Ω/sq.

The aspect ratio itself need not extend much beyond a range of 1:3 for short, fat resistors to 3:1 for long, skinny resistors to accomplish a decade of value for each paste. Herein lies the second option. Since each resistivity requires a separate screen, setup, and screen-printing operation, economics demands that resistivities be held to a minimum. If, however, the range of resistance values for the proposed circuit spans five decades, with values in each decade, a "rule" must be broken. In such cases, the designer can utilize a top-hat design, in which the aspect ratio is altered by trimming from less than unity to values of 20:1 or more.

Size itself can present yet another compromise when considered with the above factor in addition to power requirements. The power that can be dissipated in a thick film resistor can be limited by a temperature-rise requirement as determined by the packaging method, the application environment, or thermal rise due to active devices on the substrate. Manufacturers[21,22] have recommended dissipation factors as low as 15 to 100 W/in.^{2}, but resistors have been operated at dissipations up to 800 W/in.^{2}. At such dissipations the resistors are hot enough to "show colors" (approximately 800°C), and the effect is one of refiring with respect to characteristics, but the design and intended application must be considered before size and power limits are established. Minimum size for resistor design is governed by three factors.

1. The deposition of the material itself and the effect of edge flow at the conductor interface, discussed with screening parameters
2. The required area for trimming (discussed with that topic)
3. The effect of interaction at the conductor-resistor interface

Although resistors have been fabricated as small as 0.020 by 0.020 in., a generally accepted minimum dimension for length or width is 0.040 in.

Resistor termination   The value of a resistor is effected by the material with which it is terminated.[9,17,22] Loughran and Sigsbee[9] developed a profiling technique to display the interaction at the conductor-resistor interface graphically. Figure 40 is a plot from such a profile, indicating the observed resistance as a percentage of total resistance as the profile point is moved from one termination interface to the other. This technique indicates, by the slope of the curve, the resistivity of the material being probed, and reactions between materials can be detected by observing where nonlinearities occur on the plot with respect to the physical interfaces as observed on the screened and fired part.

Figure 41 is a plot of the ratio of measured to predicted resistance for ESL 3800 series as a function of resistor length. This series of curves indicates a greater interaction between Pt-Au (5800) conductors and the higher-value resistor materials than the lower-value members. It indicates a larger percentage effect on short resistors. Reaction or interaction of this type is another cause of compromise in design, since measured values for resistors with various lengths differ.

Figures 42 to 45 indicate resistivity vs. resistor length for various termination materials with several members of Du Pont's Birox 1400 series. All four of these plots are necessary to construct a composite for the resistivity family vs. a particular conductor material. Observing the effect, however, from curve to curve indicates that the reaction can cause either a net increase or decrease in measured resistivity with respect to predicted values.

Note that in these curves, the change in slope decreases as the length of the resistor increases, indicating a benefit to the designer when resistors are made as long as possible. In Fig. 46, which shows similar data for Plessey EMD series G resistors, the change in sheet resistivity vs. length is nearly linear, and the curves for

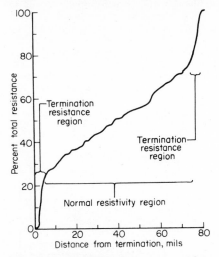

Fig. 40 Resistance profile with gold 80 by 80 mils, 100 kΩ/sq.[9]

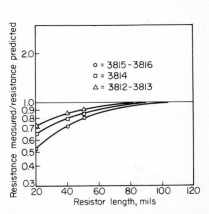

Fig. 41 Resistor length vs. sheet resistivity, for terminations ESL 5800B Pt-Au.[22]

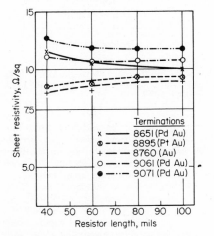

Fig. 42 Birox 1400 series resistivity vs. resistor length and termination. (*Du Pont, Bull. 1400-16, serial no. A83582, October 1972.*)

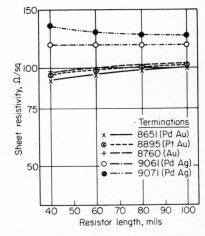

Fig. 43 Birox 1400 series resistivity vs. resistor length and termination. (*Du Pont, Bull. 1400-16, serial no. A83582, October 1972.*)

three lower-resistivity members are very nearly parallel. This indicates the same percentage reaction for each of these members.

The data presented in these figures serve to indicate a factor to be considered by the designer, and the data must be developed for each user since, as with basic resistivities, the effects will vary with processing. A further note of caution is due since some manufacturers' resistor materials are not compatible with those of other manufacturers.   Cracking and particle-to-particle separation have been observed in such instances.

The effect of termination material may also vary with termination sequence.   Prefiring resistors and posterminating with conductor material will usually result in different values than either prefiring or cofiring terminating conductors.   Each process facility must evaluate materials according to the process to be used and utilize the resulting data in the design function.

Resistor noise   Electrical noise is the spontaneous fluctuation in electric current caused by random motion of the current carriers.   Current noise in resistors is also called *excess noise,* since it is the excess over the thermal noise.   Thermal noise (Johnson noise or the brownian motion of electricity) is a function of resistance and absolute temperature.[23]

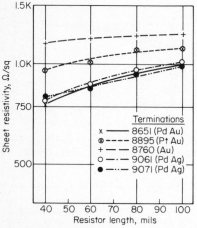

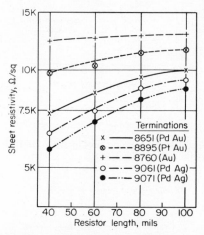

Fig. 44 Birox 1400 series resistivity vs. resistor length and termination.   (*Du Pont, Bull. 1400-16, serial no. A83582, October 1972.*)

Fig. 45 Birox 1400 series resistivity vs. resistor length and termination.   (*Du Pont, Bull. 1400-16, serial no. A83582, October 1972.*)

Unlike thermal noise, which is predictable from its mechanism and from a thermodynamic equation, no definitive relationship is known for current noise.   For discrete carbon resistors current noise varies with such parameters as materials, methods of construction, size, and shape of resistors.   Noise in thick film resistors has also been found to depend upon volume and, assuming constant thickness, upon geometry.[24] Figure 47 is a typical plot of measured Quan-Tech* noise vs. resistor area for a high-value (200 kΩ/sq) resistor paste.   The curve indicates decreasing noise with increasing resistor area.   Figure 48 shows noise-vs.-resistivity plots for Powerohm 780 series resistor material.   Curves are drawn for two resistor sizes, and again lower noise is indicated for the larger resistor.   The curves also show increasing noise with increasing resistivity.   Figure 49 shows noise vs. resistivity overlaid on a plot of TCR vs. resistivity for members of Du Pont's 7800 series resistor materials.   These curves show that the relationships are nonlinear.   Figure 50, however, drawn for ESL's 2800 series for 0.080 in. square resistors indicates a predictable straight line on the semilogarithmic plot of noise vs. resistivity.   The same general characteristic is indicated in Fig. 51 for Plessey EMD on series G resistors, where the resistor size is 0.200 by 0.100 in.

Noise also varies with processing conditions.   Figures 52 and 53 show the effect of peak firing temperature on noise for various resistivity members of ESL 7000

* Registered trademark of Quan-Tech Laboratories, Inc.

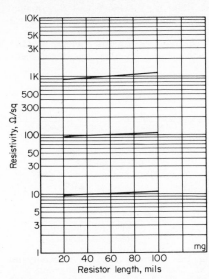

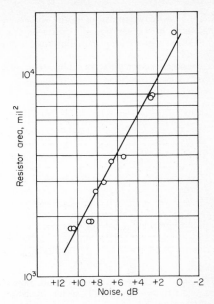

Fig. 46 Resistivity vs. resistor length for Plessey EMD series G resistors normalized to film thickness of 0.9 mil. (Profile: 915°C; 8 min above 900°C; termination: 05025; resistor width: 20 to 100 mils.[20]

Fig. 47 Noise vs. resistor area, MBI 200 kΩ/sq. (*Matthey Bishop Inc., Application Data Index EMG2.2, June 1972.*)

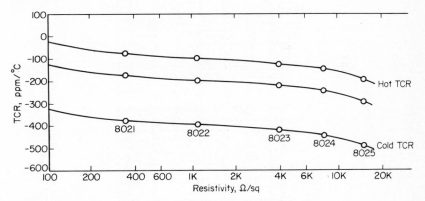

Fig. 48 Quan-Tech noise vs. resistivity for Powerohm 780 series resistor material. (*Thick Film Systems, Inc., Bull. R-780-8.*)

series and Firon. These data suggest the possibility of compromise on firing conditions depending upon the resistivity range, since noise-vs.-resistivity functions vary for these systems with firing temperature.

Screen Printing

Materials selection for thick film circuit manufacturers is a matter of choice, and the industry has adequately demonstrated that successful designs can be accomplished with all resistor materials. Design of the circuit to be screened and fired is subject to compromises previously discussed, but with consistent application of chosen guidelines, reasonably uniform results should be achieved. What remains to be described, however, are areas where consistency is most difficult to achieve and the smallest deviation can trigger distributions of values spread further than the process can correct.

Process requirements The two major requirements for the screen printing process are uniformity and predictability. To the casual observer, the requirements

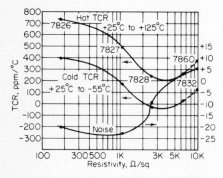

Fig. 49 TCR vs. resistivity and noise vs. resistivity for 7800 series resistor compositions.[4]

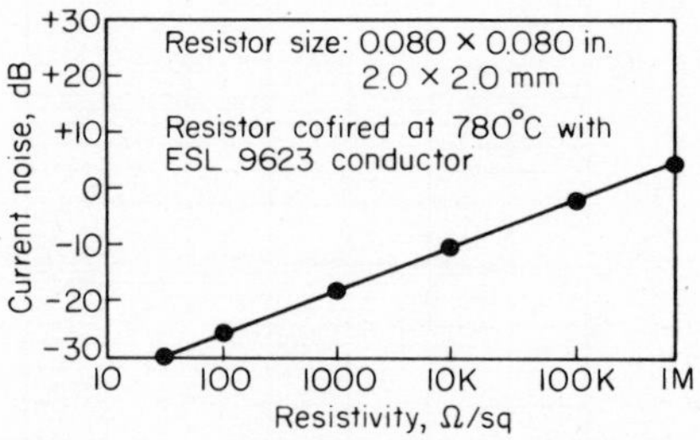

Fig. 50 Current noise vs. resistivity for ESL 2800 series. (*C. Huang and S. J. Stein, Thick Film Resistors for High Yield Processing, Electro Science Labs., Inc., Pennsauken, N.J.*)

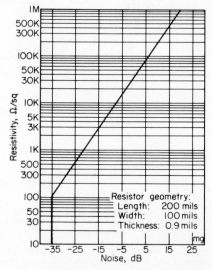

Fig. 51 Current noise vs. resistivity for Plessey EMD series G resistors.[20]

might seem obvious, but their relationship to other operations in thick film resistor manufacturing is significant enough to justify some process inefficiency to optimize their achievement.

Uniformity. In the discussion of design parameters the term sheet resistivity was used. Aside from the technical description to be found in the Glossary, this term implies a flat, sheetlike film of material. Unlike thin films, however, where the total thickness of the film is insufficient to describe irregularities, a thick film does not

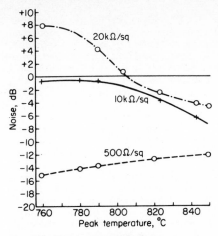

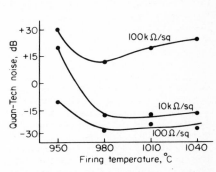

Fig. 52 Noise vs. peak firing temperature at 26°C for ESL 7000 series. (*S. J. Stein and L. Ugol, Effect of Firing Conditions on Stability and Properties of Glaze Resistors, September 1968.*)

Fig. 53 Noise vs. peak firing temperature for EMCA Firon. (*G. Lane, Designing with High Reliability Thick Film Resistors, Proc. 2d Symp. on Hybrid Microelectronics, October 1967.*)

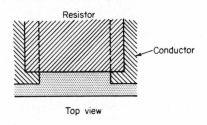

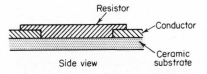

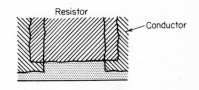

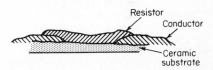

Fig. 54 Representation of an ideal thick film resistor. Ideal characteristics would be uniform resistor height; uniform conductor height; extremely sharp, well-defined conductors; extremely sharp, well-defined resistor; extremely flat ceramic substrate; TCR approaching 0 ppm; and a noise equivalency and stability factor to wire-wound resistors.[25]

Fig. 55 Typical as-fired thick film resistor. Physical characteristics: varying resistor and conductor height; conductor and resistor definition not sharply defined; ceramic substrate has camber. Electrical characteristics: TCR depending on value and paste, between ±400 ppm; noise from −35 to 20 dB; a stability ranging to that between carbon-deposited and metal-film types.[25]

approximate rectangular dimensions. Figure 54 represents an ideal thick film resistor and describes some of the desired characteristics, but the typical result of the thick film process is that in Fig. 55, which describes some of the less desirable results. With these inherent variations, the method of achieving final product uniformity has been assigned to the trimming function. However, since design parameters such as noise and coefficient TCR are functions of resistor geometry, the need for process uniformity is increased where resistor adjustment by geometry alterations, i.e., air-abrasive or laser trimming techniques, are used. For production throughput, the amount of physical material removed should be kept to a minimum. The distribution of values presented to the trimmer should then be as uniform as possible, with a mean value near the desired value. It is not sufficient to achieve this result on given lot or batch: for the manufacturing steps rightly to be called a process, this result must be achieved from lot to lot on a day-to-day basis and be predictable.

Predictability. Thick film production facilities are often really high-volume laboratories. As a resistive paste member is blended, it is sample-screened on the substrate being produced and fired. The resistor properties are then evaluated, and a go no-go decision is made with respect to the process. During the firing of this sample, the screen printer sits idle, possibly with a screen still spread with paste.

Further, having recognized success with a particular blend on a given day, many operators would repeat the sample firing step when attempting to duplicate a production lot on subsequent days. The ability to predict results can therefore be recognized as a necessity for production efficiency. It is certainly true that other steps in the process affect predictability, but absolute control of each subsequent process step will not overcome a lack of consistency in the material-deposition step.

Process variables in screen printing Several authors[26–31] have discussed the variables in the screen printing process. Hughes[26] lists 38 parameters involving the printer, paste, substrate, and screen. Kobs and Voigt[30] list 55, although a few of these might be considered as duplicates or not directly affecting the material deposit. Their investigation[30] presented a matrix of seven parameters for each of two or three parameter levels. Their procedure, however, introduces an additional variable (screen-tension variations) not identified in the initial matrix. It must also be recognized that while each of the listed factors may affect screen printing, a given setup on a given machine will establish as constants all screen parameters except tension, all squeegee parameters except speed and pressure, and all paste parameters except rheology and leave in addition the substrate-to-substrate variations. For our purposes it is proposed to classify these variables in two groups: (1) paste characteristics affecting transfer and (2) variations determined by the adjustment of the screen-printing machine.

Paste-Transfer Characteristics. Since the object of screen printing is to deposit on the substrate a precise and consistent volume of paste, and since the paste at the initiation of the process is in the screen, it is first necessary to consider those factors which affect the transfer of paste from screen to substrate. It is a two-phase operation in that the paste must first fill the cavity defined by the screen mesh and emulsion and then be transferred from the cavity to the substrate. The rheology of the paste determines its reaction to the shear forces imparted by the squeegee and subsequent recovery after removal of these stresses.

Trease and Dietz[32] investigated shear rates and viscosity as they might occur in the actual screen printing process. A curve of relative viscosity vs. time is shown in Fig. 56. Shear rates assumed by Trease and Dietz were calculated as shown in Fig. 57 using a squeegee speed of 5 in./s and assuming instantaneous and total filling of the screen cavity. This would result in a shear rate of approximately 1,000 s^{-1}; however, their work indicated that a shear rate of 100 s^{-1} more nearly correlates with observed printing results. The deviation is assumed to be due to insufficient time during the squeegee stroke for the material to achieve equilibrium. Materials are classified with respect to their response to shear rates as dilatant, newtonian, and pseudoplastic. Characteristic flow curves indicating relative response for these types of materials are shown in Fig. 58. Most thick film materials are pseudoplastic and exhibit a reduction in viscosity as the shear rate is increased. This is often confused with thixotropy, which is a reduction in viscosity with time at a constant shear rate. Trease and Dietz[32] give a mathematical model describing the rheological behavior of various systems. Figure 59 is a curve of viscosity vs. shear rate for pseudoplastic materials and shows deviations from the straight line expected from definitions of pseudoplasticity. The model, however, is described in the original text according to the relationship

$$\text{Viscosity} = \frac{B}{R^m} + N + \frac{Y}{R^{m+1}}$$

where B = base viscosity at defined shear rate of $(1\ s^{-1})$

M = exponent of pseudoplasticity or rate of reduction in viscosity with increased shear

N = viscosity at infinite shear rate

Y = plastic yield stress or minimum force to initiate flow

R = shear rate

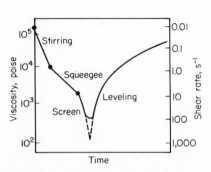

Fig. 56 Viscosity changes during screen printing.[32]

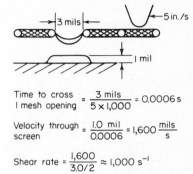

$$\text{Time to cross} \atop \text{1 mesh opening} = \frac{3\ \text{mils}}{5 \times 1,000} = 0.0006\ s$$

$$\text{Velocity through} \atop \text{screen} = \frac{1.0\ \text{mil}}{0.0006} = 1,600\ \frac{\text{mils}}{s}$$

$$\text{Shear rate} = \frac{1,600}{3.0/2} \approx 1,000\ s^{-1}$$

Fig. 57 Calculation of shear rate through a 200-mesh screen.[32]

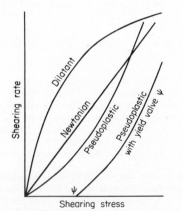

Fig. 58 Characteristic flow curves for various classes of liquids. (R. P. Anjard, *Viscometers for Thick Films, Electronic Packaging and Production, December 1971.*)

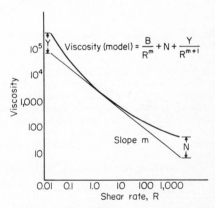

Fig. 59 Viscosity as a function of shear rate for pseudoplastic materials.[32]

Using this model, the authors established criteria for specification of pastes giving satisfactory results under the prescribed printing conditions used for the experiment:

1. Base viscosity of 6,000 to 10,000
2. Infinite shear viscosity of 20 to 60
3. Exponent of pseudoplasticity of 0.7 to 0.9

These criteria are valid only for the specific screening parameters used by the authors, but the technique could be applied to predetermine general print characteristics for any set of conditions.

Figure 60 is a plot of several members of the Powerohm 850 series indicating the index of pseudoplasticity for these materials from a shear rate close to base viscosity to a point near the correlation of viscosity as recognized in actual screen printing operations. Miller's discussion[28] of paste transfer describes experimental results of cavity volume, as affected by screen mesh and emulsion thickness, and the percentage of paste immobilized by the proximity of the emulsion sides to each other. He describes the rheological requirements as follows:

1. At screening shear rates, the paste must not flow between screen and substrate.
2. Viscosity should recover after screen separation at a rate to prevent flow beyond original screened dimensions.

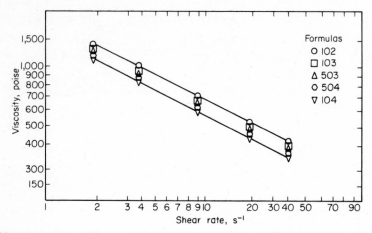

Fig. 60   Viscosity vs. shear rate, Powerohm 850 series. (*Thick Film Systems, Inc., Bull. R-850-1.*)

The settling that occurs after separation of the screen creates various cross-sectional profiles for materials with different rheological characteristics. Trease and Dietz[32] describe a stress-relaxation technique wherein the recovery of time-dependent viscosity characteristics can be determined and plotted according to relationships defined by Patton.[33] A characteristic curve is shown in Fig. 61. The study indicated that for fine-line resolution a recovery to 30 P within 15 s is required but surface imperfections (screen-mesh imprints) remain for materials with recoveries to this level in less than 30 s.

The interface between paste and substrate is the screen itself, and it could be described as part of either the transfer or the physical machine. It is included with the transfer properties in this treatment primarily because discussion of machine variables would involve the mechanical attributes of the screen.

Miller[28] describes immobilization of a percentage of paste in the cavity as being due to screen mesh (to the extent it is wet by the paste) and the shearing effect of that portion of the material remaining above the diameter of the mesh wire as the screen is withdrawn. Table 4 gives wire diameters, mesh openings, and percentage of open area for various screen mesh. This opening, however, is only for topographical areas, and the volume

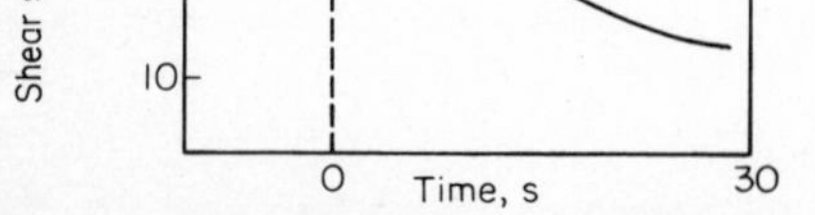

Fig. 61   Stress-relaxation technique for measuring time-dependent viscosities.[32]

$$\frac{S_0}{S_t} = e^{kt/n}$$

$$\text{or} \quad \frac{1}{n} = \frac{k\Delta \ln(S_0/S_t)}{\Delta t}$$

determinations with respect to maximum possible paste-transfer volumes are given by Herbst.[10]   He describes the notations of Fig. 62 as $y = 2r + c$, where r is the radius of the wire and c is added thickness due to the inability of mesh wire to bend to its own radius.   He then shows calculations for

1. The volume of the wire as 1.54×10^{-3} in.^{3}
2. The volume of the open area as 2.96×10^{-3} in.^{3}
3. The volume of the total cavity for 0.001 in. emulsion as 3.96×10^{-3} in.^{3}

All these calculations are for 1 by 1 in. opening in a 200-mesh screen with 0.0021-in. wire.   The results of these calculations indicate that reducing the emulsion thickness by 0.001 in. to a purely flush coating within the confines of the screen mesh would

TABLE 4   Open-Area Percentages

Mesh size	Wire diameter	Mesh opening	Percent open area
80	0.0037	0.0088	49.56
105	0.0030	0.0065	46.58
150	0.0026	0.0041	37.82
165	0.0020	0.0042	48.02
200	0.0016	0.0034	46.24
200	0.0021	0.0029	33.64
230	0.0015	0.0029	44.49
250	0.0016	0.0024	36.00
325	0.0011	0.0020	42.25
400	0.0010	0.0015	36.00

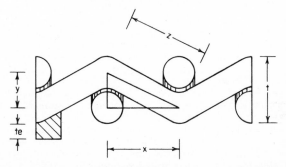

Fig. 62   Cross-sectional view, 200-mesh screen with emulsion.[10]

reduce the volume of paste by 40 percent.   A study of Fig. 62 makes it clear that total transfer is not to be expected.

Machine Parameters.   In the setup procedure for thick film screen printing, several criteria must be established.   Previously listed references[26–30] describe many variables, which become fixed for a given screen, on a given machine, used with a given paste, and a given squeegee.   Several of the parameters are subject to variation due to interaction with each other.   Squeegee pressure, for example, cannot be separated from screen tension in a printing mode where the screen must be depressed to contact the substrate.   Also the percentage of the total applied pressure necessary to depress the screen in the print area vs. that necessary to depress the screen over the total length of the squeegee blade cannot be calculated.   Screen-to-substrate distance at setup will also vary with screen tension for optimum printing conditions.   Ottaviano[31] reports screen deflections on new screen to average 66 mils at a force of 3 lb on a 3-in. squeegee.   Measurements on the same screens at various use periods disclosed an average of 83 mils.   For new screens, this reference recommends a screen-to-substrate distance of 0.039 in. Salisbury[29] found 0.040 in. to be most satis-factory and further noted that from 0.030 to 0.050 in. required additional force on the squeegee of 560 g, but he does not identify the length of the squeegee blade. Kobs and Voigt[30] used screen-separation distances of 0.040 and 0.055 in. and con-cluded that for conditions established for their matrix test, 0.055 in. offers better performance. This work referred to a screen size of 5 by 5 in. with a 3-in. squeegee length. Center-point deflection of the screen, however, changed from 0.030 to 0.034 in. for a ½-lb force. As a screen ages or stretches with use, compensation can be made by increasing the screen-to-substrate distance, but this changes the angle at which the screen peels from the substrate and introduces still another variable.

Squeegee durometer interacts with screen tension in a more subtle way. Screen deflection is normally pictured for purposes of explaining the printing mechanisms as a cross section of the screen and squeegee area in the direction of squeegee travel. In this view, screen deflection is a simple angle between screen and substrate which changes as the squeegee progresses in the print stroke. Across the screen, however, a different condition exists. Here the deflection must take the shape of an inverted trapezoid, and the definition of the deflection at the ends of the squeegee will de-pend upon squeegee hardness, or durometer. A soft squeegee (45 durometer) might have as much deflection as the screen, depending on the shape and size of the squeegee blade. For this reason, the length of the squeegee blade must be longer than the widest print area by a factor that is a combination of squeegee hardness, screen tension, and breakaway distance. Unfortunately, there are no ready formulas for the calculation of such multivariable factors, and the optimum must be deter-mined for each operation.

Screen size is an additional consideration. As the squeegee progresses, the angle between screen and substrate changes. To present minimum change over the print area, the screen should be as large as possible, since (as was discussed before) the angle will increase if substrate-to-screen distance is used to compensate for screen wear.

From this it can be concluded that the function of squeegee pressure is to ensure compression of the screen-to-substrate interface at the point of past deposition and is therefore a dependent variable determined by screen tension, screen size, squeegee length, squeegee durometer, and the setup screen-to-substrate distances. Squeegee pressure in excess of that required to perform these functions is detrimental. If suf-ficient pressure is exerted, the screen may be coined or deformed to an impression of the substrate. Excessive squeegee pressure also deforms the squeegee and changes the squeegee attack angle. The squeegee functions by imparting fluidic pressure within the bead or roll of paste, making the paste fill the cavity created by the screen-mesh openings as bounded by the emulsion. This fluidic pressure is a func-tion of both the squeegee speed and the angle between squeegee and screen. If the squeegee is deformed by excessive pressure, the angle will change and so will the pressure within the fluid that causes the cavity to be filled. This effect is the possible interaction with the last of the machine variables to be discussed here, squeegee speed.

Squeegee speed determines the force (with the angle previously discussed) on the paste, the shear rate as shown in Fig. 57, and the rate of breakaway, or peel, of the screen from the substrate. Since the rate and angle of peel both affect the shear to which the paste previously deposited in the cavity is subjected, they and the squeegee speed are related to the deposition with respect to the viscosity recovery rate discussed in the section on Paste-Transfer Characteristics.

Printing Methods The three primary techniques for screen printing will be treated separately with respect to the transfer characteristics and machine variables of the previous sections.

Off-Contact Printing. Off-contact printing is so called because of the separation between screen and substrate. Figure 63 shows relative positions of the screen, sub-strate, and squeegee with the squeegee height set to prevent depressing the screen. Indicated are a 200-mesh screen with 0.002-in. emulsion suspended 0.015-in. above a 0.025-in.-thick substrate. Screen mesh is shown rodlike rather than woven, but the relationship is given for a 0.020-in. opening in the screen. The squeegee has a 45°

angle with the screen.   In Fig. 64 the squeegee moves from left to right, forcing the bead of paste across the screen and into the cavity.   The angle between screen and substrate indicates that the squeegee has progressed somewhat to the left of the center of the screen.   To the left of the squeegee the angle between substrate and screen illustrates the snap-off, or breakaway, that causes transfer of the paste from the screen cavity to the substrate.   The ideal screening operation would permit 100 percent transfer.   As Miller[28] pointed out, however, some wetting of the screen must

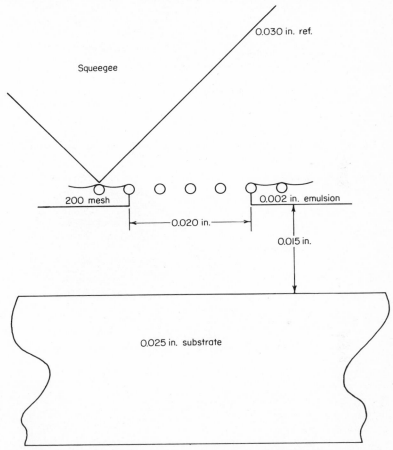

Fig. 63   Squeegee, screen, substrate relationship for off-contact screening.

take place, and the transfer cannot be complete.   In addition, the woven nature of the screen mesh itself causes paste immobilization.

Indicated here is a relatively rare but undesirable occurrence, the flow of paste under the screen ahead of the squeegee.   In this respect Fig. 64 is inconsistent in that no flow or slump is indicated in the nearly perfect transfer behind the squeegee. As the screen-to-substrate separation is increased, the angle between substrate and screen increases, changing both the opportunity for paste leakage and rate at which the screen separates from the substrate.   Variations in substrate thickness and substrate camber affect this characteristic from substrate to substrate.   To minimize this effect, the screen must be made larger with respect to the area to be printed.

Off-contact printing occurs as a "wave" or "front" action in line with maximum

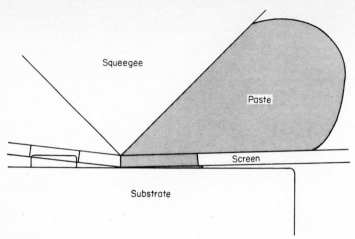

Fig. 64   Squeegee, substrate, screen relationship during print stroke.

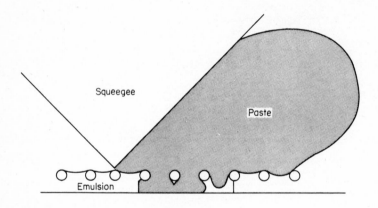

Fig. 65   Squeegee, substrate, screen relationship for contact printing.

screen depression as the squeegee moves across the screen.   Squeegee speed determines the shear rate to which the paste is subjected and (in conjunction with the
chosen squeegee angle) the fluid forces which causes the cavity to fill and the substrate to be wet.   With the instantaneous snapback of the screen characteristic of
this printing method, the paste has no chance to recover its rest viscosity, and the
tendency of substrates to stick to the screen is minimized.

Contact Printing. This method of printing is characterized by constant contact between screen and substrate for the total duration of the print stroke. In Figure 65, showing the screen, substrate, and squeegee for the same relative dimensions as in Fig. 63, the screen-substrate relationship more closely resembles a cavity, and the suspected nature of the cavity-fill operation is clear. In this printing method, all cavities or areas to be printed are filled with paste, and the screen or substrate is then withdrawn.

Two withdrawal methods are currently available. The first is a peel motion, in which the screen or the substrate is rocked or tilted away from the print position. This breakaway approximates the wavefront of off-contact printing for the paste-transfer functions. The second approach is the vertical breakaway, in which the substrate is firmly held and the screen or substrate is separated in direct vertical motion maintaining parallelism with the other member. There is (at least on large print areas) a peeling action here also, but it occurs first on the edges of the substrate or print area and progresses as a circle of decreasing size until final separation is complete.

Variations in thickness and camber from substrate to substrate modify this type of printing. If the screen is just touching for the thinnest substrate, it will be in interference contact with all thicker substrates. By the same reasoning, camber on such a setup could cause interference contact in one area and off contact in another. The effect, however, on all but the very largest substrates is still insufficient to cause screen stretching. Two prime advantages are offered by contact printing. The need for very large screens is eliminated. Since there is no intended depression of the screen, the angle between screen and substrate does not change and the screen need only be large enough to accommodate the squeegee excursion. Also pattern distortion and stretching caused by exceeding the allowable percentage elongation of the mesh wire is minimized.

The second advantage comes from the fact that the printing variables are no longer interactive. The cavity is filled by squeegee action alone, and the single variable, i.e., squeegee speed, is measurable and controllable. The paste transfer is accomplished by screen-substrate breakaway, and again this mechanical action can be measured and controlled. Screen-life improvement is marginal over that achieved with "near" off-contact (0.003 to 0.005 in.) printing.

Indicative of the print quality possible with a printing method is the distribution of the as-fired values of resistors. Most of the literature sources refer to distributions that are gaussian or normal with equal population above and below the median value. Contact printing, however, has shown the possibility of achieving a binominal[35] distribution, wherein the mean, or modal, value is very close to the minimum value and a tail exists only for upper values. This suggests near-optimum printing since the lowest value must represent the maximum volume transfer of paste.

Curved-Screen Printing. A few machines now utilize a curved screen. In their operation the squeegee remains in a position representing the line of contact between a rolling cylindrical screen or mask and the plane of the substrate surface. Very little has been written about this method of printing or the characteristics of paste transfer, but analysis of the mechanism indicates the possibility of combining the better aspects of both off-contact and contact printing. This method presents the wavefront, or sequential, print action of the off-contact mode with a constant peel angle. It eliminates the necessity of stretching the screen, permitting large print areas with relatively small mask areas. And effective squeegee speed and breakaway speed are both measurable and controllable although still inseparably interactive.

Print Drying

Function After the printing operation, the material must be dried. Depending upon the vehicle makeup and the manufacturer's recommendations, various periods of settling or slumping may be necessary before the actual drying takes place. The volatile solvents must be removed from the deposit before furnace firing. This is the primary function of the drying step, although some fixing or freezing of the solids may be accomplished during the process. From Figs. 34 and 35 it can be seen that several hours are required at room temperature to effect the same evaporation as a few minutes at elevated temperature. This rapid change in evaporation rate from room temperature to 85°C means that near-explosive expansion would occur at an early point in the furnace firing cycle.

Methods   Several methods of drying paste have been suggested,[25] including the simple hot plate of the laboratory to complex systems of rotary motion in a reflective enclosure. Suggestions have been made that the drying take place in periods up to 24 h in an atmosphere containing vapors of the solvent to prevent entrapment of solvent within the body of the resistor caused by surface "freezing." Literature from most paste manufacturers gives recommendations for drying applicable to most operations. Methods for the most part are not specified, and the user is permitted a choice so long as the conditions of time and temperature are met.

Postdry measurement   Since furnace-firing profiles require up to 1 h to accomplish, it is convenient to attempt control of the process before furnace firing, which

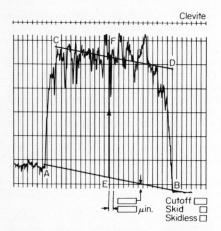

Fig. 66   Surfanalyzer trace. (*Du Pont, The Thick Film Handbook, Prod. Bull. A63326, January 1969.*)

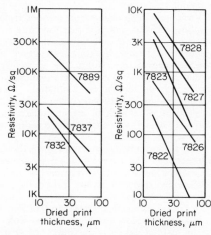

Fig. 67   Resistivity vs. dried-print thickness for 7800 series resistor compositions. (*Du Pont, The Thick Film Handbook, Prod. Bull. A80631, May 1972.*)

is made possible by data from suppliers or generated at the paste-preparation phase through correlation with dried-film thickness. This thickness is determined by a variety of devices ranging from a light section microscope to a surface-profiling device. Figure 66 represents a profile trace from a Clevite Surfanalyzer. Devices of this type are marketed by several suppliers and are quite accurate. Traces are repeatable to the extent that reversing the traverse mechanism after profiling a print will produce a trace undetectable from the original if the tracing medium is folded and subjected to inspection on a light table.

Measurement of print thickness is accomplished as shown in Fig. 66 by sketching the suspected substrate surface (line *AEB*) and approximating the average print surface (line *CFD*), then calculating the print thickness based upon the sensitivity of the vertical scale. Correlation of this measured thickness with observed results after firing can be plotted. Figure 67 indicates a correlation between dried-print thickness and resistivity for members of Du Pont's 7800 series. Similar data for the 8000 Series of Du Pont are shown in Fig. 68. Figure 69 indicates resistivity vs. film thickness for the Powerohm 850 series, while Fig. 70 indicates TCR vs. dried-film thickness for some members of the Du Pont 7800 series.

Correlation between dried and fired values of resistivity can be made from Fig. 71 data compared with those of Fig. 67. The effect of firing conditions must be considered, however, and conditions must be tightly controlled for reasons discussed in the section on furnace firing.

Firing

Firing accomplishes a twofold function with respect to resistor performance: it removes the organic binder from the dried deposit, and it sinters the frit-metal solids content into a solid structure with characteristics determined by the materials and the process.

Material Reactions *Resistor Material.* The oxidation and reduction of palladium as a function of temperature has been documented.[4, 7] Here the reaction is bilateral at some temperature experienced in the firing process for most PdO, Pd-PdO-Ag, and Pd-Ag resistor systems. Less is known, however, about the reactions occurring in some of the newer systems. Bube[15] describes experiments indicating partial reduction of ruthenium dioxide during combustion of the organic vehicle in one system. He further states that the RuO_2, in a normally fired system

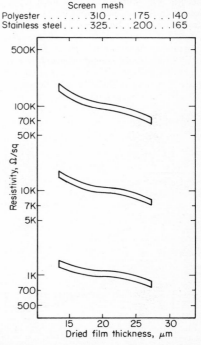

Screen mesh
Polyester . . . . 310 . . . . 175 . . . 140
Stainless steel . . . . 325 . . . . 200 . . . 165

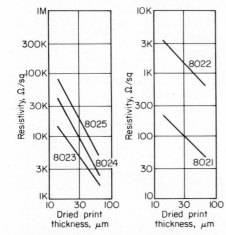

Fig. 68 Resistivity vs. dried-print thickness for 8000 series resistor compositions. (*Du Pont, The Thick Film Handbook, Prod. Bull. A80631, May 1972.*)

Fig. 69 Film-thickness effects on resistivity for Powerohm 850 series resistors (850°C peak firing temperature, 25-min firing cycle; conductor: Conductrox Pd-Ag 3412.). (*Thick Film Systems, Inc., Bull. R-850-3.*)

based upon this material, is oxygen-deficient, as evidenced by a decrease in resistance as the pressure of oxygen is increased during firing. Without specific knowledge of the content and condition of the materials used to formulate the paste, the user is not able to predict reactions and it becomes necessary to maintain constant process conditions.

Resistor-Conductor Interface. Several reactions can occur at the interface between conductors and resistors. Those affecting values of resistance through the contact-resistance mechanism are described in the section on termination effects. The reactions, however, can be affected by firing sequence. Figure 72 indicates the effect of preterminating, postterminating, and cofiring the Powerohm 850 series. The curves are shown for several conductor compositions. A ratio of observed values as great as 4:1 is indicated for postterminated material with respect to cofired material. Separation of materials is the extreme reaction but has been reported in the literature.[36]

Frit-Material Reactions. The primary reaction to furnace processing by the frit would ideally be that of sintering or melting into an amorphous mass within which particle-to-particle contact can occur. Unfortunately, glass is a complex compound classified as a supercooled liquid. The prime constituents of glass are metal oxides, and the metal of the frit oxide may react with the oxides of the conductive pigment to form an entirely foreign compound.[15] Glass is also sensitive to processing temperature. The viscosity or flow rate of various glasses for the respective temperatures are shown in Table 5. The table is not intended to be complete, nor does it indicate characteristics of particular glass frits to be found in resistor materials. It serves only to indicate relationships of temperature to viscosity and to indicate those viscosities for reference which have been defined. For example, strain point is defined as a viscosity of $10^{14.6}$ P or conversely as the minimum

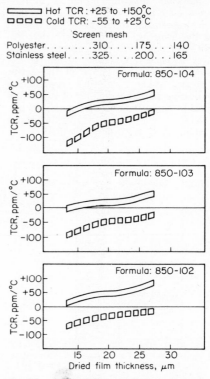

Hot TCR: +25 to +150°C
Cold TCR: −55 to +25°C

Screen mesh			
Polyester	.310	.175	.140
Stainless steel	.325	.200	.165

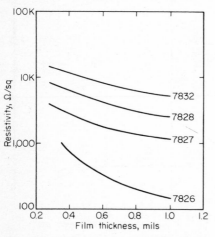

Fig. 70 Resistivity vs. fired film thickness for several Du Pont 7800 series resistor compositions.[4]

Fig. 71 Typical film-thickness effects on TCR for Powerohm 850 series resistors (850°C peak firing temperature; 25-min firing cycle; conductor: Conductrox Pd-Ag 3412.). (*Thick Film Systems, Inc., Bull. R-850-3.*)

temperature at which one can expect stresses to be relieved. Annealing point is defined as a viscosity of 10^{13} P or a temperature at which internal stress should be relieved within 15 min. From Figure 73, a plot of some of the data in Table 5, it can be seen that although the melting-point temperature of Corning 7740 is lower than that of the unidentified lead silicate, the lead silicate has lower softening and annealing points. The curves presented here represent a characteristic of the glasses themselves. The nature of the frit when combined with metal or metal oxide particles cannot be predicted. In addition, the particular frit formulations used by paste manufacturers are closely guarded secrets. Time-temperature treatment of thick film pastes is therefore not quite as simple as "time at temperature."

Furnace considerations Two primary factors determine the process for furnace firing: the time-temperature profile and the content of the atmosphere within the furnace.

Profile. The effects on resistor pastes of varying time at peak temperature or of varying peak temperature are supplied by most manufacturers for their materials. Sample or suggested profiles are available (see, for example, Fig. 4, 7, or 17); and the functions that a furnace performs on thick films in general have been described in literature.[4, 7, 25] With the temperature sensitivity displayed by some materials, requirements exist for temperature repeatability and uniformity to within a few degrees. Detailed effects of phase changes[8] and bilateral reactions[37] in the chemical makeup of the pastes have been described. The final profile determination, however, is left to the paste user, and lack of compatibility between desired profile and equipment on hand may prevent achievement of expected results. Figure 74 indicates a general profile that will meet the requirements of all paste systems listed in Table 3 (see the section on Material Selection) if the high-temperature fire section is considered to be variable with respect to both time and temperature. If such a concession is made, the fire section can be either bell-shaped or a flat-topped trapezoid, as indicated. Belt travel in the indicated profile is from left to right, and as the part proceeds, it comes first to the burnout section, where all traces of organic binder and ash are removed. This can generally be accomplished at temperatures below 480°C, but 550°C is far enough above the burnout point to ensure total decomposition and below the temperatures of the lowest softening point of known resistor materials. The dwell, or flat, at this temperature further ensures that softening will not occur until reducing atmospheric effects of hydrocarbon decomposition are minimized.

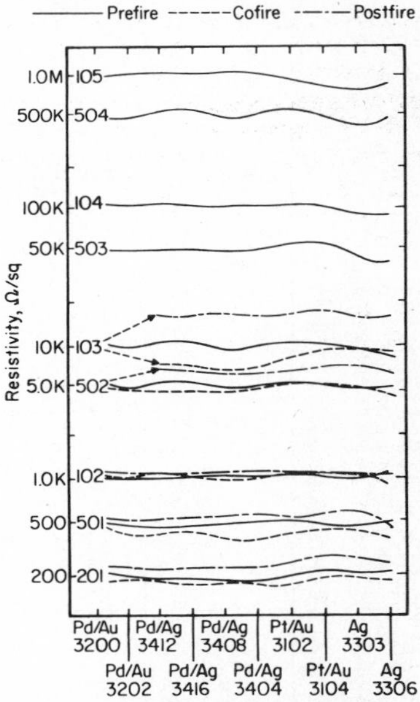

From this point, the slope of the temperature rise to peak and the time at peak temperature are a matter of material selection and process determination that may differ for each process location. As already mentioned, either bell-shaped or trapezoidal profiles can be generated by assignment of rise and fall slopes and time and temperature at the peak.

The slope of the temperature fall from

Fig. 72 Resistivity comparison of prefired, cofired, and postfired resistors of Powerohm 850 series with various conductor types. (*Thick Film Systems, Inc., Bull. R-850-5.*)

peak should generally be as fast as possible to about 600°C. This has the effect of stabilizing the chemical composition of materials as they were established at the peak temperature. This temperature is chosen because it is above the annealing-point temperature of any of the known frit materials in thick film resistors. Referring again to the general nature of glass viscosity vs. temperature in Fig. 73, we see that the increase in viscosity as temperature is reduced from 600°C is much more rapid than for temperature changes from 800 to 600°C.

From 575°C to below 375°C a time rate of change of temperature of less than 40°C/min is recommended. This rate, derived from the nature of glasses, is intended to prevent generation of internal stresses in the microstructure of the resistors. The temperatures are determined from the highest annealing- and lowest strain-point temperatures determined for glasses in present resistor systems. Chemical reactions may occur in the firing of some resistor systems, altering their composition or behavior with respect to annealing- and strain-point temperatures, but they must be determined empirically.

TABLE 5 Temperature-Viscosity Relationship for Various Borosilicate Glasses, °C

Comparative flow†	log viscosity,* P	Borosilicate* Pharmaceutical	Borosilicate* Laboratory	Lead silicate*	7040‡	7740‡	7760‡	
Water	−2							
	−1							
Olive oil	0							
Castor oil	1							
Honey	2							
	3	1399	1419	1195	1080	1240	1210	Melting point‡
Molasses in January	4							Working point
	5							
	6	932	974	760	· · · · · ·	960	· · · · · ·	Sinter solid from powder
Asphalt	7							
Plastics deformation	8	796	835	663	700	820	780	Softening point‡
	9							
	10							
	11							
	12							
	13	577	566	477	490	565	525	Annealing point‡
	14							
	15	549	530	443	450	520	480	Strain point‡

* H. H. Holscher, "The Relationship of Viscosity to Processing of Glass," Owens Illinois Technical Center.
† Properties of Selected Commercial Glasses, *Bull.* B-83, Corning Glass Works, Corning, N.Y., 1961.
‡ ASTM viscosity definitions.

Atmosphere. Both the elemental makeup and the moisture content of the atmosphere in a furnace can affect the results. Figure 75 shows relative resistivity vs. oxygen content in furnace atmosphere for Du Point 7800 series resistor paste. The curve indicates increasing resistivity with increasing oxygen content up to 40 percent. Overlayed on the value of normalized resistivity in this curve are ranges of resistance value. Since normally air contains approximately 20 percent oxygen, the data suggests that (for this series of materials at least) an oxygen-enriched atmosphere would

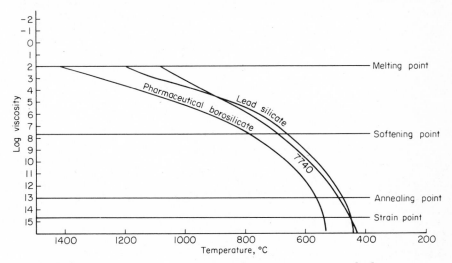

Fig. 73 Viscosity vs. temperature for various commercial glasses.

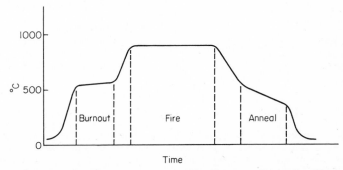

Fig. 74 Proposed all-purpose furnace profile showing functions of the various furnace zones.

improve uniformity. These data are the results of a study prompted by the observation that a furnace belt full of parts yielded lower resistivity than a sample of five or six substrates from the lot, evidence of a partially reducing or at least nonoxidizing atmosphere presumed to be the result of organic material decomposition. Forced-atmosphere systems are now installed in thick film firing furnaces. Partial reduction of some of the ruthenium oxides is possible in the normal combustion of organic vehicles. Bube[15] claims that for the system investigated the normally fired resistor contains RuO_2 that is oxygen-deficient.

Base-metal conductor systems are presently under investigation by several companies and will of course require reducing atmospheres in furnaces. Compatibility between the reducing atmospheres necessary for such conductors and present resistor materials has yet to be determined, but the above data suggest difficulty for some systems.

The effects of moisture in furnace atmosphere are much more subtle. They include (1) the nature of water as a catalyst in chemical reactions, (2) hydration and ion exchange in glass in the presence of superheated water, and (3) infrared-energy absorption by moisture in air. The first effect has not been documented for materials used in thick film resistor systems. The need to maintain constant conditions for chemical reactions demands attention; in addition, moisture in furnace atmospheres causes formation of HCl from decomposed residues of chlorinated solvents. Some glasses are soluble in water at high temperature, and all are subject to some extent to corrosion by ion exchange or hydration[38] in the presence of superheated water. Infra-

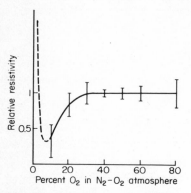

red energy emitted by the furnace surface within the muffle is not absorbed by the gases making up furnace atmospheres, but moisture in these atmospheres will absorb infrared energy. The direct effects of moisture may not be measurable, but these conditions suggest that the level of moisture content be controlled to minimum attainable levels. Other suggestions for optimum firing conditions would include atmosphere flow counter to belt travel in the burnout section and sufficient supply of fresh atmosphere in the firing section to ensure completion of desired chemical reactions. The recommendations must not overlook expansion and contraction of the gases at the furnace temperatures.

Fig. 75 Resistivity vs. atmosphere for Pd-Ag glass composition. Curve showing effect of increased oxygen content in furnace atmosphere on resistivity and distribution of values for Du Pont 7800 series paste. Resistors 0.200 in. long by 0.100 in. wide on prefired 7553 terminations. Values normalized for length, width, and average thickness. Fired 7 min in box furnace at 760°C.

Trimming

As discussed in sections on screen printing, the thick film process has sufficient inherent variables to result in distributions of resistor values too far from nominal or desired value to be used as fired for some applications. The corrective process for this condition is trimming. There are several approaches, according as the user wishes to trim up or trim down. Trimming up generally implies selective removal of resistor material; trimming down involves electrostatic or discharge methods of improving conductivity within the resistor material.

Material removal   Methods of material removal used for resistor trimming include rotary abrasive tools, air-abrasive jets, lasers, and surface-lapping techniques. Common use centers on air-abrasive and laser techniques, to which the discussion will be limited. Material-removal methods operate on the principle of reducing the cross-sectional area or increasing the effective length-to-width ratio in trimming top-hat resistors.

Air-Abrasive Trimming.   In air-abrasive trimming, very small particles of a hard material, e.g., silicon carbide, are forced through a small orifice or nozzle by high-pressure air. The mass of the particles is so low that each scratch, or "bite," is infinitesimally small. Since almost no stress is introduced at the work surface, the process is very useful for cutting hard, brittle materials without damage. These characteristics are applicable to trimming resistors where the medium containing the conducting mechanism is glass. In practice, nozzles as small as 0.006 in. can be directed at thick film resistors creating a cut width, or kerf, of approximately 0.010 in.; rectangular orifices with dimensions up to 0.180 by 0.006 in. can also be used. With such flexibility, a resistor 0.200 in. long can be trimmed for most of its length to minimize thermal hot spots. With a trim motion at 90° a top-hat resistor with brim dimensions of 0.010 in. and body of 0.030 in. can be increased in effective length by 2 sq for each 0.010 in. of travel.

Air-abrasive trimming was the first trimming method to gain wide acceptance in the hybrid industry; however, the drive for physically smaller resistors placed in closer proximity to other resistors makes an overglaze necessary to protect resistors

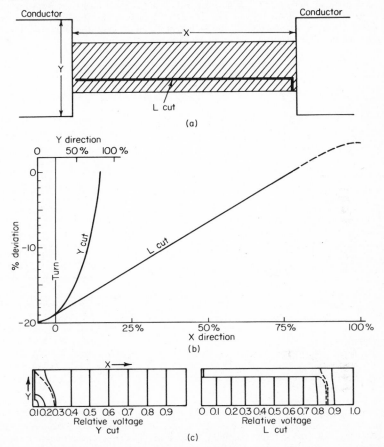

Fig. 76 (*a*) Ideal L cut. (*b*) Resistance deviation vs. length of laser cut for a 4:1 aspect ratio resistor. Resistance increases very rapidly as the cut moves across in the Y direction. The rate of change in the X direction is of most interest. (*c*) The resistance change there is very linear, decreasing to almost zero near the end of the cut. Thus, the longer the L cut, or the closer it is to the ideal cut, the finer the adjustment and the less the sensitivity to the trimming process. (*A. G. Albin and E. J. Swenson, Laser Resistance Trimming from the Measurement Point of View, Proc. 1971 Electron. Components Conf.*)

adjacent to the trimming area. With earlier resistor systems that required overglazes for physical or environmental protection, this was no disadvantage, but with the advent of resistor systems without an overglaze, laser trimming had the advantage.

Laser Trimming. The laser uses a much more complicated technique to give the same net trim action as the air-abrasive jet, i.e., the technique of the B cut (Fig. 76*a*). The laser spot is only a few mils in diameter and the kerf not much wider than the spot. If the trim is directed in the Y direction, the change in resistance as a function of distance trimmed is nonlinear (Fig. 76*b*). If the beam is then programmed to move parallel to the length of the resistor for an L cut, the deviation vs. travel is shown by the curve labeled "L cut." These reactions are further indicated by the plots of lines of equal voltage distribution in Fig. 76c. Figure 77 shows curves of resistance change vs. length of cut and sensitivity (percent $\Delta R / \Delta$ length) for various configurations of L cuts on square resistors. To achieve the greatest degree of accuracy in trimming, the final value should be approached on a trim curve of least sensitivity. The burden of laser programming then becomes the predetermination of the length of the initial Y cut to permit maximum utilization of the decrease in sensitivity with corresponding increase in the L cut. This in turn

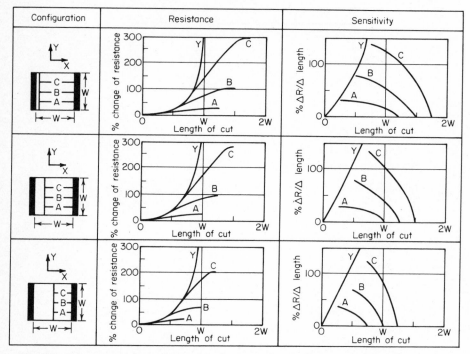

Fig. 77 Deviation and sensitivity as a function of trim distance in L cuts on square resistors. (*E. J. Swenson, G. O. Vincent, and J. C. Riley, Laser Resistor Trimming Analysis Using a Resistive Sheet Analogy, 1972 Proc. Electron. Components Conf.*)

places an additional burden on the screen-printing operation to produce uniform distributions and accurate physical placement.

The nature of laser trimmers calls for the use of computers to make effective use of the L-cut technique. With these, such techniques as sensing the edge of a resistor and trimming in the Y direction until 80 percent of desired value is reached (deviation trimming) are accomplished by programs that alternately trim and measure, as opposed to the constant sensing used with the air-abrasive technique.

Conductivity changes Trimming resistor values down from as-fired results by high-voltage discharge[39] and high electromagnetic stressing[5] has been proposed. From data presented in the references, the mechanism of resistance change appears to be one of increasing the particle-to-particle contact by eliminating dielectric microbarriers or bridging microgaps. The changes that occur are said to be permanent, and reported stability approaches that of other trimming methods. A major drawback is that sensitivity to the methods varies with respect to material, supplier, and process to the extent that one of the tested materials could not be adjusted. The technique may prove useful, however, for lot-recovery processes where as-fired values or unique design changes require selective reduction of resistor values.

Glazing

Materials for protective overglazes for resistors are covered elsewhere in this handbook. Such steps are necessary in a process when the resistor material used is subject to chemical reaction with subsequent processing or packaging environments. As mentioned in the discussion of trimming, an overglaze can be of assistance in protecting resistors close to one being trimmed by air-abrasive techniques. An unexpected benefit sometimes results from the profile used to fire these glazes. Temperatures for protective-glaze firings have customarily been maintained between 550 and 650°C, to keep them somewhat lower than the firing temperatures of the resistors and to keep refiring effects at a minimum. These temperatures, however, serve well as annealing treatments if the cooling slopes are not steep enough to reintroduce stress in the resistor glass. Some published "better drift" data on glaze-protected resistors may reflect the lack of long-term stress relief.

RELIABILITY

The reliability of thick film resistors is difficult to treat apart from the materials themselves. Most failure rates are described with respect to application or to manufacturer's test data and are difficult to compare. Failure mechanisms, on the other hand, are usually inherent in the materials.

Failure Mechanisms

Various failure mechanisms exist for thick film resistors depending upon material selection and process. They can be classified by material type, process, or failure type. We classify them here by type of failure as physical, or catastrophic, and chemical or degradation.

Physical failure mechanisms Cracks or cracking can occur in thick film resistors during processing or during application and for various reasons, but the results are not always the same. In resinate materials, where film thicknesses demand an underglaze and an overglaze, if a crack in the glaze is propagated through the resistor, it results in an open resistor and a catastrophic failure. Cracks in the overglaze of cermet resistors, however, do not necessarily have such a result. In fact, in a reducing atmosphere, e.g., the hydrogen released by some epoxies during cure, such a crack could result in negative-value drift of a resistor fabricated with a palladium oxide-based resistor paste.

Cracks in the body of cermet resistors have been reported as a result of laser trimming. [40, 41] They can be minimized by proper selection of laser operating parameters and have not yet been shown characteristically to cause open-type failures. In the sense that these cracks might cause shifts of value beyond specified tolerances and can be physically detected with a scanning electron microscope, they will be classified as catastrophic.

Degradation failures *Chemical Reaction.* Chemical reaction can occur in the form of reduction of oxide to metal in the case of PdO-Ag resistor systems. This can occur during processing, e.g., semiconductor chip or wire bonding, where slight reducing or nonoxidizing atmospheres are used. It can also be initiated during hermetic-sealing operations, which may use the same conditions. The result is a downward or negative drift in resistor value.

Reaction can also take place at the resistor-conductor interface, and although the more violent reactions occur during furnace firing, long-term-drift data on resistor materials terminated with different conductors indicates continued postprocessing reaction.

Bulk Material Drift. Many of the data published by materials suppliers describing stability of thick film materials indicates a bulk shift in resistivity. This shift invariably shows within the first 50 to 100 h of indicated test time. No explanation exists for the observed change, but initial data indicate that proper selection of the parameters of the furnace profile will eliminate it.

Reliable History

Other than drift of resistors to values beyond specified limits, failures can be detected only if they cause circuit malfunctions. In this sense most published reliability data indicate excellent results. The solid-logic technology program at IBM, for example, accumulated more than 90 billion resistor hours without a single field failure.[42] These figures indicate with 90 percent confidence a statistical failure rate of less than 0.000003 percent per 1,000 h. Load-life test data by independent sources have indicated failure rates of 0.029 percent per 10^6 h.[43] Here, as in the equipment or field test, the true members cannot be known since no failures occurred.

REFERENCES

1. D'Andrea, J. B.: private communication.
2. Hopper, R. T.: How to Apply Noble Metals to Ceramics, *Ceram. Ind.*, June 1963.
3. Place, T. M. et al.: U.S. Pat. 2,950,966, Aug. 30, 1960 (assignor Beckman Instruments, Inc.).
4. E. I. du Pont de Nemours & Co.: The Thick Film Handbook.
5. Polinski, P. W.: Stability of Thick Film Resistors under High Electromagnetic Stress, *Solid State Technol.*, May 1973.
6. Brady, Lynn J.: Relation of the Particle Size of RuO_2 in Cermet Resistor Inks to the Electrical Properties of Fired Resistors, *IEEE Trans. Parts, Mater. Packag.*, vol. PMP-6, November 1970.
7. Ilgenfritz, R. W.: Controlled Processing for Precision Thick Film Resistors, *IEEE Electron. Components Conf. 1967.*
8. Ras, U. V., and R. Swayne: Thermal Behavior of Thick Film, *Solid State Technol.*, June 1973.
9. Loughran, J. A., and R. A. Sigsbee: Termination Anomalies in Thick Film Resistors, *ISHM Conf. Proc. 1969.*
10. Herbst, D. L.: Composition of Thick Film Resistors, *IEEE Electron. Components Conf. 1971.*
11. D'Andrea, J. B.: Ceramic Composition and Article, U.S. Pat. 2,924,540, Feb. 9, 1960.
12. Deunesnil, M. E.: Resistor and Resistor Composition, U.S. Pat. 3,052,573, Sept. 4, 1962.
13. Hoffman, L. C.: Resistor Compositions, U.S. Pat. 3,207,706, Sept. 21, 1965.
14. Miller, Lewis F., and K. E. Neiser: Electrical Resistor Compositions, Elements and Method of Making Same, U.S. Pat. 3,390,104, June 25, 1968.
15. Bube, K. R.: The Effect of Prolonged Elevated Temperature Exposure on Thick Film Resistors, *ISHM Proc. 1972.*
16. Love, G.: Designing with High Reliability Thick Film Resistors, *2d Symp. Hybrid Microelectron.*, October 1967.
17. Peckinpaugh, C. J., and W. G. Profitt: Termination Interface Reaction with Non Palladium Resistors and Its Effect on Apparent Sheet Resistivity, *IEEE Electron. Components Conf. 1970.*
18. Collins, F. M., and C. F. Parks: Thallium Oxide Glaze Resistors, *IEEE Electron. Components Conf. 1967.*
19. Miller, L. E.: Glaze Resistor Paste Preparation, *IEEE Electron. Components Conf. 1970.*
20. Greenfield, M., and R. Delaney: A New Thick Film Resistor System, *NEPCON Proc., February 1973.*
21. E. I. du Pont de Nemours & Co.: The Thick Film Handbook, *Bull.* A76967. December 1971.
22. Garvin, J. B., and S. J. Stein: The Influence of Geometry and Conductive Termination on Thick Film Resistors, *EIA, IEEE Electron. Components Conf.*, 1970.
23. Van der Ziel, A.: "Noise," Prentice-Hall, Englewood Cliffs, N.J., 1956.
24. Kuo, C. Y., and H. G. Blank: The Effects of Resistor Geometry on Current Noise in Thick Film Resistors, *Proc. ISHM Hybrid Microelectron. Symp. 1968.*
25. Van Hise, J. A.: Process Variables in Thick Film Resistor Fabrication, *ISHM 1969 Symp. Proc.*

26. Hughes, D. C., Jr.: Variables Affecting Uniformity in the Screen Process Printing of Fired on Films, *IEEE, EIA Electron. Components Conf. 1967.*
27. Austin, B. M.: Thick Film Screen Printing, *Solid State Technol.*, June 1969.
28. Miller, L. F.: Paste Transfers in the Screening Process, *1969 SAE Microelectron. Packag. Conf.*
29. Salisbury, I. D.: Variables in the Screen Printing Process and Their Effect on Thick Film Resistor Deposition in Large Scale Production, *Electron. Components,* 1970.
30. Kobs, D. R., and D. R. Voigt: Parametric Dependencies in Thick Film Screening, *ISHM Conf. Proc. 1970.*
31. Ottaviano, A. V.: Repeatability in Screen Printing Hybrid Microcircuits, *ISHM Conf. Proc. 1969.*
32. Trease, R. E., and R. L. Dietz: Rheology of Pastes in Thick Film Printing, *Solid State Technol.*, January 1972.
33. Patton, T. C.: A New Method for Viscosity Measurement Using a Spring Relaxation Technique, *J. Paint Technol.*, 1966.
34. Coronis, L. H.: Fine Line Printing Revisited, *Proc. ISHM Microelectron. Symp. 1972.*
35. Waite, G. C.: Assembly and Packaging of Dual Incline Hybrids, *NEPCON Proc.*, 1973.
36. Driear, J. R.: Observations on Formation of "Cleavage" at the Interface between Glaze Resistor and Termination, *Proc. IEEE Electron. Components Conf. 1965.*
37. Burks, D. P., B. Greenstein, and J. P. Maher: Screened "Thick Film" Resistors, *Proc. IEEE Electron. Components Conf. 1967.*
38. Corning Glass Works: "Properties of Selected Commercial Glasses," Corning, N.Y., 1961.
39. Taketa, Y., and M. Haradome: Thick Film Resistor Adjustment by High Voltage Discharge, *ISHM Proc.* 1972.
40. Howard, R. T., and R. V. Allen: Characterization of Laser-trimmed Thick Film Resistors by Screening Electron Microscopy, *Proc. ISHM,* 1971.
41. Headley, R. C., M. J. Popowich, and F. J. Anders: YAG Laser Trimming of Thick Film Resistors, *Proc. Electron. Components Conf. 1973.*
42. Platz, E. F.: Reliability of Hybrid Microelectronics, *Proc. Int. Electron. Circuit Packag. Conf. 1968.*
43. Paquette, G. H.: Load Life Characteristics and Reliability of Thick Film Resistors, *ISHM Conf. Proc. 1973.*

Chapter 7

Dielectric Materials, Processing, and Controls

DONALD R. ULRICH
Space Sciences Laboratory, General
Electric Company, Philadelphia

INTRODUCTION

Although circuit manufacturers have been printing resistor-capacitor networks for over 20 years, thick film circuits as we know them today were born in the early 1960s. The first screened capacitor was reported in 1964.[1] New material concepts were introduced during the mid-1960s. During the past 6 years the increased demand for improved and versatile microcircuits has stimulated considerable research for new and improved dielectric thick films. While a large volume of information has been published on thick film resistor and conductor compositions and processing, less has been published on dielectric films.

Dielectric thick film inks can be defined as high-viscosity thixotropic pastes containing dielectric oxide powders and vitreous or glass binders suspended in an organic vehicle. Dielectric thick film inks have interchangeably been referred to as glazes, frits, pastes, and inks. From the point of view of ceramic science there are technical distinctions between these terms. Frits are powdered glasses containing no crystalline or organic materials. They are formed by pouring and quenching high-temperature melts in water or other quenching media. If a frit is applied to the surface of a ceramic substrate and heat-treated, the glass flows in situ and forms a solid vitreous film on the ceramic (if it is applied to a metal, the glass film is called an enamel).

Conventional thick film dielectrics consist of dielectric or ferroelectric oxide powders and a low-softening-glass frit intermixed with an organic vehicle which gives it a pastelike or inklike consistency. After application to the ceramic substrate by screening, the deposited film is heated to (1) burn out the organic vehicle, (2) cause softening and flow of the glass for the frit to bond to the dielectric powders and form a glazelike film, and (3) chemically bond to the ceramic substrate.

The original dielectric pastes consisted of the frit-dielectric powder combination. Other composite forms reported in the mid to late 1960s are now in use. These are based on recrystallized or glass-ceramic films, in which the paste consists essentially of a devitrifiable or recrystallizable glass frit suspended in an organic binder. The frit is specially formulated so that upon reheating the dielectric or ferroelectric crystallizes or precipitates from the glass. In another concept, in particular for high-dielectric-constant films, the paste consists of recrystallizable frit and ferroelectric powders suspended in an organic vehicle. The glass crystallizes the same ferroelectric phases as the powders and binds them in situ into a monolithic film.

In thick film technology dielectric pastes have four principal uses:
1. Capacitors
2. Crossovers and multilayers

3. Encapsulation
4. Hermetic seals and packaging

The thick film capacitors consist of a screened dielectric layer between two printed conductors. This permits the construction of capacitive devices on ceramic substrates. Dielectric crossovers provide good isolation between top and bottom conductors and isolate layers of conductors in multilevel arrays. Encapsulating dielectric films minimize the effects on resistors and capacitors of moisture and reducing atmospheres and protect these elements during such processing steps as trimming.

The processes, materials, and controls pertinent to the fabrication of thick film dielectrics are discussed in the following sections.

PROCESS SEQUENCE

The processes for screening and firing dielectric pastes have been well established on an economic basis, as well as a large-scale production basis. The basic capital equipment required for manufacturing includes a manual or automatic screening press and a closely controlled zone furnace. For some types of dielectric pastes a closely controlled ceramic furnace may suffice. The flow chart for a typical thick film process sequence is shown in Fig. 1. First, suitable substrates are selected and prepared, and the compositions are thoroughly dispersed. Bottom conductor and capacitor electrodes are printed, dried, and fired. The capacitor or crossover dielectrics are printed and dried. The crossover conductor and top electrodes are printed and dried and the dielectric-conductor combinations are co-fired. The resistor compositions are then printed, dried, fired, and trimmed. The resistor encapsulants are printed, dried, and fired. Conductive patterns are soldered and active devices, leads, and discrete components are attached. The circuit is packaged and tested.

The unit processes fundamental to the fabrication of screened circuit dielectrics are:

Substrate selection, preparation, and handling.

Paste selection, prepreparation, and handling

Screen printing

Drying and firing

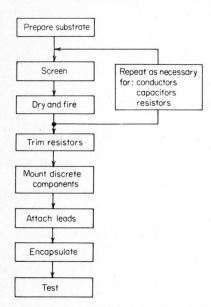

Fig. 1 Typical thick film process sequence.[2]

Substrates

The first consideration in the process sequence is the choice of substrates. These can be polycrystalline alumina, Al_2O_3; beryllia, BeO; steatites; or titanates. Ceramic substrates are used because they can withstand high temperatures during the firing step. The most common is 96 weight percent alumina ceramic; 85 percent alumina substrates are finding increased use because their electrical and mechanical properties are suitable for many applications.

When selecting a substrate, size, flatness, and surface finish should be specified. Several shapes and sizes are commercially available; the 2- to 4-in.² substrate costs less per square inch than smaller or larger sizes. Thickness tolerances of $\pm\frac{1}{2}$ percent can be obtained. Substrate flatness is important during the screening operation. A uniform camber of less than 4 mils/in. is usually required or specified. A surface finish of 20 to 40 centerline average (CLA) as fired is acceptable. Finer surface finishes are available and allow deposition of conductors with finer registration; however, a trade-off here is that the bond strength between the film and substrate is reduced. Snap-apart substrates offer an advantage where an operation involving multiple printing of a surface pattern can reduce overall manufacturing costs (Chap. 4).

Cleaning the substrates before screen printing is important in order to prevent adherence or contamination. The substrates as manufactured are usually clean and dust-free, but before use they should be solvent-degreased to remove organic contaminants, followed by a rinse in deionized water to remove inorganic impurities. A blast of dry nitrogen ensures complete evaporation of the water. Since the substrates can be abraded, handling procedures should ensure that the substrates do not come in contact with other substrates or metal parts.

Paste Handling

Prepreparation of the dielectric and electrode pastes is necessary before screen printing to restore the homogeneity of the paste system. The manufacturers thoroughly disperse the system before shipment, but some settling occurs by the time it reaches the customer. The user must redisperse the system, usually by hand, stirring with a stainless-steel spatula. This should be continued until no sediment or lumps remain and should be timed so that no further dispersion is required before production.

While a laboratory stirrer can be used, caution is advised. High stirring speeds can cause air to enter the paste system, which can alter the viscosity and change subsequent printing properties drastically. If the viscosity changes considerably, the composition should be stored on a jar or mill roller for 12 to 16 h. To determine whether the air has escaped, viscosity measurements should be taken using a Brookfield or other viscometer.

After receipt of the paste from the vendor, storage on jar rollers can prevent the solids from settling out. Rotation rates of 3 to 10 r/h are sufficient. Attention should be paid to the manufacturer's instructions. Some paste systems contain dilatant organic vehicles that exhibit increased shear and should not be placed on jar rollers. In general, the newer highly thixotropic pastes require no rolling to maintain good dispersions.

Recommended shelf-storage conditions for dielectric pastes are between 10 and 30°C and 30 to 90 percent relative humidity. The storage containers should be lightly sealed with their original covers to prevent solvent evaporation. During production some solvent will evaporate; it should be replaced by hand stirring until the original viscosity is attained. The solvent and recommended thinner are shown on the vendor's label.

Screen Printing

The screen printing process is a most important part of thick film production and must be closely controlled if good reproducibility of thick film dielectric properties are to be achieved. The screen printing process consists essentially of forcing the viscous dielectric paste through preprocessed openings in a stencil screen to deposit the required image on the ceramic substrate. The important process parameters for screen printing for the dielectric paste are

1. Squeegee stroke speed
2. Squeegee pressure
3. Breakaway height
4. Squeegee material and angle
5. Squeegee medium considerations (viscosity and particle size)
6. Screen mesh size
7. Screen tension and deflection
8. Level and rigidity of system
9. Direction of screening

When depositing the dielectric film, the optimum deposition thickness is usually about 0.0008 ± 0.0001 in. The film thickness is determined mainly by the viscosity of the formulation and the screen mesh size. For a screenable formulation with optimum deposition thickness a viscosity of $100,000 \pm 15,000$ cP as measured with a Brookfield viscometer is usually recommended. Wide variations in viscosity can cause deleterious effects. For example, when the viscosity is below about 40,000 cP, the films become thin and have poor line definition. Above 230,000 cP the formulation is unscreenable, and the films become very thick.

Screens with 105 to 325 mesh are commonly used. For optimum films, a 165-mesh screen is recommended; however, satisfactory films are deposited with a 200-mesh screen. With screen meshes finer than 250, very thin films result, which increase failure rates due to electrical shorting. Finer meshes clog more easily and must be kept thoroughly clean to obtain fine line resolution. Commercial screen-release coatings are available which prevent the composition from sticking to the screen. With screen meshes coarser than 165, there is a tendency toward voids or partial voids in the dielectric at points below the thread crossings on the screen. For example, voids are quite pronounced on an 80-mesh screen. Coarse screens yield heavy screened-on layers.

Taut screens yield good reproducibility and minimize screen and squeegee wear. The screen-substrate gap should be chosen so that the screen is in contact with the substrate only at the point directly under the squeegee blade as it passes over the screen surface. The spacing is usually 10 to 60 mils, depending on overall screen size. Too small a gap can result in poor snapback of the screen, causing thin prints and smeared or poorly defined images. With a 5 by 5 in. frame printing on 1 by 1 by 0.025 in. substrates, a 15- to 20-mil gap provides good print deposition. Squeegee pressure depends on gap and screen tension. In the 5 by 5 in. frame, a pressure in the vicinity of 1 lb per linear inch of squeegee is sufficient. The rheologies of the highly thixotropic dielectric pastes perform best at squeegee speeds of about 7 in./s.

The angle between blade and screen during printing should be selected to

1. Accommodate minor variations in substrate thickness and surface unevenness
2. Provide for full loading of the screen during printing
3. Terminate paste flow immediately above the print

Drying and Firing

After deposition by screening, the film should be given time to set to promote print leveling. A few minutes will suffice. The film should then be placed in a vacuum soak at 0.1 atm for about 10 min. The structure is then heated at 100 to 150°C for 10 to 15 min to outgas the film and remove the organic solvent. The organic solvent should be removed slowly; rapid evaporation during firing causes void and blister formation. Hot plates, single-zone ovens, or infrared lamps are used for drying. The latter have the advantage of being able to penetrate the deposited film without causing skin formation and solvent entrapment. A combination of the infrared lamp and the hot plate or oven may be utilized. Excessive drying can cause print shrinkage and peelback.

The different capacitor structures all employ glass, either as the binder or as the dielectric in glass or crystallized form. During the firing process the nonvolatile part of the organic vehicle decomposes. Firing should proceed slowly enough to permit the organic to burn or completely decompose before the glass flows.

The firing schedule and peak temperatures depend upon the film structure being employed and the materials being processed. The temperature profiles for glass crossover, dielectric–ferroelectric particle–glass composites, glass-ceramics, and ferroelectric-particle–crystallizable-glass films are discussed under Crossover Dielectrics, Composite Capacitor Compositions, Glass-Ceramic Capacitors, and Barium Titanate Glass-Ceramic Particulate Capacitors, respectively. Differential thermal analysis (DTA) is a very useful tool in profile determinations. The application and interpretation of results is discussed under Differential Thermal Analysis for Process Control.

In general, dielectric–ferroelectric particle–glass composites are fired at 1000°C. The glass-ceramics are fired over a range of temperatures depending on composition, usually from 740 to 1050°C. For more refractory ceramic systems the temperature will be higher, depending on the limitation which can be tolerated by the metal phase in the electrodes.

The firing cycle controls the capacitance density and loss factor of the dielectric materials and the adhesive of the conductor-electrode compositions. The most important part of the firing system is the ability to produce identical cycles consistently.

Dielectric films can be fired in periodic kilns or belt furnaces. A typical furnace profile for a periodic kiln is shown in Fig. 2. The profile for a typical belt furnace is shown in Fig. 3. There are two types of periodic kilns. A *box kiln* has the advantage of accommodating several batches of substrates. However, the temperature control at 1000°C is ±25°C. A *tube furnace* contains an alumina tube wound with a resistance element and has the advantage of closer temperature (±2 − 5°C) and atmosphere control. However, it accommodates smaller batches of substrates.

Good belt furnaces have at least four zones. Rates of use of 30 to 70°C/min yield good reproducibility. These furnaces give extremely good control where properties are microstructurally and stoichimetrically dependent, especially for resistor properties, such as sheet resistivity, temperature coefficient of resistance, current noise, and stability. Control can be held to within ±2°C/min for the temperature rise and decrease from peak temperature and to within ±1°C of a selected peak-temperature value. The substrate time-temperature profiles are best measured using relatively low-mass sheathed traveling thermocouples. Belt speeds are usually in the range of ½ to 6 in./min.

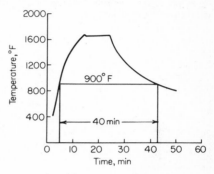

(a) Crossover dielectric

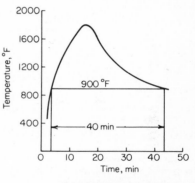

(b) Capacitor dielectric

Fig. 2 Periodic-kiln furnace profiles.[1]

PASTE FORMULATION AND PREPARATION

A wide range of dielectric pastes are commercially available, but some users prefer to formulate and blend their own pastes in order to attain specific values of control and performance. A few processing guidelines are presented for the formulation of the dielectric-powder–glass-frit type of paste.

The raw materials most commonly used in dielectric powders are barium titanate and lead zirconate titanate. The barium titanates often contain additives of strontium, calcium, or lead.

Barium Titanate Raw Materials

Purity The raw materials for formulating barium titanate dielectrics consist either of $BaTiO_3$ or of barium carbonate as the source for a BaO and anatase or rutile, TiO_2. High-purity oxide materials are available in industrial quantities and prices. Purities of over 99 percent are used and are available at 99.5 and 99.9 percent by weight. It is also possible to substitute barium hydroxide, barium nitrate, or barium oxylate for barium carbonate.

The reacted compound, barium titanate, is produced either by the solid-state thermochemical reaction of $BaCO_3$ and TiO_2 or by coprecipitation methods from solid solution. The latter product has higher purity but is more difficult to sinter because of the purity. These materials also consist of smaller-sized powders as a result of the precipitation process. Currently the precipitated barium titanate is more costly than that prepared from rutile and the carbonate. For inclusion in thick film dielectrics of the type consisting of dielectric particles suspended in a low-softening glass, the fine and ultra-fine particle sizes may be desirable. As part of these films the fine particles may have a profound effect upon the film dielectric properties because of particle-size effects.

Barium oxide and other alkaline earths, e.g., strontium or calcium oxide, are introduced as carbonates or oxylates rather than oxides, which are hygroscopic. Before mixing the dielectric formulations with organic binders for screening, the formulated powders should be calcined to decomposed the carbonates completely.

Titanium dioxide, or titania, is introduced as anatase or rutile, available in pure grades of 99.8 percent, the major impurities being aluminum, silicon, and niobium. Pigment grades, which are also used, are of smaller particle size and available in purities of 99.5 percent.

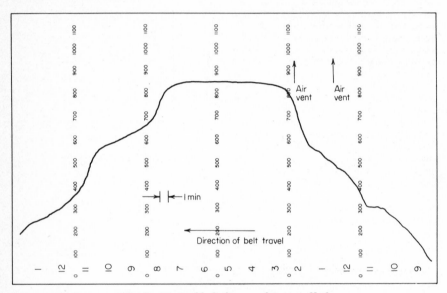

Fig. 3 Typical belt-furnace firing profile.[2]

If sulfur is present in the reacted barium titanate used as a raw material, low density and film bloating can occur. This can be eliminated during the low-temperature burnout cycle by introducing a reducing atmosphere by insufficient oxygen.

Stoichiometry and particle size The importance of purity has been emphasized. However, the stoichiometric Ba-Ti ratio is also important. A barium-rich content results in a fine-textured microstructure, while excess titanium causes the growth of crystals larger than 100 μm in the fine-grained matrix.[3]

The particle-size distributions of three ultrafine powders are shown in Fig. 4. Two of these are commercially available (C and H), and the third is for a laboratory experimental sample. The average particle size, surface area, and $BaO\text{-}TiO_2$ ratio are listed in Table 1. These materials have different processing histories and not necessarily identical purity levels. Material C, which is 99.6 percent pure $BaTiO_3$ powder, was prepared by the thermal decomposition of barium titanyl oxylate. Material H was prepared from the decomposition of barium carbonate with titania. Material M was prepared by the controlled decomposition of metal alkoxides.

Examination of the topography of the powders is important before incorporation in thick film formulations. This can best be accomplished by scanning electron microscopy, if available. If not, high-magnification optical microscopy can be employed, but it is not as informative. Material C consists of relatively large particulates composed of much smaller crystallites; this gives rise to a porous structure.

The importance of comminution before dielectric formulation is stressed here because of the agglomerate nature of the powders. In contrast, material H has a well-defined cubelike crystallite structure. No agglomeration was noticed in the ultrafine powder M.

Lead Zirconate Titanate Raw Materials

Solid-solution lead zirconate titanate, the second most commonly used ferroelectric in thick films, is available commercially. When the lead zirconate titanate is formulated from its oxide constituents, lead oxides with purity in the vicinity of 99.9 percent are available. If very small particles are preferred, the lead oxides prepared by sublimation rather than precipitation should be chosen. Zirconia, ZrO_2, is available in purities of 99.7 percent and above.

Raw-Materials Characterization

Raw materials are characterized for purity, stoichiometry, and particle size. Small, fine, and ultrafine particle sizes are desired for complete reactivity in solid-state thermochemical reaction. Impurities influence both the reactivity and the resultant dielectric and conduction properties. Stoichiometry is also important since excess major cations can increase or reduce the temperature and time of sintering of the films, affecting both the microstructures and the dielectric properties. The ignition losses of volatile impurities and foreign inclusions should be known so that allowances can be made in film formulation to preserve the stoichiometry. While dopants or additives may be only a few atom percent, the reactivity of the particles must be considered.

Particle sizes are determined by optical microscopic examination, scanning electron microscopy, the Coulter counter, Andreasen pipette, or the Sharples micromerograph. Amines can be detected by

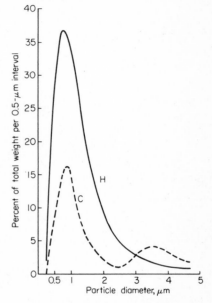

Fig. 4 Particle-size distribution for commercially available $BaTiO_3$ raw materials.[4]

TABLE 1   Characteristics of BaTiO₃ Powders[4]

	Type C	Type H	Type M
Average particle size, μm	0.8	0.75	0.01
Surface area, m^2/g	4.91	3.2	
BaO-TiO_2 ratio	0.995	1.001	1.000

ignition loss and investigated in more detail by wet-chemical methods. Cation impurities are detected with the spectograph, atomic absorption, spectroscopy, or x-ray fluorescence.

Calcination and Comminution

The dielectric properties of a thick film can be drastically affected by any lack of homogeneity. The constituents should be intimately mixed by ball milling and the loose aggregates broken up. Postcalcination mixing can only partially compensate for poor initial mixing.

After the initial mixing, the dielectric formulations are calcined in powder form to initiate thermochemical reaction among the constituents for solid-solution and compound formation, to burn off volatile impurities, to drive off CO_2 from the alkaline-earth carbonates, and to get rid of water of hydration. The initiation of the thermochemical reaction can also reduce film shrinkage during the densification firing. The calcining temperature should allow for complete reaction without densification so that postcalcination grinding is easy. For dielectric phases such as lead zirconate titanate or sodium potassium niobate which contain volatile oxides the temperature should be kept low to minimize loss. Barium titanate is commonly calcined at 1000°C for 2 h.

Postcalcination grinding is commonly carried out by ball milling. This prepares the calcined formulation for screen printing and reduces any compositional inhomogeneity which may still exist. The powders are ground to sizes of 1 to 10 μm. Larger particle sizes can result in low-density fired films having large intergranular voids.

Normally, the calcined materials are crushed to pass through a 105-mesh screen, ball-milled for 4 h in a carrying vehicle, and dried. The carrying vehicle can be deionized water (eliminate sodium contamination), tetrachloroethylene, or alcohol.

To achieve the fine particle size without contamination, the glass and ceramic can be processed in plastic-lined mills with balls or pebbles of the same composition as the material being milled. The slurry from the ball mills is wet-screened through a 400-mesh screen and the fines, i.e., less than 5μm, separated by a sedimentation process in an ultrasonic bath.

Frit Fabrication

Glass binders can be purchased or fabricated. Suitable glass frits are available from commercial suppliers. Typical formulations and component members have been discussed in other sections. If frits are to be fabricated, reagent-grade components should be weighed out, thoroughly mixed by dry blending for 1 h, and ball-milled in distilled or deionized water. After calcining in air for 1 h, the material is dried under an infrared lamp. The composition can be melted in kyanite crucibles and quenched to a glass frit in distilled water. An organic medium is used for quenching if the glass compositions contain baria since BaO can be leached by water. Usually the calcining temperature is 1000°C except for lead oxide–or bismuth oxide–containing glasses, which melt at lower temperatures. After quenching, the frit is ground until less than 1 percent of the residual is retained on a 325-mesh screen (usually about 16 h). Typically, a 1-gal ball-mill jar half full of grinding medium contains a 1,500-g charge of frit and 120 ml of water. After melting the slurry is filtered and dried at about 100°C for 16 h. The dried cake is micropulverized to break up the aggregates. Ceramic laboratory box kilns or furnaces can provide temperatures at 1500°C. If higher temperatures are required, the heating chamber can be provided with platinum-rhodium resistance wire elements or silicon carbide elements.

Paste Preparation

In general, the formulations require that the particulates of the pastes be less than 250 mesh. The processing of the formulation is as follows. First, 100 g of the batch is blended into 50 ml of distilled water. Tetrachloroethylene can also be used. The mixture is ball-milled for 4 h and the water removed. A satisfactory way to remove the water is to draw it through filter paper by vacuum-soaking the residue under an infrared lamp for 4 or 5 h. To form a screenable paste, the formulations are suspended in an organic binder. A suggested formulation of the binder is 200 ml of butyl Cellosolve acetate and 21.5 g of Ethoxyl T-10. The organic binder is prepared by adding the butyl Cellosolve acetate into a high-speed blender and sifting the ethoxyl slowly into the grinder. The mixture is left in the blender for about 10 min and filtered with a 250-mesh screen.

Organic carriers of carbitol acetate and cellulose and β-terpineol and ethyl cellulose are also utilized. The ratios of powder and organic carrier may be 75:80

and 20:25 percent by weight, respectively. The mixture can be diluted with pine oil to obtain the correct viscosity.

The screenable capacitor dielectric paste is a mixture of 100 g of the dielectric formulation and 50 ml of binder. The mixture is milled in an automatic mortar pestle for 2 h and stored in an airtight bottle.

SINTERING MECHANISMS

Ceramic Chip Capacitors

The density, strength, and dielectric properties of a capacitor dielectric are achieved through sintering. Ceramic materials which have been shaped by a variety of forming techniques are fired under controlled conditions of temperature, time, and ambient environment. Sintering is a diffusion-controlled process. During the firing process, densification occurs as a result of several mechanisms which involve the movement of material by surface or volume diffusion and glass formation as a result of chemical interactions. In multiphase ceramics these processes are often accompanied by phase changes, the decomposition of certain phases, and the formation of others. Ceramic capacitor dielectrics are densified by solid-state sintering.

Thick Film Structures

For materials of interest in thick film dielectrics, it is desirable to produce a ceramic which is a dielectric or ferroelectric powder-glass composite of known stoichiometry or a glass-ceramic structure. In the first case liquid-phase sintering processes must be employed to induce densification of the screened film. In the second case the ferroelectric material is formed as a glass and then converted into a largely crystalline film by heat treatment. The crystallites form and grow by precipitation from the glass rather than through the sintering of particles.

Densification, Shrinkage, and Pore Elimination

Compacted particulate ceramics, such as those fabricated by the cold pressing of a ceramic powder of fine particle size, contain a large fraction of pores, often over 50 percent by volume. The screened, unfired thick films contain the same volume fraction of porosity. It is probably higher in thick films due to the particulate-size to film-thickness factor.

The objective of the sintering operation is to produce a dense film by the elimination of this porosity and to preserve a uniform structure in the ceramic or film. The elimination of all the porosity while the compact is shrinking and its bulk density is increasing is a difficult operation to control, particularly for films, which are further constrained by the substrate. As an example, the mechanisms for material transport in the solid-state sintering of ceramics such as barium titanate consist of surface diffusion, lattice or volume diffusion, and evaporation-condensation mechanisms active at or near the points of contact between particles. These processes lead to the formation of interparticle necks and to the shrinkage of the compact as the pores are eliminated. The driving force for densification is the free-energy change associated with the decrease in surface area which occurs when necks form between particles and when pores shrink and are ultimately eliminated. The mechanism for shrinkage of the powder compact is the transport of lattice vacancies from the neighborhood of a pore to adjacent grain boundaries, where they are eliminated. Sintering without grain growth is desirable so that grain boundaries remain accessible to pores. Too high a temperature results in exaggerated grain growth.

Ambient Atmosphere

Ambient atmosphere is another important sintering parameter. During the final stages of densification, gas trapped in the pores must escape by diffusion to the grain boundaries, or further densification and porosity elimination cannot proceed. With the ferroelectric powders used in thick film ceramic-glass composites, firing in air or a pure oxygen environment is important to prevent rapid evaporation of constituents, which destroys the stoichiometry of the material. In cases such as the firing of lead zirconate titanate, the air or oxygen environment can be supplemented with the vapors of the more volatile constituent oxide to preserve stoichiometry. In some oxide materials it is necessary to eliminate air during the final stages of densification and use nitrogen, which does not diffuse rapidly in oxides.

Powder-Glass Composites

Solid-state sintering is not a direct mechanism for the sintering of films consisting of ferroelectric or dielectric powders and glass. Sintering takes place by liquid-phase sintering or the sintering of the ferroelectric particles in the presence of a viscous-liquid phase. The glasses will flow and bond the ferroelectric powders at temperatures which are much lower than the ferroelectric-particle sintering temperature. If the volume fractions are large, so that there is ferroelectric particle-to-particle contact and temperatures are sufficiently high, some degree of solid-state sintering will occur.

The major sintering process for ferroelectric-glass composites is through densification with the aid of the glass binder as a viscous-liquid phase. The glass serves as a bond for the dielectric particulates and for the body to the substrate and bottom electrode. The rates of shrinkage and densification depend on particle size, viscosity, and surface tension of the glass.[5] Important for control purposes are the viscosity and its rapid change with temperature and the dependency of sintering rate on particle size. If the densification process is to be controlled, the particle size must be controlled. For example, the rate of sintering is increased by a factor of 10 in changing the particle size from 10 to 1 μm. For a typical glass binder, viscosity and therefore the densification rate can change by a factor of 1,000 over an interval of 100°C. For glasses characterized by viscosity-temperature curves of this profile, temperature must be closely controlled. The viscosity of the glass can be lowered by changing composition. In general, surface tension normally does not cause problems. However, the interrelation of particle size and viscosity must also be considered. Deformation should be avoided by preventing viscosity from being too low and by providing a particle-size range such that the stresses arising from surface tension are larger than those arising from gravitational forces. On this basis the frits should be well ground and uniformly mixed.

Glass-Ceramic Process

Glass-ceramics are polycrystalline bodies or films prepared by the controlled crystallization of glasses. The presence of major amounts of crystals distinguishes glass-ceramics from glass, which is noncrystalline, or amorphous. The crystallization is accomplished by subjecting the glasses to a carefully controlled heat-treatment schedule, resulting in the nucleation and growth of crystal phases within the glass. Usually, the crystallization process can be taken almost to completion. A small amount of residual glass phase is often present.

The nucleation and crystal-growth stages of the heat-treatment schedule are shown in Fig. 5. The glass is first heated from room temperature to the nucleation temperature, the optimum nucleation temperature usually lying within the range of temperature corresponding to viscosities of 10^{11} to 10^{12} P. Heating rates between 2 and 5°C/min are employed for glass pieces, and thick films are processed at rates as high as 10°C/min.

Following the nucleation stage, the temperature of bulk glass is increased at a rate usually not exceeding 5°C/min. The slow rate allows crystal growth to occur without deforming the glass. Glass deformation will occur in the early stages when the glass phase predominates. Crystallization increases as the liquidus temperature of a predominant crystal phase is approached. The upper crystallization temperature at which maximum crystallization is achieved is usually 25 to 50°C

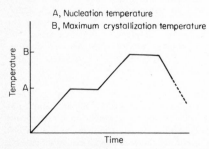

A, Nucleation temperature

B, Maximum crystallization temperature

Fig. 5 Idealized heat-treatment schedule for a glass-ceramic.[6]

below the redissolving temperature of the crystalline phase. The upper crystallization temperature is usually maintained for a period of at least 1 h and the glass-ceramic cooled to room temperature.

The cooling rate is not critical. Often glass-ceramics with high thermal expansion can be cooled at rates as high as $10°C/min$. Low-expansion composites can be cooled at higher rates. Methods to determine the optimum nucleation temperature, permissible heating rates for crystallization, the upper crystallization temperature, and the liquidus temperature have been reported.

The heat treatment converts the glass into a ceramic containing a microcrystalline microstructure. Of particular interest to thick film glass-ceramic capacitors is the fact that many high-dielectric-constant ceramic compositions can be completely dissolved in an appropriate glass and in heating be nearly completely precipitated from the glass. During heat treatment the residual glass phase will also be crystallized and converted into a ceramic material. Thus, the high-dielectric-constant crystals may be suspended in a glass or in a crystalline ceramic phase. The crystallite size of both the electronic ceramic phase and the solvent ceramic phase resulting from the devitrification of the residual glass can be controlled by varying the heat treatment. Differential thermal analyses have been made of numerous examples of this type of composition. The characteristic temperatures are known for the appearance of the electronic ceramic and for the devitrification of the remaining glass. The application to thick film processing is discussed in the sections on Glass-Ceramic Capacitors and Crossover Dielectrics.

A good example of the glass-ceramic process as applied to thick-film dielectrics is the crystallization of certain glasses containing barium titanate, $BaTiO_3$ and the feldspar $BaAl_2Si_2O_8$ as a minor phase.[7] These are obtained by the heat treatment of glasses with composition corresponding to $xBaTiO_3 + 100-xBaAl_2Si_2O_8$. The mechanism of crystallization which takes place during the gradual heating of the glass is

$$\text{Glass A} \xrightarrow[800°C]{600 \approx} BaTiO_3(c) + \text{glass B} \tag{1}$$

$$\text{Glass B} \xrightarrow[1000°C]{750 \approx} BaAl_2Si_2O_8(c) + BaTiO_3(c) \tag{2}$$

Unwanted reactions are produced by excessive heating rates. For the complete crystallization of $BaTiO_3$, step 1 should be near completion before step 2 starts. This can be attained if viscosity is kept low at the beginning of the crystallization by a moderately high heating rate and by the addition of fluorides, that is, 2 mol percent of BaF_2 or CaF_2 on an oxide basis for BaO, to suppress nucleation. However, crystallization at high viscosity is desirable to minimize surface deformation. An increase in the $BaTiO_3$ content and crystallite size occurs simultaneously during crystallization.

The sintering of thick film glass particles precedes the crystallization of the glass.[8] The particles sinter, and the glass flows into a film. The crystallization of the glass proceeds in the same manner as the bulk glass-ceramic on further heating or during subsequent heat treatment. In $BaTiO_3$ glasses, grain growth occurs through crystallization rather than by sintering of particles.[7] Grain growth depends on the extent or completeness of crystallization under any one condition rather than on the amount of $BaTiO_3$ phase produced.

BARIUM TITANATE SINGLE-PHASE FILMS

Approximating Processing Temperatures and Reaction Products

Barium titanate dielectrics are produced either by the solid-state thermochemical reaction of barium carbonate, $BaCO_3$, and titania, TiO_2 or by the sintering of reacted $BaTiO_3$ powders. The reaction of $BaCO_3$ with TiO_2 is heterogeneous. Complete reaction of these constituents is essential to produce single-phase films of $BaTiO_3$. Reference to the phase diagram for the system $BaO\text{-}TiO_2$ is useful for understanding the problems which can be encountered during and after sintering.[9] The phase diagram is shown in Fig. 6. The desired ferroelectric form, which is designated as BaO·TiO$_2$ or BaTiO$_3$, contains 50 mol percent each of BaO and TiO$_2$. BaTiO$_3$ has a melting point of 1618°C. On the TiO$_2$-rich side of BaTiO$_3$ are the incongruently melting compounds BaTi$_2$O$_5$ and BaTi$_3$O$_7$. The BaTi$_2$O$_5$ is not usually present in solid-state reactions since it is unstable below 1210°C. There is a solid-solution region of nearly 3 mol percent on the TiO$_2$-rich side before the appearance of BaTi$_3$O$_7$. This represents a perovskite with A-position vacancies and compensating oxygen vacancies.

BaTiO$_3$ has a high-temperature hexagonal form stable above 1460°C. Although the cubic phase is stable below this temperature, incomplete transition of the hexagonal to the cubic occurs with cooling, in particular, rapid cooling. Thus within

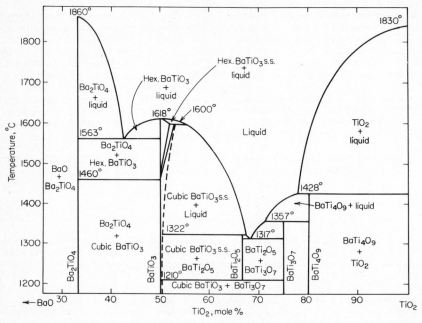

Fig. 6   Phase equilibria in the system BaO-TiO$_2$.[9]

some films the hexagonal phase will appear within the 3% solid-solution region at room temperature, as shown by the dashed line in the phase diagram.

A sintered dielectric represents a metastable condition rather than a condition of equilibrium since fast solid-state reaction and sintering schedules are followed. Reference to the phase diagrams provides a concise method for approximating sintering temperatures and reaction products. The initial reaction of BaCO$_3$ with TiO$_2$ in air starts with the formation of BaTiO$_3$ at the BaCO$_3$-TiO$_2$ grain boundaries.[10] The BaTiO$_3$ and BaCO$_3$ then react to form Ba$_2$TiO$_4$. This reaction is diffusion-controlled and proceeds until the BaCO$_3$ is depleted. Excess TiO$_2$ then reacts with Ba$_3$TiO$_4$ to form BaTiO$_3$.

For sintered capacitor disks, a sintering temperature of 1350°C is usually recommended.[10,11] Complete reaction can be obtained below 1200°C in several hours. Firing temperatures should be kept lower than 1460°C since the hexagonal phase will form and appear at room temperature.[9] This can also be caused by insufficient oxygen in the sintering atmosphere.[12]

If BaTi$_2$O$_4$ is present, the sintered film can decompose with swelling in slightly moist air since BaTi$_2$O$_4$ is hygroscopic.[13] The first indication of this mode of failure will be an increase in the loss factor with time. An immersion test in water for 24 h followed by drying the surface and checking for increases in tan δ is recommended. Below $1100°C$ the CO_2 by-product from the decomposition of $BaCO_3$ will inhibit the formation of Ba_2TiO_4.

Process incompatibility Because the barium titanate contains no volatile or low-melting oxides, it is refractory and reasonable densities (over 90 to 95 weight percent) are difficult to attain in high-purity materials. The sintering of dense, low-porosity thick films of $BaTiO_3$ is difficult for the same reasons. In addition, since the porosity content is usually higher in films because of the thickness–particulate size relationship, a glass binder is usually employed when fabricating dielectric-powder-containing composites. The introduction of a few tenths of a percent of strontium titanate, $SrTiO_3$, will enhance the density of sintered barium titanate,[14] due to small differences in ionic size between strontium and barium.

Temperatures in the vicinity of $1350°C$ are generally considered too high for the processing of pure, single-phase $BaTiO_3$ films. There are several reasons for this from a practical standpoint:

TABLE 2   X-Ray Diffraction Lines (CuK alpha radiation) for Identification of Phases in $BaTiO_3$[15]

Compound	2θ, deg	d, Å	Intensity of line relative to $BaTiO_3$ $d = 2.84$ Å	Limit of detection in $BaTiO_3$
$BaTiO_3$, tetragonal................	31.5	2.84	100	
Hexagonal......................	41.3	2.18	30	$<10\%$
Ba_2TiO_4...........................	29.3	3.05	60	4 mol %
$BaTi_3O_7$...........................	28.4	3.14	45*	4 mol %
$BaTi_4O_9$...........................	30.2	2.99		3%
TiO_2, rutile†......................	27.5	3.25		$<5\%$
$BaCO_3$.............................	23.9	3.72		$<5\%$

* This line superimposes on the CuK beta line of d spacing 2.84 Å.
† If fired over $950°C$, any other form of TiO_2 will be converted to rutile.

1. The higher processing temperature necessitates the use of the higher melting point of precious metals and alloys. Substrate screenable bottom electrode materials (other than platinum) which will sustain the temperature are not readily available.

2. The processing furnaces routinely available in a prototype or production facility for the processing of conductors, resistors, and capacitors are restricted to lower temperature limits, that is, $1250°C$.

3. The sintering temperature may not be compatible with the typical sequence of decreasing firing temperatures for the various passive elements.

X-ray diffraction X-ray diffraction spectra of the sintered thick films can be used to detect the tetragonal or hexagonal Ba_2TiO_4 or $BaTi_3O_7$.[15] These serve as a fast-turnaround process control tool since the presence of an unwanted phase can be identified quickly from the major reflection. These are listed in Table 2, with their limits of detection in ceramic barium titanate. Assuming ceramic barium titanate to have a 100 percent single-phase $BaTiO_3$ content, the limits of detection would be proportionately less in films which contain 10 to 90 percent of $BaTiO_3$ powders by weight. Since the $BaTiO_3$ raw materials used for fabricating powder-glass composites are produced by the reactions described, periodic sampling and analyses by x-ray diffraction of the raw materials should be part of the quality-control sequence.

CONCEPTS OF MICROSTRUCTURE

The properties of dielectric thick films depend on the microstructure of the films and the nature of the ferroelectric or dielectric particles used for film fabrication.

Considerable work has been done in evaluating the effects of grain size, secondary phases, and composition on the thermal, mechanical, and electrical properties of various sintered ceramic systems. Considerable work has also been done on the microstructural effects on dielectric properties of glass-ceramics. Little work has been reported on the microstructural effects on dielectric properties of thick film structures. Minimal work has been centered on the dielectric properties of particulates and their effects in glass binders. In the following sections discussion will be directed toward the following topics:

1. Properties of dielectric and ferroelectric powders
2. Property-microstructural interrelationships in ferroelectric and dielectric sintered ceramics
3. Property-microstructural interrelationships in ferroelectric and dielectric glass-ceramics
4. Property-microstructural interrelationships in thick film dielectrics

PROPERTIES OF DIELECTRIC AND FERROELECTRIC POWDERS

Particle Characterization and Thick Films

Particle size, shape, and distribution are believed to affect the dielectric properties of the final films. Little work has been reported on this important effect in films, but it has been discussed for sintered ceramics. Discussion of these effects may give some insight into the nature of particulates in films and their contribution to the apparent dielectric properties. Thus the particle characterization becomes a matter for process control in thick film manufacturing.

Ultrafine Powders and Surface Effects

Ferroelectric and dielectric powders are a major constituent of many thick film composites. Ultrafine particles are 2 to 5 μm and less in size. Several investigators have shown ferroelectric powders below this size to have a surface effect which becomes increasingly important as particle size decreases.[16-18] The characteristics of the surface-defect layer of powders play a significant role in determining the properties of the sintered ceramics.[19-25] Surface phenomena strongly affect particle-size distribution and shape, rheology in the forming operations, degasification during green-body compaction, and finally the sintering mechanism itself. All these factors are related to the microstructure of the final ceramic and hence affect all its properties. With powders of micrometer and smaller sizes, because of the increased surface-to-volume ratio, the effects due to absorbed gases and surface charges can alter the dielectric behavior of the fired ceramic. These same factors can also alter the dielectric behavior of thick film composites.

Particle Surface-Layer Effect

The presence has been postulated of surface layers which alter the properties of barium titanate, $BaTiO_3$, fine particles, the most commonly used ferroelectric in thick films.[16-18] The tetragonal deformation of the particles is progressively reduced as particle size decreases below 5 μm. Below this size, there is an increasingly important surface effect, in which the polarization that results from the tetragonal deformation is locked, probably normal to the grain surface. The locked-in deformation persists to temperatures far in excess of the usual Curie point, and the surface of the crystallites may continue to show tetragonal deformation to temperatures above 500°C. The surface layer is considered to be about 100 Å thick. There is experimental evidence that the surface layer has a higher resistivity than the bulk of the particle.[26] It has also been postulated that this surface is of the nature of a Schottky depletion layer.[18] The presence of surface layers has been invoked by several investigators to explain various properties of barium titanate single crystals.[27-36] It has been reported that surface layers may not always be observed in single crystals.[37]

Surface Phenomena and Water Vapor

Water vapor plays a significant role in determining surface phenomena of oxide powders.[38] Surface hydration plays a significant role in the surface activity of micrometer and submicrometer $BaTiO_3$ powders. It has been pointed out that the surfaces of $BaTiO_3$ powders can be characterized by evaluating the changes in ac conductivity due to controlled partial pressure of water.[39] Figure 7 shows the effect of stoichiometry on dielectric loss due to water vapor. The powders had comparable particle size (approximately 1 μm) but differing BaO-TiO_2 ratios. The two powders had nearly identical values of ac conductivity at a water partial pressure p-$_{H2O}$ of 7×10^{-6} atm and in vacuum ($\sim 10^{-6}$ Torr). However, with increased value of p-$_{H2O}$ the dielectric loss in the material with excess TiO_2 is enhanced, compared with the stoichiometric material.

Frequency-dispersion measurements of the dielectric properties indicate that a surface layer of adsorbed water is removed by as isopropanol wash of $BaTiO_3$ powders which are 0.01 μm and less in particle size.[4] Removal of surface water by isopropanol washing reduces the dissipation factor, which is important in processing thick films. Fine powders tend to absorb gaseous species because of their large surface-to-volume ratio. Varying the surface-to-volume ratios will cause differences in the degree of hydroxylation.

The effect of water vapor on the dielectric behavior of larger powders is less significant. On removal of adsorbed water, there is no significant difference in the relative dielectric constants of fine- and coarse-particle $BaTiO_3$. Removal of surface water by isopropanol washing reduces the dissipation factor, which is important in processing thick films.

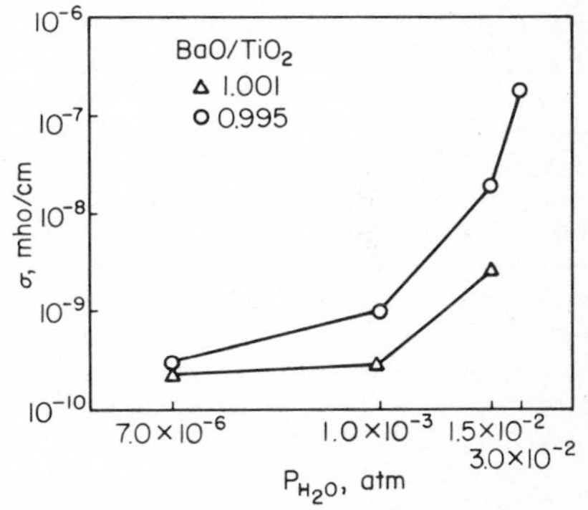

Fig. 7 Effect of stoichiometry on dielectric loss due to water vapor on $BaTiO_3$ powders at 1 kHz.[39]

MICROSTRUCTURE OF SINTERED FERROELECTRIC AND DIELECTRIC CERAMICS

Physical Microstructural Features

The electrical, magnetic, and mechanical properties depend on the microstructure of the ceramic. The important aspects of the microstructure which influence these properties include porosity, the number and distribution of phases, the size of the constituent grains, the nature of the boundaries between them, and whether the grains have a preferred orientation.

A polycrystalline ceramic is an agglomerate of individual grains, each of which may in fact be a perfect single crystal of irregular outline determined by adjacent grains. In bulk form these polycrystalline materials can be fabricated to eliminate porosity between grains completely and in many cases other secondary phases also. The feature distinguishing these materials and single crystals (whose properties for device applications are generally well known) is that they contain physical discontinuities between grains. Grain boundaries are often the controlling factor in determining the properties of the bulk material. Since adjacent grains are unlikely to have identical crystallographic orientation, there is a region of disorder between adjacent crystal lattices as they attempt to adhere or coalesce during sintering. The degree of disorder is related to the extent of misorientation and may be several atom planes wide, even in the absence of impurity phases.

Relations Between Microstructure and Electrical Properties

Grain-size requirements for high-permittivity dielectrics The relationship between microstructure and material properties should be known in order to produce the optimum microstructure for the dielectric properties required. In ceramic dielectrics, the situation is rather straightforward compared with the ceramic ferroelectrics, which are the basic materials of interest. In ceramic dielectrics, grain or crystallite boundaries are often useful imperfections, and internal pores can cause dilution effects of minor significance provided that they are not freely interconnected. Open porosity, however, enables the material to absorb moisture and increases the dielectric losses and the possibility of complete electrical breakdown.

The relations between microstructure and properties assumes a variety of forms in ferroelectrics, but in general the influence of changes in the microstructure is very pronounced.[40]   Ferroelectric barium titanate is a prime example.[41] In ceramic form it is a good ferroelectric (and suitable for piezoelectric use) only when its crystallites are large enough to consist of domains. If the crystallites are small, the ferroelectric character is completely suppressed, as are the hysteresis losses.   The material is then a very useful dielectric.

Of the various aspects of microstructure, the crystallite size and the nature of the grain boundary are the features which most critically influence the electronic processes of a ceramic material.   As a rule, pores are undesirable since they decrease domain-wall mobility, cause internal depolarization, and make the internal field inhomogeneous.

Barium titanate ferroelectrics   The ferroelectrics used in thick film dielectrics have the perovskite structure (corresponding to the mineral perovskite) and very high dielectric constants.   The best known compound in this group is barium titanate. The ferroelectrics are like ferromagnetics in that they give spontaneous polarization hysteresis, a domain structure, and wall displacement below a Curie temperature T_c. The primary polarization process arises from a displacement of the mean center of the positive ions in the crystal lattice with respect to the mean center of the negative ions.   In the dielectric or field-free state above the Curie temperature, these centers coincide.   In the ferroelectric state below the Curie temperature the centers spontaneously assume a certain separation.   In the dielectric state all perovskites are cubic; below the Curie temperature they are tetragonal or rhombohedral.   BaTiO$_3$ is tetragonal below a Curie temperature of 120 to 130°, depending on purity and particle size.[42-44]   In the spontaneously polarized state the crystal has a dielectric constant of the order of several hundred since it can be further polarized by an externally applied field.

When single crystal barium titanate cools through the Curie temperature, the cubic material becomes tetragonal by elongation along a cube axis, there being six possible directions for the polar tetragonal axis.   This causes complicated twinning patterns in the crystal, called *domain patterns;* the particular pattern formed is the result of physical imperfections, uncompensated surface changes, and stresses created at the Curie temperature.   For these reasons single-domain crystals are rarely found.

The change in shape of a crystal as it passes through the Curie temperature transition has a significant effect on the properties of the ferroelectric.   In a single crystal of a ferroelectric perovskite, the crystal is cubic on the unpolarized dielectric state above the Curie point.   In the polarized single-domain crystal below the Curie point the shape is tetragonal or rhombohedral.   For example, barium titanate is tetragonal, and the axial ratio (c/a) has changed from 1. to 1.01, a 1 percent change from the cubic structure.   This has a profound effect on polarization.   In ferroelectric lead titanate, the change is 6 percent.

In a piece of sintered barium titanate which has been cooled to below the Curie temperature, the crystallites are held tightly between their neighbors.   Spontaneous polarization can take place only if the external form of the crystallite is almost exactly preserved.   Under favorable conditions the material spontaneously satisfies this condition by forming a fine pattern of domains, the longitudinal axis of each domain differing by 90° from that of its neighbor, i.e., where there are 90° walls between the domains.[45]   Crystallites are then formed with ridges on the surface.   In the most favorable case the ridges of two neighboring crystallites fit together.   In this kind of polarization, stresses arise and have been manifested as line-broadening effects observed in x-ray diffraction photographs.   Crystallites may be too small to split into domains and often incapable of being polarized.   In their constrained position they cannot achieve the required change of shape.

Thus, in ceramics complicated domain patterns are formed to minimize the intergranular stresses which arise from the constraint imposed upon each grain by its neighbors.   If the grain size is large enough, the ceramic will have a domain structure.   Since the width of the domain is about 1 μm, the grain size must be about 10 μm for polarization to be effective.

Barium titanate capacitor dielectrics   When a ferroelectric ceramic is used as a dielectric in a capacitor, the requirements of high dielectric constant and low losses must be met.   In barium titanate these are both met by producing a microstructure with small grains, which suppresses the ferroelectric behavior.[20, 45-48]   There is no spontaneous polarization in small crystallites that are clamped tightly between their neighbors and too small to split up into domains.   This completely eliminates the principal losses in coarse-grained materials, i.e., the ferroelectric hysteresis losses. Small grains also favor a high dielectric constant.   When a crystal changes from the dielectric to the ferroelectric state, there is a stiffening of the lattice in the direction which becomes the polar axis.   The positive and negative ions become less easy to displace with respect to one another.   As a consequence the dielectric constant in coarse-grained material with grain sizes of 50 to 100 μm is relatively low.[11]   When

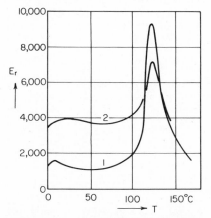

Fig. 8   The relative dielectric constant of ceramic barium titanate as a function of temperature.   Curve 1 for grains larger than 50 μm; curve 2 for grains about 1 μm.[45]

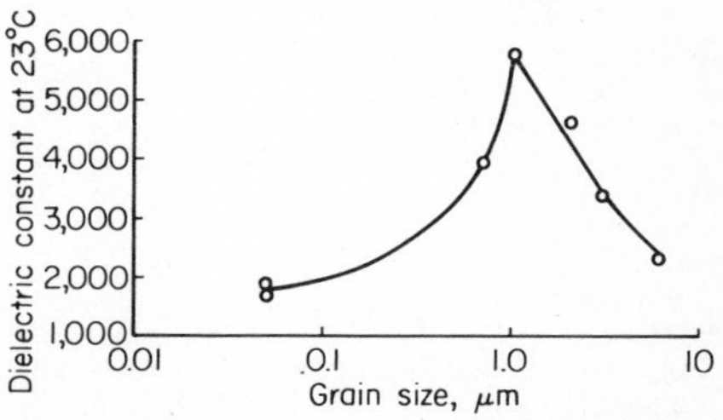

Fig. 9   Dielectric constant versus grain size for BaTiO$_3$.[49]

the ceramic is cooled, the value decreases from about 10,000 at the Curie point to 1,000 at room temperature.   In fine-grained material with grains of 1 μm the decrease is much smaller, and the final room-temperature value is 3,000 to 4,000,[45] as shown in Fig. 8.

The fine-grained material has a considerably higher dielectric constant in the temperature range below the Curie point owing to the suppression of the ferroelectric behavior.   It is therefore more suitable as a dielectric.   Figure 9 shows the effect of grain size on the relative dielectric constant of hot-pressed BaTiO$_3$.[49]   The dielectric constant increases as the grain size decreases from about 7 μm, passing through a maximum value of about 5,500 at an average grain size of 1 μm.   The increase in dielectric constant with decreasing grain size is not understood at this time.   It has been proposed by Buessem et al[46-47] that the decrease is due to an increase in the internal stress of the ceramic brought about by decreased domain twinning in the smaller grains.   Below 1 μm the dielectric constant starts to fall off.   Others have refuted these experimental results, stating that the dielectric constant continues to increase with decreasing grain size into the submicrometer range.   The room-temperature dielectric constant also varies as a function of particle size, reaching a maximum of 3,000 to 4,000 at 1 μm.   The Curie peaks are also suppressed with ultrafine-grained ceramics.   This is highly desirable since low-temperature coefficients of capacitance can be attained in high-dielectric-constant bodies over extended temperature ranges.

The grain size in ceramic barium titanate depends upon exact composition, stoichiometry, and processing conditions. [50-51] The nominal technical-grade powders are 1 to 5 μm in size. Sintered ceramics which are nearly stoichiometric ($BaOTiO_2$ ratio of 1) have grains which are approximately 20 μm; occasionally larger grains are developed. At a comparable firing temperature, grain growth is more extreme with excess Ti^{4+}; coarse-grained microstructures result with grain sizes of 50 to 100 μm or larger. Fine-textured microstructures with grain sizes of 5 to 10 μm result with excess Ba^{2+}. Firing in an oxygen atmosphere will result in large grains. The grain growth is noted even with an excess of Ba^{2+}.

Lead zirconate titanate capacitor dielectrics As in $BaTiO_3$, a decrease in grain size was found to increase the relative dielectric constant of lead zirconate–lead titanate. This is shown in Fig. 10 for 65 mol % lead zirconate and 35 mol % lead titanate with 2% Bi_2O_3. [52] The $\varepsilon/\varepsilon_0$ increase from 450 to 750 for a change of grain size is minimized. Optical-property studies on the same materials uncovered a unique optical-

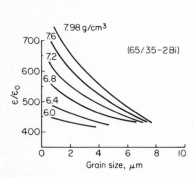

Fig. 10 Variation of relative dielectric constant with grain size for lead zirconate titanate.[52]

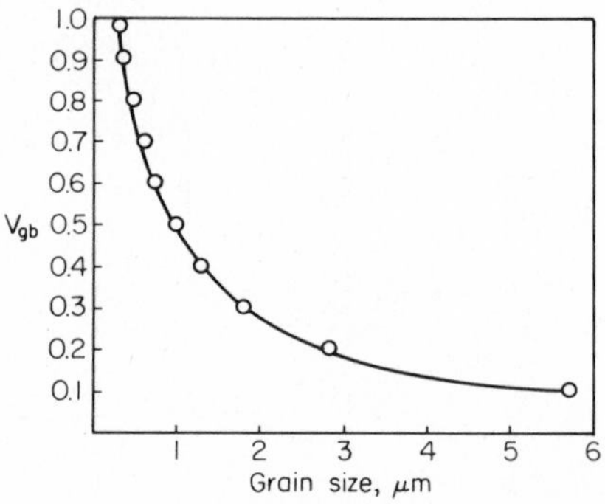

Fig. 11 Volume fraction V_{gb} of grain boundary as a function of grain size where grain-layer thickness = 0.1 μm.[54]

scattering behavior which reveal the relation of grain size and domains.[53] With grain sizes greater than 2 μm, the material behaves like a light scatterer and scatters light preferentially along the electric-dipole direction. Material with grain sizes smaller than 2 μm displays characteristics of optical retarders and does not preferentially scatter light along a given dipole direction. The unique optical scattering of this material with grain sizes of about 1 μm and less may be due to the absence of domains. Domains were observed in the large-grained material, but with the 1-μm material.

Surface-to-volume and grain-boundary-to-grain Ratios As grain size decreases to the micrometer region and below, there is a corresponding increase in the surface-to-volume and grain-boundary-to-grain ratios in particulate and sintered dielectrics, respectively. Thus, as the grain or particulate size becomes smaller, the dielectric properties increasingly assume the characteristics of surfaces and boundaries in the dielectric. Figure 11 shows the volume fraction V_{gb} of grain boundary as a function of grain size.[54] This curve assumes that the grains are spherical and that the effective grain boundary layer is 0.1 μm. The curve shows that below a critical size of 1 to 2 μm the value of $V/_{gb}$, and thus the contribution of the grain boundary to electrical properties, increases significantly. This indicates that for thick film dielectrics with average grain sizes of 2 μm and less, the film can increasingly assume the dielectric properties of the grain surfaces or boundaries, depending on the type of film structure and its microstructure.

Below 1 μm grain size the volume of the grain boundary phase becomes greater than 0.5, according to Fig. 11. Relative to the discussion of the high dielectric constants of sintered ceramics with an average grain size of 1 μm or less, this indicates that the effect of the grain boundary phase would predominate over bulk properties. On this basis, it is important to the user of ferrolectric particulate in the ferroelectric-glass type of structure to consider the size of these powders instead of the surface condition.

MICROSTRUCTURE OF GLASS-CERAMICS

Crystallite sizes in barium titanate glass-ceramics are small in comparison to those in sintered barium titanate. The crystallite sizes are in the range of 0.01 to 1.1 μm. Control of the crystallite size is possible by the crystallization process; a given heat treatment produces crystallites within a narrow size range. The mechanism of crystallization for glasses corresponding to $x\mathrm{BaTiO_3} + 100\text{-}x\mathrm{BaAl_2Si_2O_8}$, where x is the nominal barium titanate content and the $\mathrm{BaAl_2Si_2O_8}$ is the glass-forming feldspar, was discussed in the section on Glass-Ceramic Capacitors.

The electrical properties are dominated by effects due to particle size and the $\mathrm{BaTiO_3}$ content recovered by crystallization.[7] Below 0.2 μm, surface effects are

TABLE 3   Electrical Properties of Polycrystalline BaTiO$_3$[7]

Property	Material crystallized from glass			Ceramic	
Average particle size, μm	0.8	0.4	0.2	10	1
BaTiO$_3$ content, vol %	61	53	47	95	95–99
ϵ_0 at 25°C, 1 kHz	1,200	650	400	1,700	3,300
Increase of ϵ_0 with temperature, %	100	25	5	450	40–100
Decrease of ϵ_0 at 1 MHz, %	10	5	1	5	1
Tan δ at 25°C, kHz	0.025	0.018	0.011	0.025	0.010
MHz	0.045	0.020	0.015	0.01	
Dc breakdown, kV/cm	240	320	400	100	
Resistivity, Ω-cm	10^{12}	3×10^{12}	10^{13}	10^{11}	

prevalent. The dielectric properties are independent of temperature, and the ferroelectric behavior is degenerated. The properties in the range of 0.2 to 1 μm are the result of contributions due to surface and normal bulk properties. The two contributions are equal at about 1 μm where the dielectric constant is the highest of all crystallite sizes.

Table 3 shows the electrical properties of $\mathrm{BaTiO_3}$ dielectrics crystallized from glass as a function of particle size and barium titanate content. The properties improve with decreasing particle size and barium titanate content.

The variation in dielectric constant and loss tangent with temperature is shown in Fig. 12. The curves on the left-hand side are characteristic of dielectrics with a crystallite size of 0.8 μm containing different volume percentages of crystallized $\mathrm{BaTiO_3}$. The peak at the tetragonal-cubic phase transition is pronounced; it is slightly less than 120°C. Profiles on the right-hand side are for average crystallite sizes of 0.4 and 0.2 μm. They show a gradual extinction of the peak with decreasing crystallite size. The loss tangents lie between 1 and 2.5 percent for the 0.2- to 0.8-μm crystallite range and show a decrease in the vicinity of the Curie temperature.

As shown in Table 4, the microcrystalline $\mathrm{BaTiO_3}$ produced by the crystallization from glass exhibits high dielectric breakdown strength and electrical resistivity and no decade aging compared with sintered ceramics. These properties are attributed to the zero porosity of the crystallized glass and the mechanical clamping of the $\mathrm{BaTiO_3}$ crystallites within the microcrystalline matrix. Ferroelectricity decreases with crystallite size. Some degree of remanent charge and hysteresis are present down to 0.2 μm.

Microcrystalline $\mathrm{BaTiO_3}$ produced in the absence of a glass-forming compound exhibits the dielectric-constant–temperature profile (curve C) shown in Fig. 13.[55-56]

There is a near-zero temperature cofficient of capacitance up to 100°C and a broadened Curie temperature peak at 150°C. Spheres of highly disordered BaTiO₃ were produced by flame spraying BaTiO₃ powder. The particulates were melted in an oxyacetylene flame, solidified in a quenching medium, pulverized, pressed into disks, crystallized at 450 to 850°C, and sintered at 1320°C. A microstructure of 0.5-μm grains was developed. By controlling the firing cycle with a soaking period in the recrystallization-temperature range, temperature coefficients of dielectric constant be-

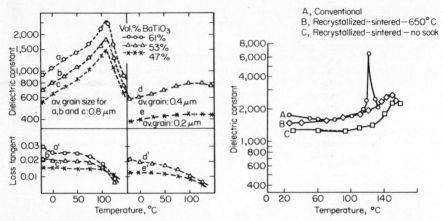

Fig. 12 Variation of dielectric constant and loss tangent with temperature, grain size, and volume percent of BaTiO₃ in glass-ceramic BaTiO₃-BaAl₂Si₂O₈ at 1 kHz.[7]

Fig. 13 Dielectric constant versus temperature for recrystallized (no glass-forming compound) and sintered BaTiO₃.[55]

TABLE 4 Large-Signal Characteristics of Polycrystalline BaTiO₃[7]

Property	Material crystallized from glass			Ceramic
Average particle size, μm	0.8	0.4	0.2	10
BaTiO₃ content, vol %	61	53	47	95
ϵ_0 at 25°C, 1 KHz, small field	1,200	650	400	1,700
Decrease of ϵ_0 with dc bias, %*	25	8	1	35
Coercive field (60 Hz), V/cm	8,000	5,000	2,500	7,400
Remanent charge, μC/cm	2.0	0.4	0.15	5.5
Loop$_{av}$/ϵ_0	1.6	1.5	1.4	3.5

* 10⁴ V/cm dc bias or 60 Hz, 2.8 × 10⁴ V/cm peak hysteresis.

came positive.[44] The largest positive temperature coefficient (curve B) was attained by soaking at 650°C for an extended time and sintering at the established sintering temperature of 1320°C. The profiles are compared with that for a solid-sintered ceramic with grain sizes of 16 to 80 μm.

THICK FILM MICROSTRUCTURES

Composites

Very little information has been published on the microstructure of thick film capacitors and crossovers. The important microstructural parameters of the ferroelectric powder-glass composites are particle size of the ferroelectric powder, porosity, and the reactivity interface of the glass and powder. The role of particle size has been discussed for particulates and sintered ceramics. The low-temperature coefficients of capacitance of the composites indicate that particle size plays a property-controlling role.

The microstructure of films containing TiO_2 and glass is shown in Fig. 14. Films with glass content larger than 50 percent by volume have their TiO_2 particles completely surrounded by glass. The glass forms a continuous network within the film. Films with glass content equal to the volume of the space between the TiO_2 grains and or smaller than 15 percent are shown in Fig. 14b and c. Porosity begins with contents of 40 percent by weight of TiO_2 and increases to 40 percent by volume for a TiO_2 concentration of 90 percent.

The microstructure of class II dielectrics based on mixtures of titanates with glasses has been discussed. [58, 59] Fine particle sizes for the titanates are necessary to overcome large boundary stresses at the glass-titanate interface. The effect of particle size on dielectric constant and temperature coefficient of capacitance has been discussed in the section on Microstructure of Sintered Ferroelectric and Dielectric Ceramics. Particle size is also known to influence the aging rate, ultrafine sizes showing little aging.

Porosity can be nearly eliminated as long as the volume fraction of glass is greater than 25 percent.[59] The residual porosity behaves like a dispersed phase, reducing the dielectric constant. At very high titanate concentrations the dielectric crazes due to differential thermal expansion between the high-volume fracture of titanate particles and the alumina substrates. The resulting cracks may be considered as a type of laminar porosity which acts as a low series capacitance and reduces the apparent dielectric constant.

With composites based on mixtures of barium strontium and barium lead titanate powder fractions in a glass frit, a dielectric-constant maximum is observed with about 10 percent by volume of glass.[60] Below 10 percent by volume there is a decrease in the dielectric constant, attributed to the porosity resulting because the amount of glass is too small to fill the voids between the titanate particles.

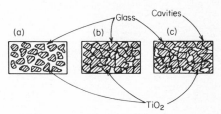

Fig. 14 Microstructure of composite films containing TiO_2 and glass.[57]

Above 10 percent an increase in the volume of glass results in an overall decrease in the dielectric constant of the composite.

Glass-Ceramics

Because the dielectric is deposited and fused as an amorphous material, porosity is expected to be negligible. The glass is crystallized at higher temperatures during the initial or subsequent heating. As a result the grain size and crystal content can be carefully controlled, and the same microstructure is developed as in the bulk components. The microstructure has considerably more effect on the electrical properties of capacitors than on crossovers.

Glass-Ceramic Particulate

The preparation and electrical properties of the ferroelectric particulate glass-ceramic type of capacitor film was described in the section on Barium Titanate Glass-Ceramic Particulate Capacitors. The microstructure consists of essentially a single-phase ferroelectric film based on 1- to 3-μm particles bonded with a crystallized glass made up of crystallites 1 μm and less in size.

CLASSIFICATION OF CAPACITORS

The capacitance ranges of glass and ceramic capacitors are compared with the basic types of capacitors in Fig. 15. The major types of ceramic and glass capacitors[61] are

1. Low-dielectric-constant glass and ceramic

2. Temperature-compensating
3. High-dielectric-constant, or general-purpose
4. Semiconducting
5. Monolithic, which are available in materials 1 to 3

Their advantages, disadvantages, and applications are listed in Table 5.

Low-Dielectric-Constant Capacitors

Low-dielectric-constant capacitors are glass or ceramic capacitors with low capacitances. However, especially with glass, these capacitors have excellent stability with time, temperature, voltage, and frequency.

Temperature-Compensating Capacitors

Temperature-compensating capacitors have TCCs available from positive to negative values and meet the requirements of the low-capacitance range. With these capacitors critical circuits can be compensated for changes in temperature. They are referred to as NPO type, which stands for negative, positive, and zero temperature coefficients of capacitance. The TCCs vary and can be controlled from +120 to

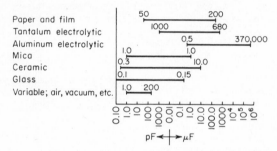

Fig. 15 Capacitance ranges of the major type of capacitors.[61]

−5,600 ppm/°C with dielectric constants of 15 to 95, low loss, and excellent stability.[63] Commercial NPO capacitors are based on magnesia-titania and barium titanate–titania compositions.

The TCC of barium titanate is controlled by several techniques, including the formation of solid solutions based on the added cation to $BaTiO_3$, the controlled additions of secondary phases, and grain-size control. The last has already been discussed. The Curie temperature capacitance peak of $BaTiO_3$ can be flattened by the addition of bismuth stannate, Ta_2O_5, Nb_2O_5, and SnO_2.

High-Dielectric-Constant Capacitors

High-dielectric-constant ceramic capacitors based on barium titanate offer a wide range of dielectric-property flexibility, i.e., dielectric constant, Curie temperature, and dielectric-constant constancy, with ambient temperature. This is accomplished through the control of microstructure, discussed elsewhere, and chemical composition. For example, the Curie temperature of $BaTiO_3$ can be controlled by the formation of solid solutions using shifter and depresser materials, as shown in Figs. 16 and 17. The tailoring of dielectric with formulations using this technique is discussed under Composite Capacitor Compositions. Lead titanate raises the Curie temperature above 120°C; strontium titanate lowers the Curie temperature. Depresser materials lower the dielectric constant. Figure 18 shows the dielectric constant as a function of temperature for several high-K dielectrics based on this approach. Table 6 shows a compilation of typical property data. It should be pointed out that the dielectric strength, voltage, and other properties degrade as the dielectric constant is increased.

TABLE 5 Advantages and Disadvantages of Ceramic Capacitors[61,62]

Capacitor type	Advantages	Disadvantages	Applications and comments
1. Low K glass and ceramics.........	Excellent stability with time, temperature, voltage, and frequency	Low K, low capacitances available	Stability and reliability dictate use
2. Temperature-compensating.........	Linear, variable TCC, low cost, and high stability	Low volumetric efficiency	Temperature-compensating; used to correct circuit changes with temperature, low capacitance needs
3. High K or general-purpose.........	Good volumetric efficiency, tailoring of properties to meet the need, low cost	High K achieved at the expense of stability	General-purpose use and hybrid microcircuitry
4. Semiconducting.........	High capacitance at low cost in a small volume	Low voltage use, low insulation resistance, very voltage-dependent, low reliability when operated at rated voltages	Conventional circuitry: high capacitance in a small size, characteristics similar to electrolytics, has a dielectric film on a semiconducting substrate
5. Monolithic, available in materials 1 to 3	Good volumetric efficiency, wide range of capacitances in small volumes	Stability is limited in the high-K materials	Microcircuitry: chip capacitors to deliver high capacitance in small volumes; usable in film circuits

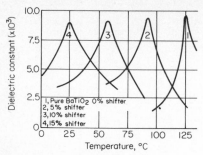

Fig. 16 The effect of shifter materials in high-K ferroelectric ceramics.[64]

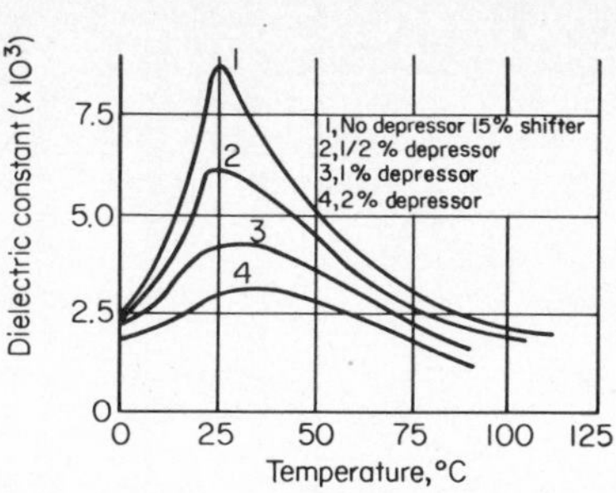

Fig. 17 The effect of depresser materials in high-K ferroelectric ceramics.[64]

Semiconducting Ceramic Capacitors

Semiconducting ceramic capacitors are based on the ability of barium titanate to be reduced to a semiconducting state and reoxidized. By reoxidizing only the surface, a thin layer of high-dielectric-constant material can be formed.[66] The barrier-layer capacitor is formed by reoxidizing the grain surfaces of the reduced $BaTiO_3$.[67] These capacitors have the advantage of high capacitance at low cost in a small volume.

Monolithic Multilayer Capacitors

Monolithic multilayer construction consists of several layers placed on top of each other to build up total capacitance.[68] For this application, the K-1200 material offers the best compromise between electrical performance and size requirements.[69]

Availability of Basic Capacitor Inks

A broad range of thick film capacitor compositions has been reported based on capacitor types 1, 2, and 3. Many are commercially available; others can be formulated and processed as dielectric films without too much difficulty. High-dielectric-constant, high-Q, and temperature-compensating thick film capacitors can be purchased. The composition, fabrication, and performance of these materials in composite, glass-ceramic, and other forms are discussed in the following sections.

DIELECTRIC TESTING

Dielectric systems in the form of parallel-plate capacitors are tested for

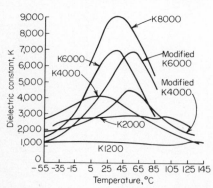

Fig. 18 Curves of dielectric constant versus temperature for high-dielectric-constant titanate dielectrics.[61, 65]

1. Capacitance
2. Dissipation factor
3. Insulation resistance
4. Breakdown voltage
5. Aging
6. Thermal shock
7. Dielectric strength
8. Temperature coefficient of capacitance
9. Frequency characteristics

Capacitance and Dissipation Factor

Dielectric constant The relative-dielectric constant is referred to as the dielectric constant and is dimensionless. Designated as K or $\varepsilon'/\varepsilon_o$, it is the ratio between the charge stored on a geometrical configuration of electroded material, i.e., disk, rectangle, or square, at a given voltage and the charge stored on a set of identical electrodes separated by vacuum. In mks notation, dielectric constant is the ratio between the permittivity ε' of the dielectric in farads per meter and that of free space $\varepsilon_o = 8.85 \times$ pF/m.

Dissipation factor The dissipation factor or dielectric loss D is calculated from the ratio of the imaginary to the real components of the specimen impedance. With alternating voltages the charge stored on a dielectric has both in-phase (real) and out-of-phase (imaginary) components. These are caused by dielectric absorption or resistive leakage. The dissipation factor is frequently referred to as the loss tangent,

TABLE 6 Typical Property Data for High-Dielectric-Constant Ceramic Materials[63]

	Material designation			
	K–0270	K–1200	K–2000	K–6000
Median dielectric constant................	280	1,275	2,150	6,300
Insulation resistance, Ω-cm $\times 10_1^0$.........	4.0	10.0	4.0	1.25
Aging, % per decade.....................	2.0	0.5	3.5	4.0
Voltage coefficient of capacitance, %, at:				
10 V dc/mil..........................	±1	±1.0	±2.0	−25.0
20 V dc/mil..........................	±1	±3.0	−4.0	−55.0
30 V dc/mil..........................	±1	−5.0	−12.0	−80.0
50 V dc/mil..........................	....	−12.0	−17.0	−85.0
Dielectric strength, V/mil, at:				
0.010-in. thickness....................	300	350	200	135
0.020-in. thickness....................	300	300	180	135
0.030-in. thickness....................	....	250	....	135
0.040-in. thickness....................	....	200	....	135
0.050-in. thickness....................	....	175	....	110
Frequency variation of dielectric properties, %, at:				
1 kHz:				
Base capacitance....................	0	0	0	0
Power factor.......................	.3	1.3	1.3	1.3
10 kHz:				
Capacitance change ΔC.............	−2.0	−0.6	−1.0	−3.0
Power factor ΔPF..................	0.4	1.7	1.35	1.2
100 kHz:				
ΔC..................................	−2.0	−2.0	−3.5	−6.0
ΔPF.................................	0.4	2.7	1.4	1.0
1 MHz:				
ΔC..................................	−10.0	−5.0	−4.0	−7.0
ΔPF.................................	.5	5.2	1.5	0.9
10 MHz:				
ΔC..................................	−12.0	−19.0	....	−8.0
ΔPF.................................	0.8	9.2	....	0.9

tan δ. In mks notation, tan δ is equal to $\varepsilon'' / \varepsilon'$, where ε'' and ε' are, respectively, the imaginary and real components of the complex permittivity, $\varepsilon^* = \varepsilon' - i\varepsilon''$.

Low-field dielectric measurements. The low-field dielectric constant K and dissipation factor of thick film dielectrics are their most important properties. With discrete ceramic components these parameters are measured on electroded disks at low fields of 1 V/mm and under. Several highly engineered instruments to measure dielectric constant and dissipation factor are available from General Radio Corporation; Boonton, Wayne and Kerr; and Hewlett-Packard, among others. These include Schering bridges, Q meters, and twin-T circuits.

The usual frequencies of measurement are 1 kHz and 1 MHz. The dielectric constant of good insulators does not vary much from direct current to microwave frequencies. However, in ferroelectrics such as barium titanate or lead zirconate titanate, there is a strong dependence of the dielectric constant on frequency. In ferroelectrics or any imperfect dielectric, interfacial polarization at low frequencies contributes to the effective dielectric constant and dissipation factor. Frequency dependence of these parameters can also arise from dipole relaxation associated with impurities or dipole wall motion in ferroelectrics.

The collection of data on dissipation factor versus frequency and temperature is as important to thick film dielectric research and batch control as dielectric constant versus these parameters. The data can easily be collected since balancing a bridge or tuning a Q meter requires the measurement of both parameters at one time. Measurements are usually made with a fired dielectric layer of 1.5 mils.

Capacitance calculations   Dielectric constant is calculated from the basic equation

$$C = \frac{K\epsilon_0 A}{t}(n-1)$$

where C = capacitance
   ϵ_0 = permittivity of free space
   A = area
   t = dielectric-film thickness
   K = relative dielectric constant
   n = number of capacitive plates or electrodes

To calculate sheet capacitance for a two-plate capacitor,

$$\frac{C}{A} = \frac{K\epsilon_0}{t}$$

where C/A is the capacitance per unit area. This can vary from as low as 1,000 pF/in.^{2} for glasses to 500,000 pF/in.^{2} or more for doped barium titanate. Since capacitance is directly proportional to the electrode area and the dielectric constant of the capacitor and is inversely proportional to the separation distance of the plates, Fig. 19 can be used to relate the K, C/A, and t relationships for designing thick film capacitors. If C, A, and t are in farads, square centimeters and centimeters, respectively, ε_0 is 8.85×10^{-14}; if C is in picofarads (pF) with the other unit in centimeters or inches, respectively, ε_0 is 0.0885 or 0.224.

Other Measurements

Insulation resistance and breakdown voltage[2]   A 100-V dc potential is applied across the capacitor at room temperature. After 1 min, the resistance is measured. The resistance is measured again after the potential is increased to 500 V. If the second resistance reading is less than three orders of magnitude lower than the 100-V value, the capacitor is considered to have passed the voltage-breakdown tests.

Dielectric strength[60]   Samples are placed individually in series with calibrated 1-MΩ resistors. A voltage bias is then applied from a constant high dc voltage supply (± 0.05 percent accuracy). The voltage across the precision resistor is monitored with a high-impedance digital voltmeter readable to 1 μV. This makes it possible to measure the dc resistance from 10^{12} Ω to breakdown.

Aging   The dielectric constant of a ferroelectric ceramic can show a change with time after firing, after any abrupt thermal change, or after the application of electrical signals or mechanical stress. These effects can be observed if the measured values of the dielectric constant are plotted as functions of the logarithm of time elapsed after the stress event. Precautions must be taken to ensure a correct zero point when aging rates are determined from short-time measurements; no errors are usually introduced over long periods of time due to errors or shifts in zero time.

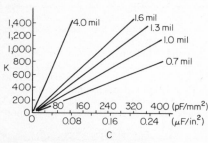

Fig. 19   Thick film capacitors: K, C, and t relationships.[70]

To determine the magnitude of the aging effect,[60] the dielectric may be held at 250°C for 5 min, withdrawn from the heating chamber, and rapidly cooled to room temperature in a desiccator. Measurement is initiated at exactly 1 h after withdrawal and completed after several readings at 1,152 h. Except for taking readings, the units are continuously stored in a desiccator, which must be supplied with fresh dessicant frequently.

Temperature coefficient of capacitance[2] Controlled temperature chambers capable of cycling from −55 to +125 or +150°C are used. Capacitance is measured at the extreme temperatures and compared to room temperature. The following equation is used:

$$\text{TCC} = \frac{(C_{T1} - C_{T2})10^6 \text{ ppm/°C}}{C_{T1}(T_2 - T_1)}$$

Quality dielectric systems yield TCCs of 1 percent or less for 100°C changes.

Thermal shock A common procedure is to place samples in a programmed temperature-control chamber. They are cycled between 0 and 100°C with a 1-min rise time, 2-min soak time, and a 2-min fall time. A 2-min lapse time completes one cycle. Capacitance and dissipation factor drift should be less than ±1 percent.

To test for cracking, the dielectrics are thermally cycled by dipping in a molten-solder bath at 220°C. After being cooled to room temperature, the substrates are dipped into a Dry Ice–acetone bath (−78°C) and permitted to warm to room temperature. After cycling the coating is treated with a dye penetrant to test for cracking.

COMPOSITE CAPACITOR COMPOSITIONS

Titanium Oxide–Glass Formulations

These are class I compositions and consist essentially of TiO_2 and glass.[59] The range of sheet capacitance is from 1,000 to 15,000 pF/in.[2] Dielectric constants up to 80 and dissipation factors of less than 0.5 percent are obtained for these films. The dissipation factor is low because the electric field is concentrated on the very low loss glass phase. For capacitors of 0.001 in. thickness, the limit of the capacitance density is slightly less than 12,000 pF/in.[2]

Trade-offs of dielectric constant versus temperature coefficient of capacitance are possible. The glass phase displays a positive temperature coefficient of the order of 250 ppm/°C; TiO_2 with its higher dielectric constant of 110 has a high negative temperature coefficient of capacitance of approximately −750 ppm/°C. Temperature coefficients of −100 ppm/°C can be obtained if a decrease in dielectric constant of 50 can be tolerated; addition of tin oxide in solid solution to the titania can improve the TCC. Films with dielectric constant of 80 contain sufficient glass to be densified in conventional belt furnaces.

The influence of baking temperature on dielectric permittivity from 850 to 950°C during 15 to 60 min is shown in Fig. 20. The influence of temperature and bake time is insignificant above 870°C. The glass content has not been reported for these curves, but based on other dielectric data reported by the authors of this work it has been assumed that the TiO_2 content is 90 percent by weight. This becomes apparent when the effect of porosity is considered since the porosity of the film depends on the concentration of TiO_2.

As the composition approaches pure TiO_2, very high sintering temperatures are required. Processing rates are tied to very slow diffusion. Controlled processes and detrimental porosity are more difficult to eliminate.

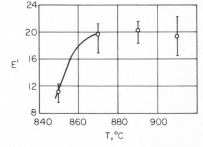

Fig. 20 Influence of the baking temperature on permittivity of TiO_2 glass composite.[57]

The high dielectric strength of the glass also gives rise to breakdown strengths exceeding 500 V.[59] Insulation resistance is a function of the volume fraction of TiO_2; values greater than 10,000 $M\Omega$ are observed for capacitor areas as large as 0.1 in.[2]. Aging is seldom a problem in most applications; however, TiO_2 glass mixtures, though lower in dielectric constant than the ferroelectric titanates, prove more satisfactory when very close tolerances are required for long periods of time.

Barium Strontium Titanate—Barium Lead Titanate—Glass Formulations

High-dielectric-constant capacitors can be based on the blending of various ferroelectric materials having different Curie points and fusing them in a glass matrix.[60] Compacts of mixtures of barium strontium titanate or barium lead titanate are sintered to form solid solutions with a specific Curie point, crushed and comminuted,

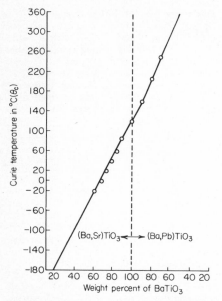

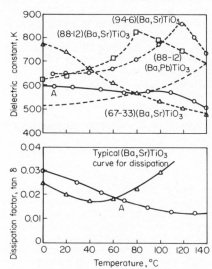

Fig. 21 The Curie point temperature in $BaTiO_3$ as a function of $SrTiO_3$ or $PbTiO_3$ additive.[60]

Fig. 22 Dielectric properties versus temperature, showing the individual components and the mixture of film A.[60]

and screened. Depending upon the dielectric constants and the thermal characterization of the dielectric within the operating range required, composites are prepared consisting of discrete granular particles containing the solid solutions separated from one another by a glass binder such as barium aluminosilicate glass, e.g., Corning 7059. The Curie point temperature shift as a function of additive is shown in Fig. 21.

The dielectric constant and dissipation factors at 1 kHz of sintered binary systems of $BaSrTiO_3$ and $BaPbTiO_3$ bodies, silk-screened capacitors of the individual bodies, and a silk-screened capacitor containing appropriate amounts of the various bodies to give a NPO capacitor film are shown in Fig. 22. The thick film formulation, designated as film A, is shown in Table 7; it will provide a dielectric constant near 600 which does not vary more than 1 percent over the range 0 to 100°C.

The dielectric-constant and dissipation-factor temperature profiles for films B and C are shown in Figs. 23 and 24. These provide thermal coefficients of −3,000 and +3,300 ppm/°C, as tabulated in Table 7. The film C type of dielectric would be suitable for the fabrication of temperature-compensating capacitors.

High-dielectric-constant capacitors can be based on the blending of various ferroelectric materials with different Curie points and fusing them in a glass matrix so that they retain their identities. Two or more ferroelectric materials with different Curie temperatures, e.g., barium titanate, lead titanate, and strontium titanate, are prepared so that the resultant composition is a mixture of discrete granular particles

TABLE 7   Compositions and TCCs of (Ba,Sr)TiO$_2$-(Ba,Pb)TiO$_2$-Glass Formulations[60]

Film	Composition, wt %		Temperature range, °C	Thermal coefficient, ppm per °C
A	94-6 (Ba, Sr)TiO$_2$ 20 88-12 (Ba, Sr)TiO$_2$ 25 67-33 (Ba, Sr)TiO$_2$ 25 88-12 (Ba, Pb)TiO$_2$ 30	93.5	0–100	NPO
	Glass (Corning 7059)	6.5		
B	67-33 (Ba, Sr)TiO$_2$ 25 74-26 (Ba, Sr)TiO$_2$ 25 81-19 (Ba, Sr)TiO$_2$ 25 88-12 (Ba, Sr)TiO$_2$ 25	93.5	25–100	−3,000
	Glass (Corning 7059)	6.5		
C	67-33 (Ba, Sr)TiO$_2$ 12.5 74-26 (Ba, Sr)TiO$_2$ 12.5 81-19 (Ba, Sr)TiO$_2$ 37.5 88-12 (Ba, Sr)TiO$_2$ 37.5	93.5	0–80	+3,300
	Glass (167 JQ)	6.5		

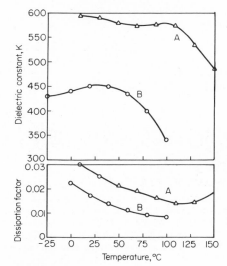
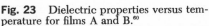
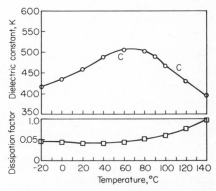

Fig. 23   Dielectric properties versus temperature for films A and B.[60]

Fig. 24   Dielectric properties versus temperature for film C.[60]

separated from one another by a glass binder. In contrast to the ferroelectric–glass-matrix thick film forms, this type of composite mixture is not characterized by a single Curie temperature with a dielectric constant maximum but by a dielectric constant that is approximately proportional to the log sum of the individual material particles. This system has an advantage where there is a requirement for an increased dielectric constant for a particular operating-temperature range. An appropriate amount of ferroelectric material having a maximum in dielectric constant in that range can be added.

Barium strontium titanate and barium strontium titanate mixtures can be prepared with dielectric constants of 500 and above. By varying the number present and the weight ratios of the sintered binary ferroelectric bodies within the composite, the thermal characteristics within a variety of desired operating temperature ranges can be controlled to give films with negative-positive-zero (NPO), positive, or negative thermal coefficients of capacitance.

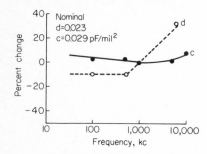

Fig. 25 Capacitance and dissipation versus frequency of film D.[1]

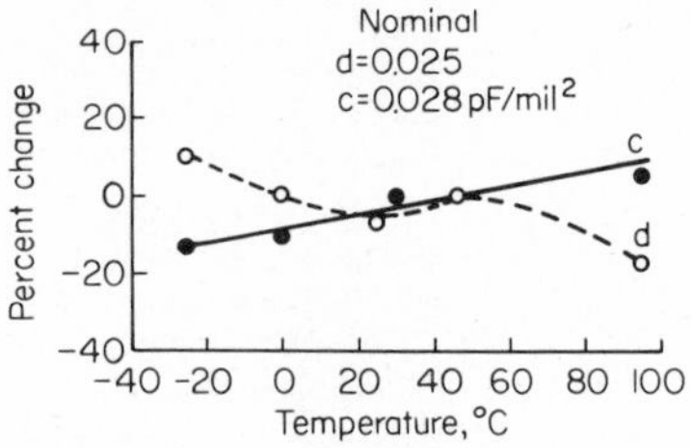

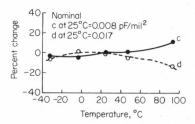

Fig. 26 Capacitance and dissipation versus temperature of film D.[1]

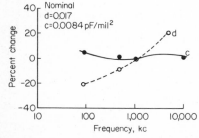

Fig. 27 Percent change in capacitance and dissipation versus frequency for film E.[1]

Fig. 28 Percent change in capacitance and dissipation versus temperature for film E.[1]

Lead Zirconate Titanate–Barium Titanate–Lead Monosilicate Glass Formulations

Dielectric films with capacitances in the range of 0.009 to 0.25 pF per 0.001-in. square and the temperature and frequency characteristics shown in Figs. 25 to 28 can be obtained based on mixtures of lead zirconate titanate with barium titanate in a lead monosilicate glass.[1] The formulations are given in Table 8. These formulations were specifically designed for 96 percent Al_2O_3 substrates and the Du Pont gold-platinum conductive film 7553.

Barium Titanate–Cadmium Borosilicate Glass

A series of high-dielectric-constant films have been reported consisting of 90 percent K9000 $BaTiO_3$ ceramic and 10 percent $CdO-B_2O_3-SiO_2$ glass.[71] Most of the data reported for this system were based on measurements made on disks consisting of 10 to 50 percent by weight of glass, the remainder being K9000 $BaTiO_3$. The effect of maximum firing temperature on the disk permittivity is shown in Fig. 29 as measured at 1 kHz and 3 V ac. Table 9 compares the dielectric constant of disks

TABLE 8   Batch Weights Lead Zirconate–Barium Titanate–Lead Monosilicate Glass Formulations[1]

Material	Formula	Weight %
	Dielectric film D	
Lead monosilicate..............	$PbO \cdot 0.67SiO_2$	65.0
Barium titanate................	$BAO \cdot TiO_2$	22.5
Lead zirconate titanate..........	$0.9281PbO \quad 0.4737TiO_2$	10.0
	$0.0719SrO \quad 0.5349ZrO_2$	
Calcium stannate...............	$1.000 \quad CaO \cdot SnO_2$	2.5
		———
		100.0
	Dielectric film E	
Lead monosilicate..............	$PbO \cdot 0.67SiO_2$	75.0
Lead zirconate titanate..........	$0.9281PbO \quad 0.4737TiO_2$	25.0
	$0.0719SrO \quad 0.5349ZrO_2$	
	1.000	———
		100.0

with screened-on capacitors fired under identical conditions. Typical firing profiles for the screened-on capacitors are shown in Fig. 30.

The temperature coefficients of the K9000-glass screened capacitors are shown in Fig. 31 for compositions containing 10 to 50 percent glass and for the $BaTiO_3$ ceramic. Glasses containing 40, 50, and 60 percent by weight of glass showed flat temperature curves. The 10 percent glass composition had a temperature coefficient of -25 to $+10$ percent from -55 to $+125°C$. The power factors increased from a low of 0.1 percent for 60 percent glass to a high of 1.9 percent for a 10 percent glass composition. The materials show good insulation resistance, most samples ranging from 10^9 to 10^{12} Ω at 200 V dc. Samples with 10 and 50 percent glass were life-tested at 20,000 V dc/cm and 85°C for 1,000 h. The capacitance of the 50 percent glass sample did not change, but the 10 percent glass samples changed 3 percent per decade of time.

Barium Titanate–Bismuth Oxide

Film formulations of barium titanate and solid solutions of barium titanate and barium stannate using bismuth oxide as the binder instead of glass have the properties shown in Table 10 and Figs. 32 to 35. The binder is varied from 8 to 50 percent of the total volume of voids in the ink.

When Bi_2O_3 is substituted for glass such as lead zirconium silicate as the binder, higher permittivities are obtained. This has been attributed to the high dielectric constant of Bi_2O_3, $\varepsilon = 40$, and its ability to react with $BaTio_3$ to form a

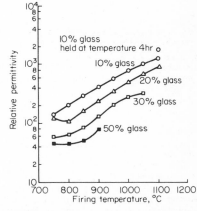

Fig. 29 Effect of maximum firing temperature on the relative permittivity of $CdO\text{-}B_2O_3\text{-}SiO_2$ glass–K9000 disks.[71]

high-dielectric-constant compound such as $Ba_2Bi_4Ti_5O_{18}$. Bismuth oxide also enhances sintering. This probably is due to the ability of Bi_2O_3 to form glasses on dissolving small amounts of other oxides. Figure 28 shows the percentage change of capacitance with firing temperature for pure $BaTiO_3$ ceramic particles bound with 10 volume percent of Bi_2O_3. The variation is smooth; the firing temperature can be used to control the

TABLE 9 Comparison of Dielectric Constant of CdO-B₂O₃-SiO₂ Glass–K9000 Disks with Screened-on Capacitors under Identical Conditions at 950°C[71]

	K	
Glass, %	Screened-on capacitors (five pieces)	Disks (five disks)
10	$331 \pm 5\%$	$560 \pm 1\%$
20	$124 \pm 5\%$	$345 \pm 1\%$

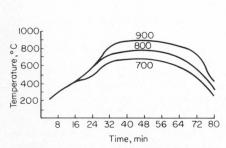

Fig. 30 Typical firing profiles for the CdO-B₂O₃-SiO₂ glass–K9000 capacitors.[71]

Fig. 31 Temperature coefficients of the CdO-B₂O₃-SiO₂ glass–K9000 screened-on capacitors.[71]

TABLE 10 Performance of BaTiO₃-Bi₂O₃ Capacitors[72]

Capacitance, pF/in.²	10,000–70,000
tan δ after aging, %	1–2
Temperature variation of capacitance (20–90°C)	Maximum deviation 5%
Stability	Typically −1% change in 500 h at 100°C off load (after initial settling period of 24 h after manufacture)
Reproducibility	Coefficients of variation, % 1.7–5.6

final capacitance value. The capacitors have stability problems since aging occurs. After manufacturing, the dielectric loss of the capacitor is high, being about 5 to 25 percent; it decreases rapidly to 2 percent or less. The loss factors obtained after the initial rapid aging are similar to those occurring in the dielectric raw materials. As with the loss factor, the major changes in capacitance occur within the first 24 h after manufacture.

The temperature variations of capacitance and dielectric loss for compositions containing 0, 8.6, and 15 volume percent of binders are plotted in Figs. 33 and 34, respectively. Figure 35 shows the variations with frequency for a formulation containing 15 volume percent Bi_2O_3. Of importance here is that the reproducibilities are similar to those for pure-glass dielectric capacitors. In the latter the dielectric remains inert, while the Bi_2O_3-$BaTiO_3$ and Bi_2O_3-$BaTiO_3$-$BaSnO_3$ solid-solution series chemical reactions were taking place between the constituents of the dielectric. This has been attributed to the high dielectric constant of Bi_2O_3, which is 40, and the ability of Bi_2O_3 to react with $BaTiO_3$ to form higher-dielectric-constant compounds

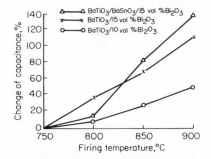

Fig. 32 Variation of capacitance with firing temperature for $BaTiO_3$-Bi_2O_3 compositions.[72]

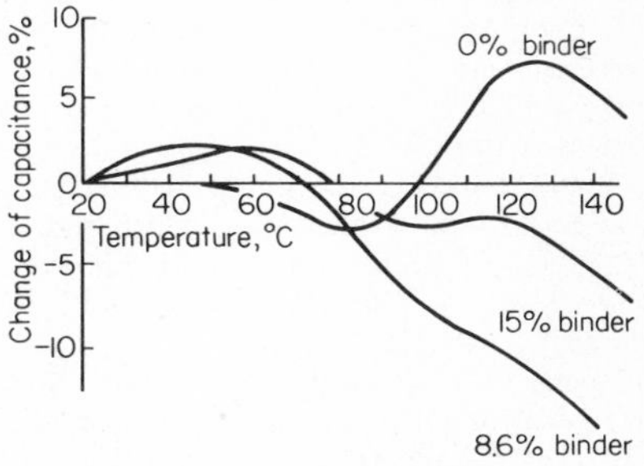

Fig. 33 Variation of capacitance with temperature as a function of Bi_2O_3 binder for $BaTiO_3$-Bi_2O_3 compositions.[72]

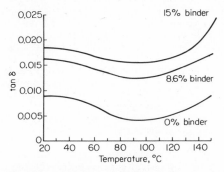

Fig. 34 Variation of dielectric loss as a function of Bi_2O_3 binder content for $BaTiO_3$-Bi_2O_3 compositions.[72]

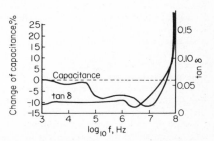

Fig. 35 Variation of capacitance and tan δ with frequency for a capacitor containing 15 vol percent Bi_2O_3.[72]

such as $Ba_2Bi_4Ti_5O_{18}$. Bismuth oxide also enhances sintering. This is probably due to the ability of Bi_2O_3 to form glasses on dissolving small amounts of other oxides. Bismuth oxide itself does not form a glass but will do so when melted with a suitable quantity of a second oxide or mixture of oxides. Up to 300-g glass batches have been reported in the borate and silicate systems, containing up to 67 and 50 mol percent of Bi_2O_3, respectively.

Other Compositions

Zinc oxide in the form of a parallel-plate capacitor was reported to be suitable for incorporation in high-speed microelectronic switching circuits.[73] Devices with apparent low-frequency dielectric constants of 3,000 and high-frequency dielectric constants of 1,700 were achieved. Zinc oxide–bismuth oxide low-Q decoupling capacitors with silk-screened film capacitance values up to 100 pF/in.[2] were reported by the same investigators.[74] However, this was with a very low (10 Ω-cm) dc resistivity. A ceramic component composed of lead zirconate–lead titanate in a silicate glass was reported to have use as a low-value dielectric ($K = 10$ to 55 with $\tan \delta = 0.0012$ to 0.0230 at 1 kHz).[75]

GLASS–CERAMIC CAPACITORS

$BaTiO_3$-$BaAl_2Si_2O_8$ Type

Concept   A family of ferroelectric glass-ceramic capacitors based on the glass-ceramic approach of Herczog[7] has been developed,[76] based on titanates and niobates, according to the dielectric-constant value required. Specific glass compositions have not been reported, although the films are based on the compositions of Herczog,[7] Herczog and Stookey,[77] Layton and Herczog,[78] and Pratt and Tarcza.[79] The thermal history of this type of material is discussed under Differential Thermal Analysis for Process Control.

Processing   The glass powders are prepared by batch melting, rapid quenching, and milling to a maximum grain size of 10 μm. They are dispersed in an organic squeegee medium, and the melting mixture is adjusted to a viscosity suitable for screening. A commercially available gold, such as Du Pont 8067, is used for the electrode material.

Successive applications are made to the substrate in the following sequence:
1. A layer of crystallizable glass such as the dielectric material
2. Bottom electrode
3. Dielectric material
4. Bottom electrode

After screening, each layer is air-dried on a heated, continuous belt attached to the screening machine. The dielectric material consists of two successfully screened layers to maintain thickness control and ensure films free from pinholes. After deposition of the top electrode the assembly is air-fired according to a schedule which peaks for several minutes at 900 to 1000°C to fuse the capacitor structure to the substrate. Volatilization of the organic screening medium takes place during the air firing.

A top layer of crystallizable glass, which is identical to the first layer deposited, is then applied. After the deposition of a final glaze of alkali-free glass, the assembly is refired on the same schedule with a longer soak at the peak to complete crystallization of the dielectric and fuse the cover glaze into a hermetic film.

The capacitor firings are performed in a continuous furnace with a stainless-steel belt. By adjusting the speed of the belt, the same furnace can be used for either of the two firing schedules. By controlling the extent of crystallization of the glass, the capacitance value can be varied by ±15 percent by adjusting the final firing cycle. This range of control is adequate to compensate for variations in dielectric thickness and electrode dimensions. It allows repeatable manufacturing of a capacitor with a ±10 percent tolerance.

Dielectric control and performance   The capacitance of the finished capacitor is determined by dielectric thickness, dielectric constant, and area between the electrodes. A primary control of dielectric constant is variation of the ratio of crystallizable to noncrystallizable constituents.

By varying the ratio of ferroelectric to network former in a single system or by changing systems (titanates or niobates) the number of process steps can be minimized while producing capacitors of several dielectric-constant values during a single firing. In this way simultaneous crystallization from the glassy state of several different compositions which have been separately screened in the required design dimensions produces dielectric constants of 400 to 1,200. To obtain dielectric constants between 20 and 400, the glass-ceramic powder can be blended with a stable, alkali-free glass and processed exactly with 100 percent glass-ceramic. The film of desired dielectric constant can be prepared by the ratio of glass to glass-ceramic mixture, as shown in Fig. 36.

The effect of temperature, applied dc voltage, time, and frequency in glass-ceramic films with a permittivity of 600 is shown in Figs. 37 to 40. The capacitor has areas of 0.002 to 0.005 in.[2] and thicknesses of 1.5 to 1.7 mils. The average dielectric breakdown is 1,200 V dc. The dissipation factor is typically 0.9 to 1.2 percent with maximum values of 1.9 percent at 1 kHz and 2 V. The insulation resistance is 10 in. or higher at 100 V dc.

BaTiO₃-Pb₂Bi₄Ti₅O₁₈ Type

The formation of a proprietary screen-printed capacitor yielding dielectric constants of 400 to 800 is based on the reaction between the paste and the electrode binder constituents.[80] The dielectric composition is said to consist of a large quantity of a high-dielectric-constant material, an inorganic reactive phase, and an organic vehicle. The fired dielectric is thought to be a solid solution formed by the reaction of the reactive phase and the electrode binder with the high-dielectric-constant phase. The phases $BaTiO_3$ and $Pb_2 Bi_4T_{51}O_{18}$ are reported.

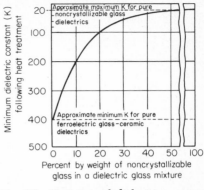

Fig. 36 Variation of dielectric constant in glass and glass-ceramic mixtures.[76]

A bottom platinum-gold electrode is fired at 760°C for 10 min. The dielectric paste is double-printed to reduce the chance of pinholes and short circuits. The dielectric thickness is controlled by the mesh of the screen. Two 200-mesh prints give about 2.2 mils of fired thickness,

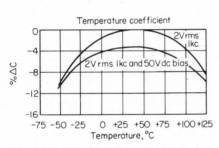

Fig. 37 Temperature coefficient of $BaTiO_3$-$BaAl_2Si_2O_8$ glass-ceramic films.[76]

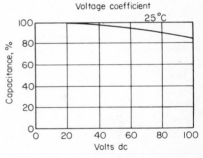

Fig. 38 Effect of applied dc voltage on $BaTiO_3$-$BaAl_2Si_2O_8$ glass-ceramic films.[76]

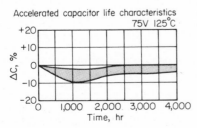

Fig. 39 Life-test results of $BaTiO_3$-$BaAl_2Si_2O_8$ glass-ceramic films.[76]

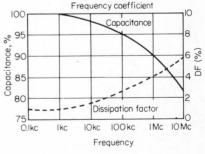

Fig. 40 Effect of frequency on capacitance and dissipation factor of $BaTiO_3$-$BaAl_2Si_2O_8$ glass-ceramic films.[76]

whereas one 80-mesh print gives about 2.0 mils. The films are fired in the range of 750 to 1050°C.

Figures 41 and 42 show the capacitance density and dielectric constant versus firing temperature. The presence of several ferroelectric phases is evidenced in Fig. 43. The capacitance and dissipation factor stay uniform with increasing frequency up to 10 MHz, where the dissipation factor starts to rise sharply. The dielectric thickness must be greater than 15 mils for maximum life even with moderate loads.

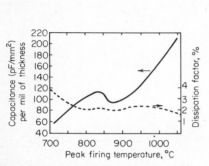

Fig. 41 Capacitance density versus firing temperature of BaTiO₃-Pb₂Bi₄Ti₅O₁₈ glass-ceramic films.[80]

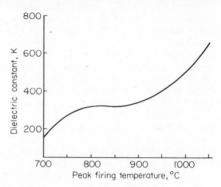

Fig. 42 Dielectric constant versus firing temperature of BaTiO₃-Bp₂Bi₄Ti₅O₁₈ glass-ceramic films.[80]

A 50-mm² capacitor generally measures in excess of 10^{12} Ω at 10 V. Under load thin dielectrics show a decrease to below 10^9 Ω at 500 to 700 h.

A reported screen-printed dielectric capacitor yielding dielectric constants of 400 to 800 is thought to be a crystalline solid solution formed by the reaction of the electrode binder constituents and a reactive phase with the high-dielectric-constant crystalline components during firing.

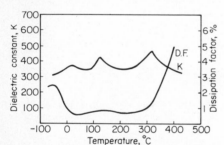

Fig. 43 Dielectric constant and dissipation factor versus temperature for BaTiO₃-Pb₂Bi₄Ti₅O₁₈.[80]

BARIUM TITANATE GLASS-CERAMIC PARTICULATE CAPACITORS

Concept

For these materials, high-dielectric barium titanate particles are mixed in a crystallizable barium titanate glass.[8, 81, 83] The final dielectric material consists of barium titanate particles bound by microcrystalline barium titanate which has recrystallized out of the glass. Therefore, the total barium content is 90 to 95 percent by weight.

Processing

Compositional ratios for two reported mixtures are listed in Table 11, which also lists the constituents of the barium titanate glass. Reagent-grade chemicals are dry-blended for 1 h and ball-milled in distilled water. The water is removed by filtering; the material is dried under an infrared lamp. After calcining in air at 1000° C for 1 h and comminuting, the composition is melted at 1450 to 1600°C and quenched to a glass frit in distilled water. The frit is milled and sifted through a 325-mesh screen.

Barium titanate power is blended with the frit, and the total composition is suspended in a binder and lubricant for screening. The binder-lubricant formulation is

200 ml of butyl Cellosolve acetate to 21.5 g of ethyl cellulose. The paste formulation consists of 100 g of the glass-frit–microcrystal mixture and 50 ml of the binder-lubricant. The suspension is blended in an automatic mortar and pestle for 2 h to form a paste suitable for screen depositing the dielectric film in place.

TABLE 11 Compositional Ratios of Barium Titanate Glass-Ceramic Particulate Mixtures and Composition of Barium Titanate Glass[81, 82]

Dielectric thick films			
	Weight %		
Component	Mixture A	Mixture B	Mixture C
Barium titanate glass...............	20	80	10
Barium titanate particulates.........	80	20	90

Barium titanate glass	
Constituent	Amount, weight %
BaO......................	54.7
BaF$_2$......................	3.2
TiO$_2$......................	24.0
Al$_2$O$_3$......................	7.9
GeO$_2$......................	2.0
SiO$_2$......................	8.2

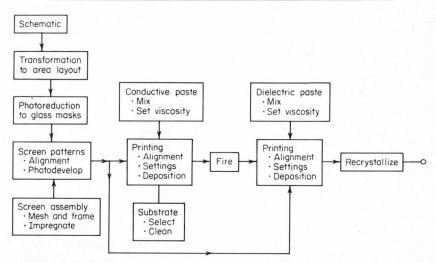

Fig. 44 Block diagram of screen circuit process for barium titanate glass-ceramic particulate films. [81–83]

Figure 44 is a block diagram showing the integration of the dielectric paste into a thick film capacitor. Conventional bottom electrodes are screened on alumina substrates. Dielectric films are screened on the bottom electrodes. The screen-deposited film is outgassed by drying in vacuum, heating at 100°C, and soaking at 425°C for

10 min to complete binder burnout. The dried mixture is heated at a rate from 5 to 20°C/min to a temperature in the range between 700°C, or the softening-point endotherm preceding crystallization, as indicated by differential thermal analysis (DTA), and 1300°C, or the first melting endotherm following the final recrystallization exotherm, as indicated by DTA. The DTA thermograms of typical pastes of this type are discussed elsewhere. The firing temperature depends on the weight ratios of the glass-frit-barium titanate powder mixture. The films are soaked at the maximum temperature for 10 to 15 min to complete recrystallization. The film dielectric is allowed to cool to room temperature with furnace shut down. The fired capacitor showing the top and bottom electrodes is illustrated in Fig. 45.

The capacitor structure can be subjected to a second heat treatment to complete the crystallization reactions. Depending on composition, these can be up to 1000 to 1050°C. Double screening of two layers is common.

|←————— l/2 in. —————→|

Fig. 45 Fired barium titanate glass-ceramic particulate film showing top and bottom electrodes.[81-83]

Dielectric Properties

The fired film capacitors have dielectric constants of 300 to 1,800 at 10^2 to 10^7 Hz and 25°C, depending upon the compositional ratio of the mixture and the firing schedule. Dissipation factors at 10^3 Hz and 25°C are in the range of 0.75 to 1.25 percent.

The dielectric properties of three thick films, A, B, and C, containing, respectively, 20, 80, and 90 percent of barium titanate particulates, are listed in Table 12. These properties are typical of all capacitors with 10 to 90 parts of barium titanate glass frit and 90 to 10 parts barium titanate powders.

The temperature and frequency dependence of the dielectric constant and dissipation factor of these films are shown in Figs. 46 to 49. The graphs of dielectric

TABLE 12   Dielectric Properties of Barium Titanate Glass-Ceramic Particulate Films at 10 kHz and 25°C[81, 82]

Film	Dielectric constant	Capacitance density, pF/cm^2	Loss factor, %
A	325 ± 10	3,850	0.75
B	615 ± 25	12,000	1.2
C	1,800	55,000	1.5

constant versus temperature in Figs. 46 and 47 show their temperature coefficients of capacitance are positive to 140°C. Frequency dependence of the dielectric constant is shown in Fig. 48. The dissipation factors are plotted as a function of temperature in Fig. 49. The dc breakdown voltage was measured at 400 V/mil.

The temperature profiles of the dielectric constant are characteristic of an ultrafine-grained microstructure. This would be expected with films, such as B, which have a continuous recrystallized microcrystalline barium titanate filler. As discussed elsewhere, these profiles would also be expected with films consisting predominantly of 1- to 3-μm barium titanate particulates.

The temperature profiles of the dissipation factors of films, such as B, which contain recrystallized barium titanate glass as the major phase behave like the pure recrystallized barium titanate glass without the barium titanate particulates. The profiles of films, such as A and C, which contain a major content of barium titanate particulates are similar to those of pure sintered barium titanate dielectrics as reported in the literature.

An indication of the ferroelectricity of these films below the Curie temperature is provided by the 60-Hz hysteresis loops of the form shown in Fig. 50. Their spontaneous polarization as a function of temperature is plotted in Fig. 51 and compared

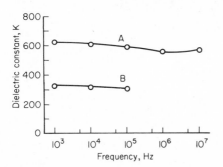

Fig. 46 Temperature dependence of the dielectric constant of films A and B at 10 kHz.[81–83]

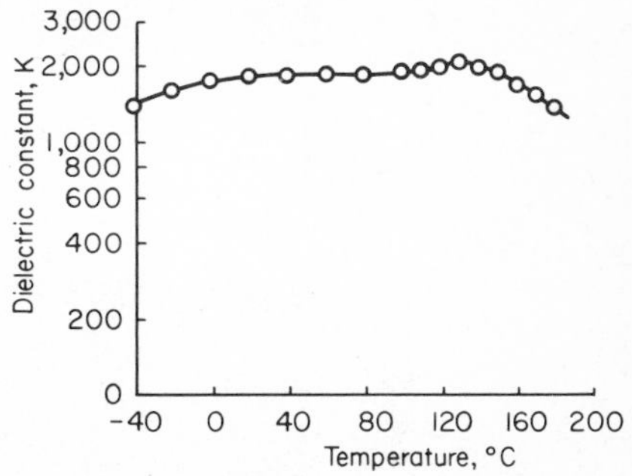

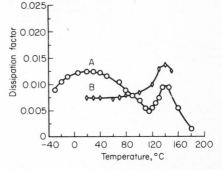

Fig. 47 Temperature dependence of the dielectric constant of film C at 10 kHz.[8]

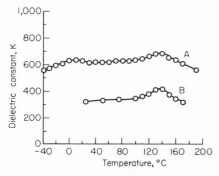

Fig. 48 Frequency dependence of the dielectric constant of films A and B at 25°C.[81–83]

Fig. 49 Temperature dependence of the dissipation factors of films A and B at 10 kHz.[81–83]

with that of a single crystal of barium titanate. For the single crystal, the loop reduces to a straight line above the Curie temperature of 120°C, indicating that the crystal is depolarized at that point. However, in thick films, spontaneous polarization, possibly arising from an internal bias or microstructural strain, is observed to nearly 200°C. This is shown for film A, with 80 percent of barium titanate particulates. Film B, with 20 percent of barium titanate particulates, is not shown. It follows the same profile as A but at half the magnitude of spontaneous polarization.

CROSSOVER DIELECTRICS

Definition and Requirements

A crossover dielectric is a low-dielectric-constant insulator capable of separating two conductor patterns through several firing steps. High-quality crossover materials

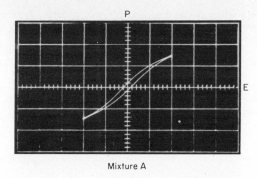

Mixture A

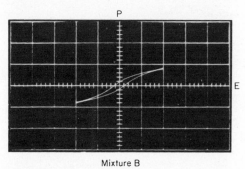

Mixture B

Fig. 50 Oscillograms (60 Hz) of polarization P versus applied field E for thick films of mixtures A and B at 25°C.[81–83]

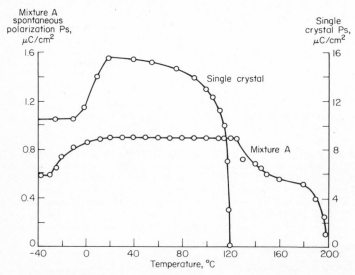

Fig. 51 Comparison of the temperature dependence of the spontaneous polarization of a single barium titanate crystal and film A.[81–83]

have been developed which give good isolation between top and bottom conductors and minimum stray capacitance.

The requirements for a crossover dielectric are

1. Control of resoftening or thermoplasticity in the top-conductor firing step
2. Low dielectric constant to prevent ac capacitance coupling between insulator circuits
3. Low electric loss or high Q values to avoid dielectric heating.
4. Minimum tendency to form pinholes
5. A low tendency to evolve gases during firing
6. High resistance to thermal-shock crazing
7. Low sensitivity to water vapor and subsequent electrical losses

DC and AC Effects

The figures of merit for the crossover are the dc leakage resistance and the dielectric constant. If the isolation is not complete, two kinds of interaction between conductors above and below the crossover can occur, the dc and the ac effect.

Dc effect[84]   If the insulation or leakage resistance of the crossover is not high, the top and bottom circuits will be connected through the leakage resistance of the crossover. As a result, the dc bias voltages for the circuits will be changed. The higher the dc leakage resistance the better the dc isolation.

Ac effect[85]   If the dielectric constant of the crossover is too high, the crossover acts as a coupling condenser between the circuits and a part of the ac signal in the conductor above the crossover goes into the conductor below it. Thus the lower the dielectric constant the better the ac isolation.

Ambient Effects on Capacitance

The dielectric constants of the crossovers may be listed as covering a range since errors in the capacitance measurements are introduced from polarization due to water absorbed by the dielectric or stray capacitance (since the crossover covers a small area).

Stability of the dielectric is of concern at ambient conditions since the electrical properties are affected by the relative humidity of the ambient. There are two effects, bulk and surface. The bulk effect arises from water molecules penetrating into the dielectric, decreasing the leakage resistance, and increasing the dielectric polarizability. The results are a decrease of leakage resistance, an increase in capacitance, and an increase of dissipation factor. The surface effect is due to adsorbed water. Protons and hydroxyl ions migrate on the surface of the dielectric, the results being the same as those of the bulk effect.

Equivalent Circuits[86]

The surface and bulk effects have been characterized by the equivalent circuits in Fig. 52. The surface and bulk leakage resistance, denoted respectively as R_1 and R_2, decrease as the ambient humidity is raised. The capacitance increases as the humidity is raised. The patterns employed to test the effect of humidity on R_1 and R_2 independently are shown in Figs. 53 and 54. To evaluate R_2 small-area electrodes were applied to both sides of a large area of dielectric; two closely spaced parallel conductors were printed on the crossover to evaluate R_1.

Advantages[87]

With the development of dielectric crossover compositions and processing compatibility with fired conductors, three-dimensional hybrid circuits have become possible. The crossover structural combinations are either

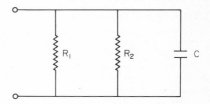

R_1: Surface leakage resistance
R_2: Bulk leakage resistance
C: Capacitance

Fig. 52 Equivalent circuit of crossover capacitor.[86]

conductor-dielectric-conductor or conductor-dielectric-resistor. The advantages of screened crossover conductor networks are
1. Low production costs
2. High reliability
3. Generally less investment than with vacuum evaporation
4. Adaptability to automatic processing
5. Increased packing density because of layered designs and use of the backside of the substrate for circuitry
6. Increased design flexibility and the ability to make complex circuits previously impractical
7. Flexibility in bonding techniques

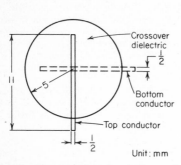

Fig. 53 Test pattern for bulk conduction in crossover capacitor.[86]

Fig. 54 Test pattern for surface conduction in crossover capacitor.[86]

Design Requirements[87]

The design parameters for crossover-conductor combinations are as follows:

Parameter	Design requirement
Capacitance, pF	0.5–5 depending on circuit
Resistance, crossover, Ω	$>10^{10}$
Conductor	0.1–1
Leakage at 40 V, nA	<0.1
Adhesion	>5 lb in 45° shear, peel
Thermal cycling, C	>20 -40 to $+85$°C
Acid resistance, S	±20 in HF·HNO$_3$·2H$_2$O
Voltage-temperature-humidity life test (crossover), h	1,000
Life test (Pt-Au pads), h	1,500, 2% resistance change

Materials and Processes

Ceramics, glasses, and glass-ceramics have been used as crossover dielectrics. Until recently glasses have been preferred since ceramic dielectrics cannot be fired in sequence or at a high enough temperature to densify the ceramics and make them nonporous; if porous, they are moisture-sensitive, and this leads to low insulation resistance between conductor lines. These problems have been eliminated with the development of glass-ceramics, which are also referred to as devitrifiable, crystallizable, or recrystallized glasses. They are superior to glass in multilayered structures since they do not reflow when heated to their original firing temperatures and have superior dimensional control.

Glass Crossovers

Problems Glass-crossover coatings permit many conductive crossovers on a single substrate and have been used in actual multilayer constructions. However, the glass dielectrics have shown some problems in application. Three firings have been required for the bottom conductor, the glass, and the top conductor. Each must be fired at a temperature which is 50 to 100°C less than the temperature of the preceding layer. If the soak temperature or firing schedule of the glass and top conductor is too long, excessive flow of the glass can occur, which distorts the geometry of the conductor lines and causes partial loss of registration. Further, diffusion may take place through the glass if the glass or top conductor is overfired. This causes shorts between the conductor layers. To minimize the formation of pin holes, the glass layer is usually double-printed. A thicker dielectric layer results, which reduces circuit-distributed capacitance.

TABLE 13   Chemical Composition of Glass Crossover Dielectric[1]

Material	Formula	Weight %
Calcia	CaO	8.0
Lead oxide	PbO	17.2
Magnesia	MgO	0.6
Soda	Na_2O	2.4
Potash	K_2O	1.7
Alumina	Al_2O_3	9.1
Boric oxide	B_2O_3	4.5
Silica	SiO_2	56.5
		100.0

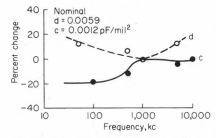

Fig. 55 Frequency profiles of capacitance and dissipation of lead aluminum calcium borosilicate glass crossover.[1]

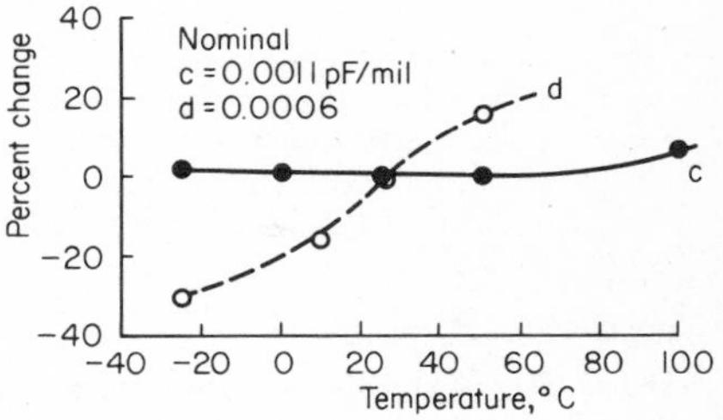

Fig. 56 Temperature profiles of capacitance and dissipation of lead aluminum calcium borosilicate glass crossover.[1]

Multilayer circuits have been constructed with considerable success using glass. However, the costs have been high since three to four separate printing and firing operations at different temperatures are required. Low yields have also presented a problem.

Lead aluminum calcium borosilicate glasses[1]   A frit which is basically a lead oxide–aluminum oxide–calcium oxide borosilicate glass (Ferro Corporation 3467) has been reported to perform in a highly reliable manner when screened into a film. Its composition (Table 13) is suitable for screening and firing on 96 percent alumina substrates with gold-platinum conductor. The film is vitrified at 890°C for 10 min.

The resulting film is clear in appearance with a linear coefficient of thermal expansion at room temperature of 7.1×10^{-6} cm/(cm)(°C) which matches that of Al_2O_3. The top conductor is processed at a peak temperature less than the vitrification temperature of the dielectric. The performance of the crossover film as a function of frequency and temperature is shown in Figs. 55 and 56. The stray capacitance between conductor is less than 0.000 pF per 0.001-in. square.

Lead borosilicate glass.[2][86]   The performance data of a crossover paste consisting of a lead borosilicate glass powder mixed with an organic vehicle are as follows:

Dielectric constant at 1 kHz...............	6–9
Dissipation factor at 1 kHz, %.............	0.5–1
Insulation resistance, Ω...................	4×10^{13} at 300 V dc (20 by 20 by 0.6 mil thick sample)
Frequency dependence of dielectric constant and dissipation factor....................	No appreciable change up to at least 1 MHz
Temperature dependence of dissipation factor.	A slowly rising function of temperature (at 70°C, DF = 1.0%)

The printed crossover is fired at a peak belt temperature of 870°C with a 6-min soak.   The top conductor is fired and printed at 760°C peak temperature.

The test patterns shown in Figs. 53 and 54 were used to simulate the effects of surface and bulk conduction on actual printed circuits.   To evaluate surface conduction, gap widths of 0.5 and 1.0 mm between the two electrodes were used, both gaps being 3 mm long.   The same electrode pattern was printed on the substrate as a reference, the reference gaps representing the surface-conduction part on the substrate.   The change of surface electrical properties to include capacitance and dissipation factor at 1 kHz and the surface leakage resistance at 20 V dc were measured for a wide range of relative humidities.   The results are shown in Table 14 and may be summarized as follows: (1) there was no significant effect on humidity up to 50 percent relative humidity; (2) capacitance, dissipation factor, and surface conduction increased as the relative humidity was raised; (3) the isolation'difference between the conductor lines on the bar substrate and crossover disappeared at 65 percent relative humidity.

The bulk conduction was measured using the test pattern shown in Fig. 53.   The electrical properties at various relative humidities are listed in Table 15.   The bulk conduction is increased drastically above 60 percent relative humidity and constant and low below 50 percent relative humidity.   At high dc voltage and fixed-relative-humidity bulk conduction is closely related to breakdown voltage.   The change of R_2, the bulk leakage resistance, at various applied voltages is shown in Fig. 57.   From the high-insulation resist and at 300 V, it can be concluded that dc isolation between the top and bottom conductor lines (20-mil width is excellent at 47 percent relative humidity.

Glass-Ceramic Crossovers

$BaO \cdot TiO_2 \cdot B_2O_3 \cdot P_2O_5 \cdot SiO_2$ Formulations[55][88]   The use of glass-ceramics for crossover dielectrics rather than glass was introduced in 1966.   A composition suitable for crossover applications is shown in Table 16.   Based on the BPO_4-SiO_2 eutectic, compositions containing up to 87 percent weight of $BaTiO_3$ can be formed into a glass, screened, and crystallized at low processing temperatures into a glass-ceramic crossover.   The optimum composition is shown in Table 16, and the range of compositions in Table 17.   Compositions of chemically pure oxides are poured at 1100 to 1550°C into preheated carbon molds or a stainless-steel quenching press and quenched into a glass or fritted in a suitable quenching media.   The glass is annealed at 550°C for 30 min after quenching to relieve stress and cooled with the furnace to ambient.   Prereaction of the constituents at 1150°C before melting and quenching may be desirable to complete all extraneous reactions.

The processing cycle of the glass is based on the differential thermal analysis (DTA) thermogram, discussed separately because of the importance of DTA as a process control tool.

Recrystallization at 740 to 830°C for soak periods of 1 h or less results in a low-loss dielectric.   The dielectric properties are plotted in Figs. 58 and 59.   The variations of the dissipation factor and dielectric constant with temperatures of the glass and recrystallized glass are shown in Fig. 58 at 10 kHz.   The dielectric constant of the glass is 14.5 and remains relatively unchanged over the temperature range from 26 to 160°C.   Recrystallization increases the dielectric constant to 22; there is a small

TABLE 14 Change of Surface Electrical Properties of Glass Crossover Dielectric due to Relative Humidity at 32.2°C[86]

Gap, mm	31% relative humidity			47% relative humidity			65% relative humidity			87% relative humidity		
	C, pF, at 1 kHz	DF, %, at 1 kHz	R_1, GΩ, at 20 V dc	C, pF, at 1 kHz	DF, %, at 1 kHz	R_1, GΩ, at 20 V dc	C, pF, at 1 kHz	DF, %, at 1 kHz	R_1, GΩ, at 20 V dc	C, pF, at 1 kHz	DF, %, at 1 kHz	R_1, GΩ, at 20 V dc
0.5	1.08	0.45	180	1.07	0.87	55	1.12	0.74	40	1.54	35.5	15
0.5*	...	0.45	140	...	0.50	120	...	2.0	62	...	...	1
1.0	1.13	0.33	160	1.13	0.25	...	1.11	0.87	80	1.50	12.5	30
1.0*	...	0.45	2,000	...	0.62	1,500	...	1.0	170	...	4.5	1

* Gap on bare substrate.

TABLE 15 Change of Bulk Electrical Properties of Glass Crossover Dielectric due to Relative Humidity at 32.2°C[86]

T, mils	31% relative humidity			47% relative humidity			65% relative humidity			87% relative humidity		
	C, pF, at 1 kHz	DF, %, at 1 kHz	R_2, GΩ, at 20 V dc	C, pF, at 1 kHz	DF, %, at 1 kHz	R_2, GΩ, at 20 V dc	C, pF, at 1 kHz	DF, %, at 1 kHz	R_2, GΩ, at 20 V dc	C, pF, at 1 kHz	DF, %, at 1 kHz	R_2, GΩ, at 20 V dc
0.6 ± 0.1	1.86	0.60	300	1.87	0.57	...	1.92	1.34	60	2.56	16.9	16

negative temperature coefficient over this temperature range.   The dissipation factor is decreased by recrystallization of the glass from 0.0005 to 0.0002.   The electrical properties as a function of frequency are plotted in Fig. 57.   The glass and recrystallized glass exhibit minima in their dissipation-factor curves; their profiles were super-imposed at 100 kHz and higher.

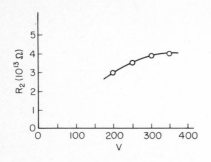

R_2: Bulk leakage resistance measured 2 min after voltage is applied

V: DC voltage applied across the crossover dielectric

Ambient condition: 23.3°C, 47% RH

Fig. 57 The change of dc bulk leakage resistance with applied voltage.[86]

To make a screenable paste of the glass, the composition is fritted, passed through a 300-mesh screen, and mixed with an organic binder consisting of 200 ml of butyl Cellosolve acetate and 21.5g of Ethoxyl T-10.   The paste consists of 100 g of glass mixed with 50 ml of binder, which had been mixed and milled in an automatic mortar and pestle for 2 h.   The paste is screened on 96 percent Al_2O_3 substrate and fired at 810 to 880°C for 3 h in a muffle furnace.   The bottom electrode is a platinum-gold paste.   The dielectric film consists of a double-screened layer which is pinhole free.

The dielectric properties of the recrystallized thick films generally follow the profiles of the bulk recrystallized glasses.[88]   The dielectric constant at 26°C and 10 kHz is 19; the dissipation factor is 0.18 percent.

$BaO \cdot PbO \cdot Al_2O_3 \cdot TiO_2 \cdot SiO_2$ formulations[89]   Glass-ceramic crossover compositions based in the system $BaO \cdot PbO \cdot Al_2O_3 \cdot TiO_2 \cdot SiO_2$ with optional additions of ZnO, PbF_2, SrO, ZrO_2, Ta_2O_5, WO_3, CdO, SnO_2, and Sb_2O_3 have been reported.   A crystallized composition and its associated dielectric properties are shown in Table 18.

To form a paste, comminuted frit is mixed with a binder consisting of 8 percent ethyl cellulose and 92 percent β-terpenol to a Brookfield viscosity of 600 P at 10

TABLE 16 $BaO \cdot TiO_2 \cdot B_2O_3 \cdot P_2O_5 \cdot SiO_2$ Crossover Composition[55]

Material	Formula	Weight %
Barium titanate..........	$BaTiO_3$	83.0
Silica....................	SiO_2	6.0
Boric oxide.............	B_2O_3	3.0
Phosphororus pentoxide...	P_2O_5	6.0
Alumina................	Al_2O_3	2.0

TABLE 17   Range of Crossover Compositions Based on $BaTiO_3$ and $0.55BPO_4$-$0.45SiO_2$ Eutectic[55]

	Composition range, weight %
Eutectic:	
$0.55BPO_4$-$0.45SiO_2$..............	7–15
Compound:	
$BaTiO_3$.......................	80–90
Al_2O_3........................	2–6

r/min. The paste is then screened through a 105-mesh screen on a 96 percent aluminum substrate containing screened platinum-gold conductors. The conductor for the BaO·PbO·Al$_2$O$_3$·TiO$_2$·SiO$_2$ crossover consists of (by weight) 55% Au, 15% Pt, 12% Bi$_2$O$_3$ 15% of the ethyl Cellosolve–β-terpineo binder and 3% glass (63.1% CdO, 16.9% B$_2$O$_3$, 12.0% SiO$_2$, 7.3% Na$_2$O). This is fired at 1050°C for 2 h or as high as the metal phase in the coductor will tolerate in order to provide good ad-

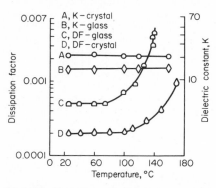

Fig. 58 Variation of dielectric constant and dissipation factor with temperature of the BaO·TiO$_2$·B$_2$O$_3$·P$_2$O$_5$·SiO$_2$ glass and recrystallized glass.[55]

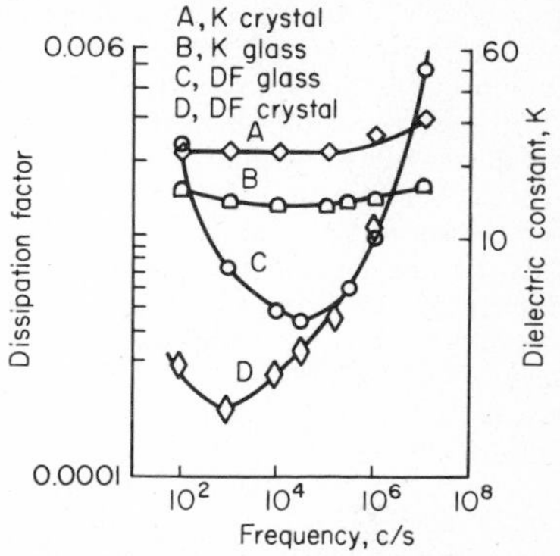

Fig. 59 Variation of dielectric constant and dissipation factor with frequency of the BaO·TiO$_2$·B$_2$O$_3$·P$_2$O$_5$·SiO$_2$ glass and recrystallized glass.[55]

TABLE 18   BaO · PbO · Al$_2$O$_3$ · TiO$_2$ · SiO$_2$ Crossover Compositions and Dielectric Properties[89]

Composition	
Component	Weight %
SiO$_2$	30
Al$_2$O$_3$	11
TiO$_2$	9
PbO	32
BaO	8
ZnO	10

Dielectric properties	
Dielectric constant at 1 kHz	13.9
Dissipation factor, at 1 MHz	580
At 5 MHz	100
Dc resistance as a crossover when capacitance varied from 7 to 265 pF, Ω	>10^9
Ac dielectric withstanding voltage on capacitance of 125 pF, V/mil	>250

hesion. The dielectric is fired to 850°C in a 10-min schedule with 5 min at peak temperature. The top conductor consists of a printed platinum-gold conductor fired at 850°C for 10 min. At this temperature there is no crossover or bottom-electrode softening and therefore no short circuiting.

This particular glass-ceramic crossover crystallizes the hexacelsian phase, BaAl$_2$Si$_2$O$_8$ and Al$_2$TiO$_5$. It contains a crystalline phase comprising less than 50 percent by weight of the glass ceramic, the preferred crystallinity being 40 percent. The mechanisms of crystallization for achieving crossover properties differ from those cited for the compositions in Tables 16 and 17, being based on partial crystallization and the crystallization of a particular compound, hexacelsian; the latter is based upon complete crystallization and a eutectic combination in which the low dielectric loss does not depend upon the crystallization of any specific compound.

Crossover Glass-Ceramic Process

During the firing of a glass-ceramic, devitrification occurs. A crystalline phase forms which is dispersed throughout the glass matrix. The crystalline material has a higher melting point than the crystalline material used. The upper use or processing temperature is determined by the melting point of the lowest-melting phase present.

With glass-ceramic capacitors, the higher dielectric constant comes with increased crystallized-phase content and microstructure optimization, at higher temperatures in the range of 800 to 1070°C. However, these factors are not as critical in crossover glass ceramics; i.e., the dielectric properties do not depend on the crystallization of a particular phase such as high-dielectric $BaTiO_3$ in capacitor compositions.

Generally the firing of glass-ceramics can be carried out over a range of temperatures depending on composition. For crossover dielectrics this can be over the range of crystallization up to the melting of the lowest-melting phase crystallized. The qualifying requirement is that the film adhere to the substrate for the firing schedule employed.

For example, the glass-ceramics reported in Tables 16 and 17 fire in the range of 810 to 830°C.[55] The crossover composition reported by Hoffman fires at 870°C.[89] The composition reported by Stein fires in the range of 900 to 950°C.[90]

The firing schedule is not as critical as for resistors. Devitrification is not as temperature- or atmosphere-sensitive as the chemical processes for resistor formation. The devitrification firing can be conducted either in a muffle furnace or conveyorized furnace.

With devitrification the glass dielectric becomes predominantly crystalline (as opposed to amorphous). The growth of crystal phases within the glass minimizes the possibility of pinholes and short circuits.

With the formation of the crystalline phase, the crossover can withstand repeated subsequent firing to the same temperature, which would be encountered when processing additional layers of conductors or dielectrics for multilayer structures. In addition, the crystalline phase tends to form a microstructure consisting of crystallites ranging in size from 1 μm down to 300 Å. The larger crystallites are formed at the higher devitrification temperatures. The crystallites are randomly oriented and form a close interlocked structure. The average crystallite size is an order of magnitude smaller than those found in sintered high-alumina substrates. As a result, they act to reinforce the film adherence and strengthen their own structure.

The conductor glazes must be compatible with the glass-ceramic dielectric. Incompatible fluxes or frits will attack the dielectric, interdiffuse, and cause bubbles and blisters. In some cases, short circuits are formed.

MULTILAYER CIRCUITRY

Several ceramic-based multilayer technologies are being developed or have been reported to provide the complex conductor interconnects needed for hybrid integrated circuits. The two major technologies are screen-printed multilayer interconnections on a single ceramic substrate[87, 91-97] and buried-wire ceramic interconnects.[98-102]

Buried-Wire Structures[98-102]

Requirements and applications A multilayer substrate is a ceramic substrate containing internally metallized wiring planes. The external and internal electrically conductive patterns are stacked in layers and separated and insulated by high-alumina ceramic. They are interconnected by conductive resins or vias with the top and bottom surfaces.

The primary application of buried-layer substrates is for monitoring and interconnecting functional groups of integrated circuits in a single package. The package designer using buried-layer structures should work toward increased integrated-circuit packaging density and shorter and fewer interconnections.   Short interconnections are made with high-conductivity buried lines at several levels.   This allows for high chip density.   Line resistance is typically 1 Ω/in. for buried conductors of 0.010-in. width, or 0.010 Ω/square.   Thus signal delay in high-speed circuits is minimized.

Multilayer substrates can be used for linear or digital circuits.   Since linear or analog circuits often require passive components which cannot be made on the silicon chip, hybrid techniques must be used.

Advantages   The advantages of buried-layer ceramic interconnects are as follows:

1. Buried metallization layers are both conductive and hermetic.

2. A large selection of air-firable thick-film components can be applied to surfaces of the multilayer boards.

3. Active-metal and evaporated thin film components are compatible.

4. Layer registration is precise.

5. The capacitance between wiring planes is low because thick dielectric layers can be used.

6. Voltage values are high.

7. Physical structure is good.

8. Versatility is possible in varying the buried layers, as opposed to varying only top-surface discretionary conduction paths to produce a given circuit.

9. Surfaces can be used in the same manner as standard aluminas.

TABLE 19   Tolerances for Buried-Layer Alumina Surfaces[100]

Dimension	in./in.
Length and width..........	±0.005
Camber...................	±0.004
Thickness.................	±0.003

10. Maximum device density is achieved by moving conductor lines from the surface to intermediate planes.

11. Reliability is improved since external wiring is minimized and higher operating speeds can be obtained through shorter conductor paths.

12. The stability of alumina permits the user to process units further without degrading the substrate or mounted devices.

13. Substrate firing in either oxidizing or reducing conditions is tolerated.

Metallization   The conductors are made of noble or refractory metals.   The alumina multilayer board with its hermetically sealed buried metallizations can be used in conventional thick film circuitry or with metallizations of the molymanganese–refractory-metal type.   The noble metals, e.g., palladium, platinum, or suitable combinations, can be specified with air-firable thick film materials.   The resistance of the buried metallization is typically 0.007 to 0.10 Ω/square.

If the buried conductors and resins on vias can tolerate exposure to air as well as reducing atmospheres at high temperatures, the multilayer structures are compatible with a wide variety of processing conditions.   For example, the use of screened, air-fired thick film conductors is a functional and inexpensive way to provide top-surface interconnection patterns in many cases.   Metallization processes other than thick film requires etching and/or plating steps, which may lead to contamination problems.   If the resins and buried conductors are noble metals, they can tolerate the high-temperature oxidizing environment of thick film processing.

Surfaces   The as-fired alumina surface is usually 20 μin. CLA which can be polished down to 4μin. CLA for use with thin films.   The routine tolerances are shown in Table 19.

Processes and design guidelines   Buried-layer structures have been described by Du Pont, American Lava, and others.   An exploded view of the Du Pont package is shown in Fig. 60.   A schematic diagram of the interconnections is shown in Fig. 61.

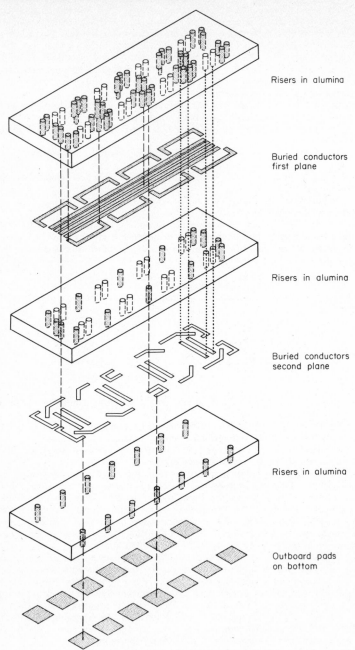

Risers in alumina

Buried conductors
first plane

Risers in alumina

Buried conductors
second plane

Risers in alumina

Outboard pads
on bottom

Fig. 60 Multilayer ceramic assembly.[101]

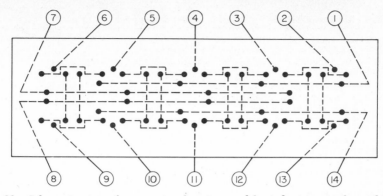

Fig. 61 Schematic view of access riser locations and buried wiring in the multilayer ceramic assembly of Fig. 60.[101]

The upper buried level carries conductors running the length of the package and has risers connecting with the top surface near each integrated-circuit position.

Design guidelines set forth by Du Pont and American Lava[101, 102] are given in Tables 20 and 21. The guidelines in Table 20 have been recommended for designing with the Du Pont Multilox° ceramic wiring structure of about 1 in. maximum dimensions.

The design parameters and tolerances for the American Lava package are shown in Table 21. The alumina is a 94 or 99.5 percent Al_2O_3 composition. Thickness of the alumina insulating layers is an important consideration, preferred thicknesses are in multiples of 0.015 in. If via holes are involved, thicknesses of 0.002 in. are suggested but many prefer 0.005 in.

Conductor widths of 10 mils or larger with 10 mils or more space between lines are recommended where possible. Maximum economy is achieved with line widths of about 15 mils and spacing of 15 mils or larger. When trade-offs must be considered, larger spacing becomes more desirable than line width. As line widths and spacings decrease, costs increase. Long parallel paths in particular increase cost.

A multilayer process, screened multilayer ceramics, combines green-ceramic technology, thick film screening technology, and a process for the protection of tungsten from oxidation during air firing.[,103] The features claimed for this process include the following: (1) the alumina and tungsten material system eliminates the structural failures in multilayer wiring imposed by thermal-expansion mismatches; (2) fine metallization up to 10 layers is possible thanks to the fine screenability on a green-state surface; and (3) there is a capability for buried capacitors as well as for thick film resistors.

A green-alumina sheet is formed with the doctor-blading technique used to form laminated ceramics. Conductive and dielectric layers are alternately printed on the flexible alumina sheet. The interconnections between layers are fabricated by screen printing subsequent conductor layers through connection vias of dielectric layers printed over the underlying first metallization. Tungsten paste is formulated as the conductor material and alumina paste as the dielectric material. The process schematic is shown in Fig. 62. The design capabilities of this system, designated SMC, are listed in Table 22.

Thick Film Multilayer Structures

In the second method alternating layers of metallic and insulating layers are serially added to a ceramic substrate by screen-printing patterns of commercially available pastelike inks. The successful fabrication of multilayers is critically dependent upon appropriate paste selection. The pastes cannot be considered on their individual merits; they must be treated as interactive components.

° Trademark of E. I. du Pont de Nemours & Co.

TABLE 20   Design Guidelines for Ceramic Wiring Structure[101]

Multilevel structure:
0.010-in.-diameter risers; 0.010-in. minimum spacings
0.010-in.-wide conductor lines on 0.020-in. centers for buried wiring
Sheet resistance ≤ 0.010 Ω/square

Top-surface conductor pattern:
0.005-in. lines on 0.010-in. centers
Sheet resistance 0.005 Ω/square
A 0.005-in.-wide conductor should not pass between two risers unless there is a spacing of 0.017 in.
Minimum overlap of 0.005 in. of a conductor over a riser where contact is required; complete coverage preferred
Minimum separation between conductor line and riser is 0.005 in. for no contact
Width of at least 0.040 in. at edge reserved for lid seal

TABLE 21   Design Guidelines for Multilayer Ceramic Substrates[102]

	Possible	Preferred
Design parameters:		
Minimum line widths, in.	0.004	0.010
Minimum space between lines, in.	0.004	0.010
Via-hole diameter (best left unspecified), in.	0.005	0.015
Via-hole centerline spacing, in.	0.010	0.030
Via-hole cover dot, in.	0.015	0.020
Via-plane insulation thickness, in.	0.002	0.005
Non Via-plane insulating thickness, in.	0.002	0.015 multiples
Maximum substrate size, in.	6 by 6	3 by 3
Tolerances:		
Line width, in.	±0.0005	±0.002
Conductor centerline spacing	±½% NLT ±0.002 in.	1% NLT ±0.005 in.
Substrate length and width	±½% NLT ±0.002 in.	1% NLT ±0.010 in.
Substrate thickness	10% NLT ±0.002 in.	10% NLT ±0.005 in.
Camber, in./in.	0.001 (ground)	0.006 NLT 0.002 in.
Through-hole diameter	½% NLT ±0.002 ni.	1% NLT ±0.005 in.
Pattern to through-hole or substrate edge	½% NLT ±0.005 in.	1% NLT ±0.010 in.
Metallized pattern resistance, Ω/square	0.007	0.015

Crossover requirements   The crossover requirements for multilayer thick film circuitry follow.

Freedom from Alkali.   This is necessary to minimize ion migration. Alkali ions such as sodium are mobile in electrical fields and will help degrade the electrical properties of the crossover.

Substrate Compatibility.   The crossover should have a thermal coefficient of expansion which is correctly matched to that of the alumina substrates to be utilized. The mismatch should be small so that on cooling the crossover is in a state of compression. The films should remain in compression over the temperature range 20 to 100°C.

Inertness to the Available Conducting Inks.   The conductor inks should not contain any frits or fluxes which will react and interdiffuse with the dielectric, causing blistering, bubbles, and possibly pinholes and short circuits. Conducting inks for multilayer structures have been discussed under Conductors.

Well-defined Softening Temperature. This permits the use of crossover layers

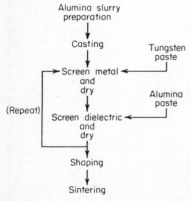

Fig. 62 Process schematic of SCM multilayer ceramic system.[103]

with different composition and firing temperatures for several planes in a single structure.

Good Flow Characteristics. The glaze should have flow characteristics in the liquid state so that on cooling a smooth layer with no pinhole results. There should be no crawling effects and a minimum meniscus of glaze at the edges of the substrate. A smooth and flat surface should be presented to the next layer of conductors. The correct temperature schedules should be adopted to control these factors.

For complex multilayer structures the use of crystallizable-glass crossovers is recommended instead of glass. The crystallized or devitrified phase have a melting point much higher than the firing temperature used or the softening temperatures of any glasses or glazes. The major advantage of using a glass-ceramic is that, once fired, the material does not resoften unless heated well above that temperature. The non-softening characteristic minimizes conductor line movement on subsequent firings and permits better circuit resolution.

Additional Requirements

Surface finish should be better than 10^{-6} in. CLA.

Crossovers should be free from gross defects such as bubbles and blisters.

TABLE 22 Screened-Multilayer-Ceramic Design Guidelines[103]

Parameter	Standard	Possible
Conductor line width, mils	8–10	4–5
Conductor clearance, mils	6–10	4–5
Via-hole diameter, mils	12–16	8
Via center spacing, mils	24–32	16
Dielectric-layer thickness, mils	1.5–2	3–4
Number of layers	3–5	10
Substrate size, in	2 by 3	5 by 5
Sheet resistivity of conductor, mΩ/square	20	10

Materials should be easily dispersed in suitable carriers and binders for screen printing.

The fired films should be easy to etch for interplane connections.

Electrical breakdown between conducting planes should not occur for the peak voltages expected during operation of the transmission lines by digital integrated circuitry.

Dielectric permittivity range should be from 5 to 500. The low part of the permittivity range is for low-capacitance crossovers. For single conductors, treated as transmission lines, the crossover should have a small permittivity and a low dielectric and resistive loss. These parameters and the crossover thickness determine the characteristic line impedance and the propagation delay time of the pulse. The high part is for decoupling between the power and ground planes.

Electrical requirements[95] Circuit-design requirements call for high-resolution conductors such as 4-mil conductor lines and spaces or voids of 8 mils or less. Electrical requirements for a multilayer depend upon its use. Connections with minimum resistance and dielectrics of minimum permittivity are two basic requirements. Logic applications are more demanding than memories since the higher operating speeds available from new silicon technologies must not be impaired by inferior interconnection performance.[104] Minimization of high parasitic capacitance preserves high signal levels and fast pulse time. Low-resistance conductors allow the use of narrow lines and wider spaces for reduced capacitance and better noise immunity. Less conductor area isolated with a lower-permittivity dielectric also reduces capacitance. A relative permittivity of 4 to 6 is preferred for high-speed use. At slower speeds 8 to 14 is useful.

Capacitance effects suggest the use of very narrow logic conductors, but their resistance is high. A compromise such as a 6-mil-wide logic conductor with less than 0.6 Ω per inch of length becomes necessary. A resistivity less than 3.5mΩ/square in the finished conductor thickness is necessary to achieve this low resistance.

Guidelines for paste selection[95] Some general guidelines based on experience can be given for paste selection.

1. Suppliers' experience and recommendations are valuable starting guides to paste selection.

2. Few suppliers have actually processed complex multilayers to learn the details and problems.

3. Combining conductor and dielectric materials from a common supplier does not assure optimum or even acceptable results.

4. Some qualification testing is essential in selecting a suitable combination of pastes for multilayer applications.

5. The pastes should be tested for undesirably high permittivity.

6. Limited resolution and via shape may restrict high-density interconnection applications.

7. Undesirably high permittivity should be avoided.

8. The paste may be incompatible with high-speed logic applications unless the permittivity is low enough.

9. Pastes may mature with porosity or show other unacceptable structural properties.

10. Lot-to-lot variability and unannounced formulation changes must be guarded against.

Processing pitfalls[95] Several pitfalls can reduce circuit yields.

Choice Conductor. Discussed in the section on Conductors, the choice of conductor can affect the permittivity slightly. For example, a dielectric with a vendor-data sheet permittivity of 10 ± 2 has a permittivity variation of 7 to 9.8 for five gold-conductor pastes from three different vendors.

Intralayer Capacitance. Conductor widths vary on and beneath the dielectric layer while their locations remain fixed.

Squeegee Motion. Fewer defects result from squeegee motion for dielectrics parallel to the conductor length. Multilayer arrays should be designed with all conductors in a given layer oriented in the same direction. Orthogonal directions in adjacent layers help reduce capacitance.

Paste Flow into Vias. Paste flow into vias can reduce their size or completely obliterate small ones. Sharp via resolution is possible only with a very stiff paste. The resulting dielectric surface can be unacceptably rough. This creates problems within the next conductor pattern. A compromise is necessary, the dielectric paste providing a reasonably smooth surface while maintaining vias of a useful size.

Dielectric-Layer Defects. Dielectric layers are susceptible to defects, particularly holes, which arise from dirt and debris, trapped air bubbles, failure to solve the screen-mesh problem, a roughened conductor surface below, irregular wetting over the conductor, and careless printing.

Dielectric-Paste Resolution Qualities. The minimum clearance between a via and an adjacent conductor must allow for via-top enlargement at the dielectric surface.

Screen Elevation at a Via Edge. This causes a thick ring of paste to surround the via. The added paste thickness tends to slump into the via and close it.

Glass-ceramic pastes Devitrified glasses or glass-ceramics are probably the best all-around crossover dielectrics for multilayers. They do not reflow when heated to their original firing temperature, a very definite advantage when other coatings are to be placed on top of the fired dielectric. The absence of pinholes is probably the most important factor in selecting a material for multilayer boards. It takes only one pinhole to make the circuit board a complete failure. For a complex multilayer circuit, the pinhole may be very difficult to locate. Glass-ceramics are virtually pinhole-free because of the workability (softening and flow properties) of the glass before nucleation and devitrification of the crystalline phases. Screening resolution is important with regard to windows or vias in the dielectric, which are used to make an interconnection between layers. Difficulties are introduced if the vias close or shrink.

Commercial-paste evaluation Several evaluations have been reported for commercial pastes. A crystallizable glass-ceramic has been evaluated and compared with glass for multilayer structures.[105] In the initial stages of firing the composition goes through the normal sintering, softening, and coalescence of single-phase glass. As the temperature is increased, crystals appear and cause a large increase in viscosity. With proper control of size and concentration, the formation of crystals can actually cause the dielectric layer to become solid. In subsequent firings the printed top con-

TABLE 23   First Comparison of Commercial Glass-Ceramic and Single-Phase Glass Crossovers[105]

	Single phase	Crystal-lizable
Hermeticity, MIL 202C, method 102A, cond. A..................	Excellent	Excellent
Surface smoothness, brush surfanalyzer, μin.......................	10	22
Top-conductor adhesion, lb/lineal in............................	20	35
Top-conductor solderability....................................	No	Yes
Capacitance, pF/mm², at 1.5 mils thickness, for 1 mm²..........	6	8
For 10 mm²..	27	32
Resistance of Pt-Au on crossover, Ω/square.....................	110	84
Resistance of silver on crossover, Ω/square.....................	15	5
Load life, 112 V dc, 85°C (capacitance, design factor, and insulation resistance)	OK to 1,100 h	OK to 1,100 h
Dielectric withstanding voltage per 1.75 mils....................	375	500

TABLE 24   Second Comparison of Commercial Glass-Ceramic and Glass Crossovers[90]

	Crystalline	Vitreous
Insulation resistance, Ω................	$>10^{13}$	$>10^{13}$
Dielectric breakdown, 2-mil film, V.....	>500	>500
Dielectric constant, 1,000 Hz...........	40–45	10
Dissipation factor, %.................	1.5	<1.0
Thermal-shock resistance, °C...........	−55 to 200	−55 to 200
Pinholes or shorts....................	Excellent	Fair
Printing resolution...................	Excellent	Good

ductors are supported by a ceramic substrate rather than a thermoplastic glass. The crystallizable material is compared with single-phase glass in Table 23.

Table 24 also compares the properties of a proprietary glass-ceramic crossover (fired at 900 to 950°C) with a glass crossover. The general compositions are not available. The differences are evidenced in the dielectric constant, printing resolution, and resistance to shorting. The dielectric constant of the glass-ceramic is 40 to 45, compared with 10 for the glass. The preliminary resolution ability to prevent pinholes is excellent for the glass-ceramics and fair to good for the glass. Insulation resistance did not deteriorate after being subject to a load of 100 V for 1000 h at 100°C.

The dielectric constant for any given capacitor varies with the conductor material used for the electrodes. The results of controlled experiments carried out with capacitors prepared with four conductors (not identified) at three different thicknesses of the same dielectric are tabulated in Table 25. The apparent dielectric constants varied by nearly a factor of 3. The variations have been attributed to the differences in composition and concentrations of glass frits and other fluxes used in the conductors rather that the metals themselves. Such conductor materials are seldom interchangeable; the capacitor-conductor combinations used to fabricate the capacitors must be completely characterized.

CONDUCTORS

Capacitor Dielectrics

The recommended procedure with high-dielectric-constant dielectrics is to prefire the bottom electrode and cofire the screened dielectric with the top electrode. Each layer should be dried after printing. Cofiring results in fewer shorts and lower dissipation factors. The dielectric should overlap the electrode pattern by at least 20 mils on all sides to eliminate surface leakage. The top electrode should fall completely within the bottom-electrode area.

TABLE 25   Variation of Apparent Dielectric Constant with Different Electrode Metals[59]

Conductor	Average dielectric thickness, mils		
	0.8	1.6	2.3
	Apparent dielectric constant		
1	170	115	75
2	60	70	65
3	70	75	80
4	90	100	80

Figure 63 shows a complete thick film circuit containing screen-printed capacitors. The geometries of the top and bottom electrodes are apparent. Also shown is a circuit containing discrete capacitor chips electroded with wire and bond techniques.

The use of silver electrodes as the top electrode should be avoided with high-K capacitors. Silver has some tendency to migrate through the dielectric during firing. Silver electrodes with no palladium are susceptible to silver migration. Shorting will occur due to thin layers and high dc voltage. Gold-containing electrodes such as gold-platinum or gold-palladium are preferred because they present no migration problem.

Closely spaced conductor leads short out by forming dendritic silver growths from one lead to another when subjected to high humidity in combination with a high-voltage gradient. Severe stress testing in rain chambers at 100 percent relative humidity of capacitors having electrodes with large quantities of silver showed consistent failure even when organic overcoating encapsulants were used.[106]

With K-500 dielectrics, the dielectric constant depends on the electrode materials as well as the firing temperature. Palladium-silver is acceptable as a bottom electrode. However, gold or platinum-gold should be used for a top electrode. The dielectric and top electrode can be cofired.

On the other hand, silver-palladium used with K-50 dielectrics gives the highest capacitance, as shown in Fig. 64. The top electrode and dielectric can be cofired. One commercial capacitor series reportedly can be used with cofired palladium-silver top electrodes.

When silver-bearing films are used as the top electrodes and the firing temperature is above 950°C, caution should be exercised since silver or silver-palladium electrodes will affect the film. Gold-platinum should be considered instead.

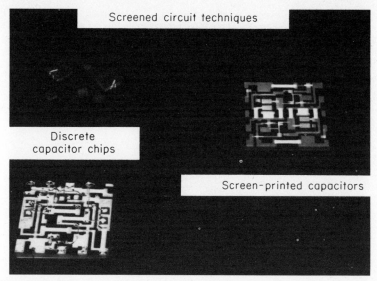

Fig. 63 Thick film circuits showing discrete capacitors and printed capacitors with printed electrodes.

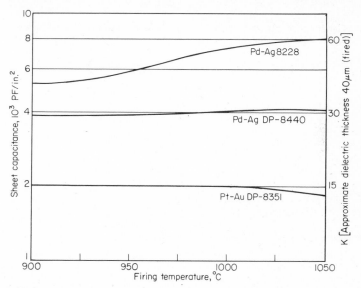

Fig. 64 Sheet capacitance versus firing temperatures for various electrode materials in a K-50 capacitor dielectric.[111]

Glass Crossover Dielectrics

Dielectric crossover compositions are compatible with most thick film conductor compositions. The bottom conductor, crossover, dielectric, and top conductor are fired separately, the crossover being double-printed. For best results the top conductor is fired separately to a peak temperature, which is considerably less than the peak temperature of the dielectric (usually 100°C or more). This is to avoid resoftening of the glass crossover, which results in lateral translation or shorting of the top conductor. Top and bottom electrodes should be of the same material if possible. In each case the vendor's literature should be checked before firing.

Reliable crossovers reportedly have been produced with Corning 167JQ and 7052 glass using silver-palladium electrodes.[106] Silver diffusion occurs during firing and causes a yellow discoloration of the glass. However, measurements of frequencies up to 150 MHz showed no open circuits, which implied that there was no significant increase in the conductivity of the glass. These glasses, as well as Kimble TM7, Corning 7059, and Drakenfeld E1313, have been shown to wet the silver-palladium electrodes.

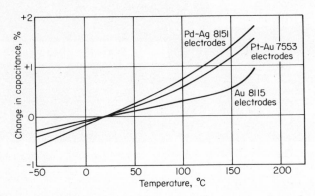

Fig. 65 Capacitance change versus temperature for various electrode materials on a glass-ceramic crossover dielectric.[112]

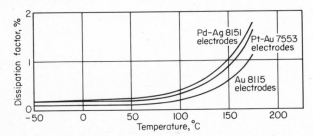

Fig. 66 Dissipation factor versus temperature for various electrode materials on a glass-ceramic crossover dielectric.[112]

Glass-Ceramic Crossovers

Crystallizable glass crossovers are compatible with most common conductor compositions. Glass-ceramics can be cofired with all conductors, including gold, platinum-gold, palladium-gold, palladium-silver, and silver. However, the change in capacitance and dissipation factor with temperature varies according to the electrodes used, as shown in Figs. 65 and 66 for palladium-silver, platinum-gold, and gold electrodes.

Conductors for Thick Film Multilayer Structures

The conductor and dielectric pastes cannot be considered on their individual merits. They must be treated as interactive components and evaluated under multilayer conditions. The assembly properties of conductors on dielectric layers can be quite different from those on alumina. [107, 108]

Gold is preferred as the multilayer conductor for the following reasons: (1) it is thermally stable when fired with multilayer dielectrics; (2) it has high electrical conductivity; and (3) it is compatible with assembly operations except for those requiring soft solders.[95] As shown in Table 26, except for silver, gold has the lowest resistivity in the narrow conductors required for multilayers.

The data in Table 26 show typical resistivity changes for the most common conductor types on alumina and the range of change in intitial resistivity resulting from covering the conductors under two separately fired prints of a multilayer dielectric. The data have been normalized to a 1-mil thickness.

The change in resistivities of the gold and gold-containing conductors under

TABLE 26   Resistivity Range of Multilayer Conductor Types[91]

Conductor type	Resistivity Ω/square at 1 mil	Change under dielectric, %
Au................	0.002–0.010	−5 to +2
Pt-Au.............	0.015–0.100	−1 to +6
Pd-Au.............	0.010–0.100	−0.5 to −0.2
Pd-Ag.............	0.010–0.050	+42 to +113
Ag................	0.002–0.010	+50 to ∞

TABLE 27   Resistance of Gold Conductors[95]

Type of gold	Thickness, mils	Conductor width, mils						
		4	5.2	6	10	14	20	25
		Resistance, Ω/in.						
Engelhard 9177............	0.49	0.82	0.56	0.47	0.26	0.19	0.13	0.11
EMCA 212B...............	0.41	0.74	0.56	0.50	0.29	0.22	0.16	0.14
282...................	0.40	0.95	0.69	0.61	0.36	0.27	0.21	0.19
Du Pont 8780............	0.53	0.99	0.70	0.60	0.36	0.26	0.21	0.18
Engelhard 1069............	0.40	1.08	0.74	0.63	0.35	0.25	0.20	0.18
ESL 8835................	0.40	1.27	0.94	0.70	0.38	0.27	0.21	0.19
Du Pont 8380............	0.51	1.27	0.85	0.74	0.43	0.32	0.25	0.23
ESL 8831................	0.40	1.35	0.97	0.82	0.45	0.33	0.24	0.22
8760................	0.38	1.49	0.99	0.83	0.46	0.33	0.24	0.22

dielectric are minimal compared with silver and palladium-silver alloys.   Silver cannot be used with most dielectrics because of its mobility during firing, which causes it to diffuse into the dielectrics.[91, 109, 110]   This results in open conductors and low insulation resistance.   Alloys of palladium and silver cause the same problems with buried conductors during multiple firings.

Conductors of platinum-gold and palladium-gold can be used when their lower conductivities can be tolerated.   These are used to overlay a gold conductor where a solderable area is required.

The resistance of gold conductors having various widths and a constant length of 400 squares is summarized in Table 27.   Measured thicknesses are also included. Figure 67 shows plots or resistivity versus conductor width.   The lowest resistivity occurs in lines 10 to 15 mils wide.   Their cross section was closest to being rectangular in shape.   As reported by Kurzweil and Loughran,[95] one commercial gold paste maintained a lower resistivity in decreased widths. This is an important consideration for low resistance in small conductor lines.

High conductivity can be obtained by increasing the thickness. For example, a 6-mil-wide conductor double-printed with the commercial paste reported in Fig. 67 was 0.9 mil thick and had a resistivity less than 1.5 MΩ/square; this is 80 percent of the conductivity of a solid-gold conductor of similar dimensions.[95]

The highest-quality narrow conductor lines result when the squeegee motion is parallel to the conductor length. Printing perpendicular to their direction broadens the conductor lines and causes an irregular feathered edge. Dielectric prints over conductors will also have fewer defects if printed in the same parallel direction. Orthogonal directions in adjacent layers help reduce capacitance.

Conductor-Dielectric Interface Porosity and Dielectric Film-Substrate Thermal-Expansion Mismatch [58, 59]

The selection of bottom electrodes compatible with the dielectric being fired influences the observed dielectric constant. Porosity is generated at the conductor-dielectric interface as a result of different sintering mechanisms, which require radically different firing schedules and peak temperatures for densification.

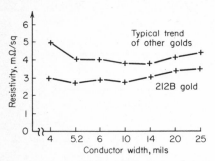

Fig. 67 Resistivity versus conductor width of various gold conductors.[95]

Reduction of the apparent dielectric constant can be caused by dispersed and laminar porosity. The reduction due to dispersed porosity should be roughly proportional to its volume fraction; however, laminar porosity resulting from differential thermal expansion with alumina substrates causes a series capacitance which drastically reduces K_m. As an example, titanate films with an expansion coefficient of 12×10^{-6} per degree Celsius compared with 7.5×10^{-6} per degree Celsius for the substrate will have an apparent dielectric constant which will be 2 to 3 times less than that of the corresponding body.

The problem of differential thermal expansion can be overcome by forming an interface which alleviates the stress, e.g., by the deposition of thick metal bottom conductors which are glass-free, and selecting glasses with low coefficients of thermal expansion (and therefore high softening temperature) to partially compensate for the titanate and thereby reduce the average coefficient for the film.

COMMERCIAL DIELECTRIC PASTES

Some commercial capacitor and crossover dielectric pastes are listed in Tables 28 and 29.

DIELECTRIC-MIXTURE RULES[58-60]

The most widely used compositions are based on mixtures of glass frits and powders of titania, barium titanate, and other ferroelectric compounds with additions of modifying oxides. The resultant dielectric constant of such mixtures is determined by the volume fractions and distributions of phases present, including porosity. During the firing of the titanate-glass composites, liquid-phase sintering occurs; the glass phase surrounds the titanate particulates and becomes continuous. As a first approximation the glass and titanate fractions behave like capacitors in series. For densified films with this type of microstructure and titanate volume fractions below 30 volume percent, the Maxwell relationship shows good agreement with experimental data:

$$K_m = \frac{\theta_c K_c(\frac{2}{3} + K_d/3K_c) + \theta_d K_d}{\theta_c(\frac{2}{3} + K_d/3K_c)} \tag{3}$$

where K_m = relative dielectric constant of mixture
θ_c = volume fraction of continuous phase
θ_d = volume fraction of dispersed phase
K_d, K_c = relative dielectric constants of phases at given temperature

For titanate volume fractions above 30 volume percent, better agreement is found with a logarithmic mixture rule

$$\log K_m = \sum_{j=1}^{N} \theta_j \log K_j \tag{4}$$

where K_m = dielectric constant of mixture
K_j = dielectric constants of individual components at given temperature
θ_j = volume fraction

The relationship is based on the assumption that there are no interactions between the various phases and no dielectric powder goes into solution with the glass frit.

Figure 68 shows experimental data points for several systems fitted with the theoretical curves for E (3) and (4). Equation (4) gives the best fit. In practice several factors complicate the comparison of data even for a simple binary system of glass and barium titanate:

1. Different sintering mechanisms require widely differing firing schedules and peak temperatures for densification (particularly at the extremes of composition).

2. Using different conductors, as a result of item 1, influences the observed dielectric constant through surface roughness, reactives affecting the chemical composition, reactives changing the fraction of the liquid phase, and generation of porosity at the conductor-dielectric interface.

An example of the use of Eq. (4) is given using barium strontium titanate and barium lead titanate, which are mixed with a glass frit such as Corning 7059 glass in the proportions shown in Table 30. The densities for the glass frit and the ferroelectric powders were measured as 2.8 and 6.5 g/cm³, respectively.

Equation (4) based on a zero glass volume fraction K_m of 1,000 is shown in Fig. 69. This is determined by extrapolating the experimental curve back to the ordinate axis. As can be seen, there is a major deviation of the experimental results from the calculated results because the ferroelectric powders go into solution in the glass frit. A modified Lichtenecker's rule which will fit the experimental data is

$$\log K_m = \theta_g \log K_g + (\theta_{DP} - \tfrac{1}{2}\theta_g) \log K_{DP} \tag{5}$$

where m = film dielectric
g = glass frit
DP = dielectric powder mixture

It is interesting to note that 10 volume percent of glass in the reported system represents optimum mixture for a maximum effective dielectric constant. Below 10 percent, porosity is introduced, which tends to decrease K_m because there is sufficient glass to fill the voids between the titanate particles.

DIFFERENTIAL THERMAL ANALYSIS FOR PROCESS CONTROL

Differential thermal analysis (DTA) is used as a process control tool for determining processing temperatures and furnace profiles. The technique has been used as a standard analytical method in chemical, mineralogical, and ceramic applications. Literature discussing the basic principles and the practical requirements for accurate and reproducible data is extensive.[113-115] Qualitative DTA has been found to be very satisfactory for the process control of thick film dielectrics. The points important for prototype or production shop use will be emphasized under the following sections: Principles of Operation, Equipment and Procedure, Process Control Variables, Interpretation of Thermogram Profiles, and Commercial Equipment

Principle of Operation

Differential thermal analysis measures the temperature difference between a sample and reference material as each are programmed through the same temperature

TABLE 28 Some Commercial Capacitor Dielectric Pastes

Material	Firing temperature, °C	Dielectric constant	Dissipation factor at 1 kHz and 25°C, %	Capacitance density, pF/in.²	Insulation resistance, Ω	Breakdown voltage, V	Capacitance change, %	Dielectric strength, V/mil	Comment
Du Pont:									
DP-8315..........	950–1050	30–60	0.5–1.5	3,000–6,000	$>10^{9a}$	>500	$<0.2^b$	...	Thickness 1.5 mils, TCC ~500 ppm/°C (−10 to 85°C)
DP-8229..........	850–1050	300–800	1–2	25,000–85,000	3×10^{10c}	...	...	...	Fired thickness should be at least 1.5 mils, flash breakdown >400 V
DP-8289..........	850–1050	~1,200	<3.5	52,000–160,000	$>10^{9d}$	...	...	...	Thickness 1.7 mils, aging ~−3%/decade-h (1,000 h, 85°C, 50 V)
Electroscience Laboratories, Inc.:									
4110..........	910–950	50 ± 10	<1.5	$10,000^e$	...	...	...	300^f	
4210..........	910–950	100 ± 20	<1.5	$20,000^e$	...	...	...	300^f	
4310..........	910–950	250 ± 50	<1.5	$50,000^e$	...	...	...	300^f	
4410..........	910–950	600 ± 100	<1.5	$120,000^e$	...	...	...	300^f	
4510..........	925–1050	1,000 ± 300	2.5–4.0	70,000–175,000	$10^9 \min^g$	...	...	...	Aging ~2–3%/decade
Electro Materials Corp. of America:									
CDP-10..........	1000	40	0.001	2,000	10^{12h}	...	...	800	TCC ± 50
CDP-30..........	1000	30	0.09	6,000	10^{12h}	...	...	500	TCC ± 30
CDP-400..........	1000	400	1.5	55,000	5×10^{11h}	...	...	500	TCC −10%
CDP-1000..........	1000	1,100	1.5	175,000	10^{12h}	...	...	500	TCC −15%
CDP-2000..........	1000	2,000	1.6	240,000	5×10^{11h}	...	...	500	TCC −18%
Matthey Bishop, Inc.:									
LS-5015..........	760	~70	<1	...	10^{8i}	>350	$+5^j$	...	Thickness 1.5 mils
LS-5106..........	760	~450	~1.5	...	10^{8i}	>450	-4^j	...	Thickness 2 mils
LS-5014..........	760	10	1	...	10^{8i}	>500	$+6^j$	...	Thickness 2 mils

[a] At 100 V, 25°C, 0.1 in.²
[d] At 100 V, 25°C.
[g] At 50 V, 25°C.
[j] From 25 to 125°C.
[b] For 500 h, 50 V, 85°C, unencapsulated.
[e] pF/(in.²)(mil).
[h] Ω/square.
[c] Ω/(cm²)(mil), 10–100 V.
[f] Rated at 50 V.
[i] At 100 V.

TABLE 29 Some Commercial Crossover Dielectric Pastes

Material	Firing temperature, °C	Dielectric constant for 1 kHz at 25°C	Dissipation factor for 1 kHz at 25°C, %	Q	Capacitance density, pF/(in.²)(mil)	Voltage breakdown, V	Insulation resistance, Ω	Comment
Electroscience Laboratories, Inc.:								
4747	800	8–10	<1.0	...	1,900	<500*	<10^{13}	
4640	780–980	40	<1.0	...	9,000	<500*	<10^{13}	
4610	875–1000	9–10	<1.0	...	2,000	<500*	<10^{13}	
Du Pont:								
8299	875–1000	10–20	<1.5	500–700†	...	>500	>10^{11}‡	Thickness 1.5–2.0 mils
8399	875–1050	10–20	<1.5	500–700†	...	>500	>10^{10}‡	Thickness 1.5–2.0 mils
Matthey Bishop, Inc.:								
LS-5524	760	10	<1	...	...	>600	>10^{10}†	Thickness 2.0 mils; capacitance change 25 to 125°C, +7%
Electro Materials Corp. of America:								
1371	900–1000	7–20	...	>2,000 >600	...	...	2 × 10^{12}	TCC 0 to +30 ppm/°C; dielectric withstanding voltage 1,000 to >700 V
Sel-Rex Company:								
9300	950	10–13	~1.0	...	...	>400	>10^{11}	
9301	850–950	7–10	<1.0	...	...	>400	>10^{11}	

* 1.5-mil film.   † At 1 MHz.   ‡ At 100 V.

range. It is a heat-averaging technique and reflects the absorption or release of heat over a finite time span, which is determined by the sample geometry, heating rate, and thermal contact of the sample with the thermocouple sensor.

A sample of the dielectric powder to be processed and a powdered reference sample which shows no heat effects in the temperature range of interest are both heated in a suitable furnace. Each powdered sample is packed around a thermocouple junction. The differential voltage readings of the two couples plotted against temperature as the general temperature of the furnace is increased indicate any heat absorption or evolution in the dielectric sample.

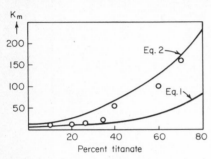

Fig. 68 Dielectric-mixture rules showing experimental data points.[58,59]

In the absence of enthalpic change during heating or cooling the temperature difference between sample and reference is constant; therefore, the net output is constant. This yields a level base line rather than a sloping one. During a chemical reaction or enthalpic change, heat is absorbed (endothermic reaction) or evolved (exothermic reaction) from the sample material. During the reaction, the rate of change deviates from that of the inert reference material.

For an endothermic reaction, an increase in heat content is required to accomplish the transformation. The temperature of the sample tends to remain constant while the heat required to complete the phase transformation is absorbed by the sample. Once the transformation is completed, the sample will heat at a faster rate until the temperature difference between it and the reference is minimized. At this point, the substance resumes heating at the programmed rate.

Equipment and Procedures

Sample holder design Since DTA is a comparative method, the heat of two samples must be symmetrical. The slightest asymmetry in the fabrication of the sample holder or furnace can produce a temperature difference even though identical samples are placed in sample and reference sides. An adjustment in thermocouple position or some other variable is usually made to cancel the effect of asymmetries. This is important during high-sensitivity and high-heating-rate runs.

The effect of change in sample conductivity is an important consideration when choosing a sample-holder design. Low-mass sample holders assure that temperature gradients within the system are reduced to a minimum. A proper shape for a sample is a cylinder heated as symmetrically as possible from the outside. The uniform

TABLE 30 Compositions for Evaluation of Lichtenecker's Rule[60]

Ferroelectric powders, vol %	Glass frit, vol %
100	0
96	4
92	8
87	13
83	17
79	21
71	29
63	37
56	44
50	50

approach of heat to the center permits the highest convenient symmetry and hence little disturbance of the peak shape due to irregularities. Other shapes are used when the cylinder may be impractical.

Many different sample-holder designs have been used, depending on the particular sample and information desired. Each design has its particular advantages and disadvantages. Commercial instrument suppliers provide sample holders designed for different purposes. The choice of the best crucible or sample holder for DTA is usually a compromise depending upon the type of reaction under study. A flat dish may cause a loss of $\triangle T$ (differential-temperature) sensitivity through radiant heat loss. In solid-solid transformations, e.g., crystal transitions, glass crystallization, of fusion phenomena, a spherical crucible might be ideal for measuring a maximum $\triangle T$ but difficult to construct. The closest practical approach to this shape is a cylinder of about a 1:1 diameter ratio.

If the crucible is firmly spot-welded to the $\triangle T$ thermocouple, a maximum $\triangle T$ signal will be transferred to the measuring and recording system.

Standard crucible materials are platinum, platinum-rhodium, alumina, and aluminum; ultrahigh-temperature sample holders usually consist of tungsten. Micro crucibles have volumes of 0.1 to 0.15 cm³, and macro sample holders have volumes up to 1 cm³. The crucibles may have a ring for achieving a homogeneous temperature field.

The heat transfer from the furnace to the sample should be as near instantaneous as possible. A thermocouple in direct physical contact with the sample and sitting in a small-mass heating block of high thermal conductivity is one of the most desirable geometries. A thermal lag results if the cell assembly has a container situated between the measuring thermocouple and the sample. In this case, slow heating rates (5°C/min) are needed to assure temperature accuracy.

Thermocouples Each sample holder should be equipped with thermocouples for $\triangle T$ measurements and for measure-

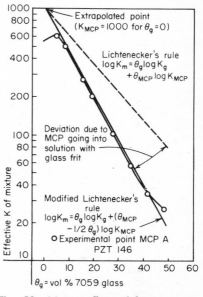

Fig. 69 Matrix effect of barium strontium titanate and barium lead titanate systems mixed with a glass frit.[60]

ments of the furnace temperature. The elements for the $\triangle T$ measurement should be designed to give broad surface contact with sample and reference. This reduces thermal resistance and increases the reproducibility of measurements. The thermocouples are placed either on the bottom of the crucible or in the center of a cylindrical crucible. With the latter type, the element can be brought protected or unprotected into the center of the substance. Rapid interchangeability and easy cleaning of the crucible are important considerations in the purchase and use of any DTA equipment.

NiCr-Cr thermocouples are used with low-temperature furnaces or for sensitive DTA in the middle-range temperature up to 600°C. Pt-Pt10%RH thermocouples are used with high-temperature furnaces. W-W25%Re are generally employed with the ultrahigh temperatures.

Temperature calibration To check the sensitivity and accuracy of a DTA apparatus, samples which have a known sharp melting or transition temperature are often used. Test runs are conducted with samples of potassium sulfate, benzoic acid, or Brazilian quartz. Of these, the behavior of quartz during DTA is well documented in the literature and the rapid alpha to beta inversion at 573°C can be used to calibrate the system. Quartz or potassium disulfate is sized in the range of −100 mesh + 200 mesh and analyzed against dried calcined alumina as a standard. A heating rate of 10°C/min is used, together with maximum amplification. A sharp endotherm at 574°C due to the inversion is observed.

The system is then allowed to cool at its natural rate. The cooling endotherm should fall within close proximity of the heating endotherm. In practice the most repeatable results have been found with Brazilian quartz. Provided that the base lines are stable, these tests can be taken to confirm the reproducibility of results.

Centering of sample holder   Centering the sample holder in the hot zone of the furnace is important for consistent results. Centering can be tested by the following procedure. Pack both the sample and reference crucibles with equal weights of −100 mesh + 200 mesh Brazilian quartz, seen at 10°C/min at maximum sensitivity (in some instruments the differential-temperature scale can be read with a precision of ±1.0 mV on the recorder printout with the maximum amplification setting), and monitor the system temperature in the sample side of the cell. In a perfectly balanced and symmetrical system the two thermal effects due to the simultaneous inversion of the two samples of material should interfere destructively and the resultant thermogram should show only an uninterrupted straight line. If the base line is fairly steady and the differential trace shows only a small discontinuity in the region of the inversion temperature, the cell is satisfactorily centered and there is no drastic imbalance in the system.

Source of reference temperature   The shape and position of curves obtained by DTA also depend upon the source of the reference temperature. The differential temperature may be plotted against the system temperature measured at one of several possible locations: (1) at the center of the reference sample, (2) at the surface of the sample material, or (3) at the center of the sample material. From consideration of these variables the most accurate and reproducible results are obtained by the third measurement. The effect of thermocouple location is very marked. For example, using Brazilian quartz as indicated above, the inversion temperature will be at 573 ± 2°C for measurement 3 and as high as 590 ± 2°C for measurement 1.

Assignment of transition temperature   Before a transition temperature is reached, the temperature at each junction of the differential thermocouple is essentially the same and no emf is generated. When a transition temperature starts in the sample, a certain amount of heat must be absorbed or given off before the sensitivity of the equipment allows it to be detected by the differential thermocouple. The first deviation of the DTA curve from the base line is more representative than the peak maximum of the start of a transition. The true transition temperature is still less and is also influenced by the rate of heating.

The nearest approach to the transition temperature is obtained by using the temperature at which the curve leaves the base line. Since this is often difficult to determine, the transition temperature can be taken as the intersection of the extension of the base line and the extension of the straight part of the adjacent side of the peak.

The shape of the DTA curve and the assignment of transition temperature depend on where the temperatures are measured. According as the differential temperature is plotted against surface temperature of the sample or the temperature of the center of the sample while the outside is being heated uniformly, the point of initial departure from the straight line or the peak of the curve respectively corresponds to the inversion temperature.

Process Control Variables

Rate of heating   Thermal transfer of the sample affects the symmetry of the peak. A sample with good thermal transfer tends to produce a symmetrical peak. Samples having poor thermal transfer produce an asymmetrical peak, which shows a gradual approach to the maximum deflection with a relatively rapid recovery of the original heating rate.

As heating rate is increased, the peak maximum is shifted to higher temperatures. A lower heating rate will lower the temperature of the reaction.

The initial transition temperature (beginning of a break) is independent or will change with change in heating rate, depending upon the sample material. For rapid transitions and those not accompanied by a loss in weight, the initial transition temperature is not greatly affected by changes in rate of heating. A high heating rate has been found desirable for a good determination of initial transition temperatures for rapid transitions.

A standard rate of heating for the determination of the thermal heating of thick film dielectric, glass, or glass-dielectric powders is 10°C/min.

Packing   Marked differences can occur if the sample and the inert material are packed differently. Differences in bulk density of the sample due to differences in packing create differences in weight, heat conductivity, and thermal diffusivity of the sample. The effect is most pronounced in the low-temperature range, where heat transfer is governed principally by conduction. The effect may disappear in the high-temperature range since the heat transfer is determined principally by radiation.

In general, differences in density of packing are the most common cause of deviations from straight base lines in temperature ranges where no reactions occur. For pronounced reactions with straight base lines tight packing is recommended. Loose packing leads to faint reactions. The effect on exothermic reactions is the same as endothermic with loose packing but less pronounced.

Particle size   Particle size has a significant effect on the shape and intensity of the inversion peak. The DTA curves of finely ground material tend to return more quickly to the base line after the peak maximum than the curves of a coarser fraction of the same material. However, in general, the effects are not observable until the particle size is smaller than 200 mesh. The technique by which the particles were fractionated also have an effect upon the DTA curves. Fractionation by sedimentation (5 to 20 μm) is a different process than the preparation of particles as small as 0.5 to 0.10μm by the grinding of coarser fractions. In addition to an intensity decrease, an endothermic reaction may be observed below 350°C which increases in intensity with longer grinding time.

The intensities of higher-temperature endothermic and exothermic peaks decrease with a decrease in particle size, and the peak position shifts to lower temperatures. The DTA curves of finely ground material tend to return to the zero position after the peak maximum faster than the curves of coarser fraction of the same material. The intensity of low-temperature peaks which may be associated with absorbed water increases as the particle size decreases.

Increasing the particle size of the sample results in strong reactions when those reactions are accompanied by changes in weight. Diffusion predominantly controls the rate of such reactions. For a given weight of material the sum of the surface area of particles increases with decreasing particle size. For example, the surface area is too small (if the particle size is greater than about 20 μm) for dehydration reactions to occur rapidly enough to yield pronounced reaction effects in DTA. For inversion reactions, however, these conditions are not applicable. Quartz consisting of 200-μm particles still perfectly exhibits the inversion from alpha to beta form at 573°C.

The degree of crystallinity of nominally crystalline particles may affect the DTA curve. Here we are referring to crystalline particles rather than particles of glass frit or crystallizable glass. The degree of crystallinity of fine and ultrafine ferroelectric particles was discussed under Properties of Dielectric and Ferroelectric Powders. These considerations are of importance in DTA thermograms of ferroelectric particles suspended in a glass frit. Poorly crystallizable materials give a broad endothermic peak of low intensity. Well-crystallized particulates show no such reaction. Poorly crystallized materials may also show a shifting of reactions to lower temperature ranges and the decrease of reaction intensity. The apparent effect of particle size on transformation temperature may actually be the effect of a lower degree of crystallinity in smaller particles.

Sample size   Sample size is an important operating parameter. Sample sizes used with dielectric powders range from 2 to 100 mg. There is a trend toward working with the smallest sample possible. The usefulness of the DTA technique can be extended and the errors inherent in the use of large samples (50 mg and above) are avoided or diminished. Small samples are preferred to permit the heat transfer from the furnace to the sample to be as nearly instantaneous as possible.

Large samples may lead to the creation of thermal gradients within the sample, with resulting loss in resolution due to overlapping of closely shaped reactions or poor definition of a reaction rate because of the simultaneous measurement of the reaction at different temperature levels. If large samples of 50 mg and above are used, slow heating rates such as ½ to 4°C/min will counter some of the disadvantages. However, because of the reduction in heating rate, $\triangle T$ values will be reduced. On the other hand, depending on the dielectric or glass powder used, too small a sample may not be representative of a production lot.

Reference substance The reference substance is as important for satisfactory DTA records as the sample. The two requirements for any reference are that it be inert and have thermal characteristics as similar as possible to those of the sample. Inertness is required so that no reaction occurs during heating. After any reaction in the sample, the thermal properties of the reaction products are usually different from those of the original material. It is not possible to find a reference material which can change its thermal properties continuously to match those of the sample as both are heated. Thus the choice of a reference material should emphasize selection of a substance whose thermal properties approach the average of those of various phases formed in the sample during heating.

Calcined alumina, Al_2O_3, is commonly used as the reference substance since its thermal characteristics are similar to most of the materials employed in thick film dielectrics. MgO is also a recommended reference. Calcined alumina should be replaced after each run or two since it becomes hygroscopic depending on its particle size. For satisfactory results, the particle-size distribution should approximate that of the sample. The use of coarse material results in too large a difference in heat conductivity between the sample and the reference. The difference results in difficulties in obtaining a straight base line.

Interpretation of Thermogram Profiles

The interpretation and evaluation of DTA thermogram profiles will be presented in terms of specific case histories. Emphasis is placed upon their use for the design and control of thick film dielectric manufacturing processes.

The following case histories, centered on glass and glass-ceramics, are discussed:

1. Crystallization of high-permittivity capacitors based on crystallizable barium titanate glass and barium titanate particulates or powders
2. Crystallization of low-loss capacitor dielectrics based on $BaAl_2Si_2O_8$.
3. Crystallization of microcrystalline barium titanate thick films based on $BaAl_2Si_2O_8$
4. Crystallization of crossover dielectrics based on BPO_4-SiO_2 eutectic compositions
5. Analysis of solder-glass seals in semiconductor packaging
6. Annealing effects in glass powders and frits.

Case 1: Crystallization of high-permittivity film capacitors based on crystallizable barium titanate glass and barium titanate particulates or powders[8, 55]   High-permittivity screenable capacitors can be fabricated by mixing high-dielectric barium titanate particles in a barium titanate glass. After firing, the barium titanate particles are bound in a dense film by smaller-grained barium titanate which has recrystallized from a crystallizable $BaTiO_3$ glass. The total barium titanate content is 90 to 95 percent by weight. The compositions were shown in Table 11 and the processing and dielectric properties discussed under Barium Titanate Glass-Ceramic Particulate Capacitors.

Four DTA profiles are shown in Figs. 70 and 71 to explain the use of DTA in the processing of the capacitor films. Crystallization of the barium titanate glass is shown in curve A. The exothermic reactions are noted. The pronounced exotherm at 800°C is identified with the crystal growth of $BaTiO_3$; the second exotherm at 980°C represents the crystallization of the non-$BaTiO_3$ glass-forming phase. This can be identified by comparison with curve B, which is the thermogram of the glass-forming phase. A single exotherm is observed at 990°C. The reaction at 1270°C represents the lowest-melting endotherm.

The DTA profiles for the BaTiO$_3$ particle-glass mixtures are shown in curves A and B of Fig. 71. Curve A shows the reactions for an 80:20 percent particle-glass mixture; curve B is for a 20:80 percent particle-glass mixture. Curve A has two exothermic peaks and one endothermic peak, which can be identified with the recrystallization and melting reactions in curve A of Fig. 70. However, there are no indications in the temperature range scanned of barium titanate powder or particulate

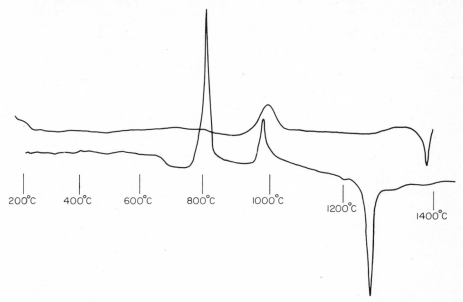

Fig. 70 DTA profiles of BaTiO$_3$ glass and the nontitanate glass-forming composition.[55, 88]

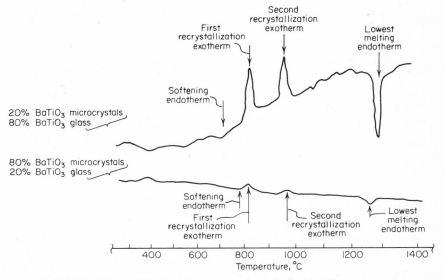

Fig. 71 DTA profiles for BaTiO$_3$-particle–BaTiO$_3$-glass film mixtures.[8]

sintering. This reaction normally is revealed as an endotherm. Curve *A* shows a profile similar to curve *B*. However, the intensities of the peaks are reduced because of the small barium titanate glass content. The absence of any barium titanate sintering endotherms over the temperature range should be noted.

The DTA profiles were obtained at rates of heating of 10 to 15°C/min.

The DTA profiles *A* and *B* are used to determine firing schedules. After completing binder burnout at (425°C for 10 min), the mixture is heated at a rate from 5 to 20°C/min to a temperature range between 700°C, or the softening-point endotherm preceding crystallization, and 1300°C, or the first melting endotherm following the final recrystallization exotherm. The specific temperature for any one composition in the mixture series lies within this range and depends upon the furnace construction and environment and the barium titanate purity, particle size, and dopants. Operable temperatures for recrystallization of the glass frit, for bond-

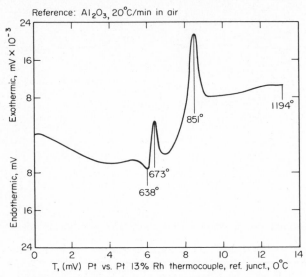

Fig. 72   DTA thermogram of a low-loss capacitor dielectric based on BaAl₂Si₂O₈.[118]

ing the barium titanate particulates and recrystallized barium titanate frit into a solid dielectric, and for bonding the film to the electroded substrate are readily obtainable from the DTA curve.

Case 2: Crystallization of low-loss capacitor dielectrics based on BaAl₂Si₂O₈[105]
Figure 72 shows the DTA thermogram of Du Pont 448-8299 crystallizable crossover. The ordinate on the figure is heat evolved or absorbed (exothermic or endothermic in microvolts and the abscissa is temperature in millivolts. The heating rate was 20°C/min with Al₃O₃ as the reference material. Two exothermic peaks are shown on the profile. The first of these, at 638°C, represents the formation of nuclei. The second represents the heat evolved when the primary phase crystallizes; this has been identified as hexacelsian, BaAl₂Si₂O₈.

A processing schedule was evolved from the DTA curve. Heat treatment at the nucleation temperature revealed by the first exotherm is not required for maximum crystallization. Raising the temperature directly to the temperature region of the second exotherm forms enough nuclei from which to grow small crystals. Soak time at lower temperatures is not necessary. Firing at 800 to 1550°C for 5 to 20 min produces the same results in terms of the size and concentration of crystals.

Case 3: Crystallization of microcrystalline barium titanate thick films based on BaAl₂Si₂O₈[7]   The DTA of the crystallization of BaTiO₃ with the feldspar BaAl₂Si₂O₈ as a minor phase by the heat treatment of glasses has been described. These compositions are presumably included in the ferroelectric glass-ceramic thick film capacitor materials discussed elsewhere.

Glass compositions corresponding to x BaTiO$_3$ + 100−xBaAl$_2$Si$_2$O$_8$, where x is the nominal barium titanate content (weight percent), were reported with stable glasses containing as much as 75 percent of BaTiO$_3$. The mechanism of crystallization was reported to be

$$\text{Glass A} \xrightarrow[800°C]{600 \approx} \text{BaTiO}_3(c) + \text{glass B}$$

$$\text{Glass B} \xrightarrow[1000°C]{750 \approx} \text{BaAl}_2\text{Si}_2\text{O}_8(c) + \text{BaTiO}_3(c)$$

The process takes place during the gradual heating of the glass.

The phase transformations associated with the mechanism of crystallization were determined from DTA. The thermal profiles are shown in Fig. 73. Curve A is the thermogram for a glass with a 74 percent nominal BaTiO$_3$ content. The heating rate was 750°C/h, or 12.5°C/min. The DTA profile shows an endotherm at 700°C, corresponding to the glass softening, two exotherms at 790 and 940°C, corresponding, respectively, to the crystallization of BaTiO$_3$ and BaAl$_2$Si$_2$O$_8$, and two exotherms at 1270 and 1380°C, which are related to the melting processes.

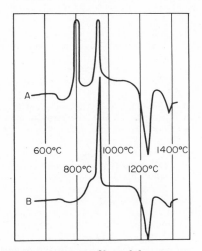

Fig. 73 DTA profiles of barium titanate–barium hexacelsian glass.[7]

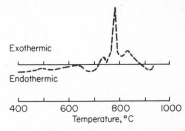

Fig. 74 DTA profile of crystallization of crossover dielectrics based on BPO$_4$-SiO$_2$ eutectic composition.[55]

Partially crystallized glass containing BaTiO$_3$ alone as a crystalline phase was used to obtain curve B. Specimens were prepared by heating for a short time at the first peak temperature (780°C) and then quenching the melt or by cooling the melt at a moderately rapid rate. Thus the first peak of curve A, which is missing on curve B, was due to the crystallization of BaTiO$_3$, and the second peak was caused by crystallization of the feldspar.

In general, the DTA profiles can serve as the basis for empirically determining the heating schedule for crystallization and fusion of the dielectric layers. The dielectric properties, i.e., dielectric constant, dissipation factor, and temperature characteristics, are influenced by the heating rate and maximum temperature used for crystallization. Minor changes in composition and heat treatment based on the DTA profile can result in an optimum manufacturing schedule. As described elsewhere, capacitor assemblies based on the titanates were fused and crystallized with schedules peaking at 900 to 1000°C, the soak time depending on whether the processing step was fusion or crystallization and on the hold time at temperature.

Case 4: Crystallization of crossover dielectrics based on BPO$_4$SiO$_2$ eutectic composition[55, 88] The DTA profile of the crossover dielectric reported in Table 17 is shown in Fig. 74. The glass was heated at 10°C/ in air with Al$_2$O$_3$ as a reference. An annealing and softening endotherm is observed in the vicinity of 700°C, followed by three exotherms of crystallization between 740 and 830°C.

A firing schedule for crystallization at 810 to 880°C was derived from the DTA curve. Although the exothermic region was between 740 and 830°C, the effects of mass and heating rate (600 and 200°C/min, respectively, for the DTA and film samples) caused the wetting and adherence and crystallization ranges to be shifted to higher temperatures. The higher temperature also results in the maximum yield of crystallization product.

The appearance of more than two exotherms indicates reaction between the glass-forming oxides and the BaO and TiO_2 during glass formation from the melt and/or interaction during crystallization. The absence of reactions as in cases 1 and 3 is indicated generally only by two exotherm reactions, one for the glass-forming systems and one for the dielectric or ferroelectric phase.

Case 5: Analysis of solder glass seals in semiconductor packaging The packaging or enclosure of semiconductor discrete devices or integrated circuits is essential for protection against degradation or unfavorable environments. Devitrifiable or crystallizable glasses, generally based on compositions in the $PbO-ZnO-B_2O_3$ system, are used for sealing. These are processed and fired in much the same way as thick film glass-ceramic dielectrics; they are similar to dielectric pastes for encapsulation, overglazing, or hermetic sealing in composition, melting temperatures, and softening-temperature range.

Two separate and detailed analyses of commercial sealing glasses have been reported which use DTA to analyze variations in different glass batches before production, characterize the crystallization characteristics of the glasses before production, investigate the behavior of the glazed parts before production, and monitor the production glazing cycles. Because of the relevance of the analyses to glass-ceramic thick film dielectric processing, the pertinent results of both studies are reported.

Study 1.[116] Solder glasses are processed and fired much in the same way as thick film dielectrics. Fritted and ground glass is mixed with a carrier vehicle and sprayed onto the alumina parts. During a first firing the binder vehicle is burnt out, and the glass is melted to the point of coalescence and wetting of the ceramic. During a second firing the glass crystallizes and forms a strong hermetic seal.

Glasses in the high-PbO region of the $PbO-Zn-B_2O_3$ system which melt at 370 to 500°C are suitable. Nominal compositions of commercial solder glasses contain 71 to 78 percent Pbo, 11 to 16 percent ZnO, 8 to 10 percent B_2O_3 and 2 percent of SiO_2 by weight and in some cases 1 percent of SnO_2 or BaO. Although the glasses have high expansion coefficients (8 to 10×10^{-6} C) relative to high-Al_2O_3 (92 to 98 percent) substrates (6.5×10^{-6} C) and lead members, the strength of the devitrified glass ceramic can usually tolerate the mismatch stress.

DTA has been used to investigate the crystallization kinetics of these glasses and to determine sealing-furnace profile design. The general form of the thermogram which characterizes the solder glasses is shown in Fig. 75.

The start of the initial endotherm A corresponds to the transformation point of the glass and a sharp increase in thermal expansion. Then endothermic peak B corresponds to the softening point of the glass. The start of the crystallization exotherm is at C. Nucleation or phase separation occurs between points A and C and gives rise to the initial thermogram at B. The exotherm D is associated more with precipitation of the primary crystalline phase than with actual crystal growth. Variations in the rate of heating have a marked effect on incipient crystallization. As an example, a change in DTA heating rate from 10 to 7°C/min reduces exotherm temperatures by about 20°C. If the recommended heating rates for manufacturing are substantially higher, they will probably exceed the limits of the sensitivity of the crystallization process to rate variations. Point F indicates the endothermic effect due to solution or melting of the crystalline phase.

DTA has been used to determine the effect of particle size upon the crystallization of glass. Solder glass is normally received through 200 mesh, which is less than 74μm. On occasion, this is too coarse for a particular use or for the accurate control of thickness required in some microseals. Within an experimental tolerance of 10°C, there is no significant change in the crystallization of glasses rescreened through a series of finer sieves. This remains true down to 400 mesh, which is less than 37μm and is the finest mechanical screen used. This has also been confirmed for finer fractions of less than 10 μm obtained by sedimentation.

Study 2.[117] The DTA curve for a typical sealing glass is shown in Fig. 76. The endotherm beginning at A is the transition temperature, at which the solid glass begins to transform into a viscous liquid. Point B is the softening point, where the glass begins to soften under its own weight. The AB part of the profile identifies the transformation range of the glass. This is a well-defined region observed with DTA on most glass-ceramics. Point C is observed only on DTA thermograms of powdered glass. It has been proposed that this slight reaction is a function of coalescence of glass particles after softening and fluid flow are obtained. The exotherm DF is the primary crystallization exotherm. At point D homogeneous nucleation begins. As the temperature increases, devitrification begins and the glassy phase is consumed. At F a rigid material is formed having a crystalline lattice with characteristic long-range order. Then endotherm at point G is the melting endotherm of the crystalline phase E with an exsolution of glass.

Beyond G other crystals may occur from the FG melt, but they are of no significance in this discussion. Point F is important since it marks the beginning of the

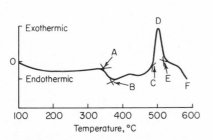

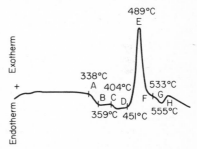

Fig. 75 DTA profile of sealing glass for study 1.[119]

Fig. 76 DTA profile of sealing glass for study 2.[120]

remelting endotherm and in this case occurs very near the devitrification zone. Sealing profiles come into the FG region and cause partial remelting with an exsolution of glass if good furnace temperature control is not maintained. Although a hermetic seal will result, the presence of glass contributes to a weaker seal. The highest content of crystallizable glass is required to create the strongest seal; this requires good control of both time and temperature.

Peak temperature position is a function of heating rate. It is a rate process that can be described with the Arrhenius equation, where rate $= Ae^{-E/kT}$. All terms have their usual meaning. Activation energies for the devitrification process can be obtained from DTA by varying the heating rates for several samples. Heating rates of 9°C/min are typical; rate reproducibility is on the order of ±1°C. The effect of heating rate on the DTA peak crystallization temperature is quite significant. A difference between 9 and 24°C/min, for example, can cause a shift of 24°C in the peak from 466 to 490°C. This is caused by mass transport; as heating rates increase, the reaction lags and occurs at higher temperatures, where the fluidity of the melt is more conducive to diffusion. Since the rate of crystal growth as well as nucleation is proportional to the fluidity of the glass, the highest practical temperature ensures maximum nucleation and subsequent maximum crystal growth rate. Nucleation can be completed at lower temperatures; however, the lower fluidity can cause some heterogeneous nucleation from surfaces initially and poor wetting of seal interfaces. This results in much weaker and less uniform seal structures.

Trial and error is the usual method of establishing the correct sealing profile. Arrhenius plots of heating rate versus reciprocal temperature for crystallization have been proposed as a possible method for obtaining supplemental information in designing furnace profiles for use with new sealing glasses or for checking present processes. If proper care is taken in obtaining the data, DTA helps furnish an estimate of the peak crystallization temperature. Extrapolation to faster heating rates should approximately indicate the temperature for maximum crystallization in the furnace at the specified rate. It has been indicated that rates of 9 to 24°C/min are typical for DTA but that sealing profile rates run closer to 100 or 150°C/min. On the plot in Fig. 77 a furnace rate of 120°C/min would show maximum crystallization to take place at about 530°C. This is very near temperatures on profiles shown to give good-quality seals for glasses.

Very high heating rates are not applicable to the plots in Fig. 77. As heating rates increase, the crystallization peak shifts upward, approaching the melting temperature of the primary crystalline phase. For heating rates of 400°C/min and above the DTA profiles would show no crystallization exotherm occurring; crystal growth rate drops off at very high heating rates. The failure to crystallize at higher temperatures has advantages for sealing glass since metal lead frames and solid discrete devices on integrated circuits can be attached.

Case 6: Annealing effects in glass powders and frits[118]   The use of glass films for crossover dielectrics, overglazes, or packaging glasses, requires the glasses to be stress-free to prevent crazing and cracking the glass. Stresses usually arise from thermal-expansion mismatched between the substrate and the glaze; however, the glasses, especially if of a proprietary nature and formulated by the user, may not have been annealed according to the proper schedule. This can be revealed by the DTA endotherm preceding the crystallization exotherms.

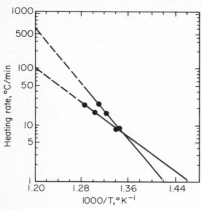

Annealing refers to the process by which internal stresses are removed. These stresses arise from unsuitable cooling conditions following any manufacturing process during which the glass has reached a sufficiently high temperature. This is usually greater than that temperature corresponding to a viscosity of 10^{12} P. The stresses can be removed only by reheating to a temperature in the annealing range which corresponds to a viscosity range of 10^{12} to 10^{14} P and controlled cooling according to a predetermined schedule. The glass for thick film pastes is usually prepared as a frit; i.e., it is rapidly quenched and shattered into small, brittle pieces by pouring the melt in water (preferably distilled to avoid the introduction of alkali-ion impurities) or other suitable quenching medium (as in the case of compositions which can be leached by water such as glasses containing barium oxide and boric oxide). After fritting, the particulate glass is not usually annealed. The glass is subjected to further size reduction by comminution with pebble mills. Each step introduces internal stresses. Thus by the time the glass is mixed and ground with dielectric powders further stress can be introduced. Incomplete annealing during processing of the glass-dielectric powder mixture through the softening range may be detected by the DTA if the stress is noticeable.

Fig. 77 Plot from Arrhenius relation for DTA heating rate versus maximum crystallization temperature for two commercial glasses, denoted by circles and squares.[120]

DTA curves showing differences in the annealing effect of glass as a function of heat treatment are shown in Fig. 78. The intensity of endothermic peaks in the glass transition range is an important indication of the thermal history of the glass. In relation to the total DTA profile, the annealing endotherm is the softening endotherm which precedes the first crystallization exotherm. Typical curves for a glass which is annealed normally, very carefully, or cooled rapidly are shown. The intensity of the peak varies with the degree of annealing. The intensity is high when the glass is well annealed but reduced or spread out if the glass is chilled. However, rapid cooling will accentuate an exothermic area preceding the endotherm.

This is a marked heat evolution at a somewhat lower temperature. Most glasses will produce these effects, depending upon the conditions of annealing; however, it is more difficult to observe in glasses with a high silica content.

Commercial Equipment

DTA equipment is readily available from a number of suppliers, including Du Pont, Tem-Pres Research Division of Carborundum Company, Edward Orton Jr. Ceramic Foundation, Mettler Instrument Corporation, and Columbia Instrumentation, among others.

Equipment is available from simple DTA setups for production control through systems employing combined thermogravimetry and DTA for research problems. It is recommended that before purchase, the specifications, features, and prices of each

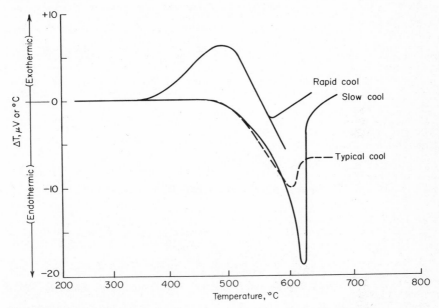

Fig. 78 DTA profiles showing differences in the annealing effect of glass as a function of heat treatment.[115, 121]

model offered be investigated in light of the requirements of the user. Specifications to be checked should include:

1. Range of temperatures available. Currently, they are $-190°$ to $1600°C$ in several ranges for standard equipment and up to $2100°C$ with interchangeable furnaces.

2. Interchangeability of furnaces.

3. Linear heating rates (0.1 to $100°C/min$).

4. Program-mode selection.

5. Heating-rate accuracy, reproducibility, and linearity.

6. Isothermal temperature control, accuracy, and drift.

7. Programmed temperature accuracy, reproducibility, linearity, and line-voltage stability.

8. Minimum detectable differential temperature ΔT.

9. Differential-temperature sensitivity ranges.

10. Data presentation.

11. Time-base ranges and accuracy.

12 Module and derivative sensitivity.

13. Base-line stability, shifts, displacement, and response sensitivity.
14. Vacuum capability.
15. Static or dynamic atmospheres.
16. Interchangeability of control console with other thermal-analysis equipment, e.g., thermal gravimetric analysis, scanning calorimetry, and thermal expansion (this consideration usually applies only to research and development procurement).

REFERENCES

1. Stermer, R. L.: *IEEE Int. Conv. Rec.*, vol. 12, pt. 9, pp. 47–53, 1964.
2. Reissing, T. C.: *Proc. IEEE*, vol. 59, no. 10, pp. 1448–1454, 1971.
3. Jaffe, B., W. R. Cook, and H. Jaffe: "Piezoelectric Ceramics," pp. 67–70, Academic, London, 1971.
4. Mountvala, A. J.: Characterization of Ceramic Materials for Microelectronic Applications, pt. III, *NASA*, CR-66817, October 1969.
5. Kingery, W. D.: "Introduction to Ceramics," pp. 380–384, Wiley, New York 1960.
6. McMillan, P. W.: "Glass-Ceramics," pp. 85–124, Academic, London, 1964.
7. Herczog, A.: *J. Am. Ceram. Soc.*, vol. 47, no. 3, pp. 107–115, 1964.
8. Ulrich, D. R.: U.S. Patent 3,649,353, Mar. 14, 1972.
9. Rase, D. E., and R. Roy: *J. Am. Ceram. Soc.*, vol. 38, pp. 102–113, 1955.
10. Templeton, L. K., and J. A. Pask: *J. Am. Ceram. Soc.*, vol. 42, pp. 212–216, 1959.
11. Brajer, E. J.: *Am. Ceram. Soc. Bull.*, vol. 36, pp. 333–336, 1957.
12. Krueger, H. H. A.: *Phys. Rev.*, vol. 93, p. 362, 1954.
13. Ref. 3, p. 62.
14. Ref. 3, p. 243.
15. Ref. 3, p. 62.
16. Anliker, M., H. R. Brugger, and W. Känzig: *Helv. Phys. Acta*, vol. 27, pp. 99–124, 1954.
17. Anliker, M., W. Känzig, and M. Peter: *Helv. Phys. Acta.*, vol. 25, pp. 474–475, 1952.
18. Känzig, W.: *Phys. Rev.*, vol. 98, pp. 549–550, 1955.
19. Kniekamp, H., and W. Heywang: *Naturwiss.*, vol. 41, p. 61, 1954.
20. Kniekamp, H., and W. Heywang: *Z. Angew. Phys.*, vol. 6, pp. 385–390, 1954.
21. Oshry, H. I.: U.S. Patent 2,695,239, Nov. 23, 1954; U.S. Patent 2,695,240, Nov. 23, 1954; U.S. Patent 2,803,553, Aug. 20, 1957.
22. Brandmayr, R. J., A. E. Brown, and A. M. Dunlap: *NASA Access.* N66-14809, *Rep.* ECOM-2614, 1965, avail. CFSTI.
23. Brown, A. E., R. J. Fisher, and S. DiVita: *Bull. Am. Ceram. Soc.*, vol. 37, p. 421, 1958.
24. Brandmayr, R. J., A. E. Brown, S. DeVita, and R. J. Fisher: U.S. Patent 2,990,602, July 4, 1961.
25. Mountvala, A. J., and D. R. Ulrich: Characterization of Dielectric Properties of Surface and Boundary Layers in $BaTiO_3$ Ceramics, *Bull. Am. Cer. Soc.*, vol. 47, no. 8, 1967.
26. Gerthsen, P., and K. Hardtl: *Z. Naturforsch*, vol. 18a, p. 423, 1963.
27. Chynoweth, A. G.: *Phys. Rev.*, vol. 102, pp. 705–714, 1956.
28. Merz, W. J.: *Prog. Dielectr.*, vol. 4, pp. 101–149.
29. Merz, W. J.: *J. Appl. Phys.*, vol. 27, pp. 938–943, 1956.
30. Fatuzzo, E., and W. J. Merz: *J. Appl. Phys.*, vol. 32, pp. 1685–1687, 1961.
31. Drougard, M. E., and R. Landauer: *J. Appl. Phys.*, vol. 30, pp. 1663–1668, 1959.
32. Brezina, B., and V. Janovec: *Czech. J. Phys.* vol. 9, p. 758, 1959.
33. Callaby, D. R.: *J. Appl. Phys.*, vol. 36, no. 9, pp. 2751–2760, 1965.
34. Callaby, D. R.: *J. Appl. Phys.*, vol. 37, no. 6, pp. 2295–2298, 1966.
35. Crawford, J. C., and R. D. Dragsdorf: *J. Appl. Phys.*, vol. 36, no. 9, pp. 2766–2771, 1965.
36. Williams, R.: *J. Phys. Chem. Solids*, vol. 36, pp. 399–405, 1965.
37. Tanaka, M., N. Kitamura, and G. Honjo: *J. Phys. Soc. Jap.*, vol. 17, pp. 1197–1198, 1962.
38. Mountvala, A. J.: Characterization of Ceramic Materials for Microelectronic Applications, *NASA Rep.* CR-6605, March 1968.
39. Mountvala, A. J.: Electrical and Magnetic Behavior of Ultrafine-Grain Ceramics,

in J. J. Burke, N. L. Reed, and V. Weiss (eds.), "Ultrafine-Grain Ceramics," p. 382, Syracuse University Press, Syracuse, N.Y., 1970.
40. Jonker, G. H., and A. L. Stuijts: *Philips Tech. Rev.,* vol. 32, no. 3–4, pp. 79–95, 1971.
41. Jonker, G. H.: *Ber. Dtsch. Keram, Ges.,* vol. 44, p. 265, 1967.
42. Ref. 3, p. 70.
43. Ref. 40, p. 90.
44. Ulrich, D. R., and E. J. Smoke: *Am. Ceram. Soc. Bull.,* vol. 44, no. 4, p. 337, 1965.
45. Jonker, G. H., and W. Noorlander: in G. H. Stewart (ed.), "Science of Ceramics," vol. 1, pp. 255–264, Academic, London, 1962.
46. Buessem, W. R., L. E. Cross, and A. K. Goswami: *J. Am. Ceram. Soc.,* vol. 49, pp. 33–36, 1966.
47. Ibid., pp. 36–39.
48. Ulrich, D. R.: unpublished data.
49. Brandmayr, R. J.: *Tech. Rep.* ECOM-2719, August 1966.
50. Kulcsar, F.: *J. Am. Ceram. Soc.,* vol. 39, pp. 13–17, 1956.
51. Ref. 3, p. 253.
52. Haertling, G. H., Improved Ceramics for Piezoelectric Devices, *West. Elec. Show Conv., Los Angeles, Aug. 23–26, 1966.*
53. Land, C. E.: Ferroelectric Ceramic Electro-optic Storage and Display Devices, *IEEE Int. Electron. Dev. Meet., Washington, D.C., Oct. 18–20, 1967.*
54. Ref. 39, p. 368.
55. Ulrich, D. R.: *Proc. Electron. Components Conf. IEEE,* pp. 17–25, 1966.
56. Ulrich, D. R., and E. J. Smoke: *J. Am. Ceram. Soc.,* vol. 49, no. 4, pp. 210–215, 1966.
57. Borek, R., B. Licanerski, and B. Rzasa: *Microelectron. Reliab.,* vol. 11, pp. 511–523, 1972.
58. Nester, H. H., and T. Cocca: *2d Symp. Hybrid Microelectron.* Western Periodicals, North Hollywood, Calif., 1967, pp. 7–12.
59. Nester, H. H., and D. Mason: *Proc. 1968 Electron. Components Conf., Washington, D.C.,* pp. 233–238.
60. Delaney, R. A., and H. D. Kaiser: *IBM J.,* vol. 11, pp. 511–519, September 1967.
61. Bratschun, W. R., A. J. Mountvala, and A. G. Pincus: Uses of Ceramics in Microelectronics, *NASA* SP-5097, 1971.
62. Socolovsky, A.: Capacitors; A Comprehensive EDN Report, *Electron. Des. News,* May 1966.
63. Dielectric Ceramics for Capacitors, American Lava Corp., Chattanooga, Tenn., *Bull.* 673.
64. Hamer, D. W.: *1968 Hybrid Microelectron. Symp.,* pp. 99–109.
65. Roup, R.: *J. Am. Ceram. Soc.,* vol. 44, no. 11, pp. 499–501, 1958.
66. Roup, R., and J. S. Kirby: U.S. Patent 2,841,508, July 1, 1958.
67. Hamer, D. W.: *Proc. Electron. Components Conf. IEEE, 1968,* pp 256–264.
68. Hamer, D. W.: *Proc. Electron. Components Conf. IEEE, 1969,* pp. 223–230.
69. Sherry, I.: *Electron. Prod.,* pp. 23–28, Mar. 15, 1970.
70. Rairden, J. C.: *Solid State Technol.,* January 1970, pp. 37–41.
71. Biggers, J. V., G. L. Marshall, and D. W. Strickler: *Solid State Technol.,* May 1970, pp. 63–66.
72. Holden J. P.: *Radio Electron. Eng.,* December 1968, pp. 381–387.
73. Delaney, R. A., and H. D. Kaiser: *J. Electrochem. Soc.,* vol. 114, no. 8, pp. 833–842, 1967.
74. Delaney, R. A., and H. D. Kaiser: *IEEE Trans. Parts, Mater. Packag.,* vol. PMP-2, no. 1/2, pp. 9–24, March/June 1966.
75. Delaney, R. A., and R. K. Spielberger: U.S. Patent 3,495,996, Feb. 17, 1970.
76. Asher, J. W., and C. R. Pratt: *Proc. 1968 Electron. Components Conf.,* pp. 239–245.
77. Herczog, A., and S. D. Stookey: U.S. Patent 3,195,030, July 13, 1965.
78. Layton, M., and A. Herczog: *J. Am. Ceram. Soc.,* vol. 50, no. 7, pp. 369–375, 1967.
79. Pratt, C. R., and W. H. Tarcza: U.S. Patent 3,267,342, Aug. 16, 1966.
80. Hoffman, L. C., and T. Nakayama: *Microelectron. Reliab.,* vol. 7, pp. 131–135, 1968.
81. Ulrich, D. R.: High Dielectric Thick Films for Screened Circuit Capacitors, *NASA Microelectron. Conf., Boston, February 1968.*

82. Ulrich, D. R.: Technical Support Package for Tech Brief 68-10542 High Dielectric Thick Films for Screened Circuit Capacitors, *NASA off. Technol. Utiliz.* PB180960, 1968.
83. Ulrich, D. R.: *Solid State Technol.*, vol. 12, p. 30, 1969.
84. Handler, P.: *Proc. IEEE*, vol. 52, pp. 1444–1447, 1969.
85. Stoller, A. I., J. A. Amick, and N. E. Wolf: *Electronics*, vol. 40. no. 6, pp. 97–105, 1967.
86. Nakayama, T., and L. C. Hoffman: *2d Symp. Hybrid Microelectron.*, October 30–31 1967, pp. 39–49.
87. Abrahms, H.: *8th Int. Electron. Circuit Packag. Symp.*, 1967 *Wescon Show*, vol. 8, pap. 2/2.
88. Ulrich, D. R.: *NASA Int. Rep.* 481, Langley Research Center, May 1966.
89. Hoffman, L. C.: U.S. Patent 3,586,522, June 22, 1971.
90. Stein, S. J.: *Proc. 1969 Electron. Components Conf.*, pp. 118–129.
91. McCormick, J. E., and D. W. Calabrese: *Proc. 1969 NEPCON*, pp. 500–506.
92. Topfer, M. L., A. H. Danis, and R. C. Heuner: A Universal Multilayer Hybrid Array, *NEPCON 1968.*
93. Fagersten, E. G.: *Proc. 1968 ISHM Symp.*, pp. 401–403.
94. Sarson, A. E.: *Microelectronics*, May, 1970, pp. 14–18.
95. Kurzweil, K., and J. Loughram: *Proc. 1973 Electron. Components Conf.*, pp. 212–219.
96. Bingham, K. C., and Y. Gurler: *Radio Electron. Eng.*, December, 1968, pp. 367–372.
97. Buck, R. H.: *UK Atom. Energy Auth. AWRE Rep.* 024/69, May 1969.
98. Theobald, R. R., M. P. Davis, and J. T. Bailey: *Proc. 1969 ISHM Symp.*, pp. 447–454.
99. Cox, J. T.: *Microelectronics*, February 1969, pp. 34–38.
100. Keller, W. R., F. E. Pirigyi, G. R. Cole, and J. P. Budd: *Proc. 1969 Electron. Components Conf.*, pp. 52–58.
101. Designing with Multilayer Ceramic Wiring Structures, E. I. Du Pont de Nemours & Co., Thick Film Technology data sheet ML-1, ser. A-66420.
102. Design Considerations for Multilayer Ceramic Substrates, American Lava Corp., data sheet dated 10/22/69.
103. Ihochi, T.: *1973 Proc. Electron. Components Conf.*, pp. 204–211.
104. Surina, J. J.: Tradeoffs in High Speed Interconnections, *NEPCON '70 Proc.*, 1970.
105. Hoffman, L. C.: Crystallizable Dielectrics, *1968 Hybrid Microelectron. Symp.*, pp. 111–117.
106. Miller, L. F.: "Thick Film Technology and Chip Joining," pp. 53–55, Gordon and Breach, New York, 1972.
107. Isaak, H. R., J. W. Kanz, and E. G. Baberacki: Development of Large Thick Film Multilayer Assemblies, *Proc. ISHM Symp., 1971.*
108. Hailes, A.: Thick Films: The Substrate-Conductor Relationship, *Electron. Packag. Prod.*, June 1972.
109. Horning, A.: Electrical Properties of a Silver Contaminated Borosilicate Glass, *Proc. ISHM Symp. 1968.*
110. Horning, A.: Diffusion of Silver in Borosilicate Glass, *1968 Proc. Electron. Components Conf.*
111. DuPont data sheet A-73482.
112. DuPont data sheet A-71765.
113. Smothers, W. J., and Y. Chiang: "Handbook of Differential Thermal Analysis," Chemical Publishing, New York, 1966.
114. Smothers, W. J., and Y. Chiang: "Differential Thermal Analysis: Theory and Practice," Chemical Publishing, New York, 1958.
115. Garn, P. D.: "Thermoanalytical Methods of Investigation," Academic, New York, 1965.
116. Forbes, D. W. A.: *Glass Technol.*, vol. 8, no. 2, pp. 32–42, 1967.
117. Ramsey, T. H.: *Ceram. Bull.*, vol. 50, no. 8, pp. 671–675, 1971.
118. Ref. 113, p. 94.

Chapter **8**

Component-Attachment Techniques and Equipment

ALBERT E. LINDEN

Honeywell, Inc., Fort Washington, Pennsylvania

INTRODUCTORY BACKGROUND

This chapter deals with the processes, procedures, materials, devices, and techniques associated with the assembly of hybrid thick film circuits. In addition, it mentions the material interfaces and compatibilities that can occur in any complex system of piece parts, conductors, resistors, insulators, and dielectrics associated with the fabrication of the thick film circuit.

Fig. 1 Servo amplifier with tantalum capacitors.

Discrete Piece Parts in Packages

Any discrete electronic piece part, whether active or passive, that can be soldered or otherwise bonded to a metallic interface can be utilized in a thick film circuit. It may not be the most efficient approach from a standpoint of packing density, cost, or volume, but nevertheless most "standard" component parts can be used in a hybrid thick film circuit. This list could be endless but would certainly include glass and plastic diodes; hermetic and plastic transistors; capacitors of all types, paper, plastic tantalum, mica, etc.; coils; transformers; reed and other types of relays; resistors, carbon, metal film, etc.

The foregoing refers essentially to the use of standard piece parts not necessarily designed for use in hybrid circuits. The message is that almost anything can be designed into a hybrid circuit and it is this versatility that has contributed to the expanding popularity of the technology. Figure 1 shows a circuit using standard tantalum capacitors in a thick film circuit. It should be noted that a conductive epoxy was used to attach the leads of the tantalum capacitors for two reasons: (1) the capacitors are heat-sensitive devices, and (2) the use of solder with the associated use of flux is to be avoided in the presence of uncased semiconductors, as seen in Fig. 1.

The foregoing discussion relates only to the use of parts not designed for use in hybrid circuits. Increasing numbers of part types are specifically designed for hybrid attachment, and these should be noted and sought out when planning any hybrid circuit.

Chip Parts

Several categories of parts are available in the general classification of *chip* devices. Care must be exercised with nomenclature, however, to avoid assuming that a term related to one device means the same related to another. A chip capacitor, for instance, has a completely different configuration from a chip transistor or integrated circuit. In this description of chip parts, we are referring only to leadless devices or *unpackaged* semiconductors. The descriptions of these chips are related only to the attachment problems they create as a result of being leadless devices and not to their electrical parameters.

Capacitor chips Monolithic ceramic chip capacitors have become increasingly popular in hybrid thick film circuits. While they are fairly expensive (especially in high production as compared to printed thick film capacitors) and require a given

TABLE 1   Chip-Capacitor Sizes

Length, in.	Width, in.	Thickness, in.
0.050	0.040	0.040
0.075	0.045	0.040
0.077	0.054	0.050
0.100	0.050	0.060
0.150	0.050	0.050
0.170	0.065	0.065
0.190	0.065	0.065
0.180	0.080	0.065
0.270	0.080	0.070
0.125	0.095	0.065
0.175	0.125	0.065
0.197	0.180	0.065
0.225	0.210	0.065
0.390	0.425	0.065
0.350	0.095	0.070
0.395	0.145	0.060
0.595	0.185	0.070
0.585	0.298	0.070

amount of labor for attachment, they nevertheless are very conservative in their use of real estate, and this, of course, is their major advantage in microminiaturization.

Chip capacitors come in a variety of sizes, which are, of course, related to their electrical values. Capacitances in picofarads range from below 10 to 10,000; although the available sizes vary with the manufacturer, they are generally as shown in Table 1.

Working voltages can vary from under 25 WVDC to over 200 WVDC—obviously a wide range of choice.

Of more importance to the assembly operations than the size or electrical characteristic of the capacitor are the end metallizations. End metallization is available in a variety of formulations, and the type should be chosen in accordance with the assembly process to be utilized. Standard electrode metallization can be obtained in Pt-Au, Pd-Ag, gold, and solder-coated. The metallization should be chosen to be compatible with the bonding process and optimized to it. If solder attachment is being used, a solder-coated metallization would be the most natural choice. Table 2 shows the type of metallization that would be used for different bonding systems.

Resistor chips The first question about resistor chips is why they should be required in a thick film circuit at all. Since one of the basic advantages of the thick film circuit is the ability to print resistors, why worry about purchasing, testing, inventorying, and attaching separate piece-part resistor chips? The reasons can be any of the following:

1. *To eliminate a separate screening.* Most of the resistors in a given circuit may fall into a category that can be screened with one paste, that is, 1,000 Ω/sq, while one resistor (or even a few) may require a higher-value paste, that is, 10,000 Ω/sq. Economics may dictate, therefore, that a chip resistor be used rather than going the route of a separate piece of artwork, a separate photoreduction, a separate screen, another printing, etc.

2. *Low-noise requirements.* Thick film resistors are noisier than thin film resistors. A circuit such as a very high gain amplifier may require low-noise resistors on the input. The only solution could conceivably be the use of a thin film chip resistor.

3. *The requirement for one or more high-value resistors.* Here again, high-value chip resistors are available in thick film, thin film, and silicon-chip configurations.

TABLE 2   Electrode Metallizations for Various Bonding Systems

Metallization	Bonding system	Results
Pt-Au..........	Pb-Sn solder	Good
	Au-Sn solder	Good
	Au-Ge solder	Good
	Conductive epoxy	Good
	Wire bonding	Good
Pd-Ag..........	Pb-Sn solder	Excellent
	Au-Sn solder	Good
	Au-Ge solder	Good
	Conductive epoxy	Good
	Wire bonding	Fair
Gold..........	Pb-Sn solder	Poor
	Au-Sn solder	Good
	Au-Ge solder	Good
	Conductive epoxy	Good
	Wire bonding	Excellent
Solder-coated....	Pb-Sn solder	Excellent
	Au-Sn solder	Good
	Au-Ge solder	Good
	Conductive epoxy	Poor
	Wire bonding	Poor

4. A need for a special thermal coefficient of resistance (TCR) may force the use of a special chip resistor in order to satisfy a special circuit requirement.

Recognizing that chip resistors may be required, it is necessary to know how best to attach them to the hybrid circuit. In general, if the chip resistor has metallized ends, it may be attached in the same manner as a chip capacitor. Referring again to Table 2, we see that a choice of end metallization is available in accordance with the attachment methods.

The thin film chip resistor will probably have only surface metallization for bonding points, and in general the chip will be bonded to the substrate with a non-conductive epoxy while the interconnections are made with wire bonds. Silicon chip resistors are interconnected in the same manner (with wire bonds), but the silicon chip can be bonded like any semiconductor dice. More will be said on these die-bonding methods later in this chapter.

LID or channel devices   Along with other ceramic chip-type devices, there is the *leadless inverted device* (LID), sometimes called the *channel carrier*. At their inception, LIDs carried only diodes and transistors due to the limitations set by the four-terminal configuration. Now channel devices are available that house integrated circuits with up to 14 terminals. The basic advantages of using LID, or channel devices, are as follows:

1. They are considerably smaller than a discrete device with leads, and yet the user does not have to concern himself with handling, bonding, and wiring a die.

2. The LID can be tested and characterized in spite of its small size. It is almost, but not quite, impossible to test a die after it has been diced from the wafer.

3. The LID can be attached to a hybrid circuit by standard solder techniques without the necessity for investment in die-bonding and wiring machines.

4. LIDs are much less apt to be damaged during handling than the unprotected die.

The metallization on the LIDs is usually a moly-manganese with Ni-Au plating. This permits the LID to be attached to substrate metallization by normal solder or epoxy attachment techniques (discussed in detail later in this chapter). It should be noted, however, that the extremely small size of the metallized pads on the LIDs and the small spacing between pads can cause considerable difficulties in the attachment processes:

1. Good cleanliness must be maintained to prevent finger oils from causing insufficient solder wetting of the attachment pads.

2. Solder splashes or epoxy spread can cause shorting between attachment pads during assembly.

3. The ability to inspect each attachment pad is obscured by the body of the LID, making inspection of the interconnections essentially impossible.

4. Not all devices (transistors, diodes, integrated circuits) are necessarily obtainable in LID form.

Semiconductor chips   Anyone working with uncased semiconductor chips must remember that he is assuming part of the responsibility of the manufacturer of the transistor, diode, or integrated circuit. He must therefore be ready to do that part of the assembly job the manufacturer does in bonding, wiring, and packaging the chip into its normal salable configuration. This includes die bonding, wire bonding, and hermetic sealing or encapsulation of the device, along with all the associated processes of inspection, cleaning, and testing. Anyone who is not ready to assume the process control and capital investment involved in doing this job had better give it a second thought.

Silicon semiconductor chips range in size from a few mils (0.015 in.) square to as large as 200 mils (0.200 in.) square. Making a eutectic die bond requires a vacuum pickup tip for each size chip it is expected to bond. A eutectic bond can be accomplished with no more equipment than a heated stage and a pair of tweezers, but this approach is highly unsuitable for complex hybrid circuits.

Chips can be purchased with gold backing or with just the bare silicon backing. In either case, the eutectic bond is made to a gold conductive pad. More will be said about the Si-Au eutectic bond later in this chapter. The bond quality depends largely on the character of the gold in the conductive pad, the temperatures involved, and the cover gas used. A good eutectic bond will show a smooth fillet on all four sides of the silicon chip.

It is more difficult to die bond to some *cermet conductors* than to others. The manufacturers of the cermet pastes generally have recommendations for their product, rating it good for die bonding, solder bonding, epoxy bonding, etc. The eutectic formation around a chip should be as small and well confined as possible since it is difficult or almost impossible to wire-bond to the Si-Au eutectic. If a wire has to be joined to the collector portion of a transistor chip, for instance, the pad should be designed to allow room for the wire bond to be made, well away from the eutectic formation.

From the standpoint of electrical characteristics, the best way to purchase semiconductor chips is in wafer form. This permits the hybrid user to characterize and test the devices in wafer form, when they can be probed, prior to dicing. This of course imposes the additional responsibility of dicing and sorting on the fabricator of the hybrid circuit. Another method of testing is to probe the substrate after partial or complete assembly. Figure 2 shows a device suitable for such testing. It contains the number of probes and the latitude of adjustment required.

Fig. 2   Probe for testing complete or partially complete substrates.   (*Micro Dynamics, Inc.*)

SOLDERING

Surface Materials

The four basic characteristics to consider when selecting a thick film conductor to be used for the solder attachment of parts are wettability, leaching, pull strength, and resistivity.   Each of these characteristics should be checked on the vendor's specification sheet before a selection is made.   The user should then check the vendor's paste in his own laboratory to assure that his equipment and processes can duplicate the vendor's specifications.   Each of the conductor characteristics is important, and a trade-off must often be made to arrive at the optimum conductor characteristic for a given circuit or process.

Wettability   Defined as the ability of solder, whether Sn-Pb or any other eutectic combination of metals, to adhere to another metallic surface, the wettability of given cermet conductors can be affected by the firing temperatures and the print thickness. Conductor materials are usually rated by the vendor's specification sheets as excellent, good, fair, or poor, with relation to their ability to be wet by 60-40 or 63-37 lead-tin solder.   In many cases, wettability can be improved by using the proper fluxes or burnishing the cermet conductor.   The proper appearance of a well-wet conductor is an unbroken mound of solder without dewetted areas, as shown in Figure 3.

Leaching   Also called *resistance to solder leaching*, this is a factor of importance when an assembly is fabricated by the solder-reflow technique.   The measurement is usually expressed in seconds of immersion in a solder pot at 225°C.   Thus Pt-Au conductors, which have a high resistance to leaching, will withstand as much as a 240-s immersion in a solder pot while Pd-Au, also a good conductor for soldering, will withstand only 80 s in a solder pot (these are approximate figures and vary with the vendor's formulation of the paste).   Each vendor has his own data sheets, or should have, and these properties must be given for an evaluation to be performed.

Pull strength   This measurement, which indicates the adhesion properties of the conductor to the ceramic substrate after firing, is sometimes expressed in pounds per inch width and can vary from a few pounds to as much as 50 lb of pull strength. Most companies screen a test pattern of conductors on which measurements of this nature can be accomplished. In general, the method for conducting an adhesion measurement consists of soldering a wire to a conductor pad of a predetermined size. The conductor pad can be 200 by 200 mils or even smaller. The important part is that various conductors be evaluated against each other using the same size test pads.

After the wire is soldered, the pulling should be done at 90° to the substrate surface, using any of the standard pull-test machines with a gage selected for its range and an indicator dial with memory. When the adhesion of the conductor is very good, instead of the conductor separating from the ceramic substrate, the solder may pull away from the conductor. When this happens, an evaluation must be made to determine whether the cause is excellent adhesion or poor solder wettability to the conductor. Firing profiles and peak furnace temperature affect both wettability and adhesion of most conductor materials.

Resistivity Typical values for this important conductor characteristic are shown in Table 3.

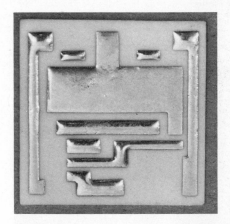

Fig. 3 Well-wet conductors. (*Electrovert Inc.*)

Surface Preparation

Cleanliness For any soldering operation, the parts involved (if a quality joint is to be achieved) must be clean. In microelectronics, the small areas involved dictate that this cleanliness be scrupulous. The most prevalent form of contamination in hybrid thick film circuits is common finger oils, which cause poor wetting, staining, and, in reflow operations, dewetting. Often these problems will be blamed on the

TABLE 3   Resistivity of Common Thick Film Conductors

Conductor type	Resistivity, Ω/sq
Gold...............	0.005
Pt-Au..............	0.1
Pd-Au..............	0.1
Pd-Ag..............	0.04
Silver..............	0.001

cermet materials, the solder, the fluxes, or the process temperatures. In general, however, the cleanliness of the surface will be the primary fault.

The best way to achieve a clean surface is not to get it dirty. When the cermet film emerges from the firing furnace, one can be sure it is clean—as clean as it will ever be—due primarily to the firing temperatures to which the material has just been exposed. All preexisting contamination will have been thoroughly burned off.

The problem now is to *keep* the surfaces clean. This can be accomplished best by having the assembly operators wear gloves or finger cots. It should be noted that even in a simple situation like this, some training is necessary. The assembly operator must be trained not to pick up "dirty" things while wearing gloves or finger cots and not to scratch his head or brush things from his face, obviously contaminating the protective garment.   For this reason a routine of glove or finger-cot changing must be established and adhered to.

Burnishing   All cermet films contain some glass, which often appears at the surface of the conductive film in the form of nodules.   It should be removed prior to assembly of devices to the thick film circuit.   Glass and other surface contamination can be removed by burnishing the surfaces.   The burnishing action makes the conductive film denser while simultaneously removing surface impurities and contamination.

The burnishing operation is simple and can be accomplished by hand or with automation, as the case demands.   The simplest implement for burnishing could be a pencil eraser.   An "ink" eraser is more efficient and is used by stroking it (with moderate pressure) across the conductive film.   An improvement on this would be a draftsman's motorized eraser machine.   Care must be exercised to prevent the eraser from burning due to the friction, as this would leave additional contamination on the conductive surface.   A preferred burnishing tool is the fiber-glass brush, which is more efficient than the ink eraser and less apt to leave contaminants on the surface of the conductive films.   Figure 4 shows a typical fiber-glass brush, and Fig. 5

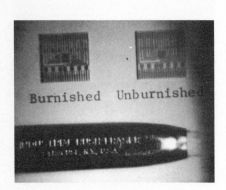

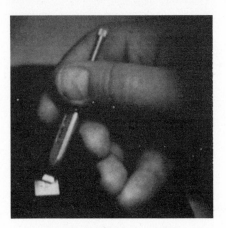

Fig. 4   Fiberglass brush.   **Fig. 5**   Burnishing process.

shows the burnishing process.   Note that the substrate is being held by a simple vacuum chuck while the operator strokes the brush across the substrate.   The stroking action should be in all four directions to assure complete burnishing.   This operation, as mentioned before, can be accomplished automatically by feeding the substrates under a revolving fiber-glass wheel.

Solvent cleaning   After burnishing the conductive surfaces, the substrate will be left with eraser or fiber-glass contamination on the surface, which must be removed by solvent cleaning.   Solvent cleaning is best done with the assistance of ultrasonics. A small ultrasonic cleaning machine of 25 to 50 W energy will suffice.   While many methods for cleaning are equally satisfactory, a simple one is outlined as follows:

1. Place the substrates to be cleaned in a beaker of trichloroethylene and put the beaker in an ultrasonic cleaner for a full 5 min.

2. Remove the substrates from the first beaker of trichloroethylene, let them drip (but not so long as to dry), and place them in a second (clean) beaker of trichloroethylene.   Put the second beaker in an ultrasonic cleaner for a full 5 min.

3. Remove the substrates from the second beaker of trichloroethylene, let them drip to remove the surplus solvent, and place them in a third beaker of methanol. Put the beaker of methanol in an ultrasonic cleaner for a full 5 min.

4. Remove the substrates from the methanol and set them under a flow of nitrogen gas until completely dry.   Although this cleaning procedure is a good one, it may be too cumbersome, time-consuming, or expensive for a low-cost commercial operation.

It can naturally be modified to suit the need.   The important point is that some cleaning process is necessary after burnishing and prior to further assembly.

Fluxing   If a flow- or wave-soldering machine is to be used for the tinning operation, a *foam fluxer* may be a part of the machine.   The foam fluxer automates the fluxing operation and makes it integral with the tinning operation, thus saving a separate process step and the additional labor that would be required.   The one problem is that a wave-solder machine results in solder pads with a buildup of solder, hindering the placement of piece parts.

Tinning   Tinning is a prerequisite to solder assembly and offers visual indication of the cleanliness and wettability of the surfaces.   Well-tinned surfaces always pro-

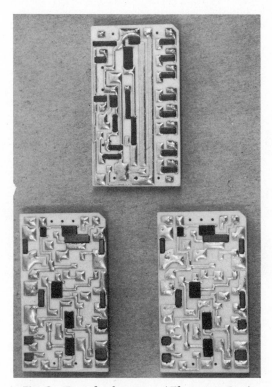

Fig. 6 Tinned substrates. (*Electrovert Inc.*)

duce good solder joints.   Poorly tinned surfaces usually do not produce acceptable solder joints.

There are two basic tinning techniques for hybrid thick film conductors, of which other techniques are modifications, namely, solder dip and solder wipe.

Solder Dip. This expression tells less about the process itself than the results obtained. If the substrate, with thick film conductors printed thereon, is fluxed and dipped in a solder pot or run through a wave-solder machine, the result should be well-wet conductors with nice solder buildup on each pad area and a proper and even meniscus around each pad.   Figure 6 shows such a tinned substrate.

A solder coating of this nature can be achieved by dipping the substrate, conductor side down, in a solder pot and removing it slowly, maintaining approximately a 60° angle with reference to the surface of the solder.   The same effect can be achieved by passing the substrates over a solder wave, as shown in Fig. 7.   It should be noted that this is a modified printed-circuit wave-solder machine, fixtured to hold the small ceramic substrates.   To the left can be seen the foam-fluxing attachment while to the right of center the substrates can be seen passing over the solder wave.

A substrate tinned in this manner is ideal for the attachment of piece parts one at a time.   But for mass production, where the piece parts must be preset and positioned on the pads, the meniscus of solder on each pad prevents easy positioning of the piece parts.   For this reason the solder-wipe method is sometimes used.

A hydro-squeegee machine can be used to solve this problem.   This machine will take a tinned substrate and cause the excess solder to be "wiped" off by the use of hot oils under high pressure.   Machines of this nature can be obtained commercially.

Solder Wipe. Tinning by the solder-wipe method is designed to leave a flat tinned surface on the conductor.   This is difficult to achieve by the solder-dip method.   The way to obtain a flat tinned surface is to wipe off the excess solder while the substrate is still hot enough to keep the solder molten.   The wiping can be accomplished with a lint-free rag, a squeegee, or a damp sponge.   All three methods can be made to work, and the optimum technique is really a matter of personal choice.   The tinned

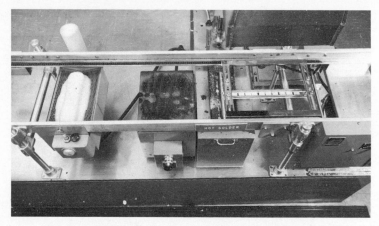

Fig. 7   Substrates being wave-soldered.   (*Electrovert Inc.*)

surface obtained from the solder-wipe technique will be flat and will readily accept the placement of piece parts.

Solder-Cream Printing. In the last few years, several solder companies have introduced solder creams which are printable by standard thick film printing techniques. These solder creams are usually proprietary formulations of lead and tin, finely ground and mixed with flux as a vehicle.   When printed through a coarse screen (200 mesh or larger), the solder cream will remain sticky long enough to apply parts, which are held in place by the stickiness.   The usual procedure is to apply all the parts to the wet solder cream and then subject the substrate to a heat source (hot plate, infrared lamp, etc.) to flow the solder and attach the parts.   Even when using solder cream, a superior job will result if both the substrates and the parts have been tinned prior to solder-cream printing and assembly.

Solders

For both tinning and assembly, common 60-40 solder is an acceptable material. The problem of thick film conductor materials leaching into the solder is one to cause concern.   Each manufacturer of thick film cermet conductor materials will give, in their data, a figure relating to the resistance to leaching.   This is usually expressed in seconds and is indicative of the length of time the material can soak in a solder pot before all the conductor material sublimates into the solder.   The sublimation rate can be lengthened by adding a small amount of silver to the solder formulation.   For instance, a formulation of 62% tin, 36% lead, and 2% silver will impede conductor sublimation considerably. It is even more effective if the conductor formulation contains silver.

Often a process requires different solder-melting temperatures so that a process sequence can be accomplished. Naturally the hottest operation would be performed first and then the next lower and so on. Table 4 shows solder combinations with their melting temperatures and comments on tarnish resistance and solderability.

It is important to find the adhesion capabilities of the solder pads to which parts are to be attached. This can be accomplished by performing solder pull tests, which take one of two forms. The most common method is to solder a wire to a 0.050 by 0.050 in. pad so that the wire is vertical to the pad. A force is then applied to the wire until the solder pad is pulled loose from the cermet or the cermet is pulled loose from the substrate. Typically, a 0.050 by 0.050 in. pad should withstand several pounds of pull before breaking loose (about 4 lb). Another method is to solder the wire parallel to the solder pad and to then apply the force 90° to the pad. This tends to peel either the solder from the cermet or the cermet from the substrate. The absolute value of the bond is less important than the repeatability of the tests, as an indication that the firing and tinning processes are under control.

TABLE 4   Common Thick Film Solders

Type	Liquid temperature, °C	Tarnish resistance	Solderability
60Sn-40Pb..........	183	Good	Excellent
60Sn-38Pb-2Ag.......	185	Good	Excellent
90Pb-10Sn..........	280	Fair	Good
95Sn-5In............	315	Fair	Excellent
100Sn..............	232	Good	Good

Reflow Assembly

In almost all cases where parts are soldered to thick film substrates, the reflow-soldering method is used. Essentially, this means that a part with a pretinned lead is soldered to a pretinned pad, without the application of additional solder. Generally, flux is used to assist the reflow process. The heat required to cause the solder to reflow can be provided in a number of ways, and the parts attachment can be accomplished singly or all together, depending upon production quantities or the quality of the work desired.

If part leads are being attached singly, it is wise to have the substrate on a preheated stage to simplify bringing an individual pad to the solder-reflow temperature. The heated stage should be about 125°C to obtain a quality solder joint with good meniscus. If mass soldering is being utilized, a preheat stage should be used to reduce the thermal shock.

Heat sources Any heat source properly designed and applied can be made to work for solder assembly. When choosing a heat source, the concerns should be efficiency, repeatability, ease of application, equipment cost, skill-level requirement, and tooling needs. The numerous heat sources available include soldering irons, resistance sources (parallel gap), power pulsing (welding power supplies), hot plates, hot gas, hot wires (nichrome), infrared, flame, laser, induction, and furnaces.

The listing mentions *types* of equipment and *sources* of heat, but this is unavoidable since a heat source without the equipment is usually meaningless and vice versa. The important thing is the method by which the heat source is implemented and its adaptability to solder assembly.

Soldering equipment The necessity for the proper equipment tailored to the specific job cannot be treated lightly. Since so many kinds of equipment are available, a careful survey should be made before funds are committed. The ease of use, the production rate, the temperatures involved, the pad sizes, and all other factors should be considered.

In most cases, when soldering to alumina thick film substrates, it is best to preheat the substrate prior to soldering. This is best accomplished by using a heated stage which will elevate the substrate to between 125 and 150°C. The task for the heat source, which is to elevate the substrate to solder-melt temperature, is considerably eased by the application of this technique.

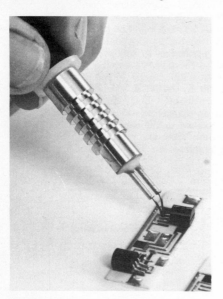

Any method of applying the additional heat necessary to reflow the solder will suffice and is merely a matter of convenience. It should be mentioned that for the solder joint to reflow properly, flux is generally required. This can be applied with a hypodermic needle, a toothpick, a pin, or any other convenient method. With mass reflow soldering (all parts on the substrate at once), the flux may assist in holding the parts in place as the substrate is passed from the preheat stage to the reflow stage. This occurs due to the stickiness of the flux.

When reflowing parts singly, a convenient tool is a soldering pencil, generally a nichrome resistance loop. A tool of this nature can be simple or made sophisticated by the addition of time controls, temperature controls, or both. Figure 8 shows a simple hand-soldering tool poised on a transistor lead. The picture is

Fig. 8 Hot-wire soldering tool. (*Browne Engineering Co.*)

posed, in that no hot stage is present to preheat the substrate, but the illustration is nevertheless clear.

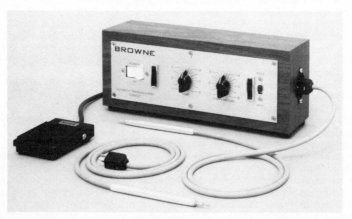

Fig. 9 Hand-soldering tool with controls. (*Browne Engineering Co.*)

Equipment for reflow soldering can become very exotic. The simple tool shown in Fig. 8 can be temperature-controlled only, with foot-peddle actuation, or it can have both temperature and time controls, as shown in Fig. 9. The next step toward sophistication is the desk-top model shown in Fig. 10. Notice that this model is using the heated stage so that the substrate can be preheated.   In the closeup of the reflow-soldering tool in Fig. 11 the resistance heating element can easily be seen.

If lead frames were to be attached to substrates by the solder-reflow method, a simple technique would be to use a heating device as shown in Fig. 12.   While this device was designed for soldering multiple-lead dual-in-line or flat-pack devices to printed-circuit boards, it can be adapted to lead assembly to substrates by merely substituting a heated stage for the pictured workholder.   It is necessary to preheat the substrate, for reasons mentioned before.

For high-production reflow soldering, the parts would of necessity have to be fixtured into position on the substrate and a method devised for applying heat to the entire substrate.   There are many ways of generating heat.   The one illustrated in Fig. 13 shows infrared heaters set over a moving belt.   By controlling the energy

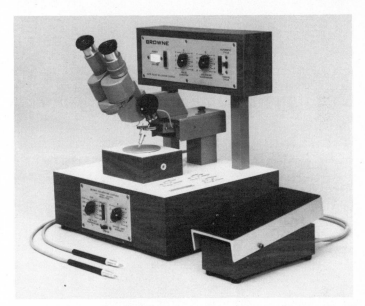

Fig. 10   Desk-console soldering equipment.   (*Browne Engineering Co.*)

setting of the infrared heaters and the belt speed, a temperature profile can be set that will preheat the substrate, reflow the solder, and slowly cool down the substrate at any desired slope.

Another technique of infrared solder reflow can be accomplished by focusing the infrared heat to a specific area.   The dwell times can be reduced to 8 to 10 s.   The application of a solid-state power supply will control the infrared energy and set the temperatures that will be seen by the substrate.   By adjusting the focus points, the diameter of the heated spot can be determined.   If the system is to be automated, an electrically driven index table can be supplied.   Such a system is shown in Fig. 14.

Substrate Bonding

As with all other hybrid operations, each assembly technique depends upon previous operations.   It must be decided during the initial assembly planning whether the substrate will be bonded to the package before or after parts assembly.   Once this is determined, the temperature tolerance of the bonding operation can be set. To clarify this statement, it should suffice to say that if parts have been bonded to the substrate with solder that melts at 183°C, it would not be possible to bond the substrate to the package with solder that melts at the same temperature (183°C) or

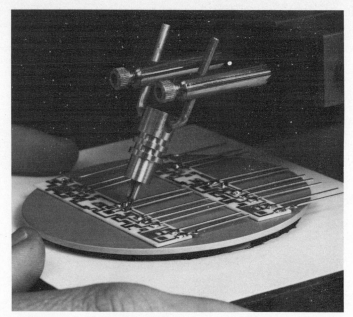

Fig. 11   Reflow-soldering tool.   (*Browne Engineering Co.*)

Fig. 12   Tool for lead frames or flat packs.   (*Wells Electronics Inc.*)

the parts would float free during the substrate-bonding process. Each elevated-temperature process must be at a lower temperature than the previous process. The choices that exist for substrate-to-package bonding are therefore limited to solder bonding and adhesive bonding.

Solder bonding There are a variety of solder alloys available, and a wide choice of melting temperatures can be selected for this task. Some of the highest temperatures used are derived from Au-Si alloys, while the lower-temperature alloys result from Pb-In mixtures. These can be selected according to the process steps and the

Fig. 13 Infrared heater for solder reflow. (*OAL Associates.*)

Fig. 14 Focused infrared soldering system. (*Spectra Instruments Inc.*)

temperature of the processes before and after the substrate-bonding step. There are basically two methods of solder bonding a substrate to a package.

1. The first method is to solder between the base of the package and the bottom of the substrate. For this purpose the bottom of the substrate must be metallized with a solderable coating. In general, the solderable coating will be selected from the same conductor materials as those used for the circuit portion of the substrate. Where the bottom of the substrate must be metallized for solder mounting, the metallization may be done while the circuit lines are being printed; i.e., the bottom of the substrate would first be printed and dried. Then the top (circuit portion) would be printed and dried; and then both the top and bottom metallization would be fired simultaneously.

After the substrate is completely assembled (or before, as the case may be), it is bonded to the package. If the back of the substrate is metallized for solder bonding, it is assumed that it will be bonded to a package with a metallized base to receive the substrate. As with reflow soldering of parts, both surfaces should first be tinned. Once this is accomplished, the bonding operation is simply a matter of heating the two parts to solder-melt temperature in the presence of flux.

2. A second method of soldering a substrate into a package is to solder it directly to the input/output pins. This requires either pins that can be bent down over corresponding pads on the substrate or holes in the substrate which will register with the pins. Naturally the holes in the substrate must be surrounded by solderable pads so that the substrate can be soldered to the pins and the input/output connections accomplished. The TO-8 package is ideal for this type of substrate attachment.

It is often necessary to bond the substrate to the base of the package in addition to soldering it to the input/output pins so that a thermal transfer will occur. This is especially important on high-power circuits, where the pins cannot be relied on for sufficient heat transfer.

Solders   Matching the solder used to the process requirement requires a fund of knowledge of available solder alloys. Melting temperatures can range from as low as 70°C (Bi-Cd) up to 837°C (silver alloys). To assist in making the proper choice, illustrations are presented in Tables 5 to 9, indicative of the exceedingly wide choice available.

A study of the tin-lead fusion diagram in Fig. 15 will explain the wide range of temperatures through which an alloy may be plastic, liquid, or solid.

It is not uncommon for a substrate assembled with solder to contain three or even four different alloys so that the process temperatures can progress properly.

Solder pastes   For high-production assembly with solder, screenable solder pastes are often excellent. These compositions consist of finely divided metal and alloy powders dispersed in thixotropic organic vehicles containing fluxes. The viscosity is such that the pastes can be screened through 105-mesh or larger screens, in the same manner that conductors, resistors, and insulators are screened; however, 80-mesh screens are recommended due to the thickness of most of the pastes. Naturally the choice of the solder paste to be used depends, as with standard alloys, on the melting temperature, compatibility with materials joined, and service temperature. Table 10 shows the choice of solder pastes available, along with their melting temperatures.

The obvious advantage of screening solder paste is that it can be applied selectively and in mass; i.e., one stroke of the squeegee prepares the entire substrate. In addition, controlled quantities can be applied, and the jigs necessary to hold or apply preforms are eliminated. Discrete parts that are to be reflowed can be held in place by the tackiness of the solder paste until the heat is applied to reflow the solder. By the selection of the screen mesh, the thickness of the solder paste and the resulting metal film can be controlled. Table 11 shows the thickness in mils that can be obtained and the resistivity of the resulting metal films.

These data were obtained by printing and flowing Formon* compositions over 50-mil-wide thick film conductor lines on an alumina substrate. The conductor was Du Pont's Pd-Ag 8151, with a fired thickness of about 0.5 mil. The sheet resistivities of the solder or braze layers were calculated from the conductance differences between tinned and bare conductor patterns.

For thinner prints, one can use 105- or 200-mesh screens. For thicker prints, which are often required, one must employ an 80-mesh screen with a metal-foil backing. Metal-foil thicknesses of 2 to 5 mils cover the usable range nicely. Solder thicknesses up to 10 mils have been produced routinely with these foil-backed screens. For some patterns, thick metal masks can also be used.

As with most solder alloys, better results may be obtained by accomplishing the reflow under inert or reducing atmospheres, which decreases the tendency of the solder to oxidize and results in better and cleaner solder joints.

As with any soldering operation that uses flux, cleaning will be required after reflow soldering with solder paste. Solvents that are normally used for flux removal will be satisfactory. A final rinse with Freon† will remove all latent films.

Solder paste has been gaining acceptance as new formulations appear and good

---

* Trade name of E. I. du Pont de Nemours & Co.
† Trademark of E. I. du Pont de Nemours & Co.

TABLE 5   Common Tin-Lead Solder Alloys

Alloy	Composition, %					Temperature at which solder becomes:			
						Plastic		Liquid	
	Sn	Pb	Ag	Sb	Cd	°C	°F	°C	°F
1	0	100	. . .	. . .	. . .	. . .	. . .	327	620
2	5	95	. . .	. . .	. . .	270	522	315	598
3	10	90	. . .	. . .	. . .	225	440	300	575
4	15	85	. . .	. . .	. . .	183	360	285	560
5	20	80	. . .	. . .	. . .	183	360	280	535
6	25	75	. . .	. . .	. . .	183	360	265	513
7	30	70	. . .	. . .	. . .	183	360	255	491
8	35	65	. . .	. . .	. . .	183	360	246	475
9	40	60	. . .	. . .	. . .	183	360	236	457
10	45	55	. . .	. . .	. . .	183	360	225	437
11	50	50	. . .	. . .	. . .	183	360	214	417
12	55	45	. . .	. . .	. . .	183	360	200	392
13*	60	40	. . .	. . .	. . .	183	360	188	370
14*	63	37	†	. . .	. . .	‡	‡	183	361
15	65	35	. . .	. . .	. . .	183	360	186	367
16	70	30	. . .	. . .	. . .	183	360	186	370
17	75	25	. . .	. . .	. . .	183	360	192	378
18	80	20	. . .	. . .	. . .	183	360	199	390
19	85	15	. . .	. . .	. . .	183	360	205	403
20	90	10	. . .	. . .	. . .	183	360	213	415
21	95	5	. . .	. . .	. . .	183	360	223	434
22	100	0	. . .	. . .	. . .	. . .	. . .	232	450
23§	95	. . .	. . .	5	. . .	232	450	240	464
24	27	70	3	. . .	. . .	180	355	253	487
25	37	60	3	. . .	. . .	180	355	233	450
26	61.5	35.5	3	. . .	. . .	180	355	. . .	. . .
27	62.5	36.1	1.4	†	. . .	‡	‡	180	355
28	. . .	97.5	2.5	†	. . .	‡	‡	305	581
29	. . .	95	5	. . .	. . .	305	581	365	685
30	95	. . .	5	. . .	. . .	221	430	295	563
31	44	55	1	. . .	. . .	177	350	210	410
32	60	34	6	. . .	. . .	177	350	304	580
33	1	97.5	1.5	†	. . .	‡	‡	309	588
34	2	95.5	2.5	. . .	. . .	300	570	305	580
35¶	95.5	. . .	3.5	. . .	1	. . .	425	. . .	430

source: Alloys Unlimited, Inc.
 * Best tensile strength.
  † Eutectic, i.e., that mixture of two or more metals which goes directly from solid to liquid (with no plastic range) upon reaching its melting point.
  ‡ No plastic range.
  § Best shear strength.
  ¶ Best creep strength.

TABLE 6 Physical Properties of Silver Solder Alloys

Alloy	Composition, %					Melting range, °F		Recommended soldering temperature, °F	Shear strength, psi		Tensile strength as-cast, psi
	Ag	Cu	Zn	Cd	Other	Solidus	Liquidus		Joint	As-cast	
36	85	...	...	...	Mn 15	1760	1778	1800	20,000	20,000	20,000
37	80	16	4	...	...	1360	1490	1500	25,000	42,000	40,000
38	75	22	3	...	...	1365	1450	1450	25,000	41,000	40,000
39	75	20	5	...	...	1350	1425	1450	25,000	42,000	40,000
40	75	...	25	...	...	1310	1325	1325	27,000	30,000	30,000
41	72	28	...	...	...	1435	1435	1450	23,000	40,000	40,000
42	70	20	10	...	...	1335	1390	1400	30,000	44,000	40,000
43	65	20	15	...	...	1280	1325	1350	33,000	50,000	30,000
44	65	28	...	...	Mn 5, Ni 2	1380	1450	1500	32,000	41,000	40,000
45	60	25	15	...	...	1260	1325	1350	30,000	50,000	45,000
46	60	30	...	...	Sn 10	1120	1370	1350	30,000	50,000	45,000
48	56	22	17	...	Sn 5	1145	1205	1250	28,000	45,000	50,000
49	54	40	5	...	Ni 1	1375*	1575	1600	30,000	50,000	50,000
50	50	15.5	16.5	18.0	...	1160	1175	1175	30,000	43,000	60,000
52	50	34	16	...	...	1275	1425	1425	30,000	46,000	45,000
53	50	15.5	15.5	16	Ni 3	1175	1450	1270	30,000	46,000	60,000
54	50	28	22	...	...	1250	1340	1350	35,000	51,000	45,000
55	45	15	16	24	...	1125	1145	1150	35,000	45,000	60,000
56	45	30	25	...	...	1250	1370	1375	33,000	47,000	55,000
59	40	18	15	27	...	1135	1205	1200	25,000	36,000	55,000

No.											
60	40	30	28	⋯	Ni 2	1240	1435	1400	37,000	46,000	50,000
61	40	36	24	⋯	⋯	1250	1400	1400	35,000	49,000	55,000
62	40	30	25	⋯	Ni 5	1250	1600	1500	40,000	48,000	50,000
63	35	26	21	18	⋯	1125	1295	1300	35,000	38,000	50,000
65	30	38	32	⋯	⋯	1370	1410	1400	45,000	50,000	60,000
66	20	45	30	5	⋯	1140	1500	1400	32,000	45,000	42,000
67	20	45	35	⋯	⋯	1430	1500	1450	35,000	40,000	55,000
68	15	80	⋯	⋯	P 5	1185	1500	1300			
69	10	52	38	⋯	Sn 8	1550	1580	1580	35,000	45,000	65,000
70	7	85	⋯	⋯	⋯	1225	1805	1550			
71	6	86.5	⋯	⋯	P 7.5	1185	1380	1300			
72	5	58	37	⋯	⋯	1575	1600	1600	33,000		
73	5	⋯	⋯	95	⋯	640	740				
74	5	⋯	16.6	⋯	Cd 78.4	480	600				
75	2	91	⋯	⋯	P 7	1185	1450	1300			

SOURCE: Alloys Unlimited, Inc.

TABLE 7 Indium-based Solders

Alloy		Composition, %					Melting point, °C	Tensile strength, psi
	In	Sn	Pb	Cd	Ag	Cu		
76	50	50	...	...	...	...	117	1720
78	74	...	...	26	...	...	123	
79	80	...	15	...	5	...	157	2550
80	90	...	...	...	10	...	230	1650
81	100	...	...	...	...	...	156.7	515
82	25	37.5	37.5	...	...	...	138	5260
83	5	...	92.6	...	2.4	...	280–285	4560
84	50	...	50	...	...	...	215	4670
85	12	70	18	...	...	...	150–174	5320
86	25	...	75	...	...	...	230	5450
87	5	...	95	...	...	...	315	4330
88	5	...	90	...	5	...	292	5730
89	14.5	...	...	...	61.5	24	630–685	

SOURCE: Alloys Unlimited, Inc.

TABLE 8 Special Low-melting Bi-Cd Alloys*

Alloy no.	Composition, %					Temperature at which alloy becomes:			
						Plastic		Liquid	
	Sn	Pb	Cd	Bi	Sb	°C	°F	°C	°F
90	20	19	13	48	...	70	158	76	168
91	18.8	31.2	...	50	...	96	205	97	207
92	5	70	...	25	...	100	212	116	240
93	50	40	...	10	...	120	248	166	330
94	23	68	9	...	...	145	293	235	455
95	51.2	30.6	18.2	†	...	‡	‡	145	293
96	50	25	25	...	...	145	293	160	320
97	18.8	31.2	...	50.1	†	‡	‡	97	207
98	...	82	2.5	14.5	1	127	261	259	498
99	25	38	...	37	...	93	200	127	260

SOURCE: Alloys Unlimited, Inc.
   * Due to the brittleness of bismuth some of the above alloys might not be available in wire form for rings.
   † Eutectic.
   ‡ No plastic range.

TABLE 9   Tin-Lead-Silver Alloys

Alloys with corresponding plastic and liquid temperatures

Tin	Lead	Silver	Antimony	Others	Temperatures (°F)
0	100				621°, 522°
5	95				597°
10	90				576°, 435°
15	85				550°, 440°
20	80				532°, 361°
25	75				514° PLASTIC RANGE 361°
30	70				491°, 361°
35	65				477°, 361°
40	60				460°, 361°
45	55				441°, 361°
50	50				421°, 361°
55	45				392°, 361°
60	40				374°, 361°
63	37				361°*
65	35				367°, 361°
70	30				378°, 361°
75	25				383°, 361°
80	20				396°, 361°
85	15				405°, 361°
90	10				418°, 361°
95	5				432°, 361°
100	0				450°
95			5		460°, 450°
35	63		2		459°, 369°
27	70	3			594°, 354°
40	57	3			543°, 354°
50	47	3			500°, 354°
61.5	35.5	3			478°, 354°
62.5	36.1	1.4			354°*
96		4			465°, 430°*
95		5			430°
	97.5	2.5			580°*
95		5			689°, 580°
.75	97.5	1.75			590°*, 203°*
16	32			52 Bi	
50	40			10 Bi	332°, 288°
51.2	30.6			18.2 Cd	288°*

Note: Bi - Bismuth   Cd - Cadmium   *Eutectic

SOURCE: Gardner Solder Company.

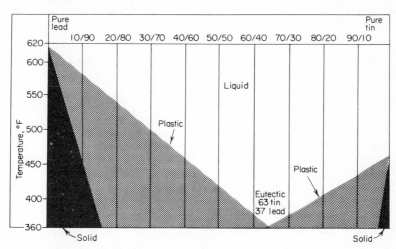

Fig. 15   Tin-lead fusion diagram.   (*Alloys Unlimited Inc.*)

results become more easily achieved. There are some minor disadvantages, including perishability of the print; i.e., the piece parts must be applied to the substrate while the paste is tacky and before it has had a chance to dry. Best results will be obtained in the reflow operation if it is accomplished before all the volatiles in the flux have had a chance to dry out.

Flip Chips for Solder Assembly

Among the special component parts developed for solder assembly is the *flip chip*, a semiconductor device (transistor, diode, or integrated circuit) which has been designed for bonding to an acceptably registered conductive pattern in a facedown configuration. Originally flip chips were bonded to a thin film circuit with ultrasonic

TABLE 10   Solder Pastes

Product	Type	Melting point, °C	Suggested flow temperature, °C
DP-8511	80 Au-20Sn	280	340–360
DP-8516	95Sn-5Sb	240	260–300
DP-8518	95Sn-5Ag	221	240–275
DP-8520	10Sn-90Pb	300	320–350
DP-8522	60Sn-40Pb	193	210–240
DP-8523	62Sn-36Pb-2Ag	189	200–230

SOURCE: E. I. du Pont de Nemours & Co.

TABLE 11   Alloy Thickness from Solder Paste

Product	Type	Thickness, mils	Resistivity, Ω/sq
DP-8511	80Au-20Sn	1.5–2.0	0.010
DP-8516	95Sn-5Sb	3.5–4.0	0.003
DP-8518	95Sn-5Ag	3.5–4.0	0.002
DP-8520	10Sn-90Pb	3.0–3.5	0.005
DP-8522	60Sn-40Pb	2.5–3.0	0.003
DP-8523	62Sn-36Pb-2Ag	2.5–3.0	0.003

SOURCE: E. I. du Pont de Nemours & Co.

or thermocompression techniques. This involved either builtup metallized pads on a very flat substrate or builtup pads on the semiconductor device. Both these techniques proved very difficult to implement, and while some flip-chip thin film fabrication still goes on utilizing ultrasonic and thermocompression bonding techniques, the overwhelming bulk of flip-chip fabrication is now accomplished by the reflowsolder process proved on over a billion thick film substrates by IBM. Considerable in-depth study and development were accomplished before the process was approved, including the effect of expansion coefficient, plastic deformation, thermal stresses, and elastic stresses on the solder joints. In the solder-reflow flip chip, the solder ball is formed on the chip and registers with a pretinned land pattern on the substrate. The volume production associated with this program permitted the development and utilization of very sophisticated automatic production equipment, which was a major factor in the success of the process. Several reports available from IBM describe the flip-chip solder-reflow process in great detail.

The chip design and manufacturing process used by IBM in their fabrication is much too complex for presentation in this chapter, but Fig. 16 shows the elements and materials involved. The figure also is indicative of the chip-fabrication complexity that requires a copper microball as part of the metallization system. It is this attention to terminal metallurgy that has made the IBM flip-chip method such a success.

Other companies also manufacture flip-chip devices for solder-reflow assembly, which can be purchased for inclusion into any hybrid thick film circuit designed for their use. The reader should be warned, however, that the IBM flip-chip devices are not generally available. Other companies, in addition to the six pictured in Fig. 17, also supply flip-chip devices. It is naturally important that the geometry of the flip chip be well specified so that an exact match with the thick film conductors can be made during the assembly process. The Electronic Industries Association (EIA) attempted to standardize bump geometry, and their recommendations (committee E1A MED 4.1) are available. Manufacturers do not necessarily follow the recommendations, however.

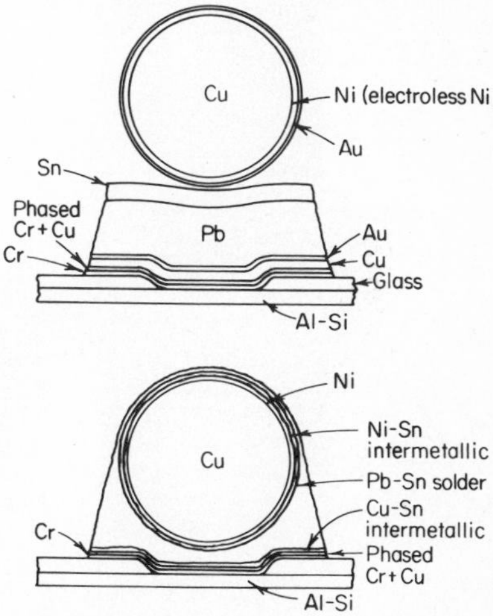

LID and Channel Devices

The LID appeared on the market several years ago as an intermediate phase between the utilization of packaged semiconductor devices and the chip-and-wire technology. The LID presented a very small, leadless semiconductor that could be soldered into a circuit without the problems and precautions associated with bare semiconductor chips.

Initially diodes and transistors were the only devices available in LID format, but as time went on, channel devices appeared which were merely 14- or 16-legged LIDs rather than the common 4-legged variety. This permitted the packaging of integrated circuits in these channel devices so that they too could be soldered into a hybrid thick film circuit with an economy of space and without leads.

Type availability problems always existed with LID devices since it was never quite

Fig. 16 SLT ball terminal before and after copper-ball soldering. (*IBM.*)

possible to get all types of semiconductor devices in the LID package. As a result, higher costs occurred whenever a special order had to be placed. Figure 18 shows a standard four-legged LID device, and Fig. 19 shows one with eight legs.

Chip Resistors

What used to be a curiosity, the chip resistor, is now a common piece part used in hybrid thick film circuits. There are numerous reasons why a designer would choose to use a chip resistor or even several chip resistors rather than printing and firing the equivalent thick film resistor. Some of these reasons follow:

1. When an exact lead-length approximation is required in a breadboard, chip resistors might be used instead of discrete leaded parts. Chip resistors can be changed easily, and circuits thereby optimized.

2. An additional printing and firing process step is avoided when one or a few resistors are out of range for a given paste.

3. It is possible to salvage a substrate that may have one open or too-high resistor.

Chip resistors can be assembled to hybrid thick film substrates by reflow soldering, conductive-epoxy or nonconductive-epoxy, and wire-bonding techniques. They are available with end metallizations capable of accepting solder (as with chip capacitors), already tinned for solder reflow operations, or with gold for chip and wire assembly. Table 12 lists the size availability and the associated wattage ratings.

Values are available in a wide range of resistances and percentages. The resistance range, for instance, of one manufacturer is from 10 Ω to 1,000 MΩ.

Soldering Chip Devices

The most obvious method for soldering chip devices to a substrate is by some means of reflow soldering. This requires that both the chip device and the substrate have been previously tinned and that solder (in general) need not be added. This, of course, is easier said than done since many difficulties can occur. Although mass soldering techniques, where the entire substrate is heated to solder-reflow temperature and all the parts are attached at once, are most common, we shall look briefly at some of the methods for soldering individual chip devices. As mentioned

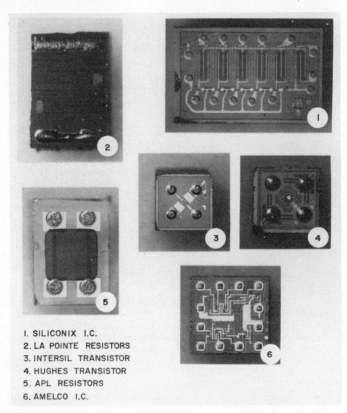

1. SILICONIX I.C.
2. LA POINTE RESISTORS
3. INTERSIL TRANSISTOR
4. HUGHES TRANSISTOR
5. APL RESISTORS
6. AMELCO I.C.

Fig. 17 Flip-chip devices from six manufacturers. (*Courtesy of The Johns Hopkins University/Applied Physics Laboratory.*)

before in this chapter, there are focused-infrared, hot-wire, and other techniques for applying the heat. The problem with a ceramic chip device is that both ends of the chip should be heated at once so that the chip can settle evenly onto its solder pads. Focused infrared can accomplish this, but with the hot-wire technique it is not as easy. A single hot wire can obviously heat only one end of the ceramic chip device and depends on the transmitted heat through the substrate and the device to heat the other end. The solution is to have two hot wires, one on each end of the chip device so that the ends can be heated equally and together. This requires separate power supplies and an elaborate mechanical arrangement, but it does accomplish the purpose. It also avoids heating one end of the chip to such a high temperature that metallization damage may result. A heated stage to elevate the temperature of the substrate is still a requirement.

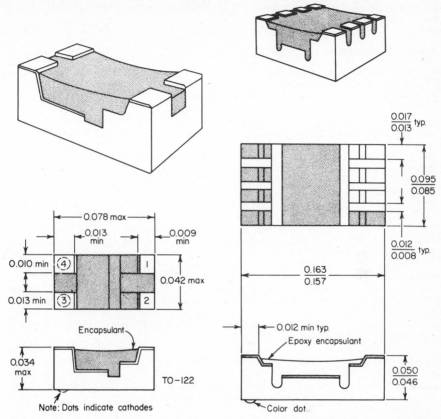

Fig. 18 Four-legged LID. (*Dickson Electronics Corporation.*)

Fig. 19 Eight-legged LID. (Shaded areas indicate gold plating. All dimensions are in inches. (*Dickson Electronics Corporation.*)

TABLE 12 Chip-Resistor Sizes and Power Ratings

Length, in.	Width, in.	Wattage
0.035	0.035	35 mW
0.050	0.050	⅛ W
0.100	0.025	100 mW
0.100	0.050	¼ W
0.100	0.100	½ W

When working with the assembly of flip-chip devices by solder-reflow methods, it is important to design the substrate metallization so that the solder is restricted from flowing away from the solder joint. This solder-runoff problem can be solved by providing a solder dam of nontinnable material, such as glass, to restrict the solder to the area of the contact. This is a very important feature in the IBM *controlled-collapse* chip-joining process. The mating areas where the chip contacts mate with the substrate conductors are surrounded by regions of nontinnable material, forming the required solder dams.

EPOXY BONDING

Epoxy bonding can generally be substituted for solder bonding, provided that the epoxy selected will withstand the circuit environments. Naturally if the epoxy will be required to perform a circuit-interconnection function in addition to the bonding function, it must also be conductive. The main advantage of epoxy bonding over solder bonding is that it is a room-temperature process. Even though most epoxy formulations must be cured at 100 to 125°C, this is still considerably lower than the 385 to 420°C required for Si-Au eutectic bonding or even 185 to 220°C required for Sn-Pb soldering. Epoxies have become popular for such things as chip bonding, capacitor bonding, LID and channel-device bonding, and package sealing. It should always be kept in mind that some military specifications prohibit the use of epoxies because of their organic content.

The choices in conductive epoxies for hybrid-circuit application are essentially two, silver epoxy and gold epoxy. Since this chapter is directly concerned with assembly processes, it is necessary to exclude the large numbers of epoxies used primarily for adhesive applications and concern ourselves with epoxies used primarily for conductive interconnection and secondarily for adhesive purposes.

There are numerous suppliers of silver and gold epoxies for hybrid thick film applications, and it would be impossible to cover them all in detail. In general, there are two-part systems, where the adhesive is mixed with a catalyst, and one-part systems, where the mixing has already been done by the vendor, which can be applied directly from the jar. In general, the two-part systems have a longer shelf life prior to mixing and can be mixed to the desired consistency by adding a thinner at the same time that the catalyst is being introduced.

Most of the epoxies require curing at elevated temperatures (although these temperatures are usually under 150°C and do not even approach eutectic-bonding temperatures), while some will cure at room temperatures. Here again the room-temperature-curing formulations have a significantly reduced shelf life. In almost all cases, the shelf life can be extended by storing the mixture in a refrigerator at approximately 40°F. Care should be exercised to prevent freezing.

There are so many vendors and formulations available that the characteristics of only a few can be given here. For complete details the vendors' specification sheets must be carefully studied. Tables 13 to 15 contain condensed information on some popular epoxy formulations.

There are numerous other epoxy types, some of which have been formulated for specific purposes rather than general bonding. In general, there are advantages to using an epoxy which has been formulated specifically for the job, as this may make application easier or in some other way simplify the use. For example, an epoxy that would be best for active-device bonding would be rather soft, while an epoxy best for passive-device bonding would be slightly thixotropic. Table 16 illustrates a group of epoxy formulations designed for specific purposes.

Another group of epoxy formulations is shown in Table 17. It should be noted that the products of each manufacturer have their own advantages and special considerations. It is difficult to select the "right" epoxy because there are so many to choose from and each process differs enough to make one formulation seem advantageous over another.

Epoxy Preparation

In all but very high production applications epoxy usage is quite wasteful, due primarily to the pot life of the preparations once the two-part systems are mixed. As a result, in certain preparations, such as gold-filled epoxies, a one-part system is preferable. The pot life in one-part systems is usually considerably longer than in two-part systems after they are mixed, i.e., months versus hours. The two-part systems will last indefinitely, however, until blended together.

In some epoxy systems an equal number of part A and part B are blended together, whereas in other systems a catalyst is added (usually in very small ratios) to the main element. Usually the catalyst is mixed by weight in order to obtain the proper ratio to a very close tolerance. Mixing small quantities of the size generally used for hybrid-assembly work requires an accurate gram scale capable of reading to 1/100 g.

TABLE 13 Properties of Eccobond 56C Conductive Epoxy

Temperature limits in use...........	-70 to $+350°$F
Lap shear strength.................	800 psi
Flexural strength..................	12,200 psi
Volume resistivity................	2×10^{-4} Ω/cm
Thermal conductivity..............	40 Btu/(h)(ft^2)(°F/in.)
Thermal expansion.................	20×10^{-6} in./(in.)(°F)
Cure temperature..................	120 to 200°F
Catalyst ratio....................	1:40 parts
Application.......................	Solder substitute, die bonding, capacitor attachment, chip attachment, etc.

SOURCE: Emerson & Cuming, Inc.

TABLE 14 Properties of Eccobond 57C Conductive Epoxy

Temperature limits in use...........	-70 to $+300°$F
Lap shear strength.................	700 psi
Flexural strength..................	10,200 psi
Volume resistivity................	0.6 mΩ/cm
Thermal conductivity..............	60 Btu/(h)(ft^2)(°F/in.)
Catalyst ratio....................	Equal parts A and B
Cure temperature..................	Room temperature to 225°F
Application.......................	Solder substitute, die bonding, capacitor attachment, chip attachment, etc.

SOURCE: Emerson & Cuming, Inc.

TABLE 15 Properties of Eccobond 58C Conductive Epoxy

Temperature limits in use...........	-65 to $+500°$F
Lap shear strength................	1,200 psi
Flexural strength..................	9,700 psi
Volume resistivity................	2 mΩ/cm
Thermal conductivity..............	70 Btu/(h)(ft)(°F/in.)
Thermal expansion.................	18×10^{-6} in./(in.)(°F)
Cure temperature..................	300 to 500°F
Solvent removal...................	170°F for 1 h
Catalyst ratio....................	One-part mixture
Application.......................	Solder substitute, die bonding, chip attachment, capacitor attachment, etc.

SOURCE: Emerson & Cuming, Inc.

It is difficult to give general mixing instructions since each vendor's product is usually accompanied by very specific instructions. These instructions, of course, should be carefully followed. For larger-quantity use *Static Mixers** are available, as are epoxy dispensers. The use of these devices, where the volume justifies, will

* **Trademark of Kenics Corp.**

TABLE 16 Epoxies for Specific Applications

EPO-TEK type	Recommended use	Number of components	Filler	Pot life, h	Curing schedule (typical) Temp, °C	Time, min	Consistency	Volume resistivity (rigid specifications), Ω-cm	Thermal conductivity, Btu/(h)(ft²)(°F/in.)	Lap shear strength, psi	Temperature range, °C Intermittent	Continuous	Shelf life, years
H20	Bonding active components	2	Silver	2	100	40	Soft thixotropic paste	0.0001–0.0005	11.5	1,000	300–400	250	1
H21	Bonding passive components	2	Silver	2	100	40	Slightly thixotropic paste	0.0006–0.0009	11.0	1,000	300–400	250	1
H22	Hermetic sealing in packages	2	Silver	2	100	40	Free-flowing paste	0.001–0.005	10.0	1,000	300–400	250	1
H23	Bonding testing and retrieving chips	2	Silver	3	100	40	Soft thixotropic paste	0.0001–0.0005	11.0	1,000	250–300	200	1
H24	Bonding substrates in packages	2	Silver	3	100	40	Low-viscosity smooth-flowing paste	0.01–0.05	10.0	1,000	300–400	250	1
H40	Bonding active components	1	Gold	…	120	45	Soft thixotropic paste	0.0001–0.0003	…	…	300–350	200	0.5
H41	Bonding passive components	1	Gold	…	120	45	Slightly thixotropic paste	0.0001–0.0005	…	…	300–350	200	0.5
H54	Coating crossovers	2	None	4	100	30	Soft thixotropic paste	5×10^{15}	…	3,100	300–400	250	1
H61	Bonding passive components	1	Thermally conductive	…	120	45	Slightly thixotropic paste	…	7.55	…	300–350	…	0.5
H74	Bonding substrates in packages	2	Thermally conductive	4	100	20	Flowable paste	2.5×10^{15}	7.60	3,500	300–400	250	1

source: Epoxy Technology, Inc.

assist in maintaining area cleanliness and assuring a more uniform end product. Figure 20 shows an epoxy-dispensing device.

Epoxy Application Techniques

The application of conductive epoxy for hybrid-assembly purposes is a skill-oriented job. A steady hand, a keen eye, and a reasonable amount of judgment are required of the operator. Depending upon the consistency of the epoxy preparation, there can be considerable danger, especially when bonding chip capacitors, that the epoxy will run under the part and cause a short circuit. Obviously the judgment exercised by the operator concerning the proper amount of epoxy to use is of paramount importance if shorts are to be prevented.

The universal tool for applying epoxy (for the purpose of hybrid thick film assembly) is still the sewing needle. It should be held in a pin vise and cleaned frequently. Not only is it a very convenient tool, it is very inexpensive and almost never wears out. After the sewing needle, the disposable hypodermic syringe is next in popularity. The 1- or 2-cm³ size is the most popular. The syringe can be used for the direct application of the epoxy only when a large enough part is being assembled. For small semiconductor chip attachment or even for capacitor attachment, it is used only to meter out the epoxy so that it can be transferred to the sewing needle for actual application. As a safety measure, the sharp point should be ground off the needle of the disposable hypodermic syringe so that the possibility of accident is reduced. This is not required; nor can it be done with a sewing needle held in a pin vise, as the very sharp point is needed for the precise epoxy application.

It is entirely possible also to screen-print epoxy, in the same manner that solder-paste formulations are printed. The two problems are (1) being able to screen-print a small enough amount of epoxy for a small semiconductor die and (2) having sufficient time to attach the piece parts after printing and before the epoxy starts to set up and cure.

The assembly of semiconductor dice with epoxy is becoming increasingly popular (except where prohibited by a military specification) because of the ease of die attach, the avoidance of large capital investment (die bonder), and the room-temperature

Fig. 20 Epoxy dispenser. (*Kenics Corporation.*)

assembly procedure. The following steps are all that are necessary for bonding dice with epoxy:

1. Clean the substrate to assure that all oils and foreign matter are removed.
2. Mix the epoxy in accordance with the manufacturer's specification.
3. Apply the required amount of epoxy to the substrate at the place where the die is to be attached (about 0.005 in. thick and one-half the die area).
4. Place the semiconductor die in position over the epoxy (check for die orientation).
5. Seat the die to the substrate surface by pressing it down with a soft implement (a small Teflon stick or a toothpick will suffice).
6. Reposition the die as necessary to comply with the assembly drawing.
7. Place the other die or epoxy-assembled devices on the substrate.
8. Cure the epoxy as directed in the manufacturer's specification sheet.

At the end of the cure cycle, the assembled substrate should be ready for wire bonding. While both thermocompression and ultrasonic wire bonding are possible

TABLE 17 Characteristics for Two Epoxy Systems

Number	Description	Base	Hardener ratio[a]	Typical tensile shear strength at 25°C, psi	Pot life[b]	Typical cure schedules		Typical resistivity, Ω-cm	Max. operating temperature, °C
						Temp., °C	Time		
E-Solder:[c]									
3102[d]	One-part paste	Epoxy	....	3,200	...	125 200	6 h 10 min	0.001	200
3021	Two-part paste	Epoxy	1 pt A, 1 pt B	2,500	30 min	27 65	24 h 3 h	0.001	150
3025	Two-part paste	Epoxy	1 pt A, 1 pt B	2,200	4–6 h	27 65	24 h 4 h	0.001	150
3022	Two-part paste	Epoxy	100 pt 3022, 8 pt no. 18 hardener	1,800	1–2 h	27 65	24 h 3 h	0.001	150
3500	Powder or pellet	Epoxy	...	2,000	...	150 125	30 min 1 h	0.001	200
3026	Two-part paste	Epoxy	100 pt 3026, 6½ pt no. 45 hardener	...	1 h	27 60	24 h 4 h	0.01	150
3044	Two-part paste	Epoxy	100 pt 3044, 8 pt no. 66 hardener	...	3–4 h	120 100	1 h 2 h	0.001	175
3205[d]	One-part paste	Gold-epoxy	....	3,200	...	125 200	6 h 10 min	0.001	200

E-Kote[a]								
40	Aerosol can	Acrylic	⋯	⋯	[g]	[g]	0.001	175
3027	Paint	Silicone	⋯	⋯	232	30 min	0.01	538
3028	Grease	Silicone	⋯	⋯	149	24 h	0.1	200
3030	Paint	Acrylic	⋯	⋯	[g]	[g]	0.001	175
3042	Paint	Acrylic	⋯	⋯	[g]	[g]	0.001	175
3202	Chalking	Vinyl	⋯	⋯	[g]	[g]	0.01	

SOURCE: Epoxy Products Company.

[a] Formulations which have a 1:1 hardener ratio can be mixed in either equal weights or equal volumes. Formulations which do not have a 1:1 ratio should be mixed by weight.

[b] Pot life is measured on a 50-g sample at 25°C.

[c] Minimum shelf life is 4 months at 77°F and 1 year at 40°F.

[d] Shelf life is 6 months at 40°F.

[e] Minimum shelf life is 6 months at 77°F in unopened container.

[f] Air dry.

[g] Nondrying.

NOTE: E-Solder and E-Kote can be thinned with acetone, cellosolve, or any lacquer thinner. Before E-Kote formulations can be baked, they must be allowed to dry tack-free at room temperature. If it is not tack-free, solvent entrapment may cause the film to blister. *Caution:* Avoid skin and eye contact. If skin contact occurs, wash with soap and warm water.

to epoxy-bonded die, the high temperatures associated with thermocompression wire bonding may damage the bond strength of certain epoxies.

Figure 21 shows a semiconductor die being bonded with epoxy. This is quite common with COS/MOS° devices and other types of semiconductors sensitive to heat. It is also popular with hybrid circuits when the large number of semiconductors to be bonded makes the total time on the heat column excessive with eutectic bonding.

Chip capacitors are quite commonly bonded with epoxy, again because it is a simple process and can be accomplished at room temperature. Since the chip capacitor is significantly heavier than a semiconductor die, enough epoxy must be used for the bonding to prevent loss of adhesion during environmental stresses, such as shock, vibration, and centrifugal force. At least the three sides of the capacitor ends should be covered, and some epoxy material should be under the surface of the capacitor mating with the substrate conductor.

When bonding large chip capacitors, thermal cycling tests should be run to assure that the differences in the thermal coefficient of expansion between the capacitor and the substrate will not become a problem.

Since compatibility of materials is always important, it is a good idea to bond to gold conductors with gold conductive epoxy and to silver-bearing conductors with

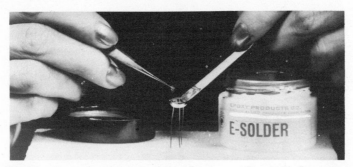

Fig. 21 Epoxy bonding of semiconductors. (*Epoxy Products Co.*)

silver epoxy. Capacitor end metallization should always be compatible whenever possible.

Mechanical dispensing systems can be quite useful if they are controllable to sufficient precision. The one shown in Fig. 22 is an air-controlled Vari-Meter† which will dispense epoxy from a syringe with a wide variation of epoxy viscosity. The air can be regulated from a fraction of 1 psi to over 50 psi to feed viscous pastes.

The use of epoxy for bonding piece parts to hybrid thick film circuits has been successful for years. It should be kept in mind, however, that the epoxy bond will never be quite as good as a eutectic-solder bond. In many cases, though, it is certainly good enough to meet the requirements.

EUTECTIC BONDING

The most common method of bonding a silicon chip device (transistor, diode, integrated circuit, capacitor, or resistor) to a hybrid thick film conductor is by means of eutectic bonding. In general the hybrid thick film conductor used is pure gold. The silicon chip forms the second material in the eutectic formation. In a Si-Au eutectic bond, no additional material is required to form the solder. The Au-Si eutectic point is at approximately 31 atomic percent silicon. This forms a eutectic with a temperature of 370°C. The die bond can be made using a gold-backed silicon die, a bare silicon die, or either type of die with a preform.

° Trademark of RCA.
† Trademark of Techon Systems, Inc.

The advantage of using a gold-backed die is that the gold backing, if deposited soon enough after wafer fabrication, will prevent the silicon from oxidizing. If the silicon has had a chance to oxidize, the Au-Si eutectic bond is difficult to form. The need to break through any silicon oxide that may have formed on the back of the chip is the reason some mechanical motion is required during the chip-bonding process. Once the oxide is broken down, the eutectic formation proceeds. A proper eutectic formation should occur around approximately 90 percent of the periphery of the chip.

When the Si-Au eutectic temperature requirements cannot be tolerated, a lower-melting preform must be used. There are many preform materials to choose from, including Au-Ge, which melts at 356°C, and Au-Sn, which melts lower, at 280°C. Table 18 lists eutectic combinations from which suitable preform materials might be chosen.

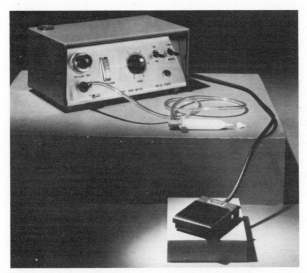

Fig. 22   Hand-held foot-controlled epoxy dispenser.   (*Techcon System, Inc.*)

Die-bonding equipment can take many forms. The simplest is a heated stage and a pair of tweezers. The die is picked up by the tweezers and assembled to the substrate, which is set on the heated stage. The substrate must be above the Si-Au eutectic temperature for this method to work. There are two basic disadvantages in using this method for hybrid thick film assembly work: (1) the substrate must be above the eutectic temperature (assuming that a lower temperature preform is not used), and (2) there is a tendency for the tweezers to damage the edges of the silicon die. This method is very skill-oriented and is unsuitable where large numbers of dice must be bonded to a single substrate. The reasons should be obvious.

If the entire substrate is held at eutectic temperature while more than one or two dice are being attached, the first die will continue an eutectic flow while the succeeding dice are being attached, and this continued eutectic flow will destroy the electrical characteristics of the semiconductors. Even if a more sophisticated die-attach machine is used that has a pickup collet and an associated vacuum system, the problem of keeping the heated stage at eutectic temperature still exists. The answer, then, is some form of selective heating. The many forms that selective heating can take include resistance heating, hot-gas jets, hot collet, and infrared. With each system the substrate must still be heated, but it can be kept well below the eutectic temperature, leaving the additional temperature increment to the selective heater. In

TABLE 18 Solder Alloys

Semalloy no.	Bi	Sn	Pb	Cd	Other	Celsius Liq.	Sol.	Fahrenheit Liq.	Sol.
						Melting point			
		Composition, %				Celsius		Fahrenheit	
	Bi	Sn	Pb	Cd	Other	Liq.	Sol.	Liq.	Sol.
2350	...	95	...	...	As 5	425	231	797	448
2345	...	...	...	...	Zn 100	421	421	787	787
2340	...	...	98	...	Zn 2	417	318	782	604
2330	...	...	...	94.5	Ag 5.5	410	343	770	649
2320	...	...	99	...	Te 1	400	326	752	618
2317	...	...	...	10	Zn 90	399	266	750	509
2315	...	...	...	95	Ag 5	393	338	740	640
2313	...	...	...	...	Zn 95 Al 5	382	382	720	720
2310	...	99.5	...	...	P 0.5	375	232	707	449
A914	...	...	...	...	Au 94 Si 6	370	370	698	698
AG5.5*	...	5.5	94.5	...	...	366	304	689	579
2300	...	...	95	...	Ag 5	363	305	685	581
2297	...	20	...	...	Au 60 Ag 20	360	360	680	680
A912	...	...	...	...	Au 75 Sb 25	360	360	680	680
A911	...	...	...	...	Au 88 Ge 12	356	356	673	673
2295	...	1	83	...	Sb 15 As 1	353	312	667	595
2292	...	60	...	...	Zn 40	343	199	645	390
2291	...	...	...	40	Zn 60	335	266	635	509
2290	...	98	...	...	As 2	330	231	626	447
2269	...	...	100	...	...	327	327	621	621
2284	...	...	...	100	...	321	321	610	610
2278	...	...	99	...	Sb 1	320	312	608	595
2270	...	...	96.5	...	Ag 3.5	317	310	603	590
2265	...	...	...	78.4	Zn 16.6 Ag 5	316	249	600	480
2262	...	...	92	...	In 8	315	305	599	580
SN5*	...	5	95	...	...	312	270	594	518
2253	...	0.75	97.5	...	Ag 1.75	310	310	590	590
AG1.5*	...	1	97.5	...	Ag 1.5	310	310	588	588
2245	...	70	...	...	Zn 30	310	199	592	390
2240	...	...	97.5	...	Ag 1.5 Se 1	309	309	589	589
2230	...	2	95.5	...	Ag 2.5	304	299	580	570
AG2.5*	...	...	97.5	...	Ag 2.5	304	304	580	580
2220	...	60	34	...	Ag 6	304	177	580	350
2218	...	10	90	...	...	302	224	576	435
2217	...	...	90	...	In 10	300	294	572	561
2215	...	...	96	...	Sb 4	299	252	570	486
SN10*	...	10	87.9	...	Ag 2.1	299	268	570	514
2213	...	...	...	90	Zn 10	299	265	570	509
2210	...	...	95	...	Sb 5	295	252	563	485
2200	...	95	...	...	Ag 5	295	221	563	430
2190	...	5	90	...	Ag 5	292	292	558	558
2188	...	...	90	...	Ag 5 In 5	290	290	554	554
2187	...	40	57	...	Ag 3	289	179	543	354
2180	...	15	85	...	...	288	227	553	437
2178	...	10	...	60	Zn 30	288	157	550	315
2175	...	...	92	...	Sb 8	286	271	552	520
2173	...	...	94	...	Sp 6	285	252	545	486
2170	...	5	92	...	Sb 3	285	239	545	463
A905	...	20	...	...	Au 80	280	280	536	536
2160	...	5	92.5	...	Ag 2.5	280	280	536	536

TABLE 18   Solder Alloys (Continued)

Semalloy no.	Composition, %					Melting point			
						Celsius		Fahrenheit	
	Bi	Sn	Pb	Cd	Other	Liq.	Sol.	Liq.	Sol.
SN20*	...	20	79	...	Sb 1	277	181	530	360
PB80*	...	20	80	...	...	277	181	530	360
2145	...	20	79	...	Sb 1	271	182	517	363
2140	100	...	...	...	...	271	271	520	520
2139	...	80	...	...	Zn 20	271	199	518	390
2137	...	10	75	...	Sb 15	268	240	514	464
2135	...	25	75	...	...	268	183	514	361
2133	...	...	91	...	Sb 9	265	252	509	486
2131	...	...	...	82.5	Zn 17.5	265	265	509	509
2138	...	...	75	...	In 25	262	255	504	491
2129	...	25	73.7	...	Sb 1.3	262	184	504	364
2128	...	50	47	...	Ag 3	260	179	500	354
2127	...	75	...	...	Zn 25	260	204	500	401
2126	14.5	...	82	2.5	Sb 1	259	127	498	261
2125	...	34	63	...	Zn 3	256	170	492	338
2120	...	5	85	...	Sb 10	256	240	493	464
PB70*	...	30	70	...	...	256	182	490	360
SN30*	...	30	68.4	...	Sb 1.6	256	182	490	360
2100	95	5	...	...	...	251	134	483	273
2098	...	...	83	17	...	248	248	478	478
2097	...	61.5	35.5	...	Ag 3	248	179	478	354
2096	...	35	65	...	...	247	183	477	361
PB65*	...	34.5	64.5	...	†	246	183	475	360
SN35*	0.2	35	63	...	Sb 1.8	246	183	475	360
2095	...	60	36	...	Ag 4	245	179	475	354
2094	...	35	63.2	...	Sb 1.8	243	184	470	365
2093	...	72	...	28	...	243	177	470	350
2092	...	38	62	...	...	242	183	468	361
SB5	...	94.8	0.2	...	Sb 5	240	232	464	450
2090	...	95	...	...	Sb 5	240	232	460	450
2080	...	97	...	...	Sb 3	238	232	460	449
2070	...	37	62	...	As 1	238	183	460	361
2060	...	40	60	...	...	238	183	460	361
SN40*	...	40	59.4	...	†	238	183	460	360
2055	...	35	63	...	Sb 2	237	187	459	369
2050	...	23	68	9	...	235	145	455	293
2040	...	37	60	...	Ag 3	232	179	450	355
2030	...	100	...	...	...	232	232	450	450
2025	...	40	58	...	Sb 2	231	186	448	365
2010	...	99	...	...	Ga 1	228	228	442	442
2006	...	99.25	...	...	Cu 0.75	227	227	441	441
2003	...	61	36	...	Ag 3	227	179	440	354
2000	48	14.5	28.5	...	Sb 9	227	103	440	217
1990	...	97.5	...	...	Ag 2.5	226	221	438	430
1994	...	45	55	...	...	225	183	437	361
1982	...	95	5	...	...	222	183	432	361
1980	...	95.5	...	1	Ag 3.5	221	218	430	425
1978	...	96.5	...	...	Ag 3.5	221	221	430	430
SN96*	...	96	...	...	Ag 4	221	221	430	430
1970	...	...	55	...	In 45	220	173	428	343

TABLE 18   Solder Alloys (Continued)

| Semalloy no. | Composition, % | | | | | Melting point | | | |
| | | | | | | Celsius | | Fahrenheit | |
	Bi	Sn	Pb	Cd	Other	Liq.	Sol.	Liq.	Sol.
1968	...	48	52	...	...	218	183	424	361
1960	...	50	50	...	...	216	182	421	361
SN50*	...	50	49.4	...	†	216	182	420	360
1956	...	...	50	...	In 50	216	216	419	419
1007	...	...	45	...	In 55	215	215	419	419
1952	...	90	10	...	...	213	183	415	361
1940	...	44	55	...	Ag 1	210	177	410	350
1938	...	85	15	...	...	205	183	403	361
1937	...	50	47	...	Sb 3	204	186	399	365
1936	...	80	20	...	...	204	183	399	361
1935	...	55	45	...	...	200	183	392	361
1934	...	92	...	...	Zn 8	200	198	392	388
1930	4	40.5	55.5	...	...	197	170	386	338
1927	...	75	25	...	...	195	183	383	361
SN70*	...	70	29.4	...	†	193	182	380	360
1923	...	70	30	...	...	192	183	377	361
SN60*	...	60	39.4	...	†	189	176	372	350
SN62*	...	62	35.4	...	Ag 2†	189	177	372	350
1920	...	60	40	...	...	188	183	370	361
1918	...	65	34	...	Sb 1	188	184	370	363
1914	...	63	37	...	...	183	183	361	361
SN63*	...	63	36.4	...	†	182	182	360	360
1890	25	60	15	...	...	180	96	356	205
1880	...	62.5	36	...	Ag 1.5	179	179	355	355
1876	...	62.5	36.1	...	As 1.4	179	179	355	355
1873	...	68	...	32	...	177	177	349	349
1870	13	40	47	...	...	176	146	349	294
1865	...	70	18	...	In 12	174	150	347	302
1860	20	30	50	...	...	173	130	343	266
1840	22	27	51	...	...	170	131	338	268
1850	40	60	...	...	...	170	138	338	281
1836	10.2	48.8	41	...	...	166	142	331	288
1830	10	50	40	...	...	166	120	330	248
1820	14	43	43	...	...	163	144	325	291
1815	...	...	...	...	In 95 Ag 5	162	160	324	319
1810	16	48	36	...	...	162	140	324	284
1790	...	50	25	25	...	160	145	320	293
1780	45	...	55	...	...	160	123	320	253
1775	25	50	25	...	...	160	144	320	293
1770	45	...	33	22	...	158	92	316	197
1765	...	...	15	...	In 80 Ag 5	157	157	315	315
1760	...	...	...	...	In 100	157	157	315	315
1742	...	...	...	...	In 99 Cu 1	153	153	307	307
1740	21	37	42	...	...	152	120	306	248
1730	...	58	...	...	In 42	145	117	292	242
1720	60	...	...	40	...	144	144	291	291

SOURCE: Semi-Alloys.
* Federal specification QQS-571d.
† Balance 0.20 to 0.50 antimony, 0.25 bismuth.

general, a cooling-gas nozzle is also incorporated to chill the eutectic as soon as the formation is complete.

Resistance Heating.   Accomplished by forcing high current at low voltage through the bonding pad, resistance heating can be illustrated schematically as shown in Fig. 23.   The probes that contact the bonding pads must be of low resistance and capable of carrying the high current.   There must also be some protection built into the system to prevent fusing of the bonding pad by too high a current.   This is not a popular method of die attachment since the control of the eutectic temperature is difficult and the process is rather slow.

Hot Gas.   This method can give excellent results and is very practical for selective heating in die bonding.   It is necessary to have (1) a well-constructed mechanical arrangement so that the gas jet is as close to the die as is possible and a method for (2) controlling the temperature of the gas jet and (3) triggering it on and off.   The simplest way to accomplish this is to heat the gas tubes by resistance heating from a low-voltage high-current power supply and trigger the gas jet by means of a foot-con-

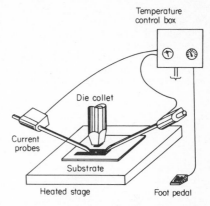

Fig. 23 Selective heating by resistance.

trolled solenoid valve.   Either nitrogen or helium gas can be used, but helium gives superior results due to its better thermal-transfer characteristics.   A schematic of the hot-gas die-attachment technique is shown in Fig. 24.

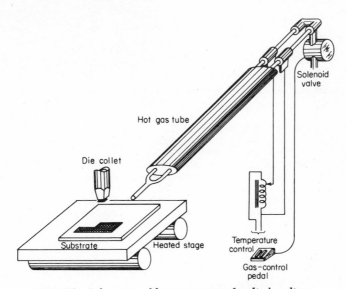

Fig. 24 Schematic of hot-gas system for die bonding.

Hot Collet.   This system of selective heating to eutectic temperature is probably the simplest method since the incremental heat is localized directly at the die to be bonded.   The collet that picks up the die to be bonded is heated to approximately 380°C.   As the die in the heated collet comes in contact with the preheated bonding pad, the die-pad interface temperature rises to the eutectic temperature necessary to cause the Au-Si eutectic flow.

Infrared.   Since this method of die bonding is rather new, the large body of experience associated with other forms of eutectic die bonding is lacking.   It is a unique technology, however, since the infrared energy can be closely controlled and even focused for maximum efficiency.   The method is implemented by having the infrared-energy source below the substrate and adjusted so that the energy is focused through the substrate at the thick film conductor pad where the die is to be bonded. Three elements can be closely controlled: (1) the focus of the infrared energy, (2) the energy level, and (3) the length of time the energy is applied.   Naturally the bonding machine must have provisions for these controls and be able to maneuver the substrate into the proper position over the infrared source.   Eye protection may also

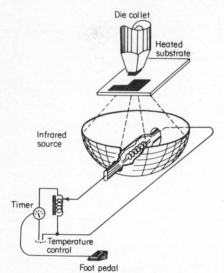

Fig. 25 Schematic of die bonding by infrared heating.

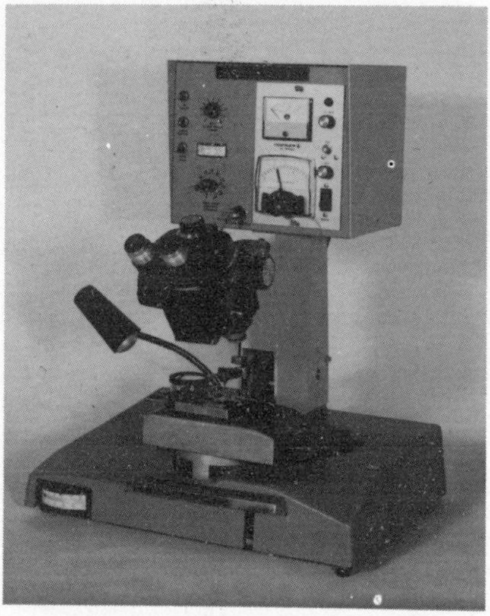

Fig. 26 Infrared die bonder. (*Kulicke and Soffa Industries, Inc.*)

be required for the operator.   As with other selective heating techniques, the substrate and the collet can also be heated to lighten the burden on the infrared source. A schematic of this technique is shown in Fig. 25.

Figure 26 shows an infrared die bonder with all the features discussed.   The complexity of the control console is indicative of the greater sophistication required for this technique.

It can be seen that there is a wide choice of die-bonding techniques.   One technique may work well for one person, whereas another may not work at all.   Each configuration should be studied independently, and the overall process flow coupled with the rework techniques must be taken into consideration before the choice is made.

Figure 27 shows a typical die bonder which is adaptable to hybrid thick film use.   This machine was designed for high-speed-production die bonding to headers or flat packs, but by tooling the substrate carrier properly, it can be used for multichip bonding to hybrid circuits.

Specifications that are important in the selection of a hybrid die bonder are as follows:

1. Ability to accept a wide variety of substrate configurations up to a minimum of 2 in. square

2. Rapid collet change capability or a multicollet holder

3. Repeatable, accurate die-placement capability
4. Work-stage temperature control and indication to ±2°C
5. Presettable energy and force settings
6. Means for selective heating of the die area being bonded
7. Ultrasonic or mechanical scrubbing with adjustment
8. Cooling-gas control
9. Inert-atmosphere blanket
10. Stereo-zoom optics
11. Variable pressure settings
12. Large stage movement (up to 2 in.)

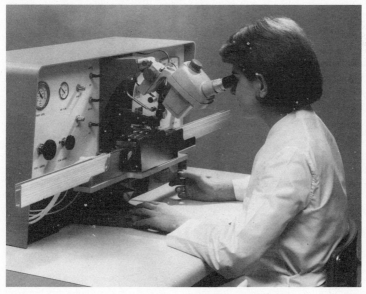

Fig. 27   Die bonder.   (*Lindberg Hevi-Duty.*)

WIRE BONDING

There are several methods for wire bonding to thick-film circuits. The most common are thermocompression ball bonding, thermocompression wedge bonding, ultrasonic bonding, and combinations of all three. All these lead-bonding techniques depend upon obtaining intimate contact between the materials to obtain an atomic interface at the connection. In general, the item most likely to prevent the formation of good bonds is contamination. This can be in the form of oxides which have formed on the bonding interfaces or chemical contamination which has formed either on the metal surfaces or on the oxides already present on the metal surfaces.

Probably the most common form of chemical contamination is finger oils accumulated because of improper precautions during handling. While the deformation of the metals that occurs during the bonding process will go a long way toward breaking the oxide layer that may have formed, it will do little to eliminate the effects of chemical contamination. One further item that can affect the formation of a good intermetallic bond is the surface irregularities of the metals involved. The surface irregularities should be such that they do not prevent intermetallic contact over too large a portion of the bond area. Anything that can be done to remove oxides or contamination from the bonding surface will improve the quality of the bond. The burnishing operation will accomplish this improvement.

Thermocompression Wire Bonding

As its name indicates, thermocompression wire bonding depends upon heat and pressure. In general, the bonding equipment contains a microscope, a heated stage, and a heated wedge or capillary that will apply pressure to the wire at the interface of the bonding surface. In addition, a wire-feed mechanism is required, as is some method for manipulation reduction and control.

Bonds can be accomplished utilizing thermocompression techniques which will exceed the wire-breaking point in strength; i.e., instead of the bond's breaking, the wire will break during a pull test.

Typical difficulties that may be experienced in wire bonding are as follows:

1. Semiconductor targets too small for the wire diameter; i.e., bonds cannot be made without shorting to adjacent metallization.

2. Improper stage or capillary temperatures. The stage should be held between 200 and 320°C while the capillary can be between 100 and 280°C. The interface temperature between the metals can be critical to a good bond.

3. Poor thermal contact between the heated column and the substrate. A warped substrate, especially if held down by vacuum, will be cool in the area where contact to the heated column is poor.

4. Wire ductility. Bending or cold working of the wire is sufficient to cause poor bonding.

5. Contamination of the wire or bonding surfaces by humidity or dust will cause difficulties in bonding.

6. Wire sticking to the capillary or wedge. The tool used for bonding should have a polished finish at least as good as 10 to 20 μin. to prevent the wire from sticking to the tool. Even with a glass capillary the wire can stick so much that the bond is broken when the tool is withdrawn.

Ball bonding requires that the small wire be fed through a quartz, tungsten carbide, or titanium thick-walled capillary tube. The capillary tube, with one end tapered to a few mils in diameter on the outside, is mounted in a suitable mechanical fixture so that it can be moved both vertically and horizontally. The horizontal positioning must be accomplished by means of precision manipulators while being observed through a microscope. Positioning accuracies on the order of 20 millionths of an inch are required.

Before bonding, a small ball is formed on the end of the gold wire by a hydrogen flame. The conductor to which the wire will be bonded is positioned on the work stage below the bonding capillary. The capillary is brought down over the bonding pad and lowered until the ball is brought into contact with the conductor, where a predetermined amount of force is applied. This deforms the ball and establishes intimate contact between the gold ball and the bonding pad. Wire-lead attachment to the other terminal can be accomplished by bonding the wire with the edge of the capillary providing bonding pressure. The capillary is then raised, and the hydrogen flame is used to cut it off while forming another ball for the next bonding operation. This leaves a pigtail, which must be pulled. Figure 28 illustrates the ball-bonding sequence.

A variation of the ball-bonding technique is wedge bonding, accomplished by thermocompression or ultrasonic techniques. There is a difference in the type of tool used in these two techniques, in that the thermocompression wedge bond relies on heat and pressure while the ultrasonic bond relies on high-frequency vibration and pressure. The tool for ultrasonic wedge bonding must be designed to grip the wire during the vibration sequence. Figure 29 shows the thermocompression wedge-bonding sequence.

Also available for thermocompression bonding is the quartz capillary, which is considerably less expensive than its metal counterpart and offers certain advantages, e.g., visibility and inertness. Figure 30 shows a heated bonding capillary designed to accept a quartz capillary tip.

The thermocompression wedge bond is capable of being used on a smaller target than the thermocompression ball bond. The ultrasonic wedge bond, however, re-

1
BALL FORMED ON WIRE

CAPILLARY TIP
(temperature range:
150 to 200°C)

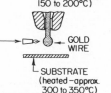

←─ GOLD
WIRE

└ SUBSTRATE
(heated–approx.
300 to 350°C)

HYDROGEN TORCH
wiper-type or stationary
pulsed ignition

2
CAPILLARY BRINGS BALL
DOWN TO SUBSTRATE
TO MAKE BALL BOND

HEAT AND
PRESSURE

PRESSURE depends on
static load (approx. 55 to
75g for 1-mil wire) and
drop load (dependent on
capillary hover height).

NOTE: Some machines are
constructed to give only a
static load, while with
other machines, one
must consider the
dynamic load due to the
drop from the capillary
hover height.

3
CAPILLARY IS
RETRACTED, ALLOWING
WIRE TO FEED OUT, AND
MOVED TO SECOND
BOND POSITION

Gaiser suggests a large loop
for wire strength. Do not strain
wire by stretching tight when
feeding out.

4
SECOND BOND

4A. TAILLESS
HEAT AND
PRESSURE

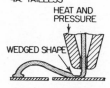

WEDGED SHAPE

4B. CONTOUR

ROUNDED

4C. NAILHEAD

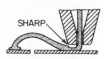

SHARP

NOTE: For highest bond
strength make two
wedge bonds. This
assures that bond is
not weakened when
the tail is removed.

METHODS USED IN TERMINATING A SEQUENCE
OF CONNECTED BONDS:

A HYDROGEN TORCH CUTS WIRE
AND FORMS NEW BALL FOR NEXT
BOND
To avoid breaking the tail and
leaving part of it attached to the
bond, use of the tailless capillary
tip is recommended.

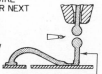

Tail removed by tweezers
or automatic tail-puller

B TAILLESS TERMINATION
The shape of the tailless tip assures
easy removal of the tail. This tip
design allows for a bonding process
and machine design in which no
tail is formed. After the second
bond is formed, the tip is raised,
allowing some wire to feed out.

Then the wire is clamped and the
capillary further retracted, breaking
the wire at the bond without
weakening the bond. Then a new
ball is formed.

Fig. 28 Ball-bonding sequence. (*Gaiser Tool Co.*)

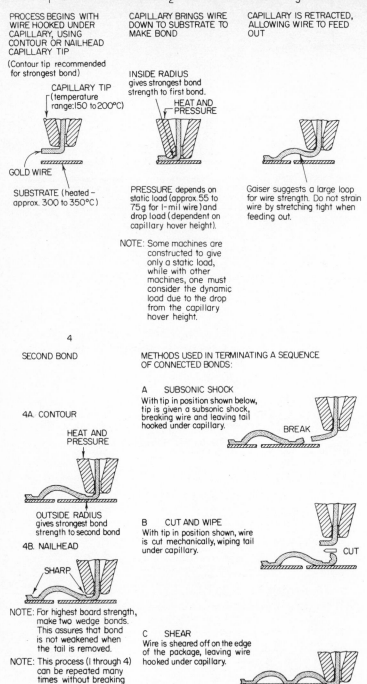

1

PROCESS BEGINS WITH
WIRE HOOKED UNDER
CAPILLARY, USING
CONTOUR OR NAILHEAD
CAPILLARY TIP

(Contour tip recommended
for strongest bond)

CAPILLARY TIP
(temperature
range:150 to 200°C)

GOLD WIRE

SUBSTRATE (heated –
approx. 300 to 350°C)

2

CAPILLARY BRINGS WIRE
DOWN TO SUBSTRATE TO
MAKE BOND

INSIDE RADIUS
gives strongest bond
strength to first bond.

HEAT AND
PRESSURE

PRESSURE depends on
static load (approx.55 to
75g for 1-mil wire)and
drop load (dependent on
capillary hover height).

NOTE: Some machines are
constructed to give
only a static load,
while with other
machines, one must
consider the dynamic
load due to the drop
from the capillary
hover height.

3

CAPILLARY IS RETRACTED,
ALLOWING WIRE TO FEED
OUT

Gaiser suggests a large loop
for wire strength. Do not strain
wire by stretching tight when
feeding out.

4

SECOND BOND

4A. CONTOUR

HEAT AND
PRESSURE

OUTSIDE RADIUS
gives strongest bond
strength to second bond

4B. NAILHEAD

SHARP

NOTE: For highest board strength,
make two wedge bonds.
This assures that bond
is not weakened when
the tail is removed.

NOTE: This process (1 through 4)
can be repeated many
times without breaking
wire.

METHODS USED IN TERMINATING A SEQUENCE
OF CONNECTED BONDS:

A   SUBSONIC SHOCK
With tip in position shown below,
tip is given a subsonic shock,
breaking wire and leaving tail
hooked under capillary.

BREAK

B   CUT AND WIPE
With tip in position shown, wire
is cut mechanically, wiping tail
under capillary.

CUT

C   SHEAR
Wire is sheared off on the edge
of the package, leaving wire
hooked under capillary.

SHEAR

Fig. 29 Thermocompression wedge-bonding sequence. (*Gaiser Tool Co.*)

quires an even larger target than the thermocompression ball bond. Typical target requirements are shown in Table 19.

In a typical thermocompression wire bonder the following parameters should be variable:

1. Heat column temperature, 25 to 350°C
2. Bonding tip or collet temperature, 25 to 350°C
3. Bonding pressure, 25 to 125 g
4. Dwell time, 1/10 to 5 s

By varying these parameters, improvements in the pull strength of both the ball and wedge bonds can be accomplished. By individual parameter variation a bonding schedule can be established similar to a weld schedule. This bonding schedule

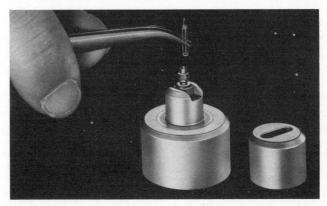

Fig. 30 Heated bonding capillary with replaceable glass tip. (*Specialty Glass Products, Inc.*)

TABLE 19 Typical Target Requirements for Ball Bonding

Wire diam, in.	Target size required, minimum dimension, in.
0.002	0.005
0.0015	0.004
0.001	0.003
0.0007	0.0025

would take into account all the referenced variables plus the wire diameter, whether the bond is being made to an aluminum (semiconductor) or gold (substrate) metallization, etc.

Ultrasonic Wire Bonding

Ultrasonic wire bonding also involves heat and pressure, but the heat is supplied by ultrasonic energy rather than by heated stages or capillaries. In addition, with aluminum wire, the ultrasonic energy and the acoustical high-frequency movement of the wire against the conductor pad breaks the refracting oxides surrounding the aluminum wire. Pressure is also used but is incidental to the effect of the ultrasonic energy. The ultrasonic vibratory energy causes a temperature rise at the wire-conductor interface that can approach 30 to 50 percent of the melting point of the metal.

One of the advantages of ultrasonic aluminum wire bonding is the absolute avoidance of *purple plague*. Since purple plague is the result of the combination of aluminum, gold, silicon, and heat, it is avoided by eliminating gold and heat.

Ultrasonic wire bonding is less commonly used with gold wire. In general, 0.002-in.-diam gold wire or larger can be ultrasonically bonded with relative ease, while the finer 0.001-in.-gold wire presents many difficulties. The ultrasonic-tool configuration is considerably different from that of the thermocompression tool. The tool design restricts the direction of wire stitching from the first bond to the second bond since the wire must be drawn directly toward the machine operation. Table 20 shows an example of an ultrasonic bonding tool along with the size availability.

Combinations of Ultrasonic and Thermocompression Wire Bonding

As wire bonding becomes more sophisticated, machines which once were very simple devices have also become more sophisticated, e.g., *ultrasonic ball bonders*. In this type of machine, the ultrasonic head is identical to the usual type except that a straight-wire capillary is used, as on a thermocompression bonder. Also included is the flame-off device necessary to form the ball on the gold wire. Whereas in straight ultrasonic gold-wire bonding it is difficult to bond gold wire of less than 0.002 in. diameter, on an ultrasonic ball bonder gold wire of 0.001 in. diameter is usually used. The differences are in the capillary design and the fact that, in general, a heated stage is used. So now we have the almost complete combination, i.e., a heated stage, a capillary-type tool, and an ultrasonic transducer. The only thing missing is the heated capillary, which becomes impractical with an ultrasonic transducer.

Pulse-heated Thermocompression Bonders

In an attempt to work with a "cold" substrate, many manufacturers of bonders have introduced *pulse bonding*, in which a welding-type power supply is used to send a pulse of current through the wire capillary, thereby heating it controllably to incandescence. This instant current provides enough heat at the interface between the wire and the conductor to effect an intermetallic thermocompression bond. Some equipment provides the pulse-heated capillary, the claim, of course, being that a lower temperature can be used at the heated stage. In the pulse bonder the flame for cutting the gold wire and forming the ball is still a necessity.

BEAM BONDING[*]

An effort to eliminate the "flying wire" has led into three additional methods for interconnecting the semiconductor device to the circuit, flip-chip bonding, spider bonding, and beam-lead bonding. Except for a special class of flip-chip bonding where the bonding pads are built up on the substrate, each of these methods requires specially fabricated semiconductor chips.

Before discussing the beam-lead bonding technique, there must be some understanding of the beam-lead device and its advantages. There are several major and some minor advantages.

The beam-lead device technology produces "sealed" junctions on the semiconductor chip. This essentially eliminates the necessity for hermetic packaging of the completed hybrid circuit. While some physical protection is still required, the need for hermeticity around the semiconductor no longer exists.

Since the die itself is not eutectically bonded to the conductor, the necessity for a separate die-bond operation is eliminated. The disadvantage here is that the thermal conductivity from the semiconductor junctions depends on the thermal path through the beams. This lessens the thermal capacity compared to that of a eutectically bonded die. As an advantage, however, the die is not touched directly during the bonding process, and additional reliability should result.

During severe shocks, both thermal and mechanical, the beams tend to isolate the die from damaging stresses that might otherwise affect the reliability of the chip. This resiliency of the beams also permits minor variations to occur in the height of the conductor patterns.

[*] For bonding processes and capillaries the following articles, which were published too late to be considered in writing this chapter, are recommended: Bonding Capillaries, *Circuits Manuf. Mag.*, December 1972, p. 26, a staff report, and F. Villella and R. Morton, Does Your Bonding Process Doom Devices?, ibid. January 1973, p. 22.

TABLE 20 Ultrasonic Bonding Tool

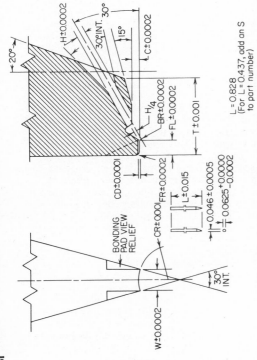

L=0.828
(For L=0.437, add an S to part number)

Part no.	Suggested wire diam, in.	Hole diam H ±0.002, in.	Foot width W ±0.0002, in.	Foot length FL ±0.0002, in.	Front radius FR ±0.0002, in.	Bond length BL ±0.0002, in.	Concavity, in.		C, in.	T ±0.002, in.	Bonding pad view reliefs
							CR	CD			
2010-13	0.0007	0.0013	0.003	0.0025	0.001	0.0015	0.010	0.0002	0.0015	0.015	Yes
2011-20	0.001	0.002	0.004	0.003	0.001	0.002	0.012	0.0003	0.0015	0.015	Yes
2012-20	0.001	0.002	0.004	0.0035	0.001	0.0025	0.012	0.0003	0.0015	0.015	Yes
2013-20	0.001	0.002	0.004	0.004	0.001	0.003	0.012	0.0003	0.0015	0.015	Yes
2020-25	0.001–0.0013	0.0025	0.005	0.0035	0.0015	0.003	0.012	0.0003	0.0015	0.018	No
2021-25	0.001–0.0015	0.0025	0.005	0.0045	0.0015	0.004	0.012	0.0003	0.0015	0.018	No
2023-35	0.0015	0.0035	0.008	0.0045	0.0015	0.003	0.012	0.0004	0.002	0.025	No
2024-35	0.0015–0.002	0.0035	0.008	0.0055	0.0015	0.004	0.012	0.0005	0.002	0.025	No
2025-35	0.0015–0.0025	0.0035	0.008	0.0065	0.0015	0.005	0.012	0.0005	0.002	0.025	No
2020-45	0.003	0.0045	0.008	0.0075	0.0015	0.006	0.012	0.0005	0.002	0.035	No

SOURCE: Gaiser Tool Co.

The bonding of the beams to the conductors is a gold-to-gold system. There are distinct advantages to a single-metal system (gold to gold) over the common wire bonding of a two-metal system (aluminum to gold). Without delving into the metallurgy involved it is sufficient to state that the single-metal interconnection system is more reliable.

Finally, the bonding of a beam-lead device (using proper equipment) is a single operation regardless of the number of beams a chip may contain. Whereas an integrated-circuit die with 14 input/output pads would require a minimum of 28 individual wire bonds, the same beam-leaded die requires only one bond operation. In addition, since a photomask operation is used, the resulting beams are very uniform with minimum tolerance variations.

While the bonds on a beam-lead die are visible for inspection, this is not true of the flip-chip device. This ability to inspect may be the deciding factor in choosing between beam lead and flip chip. If an inspection of a beam-lead device shows one or several beam bonds to be questionable, proper tooling will permit an individual beam to be rebonded without any effect on adjacent beams or adjacent components.

The technique for beam-lead bonding has gone through various stages of development over a period of several years. Initially parallel-gap welding was used to weld one beam at a time. This created problems in holding the chip and keeping it from skewing during the first couple of welds. As in all parallel-gap welding operations, problems of maintaining electrode dress, proper contact, uniform current and timing, etc., existed. In addition, parallel-gap welding could not adapt to the bonding of more than one lead at a time.

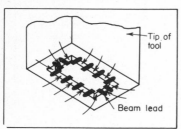

Inrush of air during vacuum holding of chip

Fig. 31 Die pickup of beam-lead device. (*Kulicke and Soffa Industries, Inc.*)

Wedge bonding utilizing the proven thermocompression bonding technique was also used. Initially thermocompression wedge bonding was accomplished by bonding one lead at a time. This led to the same skewing and holding problems that individual parallel-gap welding caused. The next obvious step was to wedge-bond all the beam leads on one side of the chip at once. This technique worked well, but still presented the problem of holding the chip during bonding.

The ultimate advance in beam-lead bonding incorporated the advantages of thermocompression wedge bonding with the efficiency of the collet used for eutectic die bonding. In this system, a square or rectangular collet, slightly larger than the beam-lead die, is used to hold the die by vacuum, and all the leads are bonded in one operation using thermocompression techniques. Figure 31 shows this technique for holding the die.

Several advantages accrue, but the most important is that holding the die during the bonding process is no longer a problem. The real problem occurs with the optical requirements for beam-lead alignment. Since the pickup and bonding collet now hides the beams, special optical arrangements must be made so that the beams can be aligned with the corresponding conductor lines on the substrate.

An improvement over the collet bonding was the *wobble head*. While the immovable collet required precise parallel alignment between itself and the substrate, the wobble head overcame this requirement and permitted a wider variation in conductor heights without the risk of a beam's failing to make an acceptable bond. Essentially, the necessity for coplanarity between the bonding tool and the substrate was no longer necessary with the wobble-head bonder. The choice then arose of whether to wobble the head or the table. The choice became dependent upon the complexity of the mechanics rather than any advantage in the bonding.

If only one beam-lead device were being bonded to a given substrate, the header or the substrate could be centered under the bonding collet and wobbled to accomplish the bond. This assumes that the axis of the device being bonded is aligned with the bonding collet. While this is possible with a single beam-leaded die on a substrate, it becomes very difficult when more than one die has to be centered under the collet.

As a result, it became more feasible to wobble the collet so that centering of the die on the stage was not necessary. With the wobble collet, many dice could now be placed on a single substrate and all bonded with the same reliability.

As semiconductor integrated circuits became more complex with medium- and large-scale integration (MSI and LSI, several thousand devices and a few dozen input/outputs on one chip) the beam-lead device becomes more advantageous. Savings in labor are significant, and the reliability, especially of the interconnections, goes up. The EIA committee JC 11.4 for uncased devices has already submitted standard designations for beam-leaded devices with 32 input/outputs.

Figure 32 shows a beam-lead bonder which utilizes the wobble-collet technique.

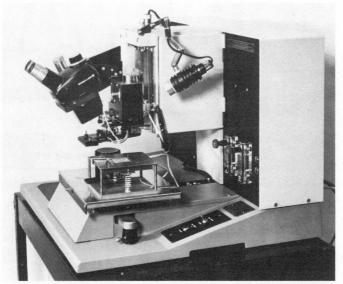

Fig. 32 Beam-lead bonder. (*Kulicke and Soffa Industries, Inc.*)

Compliant Bonding

Compliant bonding of beam-leaded dice is a technique for thermocompression bonding of all leads at the same instant, as opposed to wobble-tool bonding, which essentially bonds each beam lead separately even though all leads are bonded in one operation. In order to compensate for the variations in tolerances as a result of tool alignment, beam thickness, and conductor thickness, a compliant material, usually soft aluminum, is placed between the tool and the beam-lead device. As the bonding tool comes in contact with the beams, the compliant member begins to flow and an equal bonding force is exerted on all the beams. The compliant member flows around the beams because of the force exerted by the bonding tool (typically about 30 lb) and because of the temperature. The bonding tool is heated to approximately 450°C, and aluminum starts to flow at about 300°C. Figure 33 illustrates the principle of compliant bonding.

While there are advantages to compliant bonding over wobble-tool bonding, there are also disadvantages (as with any other process). The compliant member (aluminum foil) compensates for tolerance variations in the beam-lead device, the substrate, the stage, and the bonding tool. The compliant member, in flowing around the beams, also maintains their alignment and prevents their lateral movement, which could cause shorts during bonding.

The problems associated with compliant bonding are mainly in handling the foil. The bonding machine becomes more complex because a reel of compliant foil must be fed and aligned for each beam-lead device bonded. That portion of foil must then be transferred out of the way of the next bond. In addition, unless the compliant bonder has a means of separately picking up the beam-lead die and aligning the die and the foil, the beam-lead die must be tacked to the compliant foil as a separate operation prior to bonding.

In compliant bonding, the bonding force must be changed when a beam-lead die has a different number of beams than the one previously bonded because the

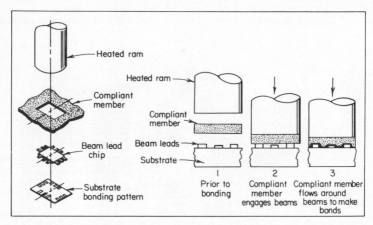

Fig. 33 Compliant bonding of beam leads. (*Kulicke and Soffa Industries, Inc.*)

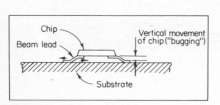

Fig. 34 Bugging of a beam-lead die. (*Kulicke and Soffa Industries, Inc.*)

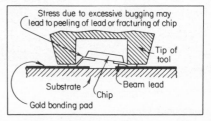

Fig. 35 Excessive bugging. (*Kulicke and Soffa Industries, Inc.*)

force is equally distributed over all the beams. In wobble-tool bonding the force can be kept constant regardless of the number of beams, since the tool only contacts a couple of beams at a time regardless of the number of beams the die may contain.

In either method of beam-lead bonding the pressure exerted by the bonding tool causes *bugging,* a condition which lifts the die away from the substrate due to the deformation of the beams during the bonding operation. Bugging of the die is advantageous since it lifts the active surface of the die away from the substrate, permitting easier final cleaning (before sealing) and less possibility of entrapment of contaminants between the die and the substrate. An example of bugging is shown in Fig. 34.

Excessive bugging could lead to stresses imparted between the beam and the die itself, causing failure of the beam-die interface or even cracking of the chip. Excessive bugging is shown in Fig. 35.

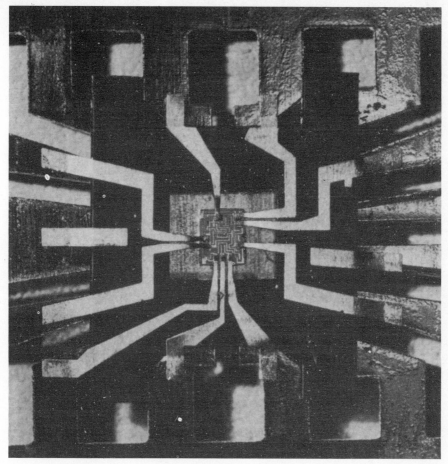

Fig. 36   A miniMod* chip.   (*General Electric Company.*)

SPECIAL INTERCONNECTION TECHNIQUES

Economics dictate chip interconnection techniques that are progressively less costly than hand wiring or even beam-lead devices.   Handling, alignment, die bonding, and wire bonding are all costly in labor and time.   As a result, new techniques continue to appear.   In general, these new techniques are tailored for complete automation, a condition incompatible with wire-bonding (flying-leads) technology.

The first of these techniques to be used on a large scale was the flip-chip solder-bonding technique by IBM.   Next came the beam-lead devices and several forms of *spider* bond devices.   The newest spider technique is the miniMod.*   In the miniMod device a silicon integrated circuit chip is bonded to plated-copper fingers suspended (laminated) in a polyimide filmstrip capable of withstanding temperatures as high as 300°C.   After the silicon chip is bonded to the fingers, it is protected with a drop of epoxy encapsulation.   The fingers can then be punched or sheared from the filmstrip and bonded to a substrate in an automatic process within 3 s.   The filmstrip (approximately 35 mm) can be reeled so that handling, bonding, and testing can all be automated.   Figure 36 shows the miniMod* chip configuration.

* Trademark of General Electric Company.

TABLE 21 Inspection Criteria

Inspection attribute	Magnification	Reject criterion
Chip and bonding to substrate:		
Chip placement on pad	30×–80×	Not 100% on cermet pad
		Hollow area visible under chip
		Part placement not per applicable drawing
Chip condition after bonding	30× min.	Cracks in active area
		Cracks in silicon over 1 mil in length and pointing toward an active area
Metallization defects on chips	80× min.	Shorting between rungs
		Scratches or voids down to substrate which reduce conductor to less than one-half design width
		Any opaque contamination, smears, or peaks which reduce distance between any metallization to less than 25% of design width
Appearance	30× min.	Melt not visible on two sides of chip
		Excessively crumbly melt
Wire bonds to chips and jumpers.	30×–80×	
Ball bonds:		
Size	...	Less than 2 times or over 4 times wire diameter
Placement	...	Less than 0.5 mil from adjacent bond or bonding area when viewed from above
Configuration	...	Bonding wire positioned on bond less than one-half wire diameter from bond edge
		Wire from ball bond perpendicular to surface of substrate for less than 0.5 mil before bending
Migration	...	Gold migration from ball bond onto chip metallization which extends over 0.5 mil from ball bond
		Silver migration onto gold wire for distance over 3 times wire diameter
		Purple plague visible beyond 0.5 mil from bond periphery
Metallization	30×–80×	
Wedge bonds		
Bond tails	...	Over 3 mils in length
Configuration	...	Less than 1.2 times or over 3 times wire diameter
		Wire angle between two wedges exceeds 15° from original run
		Less than three-fourths on pad area
Bond placement		Less than ½ mil from adjacent bond or bonding area

Inspection item	Magnification	Reject criteria
Wire leads:		
Wire size reduction by nicks, cuts, crimps, or necking down	30X–80X	Wire diameter reduced by over 25%
Proximity to other land areas or other wire leads	⋯	Less than 2 mils distance
		Crossing active unglazed land or bonding areas
Configuration	⋯	Too taut (little or no slack)
		Wire not straight when viewed from above (over 3 times diameter bend)
Epoxy-paste bonding:	10X–30X	
Capacitor-to-substrate bonding:		
Paste fillet	⋯	Cracks in cured paste of fillet
		Less than 90% of fillet on capacitor face with the 10% void on upper periphery of chip face
Excessive paste	⋯	Granular coarse appearance with flaky areas
		Paste shorting between pads or runs
		Loose paste on substrate
		Paste on resistors or conductor runs
		Paste on glass (insulation) where glass is over a resistor
		Paste over substrate edge shorting to pads or runs, except for grounds
Substrate in case	⋯	Paste on flange area of package, except where intentional grounds have been made
Wire-bond interconnect leads	30X–80X	
Lead length	⋯	Nonconformance to detail drawing
Bond configuration	10X–30X	Bond not 100% on package land area
Paste bonding of chips	10X–30X	
Fillet	⋯	Paste not visible on 100% of periphery
		Shrinking of paste away from chips or cracks in paste
Excessive paste	⋯	Paste shorting to active areas
		Loose paste on chip surface
Tack mounting of tantalum capacitors	10X–30X	
Form leads	⋯	Lead bend closer than 30 mils to part body or as recommended by part manufacturer
		Radius less than 1.5 times lead diameter
		Leads so formed that capacitor is not in parallel plane to substrate
Lead attachment	⋯	Part leads over top of bonding pad for distance less than 25 mils where epoxy-paste-bonded

	Conventional chip	Flip chip	Beam-lead device	Ceramic flip chip	Miniature package	Spider bond
Availability	Excellent	Fair	Fair	Good	Excellent	Low
Original cost	Low	High	High	Med. to high	Med. to high	Med. to high
Handling ease	Difficult	Good	Fair/good	Very good	Very good	Fair
Testing ease	Difficult	Fair	Fair	Good	Very good	Good
Comparative size	Small	Very small	Very small	Large	Large	Medium
High-frequency performance	Good	Good	Excellent	Fair	Fair	Very good

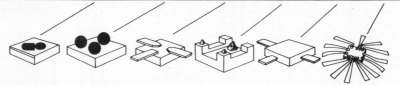

Table 37 Comparison of active devices. (*Electronic Component News.*)

INSPECTION CRITERIA

Table 21 tabulates inspection criteria (some abstracted from MIL STD 883) and defines visual and workmanship tolerances that can be reasonably expected in the assembly of thick film hybrid circuits. Microscopic examination should be performed for the inspection attributes at the listed power of magnification.

QUALITY-ASSURANCE RECORDS AND DATA BANK

Inspection results should be maintained to generate statistics to measure and control process steps and to assign priorities to corrective actions as required. These data should include part number and title, number of units submitted for acceptance, number of rejected units, rejects (or rejection causes) categorized at each process step or station.

HANDLING AND STORAGE

All substrate assemblies should be stored in a dry nitrogen atmosphere before and after inspection. Tweezers and/or finger cots should be used in handling in-process substrates. All substrates, during and after assembly processing, should be packaged in individual cushioned containers. Precautions should be taken with assembled substrates to assure that the necessary clearances are maintained to prevent chip or wire-bond damage.

Chapter **9**

Packaging and Interconnection of Assembled Circuits*

RICHARD J. CLARK

Advance Development Engineering,
General Electric Company,
Heavy Military Electronic Systems, Syracuse, New York

* The author wishes to acknowledge the information on design automation provided by his colleague F. C. Bergsten. Special acknowledgment is also given to Karin DeRegis for her assistance in typing the final manuscript.

INTRODUCTION

During the past decade the role of electronic packaging has grown from a relatively minor support function to a major area of responsibility in most engineering organizations. The product design engineer has evolved into an electronic packaging engineer. His areas of responsibility encompass the basic mechanical design areas but have broadened to include the fields of microelectronics, materials, and processes necessary to implement the newer breeds of advanced electronic hardware. In becoming increasingly competent in these fields he has himself become a "hybrid" engineer, interested in the mechanical, electrical, chemical, and physical properties of the various components making up the finished product.

This chapter will attempt to deal in some detail with the key areas of hybrid packaging. It starts at the hybrid package and follows it through the various levels of packaging to the cabinet level. Figure 1 indicates the various sections of the chapter and the material covered therein. Each section includes a brief description of the material covered in it.

Key areas of hybrid packaging The choice of the hybrid-circuit package greatly influences the rest of the system packaging design. The first section deals with various types of hybrid packages used in the industry. New varieties are constantly being introduced to meet new requirements of the semiconductor or system groups. Standardization of package size has made some headway, primarily in the dual-in-line package type. Here the 0.1-in. lead spacing on 0.3-, 0.6-, and 0.9-in. row centers is

Components	Interconnection techniques	Assemblies	Design considerations
		First level	
Integrated-circuit packages Hybrid packages Discrete components	Techniques within package covered in Chap. 8	Hybrid packaged assemblies	Key areas of hybrid packaging Trade-off factors Current status
		Second level	
Printed-wiring boards, Double- and single-sided, and Multilayer Flexible circuit board Ceramic substrates	Flow solder Reflow solder Point-to-point solder Point-to-point welding	Modules Printed-wiring board assemblies	Packaging design considerations Computer-aided design techniques Examples of hybrid packaging Future trends

Fig. 1 Chapter organization chart.

well established. The other areas of flat pack and special lead configurations show standardization only within families. The need for increasing numbers of pins and larger packages will be an important impetus to continued package development.

The section on System Packaging Design Considerations shows the diversity of concerns in the choice of a packaging system. The electrical design requirements generally make up the majority of the system specifications. Then the thermal, weight, volume, reliability, environmental, maintainability, and producibility requirements must all be met without jeopardizing the final system cost and operability.

It can be said that the name of the game is interconnection. This is true within the hybrid package, and it is equally true at the next higher levels of packaging. The various interconnection techniques available are discussed, and advantages and disadvantages of each are given.

An introduction to the subject of computer-aided design (CAD) techniques is given to acquaint the designer with the various methods of CAD presently available and the relative merits of each. Other detailed sources are available for specific programs and equipment.

Examples of hybrid microelectronic packaged assemblies are shown at various levels of assembly and for a variety of applications.

To assure that the information given in this chapter is as up to date and useful as possible, the last section covers future trends. Examples of some of the new devices, new packages, new processes, and new cooling techniques are shown. The impact of many of these developments on the growth of hybrid packaging will be great. Only by continued development of better and more sophisticated packaging techniques can we hope to keep pace with the development of new devices. The success or failure of the ultimate electronic system depends to a large extent on the competence and forcefulness of the device and system packaging engineers.

Key design factors Many design parameters must be considered when a new system is being designed. We will consider only the basic design philosophy at this time. The section on Packaging Design Considerations goes into more detail. The thick film hybrid circuit itself is only one variety of electronic circuit which when interconnected with other portions of the system performs a particular function. The key design factors which face the hybrid designer are much the same as those he faced in previous systems which used discrete transistors and integrated circuits. To many designers, the thick film hybrid circuit is factored into the system design much like any other conventional component. In fact, however, the typical thick film hybrid circuit will be larger, will consume and dissipate more power, and will perform a fairly complex function. Of course, it must perform its function better and at a lower cost to be advantageous. It must do all this while meeting many stringent requirements set down by its potential user.

Table 1 shows the key design factors which must be considered. These factors can and should be applied at all levels of packaging.

Advantages of thick film hybrid packaging The basic reasons for using thick film hybrid circuits can be listed as follows:

1. *Flexibility*. The ability to package a wide variety of active and passive chip and conventional devices having different physical characteristics into an integrated assembly.

2. *Low cost*. The inherent lower cost of the screening processes has led to the development of mass-produced thick film circuits.

3. *Reliability*. The basic thick film technique makes it possible to use proved materials and processes along with tested components finally sealed in hermetic enclosures for highest reliability.

4. *Varied technologies*. The combination of the varied devices along with various attachment and film materials allows optimum use of available technology.

5. *Reduction in size and weight*. Compared to other implementations using discrete packaged devices, the thick film hybrid is always smaller and lighter.

6. *Improvement in electrical performance*. Certain circuits where noise, speed, or other parameters are critical may benefit from the compact layouts permitted in the hybrid circuit.

Current status of thick film hybrid packaging Increasing pressure on system designers to develop new systems with greater complexities using the latest technologies at lower total system costs has led to a rapid development of thick film technology and created the need for compatible packaging technology to utilize all its advantages fully. Thick film hybrid technology is finding its way into a large segment of the commercial, industrial, and military electronics business. Since each new application must be evaluated on its own particular requirements, a variety of reasons are given for its use. The need for careful trade-off studies is also to be noted. Subsequent discussion in the section on Trade-offs will go into this in greater detail.

TABLE 1. Package Design Considerations

Chemical properties:
  Absorbtion
  Porosity
  Grain and structure
  Purity
  Inertness
Electrical properties:
  Dielectric constant
  Dissipation factor
  Resistivity
  Loss tangent
  Effect of temperature, humidity, frequency
Mechanical properties:
  Density
  Hardness
  Strength:
    Compression
    Flexibility
    Tensile
  Dimensional stability
  Hermeticity
Thermal Properties:
  Conductivity
  Specific heat
  Expansion
  Service temperature and softening point
  Thermal shock
Cost
Producibility

The design engineer has available a wide variety of packaging techniques and an in-depth technology providing constantly changing and improving materials and processes. New thick film pastes are available which will allow lines finer than 10 mils wide with 10-mil spacing to be manufactured on a regular production basis. New techniques for bonding copper directly to ceramic will provide a means for high-current-carrying capability or heat transfer in ceramic thick film packages. The variety of packages—multilead dual-in-line, single- and double-edge mount, as well as a variety of custom packages—continues to grow at a steady rate.

The techniques for interconnecting the basic thick film hybrid packaged circuit remain substantially the same, i.e., basic printed-wiring-board techniques together with a basic connector and back-plane configuration. The use of wire-wrap, Termi-point,* or soldered back planes is still the predominant method for system interconnections. Newer techniques such as Multiwire,† Infobond,‡ and special three-

* Trademark of AMP, Incorporated.
† Trademark of Photocircuits.
‡ Trademark of Inforex, Inc.

dimensional techniques are being developed and will undoubtedly see expanded use in the future.

The continued development of reliable, cost-effective thick film techniques will encourage their greater use in the automotive, television, industrial, and computer industries. There are already signs of successful implementation of microwave power modules using thick film technology.

SELECTION OF HYBRID PACKAGE

The variety of potential applications for thick film hybrids and the large number of different types of package available make it essential to understand the characteristics of each package type and how it affects the circuit to be packaged.

Table 1, package design considerations, is a useful checklist when considering a package for a particular application. In many cases trade-offs must be made to assure an optimum selection, and each characteristic can be weighted and compared.

Types of Packages

There are two basic types of packages used for hybrid circuit packaging, the peripheral-lead package and the axial-lead package. There are literally hundreds of available off-the-shelf packages for a variety of applications. Reference 23 outlines them according to family, and Fig. 2 shows a summary of standard packages. The most popular industry package configurations are the two-sided flat packages, the transistor-type axial-lead package, and the dual-in-line package. The availability of standard packages using reliable construction techniques has been a distinct advantage for the hybrid packaging field as this new generation of hybrid circuits is introduced. The broad variety of package types makes it essential to understand the basic differences in construction and choice of materials for the different types. The proper package for a given application must meet all the requirements placed upon it by the electric circuit, the manufacturing processes, and a variety of environmental conditions.

The various types of package construction are discussed with key materials and processes indicated.

Figure 3 shows a wide variety of typical hybrid packages, and Fig. 4 gives a brief description of some popular sizes, with their dimensions, substrate and device capacity, and power-handling capability. Table 2 lists a variety of hybrid packages compiled according to various physical characteristics.

Package Construction

The basic package design originally developed for single transistor packaging depends on matched-seal glass construction. The package materials shown in Fig. 5 are all used in various combinations. The package materials must be selected to operate properly in the final system and to achieve a proper balance of such important package characteristics as thermal transfer, radiation hardening, and special electrical or mechanical requirements, and they must be able to be assembled in a cost-effective manner into a final packaged hybrid circuit. In all cases, the choice of material combination is determined by process temperatures, chemical and electrical compatibility, and thermal-expansion characteristics. Figure 6 shows the thermal properties of the commonly used packaging materials and points up the need for compatible materials. The four basic types of construction for hybrid flat packages are discussed below.

All-metal construction The all-metal construction uses an Fe-Ni-Co alloy (Kovar)[*] for the metal base, body, lead frame and cover (Fig. 7). In most cases the base and package body are stitch-welded together. The lead frame is sealed with a borosilicate glass (Corning type 7052) in a high-temperature furnace (1130°C). This process is common to all matched-seal-construction packages, with modifications in fixturing to accommodate the different package-material combinations. After furnace firing the parts are descaled, cleaned, and gold-plated.

[*] Trademark of Westinghouse Corporation.

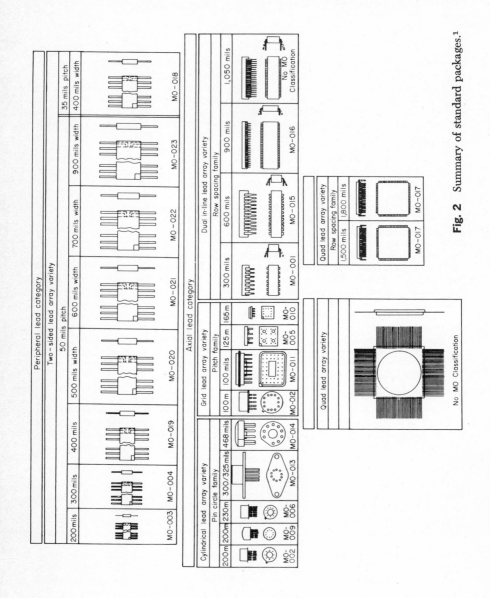

Fig. 2 Summary of standard packages.[1]

All-glass construction In this type of package a Kovar or similar alloy lead frame is sealed into an all-glass (Corning type 7052 or equivalent borosilicate glass) package. A low-cost version of this type utilizes a soft glass to form the lead seals during package fabrication. The soft-glass package uses ceramic covers to form the finished package (Fig. 8). The hard-glass version uses a molded glass body which

Fig. 3 Typical hybrid packages. (*Heavy Military Electronic Systems Department, General Electric Company.*)

Package dimensions, in.	Available leads	Substrate size, mils	Active-device density		Power dissipation in free air
			Transistors	Integrated-circuit chips	
$\frac{1}{4} \times \frac{1}{4}$	14	110 × 110	3–6	1–3	250 mW
$\frac{1}{4} \times \frac{3}{8}$	14	110 × 240	4–8	2–3	300 mW
$\frac{3}{8} \times \frac{5}{8}$	14	240 × 240	8–12	2–4	400 mW
$\frac{5}{8} \times \frac{5}{8}$	20	500 × 500	12–20	4–8	800 mW
$\frac{3}{4} \times \frac{3}{4}$	22	540 × 540	15–20	10–15	
$\frac{3}{4} \times 1$	32	540 × 820			830 mW
1×1	30	750 × 750			1.2 W
TO-5	6, 8, 10	180 diam	2–4	1	280 mW
TO-8	12 (small)	230 diam	6–8	2–3	1.0 W
TO-8	12, 16 (large)	300 diam	10–16	3–5	1.2 W

Fig. 4 Package capabilities.[2]

provides the structural body of the package as well as the seals (Fig. 9). The cover (either ceramic or metal) is then sealed with soft glass to give a low-cost package. When an alumina-ceramic cover is used, the alumina can be metallized with a Mo-Mn alloy, which forms an excellent bond with the alumina and accepts gold plating, so that a solder braze seal can be made. This provides a reliable hermetic seal for the package.

TABLE 2. Hybrid Microelectronic Packages[3]

Source	Type	Substrate size,* in.	Package size, in.	Number of leads	Lead configuration	Material	Package no.	Cover no.	Lid seal
Round substrate:									
Saegertown	TO can	0.475	TO-8	20	38 per 2 sides	Metal	K-820	C1101	Cold weld
Texas Instruments	Flat	1.50	2.125 × 2.125	154	29 per 2 sides	Ceramic	CN6035		
Saegertown	TO can	0.475	TO-8	16		Metal	K-818	C1101	Cold weld
Veritron West	TO can	0.375	0.500 diam	12		Metal	P-2435-6		Weld
		0.475	TO-8	12		Metal	P-2439-5		Weld
		0.475	TO-8	16		Metal	P-2441-7		Weld
Square substrate:									
Veritron West	TO can	0.350	TO-8	12		Metal	P-2439-5		Weld
Saegertown	TO can	0.350	TO-8			Metal	K-818	01101	Cold weld
American Lava	Flat	0.475	0.625 × 0.625	20	10 per side	Ceramic	SK-26796		Solder
Tekform Prod.	Flat	0.500	0.650 × 0.750	12/20	2 per 4 sides on base	Metal/glass	20260	20266	Weld
		0.750	0.975 × 0.975	14/20	2 per 4 sides on base	Metal/glass	20265		Weld
Coors Porc.	Flat	0.750	1.030 × 1.150	30	15 per side	Ceramic	A-1212-88-CS		Solder/ Ceramic
Tekform Prod.	Flat	0.750	0.904 × 1.048	7/1A	7 per side	Metal	70124-1	70110	Solder
		1.000	1.200 × 1.300	20	10 per side	Metal	70073 70202	70208	
		0.750	0.936 × 1.140	14	7 per side	Metal	30174	30176	
		1.000	1.335 × 1.372	30	10 per side	Metal	30101	30206	
		0.875	1.000 × 1.000	16/30/32		Metal	50260 50103 50270	With pkg.	
Isotronics	Flat	1.000	1.335 × 1.372	30	15 per side	Metal	60106	60107	Cold weld
Seal-A-Metic	Flat	1.000	1.180 × 1.180	72	18 per side	Metal	1P1155	With pkg.	Solder
Hermetic Seal	Flat	0.820	0.965 × 1.020	30		Ceramic	K1C-131	With pkg.	Solder
		1.000	1.076 × 1.356	16	8 per side	Metal	K300001-16-SC30		Solder
			1.255 × 1.255	44	22 per side	Metal	FP11231-44		Solder
Sylvania	D.I.P.	0.200	0.530 × 1.226	24	12 per side	Ceramic	7990-0055	7999-0122	Solder
American Lava	Flat	0.760	1.130 × .925	30	15 per side	Ceramic	SK-26685		Solder
Texas Instruments	Flat	0.770	1.000 × 1.000	64/80	15 per side	Ceramic	CN6180-83		Solder
Mitronics	Flat	0.800	0.960 × 1.020	30/15	16 per side	Metal/glass	HA-1752-		Solder
Texas Instruments	Flat	0.812	1.00 × 1.00	64	2 per 4 sides on base	Ceramic	CN6036		Solder
Tekform	Flat	1.000	1.150 × 1.250	20/30	22 per side	Metal/glass	20269	20270	Weld
Ceramic Metal	Flat	1.030	1.260 × 1.260	44		Ceramic	441260RW		Solder
Tekform Prod.	Flat	0.500	0.650 × 0.800	10	5 per side	Metal	70186	70184	Solder

Rectangular substrate

Manufacturer	Type	Substrate size, in.	Package size, in.	No. leads	Lead arrangement	Material	Part no.	Alt. no.	Seal
Seal-A-Metic	Flat	0.234 × 0.524	0.415 × 0.655	22	11 per side	Glass/ceramic	K1C-142-1		Solder
Coors Porcelain	Flat	0.300 × 0.400	0.520 × 0.520	16	8 per side	Ceramic	A-1212-88-CS	10-239198-16	Solder
Bendix	Flat	0.450 × 0.520	0.655 × 0.670	22	11 per side	Metal/glass	10-299122-1		Solder
Seal-A-Metic	Flat	0.460 × 0.525	0.665 × 0.665	22	11 per side	Glass/ceramic	K1C-143		Solder
Philco-Ford	Flat	0.470 × 0.500	0.640 × 0.640	20	10 per side	Metal/ceramic	FP-2082, 84	10-299198-16	Solder
Bendix	Flat	0.490 × 0.520	0.645 × 0.650	11	11 per side	Metal/glass	10-299122-2		Solder
Sprague Electric	Flat	0.500 × 0.520	0.780 × 0.780	22	11 per side	Metal/glass	FB-49		Solder
Coors Porcelain	Flat	0.500 × 0.600	0.770 × 0.770	26	13 per side	Ceramic	A-3434-1313-CS		Weld
Tekform Products	Flat	0.500 × 0.750	0.750 × 0.750	14/20	2 per 4 sides on base	Metal/glass	20261	20262	Weld
	Flat	0.500 × 1.000	0.750 × 1.150	20/30	2 per 4 sides on base	Metal/glass	20263	20264	Solder
Bendix	Flat	0.510 × 0.570	0.785 × 0.795	22	11 per side	Metal/glass	10-299113-1	10-299198-9	Cold weld
	Flat	0.530 × 0.820	0.810 × 1.035	32	16 per side	Metal/glass	10-299111-1	10-299198-8	Weld
Saegertown	Flat	0.700 × 0.800	0.900 × 1.000	16	8 per side	Metal	FPB-1000	FPC-1000	Solder
Tekform Products	Flat	0.750 × 1.000	1.050 × 1.150	20/30	2 per 4 sides on base	Metal/glass	20267	20268	Cold weld
Isotronics	D.I.P.	0.625 × 0.800	0.850 × 1.200	14		Metal	50264	With pkg.	Cold weld
		0.800 × 1.000	0.740 × 1.140	10		Metal	60100	60101	Cold weld
		0.800 × 1.600	0.740 × 1.240	20		Metal	60102	60103	Cold weld
		0.250 × 0.750	1.140 × 1.740	30		Metal	60104	60105	Weld
	Flat	0.250 × 0.870	0.500 × 0.870	14	11 per side	Metal	CD2010	With pkg.	Weld
		0.280 × 0.560	0.500 × 0.970	16	8 per side	Metal	CD2020	With pkg.	Solder
		0.390 × 0.440	0.375 × 0.625	22	11 per side	Metal	1P1030	With pkg.	Solder
		0.500 × 0.516	0.500 × 0.500	16	8 per side	Metal	1P1040	With pkg.	Solder
		0.562 × 0.600	0.625 × 0.625	20	10 per side	Metal	1P1060	With pkg.	Solder
		0.640 × 0.700	0.700 × 0.700	12	6 per side	Metal	1P1580	With pkg.	Solder
		0.500 × 1.000	0.810 × 0.810	14	7 per side	Metal	1P1590	With pkg.	Solder
		0.625 × 1.562	0.660 × 1.130	38	19 per side	Metal	1P1076	With pkg.	Solder
			0.810 × 1.700	32	16 per side	Metal	1P1600	With pkg.	Solder
Hermetic Seal	Flat	0.380 × 0.820	0.500 × 1.000	27	On 4 sides	Metal	FP11207-12		Solder
		0.430 × 0.975	0.450 × 1.125	18	9 per side	Metal	FP11193		Solder
		1.000 × 1.100	1.255 × 1.255	44	22 per side	Metal	FP11231-44		Solder
		1.000 × 2.100	1.358 × 2.256	32	16 per side	Metal	K300000-32-SC30		Solder
American Lava	Flat	0.500 × 0.700	0.750 × 0.875	28	14 per side	Ceramic	AP-645ST	AT-60927	Solder
Coors Porcelain	Flat	0.500 × 0.630	0.770 × 0.770	26	13 per side	Ceramic	A-3434-1313-CS	With pkg.	Solder
		0.790 × 0.840	0.990 × 1.100	30	15 per side	Ceramic	A-1X1-1/8-1515-CS	With pkg.	Solder
Bendix	Flat	0.510 × 0.570	0.785 × 0.795	22	11 per side	Metal/glass	10-299113-1	10-299198-9	Solder
		0.530 × 0.800	0.810 × 1.035	32	16 per side	Metal/glass	10-299111-1	10-299198-8	Solder
Saegertown	Flat	0.700 × 0.800	0.900 × 1.000	16	8 per side	Metal	FPB-1000	FPC-1000	Cold weld
Tekform Products	Flat	0.750 × 1.000	1.050 × 1.150	20/30	2 per 4 sides on base	Metal/glass	20267	20268	Weld
Mitronics	Flat	0.760 × 0.900	0.960 × 1.020	45/60	15 per side	Metal	HA-1752-( )		Solder
Tekform Products	Flat	1.000 × 1.500	1.300 × 1.800	32/42	2 per 4 sides on base	Metal/glass	20163	20164	Weld
Saegertown	Flat	1.100 × 1.200	1.300 × 1.400	24	12 per side on base	Metal	FPB-1400	FPC-1400	Cold weld
	Flat	1.100 × 2.000	1.300 × 2.200	40	20 per side on base	Metal	FPB-2200	FPC-2200	Cold weld

TABLE 2. Hybrid Microelectronic Packages[3] (Continued)

Source	Type	Substrate size,* in.	Package size, in.	Number of leads	Lead configuration	Material	Package no.	Cover no.	Lid seal
Tekform Prod...........	Flat	0.500 × 0.750	0.800 × 0.904	14	7 per side	Metal	70297	70233-030	
		0.500 × 1.000	0.800 × 1.200	20	10 per side	Metal	70187	70199-110	
		0.750 × 1.000	1.048 × 1.200	20	10 per side	Metal	70188	70200-110	
		0.750 × 1.500	1.048 × 1.703	30	15 per side	Metal	70075	70116-110	
		0.355 × 0.500	0.590 × 0.790	20	On 4 sides	Metal	20146	20147-130	
		0.355 × 0.955	0.545 × 1.145	28	On 4 sides	Metal	20258	20272-200	
		1.250 × 1.350	1.595 × 1.600	22	On 4 sides	Metal	20110	20109-150	
		0.750 × 1.500	1.140 × 1.936	40	20 per side	Metal	30175	30177	
		0.500 × 1.750	0.645 × 1.057	10		Metal	50100	With pkg.	Solder
		0.625 × 0.875	0.750 × 1.000	32/30/16	On 2 sides	Metal	50262	With pkg.	Solder
							50112		
							50113		
Seal-A-Metic...........	Flat	0.460 × 0.525	0.665 × 0.665	22	11 per side	Ceramic	K1C-143	With pkg.	Solder
U.S.E.S...............	D.I.P.	0.150 × 0.700	0.260 × 0.750	14	7 per side	Ceramic	100-0125	601-0002	Solder
	Flat	1.000 × 1.100	1.160 × 1.160	36	18 per side	Ceramic	100-0803	601-0003	Solder
		0.500 × 0.550	0.600 × 0.600	16	8 per side	Ceramic	100-0604		Solder
Sylvania...............	D.I.P.	0.750 × 0.500	0.295 × 0.715	14	7 per side	Metal	7990-0056	7999-0122	Solder

* Size is diameter for round substrates.

Metal-glass combination The use of Kovar elements, in combination with the necessary Corning type 7052 sealing glass, achieves the required hermeticity and thermal matching for the final package. A Kovar type of sealing frame is used alone or with a base plate (Fig. 10). In this type all parts can be fixtured in graphite jigs and passed through a conveyor-furnace firing cycle. The Kovar pieces are used to give flat and parallel surfaces without extra machining.

Glass-ceramic combination This type of package uses prefired ceramic elements for its basic components with sealing glass (Corning 7052) to hermetically seal at final assembly (Fig. 11). The ceramic base can be metallized in patterns for chip

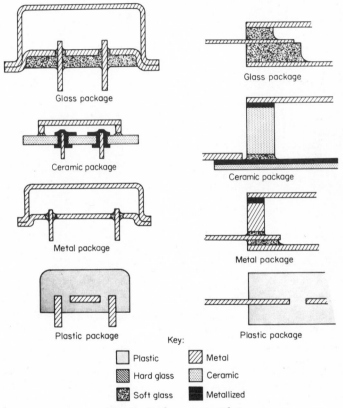

Glass package

Glass package

Ceramic package

Ceramic package

Metal package

Metal package

Plastic package

Plastic package

Key:

Plastic   Metal

Hard glass   Ceramic

Soft glass   Metallized

Fig. 5 Package materials.[1]

bonding, and the ceramic sealing flange can be metallized for convenient furnace sealing. The all-ceramic package has good dimensional stability and is useful in larger packages, such as microwave hybrid circuits, where microstrip conductors are brought out under the ceramic seal.

Transistor or axial-lead packages The same basic materials are used in this type of package, and similar sealing techniques are used. The metal base is formed to provide a mounting platform with holes through which the leads can be placed and sealed. The sealing glass is flowed into the base and around the leads to structurally support and hermetically seal them. The leads may be any length, and a typical hybrid circuit can then mount one or several ceramic substrates to the leads to form an interconnected circuit (Fig. 12a and b). Up to 20 leads are available in this type of package, and various cover heights are available to permit substrate stacking.

The resistance-weld seal makes this one of the highest-yield package types. Either nickel metal or gold-plated Kovar covers are used.

A variation of this type uses a thicker metal base for better heat transfer but

	Room-temperature thermal expansion, $10^{-7}\ °C^{-1}$	Room-temperature thermal conductivity, $(cal)(cm)/(s)(cm^2)(°C)$
Metal:		
Aluminum	230	0.52
Copper	170	0.96
Gold	140	0.74
Fe-Ni-Co alloys	50	0.04
Fe-Ni alloys	50–100	0.03
Molybdenum	50	0.32
Nickel	130	0.16
Silicon	45	0.31
Silver	190	1.00
Tungsten	45	0.38
Glass:		
Borosilicate	46	0.003
Aluminosilicate	47	0.003
Quartz	5	0.003
Lead borosilicate	90	0.003
Ceramic:		
Alumina, 94%	63	0.07
99.5%	68	0.09
Beryllia, 99.5%	70	0.60
Plastic:		
Epoxy, no fill	700	0.0004
Ceramic fill	120–250	0.001–0.02
Phenolic, no fill	800–1,000	0.0005
Filled	320–1,000	
Silicone rubber	3,000	0.0007
Teflon	800–1,000	0.0006

Fig. 6 Thermal properties of common packaging materials.[1]

requires individual glass seals for the leads and results in higher cost. The better thermal properties, however, may offset the higher cost.

Dual-in-line package The dual-in-line package is used primarily for monolithic integrated-circuit chip packaging (Fig. 13). As the large-scale integration (LSI) devices have become more available, a variety of 14-, 16-, 24-, 40-, and 50-lead packages with leads on 0.1-in centers and row spaces varying from 300 to 900 mils have been introduced to the industry. Ceramic, glass, and metal packages are available. The small mounting area available in these packages has limited their use in hybrid-package applications. They have also found use in discrete-component packages, where the dual-in-line format is needed for packaging standardization.

Plastic packages The use of plastics, such as epoxies, phenolics, and silicones,

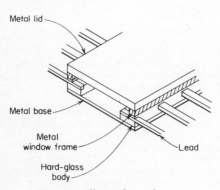

Metal lid

Metal base

Metal window frame

Hard-glass body

Lead

Fig. 7 All-metal package.[1]

as a packaging agent for hybrid microelectronic circuits has not been widely accepted in military-type equipment because they cannot be hermetically sealed and therefore their growth has been rather slow. Several varieties of 14- and 16-lead dual-in-line packages have been introduced for commercial applications (Fig. 14). Most of these were troubled with problems of humidity (moisture penetration between plastic and lead frame) and broken and lifted wire internal bonds caused by expansion differentials. As package leads grew to 40, new techniques were devised to overcome these shortcomings (Fig. 15). The package shown has a molded epoxy base around the ends of a Cu-Ni lead frame, the tips of which have been bent at right angles to be flush with the plastic surface. A lead pattern is then metallized from the tips

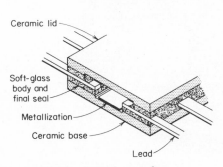

Fig. 8 Soft-glass package.[1]

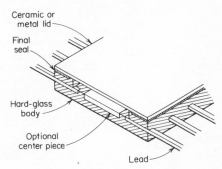

Fig. 9 Hard-glass package.[1]

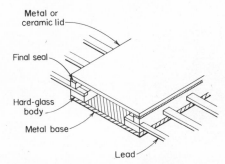

Fig. 10 Metal-glass combination package.[1]

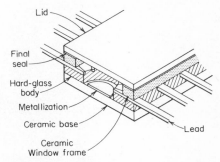

Fig. 11 Glass-ceramic combination package.[1]

of the lead frame to a chip cavity in the center of the epoxy base. A second sheet of semicured epoxy with a center opening is then molded with the epoxy base. After chip bonding a lid is epoxy-sealed over the cavity. Another variety of plastic package is the plastic axial-lead transistor package (Fig. 16).

Hermetic Sealing of Hybrid packages

Since the basic function of the hybrid package is to protect the microelectronic components within it, the final hermetic sealing of the package is very important. Semiconductor devices are very susceptible to minute quantities of contaminants, such as moisture and sodium, oxygen, and hydrogen ions. Moisture can attack the metallization on an integrated-circuit chip, causing an open circuit. Also the formation of purple plague (an Au-Al intermetallic) is accelerated by the presence of moisture. Ionic contaminants can reduce collector-base breakdown voltage in *npn* transistors, increase leakage current and parasitic capacitance in *pnp*'s, change

(a)

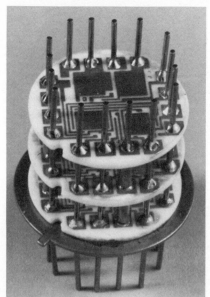

(b)

Fig. 12 (a) Single-substrate 16-lead TO-8 package; (b) multiple-substrate 12-lead TO-8 package.

low-current transistor gain, and in MOS devices cause time and bias variable changes in gate threshold voltages.

The severity of these changes is determined by the amount and type of contaminant present. Since some contaminants can exist even within a sealed package, inert gases such as dry nitrogen are used to purge the packages before sealing. Also, progress is being made in sealing chips with silicon nitride passivation so that true

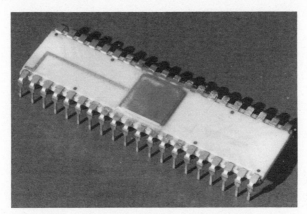

Fig. 13   Forty-lead ceramic dual-in-line package.

hermetic packages will not be required. These techniques are discussed later in the chapter.

Therefore, to minimize the effects of environmental changes on the finished package it is desirable to (1) clean the package and components thoroughly before sealing, (2) purge with a dry inert atmosphere, (3) introduce no contaminants inside the package during sealing, (4) seal at the lowest possible temperature, (5) seal for maximum hermeticity, and (6) test and screen out all defective seals.

The criteria of a good hermetic-sealing technique are shown in Table 3.

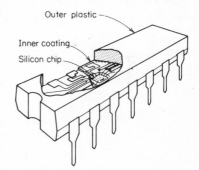

Fig. 14   Plastic dual-in-line packages.

Package-sealing Techniques

The final package seal of a thick film hybrid circuit can utilize welding, brazing, soldering, glass-frit sealing, plastic molding, and conformal coating, as shown in Table 4. All but the last two result in a hermetic seal in the 10^{-7} to 10^{-8} torr range. For many industrial and commercial applications a plastic transfer mold is used to give environmental protection to the hybrid devices. In a similar way a plastic conformal coating is used on certain hybrid circuits mounting silicon nitride–passivated beam-lead devices. These circuits, although in the strictest sense nonhermetic, have passed many hours of severe environmental testing. For military applications, however, strict compliance to hermeticity specifications is normally required.

Included in the welded seals are resistance, stitch, and cold welding. These all provide a reliable metal-to-metal seal for packages designed to be welded. Soft-soldered and brazed seals use a relatively low-melting solder to bond the other case materials. This requires special care in the choice of solders and in the use of cleaning fluxes to avoid package failure due to incomplete joints or thermal-expansion problems. The glass-frit seal is used where an insulating sealed joint is required and in low-cost applications where lower-strength seals can be tolerated.

Resistance welding   Resistance welding has been used for many years to seal

Fig. 15   Forty-lead plastic dual-in-line package.

the conventional TO series of semiconductor packages.   A special die set is required, and the package must be designed to provide a mating weldable lip, usually Kovar°

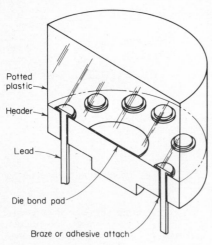

Potted
plastic

Header

Lead

Die bond pad

Braze or adhesive attach

Fig. 16   Molded-plastic transistor package.

or nickel-alloy materials (Fig. 17).   Plating may be used to assure metal cleanliness and efficient metallic melting during the welding process.   Resistance welding may be either ac or capacitor-discharge type.   The main difference between the two is that the capacitor-discharge weld energy is delivered in short welding pulses and the ac type has a longer welding cycle, resulting in more heat transfer into the package.

Parallel-seam welding   Parallel-seam welding is a fast, reliable method of sealing hybrid integrated-circuit packages. The weld is made by applying current through two conically shaped electrodes, as shown in Fig. 18.   The combination of correct pressure and current pulses along with proper package design assures good welds.   Since the weld is determined by the resistance presented by the package lid and lead frame, care must be taken to design the lead frame for low resistance relative to the lid.   This permits welding at moderate currents, 500 A peak to peak, of short duration, 20 ms. It has been found that lid thicknesses of 3 to 5 mils result in minimum tempera-

° Trademark of Westinghouse Electric Corporation.

ture rise in the package. For larger packages a thicker lid with the edges coined to a reduced thickness in the weld area is used.

Solder-sealing techniques Soft solder sealing employs Pb-Sn solders which melt from 190 to 225°C. The low temperature is the basic advantage of solder sealing. In addition, the seal is easy to repair. Soldering is a particularly attractive method of packaging thick film circuits or circuitry in which organic adhesives or plastics have been used or temperatures higher than 225°C could have detrimental effects on device performance. Care must be taken if the seal requires a flux before soldering. Most Pb-Sn solders require flux to achieve appropriate removal of oxides to obtain good surface wetting. Flux residues trapped during this process can be a major failure mechanism.

Bond strength is also a problem in solder sealing. Achieving maximum strength requires a minimum amount of solder. However, controlling the thickness of this

TABLE 3. Criteria for Good Hermetic-sealing Technique

Produce high yield of sealed product passing 10^{-8} cm³/s leak-rate requirement
Produce a strong seal which will not fail in handling, assembly into the system, or use
Seal lid to package without raising the temperature of the contents significantly
Perform the seal in an ultradry inert atmosphere
Create no molten areas of sealing material inside the package which can short-circuit to circuit parts or interact with them
Create no gases, vapors, or loose particles within the package cavity
Create minimal additional strains in the package structure which might cause it to fail
Require minimum manual labor
Readily adapt to a variety of package shapes and sizes and accommodate increasingly larger packages with minimum tooling and adjustment

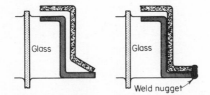

Fig. 17 Resistance-welded package.[1]

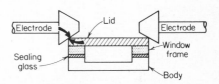

Fig. 18 Parallel-seam welding.[1]

solder seal requires elaborate precautions, thereby quickly losing its economy. The melting temperature of the solder may sometimes be a problem. Because of the temperature limitations of the solder, it cannot be used for high-temperature storage or operation. A special purpose solder seal is shown in Fig. 19.

Braze-sealing techniques Although a large number of braze materials are available for integrated-circuit package sealing, most use the low-melting gold eutectics. Included in these materials are Au-Si, Au-Ge, and Au-Sn. The respective melting points of these materials are 377, 356, and 280°C. The Au-Sn preform material is generally preferred because of its eutectic temperature, high gold content, and relatively fine grain structure. These materials give good wetting of either gold-plated piece parts or metallized gold surfaces. Although Au-Sn has the lower melting point, it is comparatively fragile and tends to crack easily under mechanical shock. Thus, it has the disadvantage of giving brittle joints. Au-Ge and Au-Si result in more ductile material and resist cracking. (See Fig. 20.)

The techniques used to braze-seal generally involve a furnace. With the furnace sealer, the package lid, the braze preform, and the package are placed in a jig. The jigs are loaded onto the belt furnace and go through the hot zone of the furnace to achieve softening, melting, and flowing of the braze materials. Usually, the package lid has a weight on it to assure even pressure during the flow of the braze material.

TABLE 4. Package-sealing Techniques

Sealing process	Package type	Repairability	Hermeticity, torr	Maximum internal package temperature rise, °C	Equipment investment	Possible problems
Resistance weld	TO-8, TO-5	No	10^{-8}	100	Moderate	Large packages difficult
Parallel-seam weld	Metal flat packs, irregular shapes, custom designs	No	10^{-8}	10–50	Moderate	Package parts should be plated
Electron-beam weld	Special packages	No	10^{-8}	10–50	High	Package must withstand 15-psi differential; expensive
Soft solder	Smaller flat packs; metallized ceramic and metal	Yes	10^{-8}	175	Low	Solder-compatibility thermal design required
Braze	Smaller flat packs; metallized ceramic and metal	No	10^{-8}	150* 300†	Moderate	Au-Sn brittle; Au-Ge more ductile
Glass-frit seal	Flat packs, ceramic, custom seals	No	10^{-8}	400–600	Moderate	Brittle, non-metallic
Cold weld	Special packages designed for cold weld	No	10^{-8}	0	Moderate	
Plastic molding	Custom standard hybrid packages	No	‡	100–150	Moderate	Mold design required; not hermetic
Conformal coating	Nonhermetic custom packages	No	‡	100–150	Low	Not repairable; not hermetic

* Peripheral.
† Furnace.
‡ Not measurable.

Control of the weight on the package influences the thickness of the final braze film and regulates the amount of material extruded. This material takes the form of little metal balls, which, if they occur, can fall off and cause trouble inside the package.

Cold welding Cold-weld sealing uses single high-pressure pulses without added heat from the seal. This forces the metal members into intimate contact to form a hermetic joint. Since the materials must be ductile, copper and aluminum are the only materials which can be utilized successfully. In many instances, the metallurgy required for a cold-weld seal can be achieved with a clad material, in which case, the copper or aluminum could be clad onto a base Kovar material to achieve the cold-weld seal (Fig. 21).

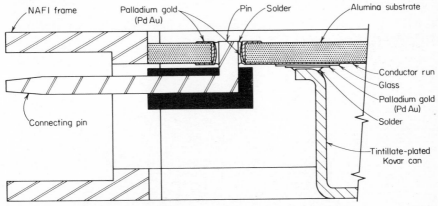

Fig. 19 Solder seal. (*Heavy Military Electronic Systems Department, General Electric Company.*)

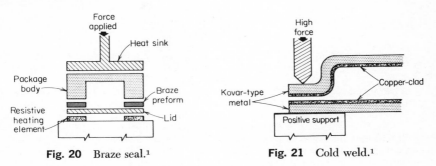

Fig. 20 Braze seal.[1] **Fig. 21** Cold weld.[1]

The main advantage of cold-weld sealing is that no thermal energy is required. Good ambient control can be achieved with the room-temperature cold-weld sealing. Thus, the cold-weld seal provides a guarantee of ambient control for surface-sensitive devices such as MOS or thin films. Also, weld flash, vaporized materials, solder imperfections, solder voids, and solder balls are all eliminated with this technique. Since this technique uses the least amount of heat, it is most attractive. A problem associated with cold weld is the requirement for a flange area in which this high force can be exerted without interfering with delicate glass-to-metal seal areas. This requirement sometimes extends the package dimensions beyond those normally associated with other types of seals.

Glass-frit sealing Glass-frit materials can be utilized with ceramic, glass, or metal packages to achieve final seals. Glass-frit sealing is generally used with a glass or ceramic body. The glass frit may be of the low-expansion high-temperature or high-expansion low-temperature type. The greater the dissimilarity in thermal expansion, the greater the permanent strain that remains. This dissimilarity is greater with lower-temperature materials. Such residual strain determines the amount of thermal or mechanical shock needed to crack this seal. The lower-expansion types do not create this problem but require high temperatures (500 to 600°C), which usually will degrade the active elements in the package.

Most of the glass frits used are vitreous; i.e., the glass does not crystallize. Vitreous glass does not give the strength available with the devitrified type, which requires a longer seal time to achieve crystallization. Optimum crystallization results from sealing times approximating 1 hr at sealing temperature. When utilizing the lower-temperature materials, it is essential to have an oxidizing atmosphere during the sealing operation to prevent the reduction of lead glasses. Again, the problems of viscous flow of glass and appropriate wetting surfaces can occur. Necessary weight is applied to get reasonable flow of the glass material and to get even coverage of the area to be sealed. Solder glasses or Pyrocerams° are typical of the devitrifying glass materials. They give a high-strength seal. The problem associated with these seals is that furnace sealing is required. The entire package must be elevated to the sealing temperature of 400 to 550°C (see Fig. 22).

Fig. 22   Large-area seal between ceramic wall and substrate using devitrifying glass sealed at 500°C (upper surface ready for solder seal.) (*General Electric Company.*)

Circuit encapsulation   Plastic encapsulation can be used to protect thick film hybrid circuits from most environmental effects at a lower cost than full hermetic sealing.

When using plastic encapsulation, it is desirable to apply an initial coating of some soft material to protect the circuits both electrically and mechanically. This material serves as a buffer between the ceramic circuit of relatively low expansion coefficient and a plastic encapsulant of much higher expansion coefficient. It also offers mechanical protection against vibration and shock damage as well as protection against moisture and ingredients of the outer shell. The materials used for this coating are usually urethanes or silicones, some of the typical properties of which are given in Table 5.

The silicones are sometimes single-component systems stabilized with acetic acid, which is objectionable because it is incompatible with some of the parts encapsulated. A preferable material is low-temperature-curing silicone rubber because the catalysts used in this system are usually innocuous. Silicone rubbers that withstand baking temperatures of 200°C or higher are reasonably moisture-resistant and will not react with subsequent outer coatings.

Urethanes can also be used either as single- or two-component systems. Among single-component systems used for noncritical purposes, solvent-containing moisture cures can be employed, although such coatings are not as impervious (nonporous) as might be desired. For more impervious coatings, phenol-blocked urethanes can be used that require baking for polymerization at about 200°C. During this baking, copious amounts of phenol are evolved. While phenol evolution is undesirable,

° Trademark of Corning Glass Company.

these urethanes are highly effective as undercoatings for such circuits as resistor networks in which a conformal coating is subsequently applied. If the high baking temperature or the evolution of phenol—or both—is objectionable, two-component systems, consisting of an isocyanate-containing prepolymer and a hydroxy-terminated curing agent, must be resorted to. These materials must be mixed as required: they have a limited pot life in the mixed state but can be cured at comparatively low temperatures.

Epoxies are generally not desirable as undercoatings. Even if they are made soft and rubbery by including copolymers, such as urethanes, in the formulation, the use of amines as curing agents constitutes a distinct hazard since any amine may reduce the palladium oxide in palladium-containing resistors if they are not initially glazed.

TABLE 5. Properties of Soft Undercoating for Circuit Encapsulation[2]

	Silicone (clear)	Urethane
Viscosity, cP*...............................	5,000	2,000
Pot life at 25°C, h.............................	2	4–6
Cure time (temperature, h(°C)	4(65)	6(65)
Water absorption in 24 h, %......................	0.1	0.5
Hardness (Shore A).............................	40	40
Expansion coefficient $\times$ 10^{-6}......................	250	150
Max. continuous-use temperature, °C..............	200	100
Dielectric constant..............................	2.7	6.0
Dissipation factor, %...........................	0.1	1.5

* cP is the ANSI abbreviation for centipoise; in the industry cps is commonly used.

TABLE 6. Properties of Typical Conformal Coatings[2]

Coating	Heat resistance, °C	Tensile strength, lb/in.	Impact strength, ft-lb/in.	Thermal expansion, μin./(in.)(°C)
Dip:				
Phenolic................	150	5,000	0.5	35
Diallyl phthallate.........	175	6,000	0.5	25
Fluidized-bed:				
Epoxy..................	225	8,000	0.5	35
Urethane...............	125	10,000	5.0	65

For economical outer coating, dip-type conformal coatings are used that are based on phenolics or dially phthalate in solution form. When properly applied, these coatings will meet stringent tests and can be made even more impervious by subsequent vacuum impregnation with wax or similar materials. One drawback of this technique is that it is suitable only for relatively large-scale production lines. For satisfactory results it is absolutely necessary that machinery be used to stir the batch, automatically control the viscosity, and provide conveyerized systems for dripping, withdrawal, and drying. Another problem is that the relatively irregular external shapes of some circuits makes automatic handling equipment unsuitable.

Another conformal technique uses a fluidized-bed coating. To apply this coating, a preheated component is immersed in a cold fluidized bed so that some of the particles adhere. Then the component is reheated to flow out the coating and cure it where necessary. Initially this process was available only for thermoplastic materials, but now some thermosets are being offered (particularly epoxies), including some relatively soft ones and some urethanes. Table 6 shows properties of various conformal coating materials. Circuit protection by conformal coating is somewhat inferior to that afforded by solution coatings because the layers, by their very nature, may contain pores, at least of the closed variety.

For final encapsulation of nonhermetically sealed circuits a rigid plastic is desired. While the final choice of material depends on the process to be used, epoxies, because of their desirable high-temperature properties and relatively low permeability, are usually selected. Epoxies can be applied in small-scale operation as liquid resins for potting of components in a premolded case or in large-scale operation more usually by transfer molding. Because epoxies have a much higher expansion coefficient than ceramic substrates and the electronic components ($50 \times 10^{-6}°\text{C}^{-1}$ for epoxy; $6.8 \times 10^{-6}°\text{C}^{-1}$ for alumina), unfilled epoxies usually cannot be applied directly over ceramic. One problem is that during cycling, moisture-leakage paths will frequently occur because the epoxy cracks, causing damage to the circuit. To overcome this problem, a buffer coating of the silicone and urethane type is usually employed. The epoxy expansion coefficient can be brought much closer to that of ceramics by using fillers. However, epoxies with fillers require higher mold pressure in transfer moldings and similar processes, so that serious damage can occur to the parts being

TABLE 7.   Properties of Typical Molding Resins[2]

	Epoxy	Phenolic	Dially phthalate
Molding temperature, °C..................	150	150	150
Molding pressure, psi.....................	1,000	2,000	1,000
Thermal expansion, in./(in.)(°C)...........	25	25	25
Water absorption in 24 h, %..............	0.05	0.3	0.3
Dielectric constant......................	4	5	5
Dissipation factor, %.....................	1.0	1.0	0.5
Maximum use temperature, °C.............	175	125	200

encapsulated. With liquid resin, this is not a problem, but because of the increased viscosity, insufficient air may be removed and moisture-leakage paths around leads can occur. Therefore, the use of unfilled or only moderately filled resins with a buffer coat underneath is preferred. Tables 7 and 8 list properties of molding and casting resins.

To obtain assemblies that will withstand temperature-cycling and humidity tests, the importance of cleanliness and suitable surface preparation, particularly of leads, to obtain optimum adhesion cannot be overstated. Many silicones and urethanes require primers, such as functional derivatives or ethoxy or methoxy esters. Even where such pretreatments are used, proper cleaning procedures are required to remove solder-flux residues, fingerprints, and the like. However, it is advisable to avoid the use of primers, because this is yet another operation and adds to the cost of the finished circuit.

Another important consideration in the selection of buffer coatings is the removability of such coatings for the replacement and addition of components. Fortunately this removal is possible with both the silicones and the urethanes, which are preferred for this kind of service. Silicones can often be removed by purely mechanical means. Where this is not completely possible, the silicones can be swelled by chlorinated solvents so that their removal becomes easy. Subsequent heating to remove the solvent from the silicone will restore the original properties. Similarly, ethanol will swell urethane coatings so that they can be removed easily; on subsequent drying, the original properties will be restored. Alternatively, urethanes can be dissolved completely with warm dimethyl sulfoxide or similar materials. Outer shells of epoxy or thermosets must be opened by purely mechanical means for repair, but this does not usually constitute a difficulty if an inner buffer coating of adequate thickness is present. This technique of plastic encapsulation can be used economically and with very high yield when the application is nonmilitary or the end environment is not likely to expose the circuit to the environment shown in Table 9. If such an environment is likely, hermetic packaging should be used. Table 10 shows the protection afforded against humidity and temperature by the various types of packages as well as the relative cost.

TABLE 8. Typical Properties of Several Common Resins[21]

Mechanical and physical

Material	Tensile strength, psi	Elonga-tion, %	Com-pression strength, psi	Impact strength, Izod, ft-lb/in. of notch	Hardness	Linear shrink-age during cure, %	Water absorp-tion, wt %
Epoxy:							
Rigid, unfilled...............	9,000	3	20,000	0.5	Rockwell M 100	0.3	0.12
Filled.....................	10,000	2	25,000	0.4	Rockwell M 110	0.1	0.07
Flexible, unfilled............	5,000	50	8,000	3.0	Shore D 50	0.9	0.38
Filled.....................	4,000	40	10,000	2.0	Shore D 65	0.6	0.32
Polyester:							
Rigid, unfilled...............	10,000	3	25,000	0.3	Rockwell M 100	2.2	0.35
Flexible, unfilled..............	1,500	100	....	7.0	Shore A 90	3.0	1.5
Silicone:							
Flexible, unfilled............	500	175	....	No break	Shore A 40	0.4	0.21
Urethane:							
Flexible, unfilled............	500	300	20,000	No break	Shore A 70	2.0	0.65

Thermal and electrical

Material	Heat-distor-tion temp., °C	Thermal shock per MIL-I-16923	Co-efficient of ther-mal ex-pansion, ppm/°C	Thermal conduc-tivity, (cal)(cm)/ (s)(cm²) (°C) × 10⁴	Dissi-pation factor*	Dielec-tric con-stant*	Volume resis-tivity,* Ω-cm	Di-electric strength,* V/mil	Arc resist-ance, s
Epoxy:									
Rigid, unfilled....	140	Fails	55	4	0.006	4.2	10^{15}	450	85
Filled.........	140	Marginal	30	15	0.02	4.7	10^{15}	450	150
Flexible, unfilled..	<RT†	Passes	100	4	0.03	3.9	10^{15}	350	120
Filled.........	<RT	Passes	70	12	0.05	4.1	3×10^{15}	130	360
Polyester:									
Rigid, unfilled....	120	Fails	75	4	0.017	3.7	10^{14}	440	125
Flexible, unfilled..	<RT	Passes	130	4	0.10	6.0	5×10^{12}	325	135
Silicone:									
Flexible, unfilled..	<RT	Passes	400	5	0.001	4.0	2×10^{15}	550	120
Urethane:									
Flexible, unfilled..	<RT	Passes	150	5	0.016	5.2	2×10^{12}	400	180

* Dissipation factor and dielectric constant are at 60 Hz and room temperature; volume resistivity is at 500 V dc; and dielectric strength is short-time.

† Room temperature.

Hermeticity Tests of Package Seals

Hermetic testing of the final package seal is required to determine the adequacy of the hermetic seal. Its hermeticity is based on the measurable leak rate determined by a variety of techniques. The various techniques are described in MIL-STD-883 and are divided into two categories, gross-leak and fine-leak tests. The relative sensitivity of these tests is shown in Fig. 23.

Gross leaks Gross leaks are detected by observing the presence of bubbles coming from a leak when the package is immersed in a hot liquid such as glycerin or polyethylene glycol. As the package warms up, the internal gases expand and, if a large leak is present, will flow into the liquid, forming bubbles. An alternate approach is to pressurize the package with 150 psig of nitrogen and then immerse it in alcohol and watch for bubbles. Another solution is to use a 90°C mixture of detergent in water. These bubble tests are highly subjective and require close observation. Another gross-leak test is the dye test (Zyglo*). The circuit packages are

TABLE 9. Typical Environmental Test Requirements[2]

Temperature exposure........	1,000 h at 125°C
Humidity cycling...........	Ten 24-h cycles from room temperature to 85°C; 90% humidity
Thermal shock.............	Three cycles from 100°C liquid into 0°C liquid
Temperature cycling........	Five cycles from room temperature to 125°C, air chamber to room-temperature chamber to −55°C chamber to room temperature

TABLE 10. Protection Afforded and Cost of 1 by 1 by 1/4 in. Packages[2]

Package type	Number of humidity cycles, room temperature to −85°C	Cost, $
Conformal coating......	10	0.02
Molded...............	20	0.05
Cast.................	50	0.20
Hermetic.............	Indefinite	2.00

immersed in a dye solution under pressure. The solution will penetrate any crack in the package, and when it is removed from the solution and wiped clean, the residual dye in the crack can be observed under ultraviolet light.

Fine leaks Fine leaks are detected by forcing helium or a radioactive tracer gas into the package after it has been placed in a pressure bomb, and then the leak rate is measured with a helium leak detector or a radioactive sensor.

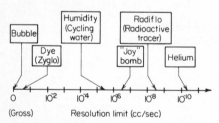

Fig. 23 Relative sensitivity of leak-testing methods.[1]

Mass-Spectrometer Leak Test. In this test the circuit packages are exposed to helium in a pressure bomb for an extended period. The helium is then pumped out and the package surfaces are washed to remove helium. If the package seal leaks, helium will have flowed into the package while it was in the pressure bomb and will be flowing out. It can be detected in two ways: (1) the package is placed in a small vacuum chamber connected to a helium-leak detector which continuously monitors the package effluent (Fig. 24), or (2) a small-bore tube is connected to the mass-spectrometer intake through a throttling valve. The valve is adjusted so that the pressure in the mass spectrometer is 10^{-5} to 10^{-4} torr. The tube acts as a sniffer to draw in the air around the suspected leaky package. The helium from the leak is mixed into the air and flows into an ionization chamber, where atoms are ionized by an electron beam and accelerated by a fixed voltage (Fig. 25).

These positive ions of known fixed energy now pass into a mass analyzer, where a magnetic field selects out ions with the charge-to-mass ratio of helium and the present energy. A very substantial increase in the test sensitivity is possible if the

* Trademark of Magnaflux Corporation.

package is filled with helium when sealed. In that case the rate of gas leakage out is greatly increased over the pressure-bomb method unless the leak is very large. Furthermore, the equipment and time to pressurize the package are no longer needed.

Helium is used as the probe gas for a number of reasons: (1) because of its low

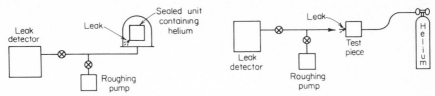

Fig. 24 Mass-spectrometer leak test using vacuum chamber.

Fig. 25 Mass-spectrometer leak test using sniffer probe.

Fig. 26 Mass spectrometer leak detector, NRC model 925. (*NRC Equipment Corp.*)

mass, its flow rate through a leak is high; (2) it is chemically inert; (3) it is easily available at a low cost; (4) fairly large partial pressures in a room are harmless; (5) it is normally present in the air at very low concentration, so that it has a very low background effect; (6) other commonly occurring gases such as nitrogen, oxygen, and possibly hydrogen have masses which differ by a large fraction so that the spectrometer resolution need not be very good to separate out the helium ions.

One variety of mass-spectrometer leak detector is shown in Fig. 26. This particular unit is equipped with an integral test station and is suitable for production testing.

SYSTEM PACKAGING DESIGN CONSIDERATIONS

The previous section on package types for hybrid circuits was primarily concerned with the portion of electronic packaging technology referred to as *microelectronic* or *hybrid-circuit* packaging, as related to the selection and design of the level 1 package. In this section we shall be looking at those areas of packaging design which relate to specific design functions, such as structural design, environmental protection, thermal management, and mechanical and electrical interconnection.

It is very important for the design engineer to understand fully the packaging requirements for a particular application as early in the design phase as possible. It is necessary to prepare some sort of rating chart for his own use in determining which of a variety of factors are important. Table 11 shows a checklist which is useful in the preparation of a packaging-method rating chart. Table 12 lists the important electronic-packaging design functions, elements, and ideal characteristics. Obviously, the final system must make engineering judgments between many of these system characteristics so that the final system is the best composite of all of the required characteristics. Table 13 lists five evaluation factors with their purpose and some of the important rating factors in each group.

In addition, the relative importance of the various factors needs to be ascertained. A typical chart (Table 14) indicates the relative order of importance of rating factors for missile and spacecraft, shipborne and military ground, airborne, and commercial equipment. This listing is, of course, subject to revision by each user and is shown only for comparison.

Trade-off Factors

The design of any electronic system, large or small, requires that certain decisions be made regarding how the system is divided into its subassemblies. The decision is usually difficult if the system is the first of a kind or little historical information is available. The complex nature of this choice and the fact that many related factors can influence it make it necessary to become familiar with these factors so that competent trade-off studies can be made to determine their relative importance to the total system requirements.

The system usually consists of individual components which are assembled into higher-level assemblies in the form of printed-wiring boards or chassis. In general, if the smallest replaceable unit is a relatively small low-cost assembly, the total number of different types of assemblies is reduced, aiding logistics and reducing manufacturing costs. If the assembly or printed-wiring board is of low enough cost, it may be considered a throwaway unit, thereby facilitating system repair. However, the inherent drawback of such a system is the proliferation of connectors and interconnection wiring required. This, in turn, tends to reduce the reliability of the system proportionately.

The direct opposite to the small, low-cost smallest replaceable unit is the large subassembly concept used in many large computers. This permits assemblies of hundreds of individual circuits mounted on a single replaceable assembly. Since these assemblies are not low-cost and are generally custom or very low quantity per system, they must be removed from the system for repair. In commercial systems, where longer periods of downtime are permitted and routine maintenance is permitted, this system-partitioning concept is acceptable. In its favor also is the increased reliability afforded by the reduced number of connectors and interconnection wiring.

Electrical versus mechanical trade-offs Overall system trade-offs should be made before any design is advanced to the point where change is too costly. Many basic trade-offs can be made in the early stages of development and design. Table 15 summarizes some of the important electrical and mechanical trade-offs for various packaging techniques.

Logic choice During any system-definition phase, trade-off studies must be carried out in such areas of microelectronic implementation as types of logic circuits and degree of logic standardization across the system, utilization of large-scale integration for digital functions, and utilization of custom-integrated circuits. A careful analysis

TABLE 11.   Checklist for Preparation of a Packaging Rating Chart[4]

Factor	Characteristic	Discussion of trade-offs
Design..........	Complexity	Do design specifications exist? Development required? Versatile enough to cover all items in system?
	Drawings	What are layout requirements? Special skills required? Drawing turnaround time? Separate manufacturing or tooling drawings needed? Can artwork be automated?
Manufacturability.	Quantity	Model, limited quantity, or large production quantity to be made (this affects tooling and manufacturing method used)? Are joints made one at a time, as in welded cordwood, or all at one time, as in printed wiring? Does quantity justify setting up a line operation, or is it possible in this design? Can design be broken into separate subassemblies? Do necessary skills and manufacturing facilities exist?
	Automation	Is design adaptable to automation? With standard equipment or especially developed equipment? How much tooling is required? Hard or soft?
	Components	Are components of nonstandard material, size, shape, or function required? What is component availability? Are weldable leads needed? How are components secured in equipment? Can components be set in with automatic assembly tools?
	Schedule	Can design be made in time available? What is parts procurement lead time? Manufacturing development required?
	Inspection and testing	Are assemblies inspectable? Does good process control make inspection unnecessary? Are built-in tests included? Is special test equipment required? What are test levels?
Repairability.....	....	Is repair done at component, module, or subassembly level? How complicated? Done by operator, repair shop, or return to manufacturer?
	In-house	Easy repair and rework at all levels during manufacture? Lower cost, increase yield, and help meet schedules?
	In use	How complicated are steps required if a component fails? Are special skills required to repair? Are complicated instructions needed? Does repair lessen environmental protection?
Maintainability...	Modularity	What is the replaceable unit level: the complete assembly or smaller subassemblies? How difficult is fault isolation? Are parts and modules easily accessible? How difficult is module removal?
	Requirements	Are any special skills or tools needed? Is required procedure evident? Is a special controlled-environment maintenance area required? How much time is required to maintain equipment? How long between maintenance operations?
Logistics........	....	What is the lowest-level subassembly throwaway cost? Is this reasonable to end use? Are hardware and assembly items standard or special? Are subassemblies multiple-use items?
Physical characteristics	Size	Consider size of overall assembly, not just small piece; remember that connectors and interconnections consume considerable volume; as a rough guide, components occupy about 5% of total volume
	Weight	Consider weight required in structure as well as electronics: include weight to communicate with the outside world; can one man handle it? A nominal density figure is 45 lb/ft^3
	Thermal	How efficient is heat transfer? Do components operate within acceptable limits? Is thermal design simple or complicated? Does design give a weight penalty?
Environmental protection	....	Does design satisfy all basic use environments–minimal or better? Does protection hurt in other areas? What is time required to get this protection? Is it compatible with schedule? How well protected from handling damage, shipping shock?
Costs...........	....	Is design time minimum? Are drawings made the least expensive way? Are raw materials and hardware items reasonable, or are costly assembly parts required? Are manufacturing costs recurring or one-time items? Are processes the least expensive? How complicated is testing?
Reliability.......	....	Are components known to have high reliability—actual or calculated? How many electrical junctions (joints) in system, and how are they made? Does the design have structural integrity? Are components properly derated, shielded, etc.? How will unit operate in its intended environment?

TABLE 12. Electronic-packaging Design Functions, Elements, and Desired Characteristics[5]

Package function	Purpose	Typical elements	Ideal characteristics
Structural support........	To combine all the electronics and their auxiliary equipment into an overall integral structure that will endure transport and operation in the field environment	Encapsulation, chassis; drawer or door structure; cabinet, rack, or panel structural members	Provides for complete component accessibility with no compromise of structural integrity Provides the isolation and transmissibility characteristics required with a minimum use of material Is adaptable to large tolerance on parts Provides efficient methods of manufacture
Environment isolation........	To protect the electronic circuits from ambient stresses (such as humidity, rf interference, dust, dynamic force inputs, rain, snow, chemical-bacteriological agents) as necessary to preclude physical damage or electrical malfunction because of these stresses	Coating and hermetic cans; wall insulation, shock and vibration isolation or absorption devices and materials, rf interference gaskets and seals; air filters, rain-resistant louvers	Provides complete isolation of the electronic components from airborne particles, organisms or chemical vapor; electromagnetic and electrostatic interference, and shock and vibration Does not affect the electrical properties of the packaged circuits Does not interface with access to the package Does not adversely affect the heat-transfer properties of the package Does not increase the power requirements of the package Isolates man-machine hazards Does not add significantly to package weight or space requirements
Heat transfer........	To transfer heat from the electronic circuits to an exterior heat sink, control component operating temperatures as required, and prevent component failures or excessive variations in circuit-performance characteristics	Thermoelectric devices; heat exchangers and heat-transfer fluids; ducts, dampers, orifices, fans, sensors and blowers; regulators, sensors and switches; valves, flow-meters, and liquid lines	Has no adverse effect on the performance of electronic components Requires no special fluids Requires no special fluid-distribution system Requires no external support equipment Requires a minimum of moving parts Requires no additional power

Mechanical interconnection......	To facilitate assembly during fabrication and package installation and to provide access to the electronics for fault isolation and repair	Card retainers, mounts, and fasteners; slides, hinges, and fasteners; pins, guides, and locating devices	Allows direct access to all components Includes no fixed connections that impede maintenance Uses identical fasteners throughout Provides strength as great as that of uninterrupted structure Does not require special tools Requires a minimum number of parts for fabrication and assembly
Electrical interconnection........	To distribute electric energy within the package	Wire molds; molded cables or loose wire-cord ties; connectors and terminals; distribution panels	All connectors and connections are coded Minimum voltage drop within connection elements Good isolation between circuits Good isolation from external noise sources Minimizes radiation of internal signals Requires no special tools for installation or separation
Display and control.............	To assure efficient, safe, and comfortable equipment use during manual operation and maintenance by proper selection of hardware terminology, and displays, and their judicious arrangement at the man-machine interface	Meters; test points and their location; visual aid; knob types and relative location; illuminated switches and panels	No specific training or skill required to operate Entirely observable and adjustable from one location All related functions grouped together Maintenance functions completely separated from operational functions All warning displays affect human senses in a way that cannot be ignored All information displays are clearly discernible but not distracting

TABLE 13. Package-Design Evaluation Factors [5]

Evaluation factor	Purpose	Rating factors
Reliability.........	Evaluates package parameters that influence the consistency of electronic-circuit performance achievable, the degree of protection provided from exterior environment extremes, the effectiveness of the cooling design in controlling electronic-component surface temperatures, the quantity of critical parts (parts in which a single failure will directly cause the equipment to become inoperable) required, the quantity of low generic failure-rate parts used	Minimum number of separable connections per circuit-component connection Components sealed, encapsulated, or coated Minimum disassembly necessary in normal support activities or transport Protection against loss of electrical components because of failure of active mechanical elements Minimum number of adjustments (mechanical or electrical) necessary in normal use Maximum allowable variation in thermal or electrical interface conditions Use of prototype testing for system-reliability data
Maintainability.....	Evaluates package parameters that determine the degree of accessibility for part replacement, facilitate fault isolation, establish special tool or test-equipment requirements, determine the level of servicing skill required, influence servicing documentation requirements, affect package spare-part stocking requirements	Grouping of components by electrical function Use of integral fault indication for basic modules Components or functional assemblies removable without interruption of permanent electrical connections Elimination of tool requirements for mechanical disassembly Use of captive mechanical fasteners Direct access to removable assemblies Identification of replaceable components
Cost.............	Evaluates package parameters that affect fabrication, development, and production, including degree of part and configuration standardization; of package modularity for ease in fabrication, assembly, and modification; of flexibility of design adaptation to alternate uses; of ability to utilize alternate fabrication processes	Use of standard components Use of developed process techniques Design for tooling and processes appropriate to schedule production quantity Use of government-furnished equipment (GFE) accessories Standardization of hardware within system Use of simplified drawing procedures

Weight............	Evaluates package parameters that affect equipment weight, including use of high-strength, lightweight materials, elimination of duplicate structure of mechanical joints, design structure to provide only that degree of strength required to provide a predetermined safety factor, use of elements required for other functions to assist in carrying structural loads	Package weight as a percentage of weight of contents Use of light alloys in structure Functional integration of structure Sizing of fasteners by strength criteria Use of heavy components as stressed members
Volume............	Evaluates package parameters that affect equipment volume, including elimination of unused space, selection of exterior shape to require least possible volume when integrated with intended installation structure, elimination of separable interconnections, use of conformally shaped components and subassemblies, minimization of circuit interconnection requirements, elimination of need for circuit-support components, such as heat sinks, blowers, special coolant flow passages, elimination of redundant components, use of components and package elements to perform multiple functions	Package volume as a percentage of volume of contents Elimination of access-space requirements (all subassemblies removable or accessible from exterior) Structural use of high-strength-to-weight materials Minimum use of separable interconnections Elimination of cooling passages or ducting

TABLE 14.   Order of Importance of Rating Factors for Four Classes of Equipment*[,4]

Missile and spacecraft	Shipborne and military ground	Airborne	Commercial
Reliability	Maintainability	Reliability	Cost
Physical characteristics	Reliability	Physical characteristics	Reliability
Environmental protection	Repairability	Environmental protection	Maintainability
Cost	Logistics	Maintainability	Repairability
Manufacturability	Cost	Cost	Logistics
Design	Manufacturability	Repairability	Manufacturability
Repairability	Design	Logistics	Design
Maintainability	Environmental protection	Manufacturability	Physical characteristics
Logistics	Physical characteristics	Design	Environmental protection

* Listed in descending order of importance.

must also be made of the many system-packaging trade-offs. These include logic choice, modular commonality, thermal management, maintainability, size and weight, manufacturability, use of automated design and fabrication techniques, and overall cost effectiveness.

Logic Types and Standardization. There exists, on the one hand, the need for optimum performance within each digital functional area which requires different kinds of logic, dependent upon speed, logic characteristics, noise immunity, and power. On the other hand, there exists the systemwide need to take advantage of common logic to reduce logistic problems both in production and in the field, to reduce cost through larger-quantity purchases, and to reduce interface problems. Interacting with these factors are those of large-scale integration (LSI), particularly in terms of what forms of logic are available in LSI and of common LSI functions which can be used throughout the system.

Large-Scale Integration (LSI). The degree of complexity that will result in optimum configuration in each digital area depends upon many factors. Some of these are related to cost and logistics, e.g., repetitiveness of the function, tooling costs, module and cabinet assembly costs, and maintainability, including fault location, repairability, and spare-parts logistics. Other factors relate to equipment-packaging design, e.g., thermal density and cooling and available number of connector terminals. These must all be considered in an overall systems study to obtain an optimum solution.

Custom Integrated Circuits. Throughout the entire electronics portion of the system, there will be a need to decide on an individual basis whether it is better to use discrete components with as many microcircuits as are available off-the-shelf to create a custom hybrid integrated circuit or to use a custom microcircuit. Although this decision will have to be made on an individual basis, some common criteria must be established, from a cost-effectiveness standpoint, concerning the relative weighting given to mean time before failure, repairability, initial cost, spare-parts logistics, and, in another direction, the forms of microcircuit technology which will provide the best performance per dollar during the anticipated production period.

Modular commonality In the digital area it is reasonable to expect that similar logic blocks can be designed and used across system functional areas. This will necessitate standardization of logic families, pin assignments, and power-supply voltages. In the analog area common functions should be sought out and the least number of different circuits used. Perhaps the largest reduction in cost can be made by reducing the number of different microcircuit package types, thereby increasing the quantities of those used and lowering their cost. Package standardization will also permit standard manufacturing techniques to be used as well as standard test equipment and procedures. The fewer the types, the fewer lead variations and process variables.

TABLE 15. Electrical versus Mechanical Trade-offs for Various Packaging Techniques[4]

Technique	Electrical characteristics	Mechanical characteristics
Conventional components on printed-circuit boards	Longest interconnection propagation delays Isolation between circuits easy Compatible with high voltage and high power High-impedance shorts in humidity High-voltage drops and ground-level changes	Maximum interconnection complexity Fast reaction time Largest size Longest leads
Welded cordwood modules	Long interconnection propagation delays Isolation easy Circuit may change during embedment Final design can be breadboarded Compatible with high voltage and high power Shielding is simple	Size large in comparison with integrated circuits Throwaway once embedded Cost high in large quantities
Hybrid cordwood modules (integrated circuits and conventional components)	Long interconnection propagation delays Isolation easy Compatible with high voltage and high power Circuit may change during embedment Final design can be breadboarded Shielding is simple	Smaller size than conventional components alone, but not so small as all-integrated circuit Closer leads make welding more difficult Compatible with standard welded cordwood techniques Throwaway cost higher than for conventional cordwood modules
Hybrid compartmentalized modules (integrated circuits and conventional components)	Maximum isolation between intermediate-frequency stages Not applicable to high power or high voltage Unnecessary for common circuits which do not require shielding Circuit may change during embedment Final design can be breadboarded	Very expensive in large quantities Not as small as an all-integrated circuit system Compatible with other module packages High throwaway cost if embedded
Etched circuits	Electrical characteristics influenced by shape and proximity Especially applicable to isolation between intermediate-frequency stages Active components must be added Limited-value resistors, capacitors, and inductors Repetitive characteristics once established	Cut-and-try design method Inexpensive in large quantities Masks necessary for each circuit Fairly large size Complex assembly

TABLE 15 Electrical versus Mechanical Trade-offs for Various Packaging Techniques[4] (Continued)

Technique	Electrical characteristics	Mechanical characteristics
TO-5 can integrated circuits on printed-circuit boards	Average interconnection propagation delay Good isolation possible High-impedance circuits marginal Mutual capacitance and crosstalk between lines likely Difficult to shield effectively	Good trade-off in size; fairly small but utilizes standard printed-circuit techniques Can be flow-soldered in large quantities Individual units can be replaced Multilayer boards required for maximum density Compatible with conventional components Assembly utilized standard printed-circuit techniques Good connection but poor heat conduction characteristics
Packaged integrated circuits on printed-circuit boards	Average interconnection propagation delay Good isolation possible Crosstalk on boards likely High-impedance circuits marginal Difficult to shield effectively	Several fastening methods available Two-dimensional fastening is much simpler than stacked flat packs Small quantities require printed-circuit artwork Spacing may be dictated by connectors Individual packages can be replaced Not compatible with three-dimensional parts Heat can be conducted away easily
Stacked flat-pack integrated-circuit modules	Average interconnection propagation delay Good isolation possible Requires welding care to avoid blowing semiconductors High-impedance circuits can be embedded Shielding easy	Throwaway once embedded Compact flat-pack subassembly Interconnections are very dense Very expensive in large quantities Compatible with other module forms and thick connectors
Thick film hybrid circuits	Film or discrete chip passive components can be used Screened on resistors can be functionally trimmed Short interconnections minimize signal loss Variety of custom active chips are available Multilayer capability	Excellent heat transfer Hermetic assemblies possible Very high density capability Low cost in high quantity Large variety of packages Leadless receptacle available Batch fabrication
Thin film circuits	Resistor and capacitor values can be higher than integrated-circuit type Isolation better than integrated circuit Active devices not usually in thin film form	Medium size Active devices added in separate operation Low cost in large quantities Long lead time: artwork required

TABLE 15.   Electrical versus Mechanical Trade-offs for Various Packaging Techniques[4] (Continued)

Technique	Electrical characteristics	Mechanical characteristics
Uncased discrete chip packages	Good isolation possible Medium high resistor and capacitor values possible Breadboard circuits possible Large-value resistors and capacitors require large area	Compatible with thick and thin film techniques Complex interconnections Costly in large quantities No complex masks needed Standard pieces allow fast reaction time
Uncased multi-integrated-circuit chip packages	Isolation between circuits possible Short leads: minimum propagation delay. Active devices easy to make Resistor and capacitor values limited Not applicable to high power or high voltage	Problems in buying and handling chips Use standard circuits—no complex masks Interconnections reduced but usually made point by point High-cost throwaway unit
Large-scale integration complex monolithic chips	Fast circuit and interconnection propagation speeds Isolation between circuits difficult Active devices easy to make Changes impossible Large-value resistors and capacitors impossible Not applicable to high power or high voltage	Interconnections built in Very small size Low yield Long development time for new masks Costly for small quantities High-cost throwaway unit
MOS (metal oxide semiconductor) devices	Slow circuit propagation speeds Parameters unstable Large fan-out No passive components necessary	Small size Heat dissipation difficult Interconnections large with respect to device Low cost on basis of small area Single diffusion process

Thermal management   As circuit density increases, assuming the elemental power stays constant, the temperature of the element will increase proportional to the density increase, all other things being equal.   However, proper circuit design and component layout can do much to reduce the effects of higher density.   With heat sinks and convection cooling provided and with careful circuit layout, a reasonable trade-off can be made to provide increased density without excessive heating.   In general, in ground-based equipment, small module size and high component density should not be of prime importance but rather a logical advantage of microcircuit use.   In calculating trade-offs it should be remembered that the predicted reliability of microcircuits is halved for every 25°C temperature rise.   When the super-flat-pack type of microcircuit package is used with bipolar or MOSFET arrays, the thermal density within the package will necessitate special heat-transfer techniques.

Maintainability   In the field, maintainability is the key to system effectiveness. By carefully partitioning the system an optimum balance can be obtained between component density, replaceable module size and weight, fault locatability, replacement cost, and modular commonality.   Each factor requires a careful weighting and trade-off evaluation during the design phase.   This should be carried out by a coordinated effort of system circuit and packaging engineers.

Manufacturability The design trade-offs which effect manufacturing are module-card standardization, number of microcircuit package types and physical differences between them, e.g., lead materials, spacing and cross section, whether cards are two-sided or multilayer, and whether plated-through or mechanical connections are made between layers. Other factors are standard grid-pattern layouts to permit use of automatic numerical-control machines and tooling and setup costs for particular techniques, such as reflow solder, wire wrap, etc. The final choice of methods will depend upon the most economical way of producing the equipment to meet the system specifications. Initial studies point to the need for computer-assisted layout and machining along with automated board processing, component placement, soldering, and testing.

Packaging Density. As previously stated, in most cases, density of a particular microelectronic assembly is a variable which has a relatively lower weighting than,

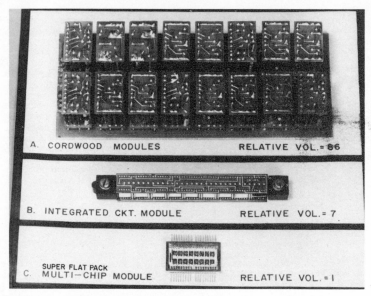

Fig. 27 Package evolution of cordwood module to large-scale integration. (*Light Military Electronics Systems Department, General Electric Company.*)

say, cost and reliability. However, it is important to understand the effects packaging techniques and microcircuit package choices have on density.

Some important things which affect density are integrated-circuit package, logic design (universal or custom), circuit complexity, pad sizes and run widths, card spacing, thermal density, and throwaway ability.

Figure 27 shows the evolution of a particular logic circuit package, its resultant size reduction, and increased component density.

INTERCONNECTION TECHNIQUES

One of the key areas of packaging design is the interconnection of the various levels of system packaging, within each level and between levels. Figure 28 shows the levels of assembly. A variety of techniques is available to the packaging engineer, starting at the level 1 interconnections, those within the basic hybrid module, and proceeding to the second level, where the hybrid module is then combined with other like packages and discrete components to form a printed-wiring board or like assembly, and finally to the interconnection of several printed-wiring boards or like assemblies into a level 3 subsystem or card-cage assembly.

Level 4 and level 5 interconnections vary from system to system but are primarily made up of cable assemblies with or without connectors. These include flat wiring, special ground and power busing systems, and other special wiring.

Many of the interconnection techniques are used in more than one level of packaging, and Table 16 shows the techniques available and the different levels where they are normally used.

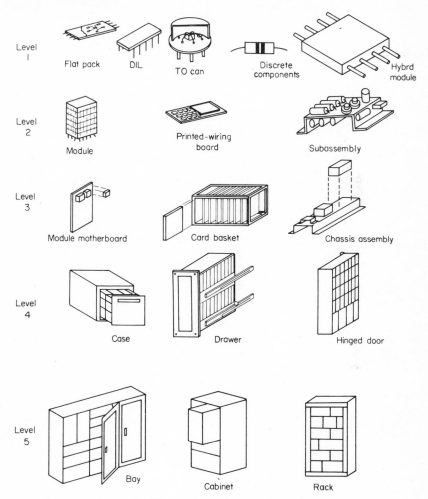

Fig. 28   Mechanical assembly levels.

Printed Wiring

Single- and double-sided printed-wiring boards are the most widely used method for interconnecting the various microelectronic and discrete components making up an electronic system. Simple layout and drafting procedures allow small quantities of printed-wiring boards to be constructed by hand in the laboratory for engineering breadboard units. After the printed-wiring-board assemblies have been assembled and tested, the final drawings can be released to an automated design area for production artwork and tooling. This design versatility has established the printed-wiring board as the prime method of interconnection. We find that even though the basic circuit functions are shrinking as a result of the increasing use of microelectronic technology, the basic component, whether packaged-device hybrid module or discrete component, usually requires a printed-wiring board of some type to mount and interconnect into a finished functional card or module.

Rigid printed-circuit boards Generically the first form of printed wiring and currently the most widely used, the rigid printed-circuit board was early recognized not only as providing a conductive wiring path but also allowing support and pro-

TABLE 16.   Interconnection Techniques Used in Different Packaging Levels

Interconnection technique	Interconnection level			
	1	2	3	4, 5
Printed wiring:				
Single-sided board		x		
Double-sided board		x		
Multilayer board		x		
Flexible board		x		
Flexible or solid board			x	
Flat cabling			x	x
Ceramic substrates:				
Double-sided	x			
Multilayer	x			
Wiring deposition:				
Plated	x			
Screened	x			
Vacuum	x			
Point-to-point wiring:				
Insulated wire, soldered		x	x	x
Welded		x	x	
Wrapped			x	x
Crimped			x	x
Bonding technique:				
Flow solder		x	x	
Reflow solder		x	x	
Welding		x	x	
Ultrasonic		x		
Opposed electrode		x		
Thermocompression		x		
Laser		x		
Electron beam		x		
Parallel gap		x		

tection of the components it connected and supplying a heat sink to aid in thermal management of the total package. The chemical methods used in producing rigid printed-circuit boards are the subtractive, or etched-foil, and the additive, or plated-up, types. The mechanical methods include stamped wiring, metal-sprayed wiring, embossed wiring, and molded wiring.

Single-Sided Printed-Circuit Boards. This type of board, with wiring on one side of the insulating substrate only, represents by far the largest volume of printed-circuit boards currently produced. Single-sided boards are used for relatively unsophisticated and simple circuitry, where circuit types and speeds do not place unusual demands on wiring electrical characteristics.

Subtractive, or Etched-Foil, Type. Illustrated by Fig. 29, the etched-foil process begins with a base laminate composed of a variety of insulators clad with copper or some other metallic foil. Following suitable cleaning and other preparation, a pattern of the desired circuit configuration is printed using a suitable negative-resist pattern for photoresist or ink resist and a positive-resist pattern if plating is to be used as an etchant resist. For the latter, gold or solder plating is applied in the nonphotoresist area. In the next step, etching, all copper not protected by the resist material is removed by the etchant. Following etching, the printed-wiring board is stripped, or subjected to removal of resist, and otherwise cleaned to ensure that no etchant remains. The board is then ready to fabricate by drilling, trimming, etc.

Plated-through processes. There are two variations of the plated-through process, both utilizing a plated conductor through the hole to make the connection. For purposes of comparison we shall call them the *conventional plated-through-hole process* and the *plating-only-through-hole process*. Both are illustrated in Fig. 30.

The conventional plated-through-hole process starts with a double-clad laminate with a series of holes placed in it corresponding to the locations where a through connection is needed (Fig. 30a). Holes are drilled and deburred, and an electroless coating of copper is applied over the entire board surface, including the holes. Copper is now electrodeposited to the exposed copper foil and sensitized walls of the hole, usually to a thickness of 0.001 in. A negative, or plating-resist, pattern is then applied, registered to both sides of the material. Resist covers all areas of foil where base copper conductor is not required; this will subsequently be etched off. The next plating step is electrodeposition of a thin

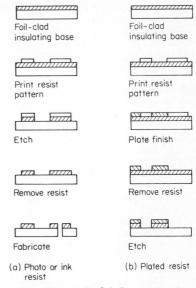

Fig. 29   Etched-foil processes.[4]

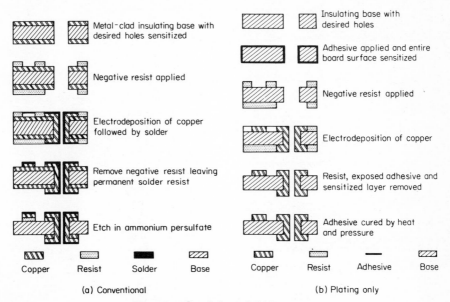

Fig. 30   Plated-through-hole process.[4]

layer of a suitable etch-resist plating, usually solder or gold. The original plating resist (photoresist) is removed, and the circuit pattern is defined by etching away the exposed copper in a suitable etchant, depending upon the type of plating resist used.

The plating-only process differs from the conventional plating-through-hole process in that no etching is required and the circuit pattern is defined at the same time the through connection is made. Holes are again drilled at desired connection points, and a thin layer of a suitable adhesive is applied, as illustrated in Fig. 30b. Electroless deposition of copper sensitizes the entire surface and holes and is followed by a "flash" electroplate.

After the registered printing of a plating-resist pattern to both sides of the board, copper is electrodeposited to the desired thickness on exposed areas. Resist is removed, unwanted electroless copper is flash-etched off, and excess adhesive is removed with appropriate solvents. The last step is the curing of the adhesive by subjecting it to a heat-and-pressure cycle.

Mechanical processes   The nonplating techniques, illustrated in Fig. 31, preceded through-hole-plating processes for double-sided boards until the reliability of the plating process was proved and the economic importance of such mass-interconnection techniques realized. While not to be construed as an integral part of the board-fabrication procedure (as represented by the plated-through hole), these mechanical interconnections, actually board-assembly techniques, serve the purpose of connecting the two sides of a double-sided board.

A simple and easily made interconnection is represented by the clinched jumper wire, illustrated in Fig. 31a. A formed, uninsulated, solid-lead wire is placed through the hole and clinched and soldered to the conductor pad on each side of the board. Part-lead wires are not normally considered as interfacial connections. Three types of eyelets are also commonly used for double-sided-board interconnection (Fig. 31b to d). Funnel-flanged eyelets are soldered to the terminal areas on the component side of the board prior to insertion of component leads. The other connection is made at assembly when the boards are dip-soldered. The funnel flange by definition has an included angle between 55 and 120°. Split-funnel-flanged eyelets differ only in the split and the fact that they need not be soldered to the terminal area on the part side of the board before insertion of part leads.

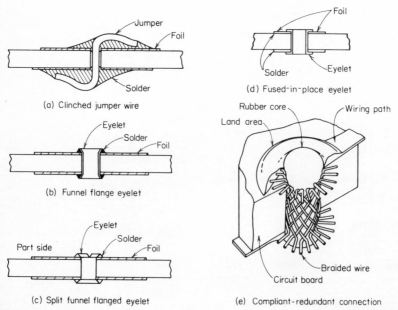

(a) Clinched jumper wire

(b) Funnel flange eyelet

(c) Split funnel flanged eyelet

(d) Fused-in-place eyelet

(e) Compliant-redundant connection

Fig. 31 (a) to (e) Mechanical interconnections.[4] [(e) *Western Electric Co.*]

Many eyelets are machine-inserted, which increases uniformity of the connection and reduces assembly cost. Figure 31e shows a recently developed interfacial through connection for double-sided printed-wiring boards called the *compliant-redundant through connection*. This technique utilizes a *sock* of multiple strands of wire braided around a rubber core. The unrestrained diameter of the rubber with braid is larger than the hole in the board; when the assembly is placed in the hole, the rubber exerts enough force on hole walls to keep solder from wetting strands within the hole, thus ensuring that the strands stay compliant after soldered assembly.

Another new technique utilizes 0.10 to 0.062 inch diameter copper pins inserted through circuit board holes and subsequently soldered to circuit runs on both sides of the circuit board. The pin is inserted, cut, and readied for connection in approximately one-quarter second by an Interface Pin Machine (Figs. 31f and g).

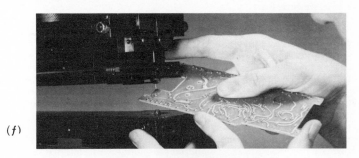

(f)

(g)

Fig. 31 (f) Interface pin being assembled to printed-wiring board using interface pin machine. (g) Interface pin machine. [(f) and (g) *Phoenix Gage Inc.*]

Originally developed by General Electric and now manufactured and sold by Phoenix Gage, Inc., Phoenix, N.Y., the Interface Pin Machine has several advantages:

1. Highly reliable and inexpensive
2. A 30:1 labor savings over hand-inserted buss wire
3. A swaging action at each connection point tightens circuit board pads, eliminates pad lift, and increases reliability
4. This same swaging action provides accurate pin projections for easier and more positive solder joints
5. Air-operated for lower operating and maintenance costs
6. Easy installation and simple operation require no change in conventional assembly methods, including automatic and dip-soldering processes; easily mounts on standard assembly bench
7. Adjustments for board thickness, pin diameter, and pin projection allow assembly of any size board
8. Special tooling is available for utilization of square copper wire, providing an extended projection on one side of the circuit board for wire wrap applications and also for various pin configurations.
9. This process is covered under Mil Specification MIL-STD-275B
10. Because of the small pin diameter and highly reliable pin connection, it can be successfully employed on high-density boards for military, industrial, or commercial use; the pins may be inserted into any material, including ceramics.

Multilayer printed-wiring boards   The problem of providing for increased wiring density required in many present electronic packaging applications could be met only by the use of more than two planes of wiring, resulting in the multilayer printed-wiring board. A multilayer printed-wiring board is a series of individual-circuit layers bonded to produce a thin, monolithic assembly with internal and external connections to each level of the circuitry determined by the system wiring diagram. As an extension of the section on terminology, this discussion of printed-wiring processes will provide definitions only, with design and application information to follow later in the chapter.

Since the processes for fabricating multilayer boards are basically extensions of methods used for single- and double-sided boards, they will be described in less detail. As identified by board-fabrication process, multilayer printed-wiring boards can be broken down into two basic types, laminated and built-up multilayer printed-wiring boards. Other than the difference between the laminated approach and the screened or deposited insulation used in the built-up approach, these methods differ only in how interlayer connections are made. As in single-side printed-wiring technology, there are countless variations of these basic multilayer processes.

Flexible Interconnects

Flexible Flat Cable.   Flexible, multiconductor, flat-wire cable is used most often with connectors at each end in the same way as conventional, round-cable wiring.

Flexible Printed Wiring.   Multiconductor, etched, flexible, flat-encapsulated wiring is generally custom-designed for a special application with regard to an electronic-systems design and connected by conventional techniques.

Either of these generic types may be piled up in the same fashion as multilayer circuits in order to obtain the required circuit density. This kind of wiring is characterized by its flexible, conforming nature and by its flat, thin shape.

Single-layer flexible wiring   Whether classified as cable or wiring wires as in the types just mentioned, the constructional features differ little. In general, however, the cable is usually supplied with an encapsulating layer of plastic insulation on the outside surfaces since the usual mode of connection is through the ends. The process outline is very similar to that for rigid single-sided boards of the etched-foil variety (Fig. 29). Except for rare instances, flexible printed wiring of the build-up or electroplating-transfer process is not fabricated or used. The manufacturing process, as illustrated in Fig. 32, is predominantly a continuous or automatic one and consists in printing an acid-resist (or plating-resist) image in the appropriate form, using either screening or photoresist methods. If the etch resist is used, the conductors are then etched. If plating resist is used, the conductors are plated with an acid-resistant metal and then etched. As explained above, an encapsulating layer may be applied and termination points made.

Double-sided and multilayer flexible wiring The etched-foil techniques are used almost exclusively for these varieties of flexible wiring, with options only in the area of interconnection between layers. For double-sided or very thin multilayer configurations, plated-through holes are rarely used. Interfacial connections in double-sided planes and interlayer connections in multilayer planes are made predominantly by the clearance-hole technique (Fig. 30) or one of the mechanical interconnection methods shown in Fig. 31.

Material and process variables The very nature of the materials used in the fabrication of printed wiring shows that changes may be expected in the printed-wiring board during its manufacture. Whether the base material is organic or inor-

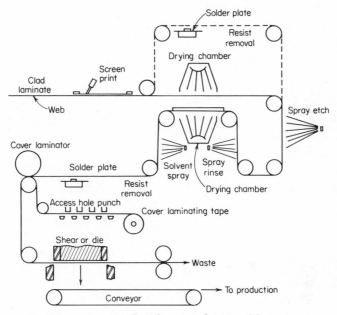

Fig. 32 Automatic, flexible, printed-wiring fabrication.[4]

ganic, changes due to thermal expansion (see Table 17), shrinkage behavior, and dimensional-stability characteristics must be known and anticipated in the initial design. In this way, printed-circuit materials are a good deal more affected by the manufacturing process than metals because the former are subjected to a variety of chemical (cleaning and etching), heat, and pressure (lamination), processes as well as regular machining, in their fabrication. Considering the physical behavior of the material, it is evident that these processes must introduce some change in the material which the engineer must consider in his initial design.

Projected use in the system The designer must, above all, be aware of the capabilities and limitations of the printed wiring he is specifying. It is the purpose of this section to outline application properties of rigid and flexible printed wiring and the way in which they can and do affect the designer's task.

Electrical Properties. The functional characteristics of the circuit and the environment to which it is subjected largely decide which insulating substrate the designer will use. Since the electrical properties of the printed wiring are integrally associated with the base-material properties, a clear understanding of these electrical properties may have a decisive effect on the design. The most important electrical properties of the printed-wiring board for dc as well as the rather low-frequency ac systems are current-carrying capacity, arc and tracking resistance, insulation resistance, and dielectric-withstanding voltage. For high-frequency applications the most important characteristics are characteristic impedance, dielectric constant, capacitance (and crosstalk) dissipation factor, and propagation delay.

Current-Carrying Capacity. The current-carrying capacity of etched-copper conductors for rigid boards is given in Figs. 33 and 34. For 1- and 2-oz conductors, allow a nominal 10 percent derating (on a current basis) to provide for normal vari-

TABLE 17.   Mean Coefficient of Linear Thermal Expansion of Selected Materials[4]

Material	Coeff., 10^{-5} in./ (in.)(°C)	Material	Coeff., 10^{-5} in./ (in.)(°C)
Thermoplastics		FR-3, epoxy paper:	
(−30 to +30°C):		Lengthwise	3
Polyethylene, 0.95 density	15	Crosswise	4
Polypropylene	10	Thickness	10
Fluorinated ethylene-		G-10, epoxy glass:	
propylene copolymer	9	Lengthwise	1.1
Polystyrene	7	Crosswise	1.5
Polycarbonate	7	Thickness	6
Polytrifluorochloroethylene	7	Ceramics (25 to 200°C):	
Polytetrafluoroethylene	5.5	Steatite	0.78
Molded thermosets		Forsterite	1.0
(−30 to +30°C):		Cordierite	0.23
Phenolic:		Mullite	0.50
Wood-flour-filled	3.7	Alumina, 99%	0.67
Asbestos-filled	2.0	Photosensitive glass	
Glass-filled	1.1	ceramic*	1.0
Diallyl phthalate:		Glasses (25 to 200°C):	
Acrylic-fiber-filled	5.5	Borosilicate (Pyrex†)	0.5
Asbestos-filled	4.0	Fused quartz	0.06
Glass-filled	3.2	Metals (25 to 100°C):	
Alkyd:		Aluminum	2.4
Asbestos-filled	3.6	Copper	1.7
Glass-filled	3.6	Silver	2.0
Laminates (0 to 60°C):		Gold	1.4
XXXP, phenolic-paper:		Solder (60-40)	2.4
Lengthwise	2	Tin	2.4
Crosswise	3	Nickel	1.3
Thickness	10	Rhodium	0.8
XXXPC, phenolic-paper:			
Lengthwise	2		
Crosswise	4		
Thickness	20		

* Fotoceram 8603, trademark of Corning Glass.
† Trademark of Corning Glass.

ations in etching methods, copper thickness, and thermal differences. Other common derating factors are 15 percent for conformally coated boards (for base material under 0.032 in. and copper over 3 oz) and 30 percent for dip-soldered boards.

Arc and Tracking Resistance. Arc resistance may be defined as the ability of an insulating substrate to resist the formation of a conducting path when subjected to electric arcs over the material surface. Track resistance is the property which enables a material to resist the formation of creepage paths under the continuous application of electrical stress. These properties are desirable only when high voltages or high-voltage pulses are expected in the packaging system. Although individual suppliers should be consulted for these values when the application warrants it, some comparative figures for arc resistance are as follows: XXXP, 5 to 10 s; G-10, 60 to 80 s; PTFE, 180 s; and G-7 (silicone), 200 to 250 s.

Insulation Resistance. Insulation resistance is propably the most important single parameter of the processed laminate from the standpoint of both design and material engineers. Insulation-resistance tests are important ways of determining raw-material properties as well as of controlling product quality since they are excellent indicators

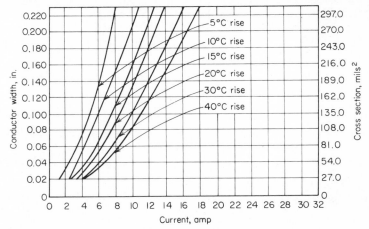

Fig. 33   Temperature rise versus current for 1-oz copper.[4]

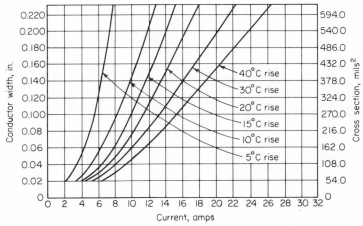

Fig. 34   Temperature rise versus current for 2-oz copper.[4]

of the propensity of the material for moisture, ions, and other contaminants commonly found in the manufacturing and handling of the printed-wiring board. The test is usually performed by making insulation-resistance tests between conductors on a layer (or between layers in the case of double-sided and multilayer boards) after a specified time in a humidity and temperature cycle. A *polarization voltage,* usually 100 V ac, is applied between conductors during the test, and resistance measurements are made immediately following the test. Typical resistance values are 1,000 MΩ per 0.001-in. separation between conductors. Among the design features which may minimize low-insulation-resistance problems are the following:

1. Keep the conductors at least 0.10 in. away from the edge of the board.

2. Machine (mill or saw) the border rather than shearing it.
3. Avoid contamination during handling and processing.
4. Employ a conformal coating.

Dielectric Withstanding Voltage.   It appears that this property, also called *flashover strength,* is dependent only on the air density and conductor spacing and seems independent of moisture content or dielectric constant of the material or conductor-surface finish.   An important factor for the design engineer to consider is the potential at which flashover occurs in the air above the conductor traces.   Table 18 may be used for a design guide in this case.   As indicated, increasing altitude lowers the insulating properties of air (causes it to ionize more readily), and, accordingly, greater spaces are called for.   Other than this, high-altitude design may call for pressurization of equipment or conformal coating or encapsulation of wiring as found in multilayer wiring.   Table 18 is not valid for alternating current above 400 Hz.

TABLE 18.   Spacing of Conductors for Various Voltages[6]

Uncoated boards sea level to 10,000 ft		Uncoated boards over 10,000 ft and coated boards		
			Minimum space, in.	
Voltage between conductors, dc or ac peak, V	Minimum space, in.	Voltage between conductors, dc or ac peak, V	Above 10,000 ft uncoated	Conformally coated all altitudes
0–150	0.025	0–50	0.025	0.010
151–300	0.050	51–100	0.060	0.015
301–500	0.100	101–170	0.125	0.020
Above 500	0.0002*	171–250	0.250	0.030
		251–500	0.500	0.060
		Above 500	0.0001*	0.0012*

* Inch per volt.

Dielectric Constant and Dissipation Factor.   The dielectric constant of an insulating material is defined as the ratio of the capacitance of a capacitor containing that particular material to the capacitance of the same electrode system with air replacing the insulation as the dielectric medium.   It may also be defined as that property of an insulation which determines the electrostatic energy stored within the solid material, i.e., the ability of the insulating material to store energy.

The dissipation factor is defined as the ratio of the equivalent series capacitive reactance of a dielectric circuit.   These values are especially important for high-frequency applications since the impedance of a printed-wiring transmission line is proportional to the frequency and the dielectric constant for most known board materials decreases with frequency.   For the lower values ordinarily found in printed-wiring materials, the dissipation factor is the practical equivalent of the power factor. Q is the quality factor, or Q factor, which is equivalent to the reciprocal of the dissipation factor and is sometimes used to rate a dielectric.   If power losses in a dielectric are to be minimized, the loss factor must be low since the mean power developed in the dielectric is proportional to the loss factor.   The interrelation of these values is given by

$$\epsilon'' = \epsilon_r \tan \delta \tag{1}$$

where ϵ_r = dielectric constant
ϵ'' = loss factor
$\tan \delta$ = loss tangent or dissipation factor

For printed-wiring substrates, a very low dissipation factor and a low dielectric constant are usually the characteristics most in demand.

Internal wiring and the two methods of achieving transmission-line capability in wiring are illustrated in Fig. 35. Microstrip may be used as a single wiring layer or the outside layer of a multilayer board. Strip transmission line may be used either as one or a multitude of wiring layers. Capacitance of strip transmission line, referring to Fig. 35c, can be approximated by[7]

$$C = 0.9 \frac{W/b}{1 - t/b} \epsilon_r \tag{2}$$

Characteristic Impedance. An important parameter for high-speed digital interconnection design, control of impedance is important to avoid signal reflections which result from the passage of fast pulses through an impedance discontinuity or a mismatch at the load end of the line. It is dependent upon the line width, dielectric thickness, and dielectric constant of the insulating medium.

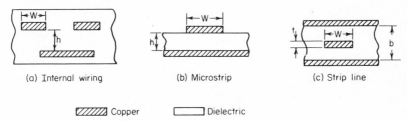

(a) Internal wiring   (b) Microstrip   (c) Strip line

▨ Copper   ☐ Dielectric

Fig. 35 Printed-wiring configurations.[4]

The general equation for characteristic impedance of any high-frequency transmission line is

$$Z_0 = \sqrt{\frac{R + jwL}{G + jwC}} \tag{3}$$

where Z_0 = impedance, Ω
    R = resistance per unit length of line
    L = inductance per unit length of line
    G = conductance per unit length of line
    C = capacitance per unit length of line

If the line is lossless, the equation is simplified:

$$Z_0 = \sqrt{\frac{L}{C}} \tag{4}$$

A specific formula for microstrip (Fig. 35b) is

$$Z_0 = \frac{h}{W} \frac{377}{\epsilon_r} \tag{5}$$

where h and W are given in the diagram and ϵ_r is the effective dielectric constant of the material (considering the effect of air). This analytical method disregards fringing effects and leadage flux and gives validity to the notion that more reliable values may be obtained by measurement. To account for the fringing effects of microstrip transmission line and if the analytical method must be used, the following formula is recommended:

$$Z_0 = \frac{h}{W} \frac{377}{\sqrt{\epsilon_r} \{1 + (2h/\pi W) [1 + \ln(\pi W/h)]\}} \tag{6}$$

Typical values of characteristic impedance values of microstrip on epoxy-glass material where the conductor is exposed to air are given in Fig. 36.

The characteristic impedance of strip transmission line (Fig. 35c) when (W/b) 0.35 is

$$Z_0 = \frac{60}{\sqrt{\epsilon_r}} \ln \frac{4b}{d_0} \tag{7}$$

where d_0, the effective wire diameter for square configuration, is $0.567W + 0.67t$.

Thermal management   The thermal design of the printed-wiring-board assembly is an important factor in keeping temperature-sensitive components at a safe operating temperature.   There are three primary modes of heat transfer; conduction, convection, and radiation.   Each of these is used to a certain extent in the cooling

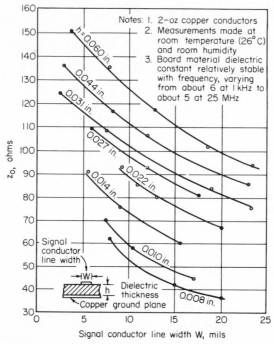

Fig. 36 Characteristic impedance versus line widths and dielectric thickness: G-10 epoxy-glass dielectric-microstrip transmission line.[7]

of any hot assembly. The majority of electronic assemblies are conductively cooled, aided in some cases by forced-air of circulated-liquid cooling. The amount of radiation cooling is related to the maximum temperature of the dissipating component and is usually a smaller portion of the total heat transfer.

The amount of heat transferred by conduction depends upon the thermal conductivity of the heat-flow path from component to heat sink. The various thermal-design techniques discussed in Chap. 8 are also applicable at the printed-wiring-board level.

Figure 37a shows a module design which has optimized the conductive-cooling mode for a standard-hardware-program module. In this case a computer program was used to determine the thermal characteristics of the module design, which were compared to laboratory tests. The previous module design was capable of dissipating up to 1 W with convection cooling. The new centerboard module was designed to increase the dissipation of the board, thus permitting cooler component operation. By designing the centerboard module with conductive strips under the flat packs it was possible to increase the heat flow to the dissipating fin area and guide areas. Increased contact pressure at the guide area also decreased the thermal resistance at

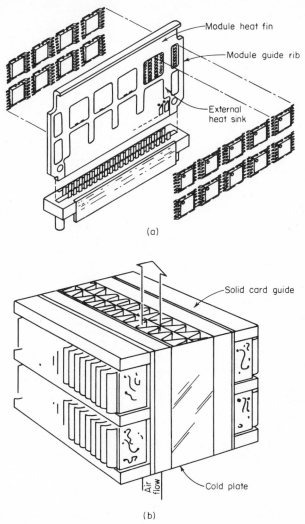

(a)

(b)

Fig. 37 (*a*) Centerboard standard-hardware-program module designed for improved conductive cooling; (*b*) additional cooling provided by air-cooled cold plate. (*General Electric Company, Ordnance Dept.*)

these points. Additional cooling was provided through an air-cooled cold plate, shown in Fig. 37*b*. The importance of choosing the proper module materials can be seen by referring to Table 19. Figure 38 shows a series of standard-hardware-program modules designed to mount various flat pack integrated circuits and hybrid integrated circuit modules dissipating up to 5 W using air and liquid cooling.

TABLE 19.   Average Thermal Conductivity K of Various Materials from 0 to 100°C (32 to 212°F)[4]

Material	Specific gravity	(Btu)(h)/(ft²)(°F/ft)*
Metals and alloys:		
Copper..................................	8.50	218–224
Aluminum.............................	2.71	117–119
Brass (70Cu-30Zn).......................	8.53	56–60
Magnesium alloy (90Mg-9Al-1Zn)............	1.81	26–28
Steel (mild).............................	7.85	26–28
Ceramics:		
Beryllium oxide.........................	2.97	104–130
Magnesium oxide........................	3.2	18–21
Aluminum oxide........................	3.7	10–15
Photosensitive glass ceramic................	2.46	1.3
Thermoplastics:		
Polyethylene...........................	0.92–0.96	0.1–0.4
Polytetrafluoroethylene....................	2.15–2.25	0.1–0.2
Molded thermosets:		
Phenolic, wood-flour-filled................	1.32–1.45	0.10–0.19
Mineral-filled..........................	1.65–1.92	0.24–0.34
Diallyl phthalate, acrylic fiber..............	1.31–1.45	0.18–0.19
Laminates, perpendicular-to-face:		
XXXP, paper-phenolic....................	1.3–1.4	0.04–0.12
G-7, silicone-glass.......................	1.6–1.8	0.07–0.17
G-10, epoxy-glass........................	1.7–1.8	0.10–0.17
PTFE glass cloth........................	2.1–2.2	0.02–0.05
Casting resins and foams:		
Epoxy, unfilled.........................	1.16	0.13–0.20
73% alumina by weight..................	....	0.82
50–55% silica by weight..................	1.6–1.7	0.29–0.53
Hollow phenolic spheres.................	0.86	0.16
Hollow glass spheres....................	0.95	0.38
Polyester, unfilled.......................	1.23	0.10–0.15
50% silica by weight....................	1.6	0.19
Polyurethane foam (10 lb/ft³)...............	0.16	0.02–0.03

* To obtain: (Btu)(h)/(ft²)(°F/in.) multiply by 12.
    (cal)(cm²)/(s)(°C/cm) multiply by 0.00413.
    W/(cm)(°C/cm) multiply by 0.0173.

SECOND-LEVEL ASSEMBLY

The packaging engineer has a variety of packaging techniques to use in mounting and interconnecting a group of discrete and hybrid integrated circuits.   Since the choice of the final package depends on system requirements, it is common practice to conduct a trade-off study by comparing the various assembly and interconnection techniques to arrive at an optimum solution for the particular application.   A thorough knowledge of the advantages and disadvantages of each technique is therefore essential.   Table 20 lists typical techniques available and shows where each can be applied.

Planar Packaging

The two basic classifications of second-level packaging are the planar and three-dimensional configurations.   The planar, the most common approach, encompasses all printed-wiring-board approaches, including single- and double-sided, plated-through or pinned, multilayer approaches, and the newer Infobond and Multiwire approaches. The planar approach has been used to mount groups of TO cans and dual-in-line packages for digital-circuit functions, as in Figs. 39 and 40.   Figure 41 shows typical functional digital printed-wiring boards including integrated circuits and large-scale integration packages used in data-processing assembly.

Fig. 38 Standard-hardware-program modules designed to mount a variety of conductively cooled hybrid packages. (*General Electric Company, Ordnance Dept.*)

The planar approach offers a wide variety of implementation techniques, as shown in Table 21. The most widely used of these have been two-sided printed-wiring boards which have plated through-holes to interconnect one connection run on one side to the other and also provide mounting holes for transistors, integrated circuits, hybrid integrated circuits, and discrete-component assemblies. The use of various sizes of printed-wiring boards allows the designer freedom in his total package concept. The use of components such as the dual-in-line integrated-circuit package with its leads on 0.1-in. grid spacings has allowed the rapid growth of the computer-assisted layout techniques. These techniques permit rapid layout of circuits and provide means of supplying manufacturing tools such as drill tapes during actual design. They also permit automatic drilling machines and component-insertion and wiring machines to be used in the factory. The standardization of this type of board layout grid has proved to be a most important decision.

Soldered-socket planar boards When it is desirable to provide for quick removal of the dual-in-line packages, a socket approach may be used. A typical board assembly is shown in Fig. 42. In this approach individual socket assemblies (Fig. 43) are assembled to the printed-wiring board like components. The boards would be completely assembled and soldered. Then the dual-in-line packages are inserted in the sockets, and the board is ready for test. This approach is used where quick board repair is desired and the individual cards are not throwaway. It is most useful in industrial applications, where the number of components in an assembly permits troubleshooting down to a particular integrated circuit or hybrid integrated circuit.

Wire-wrap-socket planar boards Another variety of socketed boards (Fig. 44) provides a wire-wrappable socket assembly with 14, 16, 24, or 36 pin contacts pressed in and soldered to a two-sided planar printed-wiring board which has printed power and ground interconnections and an edge-mounted connector. In this assembly the

TABLE 20.   Second-Level Assembly Techniques

Technique	Characteristics
Printed wiring.......	Usually used between components and small assemblies of components; may be single-sided, double-sided, or multilayer; minimum standard grid spacing is 0.100 in., although 0.050 in. is fairly readily achieved; components or modules replaceable; minimum of tools required after board is made; artwork required for board manufacture; circuit not easily changed
Soldered wire........	Point-to-point connection; with insulated wire, crossovers are permissible; most common method in use; simple tools, medium skills; spacing limited by size of wire and terminals used; with magnet wire to 0.050 in.; metallic junction; wire can be removed and reused to a limited extent; quality depending upon operator if hand-done; can be mechanized to minimize operator effect; heat may cause damage, relatively bulky; circuit easily altered
Welded wire........	May be point-to-point or matrix; usual spacing 0.100 in., but often made on 0.050-in. centers; requires special tools and skills; quality not as operator-dependent as solder; fusion joint of parent metals; permanent joints, so repair is not readily made; high-temperature operation possible; lightweight, small size
Flat cable..........	Replacement for cable harnesses; lightweight, flexible, and small; may be soldered or welded at terminals; can be designed to fit a space or bought in standard widths, conductor spacings, etc.; conductors are printed flat, or may be standard stranded wire
Solderless wrap......	Multiple turns of wire around a square or rectangular post; contact through surface distortion and high interface pressure; requires proper terminal design and special tools; may be automated; quality largely controlled by tooling; usual spacing 0.100 in., but can be performed to 0.075 in.; limited to solid wire; when removed, wire end cannot be reused
Crimped contact.....	Direct contact through deformation; use for wire terminations at lugs, pins, and removable contact elements; requires special tools, terminals, and contacts; cannot be removed and reused; joints usually preformed because of space required for tool; quality depends upon tool use; one-at-a-time connection; relatively bulky; spacings of contacts usually on 0.100-in. centers
Separable connectors.	Pressure-type contact for single or multiwire connections; wire termination to contacts may be solder, weld, crimp, or wrap; maximum accessibility and interchangeability; quality depends upon connector design and application; contacts readily available to 0.050-in. centers; assembly tends to be bulky
Screw-terminals.....	Limited application to relatively large leads which are often removed; pressure junction, subject to loosening under vibration; easiest made, no tools; quality depends upon operator; bulky

socket protrudes above the surface of the board from ¼ to ½ in. and the wire-wrap pin protrudes approximately ¾ in. from the bottom of the board. This may be objectionable for high-density applications. However, the advantage of being able to completely interconnect the assembly with point-to-point wire-wrap interconnection may outweigh the additional volume and weight. The alternative of a multilayer board assembly requires drafting or design automation time and is very difficult and expensive to modify during development. A cost study (Fig. 45) shows that for quantities under 10,000 the wire-wrapped board is less expensive. However, the use of multilayer boards is generally required for electrical reasons in applications greater than 20 MHz and switching speeds less than 2.5 ns. This is also true for applications where controlled impedance levels and crosstalk reduction are of prime importance.

Wire-wrappable panel boards   Another variety of wire-wrappable panels consists of rows of individual socket pins assembled to a printed-wiring board with power and ground planes. These are available in a variety of sizes—30-, 60-, or 90-position

Fig. 39 Printed-wiring-board assembly with TO package. (*Heavy Military Electronic Systems Department, General Electric Company.*)

Fig. 40 Printed-wiring-board assembly with dual-in-line package. (*Heavy Military Electronic Systems Department, General Electric Company.*)

Fig. 41 Typical digital printed-wiring-board assembly with dual-in-line package including integrated circuits and large-scale integration. (*Heavy Military Electronic Systems Department, General Electric Company.*)

boards or a universal board which can mount any variation of 0.1-in. grid dual-in-line packages (Fig. 46). These boards have the lowest profile of the wire-wrappable boards and are also the most expensive due to the individual custom socket. When using socketed boards in military or industrial environments, it is essential to specify gold plating on the dual-in-line package leads and at least 100 millionths of gold on the sockets. Tests[8] run by Rome Air Development Center indicate corrosion problems where dissimilar materials are used, e.g., Au-Sn.

Topside-reflow planar boards   When high density is required, a topside-reflow approach can be used, as shown in Fig. 47. Here the packages are flat packs, and the leads are formed so that they can be reflow-soldered to the mounting pads on the top surface of the printed-wiring board. This allows a maximum-density layout for the multilayer board since the mounting pads for the flat-pack packages do not extend completely through the board assembly and the other interconnecting vias need only connect adjacent layers. This approach is also helpful when ceramic printed-wiring boards are used and has been used to lower cost and increase the density of computers, as shown in Fig. 48.

New planar techniques   Some newer techniques under development are of interest when higher volumetric efficiency is required. Two of these are the Multiwire approach of Photocircuits and the Infobond system of Inforex, Inc. In both cases all the information required for manufacturing is a from-to wiring list and a component-layout drawing showing integrated-circuit positions. These techniques are primarily designed for digital integrated-circuit interconnection but are adaptable to the interconnection of a variety of hybrid integrated circuit packages.

Multiwire Process.   The process provides a four-position wiring machine (Fig. 49) which holds four identical glass-epoxy boards, each coated on one side with an

TABLE 21. Planar Printed-Wiring-Board Implementation Techniques[19]

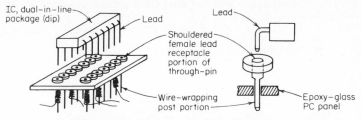

Left: Dual-in-line integrated circuits plugged into sockets (bushings, through-pins) prestaked into epoxy glass or metal panel; through-pin's bottom serves as wire-wrapping post; panel contains ground and voltage planes.   *Right:* Same concept but card-guided sockets on back panel or mother board accept printed-circuit cards with plated finger contacts.

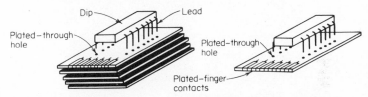

Left: Integrated circuits wave-soldered into plated-through, eyeleted, or press-fit-linered hole in a conventional multilayer board, edge-connected to outside world via finger contacts (see top right).   *Right:* Integrated circuits wave-soldered into plated-through, eyeleted, or press-fit-linered holes in conventional two-sided printed-circuit board, edge-connected to outside world via finger contacts (see top right).

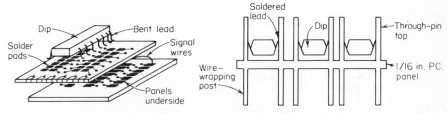

Left: Integrated circuits soldered onto pads on top side of printed-circuit planar panel; signal wires soldered to pads on panel's reverse side; panel edge-connected to two-piece connector or to card-edge receptacle staked to mother board or back plane.   *Right:* Integrated circuit solder-mounted (upside down) to tops of through-pins staked into $\frac{1}{16}$-in. epoxy-glass panels containing ground and voltage planes; bottom of through-pin serves as wire-wrapping post.

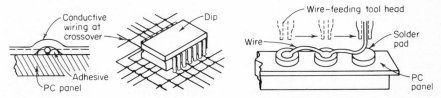

Left: Integrated circuits wave-soldered into plated through-holes; conductive wires written on insulated two-sided printed-circuit panels; panel edge-mounted to card-edge receptacle on mother board or back panel (see top right).   *Right:* Integrated circuit parallel-gap-welded or reflow-soldered into plated through-holes in two-sided printed circuit panel; wires fed in continuous path (from hollow electrode) resistance-welded to solder pads on back of panel. Panel edge-connected to outside.

adhesive layer about 5 mils thick.   The opposite sides of the board may have plated power and ground interconnections.

The machine, controlled by a punched tape, feeds a 7-mil wire onto the board and presses it into the adhesive.   A series of interconnections is made to points determined by the from-to information.   Each of the four heads is equipped with a pressure foot, cutter, and wire feeder, which sequentially feed the wire, heat the adhesive to a plastic state, press the wire into the adhesive, and cut the wire (Fig. 50).

Fig. 42 Typical printed-wiring-board assembly with integrated-circuit sockets soldered to the board.   (*Texas Instruments.*)

Fig. 43   Individual dual-in-line socket assembly.

When wiring is completed, each individual board is heat-treated to ensure complete embedment of the wire matrix.   Then the boards are drilled and the holes plated with copper to complete the interconnections (Fig. 51).   Wires can be embedded on both sides of the card and interconnected in the same manner.   The circuits are then fully tested to determine any wiring defects.   Figures 52 and 53 show the front and rear views of a typical printed-wiring-board assembly fabricated by the Multiwire process.

Infobond Process.   The Infobond process starts with a universal printed-wiring board. One side of the board has a series of pretinned pads designed to mount integrated-circuit packages.   When dual-in-line packages are used, the leads are preformed for topside mounting. The other side of the board has another series of pretinned pads interconnected to the front side by plated through-holes (Fig. 54).

The boards are mounted six at a time in a special jig mounted on the wiring-machine table (Fig. 55). Polyurethane wire (38 AWG) is fed through compression-

Fig. 44   Wire-wrap-socket planar board assembly.   (*Augat, Inc.*)

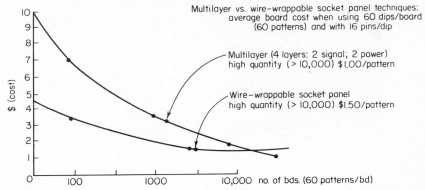

Fig. 45   Comparative costs of wire-wrapped board versus multilayer.   (*Augat, Inc.*)

bonding tips on each of the six heads. The table is numerically controlled, and the heads are moved into contact with the bonding pads and soldered. An air blast is used to cool the bonds quickly. All connections are immediately tested for bond strength and electrical continuity. If everything checks, the next series of bonds is made.   Continuity-wiring checks are made on the completed boards, and electrical-function tests are run.   Repairs are made as shown in Fig. 56.

Comparison of Processes.   The processes presented give a general overview of the major interconnection techniques being used for planar assemblies.   For any particular application a trade-off must be made between the various system require-

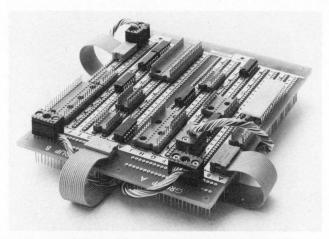

Fig. 46   Typical universal printed-wiring board with individual socket pins.   (*Augat, Inc.*)

Fig. 47   Printed-wiring-board assembly with flat-pack integrated circuits reflow-soldered to mounting pads.

ments such as density, environment, quantity, etc., and the overall system cost presented by a certain technique.   Wire wrapping of two-sided printed-wiring boards with socket terminals limits the volumetric efficiency, requires a sizable capital investment for wire-wrap machines, or necessitates outside-vendor wiring by contract. However, it does afford the greatest flexibility during the design and production cycle for making changes, and it minimizes any repair and rework costs.

The Infobond and Multiwire techniques make use of more standard printed-wiring-board techniques of assembly and soldering. They do necessitate either special equipment investment or outside-vendor contract services. The volumetric efficiency is better than the wire-wrap technique, and both processes make use of numerical-

Fig. 48 Ceramic printed-wiring board with flat packs reflow-soldered. (*Magnavox.*)

Fig. 49 Four-headed Multiwire machine lays out and embeds 7-mil wire under numerical control. (*Photocircuits.*)

control automated systems compatible with computer-assisted design programs. Repairability of both techniques is good. However, the Multiwire technique appears more difficult to change during a production run. The multilayer approach appears to offer the highest interconnect density of all the techniques and for high quantities

Fig. 50 Closeup of wire-feed mechanism of Multiwire machine. (*Photocircuits.*)

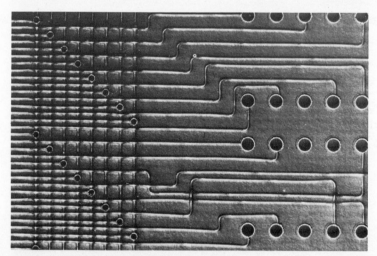

Fig. 51 Closeup of wire embedded in plastic-coated board with holes drilled and copper plated. (*Photocircuits.*)

is economical. The higher cost of artwork and tooling can then be offset by large-quantity production runs (Fig. 45). For higher-speed systems, greater than 20 MHz, the use of multilayer controlled-impedance techniques is desirable. Tables 22 and 23 summarize the advantages and disadvantages of the various techniques.

Three-Dimensional, or Volumetric, Packaging

The first impression of three-dimensional, or volumetric, packaging would be that all useful assemblies of course have three dimensions and therefore even the planar board previously discussed would be considered a volumetric form of packaging. However, for our purposes, assembled boards or modules having components primarily on one side of a two-sided board will be considered planar and all other varieties will be three-dimensional, or volumetric.   In general, three-dimensional assemblies are designed to achieve certain unique properties, e.g., high density, special shape configurations, short interconnect lengths, or unusual fabrication methods.

Volumetric packaging starts at the first-level package as shown in Fig. 57.   These stacked-substrate TO-8 assemblies house individual thick film substrates, each with its own active devices and interconnected by the package leads between levels and to the next packaging level.

Phase-Shift Driver Package.   Another example of volumetric packaging is the assembly shown in Fig. 58.   Here the thick film substrate is slid into a protective housing and environmentally sealed.

Stick Module.   A simple approach to three-dimensional packaging is shown in Fig. 59.   The *stick module,* as it is called, consists of two miniature printed-wiring

Fig. 52 Front view of printed-wiring-board assembly fabricated by Multiwire process prior to connector assembly.   (*Photocircuits.*)

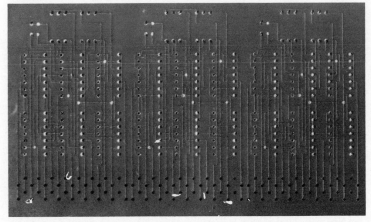

Fig. 53 Rear view of printed-wiring-board assembly fabricated by Multiwire process prior to connector assembly.   (*Photocircuits.*)

Fig. 54 Each of six bonding heads incorporates a solder tool (vertical probe, *center*), an electrical probe (*right*), a wire cutter (*left*) and a pull-tester. Pins protruding through the board serve as guides; they are not wrap posts. (*Inforex, Inc.*)

Fig. 55 Tape-controlled bonding machine wires six boards at once; cone over each tool head feeds 38-gage wire to the bonding tip. Operator can monitor the solder and test sequence through an optical viewer. (*Inforex, Inc.*)

boards mounted on opposite sides of an anodized-aluminum heat sink, in which is mounted a multipin connector. The flat-pack integrated circuits are mounted directly to the side of the heat sink and interconnected by flow solder to each of the two printed-circuit boards. Other hybrid-package mounting versions are shown in Fig.

Fig. 56 On the bottom of a finished panel, solder-bonding pads (on 125-mil grid) replace the wrap pins found on a conventional wired board, increasing circuit density to about 400 joints/in.² (100 joints/in.² with wire wrapping). (*Inforex, Inc.*)

TABLE 22. Comparison of Packaging Techniques[9]

Wire-wrappable-socket panel	Multilayer boards
Cost vs quantity: For contracted outside service to wire boards: 10 to 7½ cents per pin (both ends), about 5 to 6 cents for the pin; 3 to 4 cents for wire wrapping, or about $1.50 per pattern in large quantity; printed-circuit card costs approximately $60 prior to wire-wrapping cost (includes artwork and tooling)	Cost vs quantity: $10 to $100 per pattern in very high quantity; $30 to $60 per unassembled board; for applications requiring over 20 MHz or below 2.5 ns, multilayer boards may represent the best alternative.
Easily repaired: $200 to $300, one-time artwork charge	Not easily repaired: 3 or 4 times cost of two-sided board, depending on number of layers.
Delivery: 1 to 3 weeks	Delivery: 3 weeks to 4 months

Multiwire*	Infobond†
For applications with from 100 printed circuit cards with 15 pins/in.² to >2,000 cards with <30 pins/in.² proves the most rapid, flexible, economic approach; for <100 cards it looks less economical	Slightly higher cost per wire than multiwire but correspondingly higher interconnection density

* Trademark of Kollmorgen Corp.
† Trademark of Inforex, Inc.

60. This technique provides high density, good thermal management, easy maintenance, and good shock and vibration resistance.

Flexible-Circuitry SHP* Module.[22] The flexible-circuitry approach has been used as an alternative to a multilayer printed-wiring board, providing certain advantages in

* Navy Standard Hardware Program.

TABLE 23.   Cost and Density Comparison[9]

Process	Maximum density, dips*/in.²	Cost
Wire-wrappable-socket panel.............	2	$2–$1.50 per dip
Multilayer............................	2.5	$1–$0.75 per dip
Two-sided printed circuit...............	1	$0.25–$0.20 per dip
Multiwire............................	2	Approx. $0.10 per wire or $1.40 per dip (14 leads); cost depends on board complexity, volume, etc.
Infobond............................	2 or slightly more	Cost depends on printed-circuit-board complexity, volume, etc.

* Dual-in-line packages.

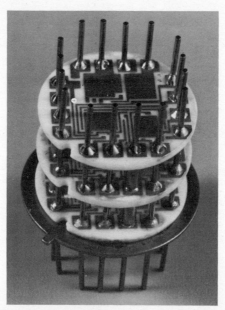

Fig. 57 Stacked substrate TO 8 package with individual thick film substrates. (*Heavy Military Electronic Systems Department, General Electric Company.*)

Fig. 58 Thick film volumetric package.   (*Heavy Military Electronic Systems Department, General Electric Company.*)

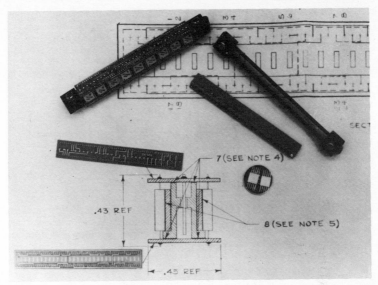

Fig. 59 Stick module. (*Light Military Electronic Systems Department, General Electric Company.*)

Fig. 60 Stick-module assembly showing individual stick module and hybrid-package type. (*Heavy Military Electronic Systems Department, General Electric Company.*)

thermal management, circuit density, reliability, size, weight, repairability, and fabrication cost. The basic module concept, shown in Fig. 61, is based on three elements.

1. A flexible dielectric copper-clad circuit board
2. Use of flat-pack packaged integrated circuits
3. A metal enclosure

The basic flexible printed-wiring board is made in a unique manner in that all the holes, both in the copper and the dielectric, are made by photoetching. This is a precise, batch process and in comparison with punching and drilling is an important factor in reducing fabrication cost. The vertical and horizontal printed-wiring runs are placed on opposite sides of the flexible printed-wiring board and interconnected by plated through-holes. A unique dummy run, shown in Fig. 62, provides adequate

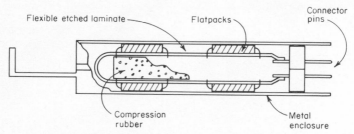

Fig. 61 Basic flexible NAFI module concept. (*Heavy Military Electronic Systems Department, General Electric Company.*)

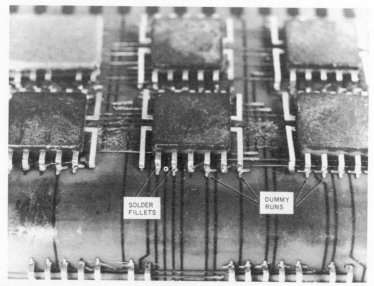

Fig. 62 Close-up of integrated-circuit flat packs with dummy runs and solder fillets. (*Heavy Military Electronic Systems Department, General Electric Company.*)

anchoring of the solder pads when a conducting line is not required and provides reliable solder bonds to the flat-pack leads and connector assembly.

The finished assembly is then folded in half, and a pressure pad of foam rubber or polyurethane is placed inside the fold; thermal grease is applied to the exposed surface of the flat packs, and the assembly is inserted into the metal enclosure (Fig. 63).

The various design techniques and concepts were carefully chosen to provide a number of benefits to the user. The thermal management of the module is assured by inverting the flat packs on the flexible circuitry with their hottest surface in direct contact with the metal housing of the module, which acts as a heat sink. This completely eliminates the circuit board as a thermal barrier.   An increase in circuit density is achieved by using the same common circuit-board area for flat-pack mounting and for circuit runs.   Plated through-holes are used for mounting the flat packs in place of space-consuming planar soldering.   The process provides a ready attainment of small-dimensional line widths, spacing, and plated through-hole diameters.

Reliability of the finished module is provided by the following factors:

1. Use of integrated circuits housed in standard flat packs

2. Attainment of favorable diameter-to-length ratio of through-holes, for good plating, even for very small-diameter holes

3. Utilization of wave soldering and consequent ready visual inspectability for good solder wetting, since the solder must flow against gravity to form a fillet around leads at the top side of the circuit (see Fig. 62)

4. Use of two-sided laminate rather than multilayer, for ready visual inspectability and electrical testing

Fig. 63   Folded circuitry partially assembled into case.   (*Heavy Military Electronic Systems Department, General Electric Company.*)

5. Use of rugged materials and avoidance of delicate ones such as thin ceramics

6. Use of fabrication techniques to which good process control can be applied, rather than dependence on operator skill

7. Use of a metal (aluminum) enclosure for its thermal and mechanical advantages

A decrease in size and weight is provided by the use of a circuit board having a total laminate thickness of less than 7 mils and the use of the same board area for flat-pack mounting and circuit runs.

Repairability is provided by good accessibility, since the folded board can be unfolded to a two-surface structure.

Package-attachment techniques   The versatility of the printed-wiring board as a means of mounting and interconnecting monolithic and hybrid integrated circuits depends on the various techniques developed for attaching these components securely —both mechanically and electrically—to the board.   These same techniques are being used as newer board materials, e.g., ceramic, are being introduced.   There are three general classes of joints to be considered:

1. Permanent, in which one or both lead ends must be broken to separate them; welded joints are of this type.

2. Semipermanent, such as soft solder joints which require special tools to separate the leads; these can be easily repaired by resoldering, permitting repair.

3. Temporary, or quick disconnect, such as plug-in socket connections.

The discussions which follow will look at some of the available welding, soldering, and friction interconnecting methods and some of the design parameters to be considered to apply them correctly.

Welding techniques Parallel-gap welding was one of the earliest techniques used to connect integrated-circuit packages to printed-wiring boards.[9] It was chosen as a highly reliable process that could be automated.

The decision to use welding depended on several system requirements. The circuit density was high, requiring very close connections and a large number of them. The circuits were so designed that components on the wiring side of the board prevented the use of wave-soldering machines. Also solder bridging between the closely spaced conductors and thermal shock to the integrated circuits posed additional hazards. Hand soldering could have been used, but because of the lack of operator control and repetitiveness of the solder joint, it was ruled out. The advantages of the welding process, therefore, were higher component and wiring density, less thermal shock, less likelihood that repair of bad components would dam-

TABLE 24. Nickel Analysis for Weldable Cladding Material[10]

Material	Percentage
Nickel*.................	99.9
Copper.................	0.008
Carbon.................	0.002
Iron...................	0.014
Silicon†................	0.003
Aluminum..............	0.003
Chromium..............	0.003
Titanium...............	0.003
Magnesium.............	0.003
Sulfur.................	0.002

* Including cobalt.
† Maximum.

age board, and elimination of operator differences because it is an automatic process that can be carefully controlled and monitored.

Design Parameters. The development of suitable board-cladding materials posed a serious problem. Several metal combinations were tried. Copper was ruled out because its high conductivity makes welding difficult. When the current flows between the electrodes on top of the integrated-circuit lead, the copper under the lead shunts the welding current, variations in copper thickness change the welding conditions, and the welding machines must be adjusted.

Kovar proved to be too resistive for the narrow run widths and subject to corrosion. A laminate of gold, nickel, and copper welded eutectically due to variations in the copper thickness. A laminate of aluminum, steel, and nickel welded well, but the steel was subject to corrosion. The final choice was high-purity nickel foil 3 mils thick, clad on both sides of a 62-mil-thick epoxy-glass board (grade G-10). It welds well, has resistivity 5 times that of copper, etches well, and has a high resistance to corrosion. The nickel analysis is shown in Table 24.

Wiring Layout. Special wiring-layout rules were generated to assure proper welding-pad areas and printed wiring runs (Fig. 64). Special feed-through techniques were developed to provide welded vias from one side of the board to the other.

Visual Criteria for Welds. The ability to quality-control the welds by visual inspection affords an important part of the overall reliability of welded joints. Figure 65 compares acceptable and unacceptable welded joints.

Repair Procedures. The recommended procedures for repair and replacement are given in Fig. 66.

Soldering

The use of soldering as the primary means of package attachment results from its successful application to a variety of electronic equipment and continued improvement in the techniques and machinery to apply the solder properly. While solder connections can be made one at a time, as with a hand soldering iron for repair purposes, the prime advantage of solder connections is that they can be made in a batch- or mass-soldering process, enabling all connections on printed-wiring-board assembly to be soldered in one operation. The large choice of solders and intermetallic com-

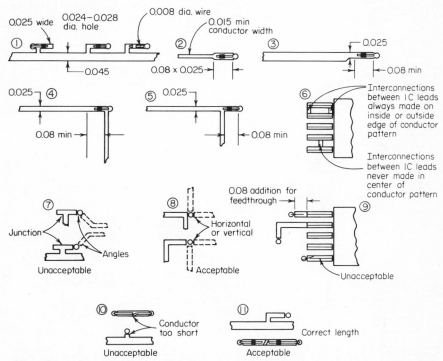

Fig. 64 Wiring-layout rules shown by example: 10 welding-pad dimensions for welding feedthrough wires to printed-wiring traces (1, 2 and 3); locations and dimensions for branch traces (4, 5 and 6); unacceptable and acceptable trace geometries on the underside of the board and at pads (7, 8 and 9); and track design for jumpers (10 and 11).

binations also proves advantageous. Figure 67 shows the melting temperature of alloys commonly used for soldering compared to those commonly used for brazing. Since the printed-wiring-board components are subjected to a large portion of the solder-bath temperature, care is taken to use the lowest-temperature solder compatible with the eventual service conditions. Brazing is used for first-level bonding and package sealing and is covered in other sections of the handbook.

Soldering fundamentals The ideal solder joint is formed when clean molten solder (Pb-Sn) is applied to a clean metallic surface, such as copper. A chemical interaction occurs, and an intermetallic compound formed of tin and copper dissolves out copper from the base-metal surface. The copper-tin compound is extremely hard and brittle in contrast to either the copper or solder base materials. It is susceptible to shock or shear forces, and the thicker the intermetallic layer, the weaker the joint. It is therefore desirable to keep the intermetallic layer to a minimum thickness by soldering at the lowest practical temperature for the shortest length of time.   The total heat applied depends on the mass of the assembly and the rate of heat flow through it.

The formation of a good solder joint depends on overcoming the metal surface contaminant or oxide layers.   Both the solder and the base metals may have coatings inherent to the material, such as copper oxide, or environmental, such as grease, oil, or dirt.   These coatings must be removed by fluxes to permit adequate melting and wetting of the solder material.

The relationship of wetting to the basic solder joint refers to the ability of the molten solder to adhere more strongly to the base material being soldered than to cohere to itself.   The phenomenon of liquid-solid wetting is best explained with the

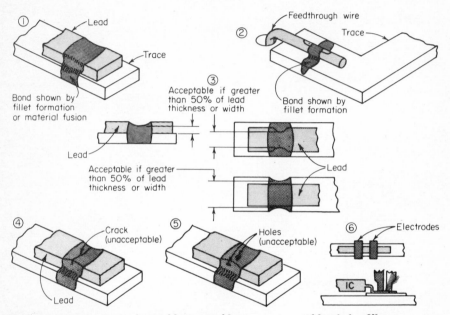

Fig. 65   Visual criteria for welds.[10]   Welds are unacceptable if the fillets are poor or off-center (1, 2 and 3) or if cracks or holes are seen (4 and 5).   Electrodes must be centered on the lead and trace (6).

aid of Figure 68, which shows in cross section a liquid droplet in equilibrium on a solid surface in a gaseous atmosphere.   At point A, the gas, liquid, and solid interfaces exist with a contact angle, θ, that results from the equilibrium of the forces associated with the various surface densities, γ_{SV} (solid-vapor interface), γ_{LS} (liquid-solid interface), and γ_{LV} (liquid-vapor interface).   It can be readily shown that the force density or surface tension (newtons/meter) is numerically equal to the surface energy density (joules/meter^2), so for convenience, the same symbols are used for both.   From the figure, equilibrium requires that

$$\gamma_{SV} = \gamma_{LS} + \gamma_{LV} \cos \theta \tag{8}$$

Since all of these surface energy forces are fixed for a given system, the contact angle, θ, is used as a measure of wetting.   If the vapor in Figure 68 is replaced by a layer of flux, the equation becomes

$$\gamma_{SF} = \gamma_{LS} + \gamma_{LF} \cos \theta \tag{9}$$

where γ_{SF} is now the interfacial tension between the solid and the flux, and γ_{LF} is the tension between the solder material and the flux.   The added flux promotes wetting if the replacement of the atmosphere by the flux results in a decreased contact angle.

The ability of a solder material to wet a surface is represented by the particular contact angle, θ, exhibited in a given system. An angle of $\theta = 180°$ represents a condition of total nonwetting. If the solder comes to equilibrium before solidifying and θ is between 90 and 180°, the wettability of the system is poor and good solder bonds cannot be obtained. Such an angle may also result from the solder "de-

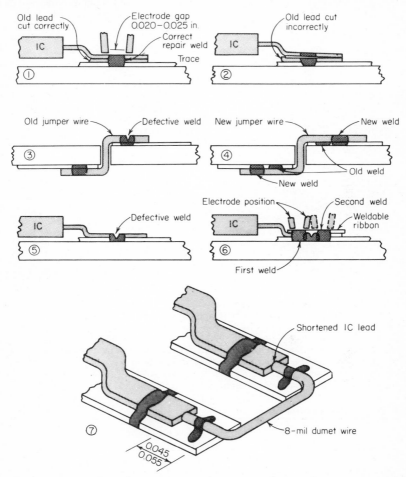

Fig. 66 Repair and replacement procedures.[10] When a flat pack is replaced, old lead must be clipped flush (1) to prevent insecure weld (2). Jumper wires are removed with flush-ground ripper and new wire welded at different weld location (3 and 4). Short length of ribbon repairs defective lead (5 and 6). Wiring modifications are made by shortening flat-pack leads and welding jumper to the welding pads on the printed-wiring traces (7).

wetting"; i.e., after the initial wetting and spreading of the solder over the surface, the force balance indicated in Equation (9) changes, with the solder tending to revert into a ball. Such a condition can be caused, for example, by dissolving thin, plated surfaces into the solder, resulting in exposure of the solder to the underlying base material, with which it may be less compatible. A satisfactory solder connection cannot be made where dewetting has occurred. With θ between approximately 75°

and 90°, partial wetting has taken place. Joints made at these angles are usually marginal. At angles of less than 75°, good wetting action has taken place.

The soldering operation is thus made successful by: (1) removal of contaminants by fluxes, (2) wetting of all joint surfaces by the solder, and (3) keeping the joint motionless while the solder freezes. Unless all of these requirements are met, the solder joint, if formed at all, will have poor mechanical and electrical properties.[14]

Fluxes The type of flux required for a particular application depends on a variety of factors. Flux is used to react chemically with the surface to be soldered, so that any oxide or other contaminant barrier must be removed, allowing the solder to wet the base material. The flux must remain active during the soldering process to prevent recontamination of the solder bond.

Fluxes are classified as *organic* or *inorganic*. Organic fluxes include the common rosin fluxes, as well as nonrosin-base fluxes. Inorganic acid fluxes, such as HCl, pro-

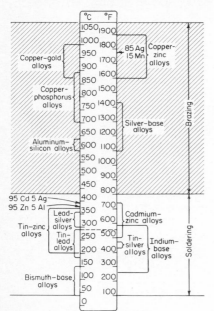

Fig. 67 Melting temperatures of alloys commonly used for soldering and brazing.[13]

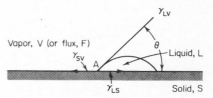

Fig. 68 Surface-tension equilibrium showing[14] contact angle.

mote fast oxide removal and are stable but are very corrosive to the base metals. Due to their corrosiveness, these water-based inorganic fluxes are not generally used in electrical assemblies. If they are used, generous washing with water must follow to remove any residue. Table 25 compares various fluxes.

TABLE 25. Flux Comparison[15]

				Results		
Liquid flux	As-sembly	Tarnish	Cost	Joint reliability	Requires cleaning	Remarks
Water-white rosin..........	Clean	Light	Average	Good	No	Requires good process and material control
	Clean	Heavy	*	Poor	No	Not recommended
	Dirty	Light	High	Medium	Optional†	Dirt might impair solderability
	Dirty	Heavy	*	Poor	Optional†	Not recommended
Activated rosin.............	Dirty	Light	Average	Very good	Recommended†	Standard practice
	Dirty	Heavy	Average	Good	Recommended†	Standard practice
Organic water-base.........	Dirty	Heavy	Low	Very good	Mandatory	Spatters excessively
Water-soluble nonspitting....	Dirty	Heavy	Low	Very good	Mandatory	Requires least amount of solderability control

* Not available.
† Depends on the specific application required for critical cases.

When resin fluxes are used, boards must be cleaned with chlorinated or fluorinated solvents to remove the nonpolar material. Care must be taken to see that board components are tolerant of the solvents used.

Solders Alloys of tin and lead are the most commonly used solders. Figure 69 shows an engineering version of the tin-lead phase diagram. Alloys ranging from 60-40 to 30-70 tin-lead are commonly used for electrical connections. The 63-37 solder has four advantages: (1) it is the fastest solder, in that it melts fastest and solidifies the quickest without going through a plastic state; (2) it has the best wetting ability (the first stage of soldering) and wets at a lower temperature than any other solder with the same ingredients; (3) it has a high fluidity to penetrate small interstices; and (4) it has a higher resistance to fatigue and corrosion, with as good ability to withstand high temperatures as any other solder.

Figure 67 shows some more common solder alloys, Sn-Pb solders, silver solders (for high-temperature use), indium solders (for intermediate-temperature use), and

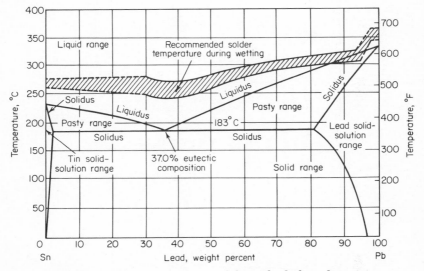

Fig. 69 An engineering version of the tin-lead phase diagram.[13]

bismuth solders (for low-temperature use). Some of the lower-melting silver-solder alloys are used to solder silver or silver-plated parts. Indium solders have been used to solder gold-plated thick film ceramic substrates to tin-plated aluminum heat-sink cases, where the dissimilar expansion coefficients would cause a regular Sn-Pb solder joint to fail under temperature cycling. The In-Pb solders offer low melting points running as low as 117°F for the 8.3% tin, 19.1% indium, 44.7% bismuth, 22.6% lead, 5.3% cadium alloy. Other alloys have a solidus as high as 243°F. See Table 26 for solder compositions.

Pure indium is also used as a solder. Its melting point is 313°F. An alloy of indium with 3 percent silver has been successfully used for fluxless soldering of thin silver surfaces. See Table 27 for properties of indium solders.

Soldering equipment The soldering process requires that several types of equipment be available to do the many different varieties of component and assembly soldering. These include different types of fluxing, preheating and soldering equipment.

Flux Application. Flux can be applied by foam, brush, or spray. The foam machine introduces air through a porous diffusing element submerged in the liquid flux. The air produces a steady stream of bubbles which are channeled into an orifice formed by vertical brushes. The bubble size and flux height depend on the

TABLE 26. Chemical Composition of Solders*[16]

Composition	Tin	Lead	Antimony	Bismuth, max	Silver	Copper, max	Iron, max	Zinc, max	Aluminum, max	Arsenic, max	Cadmium, max	Total of all others, max	Melting range, °F† Solidus	Melting range, °F† Liquidus
Sn 70	69.5–71.5	Remainder	0.20–0.50	0.25	....	0.08	0.02	0.005	0.005	.....	.....	0.080	360	380
63	62.5–63.5	Remainder	0.10–0.25	0.10–0.25	....	0.08	0.02	0.005	0.005	.....	.....	0.080	360	360
62	61.5–62.5	Remainder	0.20–0.50	0.25	1.75–2.25	0.08	0.02	0.005	0.005	.....	.....	0.080	350	372
60	59.5–61.5	Remainder	0.20–0.50	0.25	....	0.08	0.02	0.005	0.005	.....	.....	0.080	360	375
50	49.5–51.5	Remainder	0.20–0.50	0.25	....	0.08	0.02	0.005	0.005	.....	.....	0.080	360	420
40	39.5–41.5	Remainder	0.20–0.50	0.25	....	0.08	0.02	0.005	0.005	.....	.....	0.080	360	460
35	34.5–36.5	Remainder	1.60–2.00	0.25	....	0.08	0.02	0.005	0.005	.....	.....	0.080	360	475
30	29.5–31.5	Remainder	1.40–1.80	0.25	....	0.08	0.02	0.005	0.005	.....	.....	0.080	360	490
20	19.5–21.5	Remainder	0.80–1.20	0.25	....	0.08	0.02	0.005	0.005	.....	.....	0.080	360	530
10	9.00–11.00	Remainder	0.20 max	0.03	1.70–2.40	0.08	.....	0.005	0.005	.....	.....	0.10	514	570
5	4.5–5.5	Remainder	0.50 max	0.25	....	0.08	0.02	0.005	0.005	.....	.....	0.080	518	594
Sb 5	94.0 min	0.20 max	4.00–6.00	.....	....	0.08	0.08	0.030	0.030	.....	0.030	0.030	450	464
Pb 90	....	Remainder	11.0–13.0	.....	....	0.08	0.02	0.005	0.005	0.600	.....	0.080	476	478
Ag 1.5	0.75–1.25	Remainder	0.40 max	0.25	1.3–1.7	0.3	0.02	0.005	0.005	.....	.....	0.080	588	588
2.5	0.25 max	Remainder	0.40 max	0.25	2.30–2.70	0.30	0.02	0.005	0.005	.....	.....	0.030	580	580
5.5	0.25 max	Remainder	0.40 max	0.25	5.00–6.00	0.30	0.02	0.005	0.005	.....	.....	0.030	579	689

SOURCE: From Federal Specification QQ-S-571c
* Indicated in percent by weight.
† Approximately and for information only.

flux solution and the percentage of solids present. Flux solutions with a solids content of 10 to 35 percent have given good results. Viscosity must be carefully checked to prevent changes in flux height and excessive overflow.

Brush coating of flux is accomplished by either using a rotary brush in a flux-filled container or by pumping flux through a stationary brush onto the work.

TABLE 27. Properties of Indium Solders

Composition, %						Melting point, °C	Tensile strength, psi
In	Sn	Pb	Cd	Ag	Cu		
50	50	....	...	...	...	117	1,720
25	37.5	37.5	...	...	...	135–181	
75	....	....	25	...	...	123	
80	....	15	...	5	...	157	2,550
90	....	....	...	10	...	230	1,650
100	....	....	...	...	...	156.7	515
25	37.5	37.5	...	...	...	138	5,260
5	....	92.5	...	25	...	280–285	4,560
50	....	50	...	...	...	215	4,670
12	70	18	...	...	...	150–174	5,320
25	....	75	...	...	...	230	5,450
5	....	95	...	...	...	315	4,330
5	....	90	...	5	...	292	5,730
15	....	....	...	61	24	630–685	

Fig. 70 Foam Fluxer machine. (*Hollis Mfg.*)

Spray coating of flux is done by forcing air through a flux-coated screen. The screen is rotated through a flux bath and air is blown through a slot forcing the flux onto the work surface. The amount of air and the amount of flux on the screen can both be varied to provide a control on flux applied (Fig. 70).

Preheating. The printed-board assemblies are usually heated after flux application to a temperature range of between 160 and 200°F. This is done (1) to drive off excess flux solvents and to accelerate flux activation, (2) to reduce the thermal shock on the assembly when contacting the molten solder (approximately 500°F), and (3) to obtain more uniform heating of the assembly and minimize the effects of large copper areas.

Fig. 71 Preheat radiant heater. (*Hollis Mfg.*)

Fig. 72 Automatic wave-soldering system. (*Heavy Military Electronic Systems Department, General Electric Company and Hollis Mfg.*)

The boards are generally heated by built-in heat sources, such as radiant electric heaters, hot-air blowers, or infrared lamps (Fig. 71).

Wave soldering The automation of the soldering process has been successfully applied to the wave-soldering machine shown in Fig. 72. The inclined conveyor passes the board through the fluxer and then by the preheaters to the solder station. Here the inclined conveyor, combined with the wide solder wave, allows the board to be separated from the molten solder and permits the natural solder forces to pull back the excess solder gradually, virtually eliminating solder bridging or icicles.

The introduction of a tinning oil into the molten solder is done to allow lower (30 to 50°F) soldering temperatures and to further prevent formation of solder bridges on close circuitry. It also has the advantage of minimizing the formation of dross on the surface of the solder bath. Care must be taken when using oil that the boards are properly cleaned and that special precautions are taken to ensure good lead fits to board holes. There should be a maximum of 0.015 in. clearance where 360° fillets are required.

All flux residue should be cleaned from the freshly soldered boards as soon as possible. This can be accomplished by rotary brushes with solvent spray or immersion, as shown in Fig. 73.

Fig. 73 Defluxer machine for flow-soldered boards. (*Heavy Military Electronic Systems Department, General Electric Company and Baron-Blakeslee, Inc.*)

Topside-reflow soldering   When assembling flat packs to printed-wiring boards, it is sometimes desirable to use topside reflow soldering to reduce the heat flow to the flat-pack devices and permit maximum utilization of multilayer board interconnection space on buried layers. With topside soldering the leads are formed and soldered to specially prepared solder pads on the top surface of the printed-wiring board. Heat can then be applied to the assembly by hot-gas jets, as shown in Fig. 74. The temperature and size of the gas jet are controlled, as well as the speed of the workpiece under the gas jet, to assure proper solder flow. Both the component lead and the solder pads are solder-coated prior to heating.

Another technique for reflow soldering is to apply focused infrared energy to the joint that has been pretinned. The advantage of the device shown in Fig. 75 is precise control and positioning of a short energy pulse.

Parallel-gap soldering is also used where individual joints are to be made and are accessible to the soldering electrode. Figure 76 shows a typical setup for parallel-gap soldering. In step 1, the integrated-circuit flat-pack lead is positioned over the etched-circuit conductor path, which has an outer layer of fused electro-plated solder. In step 2, the electrodes are brought against the flat-pack lead with a light pressure, and a dc voltage pulse of short duration (3 to 5 ms) causes current to flow through the joint, melting the existing solder on the joint surfaces. In step 3, the soldering electrodes have been raised, and the reflowed solder has been allowed to cool, forming the interconnection.

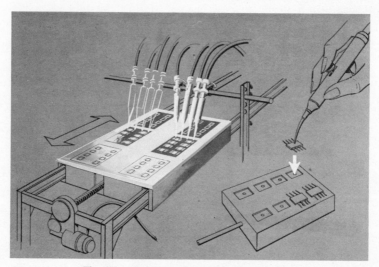

Fig. 74   Hot-gas reflow-soldering machine.

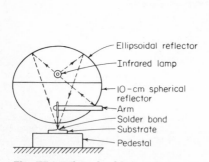

Fig. 75   Infrared soldering device.[17]

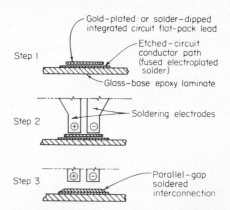

Fig. 76   Parallel-gap soldering process.[15]

Printed-Wiring-Board Connectors

The choice of the proper printed-wiring-board connector should be made early in the design phase of a printed-wiring-board assembly. The selection is usually based on a study of the operating requirements of the final system, the implementation techniques available, and the interconnection techniques used in the rest of the system. Since there are many factors to be judged and many connector types to choose from, the following list will help in this choice.

Board tolerance. More critical with card-edge connectors than with two-piece types.

Printed-circuit pad plating (if any) and base material. An important dry-circuit consideration; also influences certain types of corrosion.

Card-to-card spacing. Is the proposed card-to-card spacing compatible with programmed wiring or with available connector widths?

Card guidance. Cards mounted in awkward locations are more subject to connector damage. Card guidance may be required for locations above eye level or below waist level. Some connectors must be withdrawn in a straight line while others can be peeled apart.

Maximum voltage. Maximum dc and rms voltage can determine connector spacing and insulation types.

Current rating. Current capacity not only influences contact size and style, but minimum anticipated current in an active circuit can influence plating selection.

Contact resistance. Do not forget the total number of connections in the longest path when specifying acceptable resistance.

Intercontact capacitance. Influences spacing, materials, and design.

Termination techniques. The manufacturing facility's ability to successfully work with the selected techniques can make or break a product, especially with new techniques and high connection density.

Polarization requirements. Is mechanical polarization required to orient connection devices before mating?

Keying requirements. To prevent mismating similar connectors, several key facilities may be needed.

Color coding or circuit identification. Some cases may require color coding or circuit identification instead of (or with) keying and polarization.

Mounting methods. Are connection devices to be mounted from the front panel, with rivets, screws, friction devices or clasps, or must they be self-mounting (snap-in)?

Maintenance requirements. Maintenance needs can influence termination method, mounting, and tolerable mating forces.

Life expectancy of a product. Avoid underspecifying.

Mating-unmating frequency. For very few mating cycles, low-cost high-pressure connectors can be used.

Inserting and withdrawal forces. Connectors requiring 10 lb of force can engage and disengage readily by hand. Higher forces may require mechanical aids.

Number of contacts and spacing. Number of contacts affects insertion and withdrawal forces; spacing affects production speed.

Operating temperatures. Operating temperatures can dictate selection of materials (both metal and plastic in connector).

Humidity conditions. Corrosion (especially galvanic) is affected by humidity; moisture absorption can also change electrical properties of the insulation.

Barometric pressure. At high altitudes, outgassing of plastic contaminates sealed compartments.

Atmospheric contaminants. The major contaminants present can determine metals, plastics, and plating required.

Shock and vibration. The plane and amplitude of expected shock or vibration affect connector design. A resonantly vibrating member can destroy electrical continuity.

Storage conditions. Will warehousing or nonoperating environment be the same as operating conditions? Shipping methods and conditions must be considered too.

Projected storage time. Excessively long storage time permits films to form on some metal surfaces. Some insulating plastics can be adversely affected as well. Naturally, storage conditions are a factor here.

Projected usage. Are all connection devices required simultaneously, or can orders be delivered in scheduled lots to avoid rush-job pitfalls? Know the production scheduling.

Approvals required (UL, MIL, CAS). This could impose a completely new set of specifications. Is there a conflict between component specifications and those covering the finished product?

Wire. What is AWG size or diameter, number of strands, material, insulation, diameter, and type?

Manufacturing methods. Heavy-duty or specially constructed connection devices may be required to withstand production processes.

Impedance. Normally the circuit involved determines the required connector impedance, but not all printed-circuit connectors can provide matched-impedance performance.

Sealing. Is environmental or hermetic sealing required; must there be a card-to-connector seal as well?

Closure forces. Closing forces are so great on some highly dense connections that structural support may be required for the panel.

Mating conditions. Manual mating or unmating under difficult conditions demands special consideration.

Tolerances on mounting structure. In blind-mating applications tolerances on both mounting structures can compound with connector tolerances to result in mismating or card damage.

Retention mechanism. Mounting position and mating conditions help choose between locking cams, ball detents, spring-actuated locks, or jackscrews.

Connector availability When a specification for a particular connector is finally drawn up, the next requirement is to match the requirements with an available connector design. Since the charge for tooling a completely new connector may run in excess of $10,000, it is desirable to use an existing design. The information included in Tables 28 to 33 will help select a connector from the list shown in Table 34. Table 35 is used as a representative listing of manufacturers for those types shown on the availability lists. Since new connectors are being designed on a continuing basis, the list is not complete and available literature and catalogs should also be consulted.

CERAMIC SUBSTRATES

The increased usage of complex functional integrated circuits has required the development of a newer and more sophisticated approach to the interconnection of these devices. The relatively small size of the flat-pack integrated-circuit package led to the development of new substrate technology to make maximum use of the small lead configuration and spacing. Ceramic technology was a natural choice for these new interconnecting substrates. Figure 77 shows a typical NAFI hybrid assembly which utilizes an alumina-ceramic substrate to interconnect the packaged integrated-circuit devices and discrete components.

The most common ceramic substrate materials are the high-oxide alumina-ceramic substrates with from 94 to 99.9 percent aluminum oxide. The other most common ceramic substrate material is beryllia, with approximately 99.5 percent oxide content. The other constituents are generally other oxides, such as magnesia, silica, and calcia, which act as inhibitors and fluxing agents. These ingredients are added to improve and control fabrication and firing characteristics. They also control the size of oxide crystals and bond them together. The addition of various quantities of these and other ingredients is made to change the physical properties of the ceramic material. In general, the higher the alumina content, the better the physical properties of the material. Hardness, compressive strength, dielectric strength, and chemical resistance are all much improved in the 99.9 percent alumina ceramic.

Fabrication Techniques

After the alumina powder is mixed with the fluxing agents and inhibitors, it is compacted together at high pressures by one of several methods—dry pressing, tape casting, isostatic forming, or hot pressing. The first two processes are commonly used for the majority of substrates. The last two processes are used to fabricate special high-density substrates.

Dry pressing This is the forming method used most generally for smaller, high-precision substrates in the thickness range of 10 to 50 mils. It consists routinely of compacting the granulated ceramic-powder dry mix in rigid die molds under

TABLE 28.   Connector Code Chart[11]

"H" "L" "W" →←Contact spacing **Type IA**	"H" "L" "W" →←Contact spacing **Type IB**	"H" "L" "W" →←Contact spacing **Type IIA**
"H" "L" "W" →←Contact spacing **Type IIB**	"H" "L" "W" →←Contact spacing **Type IIIA**	"H" "L" "W" →←Contact spacing **Type IIIB**
"W" "L" "H" →←Contact spacing **Type IIIC**		
"H" "L" "W" →←Contact spacing **Type IIID**	"W" "L" "H" →←Contact spacing **Type IV**	"H" "L" "W" →←Contact spacing **Type V**
"H" "L" "W" →←Contact spacing **Type VI**		
"H" "L" "W" →←Contact spacing **Type VII**	"H" "L" "W" →←Contact spacing **Type VIII**	"H" "L" Contact spacing **Type IX**
"H" "L" "W" Contact spacing **Type X**		
"L" "W" "H" →←Contact spacing **Type XI**	"L" "W" "H" →←Contact spacing **Type XII**	"H" "L" "W" →←Contact spacing **Type XIII**
"L" "W" "H" →←Contact spacing **Type XIV**		
"L" "W" "H" →←Contact spacing **Type XV**	"L" "W" "H" →←Contact spacing **Type XVI**	"L" "W" "H" Contact spacing **Type XVII**
"L" "W" "H" Contact spacing **Type XVIII**		
"H" "L" "W" →←Contact spacing **Type XIX**	"L" "W" "H" →←Contact spacing **Type XX**	→←Contact spacing "H" "L" "W" **Type XXI**
"L" "W" "H" →←Contact spacing **Type XXII**		

pressure and sintering it by firing.   Precise allowances must be made for dimensional shrinking after firing.

Major process steps are as follows:

1. Analyze samples, then grade the raw materials, chiefly for particle size and purity.

2. Vibrate the individual hoppers of original-composition alumina powders and of additive ingredients in powder form (such as calcia, magnesia, silica, etc.).

3. Prepare the cone-blended and carefully weighed formulation of dry mix, usually held proprietary, both for ratio of constituents and particle-size distributions.

4. Add binder, water, coloration, etc., and ball-mill it as a slurry.

5. Granulate by forcing the slurry through a spray nozzle and processing it through a mesh screen and a magnetic separator into drying droplets, thus forming discrete, uniform granules.

6. Issue a tagged lot of granulated powder mix to the presses to fill properly shaped mold cavities.

7. Wipe off the excess powder, either by hand using a spatula or by automatic moving-blade equipment, before applying a stroke of the press.

TABLE 29. Contact Code Chart[11]

	FR	FB	TF	SO
	"Ribbon style" formed from flat stock; can be bifurcated	"Bellows style" formed from flat stock; usually bifurcated	"Tuning fork" punched from sheet stock	"Socket" machined or formed from rod or sheet stock
PI	**RS-1**	**FW**	**PS**	**RS-2**
"Pin" machined from rod stock	"Right angle" formed from crowned square stock	"Wire style" formed from precision wire; cannot be bifurcated	"Adapter" punched from sheet metal stock	"180° stacking" formed from crowned square stock
TF-1	**WE**	**RS-3**	**TR**	**HM**
"Modified tuning fork" formed from sheet stock	"Wedge-male" usually punched and formed from sheet metal stock	"Parallel stacking" formed from crowned square stock	"Torsion type" formed from flat stock	"Hermaphrodite" usually punched and formed from sheet metal stock

TABLE 30. Contact Termination[11]

SC	Solder cup	TR	Tubular rivet
SE	Solder eyelet	TT	Taper tab
SD	Solder dip	WW	Wire wrap
ST	Slotted tongue	SO	Stand-off
CR	Crimp	CRM	Crimp removable
TP	Taper pin	WT	Weld tab
TS	Threaded stub		

TABLE 31. Contact Plating[11]

Au	Gold	Rh	Rhodium
Ag	Silver	EL	Electraloy
Ni	Nickel	Sn	Tin
Cu	Copper		

TABLE 32. Connector Mountings[11]

CH	Clearance hole	GP	Guide pins
FB	Float bushing	RI	Riveted
TS	Threaded stud	ST	Staked
SM	Solder mount	TI	Threaded insert
MB	Mounting bracket		

TABLE 33.   Connector Materials Code[11]

GDI-30F	Glass-fiber filler, flame-retardant, diallyl phthalate MIL-M-19833	MFG	Phenolic type MFG MIL-M-14
GDI-30	Glass fiber general-purpose diallyl phthalate MIL-M-19833	MME	Mineral filler, arc- and flame-resistant, MIL-M-14F
GDI-37	Glass-fiber filler, high-heat-resistant diallyl phthalate MIL-M-19833	NY	Nylon MIL-P-17091
		Poly	Polycarbonate resin Lexan†
GFN	Glass-filled nylon with epoxy alloy	PHE	Phenolic
		SDG	Short-fiber filler, diallyl phthalate, general-purpose MIL-M-14F
Lucite*	Acrylic plastic MIL-P-5425A		
MAI-60	Glass-fiber filler, flame-resistant, alkyd MIL-M-14	SDG-F	Glass-fiber filler, flame-resistant, diallyl phthalate MIL-M-14F
MDG	Mineral filler, diallyl phthalate general-purpose MIL-M-14F	SDI-5	Orlon-filled diallyl phthalate, MIL-M-14
		XXX-P	Paper-base phenolic MIL-P-3115
MFE	Mineral filler, best electrical properties, MIL-M-14F		

* Trademark of E. I. du Pont de Nemours & Co.
† Trademark of General Electric Company.

8. Actuate the press, which will complete the forming operation.   NOTE: Better machines today have either a *floating die* or provisions for otherwise equalizing the pressure also from the bottom half upward, to prevent bridging in the green ceramic, i.e., denser compacting below and close by the upper die half.

9. Load compacted substrates in the green state onto refractory setter slabs and stack them.

10. Process the substrates on setters through the pyrometrically controlled firing cycle.   Sintering takes place either in preprogrammed *periodic kilns* or in straight-through *tunnel kilns.*

11. Subject the finished, hard-fired substrates to verification of compaction density and to dye-check inspection for hairline cracks, then to ultrasonic cleaning.

Supplementary operations may include tumbling, grinding, lapping, polishing, drilling, machining, sawing, coating, metallizing, and others.   Note that hole patterns as specified on the drawings are usually molded in during dry pressing, with the help of pins around which the free-flowing powder settles prior to compacting.

Tape casting or slip casting   The *tape* or *slip* refers to the thin, tapelike appearance of the material after effortless compaction via evaporation of the suspension medium.   This method is a natural means of making essentially two-dimensional, large, flat substrates of all shapes almost devoid of internal stresses, as well as the layers for laminated monolithic chip capacitors.

Major process steps, essentially the same for alumina as for barium titanate, forsterite, and other similar ceramics, are:

1. Vibrate the cone-blended raw material, an original powder-mix composition of desired purity, particle size, shape, additives, etc.

2. Add plasticizers and binders to make it into a slurry and ball-mill.

3. Add liquid, organic vehicle for easy pouring; then de-aerate it.

4. Strain through sieve to minimize agglomerates and contaminants.

5. Cast the liquid *slip,* pouring it at a uniform rate onto a long, clean, glass-plate surface; or pour it onto an endless, smooth plastic carrier belt.

6. Carefully control the variables influencing the wet-cast slip (film) thickness, namely, the viscosity of the slip, the hydrostatic head of its reservoir, and the rate of speed at which it is cast.

7. Activate the travel of a preset thickness-control *doctor blade* along the length of the glass plate, removing the excess and leveling the ceramic slip to the desired thinness; or, alternately, let the slip pass on a moving, smooth surface under a stationary preset doctor blade.

TABLE 34.   Printed-Circuit-Connector Availability Chart[11]

Connector type (refer to Table 28)	For single- or double-sided board*	Recommended nominal board thickness, in.	Number of contacts available	Contact spacing center to center, in.	Length L as function of number of contacts	Height H, in.	Width W, in.	Contact type (refer to Table 29)	Contact termination (refer to Table 30)
XIX	S	⅛	6, 10, 12, 15, 18, 22	0.156	1.785–4.320	0.458	0.340	Spade	SD
IIIA	S, D	¹⁄₁₆, ³⁄₃₂, ⅛	7, 12	0.100, 0.156	1.18, 2.31	0.17, 0.25	0.19, 0.31	PI	SE, SD, TP, SC
V	S, D	¹⁄₁₆, ³⁄₃₂, ⅛	11, 15, 23, 37	0.100	1.70–4.30	0.61	0.36	PI	SE, SD, TP, SC
V	S, D	¹⁄₁₆, ³⁄₃₂, ⅛	45	0.050	2.90	0.67	0.36	PI	SE, SD, TP, SC
IIID	S, D	¹⁄₁₆, ³⁄₃₂, ⅛	50	0.100	4.50	0.51	0.46	PI	SE, SD, TP, SC
V	S, D	¹⁄₁₆, ³⁄₃₂, ⅛	7, 11, 15, 19, 25	0.125	1.45–3.70	0.45	0.36	PI	SE, SD, TP, SC
XIX	S, D	¹⁄₁₆, ³⁄₃₂, ⅛	10, 12	0.200	2.80–3.20	0.35	0.31	PI	SE, SD, TP, SC
	S, D	¹⁄₃₂, ¹⁄₁₆	39, 59	0.100	3.280, 4.280	0.529	0.386	SD, CR	SO, PI
	S, D	¹⁄₃₂, ¹⁄₁₆	28, 30, 60, 80	0.100	890–1.590	1.040	0.410	SD, CR	SO, PI
1B	D	¹⁄₁₆	15, 22, 29, 30, 43, 50	0.100	2.600–5.400	0.835	0.500	FR	CRM
1B	D	¹⁄₁₆	8, 15, 18, 22, 25, 30, 36, 43	0.156	2.096–7.600	0.670–0.703	0.500	FB	CRM
1B	S, D	¹⁄₁₆	15, 22, 30, 36, 43	0.156	3.190–7.555	0.830	0.440	FR	WW
XIII	S, D	¹⁄₁₆, ³⁄₃₂	17, 23, 28, 41	0.150	1.550–3.350	0.260	0.250	PI, SO	SD
XIII	S, D	¹⁄₁₆, ³⁄₃₂	17, 25, 33, 41	0.100	1.150–2.350	0.260	0.250	PI, SO	SD
XII	M	¹⁄₁₆, ³⁄₃₂	13, 25, 37, 49, 61, 92	0.100	0.800–2.400	0.350	0.345	PI, SO	SD
1A	S	¹⁄₃₂	20	0.100	2.500	0.250	0.200	FB	SD
1B	D	¹⁄₁₆	12, 20, 30, 40, 50	0.050	0.850–1.800	0.488	0.281	FB	SD, SC, WW, WT
XIII	S, D	¹⁄₁₆	23	0.100	3.300	0.400	0.400	PI	SD
XV	D	¹⁄₁₆	48	0.125	3.687	1.093, 1.812	0.343	FR	WW
IB	D	¹⁄₁₆	80	0.125	5.805	0.548	0.360	FR	WW
XV	D	¹⁄₁₆	30, 60	0.125	4.360–4.750	0.250	0.690, 2.000	FR	WW
IB	S, D	¹⁄₁₆	30, 44	0.156	3.250, 4.345	0.672	390	FR	WW
1B	D	¹⁄₁₆	15, 50	0.156	3.188, 4.748	0.730	0.360	FR	WW
IA	S, D	¹⁄₁₆	6–50	0.156	1.785, 4.906	0.458	0.340	FB	SE
IA	D	¹⁄₃₂	10, 30, 50, 64	0.050	1.68–3.88	0.42	0.25	FB	SD or welding
IA	D	¹⁄₁₆	10, 14, 20, 40, 50, 64	0.050	1.22–3.92	0.43	0.33	FB	SD or welding
IA	S, D	¹⁄₁₆	10, 15, 19, 22, 26, 35, 40, 50	0.100	1.88–5.00	0.42–0.61	0.36–0.41	FB	SE, SD, WW
IA	S, D	¹⁄₁₆	6, 10, 15, 18, 20, 22, 28, 30, 50	0.125	1.55–7.055	0.42–0.61	0.375	FB	SE, SD, WW
IA	S, D	¹⁄₁₆	6, 10, 12, 15, 18, 22, 28, 30, 36, 43	0.156	1.78–7.63	0.44–0.47	0.33–0.50	FB	SE, SD, WW
IA	S, D	¹⁄₁₆	18, 28, 22, 43	0.200	4.60–9.60	0.875	0.50	FB	SE, SD, WW

Standard contact plating (refer to Table 31)†	Type of connector mounting (refer to Table 32)	Dielectric material (refer to Table 33)	Dielectric withstanding voltage sea level, rms	Dielectric withstanding voltage 70,000 ft, rms	Nominal current rating, A	Military Specification for vibration and shock‡	Max. ambient temp. for use, C°	Manufacturer	
								Catalog series number	Name and location (refer to Table 35)
Cu, Au	CH	MDG, SDG-F	1,800	450	5	C-21097	125	133, 233	AMPH
Au	CH	SDG-F	1,500	375	5.0	C-8384	125	TBHR	ARM
Ag, Au	SM, TS GP, RI	MME, GDI-30F, SDI-5	1,500	375	7.5	C-8384	125	DEP	ARM
Ag, Au	SM, TS, GP, RI	SDI-5	1,000	300	3.0	C-8384	125	DEP	ARM
Ag, Au	SM, TS, GP, RI	GDI-30, GDI-30F	1,500	375	7.5	C-8384	125	LP	ARM
Ag, Au	SM, TS, GP, RI	GDI-30, GDI-30F	2,250	500	7.5	C-8384	125	EP	ARM
Ag, Au	SM, TS, GP	GDI-30, GDI-30F	2,700	560	7.5	...	125	CP	ARM
Ag, Au	CH	SDG-F	1,300	325	5.0	STD-202	125	PCB min.	BEND
Ag, Au	CH, TI	GFE	1,300	325	3.0	STD-202	125	PCB min.	BEND
Ag, Au	CH	GDI-30F, Poly	800	...	3	STD-202	125, 105	PBD, PB	BUR
Ag, Au	CH	GDI-30F	1,800 .	...	5	C-21097	125	PC	BUR
Ag, Au	CH	Poly	1,500	...	5	STD-202, 9810	105	PWC	BUR
Ag, Cu	RI	SDG-F	1,000	...	3	STD-202	125	UPC2A	BUR
Ag, Cu	RI	SDG-F	1,000	...	3	STD-202	125	UPC2A	BUR
Ag, Cu	RI	SDG-F	1,000	...	3	STD-202	...	UPC3B	BUR
Au	CH, SM	SDG-F	500	1,660	0.5	C-21079-1	125	Mini-Tyke	CINCH
Au	CH, SM	SDG-F	500	1,660	0.5	C-21079	125	Tykon	CINCH
Au	TS	MME	...	...	...	...	...	...	CINCH
Au	TI	SDG-F	830	200	5.0	C-21079	125	...	CINCH
Ni, Au	CH	SDG-F	830	200	5.0	...	125	...	CINCH
Au	RI	GFN	1,800	450	3.0	202-C	125	...	CINCH
Au	CH	SDG-F	830	200	5.0	202-C	125	...	CINCH
Ni, Au	GH	PHE	2,000	...	3.0	...	...	...	CINCH
Au	CH, FB, TI	SDG-F	830	200	5.0	202-C	125	...	CINCH
Au	CH	GDI-30, GDI-30F	900	300	1	C-21097	125	600-2	CON
Au	CH	GDI-30, GDI-30F	900	300	1	C-21097	125	600-6	CON
Au	CH, FB, TI	GDI-30, GDI-30F	1,500, 1,400	360, 240	3	C-21097	125	600-121, 600-100	CON
Au	CH, FB, TI	GDI-30, GDI-30F	1,400, 1,350	350, 200	3	C-21097	125	600-128, 600-125	CON
Au	CH, FB, TI	GDI-30, GDI-30F	1,665	430	5	C-21097	125	600-11, 600-156	CON
Au	CH, FB, TI	GDI-30, GDI-30F	2,100	450	5	C-21097	125	600-83	CON

TABLE 34. Printed-Circuit-Connector Availability Chart[11] (Continued)

Connector type (refer to Table 28)	For single- or double-sided board*	Recommended nominal board thickness, in.	Number of contacts available	Contact spacing center to center, in.	Length L as function of number of contacts	Height H, in.	Width W, in.	Contact type (refer to Table 29)	Contact termination (refer to Table 30)
IA	S	0.054–0.070	8, 10, 12, 15, 18, 22	0.156	2.11–4.30	0.458	0.340	TF	SE, SD, WW
IIID	S, D	Max. 0.115	7, 15, 19, 25	0.125	1.45–3.70	0.453	0.360	PI	SD
V	S, D	Max. 0.160	7, 15, 19, 25	0.125	1.45–3.70	0.453	0.360	PI	SD
IV	S, D	Max. 0.115	7, 11, 15, 19, 23	0.156	2.097–4.596	0.750	0.468	PI	SD
IB	S, D	$\frac{1}{32}$, $\frac{1}{16}$	8, 16, 20, 25, 32, 50, 64	0.050	0.950–3.750	0.270	0.158, 0.190	FW	SD
V	S, D	Max. 0.124	50	0.200	4.50	0.460	0.360	PI	SD
X	S, D	$\frac{1}{16}$	64	0.050	3.425	0.250	0.250	FR	......
IIIA	S	$\frac{1}{16}$	15	0.156	2.325	0.250	0.300	FI	SD
XVI	S	$\frac{1}{16}$	34, 40, 56, 60	0.100	2.050–3.250	0.180	0.556	FR	SD
IA	S, D	$\frac{1}{16}$	6, 10, 12, 15, 18, 20, 22, 30, 36, 44	0.156	1.843–4.343	0.458	0.343	FB	SE, SD, TP, TT, WW
IA	S	$\frac{1}{16}$	31, 41	0.075	2.050–3.937	0.258–0.439	0.281–0.463	HM	SD, SE
V	S, D	$\frac{1}{16}$	2–152	0.050	0.490–7.790	0.340	0.245	HM	SD, SE
XIV	S	$\frac{1}{16}$, $\frac{3}{32}$	17, 23, 29, 35, 41, 47	0.100	2.386–5.620	0.453–0.828	0.594	HM	SD, TT, WW, ST, SE
XV	S, D	$\frac{1}{16}$	25, 36, 50, 72, 100	0.100	3.370–5.870	0.920	0.292	FR	WW
XV	S, D	$\frac{1}{16}$	20, 30, 40, 60, 80	0.125	2.747–5.247	0.720	0.245	FR	WW
IA	D	$\frac{1}{16}$	10–20, 20–40, 30–60, 40–80	0.050	1.176, 1.676, 2.276, 2.826	0.450	0.250	FB	SD, WT
IA	D	$\frac{1}{16}$	Multiples of 2	0.125	Various	0.460	0.278	FB	WW
IA	D	$\frac{1}{16}$	18	0.150	3.562	0.625	0.375	FB	WW
IA	S	$\frac{1}{16}$	28	0.156	5.343	0.468	0.437	FB	SE
IA	D	$\frac{1}{16}$	15–30, 18–36, 22–44	0.156	3.25, 3.79, 4.34	0.406	0.328	FB	SE, SD
IA	D	$\frac{1}{16}$	23–46	0.200	6.050	4.425	0.500	FB	WW
IA	D	$\frac{1}{16}$	12, 20, 36, 56, 72, 96	0.156	1.839–7.600	0.670	0.328–0.500	FB	SE, SD, TP, TR, WW
IA	D	$\frac{1}{16}$	10, 20, 30, 40	0.050	1.200–3.400	0.300	0.250	FR	SD
IB	D	$\frac{1}{16}$	44, 50, 60, 72, 86, 96	0.100	3.158–5.758	0.625	0.365	FB	WW
XX	S, D	$\frac{1}{16}$, $\frac{3}{32}$	2–130	0.100	0.100 contact	0.245	0.315	Square socket	SD
XXI	S, D	$\frac{1}{16}$, $\frac{3}{32}$	64	0.100	3.69	0.650	0.355	Square pin	SD
XXII	S, D	$\frac{1}{16}$, $\frac{3}{32}$	64	0.100	3.69	0.614	0.378	Square socket	WW
XX	S, D	$\frac{1}{16}$, $\frac{3}{32}$	2–86	0.150	0.150	0.295	0.315	Square socket	SD
IA	S	$\frac{1}{16}$	28	0.1875	2.688	0.312	0.438	SO	SE
IA	S, D	$\frac{1}{16}$	30, 60	0.156	5.630	$1\frac{7}{32}$	$\frac{45}{64}$	FR	SE, SD

Standard contact plating (refer to Table 31)†	Type of connector mounting (refer to Table 32)	Dielectric material (refer to Table 33)	Dielectric withstanding voltage sea level, rms	Dielectric withstanding voltage 70,000 ft, rms	Nominal current rating, A	Military Specification for vibration and shock‡	Max. ambient temp. for use, C°	Manufacturer	
								Catalog series number	Name and location (refer to Table 35)
Au	CH, FB, TI	MFG	1,800	450	5	C-21097	125	EBT-156	DALE
Ag, Au	GP, TS	GDI-30F	3,600	975	7.5	C-8384	125	300PE	DALE
Ag, Au	GP, TS	GDI-30F	3,600	975	7.5	C-8384	125	300PR	DALE
Ag, Au	CH	MDG, SDG-F	2,500	600	7.5	C-8384, C-21097	23?	320	DALE
Au	CH	MFH	1,200	600	0.5	202-C	125	EBTL	DALE
Au, Ag	GP	GDI-30F	2,500	600	7.5	C-8384	125	315PR	DALE
Au	FB	NY	1,000	250	. . .	. . .	125	Bal-Con	EBY
Ag	SM	NY	1,000	250	5	. . .	125	. . .	EBY
Au	RI	SDG-F	1,000	250	3	STD-202	200	Ra-Con	EBY
Ni, Au	CH, TI, FB	SDG-F	2,000	450	5	C-21097	150	6007	ELC
Ni, Au	TS, CH	SDG-F	1,350	325	5	C-55302	150	8114	ELC
Ni, Au	MB	SDG-F	1,000	450	5	C-21097	150	8218	ELC
Ni, Au	CH	SDG-F	2,000	450	10	C-21097	150	7008	ELC
Ni, Au	CH, FB	GFN	1,800	450	3	E-19600	105	6309	ELC
Ni, Au	FB	GFN	1,800	450	6	E-19600	105	6313	ELC
Cu, Ni, Au	CH	SDG-F	1,000	350	1	C-21097	177	A-2289	FAB
Cu, Ni, Au	Various	GFN	1,800	450	5	STD-202	177	501401	FAB
Cu, Ni, Au	CH	SDG-F	2,100	580	5	STD-202	177	200271	FAB
Cu, Au	CH	MAI-60	3,200	1,800	5	STD-202	177	2274	FAB
Cu, Ni, Au	CH, FB, TI	SDG-F	3,200	1,800	5	C-21097	177	200242, 200243, 200244	FAB
Cu, Ni, Au	Card guide	Poly	3,200	1,800	5	STD-202	177	20015	FAB
Ag, Au	CH, TS, FB	GDI-30, SDG-F	2,500, 1,000	600, 230	5.0	C-21097	177	ERS	HCD
Au	CH	SDG-F	1,000	375	1.0	C-21097B	125	CPS	HCD
Ni, Au	CH, TI	Noryl	1,500	375	3.0	C-21097	125	EC4	ITTC
Ni, Au	SM	GDI-30F	1,500	375	3.0	C-20197	125	GO8	ITTC
Ni, Au	CH	Poly	1,500	375	3.0	C-21097	125	GO7	ITTC
Ni, Au	CH	Poly	1,500	375	3.0	C-21097	125	GO7	ITTC
Ni, Au	SM	GDI-30F	1,500	375	3.0	B-21097125	. . .	UB	ITTC
Ag	SM	NY	660	250	3.0	. . .	65	126-0110	JOHN
Ni, Au	CH	GDI-37, MFG, PHE	. . .	. . .	. . .	. . .	150–300	EHM	LMC

TABLE 34. Printed-Circuit-Connector Availability Chart[11] (Continued)

Connector type (refer to Table 28)	For single- or double-sided board*	Recommended nominal board thickness, in.	Number of contacts available	Contact spacing center to center, in.	Length L as function of number of contacts	Height H, in.	Width W, in.	Contact type (refer to Table 29)	Contact termination (refer to Table 30)
XXII	D	$\frac{1}{16}$	60	0.156	$5\frac{61}{64}$	$1\frac{7}{32}$	$\frac{45}{64}$	FR	SD, WT
XV	S, D	0.047, 0.062	6–X	0.125	Various	0.687	0.250	FR	WW
XVI	S, D	0.062–0.125	1–X	0.200	Various	0.156	0.453	TF	WT, WW, SD
XVI	S, D	0.062–0.125	1–X	0.100–0.150	Various	0.097–0.147	0.390	TF	WT, WW, SD
IA	S, D	0.062	6, 10, 15, 18, 22, 28, 36, 43	0.156	1.839–6.540	0.406	0.328–0.500	FB	SE, SD
IA	D	$\frac{1}{16}$	20–130	0.100	1.925–7.60	0.437–0.500	0.500	FB	SE, SD
XV	D	$\frac{1}{16}$	76	0.125	5.308	0.200	0.343	FR	WW
XVI	S, D	0.062	I–X	0.100–0.150	Various	0.097–0.147	0.400	WE, TF	SD, WW, WT
XX	D	0.062	2–130	0.100	0.100–6.540	0.238	0.320	SO	SE
IA	D	0.062	36–130	0.100	2.635–6.835	0.610	0.370	FB	WW
IA	S, D	0.062	6, 8, 10, 12, 15, 18, 22, 24 30, 36, 44	0.156	1.785–4.597	0.468	0.340	FR	SE, SD, WW, TT
IA	S, D	0.062	6, 8, 10, 12, 15, 18, 22, 24, 30, 36, 44, 48	0.156	1.785–4.597	0.468	0.340	TF	SE, SD, WW, ST
IA	D	0.062	60, 80	0.125	4.555, 5.805	0.610	0.360	FB	WW
IIA	S, D	0.062–0.093	6, 8, 10, 12, 15, 18, 22, 24	0.156	1.785–4.597	0.450	0.340	Wedge male	SE, DS
IIIA	S, D	0.062–0.250	10, 15	0.156	1.546, 2.328	0.312	0.312	PI	DS
IIB	S, D	0.062–0.125	6, 8, 10, 12, 15, 18, 22, 24	0.156	1.785–4.597	0.468	0.340	Wedge male	SE, SD, WW
IA	S	$\frac{1}{16}$	6, 9, 12, 15, 18, 22, 24	0.156	1.704–4.635	0.600	0.590	FR, FB	CR
IA	S	$\frac{1}{16}$	6, 9, 12, 15, 18, 21, 22, 24	0.156	1.706–4.512	0.671	0.296	FR	CR, SE
IA	S	0.062	6–50	0.100, 0.125	1.435–7.055	0.550	0.370	FR, FB	WW
IB	S, D	0.062	6–50, 12–100	0.100 × 0.200, 0.125 × 0.250	1.435–7.055	0.550	0.370	FR, FB	WW
XIV, XV	S, D	0.062	6–50, 12–100	0.100 × 0.200, 0.125 × 0.250	1.435–7.055	2.060	0.370	FR, FB	WW
IA	S	$\frac{1}{16}$	6, 10, 15, 18, 22	0.156	1.781–4.281	0.438	0.343	TF	SE
IA	D	$\frac{1}{16}$, $\frac{3}{32}$	20, 30, 36, 44	0.156	1.781–4.281	0.438	0.343–0.406	FR	SE, WW, TT
IV	S, D	$\frac{1}{16}$, $\frac{1}{8}$, $\frac{1}{4}$	7, 11, 15, 19, 23, 32	0.156	2.097–6.093	0.750	0.368–0.406	PI	SD
IV	S, D	$\frac{1}{16}$, $\frac{1}{8}$, $\frac{1}{4}$	11, 17, 23, 29, 35	0.100	2.097–4.596	0.800	0.604	PI	SD
IB	D	0.062	132	0.050	4.020	0.867	0.33	FB	SD
IB	D	0.062	100	0.100	5.875	0.74	0.40	FB	SE, SD

Standard contact plating (refer to Table 31)†	Type of connector mounting (refer to Table 32)	Dielectric material (refer to Table 33)	Dielectric with-standing voltage sea level, rms	Dielectric with-standing voltage 70,000 ft, rms	Nom-inal cur-rent rating, A	Military Specification for vibration and shock‡	Max. ambient temp. for use, C°	Manufacturer	
								Catalog series number	Name and location (refer to Table 35)
Per cus-tomer spec.	CH	GDI-30F, MFG, MAI-60, PHE, etc.	...	...	...	...	150–350	LMC 2289	LMC
Ni, Au	MB	Poly	1,300	...	5	STD-202	105	Cardex 125	MAL
Ni, Au	MB	NY	2,500	...	5	STD-202	95	WASP	MAL
Ni, Au	MB	NY	1,000–1,500	300–600	3	STD-202	95	MINI, WASP	MAL
Ag, Au Ni, Ag	CH, FB, TI	GDI30F	...	...	...	C-21097	125	SO14	MAS
Ag, Au	CH, FB, TI	GDI30F	120	250	3	STD-202	125	008	MAS
Ni, Au	TI	GFN	1,200	200	3	STD-202	95	...	MAS
Ni, Au	MB, CH	NY	1,000–1,500	300–600	3	STD-202	95		MAS
Ni, Au	...	PHE	...	...	3	STD-202	125	000406	MAS
Ni, Au	CH, TI	GD-130	...	3	...	C-21097	125	008	MAS
Cu, Au	CH, FB, TI, MB	SDG-F	1,800	450	5.0	C-21097	125	70 ser.	MET
Cu, Au	CH, FB, TI, MB	SDG-F	1,800	450	5.0	C-21097	125	90 ser.	MET
Ni, Au	CH	SDG-F	1,500	...	3.0	C-21097	125	186 ser.	MET
Cu, Au	CH	SDG-F	1,800	450	5.0	...	125	65 ser.	MET
Cu, Au	SD	MFG, SDG-F	1,800	450	5.0	...	125	50 ser.	MET
Cu, Au	CH, FB, TI, MB	SDG-F	1,800	450	5.0	C-21097	125	60, 61 ser.	MET
Sn	CH	NY	...	...	5	...	105	1800	MOLE
Sn	CH	NY	...	...	5	...	105	1796	MOLE
Sn, Pb, Au	CH, TI	SDG-F	1,800	450	3.0	C-21097, STD-202	125	P-101	SYL
Sn, Pb, Au	CH, TI	SDG-F	1,800	450	3.0	C-21097, STD-202	125	P-101	SYL
Sn, Pb, Au	CH, TI	SDG-F	1,800	450	3.0	C-21097, STD-202	125	P-101	SYL
Cu, Au	CH, FB, TS	GDI-30F, SDG-F	2,400	600	5.0	C-21097B	232	UPCR	USC
Cu, Au, Ag	CH, FB, TS	GDI-30F, SDG-F	2,400	600	5.0	C-21097B	232	UPCR-D, UPCR93-D	USC
Cu, Au	CH	GDI-30F, SDG-F	2,400	600	7.5	C-55302A	232	UPCC-M	USC
Cu, Au	CH	GDI-30F, SDG-F	2,265	565	7.5	C-55302A	232	UPCC-SGM	USC
Au	CH	SDG	375	275	0.5	C-21097	400	L ser.	VIK
Au	CH, FB, TS	SDG	650	275	3	C-21097	400	C	VIK

TABLE 34. Printed-Circuit-Connector Availability Chart[11] (Continued)

Connector type (refer to Table 28)	For single- or double-sided board*	Recommended nominal board thickness, in.	Number of contacts available	Contact spacing center to center, in.	Length L as function of number of contacts	Height H, in.	Width W, in.	Contact type (refer to Table 29)	Contact termination (refer to Table 30)
VII	S, D	0.062, 0.093	120	0.100	5.3	...	...	TF	WW
IB	D	0.062	20, 25, 30, 40, 50	0.050–0.700	1.75–5.950	0.260–0.610	0.250	FW, FR	SD, WW
IB	D	0.062	28, 31, 40, 49, 50	0.125	4.300–7.050	0.610	0.360	FR	WW
IA	S	0.062	6, 15, 18, 22, 28	0.156	1.864–5.284	0.440	0.343–0.500	FR	SE, SD
IB	D	0.062	10, 15, 18, 22, 36, 43	0.156	2.488–7.631	0.440	0.343–0.500	FR	SE, SD
IA & IB	S, D	0.062	6, 10, 12, 15, 18, 22, 28, 30	0.156	1.780–5.450	0.560	0.440	FR	SE, SD, TP, WW
IV	S, D	0.062, 0.125, 0.250	7, 11, 15, 19, 23, 24, 35	0.100–0.156	2.094–4.594	0.750	0.469	PI	SD

* KEY: S = single; D = double; M = multilayer.
† First plating given is underplate.
‡ The prefix MIL, omitted for space considerations, should be read with these numbers.

TABLE 35. Printed-Circuit-Connector Manufacturers[11]

AIR	Airborn Inc., P.O. Box 20232, Dallas, Tex. 75220
AMP	Amp Inc., P.O. Box 3608, Harrisburg, Pa. 17111
AMPH	Amphenol Industrial Div., Bunker-Ramo Corp., Chicago, Ill.
ARM	Armel Electronics, North Bergen, N.J. 07047
	Becon Connectors, Brown Engineering, Huntsville, Ala. 35804
BEND	Bendix Corp., Electrical Components Div., Sidney, N.Y. 13838
BUR	Burndy Corp., Norwalk, Conn. 06852
	Cambridge Thermionic Corp., Cambridge, Mass. 02138
CINCH	Cinch Mfg. Co., Div. of TRW, Inc, Elk Grove Village, Ill. 60007
CON	Continental Connector Corp., Woodside, N.Y. 11377
DALE	Dale Elex Inc., P.O. Box 180, Yankton, S.D. 57078
EBY	Hugh H. Eby Co., Philadelphia, Pa. 19144
ELC	Elco Corp., Willow Grove, Pa. 19090
FAB	Fabri-Tek Inc., National Connector Div., Minneapolis, Minn. 55428
HCD	Hughes Connecting Devices, Newport Beach, Calif. 92660
ITTC	Itt Cannon Electric, PC Connector Operation, Santa Ana, Calif. 92705
	Javex Elex, Cherry Valley, Calif. 92223
JOHN	E. F. Johnson Co., Waseco, Minn. 56093
LMC	Loranger Mfg Co., Warren, Pa. 16365
MAL	Malco Mfg. Co. Inc., Chicago, Ill. 60650
MAS	Masterite Industries, Torrance, Calif. 90505
MET	Methode Elex Inc, Chicago, Ill. 60656
MOLE	Molex Products Co., Downers Grove, Ill. 60515
	G. T. Schjeldahl, Box 170, Northfield, Minn. 55057
	Specialty Fastener, Englewood, N.J. 07631
SYL	Sylvania Electric Products Inc., Parts Div., P.O. Box 129, Warren, Pa. 16365
	Transitron Elex Corp., Wakefield, Mass. 01880
	The Ucinite Co., Newtonville, Mass. 02160
USC	U.S. Components, Bronx, N.Y. 10462
	Viking Elex Inc., Hudson, Wis. 54016
VIK	Viking Industries Inc., Chatsworth, Calif. 91311
WIN	Winchester Elex Inc., Div. of Litton Precision Products Inc, Oakville, Conn. 06779

Standard contact plating (refer to Table 31)†	Type of connector mounting (refer to Table 32)	Dielectric material (refer to Table 33)	Dielectric with-standing voltage sea level, rms	Dielectric with-standing voltage 70,000 ft, rms	Nom-inal cur-rent rating, A	Military Specification for vibration and shock‡	Max. ambient temp. for use, C°	Manufacturer	
								Catalog series number	Name and location (refer to Table 35)
Au	CH, FB,	SDG	...	...	...	5400	400	...	VIK
Cu, Au, Ni	CH, TI	SDG-F	1,400	450	1.0	C-21097	150	PCM	WIN
Ni, Au	CH, TI	SDG-F	1,400	440	3.0	C-21097	150	HW	WIN
Cu, Au	CH, FB, TI	SDG-F	2,100	580	7.5	C-21097	150	HB	WIN
Cu, Au	CH, FB, TI	SDG-F	1,500	540	7.5	C-21097	150	HBD	WIN
Sn, Ni, Au	CH	MFG	2,400	750	5.0	C-21097	125	8B	WIN
Ni, Au	CH	MDG	2,500	600	5.0	NAS713	150	W	WIN

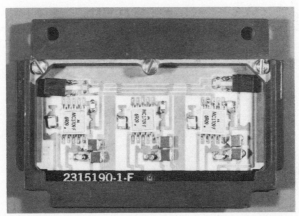

Fig. 77 NAFI module with ceramic substrate mounting package integrated-circuit devices. (*Heavy Military Electronic Systems Department, General Electric Company.*)

8. Air-dry the ceramic layer on the glass bar; then strip it off. The organic *release agent* present will burn off afterward in firing. NOTE: In the force-dried process, heated air is usually used.

9. Lay the plasticized ceramic tape between papers and smooth it out.

10. Inspect the tape for pinholes, surface protrusions, etc., over a *light table*.

At this point there is a bifurcation. If the tape is thick enough to be used for substrates as is, the next step entails slitting to width and punching it to size. A blanking die is used to punch them out separately, including their holes, for storing in the green state. The wafers are held in readiness for later possible machining and high-temperature firing, followed by the usual cleaning.

Alternatively, if the tape thickness was held close to, say 5 mils, the next process step calls for laminating the stored cut or die-punched thin wafers in a laminating press to obtain a monolithic substrate of the desired thickness in the green state. The usual high-temperature firing follows, along with any supplemental surface preparation and/or metallizing, as well as tumbling and cleaning. In automated handling, the tape is transferred onto a continuous belt and through a drying oven until it assumes a pastrylike, plastic characteristic. Then it is rolled up in its green state for later stamping.

Note that one side of the tape cast substrates, the one in contact with the glass at casting, has a very smooth surface texture, as-fired finishes of 4 to 5 in. being somewhat common.

Further, note that here, too, allowances are made in the prefired state for subsequent shrinkage of the vitrified ceramic to finished dimensions.

Isostatic forming In this cold-forming process, the dry powder mix is loaded into heavy rubber bags serving as flexible forming dies, which may or may not contain one or more internally placed mandrels. The rubber bag is lowered into a water-filled well on the shop floor which acts as a pressure vessel.

Hydrostatic pressure of 12,000 psi or more is applied to the water, which transmits the pressure evenly and compacts the encased ceramic powder uniformly into a chalky, cylindrical body. This is sliced into flat substrates. These are fired, resulting in hard, dense substrates with a truly homogeneous microstructure.

Hot pressing The hot-pressing procedure produces alpha-alumina from the original forms in situ. Primarily used not so much for substrates, this process will yield ultrafine microstructured, polycrystalline compacts of practically pore-free high-density structure. The hot-pressing procedure involves application of high-wattage electric energy in the form of heat against a hot-pressing graphite die assembly mounted in a hydraulic press. Inside the die's mold cavity, powder material has been compacted previously. During hot pressing, the mold is surrounded by powdered magnesia Fiberfrax* (Al-Si fiber) for thermal insulation. Uniform temperatures are obtained over the full length of the specimen.

Hot-pressing temperatures of 1500°C and higher are common. In effect, compacting and sintering are often combined to save time. In normal sintering, which follows the above three cold-forming processes, some degree of porosity is a way of life, and densification is accompanied by undesirable grain growth in the ceramic. In the hot-pressing process, grain growth in the microstructure of the ceramic is not a necessary condition for densification to take place. Increasing the bulk density by hot pressing affords a means of increasing not only the mechanical strength of the product but also its thermal-shock resistance, resistance to fracture due to residual strains, and other useful characteristics. Upon cooling and removal of the specimen, it can be sliced and shaped, using wafering techniques.

The translucent alumina material Lucalox† is also made by hot pressing. Special precautions are taken to offset contamination of the product's outer surfaces from contact with the mold.

A variety of ceramic substrates are shown in Fig. 78. The electrical and physical characteristics of two typical alumina formulations and a typical beryllia formulation are shown in Table 36.

Surface finish Ceramic surface finishes are either natural or applied. Natural finished are termed *as-fired* which means that no smoothing operations are performed on the ceramic after it is sintered. Applied finishes are those obtained by grinding, lapping, polishing, tumbling, or glazing as-fired ceramics.

To a large extent the surface finish possible on a given ceramic part is determined by its size, shape, method of manufacture, and composition. As-fired surface finishes on electronic-circuit substrates made from one Coors alumina ceramic, for example, are under 10 µin. As-fired finishes on other parts made from other Coors alumina ceramics range from 15 to 65 µin. A finish of less than 1 µin. can be obtained on several Coors alumina ceramics by lapping, 3 to 10 µin. being the normal surface-finish range for other lapped bodies. Glazing provides very smooth surfaces, finishes of 1 µin. or better being common.

Figure 79 shows the comparison of the surface-analyzer traces of as-fired, ground, and polished surfaces.

---

* Trademark of Carborundum Co.
† Trademark of General Electric Company.

Electrical properties   The most important factors which make ceramic so desirable for electronic applications are its electrical properties of high resistivity, high dielectric strength, and low loss factor, as shown in Table 37.   Alumina and beryllia ceramics are readily metallized, making them ideal for ceramic-to-metal assemblies, components packages, and circuit substrates.

Alumina ceramics are excellent electrical insulators with volume resistivities greater than 10^{14} even at 300°C.   Dielectric constants of alumina ceramic vary from 8 to 10 at room temperature.   Beryllia ceramics are in the 6.5 to 7 range, depending on purity and composition.

It is generally desirable to keep the dielectric constant of ceramic materials used

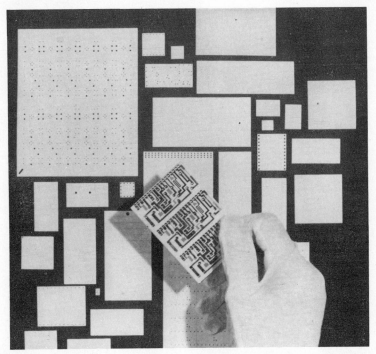

Fig. 78   A variety of ceramic substrates for thick and thin film applications.   Sizes range from 4 by 4 in. down to ¼ by ¼ in.   (*Coors Porcelain.*)

in signal transmission as low as possible.   This minimizes signal-transmission delay since

$$\frac{t_D}{l} = \frac{\sqrt{\epsilon_r}}{c} \qquad (10)$$

where t_D = delay
  ϵ_r = dielectric constant
  l = length of signal path
  c = speed of light

Table 37 lists the propagation delay in nanoseconds (10^{-9} s) per inch for selected materials.

Multilayer Substrates

The requirements of modern electronics circuitry for higher circuit density, higher operating frequencies, and low signal losses has provided the necessary development

TABLE 36. Properties of Ceramics

Property	Test	ADS-96F nominally 96% Al_2O_3 for standard thick film, circuits	ADS-995 nominally 99.5% Al_2O_3 for precision thick film circuits, coarse thin film circuits	BO-995-2 nominally 99.5% BeO
Specific gravity	ASTM C 20-46	3.75	3.90	2.90
Water absorption	ASTM C 373-56	None	None	None
Permeability	...	Gastight	Gastight	Gastight
Flexural strength (typical, 70°F), psi	ACMA TEST #3	60,000	68,000	40,000
Compressive strength (typical, 70°), psi	ASTM C 528-63T	>300,000	>400,000	310,000
Hardness (typical), Rockwell 45N	ASTM E 18-67	79	84	67
KNOOP 1 kg	...	1,200	1,400	
Color	...	White	White	White
Surface finish, as-fired, μin. (CLA)	Profilometer (0.030 in. cutoff normally used)	<25	<10	<25
Maximum temperature for no-load conditions, °C	...	1700	1750	1850
°F	...	3100	3180	3360
Thermal conductivity, (cal)(cm²)/(s)(cm)(°C):				
20°C	...	0.063	0.075	0.67
100°C	...	0.048	0.065	0.48
400°C	...	0.029	0.028	0.20
Thermal coefficient of expansion, per °C:				
25–200°C	ASTM C 372-56	5.9×10^{-6}	6.0×10^{-6}	2.4×10^{-6}
25–500°C		7.1×10^{-6}	7.3×10^{-6}	7.7×10^{-6}
25–800°C		7.8×10^{-6}	7.9×10^{-6}	8.5×10^{-6}
25–1000°C		8.1×10^{-6}	...	8.9×10^{-6}
Volume resistivity, Ω-cm:				
25°C	ASTM D 1829-66	$>10^{14}$	$>10^{14}$	$>10^{17}$
300°C		2.0×10^{12}	$>10^{13}$	$>10^{15}$
700°C		1.6×10^{7}	1.0×10^{9}	1.5×10^{10}
TE value (temperature at which resistivity is average 1 MΩ-cm), °C	ASTM D 1829-66	925	1100	1240
Dielectric strength, V/mil (60-c, ac, average rms):				
0.250 in. thick	ASTM D 116–65	225	225	260
0.025 in. thick		500	550	610
Dielectric constant at 25°C (relative permittivity):				
1 kHz	ASTM D 150-65T	9.7	10.3	6.7
1 MHz		9.7	10.3	6.6
100 MHz		9.7	10.3	6.6
Loss tangent, (dissipation factor):				
1 kHz	ASTM D 150-65T	0.0010	0.0002	0.001
1 MHz		0.0004	0.0001	0.0002
100 MHz		0.0004	0.0001	0.0002
Loss factor:				
1 kHz	ASTM D 150-65T	0.009	0.002	0.007
1 MHz		0.004	0.001	0.001
100 MHz		0.004	0.001	0.001

SOURCE: Coors Porcelain Co.

impetus for multilayer ceramic substrates. The earliest two-layer thick film screen-printed ceramic substrates were made by using both sides of the substrate for conductive patterns and interconnecting them with conductive vias through the substrate. In the IBM SLT* module (Fig. 80) metal pins are used to interconnect the two sides.

Screen-printed multilayer substrates While this is an effective means of packaging and interconnection, it has several shortcomings when compared to a "true" multilayer thick film circuit with two distinct levels of conductors separated by a layer of insulating dielectric. Early techniques simply used flying wires to cross over a conductor. Early attempts at screen-printing conductors first and then firing, and

* Solid Logic Technology, IBM Corporation.

TABLE 37. Propagation Delay for Selected Materials

	Dielectric constant ϵ_r	t_D/l, ns/in.
Alumina, 99.9%	10.1	0.270
85%	8.2	0.242
Spinel	8.5	0.246
Beryllia, 99.5%	6.8	0.220
Pyroceram*	5.6	0.200
Borosilicate glass	4.1	0.170
Silicone-glass laminate	4.0	0.169
Fuxes silicon glass	3.8	0.165
Kapton†	3.5	0.159
Mylar†	3.25	0.152
Parylene	3.00	0.142
Polyphenylene oxide (PPO)‡	2.55	0.135

* Trademark of Corning Glass Works.
† Trademark of E. I. du Pont de Nemours & Co.
‡ Trademark of General Electric Company.

then printing and firing a dielectric material, followed by a subsequent screening and firing of another conductor layer, produced some of the first multilayer ceramic substrates. The typical material had a dielectric constant of 7 to 8 and a loss tangent of 0.006. Interconnection between levels was made by leaving windows in the dielectric layer through which the subsequent conductor layer was screened. These efforts, though crude, were satisfactory for some applications, as shown in Fig. 81. However, as larger areas of crossover were used, it was found that during subsequent firings the dielectric glass would soften and permit lateral movement of the upper-level conductors. Since the top conductors were also used for mounting pads for ribbon-leaded devices, the misalignment of pads posed a serious bonding problem.

A new type of crystallizable glass dielectric was created to overcome this difficulty. Du Point DP8299 represents this type of dielectric; it exists initially as a vitreous material which is converted into a mixture of vitreous and crystalline phases during firing. The crystalline phase is distributed throughout the mixture and prevents resoftening of the dielectric during subsequent firing at higher temperatures than the original 850°C. Large (up to 4 by 6 in.) ceramic substrates can be fabricated using this technique. Various conductive and resistive pastes can be used to provide a printed circuit on the substrate.

The screen printing process can also be used to make multilayer substrates for

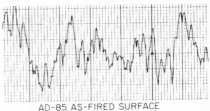

AD-85 AS-FIRED SURFACE
FINISH: Approximately 55 μin. AA

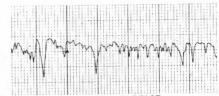

AD-85 GROUND SURFACE
FINISH: Approximately 25 μin. AA

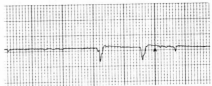

AD-85 POLISHED SURFACE
FINISH: Approximately 10 μin. AA

Fig. 79 Surface analyzer traces of typical as-fired, ground, and polished surfaces. (*Coors Porcelain.*)

Fig. 80   IBM Solid Logic Technology module.   (*IBM.*)

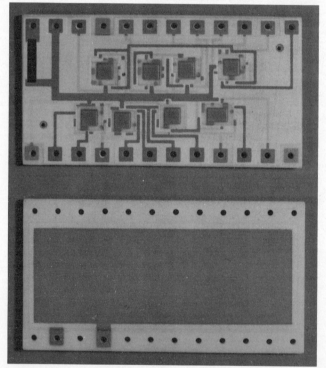

Fig. 81   Screen-printed multilayer substrate.   (*General Electric Company.*)

mounting integrated-circuit chip devices.   In cases where very fine definition lines are needed (2-mil lines and 2-mil spaces) etched metal masks can be used in place of the more conventional screens, which at the very best produce 4-mil lines and 4-mil spaces.

Metal-mask screens   Two types of metal screens are available, indirect and direct. The indirect metal mask is formed by bonding a pre-etched foil to a stainless-steel mesh which has been assembled to a screen frame.   Usually the foil thickness is 0.001 in., although some applications have required 0.002 in.   The advantage of the metal mask over direct emulsion is close control of the paste deposited and resistance to pattern breakdown. It also is capable of reproducing intricate patterns with well-defined lines. Fine-line printing requires metal masks.

Tolerances with indirect metal masks can be held to the diameter of two wires of the mesh. This is required because the process is indirect, whereby contact of the mask to the mesh results in an immediate bond. Consequently, one cannot have the latitude in lining up the pattern under the microscope that is possible with a direct emulsion process.

Deposition is controlled because the etched foil is of a specific thickness and wire diameter can be easily predetermined. This means more repeatable ink deposit is possible than with direct emulsion screens.

Pattern breakdown is forestalled, in that an image formed by metal will withstand abrasion and the organic solvents of the pastes far better than one formed with emulsion. It should be remembered, however, that in applications where overstressing of the wire is the determining factor of screen life, this will not be improved with indirect metal masks.

Direct metal masks exhibit all the qualities of indirect masks and offer several more. The direct process involves etching the mesh and the pattern from opposite sides of a metal foil. These can be perfectly registered before etching. Mesh interference is far less of a problem than with an indirect approach. If a grid with a standard number of lines per inch is used, direct masks can be fabricated reasonably quickly and at a moderate price. If, however, a pattern is so intricate, complex, or critical that a stock grid will offer intolerable interference, a special grid may be required. This has been done successfully but obviously cannot be offered as inexpensively as a stock grid register.

An important advantage the direct mask affords over all other types of printing screens is potentially longer screen life, because a solid piece of metal supports the image and there is far less stretching than with conventional screening material. Since the largest single cause of deterioration is overstressing of the wire, the direct mask has a definite advantage. It should also be emphasized here that if they are properly fastened and tensioned to the frame, direct metal masks can be successfully printed with conventional off-contact screening equipment and techniques. The mask will deflect and snap back if properly used.

DESIGN AUTOMATION

The simplest version of the artwork process in the thick film technique starts at the drawing board, with the designer converting the schematic to a layout at some scale ranging from 10× to 500×. The designer converts a resistor, for example, into a physical model with a given length and width to achieve the required resistance values. These resistors, depending on their size and the resistance value of the thick film paste screened onto a substrate, take on the approximate given value. The designer works on a grid (printed on a stable material) which defines his substrate (circuit-board) configuration. He can then make paper models of his required elements and move them around his board area until he finally has a satisfactory layout. He then proceeds to satisfy his schematic requirements by completing the interconnection between components and the outside world. After the physical drawing is complete, he inks or tapes over his circuit pattern to provide a photo image for the camera which is carefully reduced to the proper scale. With its greater and more consistent density a taped drawing provides a better camera image but still provides poor line definition and, when subjected to heat and humidity, tends to curl and distort the stable medium on which the tape is affixed.

When individual masks are required for each layer of paste, i.e., resistor, conductor, insulation, etc., registration of the various layers of artwork masks can be a problem. Photographic reversals, i.e., negative or positive images, are made available for transfer to the production screens. Even with the best available controls, line definitions are relatively poor, and component sizes lack the precision required for most of today's processes. The flow chart to the artwork mask is shown in Fig. 82.

Several other approaches for arriving at a production screen can be taken to provide better line definition and tolerances. For these applications, peel coat and Scribe coat° materials are used.   The material is a laminate on a stable base, and the outlines of devices are scribed or phototransferred to the material, whereupon device topology is peeled out to produce a photonegative image of the production mask (similar to the image in Fig. 83).   The flow chart is shown in Fig. 84.

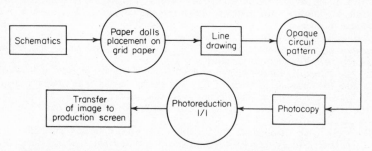

Fig. 82   Artwork flow chart for thick film mask by manual layout.

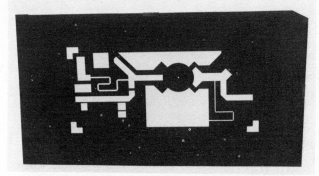

Fig. 83   Photonegative image of production mask.

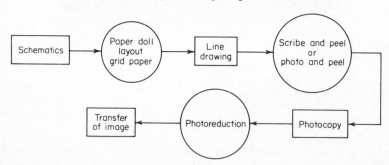

Fig. 84   Flow chart for artwork produced by manual scribe and peel techniques.

When close tolerances are required, such as strip-line design, precision results can be obtained by scribing Rubylith† (Scribe coat) and using the coordinatograph (a precision table with tolerances to ±0.0005 in.) as the scribing device (Fig. 85).   Use of the coordinatograph, a diamond stylus, and rubylith produces an excellent mask, and

° Trademark of Keuffel & Esser Co.
† Trademark Ulano Products Co., Inc.

precision control of cameras and developers results in excellent masks. The flow shown in Fig. 86 then appears as one of the better solutions to the production of artwork masks.

The time to process material by the coordinatograph may be too slow for heavy production loads, and computer techniques can then be applied. One of the interim steps to get quick artwork is to define circuit shapes to the computer and then draw them out as outlines on the Cal Comp plotter. This information is then transferred

Fig. 85   Coordinatograph machine for precision cutting of artwork masters.

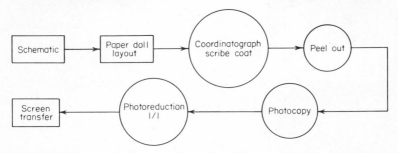

Fig. 86   Flow chart for artwork produced by coordinatograph machine.

to peel coat, and masks are made as before. Although the method is fast, it has several deficiencies. Since line definitions are poor for phototransfer, and since peel coat requires close control to avoid growth and give fair definitions, this approach is used only for laboratory models. The flow for the Cal Comp is shown in Fig. 87.

When numerically controlled precision drafting tables are available, people can handle the load going into the coordinatograph. A great deal of work has gone into the development of routines which allow these tables to be used effectively. Their accuracies run from ±0.010 to ±0.0005 in. Plotters such as the Gerber (Fig. 88) are used to scribe rubylith and with the development of the optical exposure head are able to expose images directly onto film. This approach solves the capacity problem and the use of 20× or more in cutting or photo work solves the accuracy problem. The development of special apertures also handles the precision requests and allows some mask work to be performed at 1:1.

Other devices, such as the David Mann pattern generator, are able to add step-and-repeat capabilities and through photo means get significantly more precise high-density images for large- and medium-scale integration masks (Fig. 89). Gyrex also produces a photo-head machine capable of producing highly accurate masks, much as

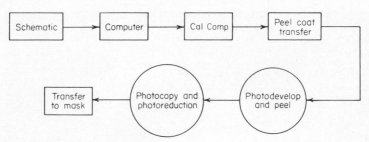

Fig. 87   Flow chart for artwork produced by Cal Comp plotter.

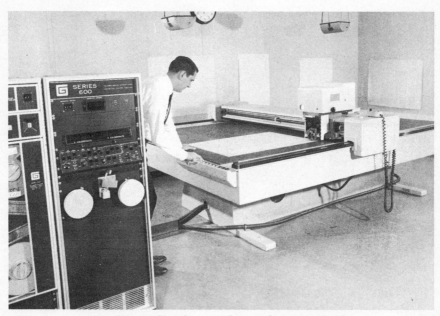

Fig. 88   Gerber plotter with optical exposure head.

the Gerber machines do, but has the same size limitations to contend with. These processes use as their input a type of digitized layout, and although we now have a few more steps in the process, the time cycles to artwork are appreciably reduced. The flow generally appears as in Fig. 90.

David Mann or Gyrex processing looks much the same, but in the Mann generator and the Mann step and repeat, we now have precision control for step and repeat to handle a great many circuit patterns for production masks. The coordinatograph and the Gerber plotters continue to handle the majority of thick film and thin film artworks. The Gerber also furnishes the capability for some weird and exotic shapes in microstripline, shapes that are virtually impossible to cut in rubylith on the coordinatograph. The optical exposure head is the best available tool for the majority of designs and also allows people using special software techniques the option of producing 1:1 artwork on the Gerber, thus eliminating the photoreduction step and providing a master pattern that can be easily updated without starting back at ground zero, i.e., adding a pull-up thick film resistor to a circuit. The 15-min update routine is a real boon to experimental and pilot-plant work.

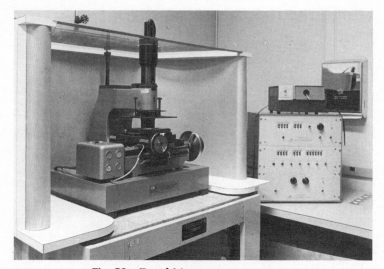

Fig. 89   David Mann pattern generator.

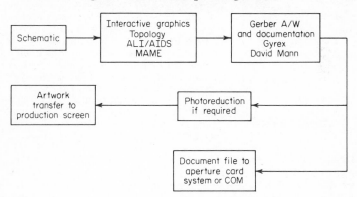

Fig. 90   Flow chart for artwork produced by the David Mann or Gyrex pattern generators.

During these periods of evolutionary artwork development it was apparent that as densities went up, the initial layout required too much time; several automated approaches were taken using simplified models to arrive at a topological layout, a use of standard library cells and sophisticated interconnection algorithms, such as AIDS*, and finally the use of interactive graphics as a layout tool. Interactive graphics now exist as the best technique covering both worlds, combining the layout process

* Advanced Interconnection Design System.

with the advent of artwork and documentation drivers for numerically controlled plotters.

We will show in capsule form what these computer-aided designs are and how they affect the artwork mask process. First there is topology, tuned mostly to the thick and thin film processes. It requires a simple mechanical model of circuit elements that embody design rules; i.e., no conductors run through the body of a thick film resistor (as shown in Fig. 91). Coupled with a NODAL equation description of the schematic and a definition of a desired substrate size, the computer program places the elements and proceeds to interconnect them in accordance with the schematic with the specific intent of minimizing crossovers. The substrate is allowed to grow, and after all elements are placed, the computer proceeds to modify the arrangements and shrink the land area to the desired size. The program produces very satisfactory paper layouts in a quick-turnaround mode and is extremely simple to use. The designer then transfers this output to line drawings for the layers and trims up or optimizes his layout. Artwork is then processed by the Gerber. Although modern versions of the topology program provide Cal Comp drawings, tie locations of exit wires down to a specific XY location, and even automatically produces artwork, the costs of these versions is higher; since, more importantly, the designer should have the opportunity to interact with the design, interactive graphics becomes a natural choice.

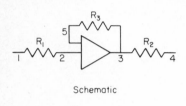

Schematic

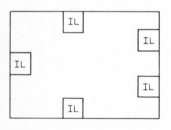

Model

Substrate

Fig. 91 Mechanical model of circuit elements.

Topology gives only a single-layer view of a substrate, and computer printouts, although fine, do not give all we want. When for example, we want to put 17 to 20 chips on an area of 0.8 by 1.2 in., it becomes obvious that more sophistication is needed. We have to accommodate many layers for interconnection and the required feedthroughs. We also have to define the pin arrangements to enter and exit to the outside world. AIDS (Advanced Interconnection Design System) is a computer program which allows us to define the schematic or logic diagram, provides for a precise definition of the chips, allows absolute location assignments to the outside world, allows a variable screen grid (325,400, etc.) for production, and provides the engineering analysis (i.e., completeness, etc.) of our circuits. Some samples of the artwork and computer printouts are shown in Figs. 92 and 93. The computer places our chips for us, interconnects them, and gives the layout very rapidly. Occasionally it gives us a nonoptimum layout, i.e., too many layers, element could have been better placed here, etc., or fails to connect 10 out of 250 wires, and this requires interaction by the designer. Again the desirability of interaction of the designer is a key factor. The program also gives us the opportunity of accessing a library of elements (similar to MACE and MAME programs, integrated-circuit design programs) and thus reduces modeling time. As input coding becomes difficult, a simple input-coding language and error-checking routines are used to further increase operating speed. The flows for integrated-circuit mask design are shown in Fig. 94 and allow turnaround time in the schedule in case operations have to be repeated because of inability to interact at the proper moment with the design.

Interactive Graphics

Interactive graphics uses a minicomputer tied to a cathode-ray tube and input/output devices such as a Rand tablet and stylus or light pen, function keyboard, teletype, plotter, digitizer, or other types of device which allow the user to input and output his results rapidly, see them graphically displayed, and then modify his data to update or modify what he sees (Fig. 95). In effect, through interactive graphics we have provided the designers with appropriate software, library devices, a pseudo drawing tablet and a high-speed electronic pencil. Appropriate software and instant hard-copy devices allow the engineer, the designer, the illustrator, factory planner, or others to determine, through interactive interplay with the computer, that they have now reached their desired result and thus produce hard copy, a MIL Standard drawing on a numerically controlled plotter, a process sheet, or even a maintenance document manual.

The design process for integrated circuits now goes something like this. In its simplest form a designer creates his own design library on line. In effect he takes his electric pencil,

Fig. 93 Computer printouts.

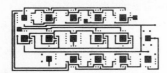

Fig. 92 Computer-generated artwork.

touches two points on the Rand tablet or cathode-ray tube screen, and thus defines a resistor model. An operational amplifier would require twelve taps of the pencil, two for the major box defining the chip area and two each for the five pad areas. Reducing these library elements to a convenient size on the side of the screen, he then taps his pencil twice for the land area of his substrate. Touching his library elements with the pencil he now places them on the substrate (this does

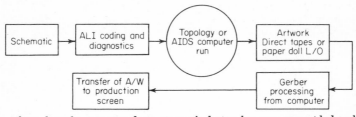

Fig. 94 Flow chart for integrated-circuit mask design by computer-aided techniques.

not destroy the library, which can be used over and over) and then interconnects the elements by tapping his pencil for each break point in the conductor. The designer may operate one layer or up to seventeen layers simultaneously or in groups of his choosing. Thus layouts can exist as layer artwork and as a composite on separate layers simultaneously. When he has satisfied the schematic, the designer pushes the button for hard copy and gets an assembly diagram, artwork, etc., delivered to him.

If the design is complex, he accesses the library models, calls on the topology program, and has displayed a paper-doll line drawing on the screen. He now can

Fig. 95 Interactive graphics machine. (*Applicon.*)

move elements as he sees fit to optimize his circuit, separate his mask layers by tapping his pencil, and expand to artwork and documentation.

Interactive Graphics, which is the natural front end to numerically controlled plotters and artwork generation extends into normal printed-circuit design, schematics, technical publications, maintenance manuals, computer-aided circuit analysis, data management, data retrieval, and even into human factor simulation and on-line programming.

APPLICATIONS OF THICK FILM HYBRID CIRCUITS

The use of thick film hybrid technology in a variety of military and space applications over the past 10 years has stimulated a much more rapid development of the technology than if it had been left to the commercial market to develop. The initial requirements for a highly sophisticated technology were generated by these applications as a result of their need for compact, lightweight, and reliable equipment. The following examples of thick film hybrid technology are representative of the various types of hybrid packaging used in the industry.

Figure 96 shows a variety of thick film circuits, most of which were designed and built at the General Electric Hybrid Integrated Circuits Facility of the Heavy Military Electronics Systems Department, Syracuse, N.Y. The upper module is a power amplifier circuit used in a bridge circuit to provide a 4-kW pulsed output to a sonar transducer. The circuit is constructed using a beryllium oxide substrate with the active diodes and power transistors hermetically sealed after assembly and test. Large ribbon leads are provided for the high current, and the substrate is directly mounted to a water-cooled heat sink to assure low transistor-junction temperatures.

The large custom dual-in-line in the next row is an active filter consisting of three lowpass stages which are functionally trimmed to provide accurate frequency breaks. The trimming of the resistors is done by an automatic air-abrasive machine. After testing, the unit is sealed with a glass-sealed metal cover.

The TO-8 package on the left is a data norm circuit built with conventional thick film techniques using two stacked alumina substrates. The chips are interconnected

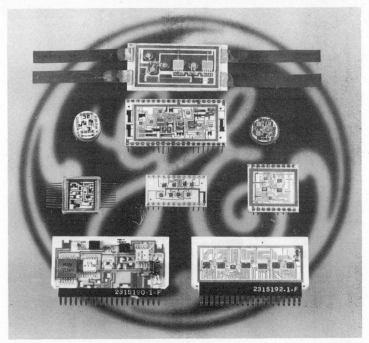

Fig. 96 Examples of thick film hybrid packaged modules. (*Heavy Military Electronic Systems Department, General Electric Company.*)

by chip and wire techniques, and the unit is sealed with a cap weld after test. The TO-8 package on the right is a phased array phase shifter driver circuit providing a 5-A pulsed output to the ferrite core. It is hermetically sealed with a cap weld after test.

The flat-pack hybrid package in the next row consists of a thick film alumina substrate with active devices interconnected with thermocompression-bonded gold wire.

The dual-in-line configuration to the right is a logic adder module made at the General Electric Corporate Research and Development Center. It consists of an alumina substrate on which are sequentially screened and fired three conductor layers and two dielectric layers plus a resistor pattern. The rear of the substrate has screened on it a large glass-encapsulated capacitor. A total of 8 integrated circuits are mounted and wire-bonded to the substrate. The module is then completed by soldering the pins and encapsulating the module.

The next module is a 4-bit adder-subtracter module fabricated on a 1 by 1 in. alumina substrate with an integral glass seal for final hermetic sealing. The package leads are soldered in place after component assembly.

The two bottom modules are standard hardware-program modules. The one on the left, a preamplifier for a sonar transducer, utilizes packaged integrated circuits soldered to the thick film alumina substrate. Therefore no external package seal is required. The module on the right is a standard-hardware-program AD-2 module with integrated circuit chips bonded to the thick film gold conductor on the alumina substrate. The chips are interconnected by thermocompression bonding, and the final seal is provided by the glass seal ring on the substrate bonded to a Kovar cover.

Standard-hardware-program beam-lead module The module shown in Fig. 97 consists of a thick film dielectric-conductor multilayer on an alumina substrate mounting beam-leaded chips bonded to gold bonding pads. Interconnections are provided by gold thick film conductors and gold-filled vias between layers. The finished module is sealed by soldering a cover to the sealing ring around the periphery of the substrate.

PEPE module   The engineer in Fig. 98 is holding a Honeywell parallel-element processing ensemble (PEPE) processing element. The PEPE system uses up to 1,000 of these elements. The closeup view in Fig. 99 shows the beam-leaded large-scale-integration devices bonded to a multilayer ceramic substrate. The assembly shown contains the electronics normally housed in large wire-wrapped printed-wiring-board assemblies. A major size reduction is possible with an increase in operational capability.

General Electric hybridized regulator   The use of thick film hybrid techniques allowed a 50 percent reduction in board size for the custom amplitude-regulator chassis (shown in Fig. 100) for a sonar transmitter. The hermetically sealed flat-

Fig. 97   Standard-hardware-program module with beam-leaded devices. (*General Electric Company.*)

pack package in the lower left corner of the hybrid version houses a large portion of the transistorized circuitry.

Turret control hybrid amplifiers   The development of thick film amplifiers has made it possible to incorporate these units in ruggedized military equipment, as shown in Fig. 101. As new functions are available, they are used along with standard power transistors. Special care is taken to provide conductive heat paths to the metal housing to provide optimum device cooling.

Bunker Ramo BR-1018 computer   The unit is modularly constructed using a three-dimensional planar coaxial micropackaging technique. The construction techniques are shown in Fig. 102a. The CPU module shown consists of six stacked wafers, each mounting up to nine large-scale-integration chips, which contain a total of 10,000 gates. The technique permits extremely dense packaging with either leaded flat packages or beam-leaded devices mounted on the component wafers. Heat transfer is provided by specially prepared Be-Cu heat sinks interleaved with the circuit wafers, providing good thermal paths throughout the package. All connections between layers are coaxial and use high-pressure contacts. The resultant system has controlled impedance throughout.

Hybrid logic module   The logic module shown in Fig. 102b is part of a commercial computer. The substrate consists of a base alumina with two gold conductor-pattern layers separated by a 1.5 mill dielectric layer. The 5-mil conductors are

Fig. 98 Honeywell PEPE compared to conventional digital integrated-circuit packaging. (*Honeywell.*)

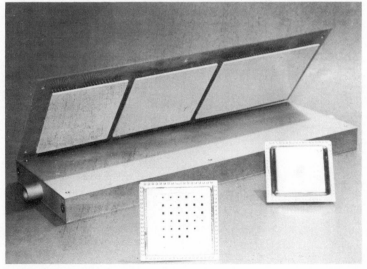

Fig. 99 PEPE module with beam-leaded large-scale-integration devices bonded to multilayer ceramic substrate. (*Honeywell.*)

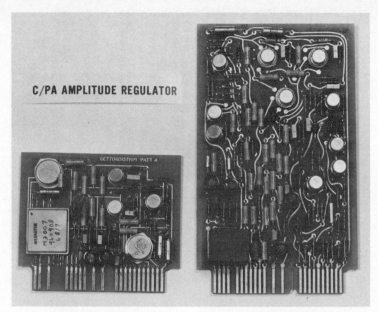

Fig. 100 Thick film hybridized regulator module showing 50 percent area reduction. (*Heavy Military Electronic Systems Department, General Electric Company.*)

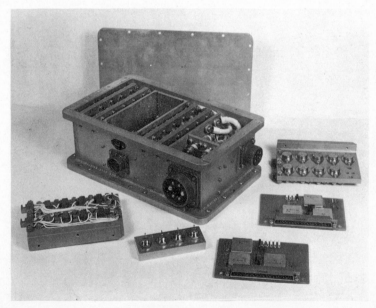

Fig. 101 Turret control unit with hybrid amplifier modules. (*Armament Department, General Electric Company.*)

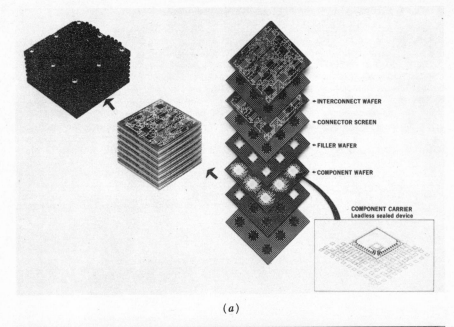

(a)

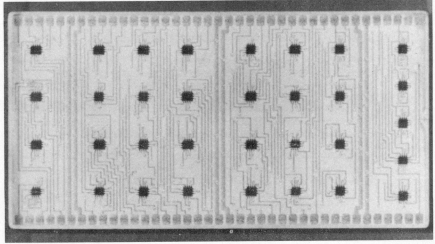

(b)

Fig. 102 (*a*) Planar coaxial micropackaging system used to construct BR 1018 computer. (*Bunker Ramo.*) (*b*) Thick film hybrid module with beam-lead chips mounted on multilayer substrate. (*Corporate Research and Development, General Electric Company.*)

laid out on a 15-mil grid. The grid is reduced to 4-mil lines on 10-mil centers for the beam-lead bonding pads. A total of 33 beam-leaded T²L integrated circuits are mounted. There are over four hundred 12-mil square vias in the dielectric layer. Six ceramic capacitors are also bonded to the substrate metallization. The process has the advantage of low tooling costs and low-cost production.

FUTURE TRENDS IN THICK FILM HYBRID PACKAGING

The growth of thick film hybrids has been stimulated by a variety of factors. The relatively low cost of the initial equipment investment and subsequent lower manufacturing costs gave thick film hybrids their initial impetus. As the economic factors encouraged its growth, users found that thick film technology offered many circuit and packaging benefits as well as manufacturing advantages. The ability to adjust component values after assembly and to provide efficient heat transfer through substrates, while at the same time providing an integral hermetic seal, using glass or solder screened on sealing rings, are only a few of these advantages.

As we look into the future, we can see several areas of importance to the continued growth of hybrid circuits (Table 38). The first of these is the increase in the varieties of integrated-circuit chips and packages being offered in the industry. Many

(c)

Fig. 102   (c) Thick film hybrid assembly using solder-bumped chips reflow-soldered to substrate metallization. (*Heavy Military Electronic Systems Department, General Electric Company.*)

new types of beam-leaded and bumped chips are becoming available which give the designer more flexibility in his substrate design and layout. The circuit shown in Fig. 102c, is fabricated using solder-bumped chips reflow-soldered to a solder-plated bonding pad. The chips can be located to the substrate pads, and all bonds are made simultaneously.

Another area that is developing rapidly is multilayer ceramic substrates. The screened-on variety has the advantage of low tooling costs and low-cost production when small quantities (under 10,000) are required. The ability for users to get quick turnaround on low-cost ($20 to $30) screens makes this technique attractive. However, the disadvantages of registration difficulties, larger required substrate areas due to via designs, and general nonplanarity of the top substrate surface should be considered. The monolithic ceramic multilayer substrate overcomes all these disadvantages but requires a much higher initial tooling investment ($7,000 to $10,000), and normally only a higher required production rate can justify its use. The resulting substrates can have better line resolution, higher circuit density, and excellent thermal-shock properties. Both techniques will continue to grow in use, and this will effect further cost reductions. New screenable ceramics and protective coatings for metallizations will find increased applications.

The increased popularity of plastic packages is due to continued testing and evaluation results which indicate their adequacy for many commercial and industrial applications. The new edge-mount package (Fig. 103) was designed to permit easier assemblies and replacement of the larger large-scale-integration packages.

TABLE 38. Important Thick Film Hybrid Growth Areas

1. Availability of new chip devices in beam-lead and bumped-chip format
2. Ceramic multilayer techniques
   a. Improved screening materials with better line definition
   b. Screenable ceramic dielectrics
3. New packages
   a. Plastic
   b. Edge mount
4. New substrate materials
   a. Flexible plastic
   b. Copper on ceramic
   c. Beryllia ceramic
5. Improved thermal design and implementation
   a. Thermal-analysis computer programs
   b. More efficient heat sinks (heat pipes)
6. Improved automation methods for circuit layout and production tools
7. Broader areas of application
   a. Commercial, automotive, and industrial
   b. Microwave, medical

Another trend is shown in Fig. 104. This new pluggable connector mounts leadless substrates with edge-contact pads. Another model is designed to handle top or bottom contacts.

Substrate material trends show the use of beryllia substrates or carriers becoming more common as higher-powered microwave devices are being developed and designed into equipment. Figure 105 shows a microwave power module with beryllia carriers mounting the power transistors inserted into the alumina substrate and bonded-directly to the metal heat sink.

Flexible-lead frames have been developed with polyimide as the flexible plastic base and various metallizations applied to its surface by selective plating or plate and etch techniques. (Figure 106 shows a lead frame for interconnecting a high-frequency power transistor to its circuit. It consists of a 2-mil polyimide carrier with 1-mil thick leads etched to a particular lead pattern. The polyimide is also selectively etched to provide relief areas.

The broader use of thermal-analysis computer programs is enabling more

Fig. 103 Edge-mount packages. (*Metalized Ceramics Corp.*)

optimized packaging techniques to be employed in mounting and interconnecting hybrid packages, both within the package and in the next higher levels of packaging. The ability to measure steady-state and transient heating effects in various types of solid state modules enables the designer to allow proper safety factors in his circuit design and assures higher reliability.

New thermal-management techniques are becoming necessary as device power density continues to rise and as requirements for packaging larger numbers of components in a given volume become necessary. One of the thermal-control devices finding increased use is the heat pipe. Figure 107 shows a schematic drawing of a typical form of heat pipe. It is a simple structure which has very high thermal conductance. A heat source is located at the evaporator end of the heat pipe, and

Fig. 104   Pluggable connector for leadless substrates. (*Amp, Inc.*)

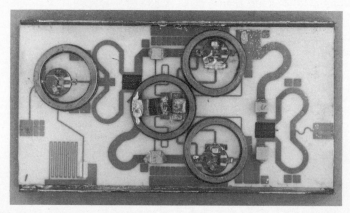

Fig. 105   Microwave power module with beryllia carriers mounting high-frequency power transistors. (*Heavy Military Electronic Systems Department, General Electric Company.*)

the condenser end is usually equipped with fins for air cooling or fastened directly to a fluid-cooled heat sink. It can transfer heat from the evaporator end to the condenser end with an efficiency exceeding 90 percent.

A typical construction would consist of a thin-walled stainless-steel tube containing a wick of porous material, such as stainless-steel mesh, in contact with the inner wall, filled with a fluid, such as water, and sealed at both ends. The choice of fluid depends on the temperature of application, water being a good choice between 50 and 200°C.

As heat is applied to the evaporator end, the water vapor produced is forced by the increasing vapor pressure to the condenser end, where it is cooled by giving up the latent heat of condensation. It then returns to the evaporator through the wick by capillary action.

The wick is essential for zero gravity or where it must work against gravity appli-

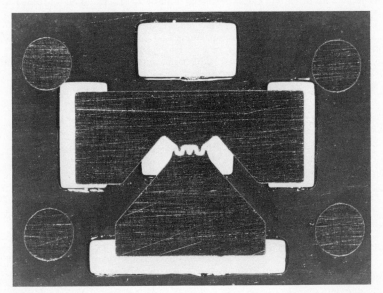

Fig. 106 Gold on polyimide lead frame. (*Buckbee Mears, General Electric Company.*)

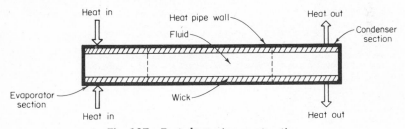

Fig. 107 Basic heat-pipe construction.

cations. It also aids liquid distribution over the evaporator surface. An excellent treatise[20] on heat pipes gives more detailed information

Figure 108 shows a large 6 by 18 by ¼ in. copper heat pipe with a fin structure attached, designed to cool an array of ceramic modules dissipating up to 250 W.

As the use of thick film hybrids continues to increase and the number of high-quantity commercial applications grows, the need for effective application of automated production and assembly methods will be essential. The need for standards within the industry will become more critical as the use of automated processes becomes more extensive within a facility. Cost effectiveness of the finished hybrid module will depend to a large extent on the basic design and process decisions made along the way. The key factor in the success of hybrid circuits has been their great flexibility.   Different technologies can be married to form a function impossible to achieve in any one by itself.   Automation and standardization will be most successful where they are carefully applied to certain large-volume circuits, where the processes and materials are optimized for their use.

Computer-assisted design of hybrid substrate layout and interconnection will continue to be improved and applied more economically to all phases of hybrid design. The newer interactive graphics facilities will be made more efficient and capable of handling more sophisticated design problems.   Many of the trade-off decisions which must be made before and during a system design will be handled by computers acting on information stored and actively fed by the design team.

The impetus behind all these improvements and the steady growth of the hybrid technology is the continuing flow of new microelectronic circuits and new materials for fabricating and packaging them.   The many potential products waiting to be hybridized provide the constant necessary pressure for the continued growth of hybrid microelectronics.

Fig. 108 Flat heat pipe with attached fin structure.   (*Noren Products and General Electric Company.*)

REFERENCES

1. Integrated Circuit Engineering Corp.: "Integrated Circuit Packages and Packaging Techniques," Phoenix, Ariz., 1970.
2. Topfer, M.: "Thick Film Microelectronics," Van Nostrand Reinhold, New York, 1971.
3. Brown, C. E., and H. G. Stech: Hybrid Microelectronic Packages: A Compilation of Various Manufacturers' Types, *Solid State Technol.*, August 1971.
4. Harper, C. A.: "Handbook of Electronic Packaging," McGraw-Hill, New York, 1970.
5. Nixen, D.: *EDN*, 1969.
6. MIL-STD-275 Printed Wiring Standards.
7. Geshner, R. A., and G. Messner: "IPC Multilayer Printed Circuit Boards Technical Manual," Institute of Printed Circuits.
8. McCormick, J.: Rome Air Development Center, Rome, N.Y., Internal Test Report.
9. *Circuits Manuf.*, June 1971, p. 27.
10. Jarosik, N. A., and H. Shapiro: Parallel Gap Welding Techniques and Precautions, in G. Sideris (ed.), "Microelectronic Packaging," McGraw-Hill, New York, 1968.
11. Institute of Printed Circuits, Inc., Connector Availability Committee: *Circuits Manuf.*, December 1970.
12. Sekeley, G.: Alumina Substrates Reliability Criteria, *NEPCON Proc., 1969.*
13. Manko, H. H.:"Solders and Soldering," McGraw-Hill, New York, 1964.
14. Manko, H. H.: Mechanism of Wetting during Solder Joint Formation, *ASTME Pap. Soldering* (STP319), 1963.
15. Coombs, C. F., Jr.: "Printed Circuits Handbook," McGraw-Hill, New York, 1967.
16. Federal Specification QQ-5-571C.

17. Bell Telephone Laboratories: Englewood Cliffs, N.J., "Physical Design of Electronic Systems," vol. 3, Prentice-Hall, 1971.
18. Wegner, Howard W.: "Reliable Mass Soldering Techniques," Hollis Engineering.
19. *Circuits Manuf.*, June 1971.
20. International Research and Development Co. Ltd., Heat Pipes: 1971.
21. Hinkley, J. R.: Resins for Packaging Electronic Assemblies, *Electro-Technol.*, June 1965.
22. Isaacson, H. M.: Flexible Circuitry Provides Many Benefits, *NEPCON Proc.*, *1971.*
23. Electronics Industries Association: *Bull* 13, Microelectronic Package Outlines, 1970.

Glossary of Terms

acceptable quality level (AQL). (1) The lowest quality level a supplier is permitted to present continually for acceptance. (2) The maximum percentage of defects or number of defective parts considered as an acceptable average for a given process or technique.

actinic (light). Radiation which causes chemical changes, e.g., the effect of light on photographic emulsions. Blue and ultraviolet are the most actinic regions of the spectrum.

active devices. Parts of a circuit that are capable of amplification, usually silicon semiconductor devices. Transistors, for instance, are active devices. Components that cannot amplify are passive, e.g., resistors and capacitors.

active element. An element in which input signal energy is converted into output signal energy by interaction with the energy from one or more auxiliary sources.

active trimming. Method of adjusting the circuit function under operating conditions.

adhesion. The ability of a conductor or insulator material to withstand a pull force attempting to separate it from the substrate.

air-abrasive. Method of trimming resistors abrasively using alumina powder propelled by air pressure.

alloy. (1) a solid-state single-phase solution of two or more metals. (2) To melt or make an alloy.

alumina. Aluminum oxide, Al_2O_3; alumina substrates are made of formulations that are primarily alumina.

angle of attack. The angle between the squeegee face of a thick film printer and the plane of the screen.

angstrom unit (Å) Unit of length equivalent to 10^{-10} m or 10^{-8} cm. There are 10,000 Å per micrometer.

anti-halation backing. The coating on the rear surface of a film or glass plate to reduce the reflection of light back into the areas of the emulsion layer which are to be unexposed.

artwork. The original scaled-up image, usually prepared on a film support, which serves as a starting point for a screen or mask.

aspect ratio. The ratio between the length of a film resistor and its width; the number of squares in a thick film resistor.

back bonding. The conventional way of bonding active chips to the substrate. The

1

physical attachment is made using the back of the chip, leaving the face, with its circuitry, up. The opposite of back bonding is face bonding.

back lighting. Illumination of an image by transmitting light through the artwork mounted in the camera copyboard.

ball bond. A bond made with the end of a gold wire on which a ball has been formed by melting the gold wire with a flame.

ball mill. A cylindrical jar filled with ceramic balls that is rotated to mix and grind ceramic powders. Intimate mixing and very fine grinding are possible.

barium titanate (BaTiO₃). The basic raw material used to make high-dielectric-constant ceramic capacitors; used also in high-K thick film ceramic pastes.

beam lead. An active chip that can be face-bonded where the connections between the chip and the substrate are made by beams of plated metal that protrude.

beam-lead chip. A chip employing electrical terminations in the form of tabs extending beyond the edge of the chip for direct bonding to a mounting substrate.

beryllia. Beryllium oxide, BeO; a substrate material used where extremely good heat conductivity is desired. Beryllia substrates are excellent electrical insulators but conduct heat better than most metals.

bifurcated squeegee. A two-part squeegee used in a screen printer so that the thick film composition fluid can be pumped down through the squeegee to provide a continuous supply of fluid. This arrangement is especially useful for a highly pseudoplastic or thixotropic fluid used for high-resolution printing.

bimetal mask. A mask formed by different metals combined by electroforming or cladding. Apertures are selectively etched through one metal to form an image in a second metal.

binders. Materials added to thick film compositions, unfired substrates, etc., to give sufficient strength for prefire handling.

bipolar. Basically, a device in which both majority and minority carriers are present. In connection with integrated circuits this term describes a specific type of construction. Bipolar and MOS (see MOS) are the two most common types of integrated circuit construction.

blistering. The development during firing of enclosed or broken macroscopic vesicles or bubbles in a body or in a glaze or other coating.

block. To plug up an open mesh in a screen.

body. The structural portion of a ceramic particle or the material or mixture from which it is made.

bolting cloth. A lighter grade of high-strength, square-mesh wire cloth used for making stencil screens.

breakaway. The screen-to-substrate separation required during off-contact screen printing; also known as the snap-off distance.

bump contacts (ball contacts, raised pads, pedestals). Contacting pads which rise substantially above the surface level of the chip; also raised pads on the substrate which contact the flat land areas of the chip.

buried layer. A distinguishable region introduced under a semiconductor circuit element, e.g., under the collector region of a transistor to reduce the series collector resistance.

burnishing. Smoothing the conductor materials, after printing and firing, with a fiber-glass brush, eraser, or other abrasive material.

camber. The amount of overall warpage present in a substrate; determined by measuring the difference between actual thickness and the separation required of two parallel plates to enable the substrate to fit.

capacitance density. The amount of capacitance available per unit area in picofarads per square inch in thick film capacitors.

capacitor. A device that can store an electrical charge when voltage is applied; its impedance is inversely proportional to the frequency of the voltage impressed; i.e., it offers little resistance or impedance to high frequencies but much to low frequencies.

casting. Ceramic forming process in which a body slip is introduced into a porous mold, which absorbs sufficient water from the slip to produce a semirigid article.

cavity. The aperture which determines the print geometry and influences the print thickness of a metal mask.

centerline-to-centerline spacing. The distance between two or more adjacent centerlines; usually used for specifying step-and-repeat arrays.

ceramic. Inorganic claylike material molded and formed by high-temperature processes.

cermet A combination of powdered precious-metal alloys and inorganic material such as alumina. Used in manufacturing resistors, capacitors, and other components for high-temperature applications.

channel. A region of semiconductor material in which current flow is influenced by a transverse electric field. Physically a channel may be an inversion layer, a diffused layer, or bulk material. The type of channel is determined by the type of majority carriers during conduction, that is p- or n-channel.

channel frame. A screen-printing frame with adjustable bars into which the mesh is mounted for tensioning; also known as a floating-bar frame.

chase. Screen frame.

chip (die). A leadless discrete part. Generally a capacitor or resistor chip; an uncased semiconductor die; the words chip and die are often interchanged.

C-MOS (complementary MOS). An MOS integrated-circuit device involving both p- and n-channel MOSFETS (see MOSFET). The technique increases logic speed but requires additional processing steps that reduce circuit density and raise cost per function.

cofiring. A process whereby two thick film compositions are printed and dried one after the other and then fired at the same time.

collet. A die pickup collet is a truncated tool designed to pick up and hold a semiconductor die, generally by means of vacuum.

compatibility (of thick film compositions). Ability of compositions to be processed together; chemical compatibility with other compositions used in the manufacture of thick film microcircuits.

conductivity. The ability of a material to conduct electricity; the reciprocal of resistivity.

conductor fluid. A thick film conductor composition complete with metal, frit, and organic vehicle ready for printing; also frequently referred to as conductor inks or pastes.

conductors. Materials that conduct electricity easily, i.e., have a low resistivity (10^{-1} Ω-cm).

contact printing. A method of screen printing where the screen is almost (within a few mils) in contact with the substrate. Used for precision printing or with nonflexible screens.

contact resistance. (1) In electronic elements, such as capacitors or resistors, the equivalent resistance between the conducting electrode and the body of the capacitor or of the resistor. (2) In thick films, the equivalent resistance between the conductor pattern and the resistor pattern.

coordinatograph. A machine for guiding a tool to cut on coordinate axes with high precision (used to generate artwork).

copyboard. The part of a process camera used to hold the artwork in place.

core. The supporting layer or center layer of material of an electroformed metal mask, usually beryllium copper.

corner mark. An L-shaped registration pattern used to define or locate the outline of a ceramic substrate.

crazing. Fine-line cracking which occurs in fired glazes or other ceramic coatings.

crossover. In a multilayer thick film circuit, a place where two conductors cross without making electrical contact. It consists of the two conductors and an insulating material separating them.

CTL (complementary transistor logic). A logic form based on the use of both npn and pnp transistors in emitter-coupled OR gates. Resembles emitter-coupled logic but is nonsaturating logic and therefore very fast.

cut and strip. Method of producing artwork by cutting the pattern and stripping away the unwanted areas of a two-layer system.

definition. The sharpness of a screen-printed pattern; the exactness with which a pattern is printed.

dewetting. The act of solder flowing away from the soldered surface during re-heating subsequent to initial soldering

dicing. Separating a semiconductor wafer into individual dice.

die (pl. dice; see also chip). The individual semiconductor element or integrated circuit after it has been cut or separated out of the processed semiconductor wafer, distinct from a completely packaged or encapsulated integrated circuit with leads attached. A semiconductor chip.

die bond Attachment of the circuit die to a hybrid or package substrate. The attachment serves as a mechanical support, thermal path, and sometimes electrical contact.

dielectric constant. The ability of a material to store discharge when used as a capacitor dielectric. It is the ratio between the charge that would be stored with free space as the dielectric material in question. Same as relative permittivity.

dielectrics. Materials that do not conduct electricity. Generally, dielectric refers to materials that are to be used as capacitors whereas insulator refers to materials that are primarily electrical insulators. The materials can serve as either or both, and the terms have essentially the same meaning.

diffused layer The region of semiconductor into which impurity dopants have been diffused to a concentration of at least the background concentration. The region is often delineated by a pn junction.

DIP (dual in-line package). The standard DIP is a molded plastic package about $\frac{3}{4}$ in long and $\frac{1}{3}$ in wide, with two rows of pins spaced on 0.1-in. centers. This package is more popular than the flat pack or transistor-outline can for industrial use because it is relatively inexpensive and is easily dip-soldered into printed-circuit boards.

direct emulsion. A sensitized-liquid coating for producing stencil screens. Material is applied to the screen mesh before exposure and development.

direct-emulsion screen. A screen whose emulsion is applied by painting directly onto the screen, as opposed to the indirect-emulsion type.

discrete (as applied to components used in thick film circuits). Elements that are added separately are discrete elements (or devices), as opposed to those made by screen-printing methods.

dissipation factor. (1) At high frequency, the ratio of the effective series resistance of a capacitor to the capacitive reactance; given by the product of the angular frequency, the series resistance, and the capacitance. (2) At low frequency, the ratio of the capacitive reactance to the effective parallel resistance; given by the inverse of the product of the angular frequency, the parallel resistance, and the capacitance.

distributed circuit. A film circuit in which the effective components are not easily recognizable in discrete form. An example is a distributed RC circuit composed of a film resistor deposited on a dielectric film which lies above a ground-plane electrode.

distributed element. The physical realization of an element incorporating more than one primary electrical characteristic (resistance, capacitance, inductance, gain, etc.) dispersed along the length of the element.

drift. Permanent change in value of a capacitor or resistor over a period of time because of the effects of temperature, aging, humidity, etc.

drying. Removing by evaporation, uncombined water or other volatile substances from a ceramic raw material or product, usually by low-temperature heating.

DTL (diode-transistor logic). An early but still widely used type of bipolar logic. The basic logic circuit is a NAND gate, consisting of input diodes for the AND function and a transistor inverter.

dynamic testing. Active elements can be tested in two ways: (1) static, where only dc tests are made, and (2) dynamic, where ac reactions (especially high frequency) are evaluated.

ECL (emitter-coupled logic). A nonsaturating form of bipolar logic in which the emitters of the input logic transistors are coupled to the emitter of a reference transistor. Cost and applications are comparable to complementary transistor logic. MECL (Motorola trade mark).

edge definition. The relative sharpness of an aperture in a stencil screen or metal mask.

electroformed mask. Usually a trimetal structure in which a core material is selectively electroplated on both sides with a dissimilar metal. Then the core is etched back to form a mesh pattern on one side and a stencil cavity on the reverse.

element. A topologically distinguishable part of a microcircuit which contributes directly to its electrical characteristics.

emulsion. The organic material used to coat and/or plug up the mesh of a screen.

encapsulate Sealing up or covering an element or circuit for mechanical and environmental protection.

epitaxy. Deposition of a monocrystalline layer of material onto a substrate material so that the layer thus formed has the same crystal orientation as the substrate, e.g., silicon on silicon or silicon on sapphire.

etch. To remove image areas physically or chemically.

etch factor. The ratio of the depth of etch to the lateral etch or undercut distance.

etched metal mask. A mask formed by etching apertures through a solid metal protected by a photoresist.

eutectic. The point of maximum fusibility of a combination of materials. The eutectic temperature of a system (if one exists) is always lower than the melting point of any of the individual components of the system. The eutectic composition is that particular composition where the eutectic occurs. The system goes from totally molten to totally solid without going through a slushy range (where it is partially molten and partially solid) at the eutectic composition.

exposure. Product of light intensity and time; the total amount of light striking a given area of photographic material.

f **numbers.** The focal length divided by the lens opening.

face bond. A bond directly between a chip bonding pad and a mounting substrate for the purpose of making electrical contact; commonly used for beam leads and flip chips.

facedown chip. A chip intended for mounting with the electrical terminations on the side attached to the mounting substrate.

faceup chip. A chip intended for mounting with the electrical terminations on the side opposite that attached to the mounting substrate.

fan in. The number of inputs available to a specific logic stage or function.

fan out. The number of stages that can be driven by a circuit output.

FET (field-effect transistor). A voltage-controlled transistor (see IGFET, JFET, and MOSFET) analogous to a vacuum-tube triode.

film integrated circuit. An integrated circuit consisting of elements which are films formed in situ upon an insulating substrate.

film stress. The compressive or tensile forces appearing in a film, e.g., internal film stress, the intrinsic stress of a film related to its mechanical structure and deposition parameters, or induced film stress, the component of film stress related to an external force such as mismatched mechanical properties of the substrate.

fire. To heat a thick film circuit so that the resistors, conductors, capacitors, etc., are transformed into their final form.

flatness. The long-range deviation from planarity of a film surface, measured in mils per inch.

flip chip. A chip having bumped terminations spaced around the device and intended for facedown mounting.

flip-chip attachment. A method of attaching a thin film active device to a thick film circuit where the device is flipped so that the connecting conductor pads on the bottom surface of the active device are set on top of the solder-coated conducting pads on the top surface of the thick film circuit and bonded by reflowing the solder.

flood bar.   A bar or other device on a screen/printing device that will drag pastes back to the starting point after the squeegee has made a printing stroke.   The flood stroke returns the paste without pushing it through the meshes, so-that it does not print but returns the paste supply to be ready for the next print.

flux.   In soldering, a material that chemically attacks surface oxides and tarnishes so that the molten solder can wet the surfaces to be soldered.

frame.   A device used to support a screen mesh or metal mask before mounting in a printing machine.   Most frames for thick film work are made of metal and are either square or rectangular.

frit.   Melted-glass composition, ground up and used in thick film compositions as the portion of the composition that melts upon firing to give adhesion to the substrate and hold the composition together.

front lighting.   Illumination of an artwork by means of light reflection off the front surface of the pattern.

gas blanket.   An atmosphere of forming gas flowing over a heated integrated-circuit chip or a substrate during bonding which keeps the metallization from oxidizing.

gate equivalent circuit.   A basic unit of measure of relative digital-circuit complexity.   The number of gate equivalent circuits is that number of individual logic gates that would have to be interconnected to perform the same function.

glaze.   The glassy coating applied to the surface of a formed article or the material or mixture from which the coating is made.

halation.   The spreading of light outside the intended area of exposure by reflection from the rear surface of the transparent base supporting the emulsion to be exposed.

hand cut.   Artwork which has been prepared without the use of a drafting machine.

hermetic seal.   A mechanical or physical closure which is impervious to moisture or gas, including air.   Usually pertains to an envelope or enclosure containing electronic components or parts or to a header.

HIC (hybrid integrated circuit).   In general, any integrated circuit that is not mono-lithic is hybrid.   Hybrid means that the circuit elements are made by two or more different technologies.   A typical HIC consists of semiconductor chips and capacitors attached to a ceramic substrate carrying printed resistors and inter-connections that are vacuum-evaporated or chemically formed.

hot zone.   The part of a continuous furnace or kiln that is held at maximum temperature.   Other zones are the preheat zone and cooling zone.

hybrid microcircuit.   A microcircuit consisting of elements which are a combination of the film circuit type and the semiconductor types or a combination of one or both of the types with discrete parts.

IGFET (insulated-gate field-effect transistor).   Most IGFETS made today are MOSFETs.   In general, an IGFET refers to any type of FET except a JFET.

image.   The sum of all geometric forms appearing in the functional pattern area. All images are identified by a reference or by actual dimension.

indirect emulsion.   A sensitized-film coating used for producing stencil screens. Materials are exposed and developed before application to the screen mesh.

indirect-direct emulsion.   A combination film and sensitized-liquid system for pro-ducing stencil screens.   Materials are applied to the screen mesh before exposure and development.

indirect-emulsion screen.   A screen whose emulsion is a separate sheet or film of material, attached by pressing into the mesh of the screen (as opposed to the direct-emulsion type).

ink.   Synonymous with composition and paste when relating to screenable thick film materials.

integrated circuit.   A microcircuit consisting of interconnected elements inseparably associated and formed in situ on or within a single substrate to perform an electronic-circuit function.

interface. The borderline region between two different thick film materials, e.g., the region where a thick film resistor composition and its connecting conductor composition meet and intermingle, react, etc.

JFET (junction field-effect transistor). A JFET consists of a gate region diffused into a channel region. When a control voltage is applied to the gate, the channel between the source and drain is depleted or enhanced by enlargement of the pn junction. Current cannot flow when the channel is pinched off.

kiln. Furnace used in ceramic firing.

lateral reversal. Image is reversed left to right; mirror image.

leaching. In soldering, the dissolving (alloying) of the material to be soldered into the molten solder.

leveling. The settling or smoothing out that takes place after a pattern is screen-printed. Immediately after printing, the mesh marks of the screen are visible. With compositions that have good leveling characteristics, these marks disappear rapidly.

line definition. Sharp, clean screen-printed lines. Capability is often described in terms of the line width and spacing that is possible, e.g., 3 by 3, indicating line-definition capabilities of 3-mil lines and 3-mil spaces.

linear scaling of resistors. The ideal situation where the resistance of a fired thick film resistor is directly proportional to its length. In practice this occurs only when resistors are long enough for conductor-resistor termination effects to be minor. How close a resistor composition comes to linear scaling on a given thick film conductor is a measure of how close the conductor comes to giving no end effects.

LSI (large-scale integration). Usually refers to monolithic digital integrated circuits with a complexity of typically 100 or more gates or gate-equivalent circuits, e.g., MOS read-only memories. Each manufacturer has a number of gates per chip that he uses to define LSI. LSI sometimes describes hybrid integrated circuits built with a number of MSI or LSI chips.

lunometer. A device used for determining the mesh count of a screen fabric.

mask (screens). A screen made not from wire or nylon thread but from a solid sheet of metal in which holes have been etched in the desired pattern. Useful for precision and/or fine printing.

MECL. See ECL.

meniscus. In a solder joint, the minimum angle at which the solder tapers from the joint to the flat area of the conductor.

mesh. Number of openings per lineal inch, measured from the center of any wire.

mesh porosity. The amount of open area in a mesh versus the amount of closed area. Also expressed as percentage of open area.

mesh size. The number of openings per inch in a screen. A 200-mesh screen has 200 openings per linear inch, 40,000 openings per square inch.

microcircuit. A small circuit having a high equivalent circuit-element density, which is considered as a single part composed of interconnected elements on or within a single substrate to perform an electronic-circuit function. (This excludes printed wiring boards, circuit-card assemblies, and modules composed exclusively of discrete electronic parts.)

microelectronics. That area of electronic technology associated with or applied to the realization of electronic systems from extremely small electronic parts or elements.

microinch (μin). One-millionth inch (10^{-6} in).

micrometer (μm). One-millionth meter (10^{-6} m).

mil. One-thousandth inch (10^{-3} in.), equivalent to 25 μm.

mill. (1) A machine for grinding or mixing material such as a ball mill, a paint mill, etc. (2) Grinding or mixing a material, e.g., milling a thick film composition.

monofilament mesh. Woven material with single-stranded threads.

monolithic integrated circuits. An integrated circuit consisting of elements formed in situ on or within a semiconductor substrate with at least one of the elements formed within the substrate.

MOS (metal-oxide-semiconductor). A method of construction distinguishes this type of integrated circuit from bipolar. MOS integrated circuits are slower than bipolar integrated circuits but have the advantage of high circuit density and low cost.

MOSFET (metal-oxide semiconductor field-effect transistor). The basic element of MOS integrated circuits. A FET consists of diffused source and drain regions on either side of a p- or n-channel region, plus a gate electrode insulated from the channel by silicon oxide.

MSI (medium-scale integration). Monolithic integrated circuits having typical complexities of 10 to 100 gates. Examples include complex gates, e.g., the quad exclusive-OR), counters, dividers, shift registers, and comparators.

multichip microcircuits. A microcircuit consisting of elements formed on or within two or more semiconductor chips which are separately attached to a substrate.

multifilament mesh. Woven material with multiple-stranded threads.

multilayer substrates. Substrates that have buried conductors so that complex circuitry can be handled. This does the same thing as a multilayer thick film circuit but with improved performance in some respects. Assembled using processes similar to those used in multilayer ceramic capacitors.

negative. A developed photographic image where light areas represent the dark areas of the original and dark areas represent the light areas of the original; opaque and clear reversal of the original.

negative artwork. An enlarged, scaled pattern in which light areas represent the image which is to be dark on the final pattern. Negative artworks are usually prepared when using cut-and-strip film.

off-contact (screen printing). The opposite of contact printing in that the printer is set up with a space between the screen and the substrate.

orthochromatic. Sensitive to ultraviolet and all colors except red.

overglaze. A glass coating over another component or element, normally to give physical or electrical protection.

pad. The metallized area on a substrate or on the face of an integrated circuit used for making electrical connections.

parasitic. An inductive, resistive, or capacitive contribution to a circuit that arises from the circuit configuration, as opposed to the design values of deliberately introduced components. Examples are the inductance of conductors, the resistance of interconnections and contacts, and the capacitance between conductors or between regions with a component.

paste. Synonymous with composition and ink when relating to screenable thick film materials.

passivation. The formation of an insulating layer directly over the semiconductor surface to protect the surface from contaminants, moisture, or particles. Usually an oxide of the semiconductor is used; however, deposition of other materials is also used.

passive element. An element which is not active.

peel strength (peel test). A measure of adhesion between a conductor and the substrate. The test is performed by pulling or peeling the conductor off the substrate and observing the force required. Units are ounces per mil or pounds per inch of conductor width.

photosensitive. Sensitive to light.

plain weave. Mesh pattern in which the filaments are woven with one over, one under.

planar (device and process). A type of semiconductor device and the process technology used to fabricate it, in which all the *pn* junctions terminate at approximately the same geometric plane on the surface of the semiconductors.

porosity, apparent. The relationship of open pore space and bulk volume expressed in percent.

positive. A photographic reproduction whose light and dark areas correspond to those of the original.

positive artwork. An enlarged, scaled pattern in which the dark areas represent the image which is dark on the final pattern; usually prepared by taping or inking.

potentiometer. A variable resistor.

pot life. The length of time a two-part epoxy system remains useful, usually measured in hours; it can be extended by refrigeration.

prefire. To fire one thick film composition, e.g., a conductor, before printing a second thick film composition, e.g., a resistor.

pressing dry. Forming ceramic ware in dies from powdered or granular material by direct pressure.

process, wet (slip process). Method of preparing a ceramic body wherein the constituents are blended in a sufficiently liquid form to produce a suspension for use as is or in subsequent processing.

profile (firing). A graph of time versus temperature or of position, in a continuous thick film furnace, versus temperature.

pseudoplastic. A characteristic of a fluid whereby its viscosity decreases as the shear rate is increased. This does not imply a change in behavior with time. This type of behavior is often obtained when solids are dispersed in organic vehicles, e.g., thick film compositions.

pyrolyzed. Burned; a material that has gained its final form by the action of heat is said to be pyrolyzed.

random mesh registration. A mask or screen in which the mesh holes are not positioned in any specific location relative to the cavity pattern.

reduction ratio. The ratio of the measurements of an original to the measurements of a reduced image of the original, expressed as $10\times$, $20\times$, etc.

registration. The accuracy of concentricity or relative position of all patterns on any mask with the corresponding patterns of any other mask of a given device series when properly superimposed.

registration marks (alignment marks; fiducial marks). The marks on a wafer or substrate which are used for aligning successive processing masks.

resistance. The property of an electric circuit which determines, for a given current, the rate at which electric energy is converted into heat; has a value such that the current squared multiplied by the resistance gives the power converted.

resistor. A device that offers resistance to the flow of electric current in accordance with Ohm's law: $R = E/I$.

resistor termination. A thick film conductor pad which overlaps and makes contact with a thick film resistor area.

resolution. The degree of fineness or detail of a screen-printed pattern; ability of photographic materials to reproduce fine materials.

rheology. The science dealing with deformation and flow of matter.

roll (mill). To disperse solid powder in an organic vehicle by passing the material through closely set steel or ceramic rolls.

roughness. The microscopic peak-to-valley distances of film-surface protuberances and depressions, measured in angstroms.

RTL (resistor-transistor logic). Form of bipolar NOR logic.

Rubylith. Ulano Company trademark for a popular masking film consisting of a transparent Mylar (Du Pont) sheet with a strippable red film coating. Used in preparing integrated-circuit artwork.

sawtooth. Small projections on the edge of a screen-printed pattern caused by fillets of emulsion left in the screen mesh and the mesh itself.

screen stencil. The screen used in thick film printing.

screen tension. The tautness of a mounted woven mesh expressed in mils per unit of weight.

sensitizer. Material used to activate a photographic emulsion to create chemical changes in that emulsion.

shear rate. The relative rate of flow or movement (of viscous fluids).

sheet resistivity. The resistance of a film material with the same length and width, i.e., one square. The resistivity of thick film resistor compositions is expressed in terms of sheet resistivity in ohms per square.

shelf life. The length of time a one- or two-part epoxy system can be stored before use, usually measured in months.

silver (Ag) migration. The growth of Ag crystals between two Ag conductor areas a few mils apart in a thick film circuit when a dc voltage is applied over a long period under conditions of high humidity. Alloying the Ag with a small quantity of a higher-melting metal such as Pd greatly reduces this tendency.

sinter. A ceramic material or mixture fired to less than complete fusion but resulting in a coherent mass.

skin depth. At very high frequencies, current travels only on the surface layer of a conductor, not uniformly throughout; this region of current flow is called the skin.

slump. A spreading of printed thick film composition after screen printing but before drying. Too much slumping results in loss of definition.

slurry. A thick mixture of liquid and solids, the solids being in suspension in the liquid.

snap-off distance. The screen-printer distance setting between the bottom of the screen and top of the substrate.

soak time. The length of time a ceramic material, e.g., substrate or thick film composition is held at the peak temperature of the firing cycle.

solderability or solder-wetting capability. A measure of how readily or how quickly a thick film conductor is wet with molten solder. The contact angle between the solder and the conductor surface may be taken as a quantitative measure of this property.

solids determination. A method of measuring the inorganic weight fraction of a thick film composition by heating the composition to temperatures of 750 to 1050°C to burn off the organic fraction and weighing the resulting material.

specific capacitance (C/A). The capacitance per unit area.

specific mesh registration. A mask or screen in which the mesh holes are carefully aligned to correspond with the apertures of the stencil cavity.

squeegee. The part of a screen printer that pushes the composition across the screen and through the mesh onto the substrate.

stair stepping. Irregular edge definition of a screen-printed pattern caused by interference between the mesh openings and the stencil cavity.

steatite. A ceramic consisting primarily of magnesium silicate popularly used as an electrical insulator; rarely used as a thick film substrate.

stencil. The emulsion or metal layer in which the image apertures are reproduced.

step and repeat. The method of dimensionally positioning multiples of the same or intermixed functional patterns over a given area.

step exposure. The technique whereby a series of exposures are made to determine the optimum amount of time and distance required for exposing a photosensitive material with any given light source.

stitch bond. A thermocompression bond in which a capillary tube is used for feeding the wire and forming the bond.

substrate (of a microcircuit or integrated circuit). The supporting material upon or within which the elements of a microcircuit or integrated circuit are fabricated or attached. Alumina is commonly used.

swimming. Lateral shifting of a thick film conductor pattern on molten-glass crossover patterns.

TC. Thermocompression bonding or temperature coefficient (resistor and capacitors).

TCC Temperature coefficient of capacitance.

TCR   Temperature coefficient of resistance.

thermal aged adhesion.   The adhesion of a thick film conductor after aging at an elevated temperature, for example, 50 to 500 hours at 125 to 150°C, with leads soldered. This test is intended to simulate the effects of aging leaded conductors at the lower normal operating temperatures for longer periods, possibly years.

thermocompression bonding.   A method of wire-bonding integrated circuits in which temperature and pressure are used to obtain the bond between wire and pad. Gold wires are often bonded to thick film conductors and integrated circuits in this manner.

thermocouple.   A device that generates voltage in accordance with temperature difference observed between one end and the other. It consists of two lengths of wires of different composition and is favored for measuring high temperatures.

thick film circuit.   A circuit fabricated utilizing a screen-printing technology associated with metallized or insulating inks as the printing materials.

thin film.   A film deposited by vacuum methods, commonly less than 20,000 Å thick. The basic process used in making thin film integrated circuits is vacuum deposition.

thixotropic.   The tendency of a fluid to decrease in viscosity as the time of exposure to a given shear rate increases. The shear-stress–versus–shear-strain rate curve of a thixotropic material should show a hysteresis loop. A purely pseudoplastic material will not give a hysteresis loop, as this property is not time-dependent. Most thick film pastes or inks are thixotropic.

TO can (transistor-outline metal-can package).   In integrated circuits, a type of package resembling a transistor can but generally larger and with more leads. The pins are arranged in a circle in the base.

trim (resistors).   (1) To change value from as fired to the final desired value, usually by removing parts of the body of the resistor. (2) To make adjustments in resistors, coils, capacitors (often by adding or replacing) to bring electrical performance into exact agreement with specifications.

TTL (transistor-transistor logic).   A form of bipolar logic similar to diode-transistor logic except that the input diodes are replaced by multiple-emitter transistors.

tweak.   A small amount of trimming.

twill weave.   Mesh pattern in which the filaments are woven with one over, two under.

ultrasonic bonding.   A method of bonding wire to integrated circuits and thick film conductors that uses ultrasonic energy to effect the metallurgical bond between the metal of the wire and the pads. Aluminum wire is commonly used.

VCC.   Voltage coefficient of capacitance.

VCR.   Voltage coefficient of resistance.

vehicle.   The organic system in the composition.

viscosimeter.   A device that measures viscosity. Not all viscosimeters are capable of measuring viscosity under conditions of varying shear rates. This is of prime importance with thick film compositions.

viscosity.   The fluidity of material, or the rate of flow versus pressure. The unit of viscosity measurement is the poise, more commonly centipoise.

vitreous.   Glassy; as used in ceramic technology, it indicates fired characteristics approaching a glassy state but not necessarily totally glassy.

vitrification.   The progressive reduction in porosity of a ceramic as a result of heat treatment or other process.

warp.   Wires or threads running parallel to the length of the roll.

weft.   Wires or threads running across the width or perpendicular to the length of the roll; also known as shute, fill, or woof.

wedge bond.   A thermocompression bond in which a wedge-shaped tool is used to apply pressure to the wire being attached.

wire bond.   The attachment between a wire and the microcircuit-chip bonding pad or package terminal.

Index

Index